Towards Radical Regeneration

Christoph Gengnagel · Olivier Baverel ·
Giovanni Betti · Mariana Popescu ·
Mette Ramsgaard Thomsen ·
Jan Wurm
Editors

Towards Radical Regeneration

Design Modelling Symposium Berlin 2022

Springer

Editors
Christoph Gengnagel
Institute of Architecture and Urban Planning,
Department for Structural Design
and Engineering, KET
Berlin University of the Arts
Berlin, Berlin, Germany

Giovanni Betti
Architectural Design for Building
Performance
University of California, Berkeley
Berkeley, CA, USA

Mette Ramsgaard Thomsen
CITA
The Royal Danish Academy of Fine Arts
Copenhagen, Denmark

Olivier Baverel
Laboratoire Navier/GSA
School of Architecture of Grenoble
and Ecole des Ponts Paristech
Grenoble, France

Mariana Popescu
Civil Engineering and Geosciences
Delft University of Technology
Delft, The Netherlands

Jan Wurm
Faculty of Architecture, Regenerative
Architecture and Biofabrication
KU Leuven
Leuven, Belgium

ISBN 978-3-031-13251-3 ISBN 978-3-031-13249-0 (eBook)
https://doi.org/10.1007/978-3-031-13249-0

This Springer imprint is published by the registered company Springer Nature Switzerland AG
The registered company address is: Gewerbestrasse 11, 6330 Cham, Switzerland

Introduction

"*Technology is the answer, but what is the question?*" famously asked Cedric Price to an audience of architects and engineers in 1966.

In the age of the growing inequalities, climate crisis, and ecosystem collapse, the questions could not be clearer for researchers, educators, and practitioners in the Architecture Engineering Construction (AEC) fields, nor the answers more urgently needed. The certainty that technology alone can be the answer is rapidly waning.

Since 1950, the world has been undergoing what some have dubbed the Great Acceleration [1]. The Great Acceleration is the steady and approximately simultaneous increase in growth rates across a wide range of indicators of human activity. The indicators are many and varied. Whether population, real GDP, primary energy use, water use, transportation, telecommunications, or inequality [2], all areas of production and consumption have undergone a near exponential growth in the last 70 years. The growth of the socio-economic indicators mentioned above is paralleled by the spike registered in the Earth System indicators such as nitrous oxide, methane, stratospheric ozone, ocean acidification, tropical forest loss, domesticated land, and terrestrial biosphere degradation. This pressure on the planetary boundaries is directly associated with the risk of abrupt environmental change [3].

In the public perception, two closely related parameters represent this development: the increase in atmospheric carbon dioxide concentration and surface temperature. But the steady, unchecked increase in almost all indicators of human activity affects far more than these easily perceived parameters. It amounts to a change of the entire planetary metabolism. As scientists from the Weizmann Institute of Science reported in 2020, global human-made mass now exceeds all living biomass [4]. The rise in our activities is radically changing the material composition of the Earth surface, and—unsurprisingly—construction materials, especially concrete, play a very significant role. The combined weight of all buildings and infrastructure ($\sim$1100 Gt) far exceeds the mass of all trees and shrubs worldwide ($\sim$900 Gt) [4], and concrete is the most used substance worldwide, second only to water [5].

This rapid and uncontrolled acceleration of human activity and output is currently running through all planetary boundaries causing the current ecological crisis. But while the crisis is a truly planetary and global one, the responsibilities are unevenly spread. As it is demonstrated in Hickel et al. [6], the so-called high-income countries are the major culprits for overshooting Earth's capacity and the planetary boundaries. This relationship bears little relation to actual population numbers and is instead much more related to choices and expectations about lifestyle and comfort.

To avoid climate catastrophe, we need to institute a break in the unsustainable paradigm fuelled by uncontrolled growth. The responsibilities of the built environment are large, but they are matched by the potential for the built environment to become a central part of the solution.

The UN 2020 Global Status Report for Buildings and Construction note that CO_2 emissions from the building sector are the highest ever recorded at almost 10 $GtCO_2$[7]. Currently, the built environment is responsible for about 40% of global GHG emissions [7] and more than 50% of developed countries waste [8].

While current efforts to minimize material usage, improve efficiency, and generally "being less bad" are important and necessary (productivity in the building sector has been stagnant in the past 30 years), they are far from being enough. Beyond the pursuit of efficiency, the built environment can become a key part of the solution to the climate equation only if it moves from an extractive logic to a regenerative one. Not only the built environment should minimize its own emissions, but the construction of new buildings and the maintenance of existing ones has to become a planetary carbon storage mechanism, an incentive to reforestation, and a driver for biodiversity restoration [9]. To enable this shift, many current paradigms need to be challenged and updated, particularly the current extractive material practice. As we have seen, the built environment literally weighs more on the planet than all biomass combined, and its constant creation and destruction are responsible for about half of the global waste by mass. Those are clearly unsustainable practices that need to be overcome through the implementation of circular economy principles and ecosystem thinking.

All of this puts us firmly within the reality of living, designing, and building in a new era: the Anthropocene. As designers and creators of the built environment, we need to ask ourselves: How can we act in the Anthropocene to create an era where both people and a stable, biodiverse, and plentiful planet can prosper?

A prosperous, equitable, and biodiverse Anthropocene can be realized only by shifting the paradigm from one of linear extraction and disposal to one of cyclic regeneration and replenishment. To achieve this, we need to build the infrastructures by which to truly rethink the end-of-life of our built environment. A circular economy must integrate upcycling across the design chain; from ideation to end-of-use so as to allow a strong retention of value as materials cycle through our socio-ecological networks. It demands new alliances between the technological and the biological, between high-tech and no-tech. In this worldview, design must be rethought as an active process partaking in both global and local cycles of production that integrate end-of-life scenarios to support a finite planet in balance.

Recent advances in AEC industries hold the promise to help us trace new paths, but, at the same time, they hold the threat of turbo-charging existing destructive paradigms. Our collective practices are tantamount to many uncontrolled planetary experiments, deploying technologies and practices at unprecedented scales first and worrying about the consequences later. We urgently need to negotiate a myriad of global and hyper-local equilibria between the built and natural environment. By synthesizing artistic, theoretical, and technological perspectives and positions, the Design Modelling Symposium 2022 "Towards Radical Regeneration" aims to challenge innovative digital techniques in the design, materialization, operation, and assessment of our built environment, examining them in relation to their impact on the future of societies and the environment. In this context, the digitalization of design and construction seems to offer multiple routes to identify the material, social, ethical, ecological, and other dimensions of new, necessary, and radically measured equilibria.

The conference is structured around four thematic areas that explore domains in which new paradigms need to emerge, from the tools that enable design, simulation, and materialization to new ways of approaching the relationship between built and unbuilt environment. Each area is cantered around four polarities and creative tensions:

Design between Human and Machine Intelligence
Design with Digital and Physical Realities
Design for Biosphere and Technosphere
Design for Humans and Non-Humans

The contributions in each area represent a collection of experiences and ongoing research in some of the crucial fields of interaction between society and environment, between natural and artificial, and between biological and technological.

In Area A, **Design between Human and Machine Intelligence**, we aim to investigate how the combination of a maturing cloud infrastructure and growing AI capabilities can help redefine a new design paradigm, one where collaboration between human and machine intelligence brings about new ways of conceptualizing and quantifying design. This is a relatively new arena, but one already where the community that for years has coalesced around ideas of computational design is now growing into a more mature and robust ecosystem. Spanning between industry and academia, we see a growing cluster of digital tools, often developed by new players in the AEC software industry often spun off in successful start-ups. They promise greater efficiencies for architects, engineers, and developers. They attempt to offer both vertically and horizontally integrated solutions and foreshadow radically new business models. We strive to engage critically with this community and understand whether there is a paradigm shift behind the technological prowess that will enable a more sustainable built environment or is it an AI-powered turbo-charge to business as usual. Given the rise and dominance of AI and computing in our daily lives, this might well be one of the deciding questions for the transition of practice towards a sustainable future.

In Area B, **Design with Digital and Physical Realities**, we explore the possibilities brought about by the growth of Extended Reality (XR). VR, AR, and MR are gathering ever-increasing interest in the AEC community. Future architects—and many contemporary ones already—might concern themselves more with the design of virtual worlds than of physical ones. Can we harness those energies and technologies to solve some of the very real problems described above? Those technologies hold the promise of creating new design methodologies and enable a new craftsmanship. At the same time, the extension of physical reality with digital environments and overlays offers a unique chance to observe and understand our biases and untold assumptions about what answers we search for in the real world. In this grouping, we try to survey some of the potentially more impactful application areas in the AEC industry, exploring the blurred line between virtual and physical, gamification and production.

Area C, **Design for Biosphere and Technosphere**, concerns itself with the shift from linear production models to cyclic ones. In this view, the biosphere is not a place to extract resources and—eventually—mitigate impacts, but it is an active and vital part of the resource cycle. Buildings may have become much more energy efficient in recent years, but existing structures keep on being routinely demolished to make room for new ones and construction sites have hardly changed. Resource depletion, fragile supply chains, and embodied impacts have not been addressed. Computation is a key enabler to understand and direct resource flows to transform from a linear to circular economy for the built environment. Tracking and understanding materials and their impacts across their entire life cycle can enable new ways of interacting with nature, using and reinventing natural materials, and fully exploring the circular potential of the built environment.

Area D, **Design for Humans and Non-Humans**.

Once we accept that we now live in the Anthropocene, we accept that every environment is human made to some degree. This calls for a radical augmentation of the "client base" for the AEC industries. We can no longer assume that we are designing and building only for one species, Homo Sapiens. Given that our collective actions have the power to alter ecosystems on a planetary scale, we need to be designing for all the living creatures of those ecosystems. This might sometimes take the form of willingly refusing to build. Mammals, plants, insects, fungi, microbes, and whole ecosystems become our clients and allies as we recognize their need to interact with the outcomes of our design and building and their importance in creating a livable built environment. As such, we need to move to a regenerative design practice, where building contributes to restoring the health of our planet. Spatial design can use computation to understand, predict, and cater for the needs and behaviours of its human and non-human inhabitants. Can we move beyond the current exploitative or destructive models towards a symbiotic approach, and can digital tools help design healthy environments of cohabitation on our planet for all life forms to prosper and flourish?

We believe that the thoughtful and diverse contributions contained in this book can offer a glimpse of tomorrow's best practices in today's research and experiments. The Design Modelling Symposium is a community born 14 years ago around the impact of digital technology on design practice. We hope this eighth edition of the DMS will make a significant contribution in shifting the axis of the conversation within our community and beyond towards a new, much-needed paradigm and that the reader will find paths leading Towards Radical Regeneration.

Olivier Baverel
Giovanni Betti
Christoph Gengnagel
Mariana Popescu
Mette Ramsgaard Thomsen
Jan Wurm

References

1. Steffen, W., Broadgate, W., Deutsch, L., Gaffney, O., Ludwig, C., The Trajectory of the Anthropocene: The Great Acceleration, The Anthropocene Review 2, no. 1 81–98 (2015). https://doi.org/10.1177/2053019614564785
2. WIDER, UNU. Trends In Global Inequality – A Comprehensive Approach, Research Brief 2021/3 Helsinki: UNU-WIDER, 2021
3. Rockström, J., Steffen, W., Noone, K. et al. A safe operating space for humanity. Nature 461, 472–475 (2009). https://doi.org/10.1038/461472a
4. Elhacham, E., Ben-Uri, L., Grozovski, J. et al. Global human-made mass exceeds all living biomass. Nature 588, 442–444 (2020). https://doi.org/10.1038/s41586-020-3010-5
5. Gagg, C., Cement and concrete as an engineering material: An historic appraisal and case study analysis, Engineering Failure Analysis, Volume 40, 114-140 (2014) ISSN 1350-6307, https://doi.org/10.1016/j.engfailanal.2014.02.004.
6. Hickel, J., O'Neill, D.W., Fanning, A.L., Zoomkawala, H., National responsibility for ecological breakdown: a fair-shares assessment of resource use, 1970–2017, The Lancet Planetary Health, Volume 6, Issue 4, e342-e349 (2022) ISSN 2542-5196, https://doi.org/10.1016/S2542-5196(22)00044-4.
7. United Nations Environment Programme (2020). 2020 Global Status Report for Buildings and Construction: Towards a Zero-emission, Efficient and Resilient Buildings and Construction Sector. Nairobi
8. Statistsches Bundesamt (Destatis). (2022). Waste Generation 2020. Retrieved from https://www.destatis.de/EN/Themes/Society-Environment/Environment/_Graphic/_Interactive/waste-management-quantity.html?nn=23400
9. Churkina, G., Organschi, A., Reyer, C.P.O. et al. Buildings as a global carbon sink. Nat Sustain 3, 269–276 (2020). https://doi.org/10.1038/s41893-019-0462-4

Organization

We would like to thank the members of the Scientific Committee of the Design Modelling Symposium Berlin 2022 for their efforts in reviewing and guidance to the authors of the published contributions.

Sigrid Adriaenssens	Princeton University, USA
Aleksandra Anna Apolinarska	ETH Zurich, Switzerland
Jane Burry	Swinburne University of Technology, Australia
Neil Burford	Newcastle University, Great Britain
Jean-Francois Caron	Ecole des Ponts Paris Tech, France
Jeroen Coenders	White Lioness Technologies Delft, Netherlands
Naicu Dragos	B+G Ingenieure Berlin, Germany /Greece
Philipp Eversmann	University of Kassel, Germany
Billie Faircloth	Kierantimberlake Philadelphia, USA
Isak Worre Foged	The Royal Danish Academy, Aarhus, Denmark
M. Hank Haeusler	UNSW, Sydney, Australia
Christian Hartz	TU Dortmund, Germany
Markus Hudert	Aarhus University, Denmark
Axel Kilian	MIT Cambridge, USA
Jan Knippers	University of Stuttgart, Germany
Toni Kotnik	Aalto University, Finland
Riccardo La Magna	University of Karlsruhe, Germany
Julian Lienhard	Universität Kassel, Germany
Rupert Maleczek	University of Innsbruck, Austria
Achim Menges	University of Stuttgart, Germany
Romain Mesnil	Ecole des Ponts Paris Tech, France
Carol Monticelli	Polytechnic University of Milano, Italy
Julien Nembrini	University of Freiburg, Switzerland
Paul Nicholas	Royal Danish Academy, Copenhagen, Denmark
Stefan Peters	Graz University of Technology, Austria
Clemens Preisinger	University of Applied Arts Vienna, Austria

Bob Sheil	UCL, Great Britain
Paul Shepherd	University of Bath, Great Britain
Martin Tamke	Royal Danish Academy, Copenhagen, Denmark
Oliver Tessmann	University Darmstadt, Germany
Tom Van Mele	ETH Zurich, Switzerland
Agnes Weilandt	B+G Ingenieure Frankfurt/Main, Germany
Roland Wüchner	TU Braunschweig, Germany

Contents

Design between Human and Machine Intelligence

A Flexible Reinforcement Learning Framework to Implement Cradle-to-Cradle in Early Design Stages

Diego Apellániz[1(✉)], Björn Pettersson[2], and Christoph Gengnagel[1]

[1] Department of Structural Design and Technology (KET), Berlin University of the Arts (UdK), Hardenbergstr. 33, 10623 Berlin, Germany
d.apellaniz@udk-berlin.de

[2] Royal Danish Academy, Architecture Design Conservation, Philip de Langes Allé 10, 1435 Copenhagen K, Denmark

Abstract. Reinforcement Learning (RL) is a paradigm in Machine Learning (ML), along with Supervised Learning and Unsupervised Learning, that aims to create Artificial Intelligence (AI) agents that can take decisions in complex and uncertain environments, with the goal of maximizing their long-term benefit. Although it has not gained as much research interest in the AEC industry in recent years as other ML and optimization techniques, RL has been responsible for recent major scientific breakthroughs, such as Deep Mind's AlphaGo and AlphaFold algorithms. However, due the singularity of the problems and challenges of the AEC industry in contrast to the reduced number of benchmark environments and games in which new RL algorithms are commonly tested, little progress has been noticed so far towards the implementation of RL in this sector.

This paper presents the development of the new Grasshopper plugin "Pug" to implement RL in Grasshopper in order to serve as a flexible framework to efficiently tackle diverse optimization problems in architecture with special focus on cradle-to-cradle problems based on material circularity. The components of the plugin are introduced, the workflows and principles to train AI agents in Grasshopper are explained and components related to material circularity are presented too. This new plugin is used to solve two RL problems related to the circularity and reuse of steel, timber and bamboo elements. The results are discussed and compared to traditional computational approaches such as genetic algorithms and heuristic rules.

Keywords: Life-cycle assessment · Cradle-to-Cradle · Machine Learning

1 Introduction

Reinforcement Learning (RL) is a paradigm in Machine Learning (ML), along with Supervised Learning and Unsupervised Learning, that aims to create Artificial Intelligence (AI) agents that can take decisions in complex and uncertain environments, with the goal of maximizing their long-term benefit (Bilgin 2020). Although it has not gained

C. Gengnagel et al. (Eds.): DMS 2022, *Towards Radical Regeneration*, pp. 3–12, 2023.
https://doi.org/10.1007/978-3-031-13249-0_1

as much research interest in the AEC industry in recent years as other ML and optimization techniques (Darko et al. 2020), RL has been responsible for recent major scientific breakthroughs, such as Deep Mind's AlphaGo (Silver et al. 2017) and AlphaFold (Jumper et al. 2021) algorithms.

Similarly to these breakthroughs in the scientific world, RL has also a great potential to help reduce the carbon footprint of our built environment. Our built environment is currently responsible for 38% of the global CO2 emissions, therefore it seems like another challenge worth being tackled with RL. However, the distinctive feature of the AEC industry, and arguably the biggest challenge for the implementation of RL, is the singularity of its problems and challenges in contrast to the reduced number of benchmark environments and games in which new RL algorithms are tested (Medium, 2022a).

Parametric design and the software Grasshopper are the most preferred approach within the AEC industry to effectively face this diversity of optimization problems in architecture (Cichocka et al. 2017), however, the Grasshopper plugin "Owl" is the only existing tool that provides a limited implementation of RL in Grasshopper in the form of a Q-Table algorithm (Zwierzycki 2022). Other current approaches for implementing RL in Grasshopper involve linking Grasshopper and external python scripts (Belousov et al., 2022) to outsource the use of neural networks to python due to the incompatibility of the python-based ML libraries "TensorFlow" and "PyTorch" with the.NET application Grasshopper. However, the necessity to write and link external python scripts may certainly discourage numerous potential users of RL that are not programming experts and are just used to the visual programming environment of Grasshopper.

This paper presents the development of a new Grasshopper plugin to implement RL in Grasshopper in order to serve as a flexible framework to efficiently tackle diverse optimization problems in architecture with special focus on the reduction of the carbon footprint of our built environment. Section 2 enumerates the components of the plugin and explains the workflows and principles to train AI agents in Grasshopper. Components to model problems related to material circularity are presented too. Section 3 shows the results of applying this plugin to two RL problems related to carbon footprint reduction. The results are compared to conventional computational approaches such as heuristic rules or genetic algorithms (GA).

2 Methodology

The new Grasshopper plugin "Pug" (Apellaniz and Pettersson 2022) developed by the authors references the library "TensorFlow.NET" (GitHub 2022) to implement neural network functionalities in Grasshopper while overcoming the incompatibility of the original "TensorFlow" python library. Its Grasshopper components (see Fig. 1) combine themselves to allow the user to apply a deep Q-network (DQN) (Mnih et al. 2015) or a Monte Carlo Tree-Search (MCTS) (Medium, 2022b) algorithm to solve optimization problems entirely defined in Grasshopper, although it also includes a predefined RL problem based on OpenAI's CartPole environment to help calibrate the hyperparameters of AI agents before using them in more complex problems. Last but not least, the plugin includes components to define stocks of recycled building materials and select the possible matches for a certain structural member, so the AI agents might be able to find an optimal structure configuration based on these material stocks.

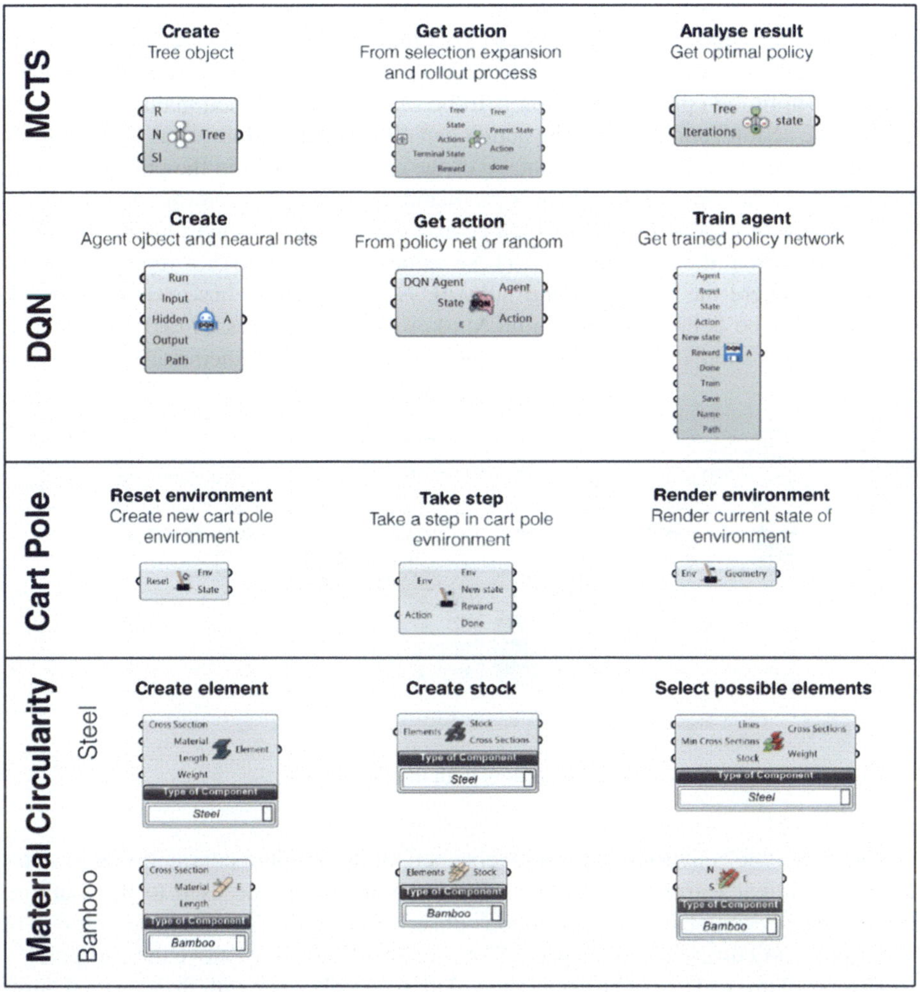

Fig. 1. Components of the new Grasshopper Plugin "Pug". It is freely accessible at www.food4rhino.com/en/app/pug.

2.1 Deep Q-Network

The DQN algorithm is a deep learning-based approach in which an AI agent is trained to be able to predict the long-term rewards of a certain environment. They achieve this by approximating the Q-values of the Bellman equation (see Eq. 1) with the agent's neural network that is trained from interactions with the environment.

$$Q(s, a) = r(s, a) + \gamma \cdot max_a Q(s', a) \tag{1}$$

When applying a DQN to solve a RL problem in Grasshopper (see diagram workflow in Fig. 2), the first step is to initialise the replay memory and the neural network. When specifying the sizes of the layers within the neural network, it is important that the input layer has the same number of units as the observation vector of the RL problem and the output layer the same as its action vector. This functionality is provided by the "create agent" component, which can generally be placed outside the training loop. For each time step within the episode an action is to be selected. This functionality is provided by the "take action" component. The action can either be a random value exploring the environment or greedily taking advantage of the experience within the policy network. In this case the action with the highest predicted discounted long-term reward "Q", is chosen. The action "a" selected can then be executed within the environment, resulting in a reward "r" and the next state "s". These values are plug into the "replay memory" component, where the agent's neural network is trained from these experiences. RL problems must be defined as sequential Markov decision processes (MDP), which is achieved in Grasshopper by combining "Pug" with the plugin "Anemone".

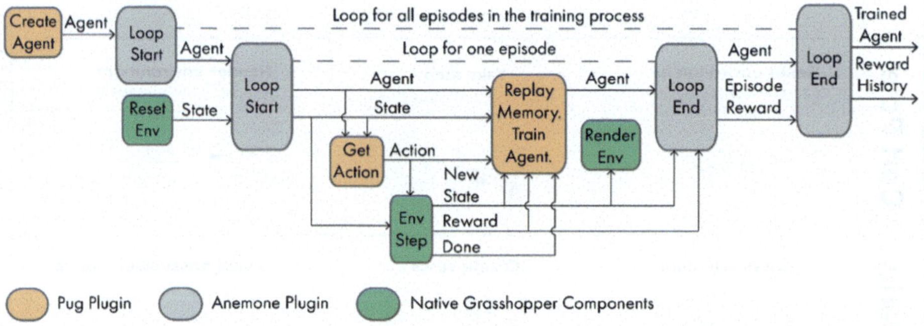

Fig. 2. Workflow with the DQN agent.

The "CartPole" environment is a reverse pendulum problem commonly used to try out reinforcement learning algorithms. The DQN agent can be used to this purpose because the environment has a continuous observation space (position and velocity of the cartpole) and discrete action space (0 or 1 to decide where to move the cart pole).

This problem, which can be easily modelled in Grasshopper with the provided built-in components of "Pug", is used to compare the DQN agent with the existing Grasshopper plugin "Owl". The "Owl" implementation is a q-leaning solution where the q values for every state action pair experienced by the agent are stored in a q-table. A discretization of the observations is therefore required. Probably due to this reason, the DQN agent is the only one that achieves to master this environment, as the training results in Fig. 3 show. However "Owl"s solution has been demonstrated as successful in other problems with discrete observation spaces. This underlines the importance of having a wide range of reinforcement learning solutions available, to enable learning within different environments.

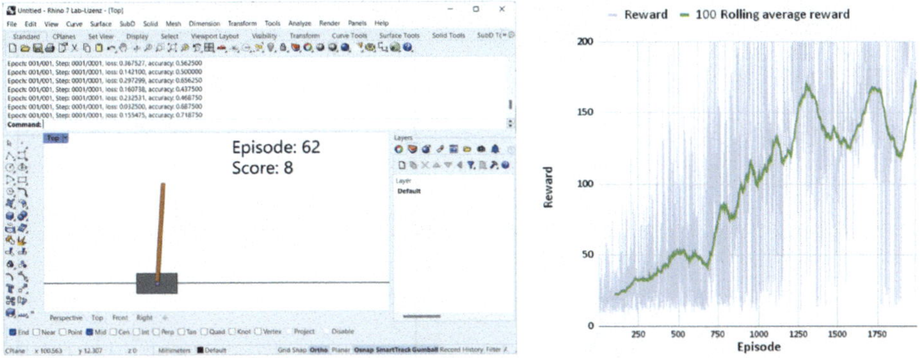

Fig. 3. Results of the learning process of the DQN agent in the Rhino viewport. After a certain number of episodes, the DQN agent learns to keep the pole balanced.

Once a DQN agent is trained and it learns to generalize the observations of a particular problem, it can be used to solve different versions of that problem within a minimum execution time, which is a very powerful feature as it is shown in Sect. 3.1.

2.2 Monte Carlo Tree-Search

The MCTS algorithm can also be applied to solve sequential problems defined as MDP. An obvious disadvantage in comparison to the DQN algorithm is that, in its basic version, this algorithm does not make use of neural networks, which means if the parameters of the environment are changed (e.g. a new labyrinth would be provided different to the one showed in Fig. 4b), the agent has to be trained again from scratch, because it does not understand generalizations. Furthermore, it is only applicable to problems with discrete observation spaces. However, a very powerful advantage is the fact that it can consider a different number of possible actions in each step, and they can therefore be used for problems with relatively large action spaces, such as many board games or the problem presented in Sect. 3.2.

The training process of the decision tree consists of the steps displayed in Fig. 4a. First, the tree is navigated downstream from the root until a not fully expanded node is selected. Secondly, a new children node will be expanded from this node. Thirdly, a simulation takes place in which successive nodes are discovered through random actions until a terminal state is reached. Finally, the reward of this terminal state is computed and backpropagated to the upstream nodes to decide which ones are more worthy of exploring during the next selection process.

Regarding the workflow with “Pug”, the selection and backpropagation processes exclusively involve already explored nodes or states, so they take place invisibly inside of the MCTS component. The expansion and simulation processes involve the exploration of new nodes, which are discovered by modelling the environment in Grasshopper as the workflow in Fig. 5 shows.

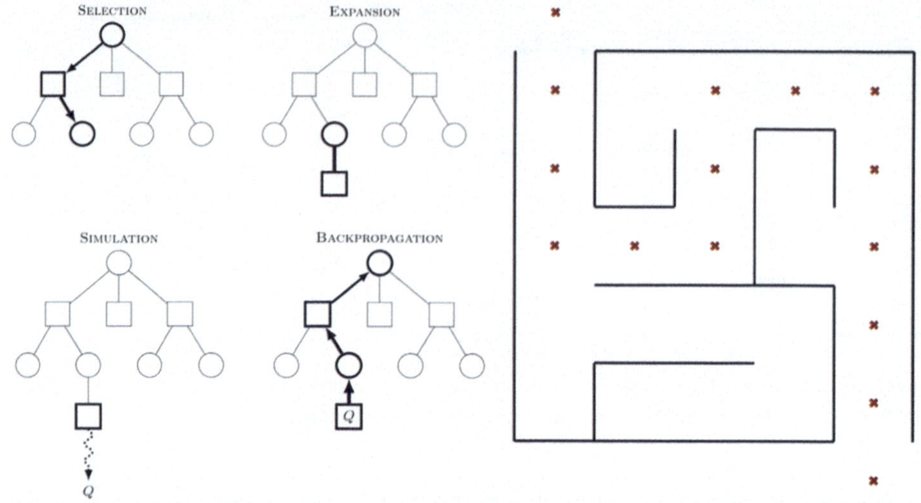

(a) (b)

Fig. 4. a) Diagram of the steps of the MCTS algorithm (created by Rmoss92 and distributed under a CC BY-SA 4.0 license) and b) Example with a labyrinth environment solved with the MCTS algorithm of "Pug" in Grasshopper.

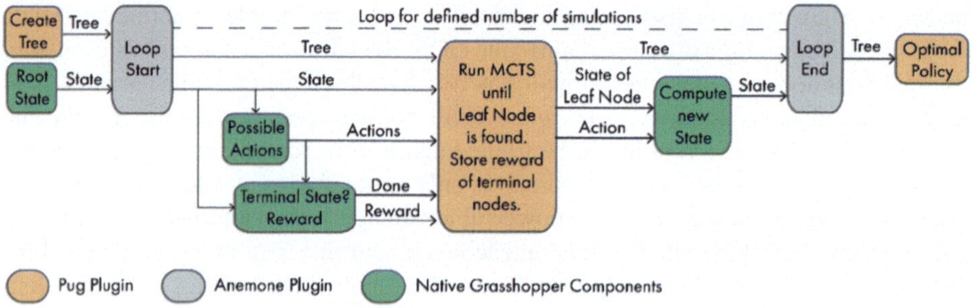

Fig. 5. Workflow with the MCTS agent.

2.3 Cradle-to-Cradle

These components help solve architectural problems involving the re-use of recycled elements. The first step would be the definition of these elements. Then they are compiled into a stock object. Finally, the potential matched elements for a certain member of the designed structure are selected from this stock according to geometric and structural requirements. Optionally, arbitrary weights can be assigned to the stock elements so the algorithm selects the less expensive ones first.

The selection requirements of steel elements are certainly straightforward. The selected elements need to be longer than the structural member and the cross section needs to be bigger than the provided minimum one, which should be determine from a static analysis as shown in Sect. 3.1. Since this check is not computationally expensive, all possible permutations are calculated and the one resulting in a minimum arbitrary weight is the final output of the Grasshopper component. Timber and other homogeneous materials with similar selection requirements can be analysed as well.

The selection requirements of bamboo poles provided with the Grasshopper components are slightly more complex. They are related to bamboo structures with bolted connections. In these cases, the connections must be located close to the nodes of the bamboo poles, because bamboo is stronger at these locations. This requirement can single-handedly strongly influence the selection process of bamboo poles, as it was the case with the structure defined in Sect. 3.2. Since this check is more computationally expensive in comparison to steel structures, all potential matches are output from the Grasshopper component and the user has to specify which bamboo pole will be used for each structural member.

3 Results

3.1 Adaptative Low Carbon Truss Structures

The aim of this challenge is to train a DQN agent that learns to produce the optimal truss configuration for a given stock of materials. The configuration of truss structures is defined by a sequence of binary actions, so that 0 implies that the next node will be closely placed and 1 that it will be widely spaced from the previous node (see Fig. 6). The reward function is inversely proportional to the resulting Global Warming Potential.

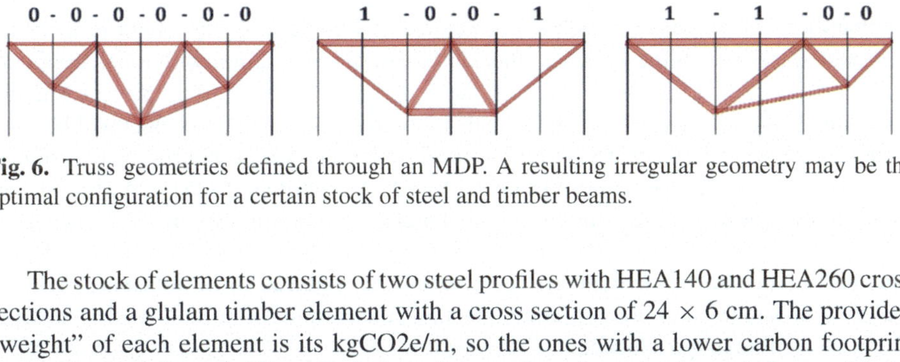

Fig. 6. Truss geometries defined through an MDP. A resulting irregular geometry may be the optimal configuration for a certain stock of steel and timber beams.

The stock of elements consists of two steel profiles with HEA140 and HEA260 cross sections and a glulam timber element with a cross section of 24 × 6 cm. The provided "weight" of each element is its kgCO2e/m, so the ones with a lower carbon footprint are preferably selected. The calculation of the carbon footprint is carried out with the Grasshopper plugin of One Click LCA (Apellaniz et al. 2021). The structural analysis that determines the minimum cross section for each member is carried out with the Karamba3d and Beaver plugins.

Table 1 shows the results for geometric configurations based on two different sequences of actions. A case with an assumed infinite length of the stock elements as well as one with limited lengths are included, which shows that the optimal truss configuration that results in a minimum carbon footprint strongly depends on the stock of available and preferably recycled elements.

Table 1. Global warming potential for truss structure configurations resulting from different sequence of actions. The DQN agent finds the optimal configuration for a given stock of steel and timber elements.

Sequence of actions	1 – 0 – 0 – 1	0 – 0 – 0 – 0 – 0 – 0
Unlimited Stock	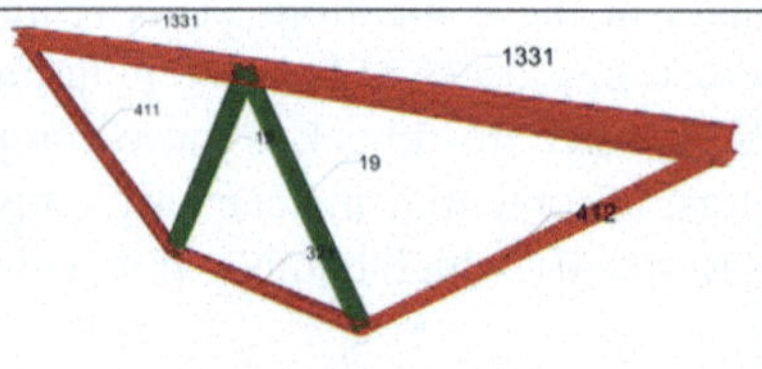 GWP = 3844 kgCO2e	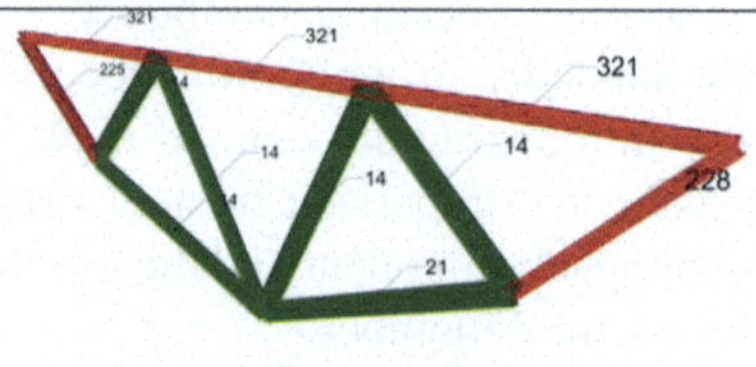 GWP = 1529 kgCO2e
Limited Stock	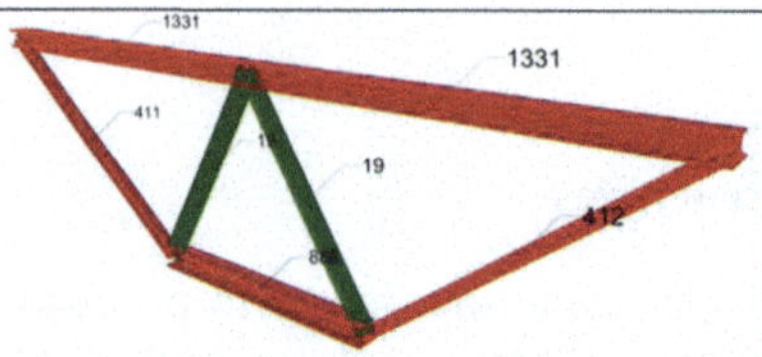 GWP = 4410 kgCO2e	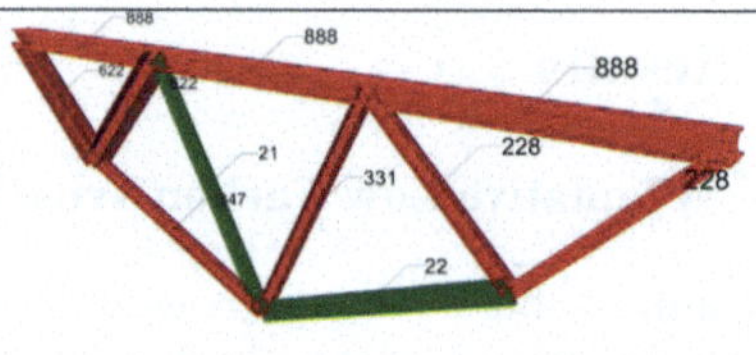 GWP = 5081 kgCO2e

3.2 Bamboo Hall in Maya Kaqchikel

The bamboo hall of Maya Kaqchikel (Guatemala) was designed and built in 2019 by a combined team from the UdK and the TU Berlin in cooperation with local architects and construction workers from Guatemala. A major challenge in this project was the selection of suitable bamboo poles for each structural element. This project will be used as case-study to train a MCTS algorithm to optimize the selection process of bamboo poles so it can be used for future projects. Firstly, an automatic selection process can considerably speed up the construction of bamboo structures. Secondly, if the agent reduces the number of required bamboo poles by using them several times for different structural elements, fewer bamboo poles would be required to be transported to the construction site from far away plantations; thus, the carbon footprint of such structures would be significantly reduced.

The problem is formulated as a MDP so in each step a bamboo pole is selected for each structural member. The stock of bamboo poles is defined with actual data from the bamboo poles of the construction site in Guatemala. The selection process is carried out with the Grasshopper components of "Pug" as explained in Sect. 2.3. The shorter the non-used remaining pole in a certain step, the bigger the score. Since this decision process has a fixed number of steps, unlike the problem described in Sect. 3.1, a GA can be also used in which the genes or parameters are the index of each bamboo pole used for each structural element. Finally, a heuristic approach is also possible in which the shortest compatible pole is selected at each step. The results of this comparison can be seen below and in Fig. 7:

- MCTS agent (10 min execution time): score 6.99 – all bamboo poles are assigned and used multiple times (see results in Fig. 6).
- MCTS agent (1 min execution time): score 3.89.
- GA Galapagos (10 min execution time): score 3.79.
- Heuristic approach: score 3.30.

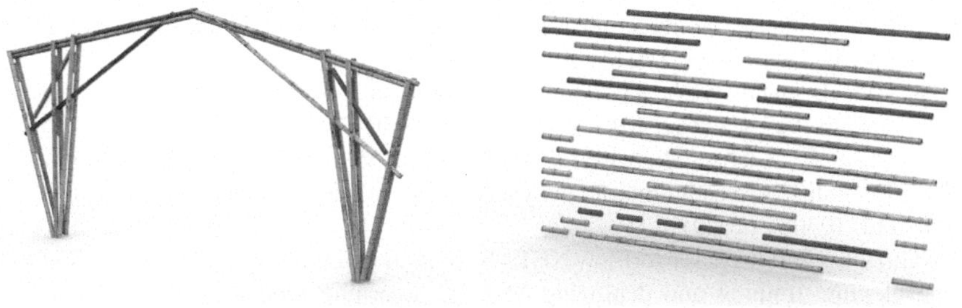

Fig. 7. Allocation of bamboo culms throughout all the members of a bamboo frame based on the optimal policy of the trained MCTS agent. The geometry of the structure and the bamboo poles are based on the bamboo hall in Maya Kaqchikel.

The MCTS agent has proven to be the best approach to this problem. The lower score of the GA is probably because the genes or parameters of the GA must consider all bamboo poles at each step and cannot be iteratively constrained to the compatible poles, which might result in impossible solutions and therefore the convergence of the learning process is much more difficult. The heuristic approach is not very effective either. Since it disregards the long-term reward, it runs out of compatible poles by using the apparently most appropriate ones at each step. Furthermore, a MCTS agent with a short training time is also able to yield sub-optimal results.

4 Conclusion

The provided case-studies have shown the potential of RL to tackle computational problems related to the AEC industry and to the reduction of the carbon footprint of our built environment. RL offers new design possibilities compared to traditional optimization approaches. In Sect. 3.1, geometries are generated through action sequences of variable lengths. The definition of the geometry is relatively basic, but it shows that there is no constrain to define problems based on a fixed number of parameters. In Sect. 3.4 reinforcement learning shows big potential for combinatorial problems with constrained possible solutions.

The only requirement for these problems is that they must be defined as sequential MDP, which can be achieved in a flexible approach with the countless possibilities of the Grasshopper environment. The AI agents included in the presented plugin "Pug" can be used to solve such problems. This tool is freely accessible and it will be further developed, so the reader is encouraged to use it for new challenges and hopefully help reduce the carbon footprint of our built environment.

References

Apellaniz, D., Pettersson, B.: Plugin "Pug" for Grasshopper. Available at: https://www.food4rhino.com/en/app/pug (2022)

Apellaniz, D., Pasanen, P., Gengnagel, C.: A holistic and parametric approach for life cycle assessment in the early design stages, SimAUD 2021 (2021)

Bilgin, E.: Mastering Reinforcement Learning with Python. Packt Publishing, Limited, [S.l.] (2020)

Belousov, B., et al.: Robotic architectural assembly with tactile skills: Simulation and optimization. Autom. Constr. **133**, 104006 (2022)

Cichocka, J.M., Browne, W.N., Rodriguez, E.: Optimization in the architectural practice. An International Survey. Caadria, 2017, April, 387–397 (2017)

Darko, A., Chan, A., Adabre, M., Edwards, D., Hosseini, M., Ameyaw, E.: Artificial intelligence in the AEC industry: Scientometric analysis and visualization of research activities. Autom. Constr. **112**, 103081 (2020)

GitHub: GitHub - SciSharp/TensorFlow.NET: .NET Standard bindings for Google's TensorFlow for developing, training and deploying Machine Learning models in C# and F#. [online] Available at: https://github.com/SciSharp/TensorFlow.NET (2022)

Jumper, J., et al.: Highly accurate protein structure prediction with AlphaFold. Nature **596**(7873), 583–589 (2021)

Medium: Reinforcement learning: deep Q-learning with Atari games. [online] Available at: https://medium.com/nerd-for-tech/reinforcement-learning-deep-q-learning-with-atari-games-63f5242440b1 (2022a)

Medium: Monte Carlo tree search in reinforcement learning. [online] Available at: https://towardsdatascience.com/monte-carlo-tree-search-in-reinforcement-learning-b97d3e743d0f (2022b)

Mnih, V., et al.: Human-level control through deep reinforcement learning. Nature **518**(7540), 529–533 (2015)

Silver, D., et al.: Mastering the game of Go without human knowledge. Nature **550**(7676), 354–359 (2017)

Zwierzycki, M.: Plugin "Owl" for Grasshopper. Available at: https://www.food4rhino.com/en/app/owl (2022)

Hybrid Immediacy: Designing with Artificial Neural Networks Through Physical Concept Modelling

Mathias Bank[1](✉), Viktoria Sandor[1], Robby Kraft[1], Stephan Antholzer[2], Martin Berger[2], Tilman Fabini[1], Balint Kovacs[3], Tobias Hell[2], Stefan Rutzinger[1], and Kristina Schinegger[1]

[1] Department of Design, i.sd | Structure and Design, University of Innsbruck, Innsbruck, Austria
{mathias.bank-stigsen,viktoria.sandor}@uibk.ac.at

[2] Department of Mathematics, University of Innsbruck, Innsbruck, Austria

[3] Center for Geometry and Computational Design, Vienna University of Technology, Vienna, Austria

Abstract. In digital design practice, the connection and feedback between physical and digital modelling is receiving increasing attention and is seen as a source of creativity and design innovation. The authors present a workflow that supports real-time design collaboration between human and machine intelligence through physical model building. The proposed framework is investigated through a case study, where we test the direct connectivity of physical and digital modelling environments with the integration of artificial neural networks. By combining 3D capturing tools and machine learning algorithms, the research creates an instant feedback loop between human and machine, introducing a hybrid immediacy that puts physical model building back at the centre of the digitally focused design process. By fusing physical models and digital workflows, the research aims to create interactivity between data, material and designer already at the early stage of the design.

Keywords: Computational immediacy · Early design phase · ANN · Physical model building · Hybrid design tool · 3D capturing · Machine intelligence

1 Introduction

Modelling is the core technique in any architectural design process. A model materialises design intentions and "objectifies" them by embedding design knowledge in the object (Oxman 2008). Herbert Stachowiak defined the general nature of models through key characteristics: They are illustrations of content that they are abstracting or "abbreviating" in order to record only the aspects that are relevant to their user. They are designed for a special purpose or task and are therefore evaluated according to their "usefulness" for the model-maker (Stachowiak 1973).

The original version of this chapter was previously published non-open access. The correction to this chapter can be found at https://doi.org/10.1007/978-3-031-13249-0_48

C. Gengnagel et al. (Eds.): DMS 2022, Towards Radical Regeneration, pp. 13–23, 2023.
https://doi.org/10.1007/978-3-031-13249-0_2

The “concept model” has a very unique role in the design process, it does not illustrate a design outcome but the initial and essential design intention. It is a tool in itself for materialising and communicating design ideas that keeps a purposeful ambiguity in order to leave space for imagination and further development[1]. According to Vera Bühlmann “models maintain a relation with ideas, and seek to sustain and communicate their power”; they do not determine the concept but rather enrich it through a “surplus capacity” (Bühlmann 2013). In the research, the concept model is an instrument for communication and exploration and in that sense a medium and a tool at the same time.

The feedback between physical and digital modelling has received heightened attention as a source of design innovation (Stavrić et al. 2013). The integration of the physical model into digital workflows opens the door for interactivity between data, material and designer (Thomsen and Tamke 2012). A common interest in many physical-digital experimentations is to increase creative capacity and immediate/intuitive control of the process without the need to explicitly define underlying geometrical rules of design objects as Mario Carpo suggested designers might directly „use“ chunks of scanned objects (Carpo 2017). Combined with artificial neural networks and growing archives of 3D point clouds of objects this vision could reach a further dimension - designers will not simply sample chunks, but they will be able to learn from spatial objects, deduce features and apply them to architectural designs without remodelling.

1.1 Related Work

Due to the continued introduction of new 3D capturing tools and scanning devices, there are promising opportunities to re-integrate physical design processes into digital design workflows. In the field of Reverse Engineering (RE), scanning technologies are used to build digital models from physical objects (Hsieh 2015). Although scanning technologies have been around for decades, their performance has improved significantly in recent years. After Microsoft developed Kinect in 2010 a much more affordable scanning device than LIDAR could offer, there has been a growing interest in its use in architectural modelling. While some research projects were focusing on the materialisation of the sensed point clouds in digital environments (Hsieh 2015), others utilised Kinect devices as navigation tools in VR environments allowing immediate ways of connecting the physical to the digital (Souza et al. 2011).

While connecting the physical with the digital through scanning devices is crucial for hybrid concept modelling, the project aims additionally to introduce artificial neural networks as interpreters in the design process. Some of the earliest attempts to implement AI as a participatory system in the early stage of design are the projects by John and Jane Frazer presented in the book An Evolutionary Architecture (Frazer 1995). In their work, the role of AI is to recognize patterns and react to the design suggestions, generating a responsive design loop. A more contemporary take on this approach is developed by SPACEMAKER where immediacy and real-time feedback become a key feature of human-machine collaborative design (Jeffrey et al. 2020). Although this project situates the process entirely in the digital space, the input of ANNs can be channelled in various

[1] A famous example of such a concept model is the digital model for the Moebius house by UN Studio. See image in (Hirschberg et al., 2020).

captured formats of the physical space. Just like the project, Deep Perception by Fernando Salcedo (Leach 2021) showcases through real-time cameras how captured folded textiles can be directly connected to a trained network.

1.2 Objectives

The authors present a workflow that supports real-time design collaboration between human and machine intelligence through physical model building. In the tested use case the immediate connectivity of the physical and digital modelling environments is challenged through artificial neural networks. Kinects are used as 3D capturing devices and a machine learning network capable of processing 3D point cloud data directly from the Kinects is established.

2 Conceptual Overview

The following workflow (Fig. 1) outlines a real-time design collaboration between human and machine intelligence through physical model building. The proposed workflow is centred around a physical installation consisting of a large black table, two Kinects, ring lights, a screen, an Arduino control panel and a computer (Fig. 2) - and is organised into four interlinked modules:

- Build
- Capture
- Machine Learning
- Post-processing

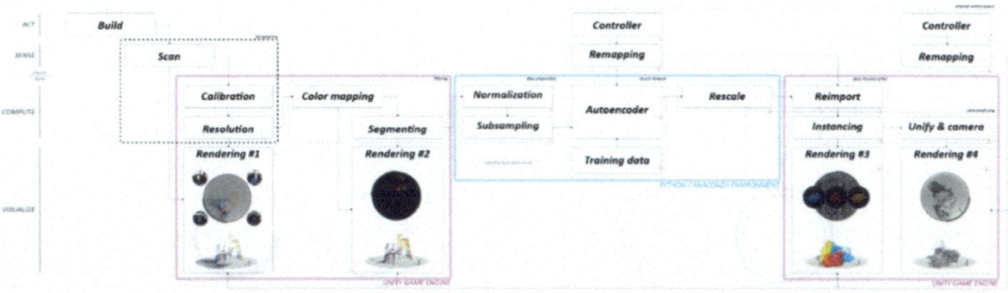

Fig. 1. Diagram of the implemented workflow. From left to right, the build, capture, machine learning and post-processing modules.

The game engine Unity is used as the platform for connecting the components in real-time while allowing users to switch between four different visualisation modes to see the immediate outcome of each step.

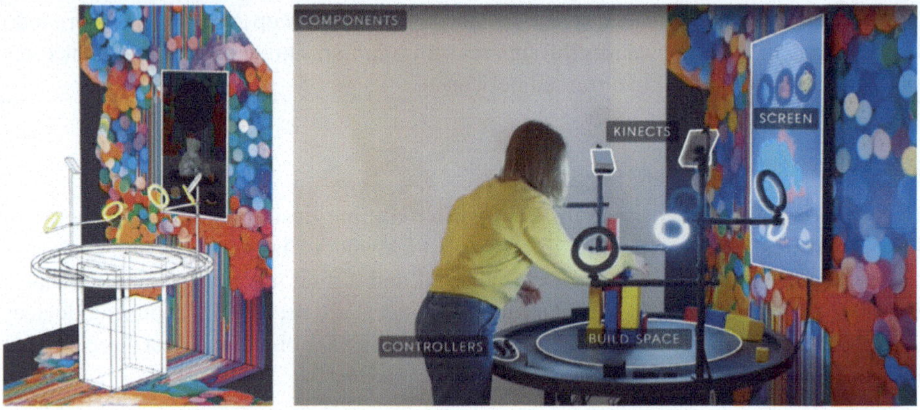

Fig. 2. Illustration and image of the physical installation depicting the main components: A large black table, two Kinects, ring lights, a screen, an Arduino control panel and a computer.

2.1 Build/Physical Setup

The build module is centred around a custom table (Fig. 2) where users can assemble architectural compositions from physical blocks in different colours. To test the workflow, three different colours were selected and applied to a limited number of different building blocks. Each colour represented a unique geometric group, which, through 3D capturing, could be manipulated by an artificial neural network on screen. A control panel embedded in the custom table was used to control the interpolation factor and target, while an additional three potentiometers controlled the scale, rotation and selection of the four visualisation modes (Fig. 3).

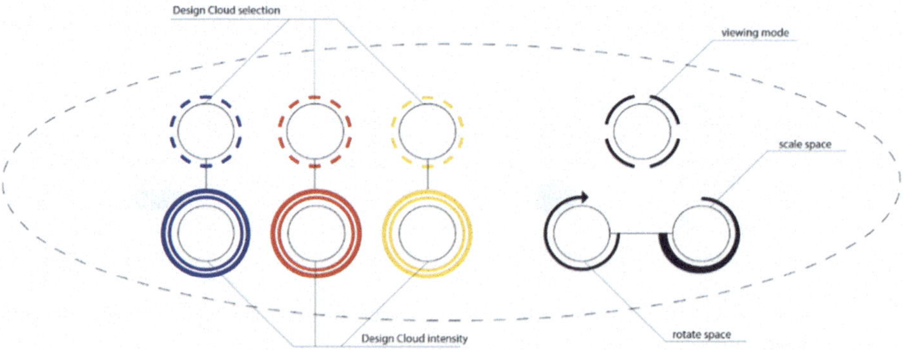

Fig. 3. The graphic depicts the control panel embedded within the custom table.

2.2 Capture

The system of capturing and processing the points required accommodating the Microsoft Azure Kinect and its SDK. A custom application was built using Unity because the Azure Kinect SDK's C# bindings interfaced seamlessly with Unity's C# front end-user code, and Unity supports custom shaders, including compute shaders, which allowed us to make heavy optimizations during the point cloud processing phase.

Homography

We ran a simple homography to virtually align the Kinect cameras in a shared 3D space. Our homography was a custom algorithm that required one scalene triangle to be placed in the centre of the table. The triangle was detected using Sobol edge detection (Kanopoulos et al. 1988), and the Kinect's depth camera was used to bring these edges into 3D. The triangles were then aligned, resulting in one 4×4 position and orientation matrix for each camera.

Point Clouds

Each Kinect device comes with one depth camera and one colour camera. A compute shader converts these raw Kinect images into a point cloud. First, each depth camera pixel is treated as a magnitude that scales its corresponding 3D vector (precomputed in a lookup table) and then this resulting point is transformed by the camera's homography matrix so that each camera's point clouds appear in the same 3D space. This is the first visualisation mode, the "raw Kinect point cloud", where each point in the cloud contains the colour as seen by the colour camera.

Colour Segmentation

The next step uses another compute shader to filter the points, both according to colour and position. The position filtering is simple: we crop points that lie outside a vertically-aligned cylinder with an infinite height and a specified radius that matches the circular installation table in physical space. The colour filtering works by analysing the colours according to the HSV values. First, the user provides a set of reference colours and $\pm$ ranges on all three HSV dimensions which define a bounding box with the reference colour at its centre. The shader rejects points based on their relations to the bounding boxes. The colours that match are categorised and coloured according to their reference colour. This is the second visualisation mode, the "segmented colour clouds". Additionally, these segmented colour point cloud arrays contain the data which is sent to the machine learning application.

2.3 Machine Learning

An artificial neural network is used to transform the Kinect scans of the physically assembled architectural compositions through a user-selected design point cloud. In the workflow, machine learning performs the following steps. We get a preprocessed point cloud from a Kinect scan and a selection of a design point cloud that was chosen by the user together with an interpolation factor. First, we cluster the point clouds into equal-sized clusters with ($n = 2048$) points and then apply the encoder to each of them.

The next step is to interpolate between each cluster from the Kinect point cloud and the design point cloud and then apply the decoder to get new (clustered) point clouds. Finally, the individual clusters are reassembled into one big point cloud. This process can also be applied to multiple point clouds simultaneously. In our case, the setup processes three separate instances at a time which encode the three colours of the used building blocks.

Training Data

A collection of more than 10000 3D point clouds were assembled to provide the training data for the machine learning network. Each point cloud was part of a category of geometric structures (Fig. 4) from which the design point clouds were selected. The intent with the used training data, was to apply the geometric structures of the selected design point clouds within the captured Kinect point clouds - enhancing the appearance of physically assembled compositions on the table.

Fig. 4. Training data and a subset of data from three categories From left to right: Basic volumes, orthographic planes and lattices.

Autoencoder

Our approach is based on the FoldingNet introduced by Yang et al. (2018). However, we make some adjustments that yield better results for our applications. First, the approach uses a 3D grid for the fixed folding input instead of a 2D grid and secondly an extended loss function was implemented, resulting in the biggest improvement. Namely, we use the Chamfer distance (as in the original) together with the Wasserstein (or earth-mover) distance which we calculate due to Feydy et al. (2019) (Fig. 5).

Thus we get a neural network that yields good abstract representations of point clouds. Since these are just vectors we can process them however we want and then use the decoder part to reconstruct a new point cloud. To be more precise we linearly interpolate between two points i.e. two outputs of the encoder and then apply the decoder yielding a new point cloud, resulting in a mix of the two original point clouds.

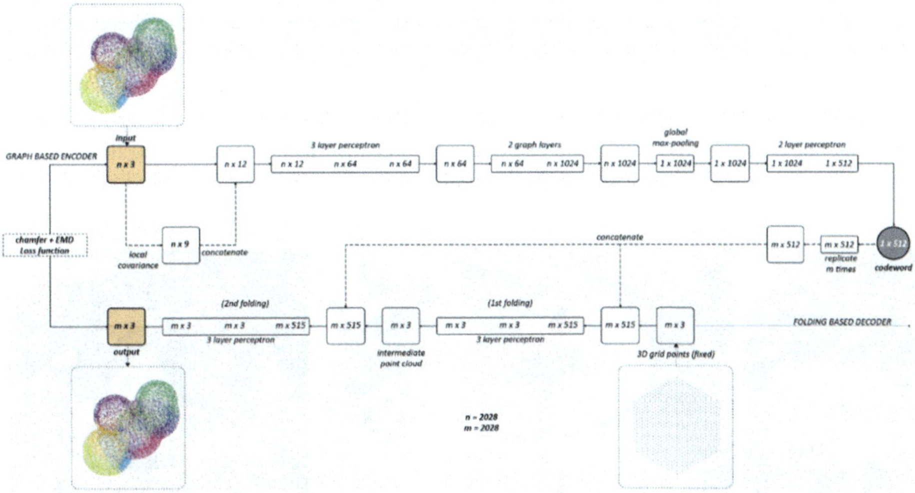

Fig. 5. FoldingNet autoencoder structure based on Yang et al. (2018).

2.4 Post-processing

After the Kinect data has gone through the artificial neural network, the three transformed points clouds are streamed back into Unity for post-processing. Here the representation of the points can be enhanced to better visualise the potential within the user-assembled concept model. This allows for the implementation of a variety of different strategies and algorithms to be applied according to the design use case.

3 Performance

The case study of the research is a hybrid instrument, a concept modelling station, tested in two different environments. The first, by being an interactive exhibition space, gives access to a wider audience for hybrid collaborations, while the second implements the workflow of the modelling station in a professional design context.

3.1 Exhibited Station

By placing the concept modelling station within an exhibition space (Fig. 6), we aimed to test its durability and user-friendliness with a non-professional audience. The station was exhibited in a gallery for three months, where it operated six days a week, ten hours a day. The age range of the visitors was very diverse, with guided tours from pre-school children to university staff.

The interactive aspect of the installation was very popular in this setup. The building blocks were in constant use, while the circular shape of the table allowed up to 4–5 users to collaborate on one model. Although over the three months the Kinect devices had to be recalibrated due to the movement of the drafting table to which they were attached, the real-time connection between the Kinect and Autoencoder was never broken. The network processed and generated three point clouds simultaneously every three seconds. While the number of navigation options via the control panel was quite limited compared to any digital 3D environment (turn-table and scale function), visitors often found it difficult to grasp the buttons' functionalities. In terms of "immediacy", the hybrid station met our expectations, but when we applied a higher interpolation factor to the scanned clouds, the resulting distortions made the real-time representation less obvious.

Fig. 6. Three images from the exhibition, where the concept modelling station was tested by a non-professional audience.

3.2 Tower Design

To test the workflow and the modelling station in an architectural design context, a brief for a tower design was defined. This more professional format of the study aims to explore the benefits of a collaborative, yet hybrid design approach at the early stages of the design. The design process starts with composing coloured wooden blocks on the table. In the example (Fig. 7), the colours are used to define different geometric characteristics in the compositions. As the blocks are placed on the table, the displayed point clouds give an alternative view of the composition. Although the building blocks and the table size limit the scale of the model, the scalability of the digital model allows for spatial interpretations regardless of the physical size of the blocks. The designers can select and assign specific 3D patterns, porosities, etc. to each of the assembled colour compositions on the table. The autoencoder (ML), controlled by the user modifies the compositions through the interpolation of selected features. The resulting Machine-interpreted point clouds are displayed on the screen in real-time.

During the collaborative tower sketching process a variety of the different compositions - both the original and the machine interpreted - were frequently saved in.xyz point cloud format. This opened up the integration into existing design software for further processing of promising concepts.

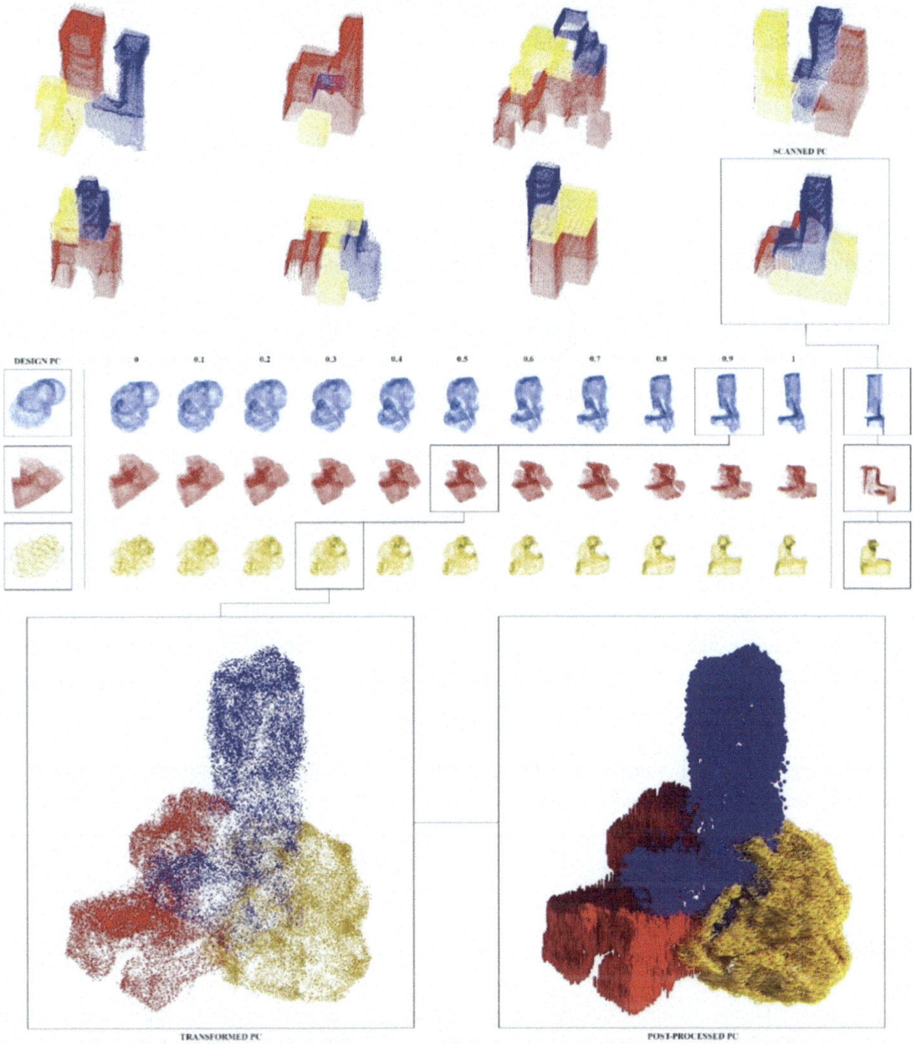

Fig. 7. Top: Variation of scanned concepts for a tower design. Middle: Transformation toward selected design point cloud via autoencoder. Bottom left: Outcome of the transformation. Bottom right: Post-processed point cloud in Unity. The depicted process is running in real-time and is responsive.

4 Conclusion

In this paper, we present a workflow that facilitates a real-time interaction with an artificial neural network through physical modelling. The user-assembled physical objects allows for an almost instant design dialogue with the trained data of the neural network through the support of capturing devices and point cloud notation. This establishes an immediate feedback loop between human and machine intelligence, introducing a hybrid immediacy that reintegrates physical model building at the centre of a digital-focused design process. The work shows the potential of using Kinects to capture the physical boundary of a model, while simultaneously using machine learning to apply selected geometric structures, and 3D patterns to it. The immediacy of the setup creates an intuitive way to physically search for conceptual design ideas without the need to remodel design features digitally.

However, during this research, we have encountered a number of limitations that need to be acknowledged. To precisely capture a physical composition from all angles, two Kinects provide a limited field of view, resulting in several unscanned areas when the physical models get more complex. This could be remedied by more Kinects but would result in an increased performance cost. Concerning the developed artificial neural network, the challenges are equally conceptual and technical. Though the integration of the network with the scanned boundary compositions works very well, the conceptual idea of blending a selected geometric structure into it was difficult to execute technically. Partially, this is a result of the inherent dilemmas within computing on large datasets, where the network gets biassed towards the most common data.

Nevertheless, the work is still in its early phases, and further developments to the demonstrated workflow could minimise the listed limitations and significantly expand on its useability within a design scenario. Additionally, we plan to expand the interactions with the produced concept models with the use of a mixed reality (MR) 3D sketching application. The sketching app combines a familiar stylus-on-tablet input and interaction paradigm with MR capabilities to allow sketching directly in 3D space, thus providing additional tools for amending and refining the concept models in real-time.

Acknowledgements. This work was funded by the Austrian Science Fund (FWF) project F77 (SFB "Advanced Computational Design").

References

Bühlmann, V.: Die Empörung des Modells // models, outraged" – Zhdk conference "David and Goliath – Models between art and architecture. https://monasandnomos.org/2013/11/19/die-emporung-des-modells-models-outraged-abstract-for-my-paper-at-the-zhdk-conference-david-and-goliath-models-between-art-and-architecture/ (2013)

Carpo, M.: The Second Digital Turn: Design Beyond Intelligence. The MIT Press, Cambridge (2017)

Frazer, J.: An Evolutionary Architecture. Architectural Association (1995)

Hirschberg, U., Hovestadt, L., Fritz, O. (eds.): Atlas of Digital Architecture: Terminology, Concepts, Methods, Tools, Examples, Phenomena. Birkhauser (2020)

Hsieh, C.T.: A new kinect-based scanning system and its application. Appl. Mech. Mater. **764–765**, 1375–1379 (2015). https://doi.org/10.4028/www.scientific.net/AMM.764-765.1375

Jeffrey, L., Hakon, D., Hakon, F., Stanislas, C.: (2020)

Kanopoulos, N., Vasanthavada, N., Baker, R.L.: Design of an image edge detection filter using the Sobel operator. IEEE J. Solid-State Circuits **23**(2), 358–367 (1988). https://doi.org/10.1109/4.996

Leach, N.: Architecture in the Age of Artificial Intelligence: An introduction to AI for architects. Bloomsbury Publishing Plc (2021). https://doi.org/10.5040/9781350165557
Oxman, R.: Digital architecture as a challenge for design pedagogy: theory, knowledge, models and medium. Des. Stud. **29**, 99–120 (2008). https://doi.org/10.1016/j.destud.2007.12.003
Souza, L., Pathirana, I., Mcmeel, D., Amor, R.: Kinect to Architecture (2011)
Stachowiak, H.: Allgemeine Modelltheorie. Springer, Vienna (1973). https://doi.org/10.1007/978-3-7091-8327-4
Stavrić, M., Sidanin, P., Tepavčević, B.: Architectural Scale Models in the Digital Age: Design, Representation and Manufacturing. Springer (2013)
Thomsen, M.R., Tamke, M.: The active model: a calibration of material intent. In: Persistent Modelling: Extending the Role of Architectural Representation, 1st edn. Routledge (2012)

Structural Form-Finding Enhanced by Graph Neural Networks

Lazlo Bleker[1](✉), Rafael Pastrana[2], Patrick Ole Ohlbrock[3], and Pierluigi D'Acunto[1]

[1] School of Engineering and Design, Professorship of Structural Design, Technical University Munich, Arcisstr. 21 80333, Munich, Germany
Lazlo.bleker@tum.de

[2] Princeton University, Form Finding Lab, Princeton, USA

[3] Institute of Technology in Architecture, Chair of Structural Design, ETH Zurich, Zurich, Switzerland

Abstract. Computational form-finding methods hold great potential for enabling resource-efficient structural design. In this context, the Combinatorial Equilibrium Modelling (CEM) allows the design of cross-typological tension-compression structures starting from an input topology diagram in the form of a graph. This paper presents an AI-assisted design workflow in which the graph modelling process required by the CEM is simplified through the application of a Graph Neural Network (GNN). To this end, a GNN model is used for the automatic labelling of edges of unlabelled topology diagrams. A synthetic topology diagram data generator is developed to produce training data for the GNN model. The trained GNN is tested on a dataset of typical bridge topologies based on real structures. The experiments show that the trained GNN generalises well to unseen synthetic data and data from real structures similar to the synthetic data. Hence, further developments of the GNN model have the potential to make the proposed design workflow a valuable tool for the conceptual design of structures.

Keywords: Structural form-finding · Combinatorial Equilibrium Modelling · Equilibrium-based design · Machine learning · Graph Neural Networks

1 Introduction

A substantial proportion of the embodied carbon of buildings and infrastructures is attributable to their supporting structures. Therefore, it is of great interest to minimise the material use of such load-bearing systems (Kaethner and Burridge 2012). The potential impact of designers on the outcome of the design process, thus including the embodied carbon of buildings, is the highest in the conceptual design phase, and it steadily decreases as the design progresses (Paulson 1976). The field of structural form-finding deals with methods that allow the transformation of form and forces to generate and explore efficient structural geometries, making it particularly suited for the conceptual design stage. Thus, applying structural form-finding holds great potential for decreasing the embodied carbon of new buildings and infrastructures. However, manually exploring the entire design space of solutions offered by form-finding methods is often infeasible.

C. Gengnagel et al. (Eds.): DMS 2022, *Towards Radical Regeneration*, pp. 24–35, 2023.
https://doi.org/10.1007/978-3-031-13249-0_3

Employing the ability of machine learning to extract information from large amounts of data could increase the potential impact of structural form-finding on the design process even more. Machine learning has already been applied to parametric structural exploration (Danhaive and Mueller 2021) as well as structural design through form-finding (Ochoa et al. 2021). This paper introduces a novel conceptual structural design workflow that relies on the use of Graph Neural Networks (GNN) to enhance the structural form-finding process.

1.1 Combinatorial Equilibrium Modelling (CEM)

Form-finding approaches such as the force density method (FDM), are tailored for structures in pure compression or tension (Vassart and Motro 1999). On the contrary, the Combinatorial Equilibrium Modelling (CEM) (Ohlbrock and D'Acunto 2020), a form-finding method based on vector-based graphic statics (D'Acunto et al. 2019) and graph theory, is best-suited for mixed tension-compression structures. The primary input for the CEM is the *topology diagram*, a graph representing the structure's topology (Fig. 1). A graph is a mathematical structure consisting of a set of *nodes* interconnected by *edges*. In a CEM topology diagram, nodes contain embedded information about the presence of supports or applied external forces. Edges are labelled either as *trail edges* (Fig. 1 - purple lines) or *deviation edges* (Fig. 1 - orange lines), where the former embed metric data related to their length and the latter related to their signed internal force magnitude.

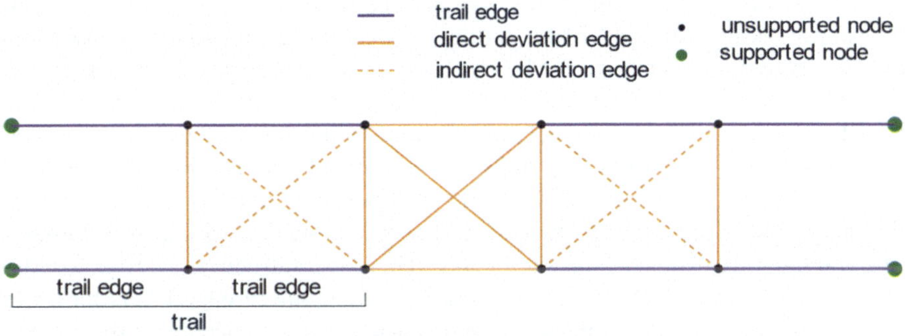

Fig. 1. Example of a CEM topology diagram and its constituting elements.

Creating a valid topology diagram that can be processed by the CEM requires the user to be knowledgeable about the correct labelling of the edges and nodes (Ohlbrock and D'Acunto 2020). As shown in Fig. 1, trail edges are connected end-to-end, forming chains of multiple edges (*trails*). For a topology diagram to be valid, every node must be part of exactly one trail, and each trail must contain exactly one *supported node* at one of its ends. A deviation edge connects a node of one trail to any node of another trail. Within deviation edges one can distinguish *direct deviation edges* and *indirect deviation edges*. Direct deviation edges connect nodes with an equal *topological distance* – the number of trail edges separating two nodes measured along a trail – to the supported nodes of their respective trails. Indirect deviation edges connect nodes with an *unequal* topological distance to their corresponding supported nodes. Manipulating the topology diagram and its embedded metric information allows generating a large cross-typological design space of efficient structural geometries.

1.2 Graph Neural Networks (GNNs)

Graph Neural Networks (GNNs) are a type of neural network that can operate on graphs (Zhou et al., 2020), such as the topology diagram of the CEM. Nodes and edges within a graph may contain embedded information, referred to as node- and edge features, respectively. GNNs employ *convolutional layers* similar to those found in deep learning. During a single forward pass of the network, each convolutional layer updates the node embedding as a function of its *neighbourhood* i.e., all directly connected nodes. GNNs have been applied to a wide range of application domains in which data can be represented as graphs such as molecules (Kearnes et al. 2016), social networks (Wu et al. 2019), and scene representations (Qi et al. 2018). In structural design, GNNs have been used as a surrogate model for truss FEM analysis (Whalen and Mueller 2021), the design of cross-section sizes (Chang and Cheng 2020) as well as the form-finding of funicular shell structures (Tam et al. 2021).

1.3 Objective: Enhancing CEM with GNN

The efficacy of the CEM is often held back in practice by the somewhat unintuitive nature of operating at the level of the topology diagram. One of the most challenging tasks when working with the CEM is setting up a valid topology diagram based on an initial design concept and correctly labelling the edges as trail and deviation edges to allow direct control of the CEM model during the form-finding process. A brute force approach to finding a valid edge label mapping rapidly becomes infeasible as this would require evaluating 2^e possible mappings, where e equals the number of edges in the topology diagram. GNNs potentially present an opportunity to overcome this challenge of finding a valid edge mapping and as such supplementing the lacking human intuition for the graph setting with machine intelligence. This paper introduces a design workflow that uses a supervised GNN model to determine a mapping of edge labels for a CEM topology that is appropriate for the form-finding task at hand. In particular, the presented experiment focuses on the design of bridge structures.

2 Methodology

2.1 Machine Learning-Based Design Workflow

The proposed design workflow allows translating rough design ideas into actual structures that are ensured to be in equilibrium. Figure 2 shows a possible manifestation of this human-machine collaborative process enabled by the AI model presented in this paper. The process consists of four separate steps.

Wireframe Sketch (A). In the first step, the human designer sketches a wireframe model of the desired structure, including the supports. Creating this wireframe sketch requires minimal knowledge of statics as it does not need to represent a structure in equilibrium, although it could very well be inspired by existing structural typologies.

Unlabelled Topology Diagram (B). In the second step, the wireframe sketch is automatically converted to a graph as an unlabelled CEM topology diagram. Every line segment in the wireframe sketch is translated to an edge in the graph. Points of intersection of line segments in the wireframe model are converted into nodes in the graph. Nodes are labelled as either supported or unsupported. Figure 2b displays an example of a topology diagram visualised as a radial graph. The choice of how the graph is visualised on the 2D plane does not play a role in the outcome of the design workflow.

Labelled Topology Diagram (C). In the third step, a GNN model is trained with a synthetic dataset of topology diagrams generated from a generic ground topology. The GNN processes the unlabelled graph representing the topology diagram and outputs a binary edge label for each edge of the graph. The labels produced for each edge by the GNN correspond to one of two classes being either a trail edge, or a deviation edge. Combining these edge labels with the previously unlabelled topology diagram results in a labelled topology diagram.

Structural Model (D). Provided that the labelling produced by the GNN is a valid CEM topology diagram (all edges are labelled correctly), this now complete topology diagram can be used for the CEM to produce a structural model in equilibrium containing both members in compression (blue) and tension (red). In this final step, the human designer is free to choose and modify the metric parameters of the model: lengths of trail edges and signed internal force magnitudes of deviation edges as well as points of application, magnitude, and direction of external loads.

2.2 Graph Neural Network for CEM Label Prediction

A synthetic dataset of 50,000 valid labelled topology diagrams is created for training the GNN. Topology diagrams are based on a parametrically generated ground topology, the size of which is defined according to two variables:

- the number of trails N_T (randomly sampled even integer between 2 and 14);
- the number of trail edges N_E per trail (randomly sampled integer between 1 and 15).

Four different sets of edge types are then activated in the ground topology (Fig. 3). Firstly, all possible trail edges (purple continuous lines) are activated to form a set of trails with an equal number of trail edges. Secondly, a random subset of all possible pairs of trails is selected; based on this selection, a subset p_d (between 0 and 1) of all possible direct deviation edges is activated. Thirdly, a random subset p_c of all possible direct centre deviation edges is activated (orange dotted lines); direct centre deviation edges are defined as the direct deviation edges at the centre of the diagram connecting

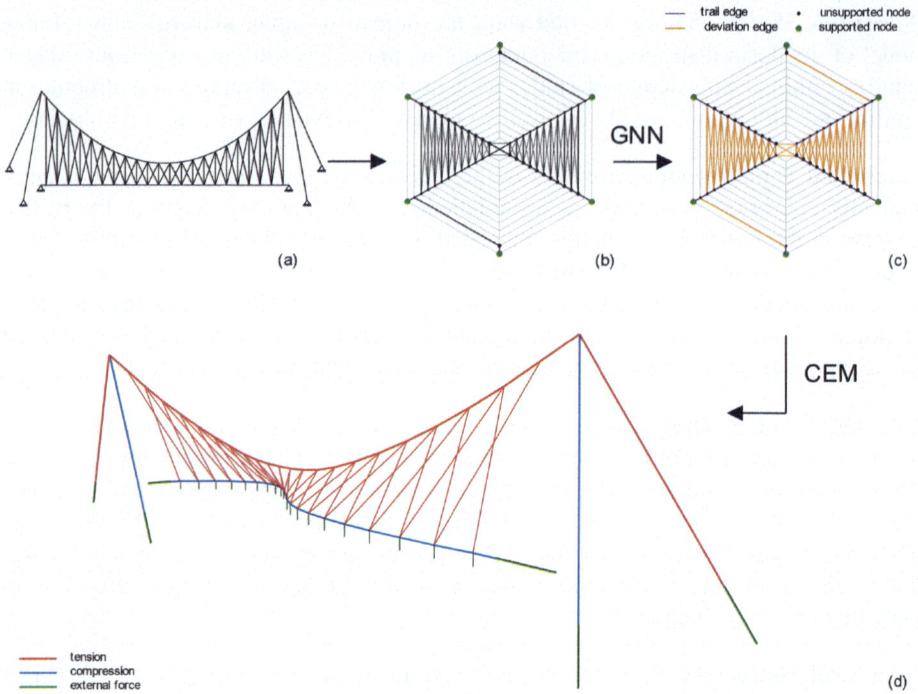

Fig. 2. Proposed design workflow: (a) Wireframe model sketched by the human designer. (b) Unlabelled topology diagram automatically generated from the sketch. (c) Valid labelled topology diagram produced by the GNN. (d) Structural model in equilibrium created by the CEM according to the designer's input of metric parameters.

the unsupported extremes of opposite trails. Finally, a similarly drawn random subset p_i of indirect deviation edges is activated (orange dashed lines).

Node embeddings are one-hot encodings indicating the class of the node being either a supported node or an unsupported node. Every edge includes a target label for training, as either trail edge or deviation edge. Figure 4 shows a set of topology diagrams randomly produced by the synthetic data generator.

Figure 5 shows the structure of the graph neural network used for the prediction of the edge labels. The model has been implemented in Python with Pytorch Geometric (Matthias and Lenssen 2019). Because any edge can belong to one of 2 classes (trail edge or deviation edge), binary cross-entropy loss is used for the loss function. During training, models are evaluated using the Matthews Correlation Coefficient (MCC), which is suitable for the imbalanced number of trail- and deviation edges in the dataset as well as independent of which of both classes is defined as the positive class (Chicco 2017). The MCC ranges from -1 to 1, where 1 represents a perfect classifier and -1 a perfect misclassifier. Through a hyperparameter study, it has been found that for accurate edge label prediction the number of convolutional layers should be equal to or larger than the maximum number of trail edges per trail N_E in the training dataset (Fig. 6). This can be explained by the fact that if the number of convolutional layers is less than N_E, information about the existence of the supported node does not reach every node of the graph. The generated dataset contains graphs with up to $N_E = 15$ trail edges per trail; therefore, the main part of the GNN consists of 15 sets of a TransformerConv (Shi et al. 2020) convolutional layer followed by a Rectified Linear Unit (ReLU) activation function. The size of the input sample of the first convolutional layer is necessarily equal to the number of node features – which are one-hot encodings of the two possible node classes – and is therefore equal to two. The size of input and output samples for consecutive convolutional layers is set to increase linearly up until 300 at the interface between layer 8 and 9, after which it is kept constant for the following layers.

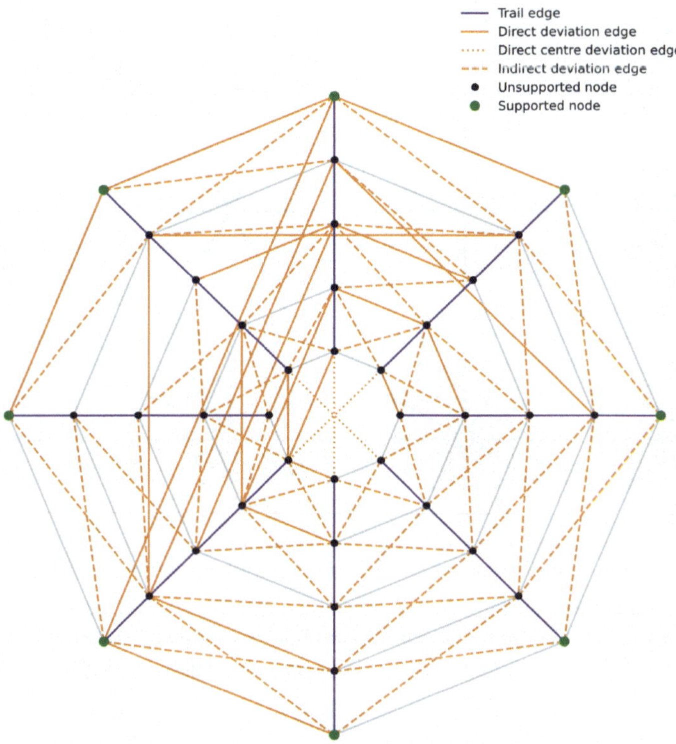

Fig. 3. Representative topology diagram with different groups of activated edges of the synthetic generator indicated. Non-activated edges of the ground topology are shown in grey.

As shown by the hyperparameter study (Fig. 7), the MCC increases steadily for maximum output sizes up to around 300 and no longer rises significantly at larger output sizes. Because creating an edge label mapping for the full graph is an edge-level task, the convolutional layers are followed by an edge pooling layer based on the dot product of adjacent node embeddings. Finally, these edge embeddings are subjected to a dropout layer ($p = 0.2$) and followed by a linear classifier with an output vector of length two. The *argmax* of this vector predicts the class to which each respective edge belongs.

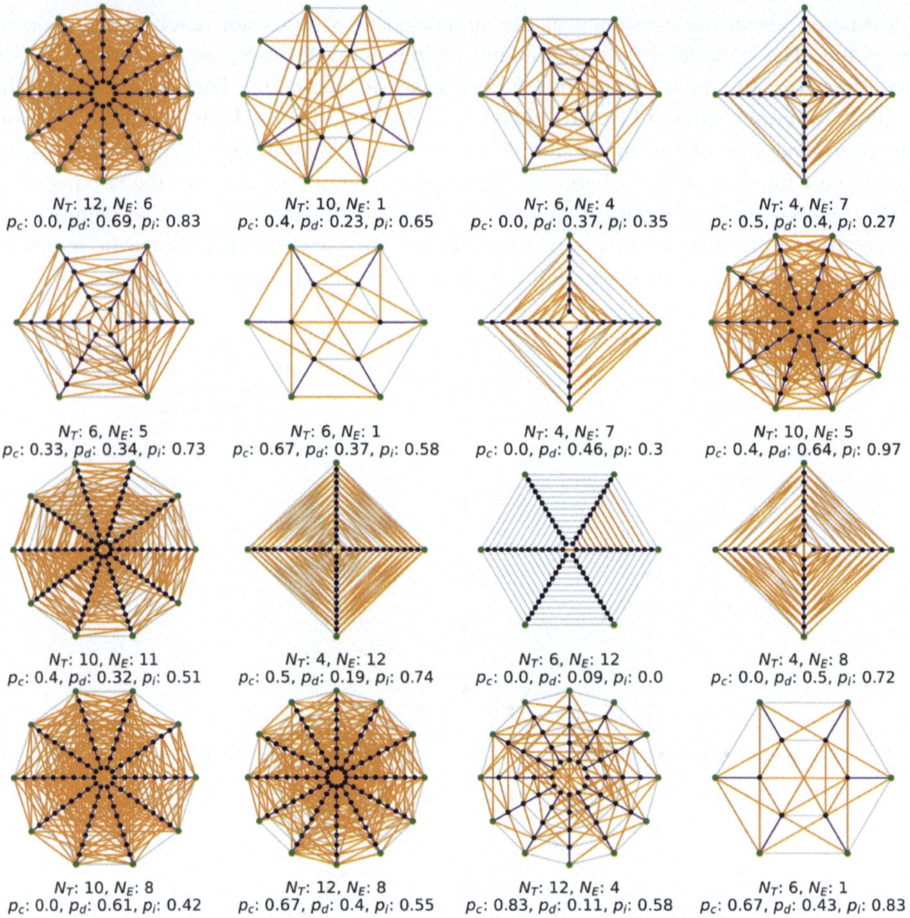

Fig. 4. 16 random graphs generated by the synthetic data generator. N_T = number of trails, N_E = number of trail edges per trail, p_c = fraction of activated centre deviation edges, p_d = fraction of activated direct deviation edges, p_i = fraction of activated indirect deviation edges.

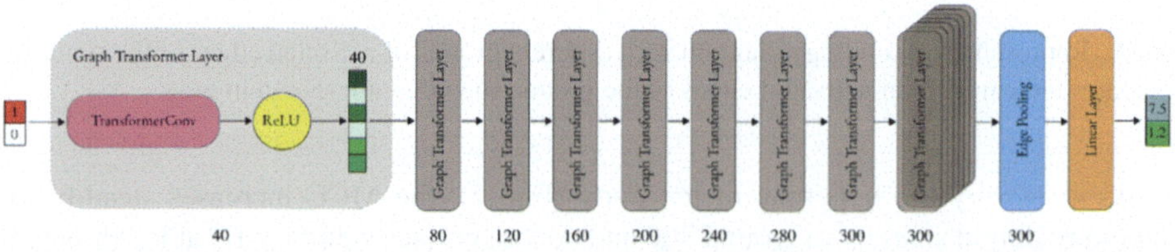

Fig. 5. The CEM edge classifier consists of a 15-layer graph neural network (GNN) followed by an edge pooling- and a linear layer.

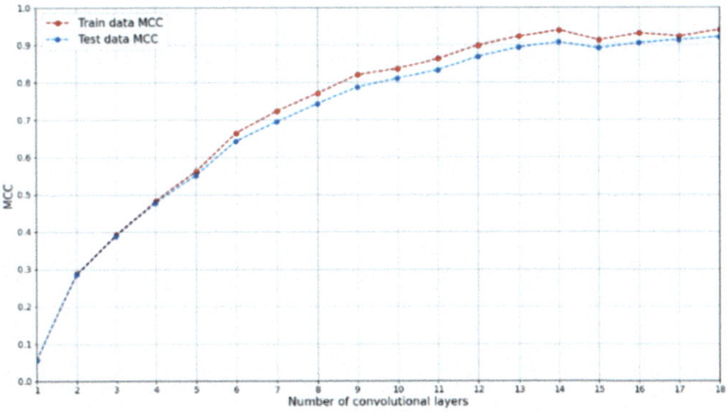

Fig. 6. Influence of the number of convolutional layers on the Matthews Correlation Coefficient (MCC) related to the edge labelling.

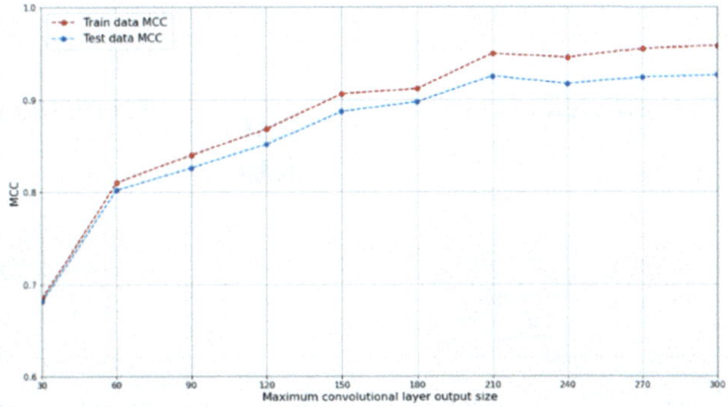

Fig. 7. Influence of maximum convolutional layer output size on the Matthews Correlation Coefficient (MCC) related to the edge labelling.

The GNN model is trained on 70% of the synthetic dataset (35,000 graphs; 14.6 million edges) and continuously tested on 15% (7,500 graphs; 3.1 million edges) for 150 epochs. Figure 8 shows the edge-labelling MCC history for both these training and testing parts of the dataset. At the end of training, the train data MCC is 0.91, and the test data MCC is 0.89, at which point it no longer significantly improves. The remaining 15% of the dataset is used as a validation set and also reaches an MCC of 0.89. It is worth noting that the number of potential graphs that the synthetic generator can produce is much larger than the number of graphs included in the training dataset. Despite the relatively modest training dataset size, the MCC on the test dataset is comparable with the MCC on the train dataset. This indicates that the trained GNN model generalises to unseen synthetic data very well. Figure 9 shows the confusion matrices for the train and test datasets for the trained GNN model. From the test data matrix, it can infer that 91.5% of unseen trail edges and 97.3% of unseen deviation edges are labelled correctly by the trained GNN model.

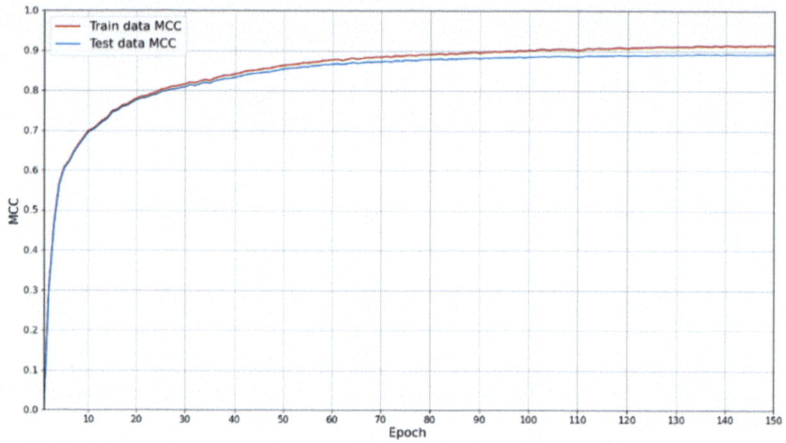

Fig. 8. Training and testing Matthews Correlation Coefficient (MCC) history on a synthetic dataset of 50,000 topology diagrams using the Adam optimizer with a learning rate of 0.001.

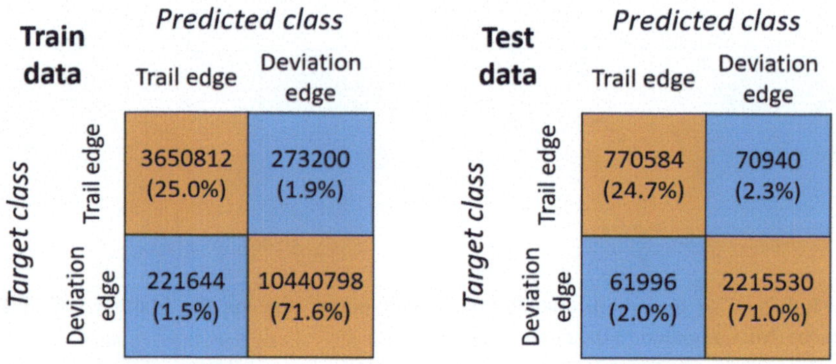

Fig. 9. Confusion matrices related to edge labelling for the train data (left) and test data (right) at the end of training.

3 Implementation and Results

To test the applicability of the proposed framework to real design scenarios, the presented model is tested on structural typologies from an available dataset of roughly 1,600 CEM topology diagrams representing different typologies of 2D and 3D bridge structures (Pastrana et al. 2021). Only those topology diagrams of the dataset that the synthetic data generator could theoretically generate are used for the test. This translates to filtering out any topologies containing indirect deviation edges that do not belong to the class of indirect deviation edges present in the ground topology of the generator, as well as any topologies containing trails of unequal lengths. After applying this filter, 173 topology diagrams of various sizes are left, spreading across eight different bridge typologies, including 2D and 3D structures for each topology. The per-typology MCC obtained from applying the GNN model trained on the synthetic dataset to these realistic cases is displayed in Fig. 10. An MCC similar to that of the synthetic test is obtained for six typologies, while a perfect edge-level classification is obtained for four of these six typologies. It is worth noting that for the GNN model to be useful, all the edges of one graph must be labelled correctly. In this respect, Fig. 10 also shows the graph-level accuracy per typology. As expected, it can be observed that for the classes for which low MCC values are obtained, the graph-level accuracy is zero. This can be explained by the fact that even a single mistake in the edge labelling renders an entire graph invalid. On the contrary, the four perfectly labelled typologies hold 100% accuracy at the graph level.

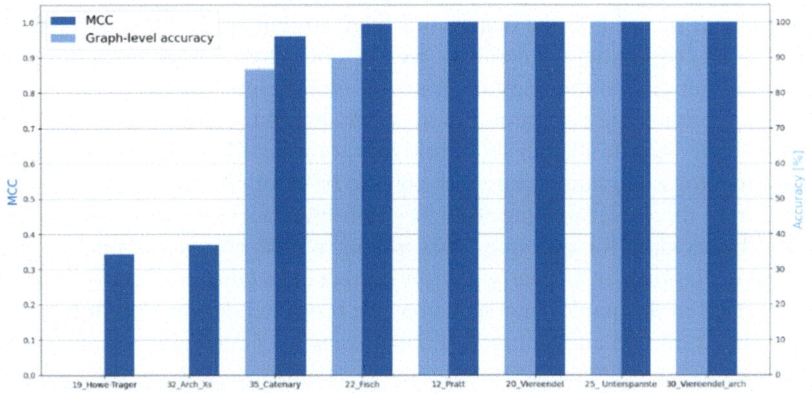

Fig. 10. Matthews Correlation Coefficient (MCC) and graph-level accuracy per typology of data from a dataset of bridge structures (Pastrana et al. 2021).

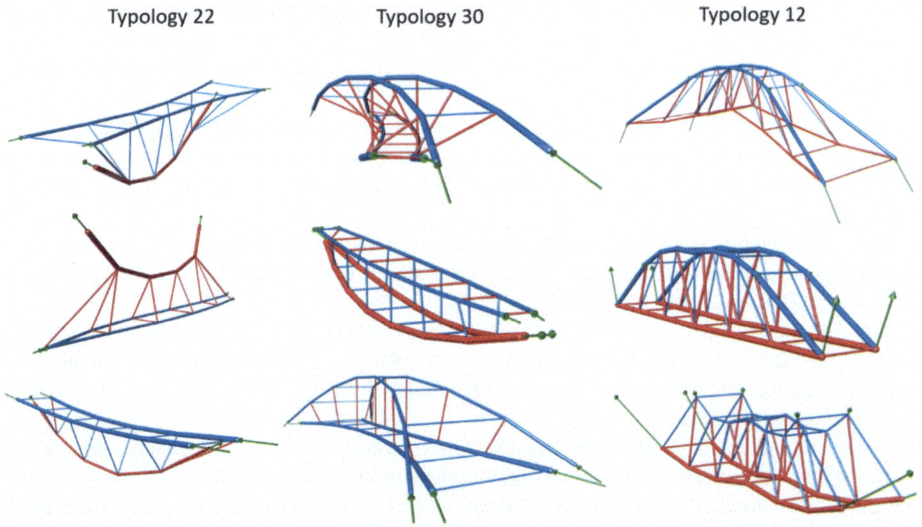

Fig. 11. Three different design variations for three typologies from the bridge dataset generated based on valid CEM topology diagrams labelled by the GNN. The metric parameters for the form-finding are defined by the human designer.

Figure 11 shows a possible outcome of the design workflow presented in this paper. Given three valid CEM topology diagrams labelled by the GNN taken from the bridge dataset, different design variations are generated via the CEM according to various metric parameters defined by the human designer.

4 Conclusion

This paper presented an AI-assisted design workflow for structural form-finding that allows simplifying the set-up of a valid Combinatorial Equilibrium Modelling (CEM) topology diagram through a Graph Neural Network (GNN). This workflow was tested in an experiment focused on the design of bridge structures. The trained GNN model developed in this experiment generalises well not only to topology diagrams from the testing dataset but also to topology diagrams of real bridge structures similar to those produced by the synthetic generator. Future work should focus on expanding the GNN training dataset to include a larger set of bridge typologies. Moreover, considering that a valid CEM topology diagram requires all edges to be labelled correctly, translating high MCC values on the edge labelling to high graph-level accuracy remains challenging. This could be tackled, for example, by applying a post-processing filter to produce valid topology diagrams from nearly entirely correct edge label predictions by the GNN model presented in this paper. In conclusion, the proposed design workflow represents a step towards a new human-machine collaborative way of designing structures through form-finding.

References

Chang, K., Cheng, C.: Learning to simulate and design for structural engineering. In: Proceedings of the 37th International Conference on Machine Learning, pp. 1426–1436 (2020)

Chicco, D.: Ten quick tips for machine learning in computational biology. BioData Min. **10**, 35 (2017)

D'Acunto, P., Jasienski, J.-P., Ole Ohlbrock, P., Fivet, C., Schwartz, J., Zastavni, D.: Vector-based 3D graphic statics: a framework for the design of spatial structures based on the relation between form and forces. Int. J. Solids Struct. **167**, 58–70 (2019). https://doi.org/10.1016/j.ijsolstr.2019.02.008

Danhaive, R., Mueller, C.T.: Design subspace learning: Structural design space exploration using performance-conditioned generative modeling. Autom. Constr. **127**, 103664 (2021). https://doi.org/10.1016/j.autcon.2021.103664

Kaethner, S., Burridge, J.: Embodied CO2 of structural frames. Struct. Eng. **90**(5), 33–44 (2012)

Kearnes, S., McCloskey, K., Berndl, M., Pande, V., Riley, P.: Molecular graph convolutions: moving beyond fingerprints. J. Comput. Aided Mol. Des. **30**(8), 595–608 (2016). https://doi.org/10.1007/s10822-016-9938-8

Matthias, F.; Lenssen, J.: Fast Graph Representation Learning with PyTorch Geometric. ICLR Workshop on Representation Learning on Graphs and Manifolds (2019)

Ochoa, K.S., Ohlbrock, P.O., D'Acunto, P., Moosavi, V.: Beyond typologies, beyond optimization: exploring novel structural forms at the interface of human and machine intelligence. Int. J. Archit. Comput. **19**(3), 466–490 (2021)

Ohlbrock, P.O., D'Acunto, P.: A computer-aided approach to equilibrium design based on graphic statics and combinatorial variations. Comput. Aided Des. **121**, 102802 (2020). https://doi.org/10.1016/j.cad.2019.102802

Pastrana, R., Skepasts, M., Parascho, S.: The CEM Framework Bridges Dataset: A dataset of bridge topologies, forms and forces. https://github.com/arpastrana/cem_dataset (2021)

Qi, S., Wang, W., Jia, B., Shen, J., Zhu, S.-C.: Learning human-object interactions by graph parsing neural networks. In: Ferrari, V., Hebert, M., Sminchisescu, C., Weiss, Y. (eds.) ECCV 2018. LNCS, vol. 11213, pp. 407–423. Springer, Cham (2018). https://doi.org/10.1007/978-3-030-01240-3_25

Shi, Y., Huang, Z., Feng, S., Zhong, H., Wang, W., Sun, Y.: Masked label prediction: unified message passing model for semi-supervised classification. In: Proceedings of the Thirtieth International Joint Conference on Artificial Intelligence, pp. 1548–1554 (2020)

Tam, K.M., Moosavi, V., Mele, T., Block, P.: Towards trans-topological design exploration of reticulated equilibrium shell structures with graph convolution networks. In: Proceedings of the IASS Annual Symposium 2020/21, pp. 784–796 (2021)

Vassart, N., Motro, R.: Multiparametered formfinding method: application to tensegrity systems. Int. J. Space Struct. **14**(2), 147–154 (1999)

Whalen, E., Mueller, C.: Toward reusable surrogate models: graph-based transfer learning on trusses. ASME. J. Mech. Des. **144**(2), 021704 (2021)

Wu, Q., et al.: Dual graph attention networks for deep latent representation of multifaceted social effects in recommender systems. In: The World Wide Web Conference May 2019, pp. 2091–2102 (2019)

Zhou, J., et al.: Graph neural networks: a review of methods and applications. AI Open **1**, 57–81 (2020)

A Digital, Networked, Adaptive Toolset to Capture and Distribute Organisational Knowledge Speckle CI

Felix Deiters[1(✉)], Giovanni Betti[2], and Christoph Gengnagel[1]

[1] Department of Structural Design and Technology (KET), Berlin University of the Arts (UdK), Hardenbergstr. 33, 10623 Berlin, Germany
felixdeiters@udk-berlin.de

[2] Center for the Built Environment, University of California, Berkeley, CA, USA

Abstract. This work explores the potential of networked, composable tools as a means to capture and distribute organisational knowledge. To solve a design problem, teams will draw from a collective body of formalised, implicit, and tacit knowledge and synthesise it into a proposal. We argue that tooling can play a part in meeting new challenges for the discipline as well as broader trends in knowledge work.

Drawing from the concepts of *Version Control* and *Continuous Integration,* a widely adopted collaboration strategy in software development making heavy use of automation, this work proposes an automation platform for *AEC* workflows as a demonstrator of such a networked tool. The prototype is implemented using *Speckle*, a data platform for *AEC*. As a case study, three automation workflows are implemented and their potential to improve planning accuracy and capture and disseminate intra- and inter-organisational knowledge is reflected on. We argue this will ultimately require a renegotiation of the relationship between planners and their tools and we speculate about a culture of toolmaking as a means to capture and evolve organisational knowledge.

Keywords: Organisational knowledge · Continuous integration · Automation · Collaboration · Toolmaking

1 Introduction

Architecture, and the built environment in general, share the paradoxical quality of being both accelerator and key to many of the overlapping and mutually reinforcing crises unfolding around us. In order to unlock its potential to store carbon rather than emit greenhouse gases, preserve biodiversity rather than devour habitats, and foster community rather than divide and displace vulnerable demographics, planners need to radically transform the way we design, build, operate, and recycle our built environment. Materialising this new sustainable, robust, and resilient architecture calls for new levels of precision while facilitating deliberate experimentation, the integration of unfamiliar

C. Gengnagel et al. (Eds.): DMS 2022, *Towards Radical Regeneration*, pp. 36–45, 2023.
https://doi.org/10.1007/978-3-031-13249-0_4

and rapidly evolving areas of knowledge as well as the rediscovery and cultivation of old ones, and the navigation of unprecedented complexity in order to arrive at something simple (Fowles 2021).

This collective learning process is interacting with and accelerating broader trends that affect not only the architecture, engineering, and construction industry, *"AEC"*, but knowledge work in general: 1. The drive to increase efficiencies, 2. The emergence of new actors and business models, 3. The move away from bespoke services, and, finally, 4. The decomposition and 5. Routinisation of professional work (Susskind Susskind 2015).

This reorganisation of work, combined with the demand to rapidly create and integrate new knowledge calls for a new generation of highly specialised, networked tools that can be composed in vertically integrated (Fano and Davis 2020), fluid workflows. A design process that emphasises as result not only the built artefact but also the production of knowledge of how to improve future iterations of itself can leverage tool making as a means to store this knowledge.

To design means to integrate a plethora of factors, often contradictory, into a coherent whole. Design problems are *wicked* problems. They can not be solved linearly, but only iteratively (Stefanescu 2020). Consequently, design solutions, or partial solutions to a subset of the problem, are usually approximate solutions. This requires the implementation of analytics to continuously assess when a proposed solution nears and eventually passes a minimum threshold to be considered acceptable. To solve a design problem, teams will draw from a collective body of formalised, implicit, and tacit knowledge and synthesise it into a proposal. Because design and planning is a highly functionally differentiated discipline, individual learnings run the risk of remaining siloed, especially since collaborators are spread across organisations. Note that the term organisation can apply at various scales here: Firms can be understood to be made up of sufficiently differentiated sub-organisations, joint ventures in turn also form organisations themselves.

Tools, especially digital tools, encapsulate knowledge and make it accessible to a broader user base (Witt 2010). Tools have thus the potential to capture and distribute organisational knowledge. But tools also encroach on the agency of their user. This happens explicitly, for example when vendors of tools intentionally limit their interoperability, as well as implicitly through nudges (Thaler and Sunstein 2008). That's why these new tools need to possess certain characteristics: They need to be scrutable, hackable, and composable (Witt 2010).

2 Methodology

To demonstrate the potential of networked adaptive tools as a means of capturing and distributing organisational knowledge, a prototype automation platform is developed and tested following a trigger-action paradigm. As a case study, three automation workflows are implemented and their potential to improve planning accuracy and capture and disseminate organisational knowledge is reflected on.

2.1 Concepts

Speckle. The prototype is implemented using *Speckle,* a data platform for *AEC* that enables sending and receiving data between a variety of authoring applications. Speckle achieves this by providing plugins, "Connectors", that translate application specific model data to a common, highly malleable data format (Stefanescu 2020). Each update to a set of data is sent to a *Speckle* server, along with metadata such as the author, authoring application, and a message describing the change. This allows collaborators to understand the evolution of a digital model over time. *Speckle* organises data in streams, branches, and commits. A stream can hold one or more branches, with each branch being defined by a series of updates to the data, called "commits".

Version Control. These features closely mirror those of *Version Control Systems* used in software development. *Version Control Systems* allow multiple collaborators to make and track changes to a joint codebase, which will usually be hosted on a private server or popular platforms such as *Github.com* or *Gitlab.com.*

Continuous Integration. To minimise the risk and impact of merge conflicts, when conflicting changes are introduced at the same time, updates should be incremental and frequent. Modern software development practices leverage automation to facilitate the *Continuous Integration* of changes. For example, an automation will check whether new code adheres to formatting standards before allowing it to be merged, or it will run a suite of integration tests to ensure no functionality was accidentally broken with the update.

Continuous Integration is often used alongside *Continuous Delivery* or *Continuous Deployment* which means to continuously compile code into a piece of software, and to distribute, and / or deploy it automatically. There is substantial overlap between the two practices which is why one will frequently find both simply referred to as *CI/CD.*

A *CI* system will allow to define a set of automations and allow those to be triggered at specific events, such as when new code is to be integrated. Automations will be composed of individual actions that are executed sequentially or concurrently (Fig. 1). The terminology differs between *CI* platforms, we will use the terms "workflow" and "action" from here on. *CI* platforms will provide a set of actions, but also allow the development of new ones. These can be made available to others, effectively sharing the knowledge encapsulated in them.

Workflows and triggers will usually be declared by adding configuration files to the code repository. The fact that workflows can be composed from existing actions and declared in a simple text file makes it very easy to incrementally adopt *CI/CD* practices and it is common for programmers to use them even in small personal projects.

2.2 Speckle CI

We argue that the simplicity of the trigger-action paradigm, together with its incremental adaptability make a *CI*-inspired automation platform an ideal candidate to demonstrate the potential of networked tools in *AEC*. Built on top of *Speckle*, being "polyglot" and itself incrementally adoptable, it constitutes, in our view, a sort of "Goldilocks path" towards a future of networked, composable tools.

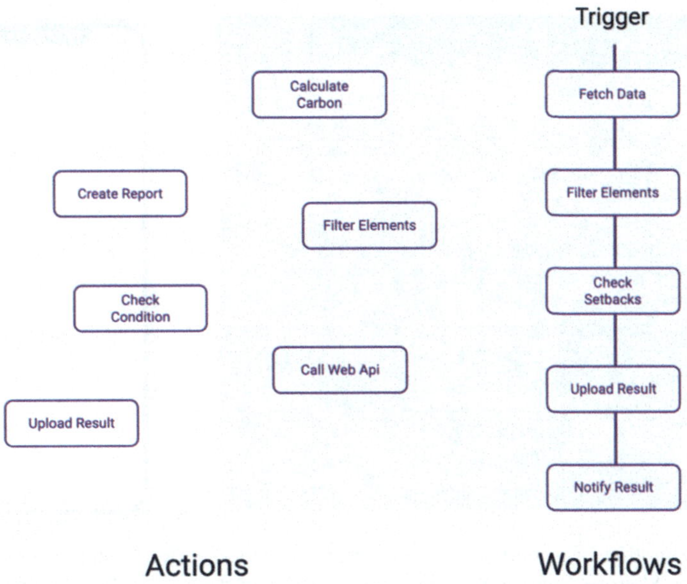

Fig. 1. *Continuous Integration* systems allow to define automations by composing actions into workflows, which are then triggered by specific events.

While *Speckle CI* draws inspiration from established *CI* platforms, key aspects were adapted to *AEC* workflows and practitioners. Based on the theory that the adoption of the automation of higher-level design tasks is often hindered by a lacking user experience (Heumann and Davis 2019), a particular emphasis was put on making the user interface as easy to use as possible. For example, the creation of workflows is done entirely via the user interface as opposed to configuration files in other solutions. While those can encode very sophisticated workflows, they might be difficult to use for practitioners unfamiliar with scripting or programming. After weighing the familiarity of visual scripting editors such as *Grasshopper* or *Dynamo*, a user interface paradigm inspired by the iOS *Shortcuts* app was selected for its simplicity (Fig. 2).

Upon login, a user will be presented with a list of the workflows they have previously set up (Fig. 3). Workflows will be sorted by the time they were last triggered and will be displayed along additional information such as whether the last run failed or succeeded, and the option to edit a workflow or create a new one.

Workflows are created or updated in the workflow editor. To keep the mental model as lean as possible, *Speckle CI* matches *Speckle's* ontology with a stream as the top hierarchy. That's why each workflow is assigned to exactly one stream, while a stream can have multiple workflows configured for it.

The editor is organised in three sections: Triggers, conditions, and the actions to be performed. After specifying a name and *Speckle* stream, users are presented with the choice to chose specific triggers on that stream, such as when a new commit is created or deleted, and conditions that will restrict a workflow from being triggered unless they are met. For example, a user could set a workflow for *Stream A* to be triggered when a new

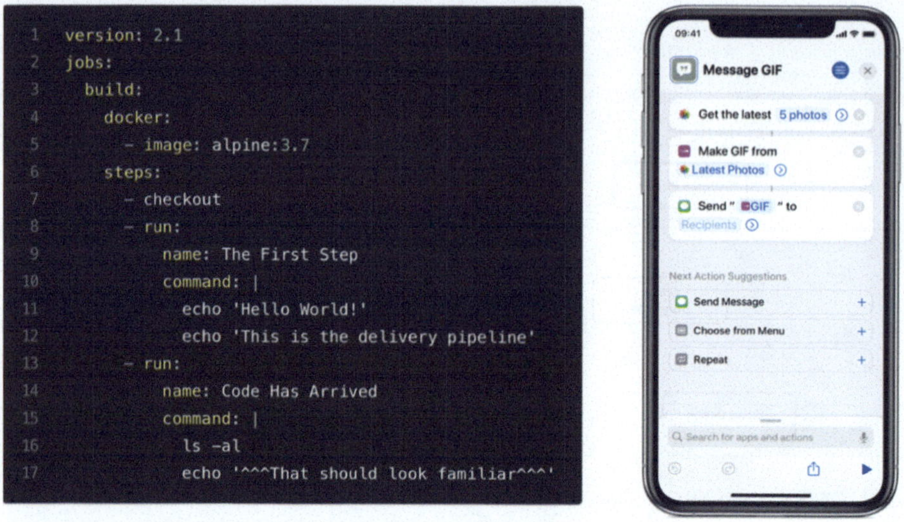

Fig. 2. Configuration of a workflow in the *CircleCI* platform. (left) compared to the simpler interface of the *Shortcuts* app on *iOS* (right) © Circle Internet Services, Inc; Apple Inc.

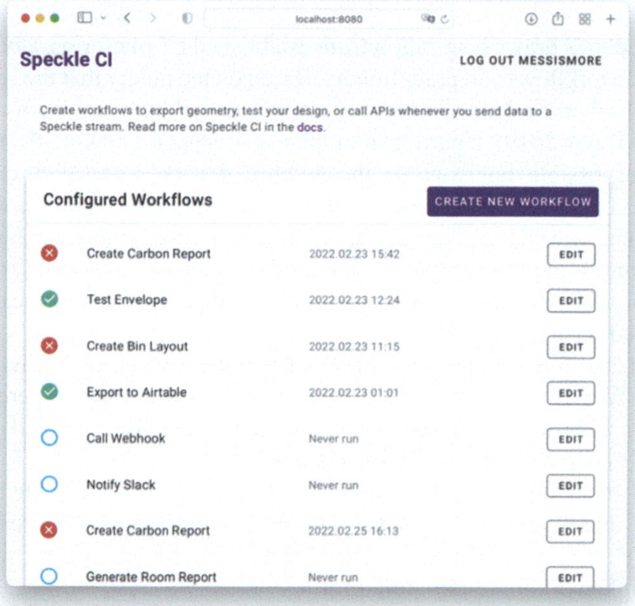

Fig. 3. The main view of the web app lists all configured workflows at a glance. Icons communicate the success or failure of the most recent run of a workflow.

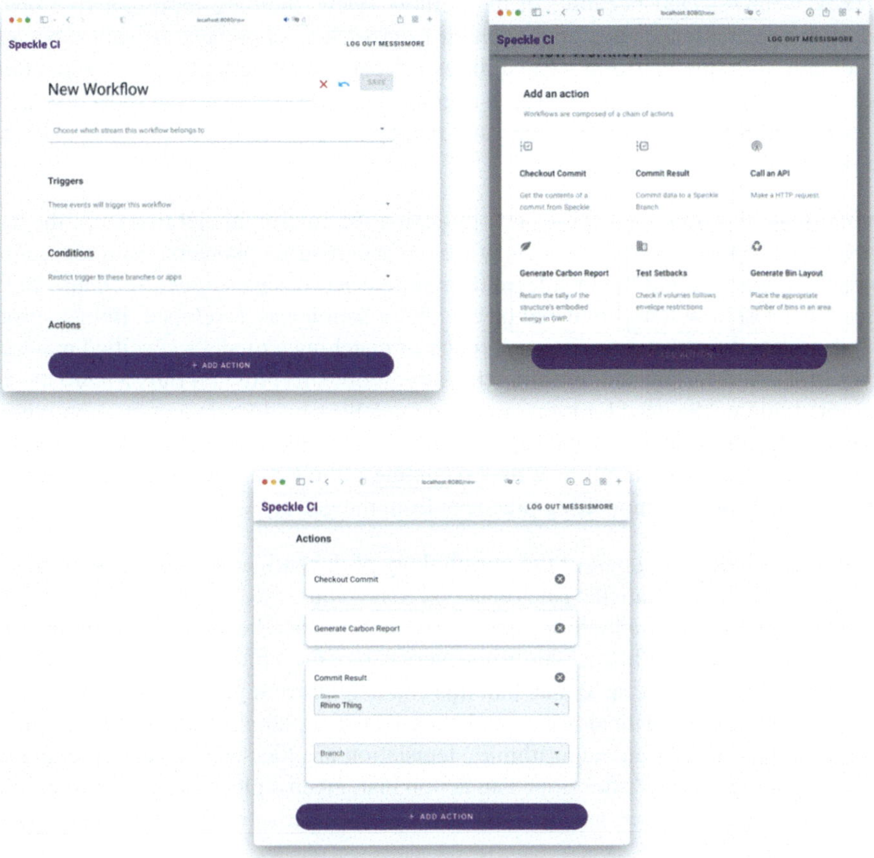

Fig. 4. Flow to create a new workflow. Actions are selected from the *Actions Store* and may offer additional settings.

commit is created, updated, or deleted, but only when the commit is made to *Branch B* and authored by *Application C*.

A prominent button opens the *Actions Store*, where the user can pick the individual actions to compose the workflow from (Fig. 4). Actions will appear in the editor as a sequence of blocks that may offer additional inputs and configuration options. For example, the action to commit data to Speckle allows the user to specify a branch.

2.3 Case Studies

Three workflows were implemented to explore the potential of this approach.

1. Carbon Calculator
2. Setback Checker
3. Bin Layout Tool

They were selected because, in the authors' view, they are representative of the kind of design challenges that could benefit from such an approach, as well as because they demonstrate the platform's capabilities to generate numerical (carbon accounting reports), spatial and geometric (setback checker), or hybrid (bin layout tool) results.

Each workflow's setup, triggering, and results were documented in videos, to serve as a reference for future user testing.

Carbon Calculator. Quantifying and optimising the environmental impact of the built environment constitutes a relatively new area of expertise for planners. A project's environmental impact should be considered from the earliest design stages (Apellániz 2021). To test this workflow, a minimal carbon calculator action was developed. Being a prototype, it applies the rather simplistic approach of matching a model's specified materials to their embodied carbon and returning the tally. While this provides only a heuristic and does not eliminate the need for more accurate modelling by domain experts, continuous, immediate feedback about how design changes impact the performance of a proposed building can help instil an intuitive understanding of how measures relate to outcomes and encourages precise modelling practices from the get-go.

Setback Checker. A big part of the morphology of the built environment is formed by building codes. To translate the rules encoded in law into software means to capture the knowledge necessary to adhere to them. Code compliance becomes thus to an extent automatically testable. While certainly not every part of the building code can be captured and tested in this way, we speculate that this can help prevent planning mistakes, direct planners' attention to critical areas, and thus increase planning precision. A simple, visual example of such an "algorithmic" legislation are the setback rules specified in Berlin's building code. We developed an action that, given a plot's perimeter, a building volume, and adjacent streets and buildings, will return whether or not it violates envelope restrictions (Fig. 5).

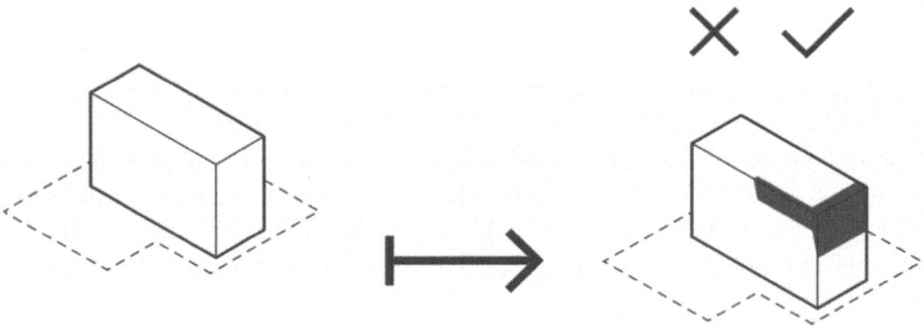

Fig. 5. The setback checker action tests whether a given volume violates envelope restrictions.

Bin Layout Tool. Planning involves many mundane, tedious, and time-consuming tasks that nevertheless require practitioners to draw from their experience. Organisations will usually develop informal best practices and conventions, and these can often be encoded

(Heumann and Davis 2019). One such example is the configuration of bins for residential buildings, because it combines hard rules, such as the required number of bins per unit, with soft, empirical rules, such as their preferred layout. As a third case study, such a tool was made available as an action in *Speckle CI*. Provided with an area and the number of residential units, the action will return whether the required number of bins will fit, and, if so, a generated layout.

3 Results

The case studies show that the abstractions and paradigms of *Continuous Integration* lend themselves well to *AEC* workflows, although further validation through user testing and the application in a real-world test-case are still to be carried out. The three case studies are, in the authors' view, representative of a whole range of domains where the application of cloud-based automation can improve planning precision.

To apply the practice of *Continuous Integration* to *AEC*, some aspects of the implementation need to be adapted. The model of a workflow was simplified and all functionality was made available via a graphical user interface. We found that the simplicity of the interface paradigm did not hinder the implementation of our case studies.

While the development of new actions requires some programming expertise and an understanding of *Speckle's* data model, the abstraction of composing actions into workflows allows for the creation and adaptation of automations by non-expert users.

Because *Speckle CI* is a web app rather than desktop software, initial setup is easy as no installation is required. This makes *Speckle CI* more approachable compared to other tools such as *Grasshopper* that are sometimes used to achieve similar goals. Leveraging *Speckle's* authentication mechanisms lowers the barrier further because users can simply start to use *Speckle CI* with their existing *Speckle* accounts.

Since *Speckle CI* is a web application, collaboration is also facilitated. Running automations locally on each staff member's computer would require their execution environment to be configured identically. This synchronisation is tedious and error prone. With *Speckle CI*, workflows are instead executed on a web server, a kind of "cloud" that contains the authoritative configuration and delivers reproducible results.

Speckle's wide range of available connectors to other software make it incrementally adoptable. Accordingly, existing workflows can be incrementally moved to *Speckle CI*. This constitutes a key advantage, as the adoption of automation tools is often hindered by their failure to integrate into existing design workflows (Heumann and Davis 2019).

4 Discussion and Next Steps

The method ultimately requires a renegotiation of the relationship between architects and their tools. Firms will usually employ teams of computational design specialists to solve specific problems for one-off projects many times over, even if these problems are very similar in nature. A platform such as *Speckle CI* allows them to shift their roles towards a new kind of toolmaker. It is positioned to herald a new culture of toolmaking as a means of capturing and developing organisational knowledge.

Developing actions, composable blocks of functionality that can be integrated into the design process by the members of other teams, allows their expertise to be more effectively disseminated in the organisation, and, through the adoption of open source methodologies, even throughout the discipline as a whole. Newly developed actions can be immediately deployed to a whole range of projects. Their effort scales.

Because their work can be more effectively reused, they can now better justify applying engineering practices to toolmaking (Davis 2013), applying lessons from design process iterations to improve functionality, thereby improving future iterations of the design process.

Tool and process share a reciprocal relationship. Actions developed to address a specific project need become available in future projects to be deployed and built upon. Their inclusion in a firm's toolkit will then represent part of its organisational design expertise and therefore make the application of this particular approach in future projects more likely, shaping the organisation's design culture.

The composition of actions into workflows can be regarded as a form of toolmaking, too. Hereby, too, is knowledge encoded, although to a lesser extent than by the development of new actions. A mechanism to share workflows between projects and users should be assessed going forward, as this will make the knowledge captured in them more easily accessible and also improve the user experience of the platform. Such a shared workflow would be scrutable, and, more importantly, adaptable by other users.

Speckle CI should be regarded as a step towards a future of networked tools and as an enabler of a new culture of toolmaking. While this work focussed on the appropriateness and applicability of *Continuous Integration* to *AEC* workflows, we suggest that the obvious next steps will be to perform user-testing, expand the set of actions, investigate a solid sharing mechanism for workflows, and apply the findings of recent research to build a more flexible and powerful backend to orchestrate the execution of workflows (Wanderley Barbosa 2022).

References

Apellániz, D.: A holistic and parametric approach for life cycle assessment in the early design stages. In: Symposium on Simulation for Architecture and Urban Design SimAUD (2021)

Davis, D.: Modelled on Software Engineering: Flexible Parametric Models in the Practice of Architecture. RMIT University, Melbourne (2013)

Fano, D., Davis, D.: New Models of Building: The Business of Technology. In: Shelden, D. (ed.) Architectural Design 90(2), pp. 32–39. John Wiley & Sons Inc, Hoboken, NJ (2020)

Fowles, E.: Make low-tech our mantra. In: RIBA journal: the journal of the Royal Institute of British Architects, August 2021, pp. 36–40. RIBA, London (2021)

Heumann, A., Davis, D.: Humanizing architectural automation: a case study in office layouts. In: Gengnagel, C., et al., (eds.) Proceedings of the Design Modelling Symposium, Berlin 2019. Springer, Cham, Switzerland (2019)

Stefanescu, D.: Alternate Means of Digital Design Communication. UCL, London (2020)

Susskind, R., Susskind, D.: The Future of Professions. How Technology Will Transform the Work of Human Experts. Oxford University Press, Oxford (2015)

Thaler, R.H., Sunstein, C.R.: Nudge. Improving Decisions About Health, Wealth, and Happiness. Yale University Press, New Haven & London (2008)

Wanderley Barbosa, V.: Computational Design Workflows Orchestration Framework. DTU, Kongens Lyngby (2022)

Witt, A.J.: A machine epistemology in architecture. Encapsulated knowledge and the instrumentation of design. In: Sowa, A., et al., (eds.) Candide. Journal for Architectural Knowledge no. 03 (December), pp. 37–88. Hantje Cantz, Ostfildern (2010)

A Design Model for a (Grid)shell Based on a Triply Orthogonal System of Surfaces

Aly Abdelmagid[1], Ahmed Elshafei[1](✉), Mohammad Mansouri[2], and Ahmed Hussein[1]

[1] Laboratoire GSA – Géométrie Structure Architecture (École nationale supérieure d'architecture Paris-Malaquais), 14 rue Bonaparte, 75006 Paris, France
a.el-shafei@hotmail.fr

[2] Laboratoire LATCH- Conception, Territoire Histoire et Matérialité (École nationale supérieure d'architecture et de paysage de Lille), 2 Rue Verte, 59650 Villeneuve-d'Ascq, France

Abstract. We present a design model for a (grid)shell that is an assembly of 3D components ('rational-voxels') fabricated from planar/developable faces. This rationalization was achieved thanks to the geometric properties of principal patches arising from a Triply Orthogonal system of Surfaces (TOS). By using such a system we were able to generate a curvilinear coordinate system where the coordinate lines are principal curves on the respective surfaces in the TOS. Next, generate 3D components (voxels) where each voxel is a curvilinear cube where its sides are principal patches, and its edges are principal curves obtained by intersecting two neighboring surfaces from each of the three families in the TOS. These voxels are then rationalized into rational-voxels having planar/developable faces and straight/planar edges. The design model allows for five degrees of design freedom for choosing: (1) the shell-slice type in the TOS, (2) the shell-slice thickness, (3) the voxel-assembly method, (4) the rational-voxel type and (5) being either a solid or a hollow voxel-assembly. A design to build process of a large scale pavilion is presented as a demonstration of the proposed design model.

Keywords: Principal network · Triply orthogonal surfaces · (Grid)shell

1 Introduction

In this work we present a design model for a (grid)shell composed from an assembly of 3D components fabricated from planar and/or developable faces. Our approach rests fundamentally on the so-called *Triply Orthogonal system of Surfaces* (TOS) giving rise to explicit parameterizations of three families of surfaces by *principal curves*. These principal-curves parameterizations are orthogonal conjugate networks, which make them optimal in the discretization of doubly-curved surfaces using planar and developable pieces. Using a TOS thus provide us a three-dimensional *principal network* in the sense of a curvilinear coordinate system which will be used to define the 3D component (or voxel) of the assembly forming the (grid)shell. The six sides of a voxel are given (by pieces of) surfaces from the three families in the TOS, hence are principal patches. There follows that we can transform our voxel into a rational-voxel whose sides are

C. Gengnagel et al. (Eds.): DMS 2022, *Towards Radical Regeneration*, pp. 46–60, 2023.
https://doi.org/10.1007/978-3-031-13249-0_5

either planar or developable, allowing for a feasible construction by simple methods. In Sect. 2, we present the geometry of principal networks and the TOS, in Sect. 3, we present the design model and degrees of design freedom, in Sect. 4, we present a realized (grid)shell based on our design model.

2 Geometry

We will now briefly recall few facts from differential geometry (cf. [4] and [5]).

2.1 Principal Patches

Consider a surface S given by a smooth parameterization $X(u,v)$ where the domain of (u,v) is an open set in the plane and the coordinate lines form a network on S. We will thus refer to $X(u,v)$ as a patch or as a network. We denote by X_u, X_v the partial derivatives generating the tangent plane T_pS and by N the normal vector. The first and second fundamental forms are denoted by I, II, the shape operator is given by $\mathrm{S} = II/I$.

Let a,b be tangent vectors in T_pS then their directions are called

$$\begin{cases} \text{Orthogonal if } \langle a, b\rangle = 0 \text{ that is } a^T \cdot I \cdot b = 0 \\ \text{Conjugate if } \langle a, S(b)\rangle = 0 \text{ that is } a^T \cdot II \cdot b = 0. \end{cases}$$

The eigenvalues of the shape operator are the *principal curvatures* k_1, k_2 and the corresponding (necessarily orthogonal) eigenvectors define the *principal directions.*

- A curve in S is a *principal curve* if at every point it is tangent to a principal direction that is, it is an integral curve for that principal direction field.
- A patch $X(u,v)$ on S is a *principal patch (network)* if its coordinate lines are principal curves, that is the directions of X_u, X_v are both orthogonal and conjugate.
- A mesh on S is a *principal mesh* if it is constituted of quadrilateral faces obtained by joining the intersection points of a principal network.

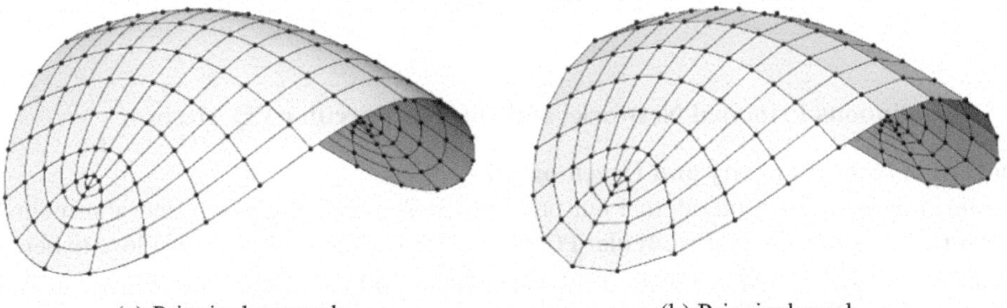

(a) Principal network. (b) Principal mesh.

Fig. 1. a) Principal network. (b) Principal mesh.

2.2 Geometric Properties of Principal Networks

Let *X(U,v)* be a principal patch (network), that is X_u, X_v are principal directions (orthogonal and conjugate). Orthogonality is clear as seen in Fig. 1(a), as for conjugation, we recall that conjugate directions *a* and *b* are parallel to the edges of a parallelogram tangent to the DUpin's indicatrix, as seen in Fig. 1(B). Naturally, there is an infinity of such parallelograms and if we impose orthogonality between *a, b* then we have a rectangle with edges parallel to the principal directions (Fig. 2).

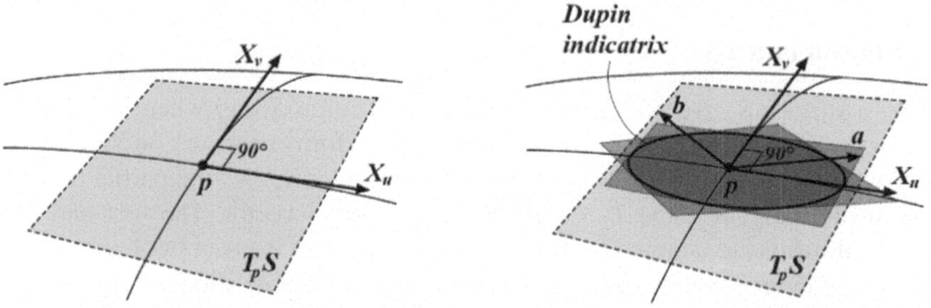

(a) Orthogonality property of X_u, X_v.   (b) Conjugation property of X_u, X_v.

Fig. 2. A) Orthogonality property of X_u, X_v. (b) Conjugation property of X_u, X_v.

We now define two important types of ruled surfaces or 'strips' that will be useful (Fig. 3).

The *Normal strips* for a fixed u_o, v_o given by:

$$\begin{cases} N_1(s,u) = X(u, v_o) + sN(u, v_o) \\ N_2(s,v) = X(u_o, v) + sN(u_o, v). \end{cases}$$

The *'Tangent' strips* for a fixed u_o, v_o given by:

$$T_1(s,u) = X(u, v_o) + s(X(u, v_o + \varepsilon) - X(u, v_o))$$
$$T_2(s,v) = X(u_o, v) + s(X(u_o + \varepsilon, v) - X(u_o, v)).$$

2.3 Continuous Principal Networks and Discrete Circular PQ Meshes

Deviation from Planarity and Developability

It is important to point out that our principal meshes (arising from generic principal networks) are not technically circular PQ meshes. However, they are very close to being so, in the sense that if Q is a quad from a principal mesh it deviates from planarity by a very small error (see [3]) as seen in Fig. 4.

Hence, non-linear optimization problems of approximating surfaces by circular PQ meshes, see [2, 7] and [9] are much more likely to work if initiated from principal networks. Another (somehow inverted way) of looking at the deviation from planarity

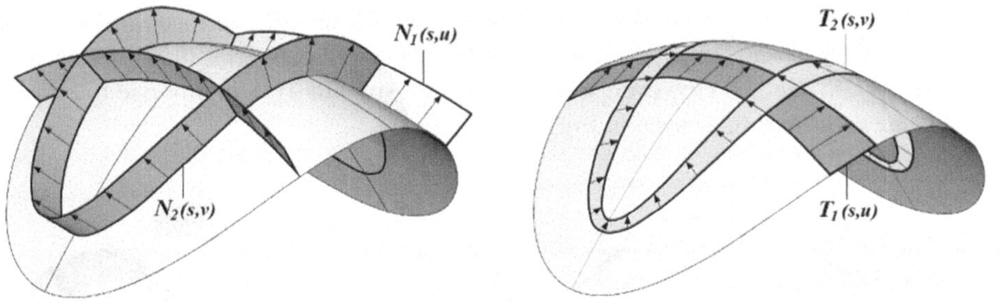

(a) Normal strips. (b) Tangent strips.

Fig. 3. a) Normal strips. (b) Tangent strips.

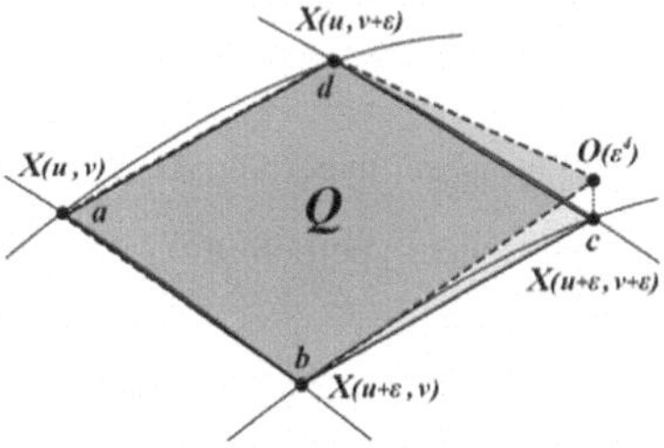

Fig. 4. Q deviate from planarity by $O(\varepsilon^4)$ as opposed to $O(\varepsilon^3)$ in other networks.

of principal meshes, which also shows the deviation from developability of the tangent strips $T_1(s,u)$, $T_2(s,v)$ is that:

$$\begin{array}{lcl} \text{Circular PQ meshes} & \xrightarrow[\text{refinement}]{\text{at the limit of}} & \text{Principal networks} \\ \text{Sequence of PQ faces} & & \text{Developable strips} \end{array}$$

Error Tolerance in our 3D Models

In our approach we start by principal networks hence the normal strips $N_1(s,u)$, $N_2(s,v)$ are truly developable. Next, we produce discrete principal meshes (hence almost planar) and semi-discrete tangent strips (hence almost developable), as explained above. However, the errors are so small, that if we consider the realization of the model using *Rhinoceros* with its tolerance set to 0.001. Then, planarity of the principal meshes and the developability of the tangent strips in question are completely within this tolerance. Thus for all purposes of our architectural application:

- The principal meshes will be considered planar
- The tangent strips $T_1(s,u)$, $T_2(s,v)$ will be considered developable.

2.4 Triply Orthogonal System of Surfaces

As mentioned above the *Triply Orthogonal system of Surfaces* (TOS) are systems consisting of three families of surfaces that are mutually orthogonal. A theorem of Dupin asserts that given such system then the curves of intersection of the surfaces are principal curves on each surface, see [6]. Similar to discrete principal networks, there is a discrete version of TOS, see [1]. However, we will focus on smooth TOS giving rise to *elliptic curvilinear coordinates,* as seen in Fig. 5.

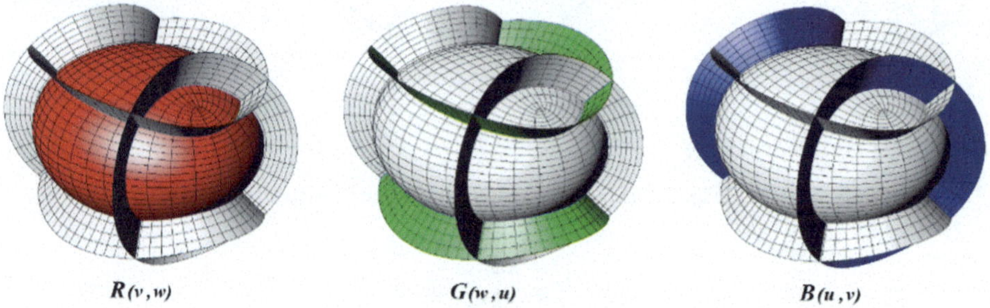

(a) Ellipsoids R. (b) One-sheet hyperboloids G. (c) Two-sheet hyperboloids B.

Fig. 5. a) Ellipsoids R. (b) One-sheet hyperboloids G. (c) Two-sheet hyperboloids B.

Explicit Parametric Definition of the TOS

For *a, b, c* real suitably bounding the parameters *u, v, w* we obtain the patches:

$$X(u, v, w) = \left\{ \pm\sqrt{\frac{(a-u)(a-v)(a-w)}{(a-b)(a-c)}}, \pm\sqrt{\frac{(b-u)(b-v)(b-w)}{(b-a)(b-c)}}, \pm\sqrt{\frac{(c-u)(c-v)(c-w)}{(b-c)(a-c)}} \right\}$$

parameterizing the elliptic TOS (R, G, B) where R are ellipsoids, G are one-sheeted hyperboloids and B are two-sheeted hyperboloids respectively, by principal patches:

$$\begin{cases} R(v, w) = X(u_o, v, w) \\ G(w, u) = X(u, v_o, w) \\ B(u, v) = X(u, v, w_o) \end{cases}$$

where u in $[-c,c]$, v is in $[c,b]$ and w in $[b,a]$.

Important Advantages of the TOS

(1) We obtain an explicit expression for the parameterizations of the surfaces in question therefore, bypassing the need for solving the differential equation of principal curves in order to obtain the principal networks.
(2) Exploring the third dimension through the elliptic curvilinear coordinates and the generation of the 3D component (voxel), as seen in Fig. 6.

(3) The TOS gives us a way to simultaneously have caps-surfaces that are principal patches with varying thickness all-the-while maintaining that the lateral-surfaces are developable. This is done by having the cap-surfaces given by two successive surfaces from one family in the TOS and the lateral-surfaces developable tangent strips on surfaces from the remaining two families, as seen in Fig. 7(b). This gives a larger variety of shell designs (of varying thickness) with the geometric advantages of constant-thickness shells. Note that, the principal patches cap-surfaces used here are not *parallel* neither are their principal meshes, by contrast to the *parallel meshes* addressed in [8].

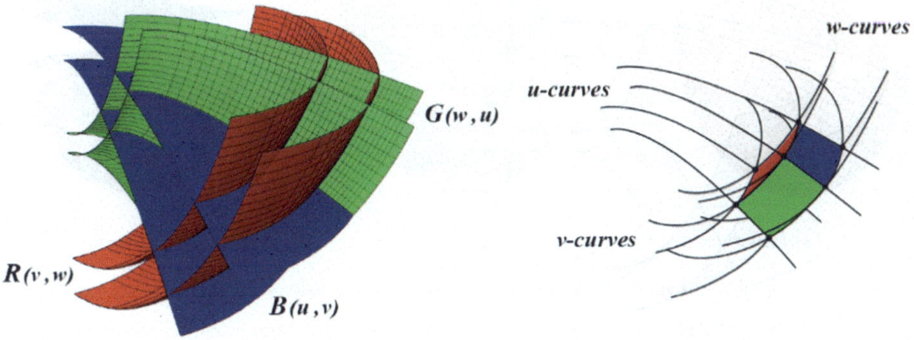

(a) Principal patches *R,G,B*. (b) Elliptic curvilinear coordinate lines *u,v,w*.

Fig. 6. a) Principal patches *R, G, B*. (b) Elliptic curvilinear coordinate lines *u, v, w*.

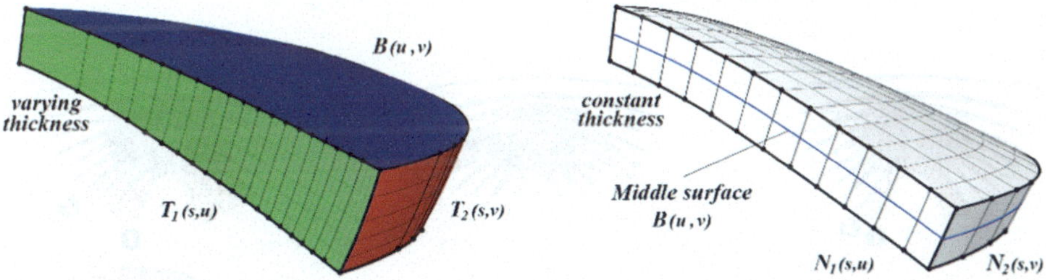

(a) Constant thickness using offset. (b) Varying thickness using TOS.

Fig. 7. a) Constant thickness using offset. (b) Varying thickness using TOS.

3 Design Model

Let us outline the five degrees of design freedom constituting the design model.

3.1 Choice of Shell-Slice Type

The first degree of design-freedom is in the choice of the leaves determining the caps-surfaces of the shell-slice. More precisely, by fixing one curvilinear direction and considering the continuum of leaves given by the family of surfaces orthogonal to that direction as seen in the Fig. 8.

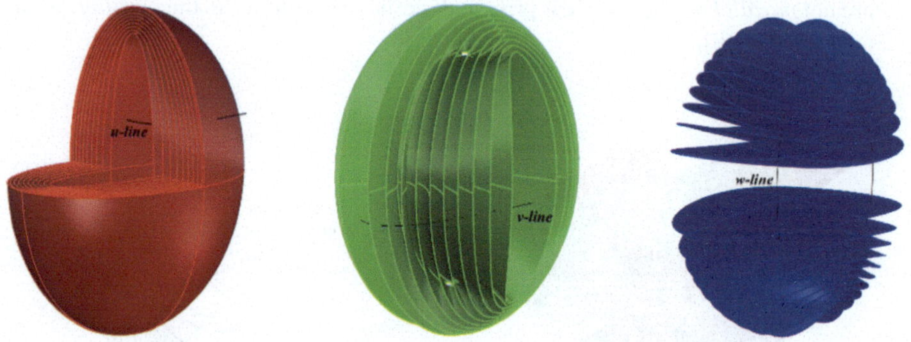

(a) *R*-leaves. (b) *G*-leaves. (c) *B*-leaves.

Fig. 8. a) *R*-leaves. (b) *G*-leaves. (c) *B*-leaves.

3.2 Choice of Shell-Slice Thickness

The second degree of design-freedom is in the choice of the specific two leaves determining the caps-surfaces of the shell-slice, as seen in Fig. 9.

(a) Highly curved thick. (b) Medium curved thin. (c) Highly curved very thick.

Fig. 9. a) Highly curved thick. (b) Medium curved thin. (c) Highly curved very thick.

3.3 Choice of Voxel Assembly

Consider the voxel bounded by two *R*-leaves, two *G*-leaves and two *B*-leaves as seen in Fig. 10(a). This voxel has (doubly-curved) faces which are principal patches and edges which are principal curves. The third degree of design-freedom is in the way we can choose a configuration of these voxels, for example, the chosen shell-slice can be seen as a two-dimensional assembly of these voxels, as seen in Fig. 10(b). This generalizes the assemblies on based revolution surfaces and normal offsets, seen in [10].

(a) One voxel. (b) Two-dimensional assembly. (c) One-dimensional assembly.

Fig. 10. a) One voxel. (b) Two-dimensional assembly. (c) One-dimensional assembly.

Three-Dimensional Assembly of Voxels
It is important to note that by construction of the TOS, this degree of design-freedom can be extended in three directions in the sense of three-dimensional assemblies, as seen in Fig. 11(b).

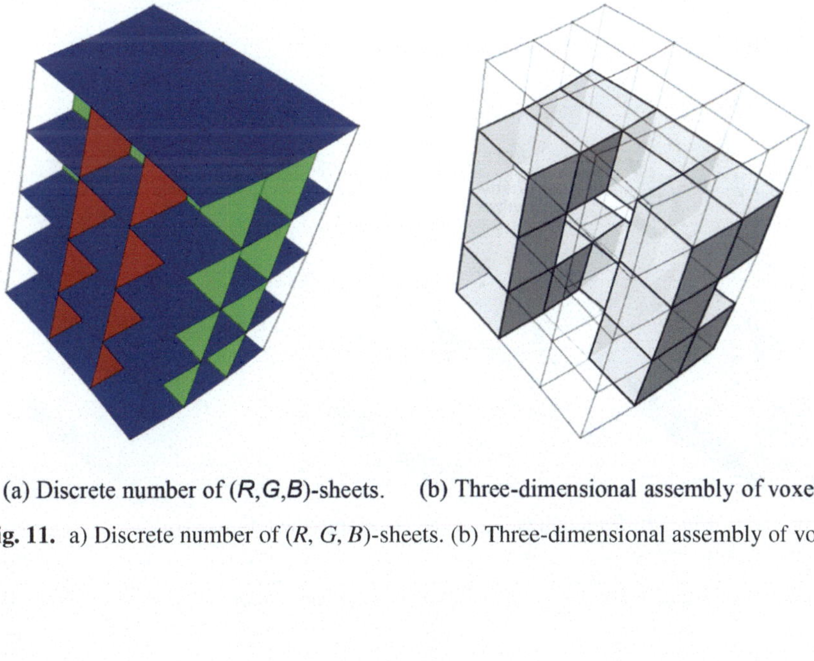

(a) Discrete number of (*R*,*G*,*B*)-sheets. (b) Three-dimensional assembly of voxels.

Fig. 11. a) Discrete number of (*R*, *G*, *B*)-sheets. (b) Three-dimensional assembly of voxels.

3.4 Choice of Rational-Voxel Type

Since a voxel has six sides all of which are principal patches then from discussion above, we can then rationalize the voxel into one whose faces are either planar or developable. The resulting rationalization of the voxel is a what we call the 'rational-voxel' and the choice of planar and/or developable faces in the voxel is the fourth degree of design-freedom, as seen in Fig. 12. We will call a rational-voxel with:

(1) Only planar faces a *P-voxel*, its assembly is called a *P-assembly*.
(2) Only developable faces a *D-voxel*, its assembly is called a *D-assembly*.
(3) Both planar and developable faces a *PD-voxel*, its assembly is a *PD-assembly*.

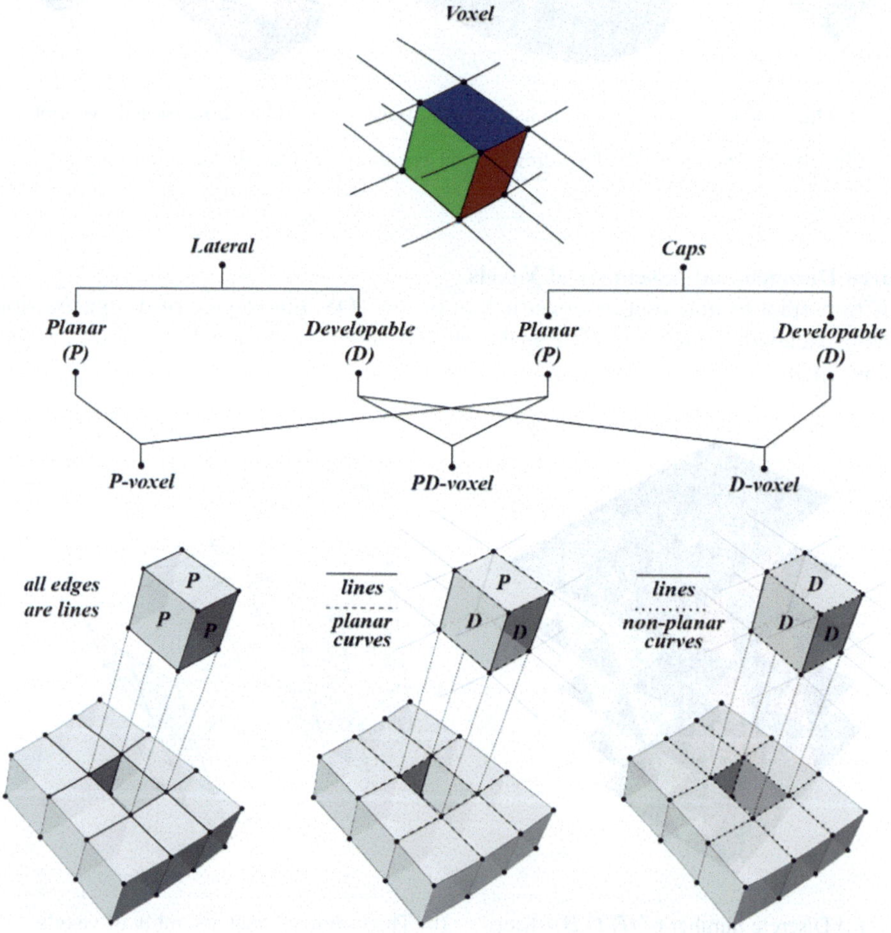

Fig. 12. Rationalization of the voxel into the rational-voxel.

Note that the choice of planarity or developability of the faces of the rational-voxel determines its continuity with its neighboring voxel, as seen in Fig. 13. In particular, the case of a D-assembly the results in series of developable strips amenable to feasible fabrication techniques seen for instance in [11].

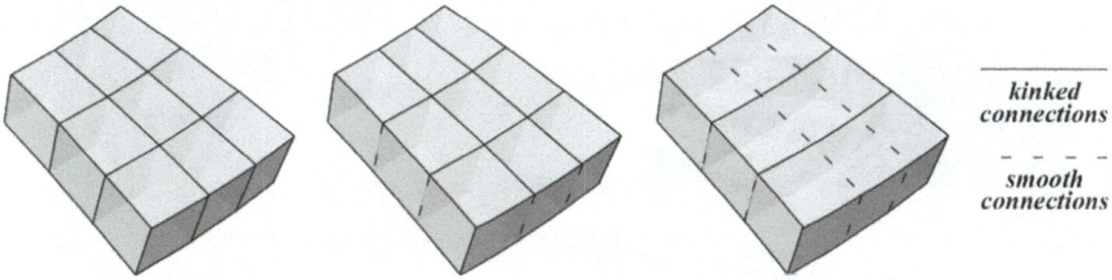

Fig. 13. P, PD and D-assemblies with the continuity with neighbors indicated.

3.5 Choice of Solid or Hollow Voxel-Assembly

The fifth degree of design-freedom is the choice of solid (shell) or hollow (grid-shell), as seen in Fig. 14. Illustrating the some of the design possibilities arising from the five degrees of design-freedom.

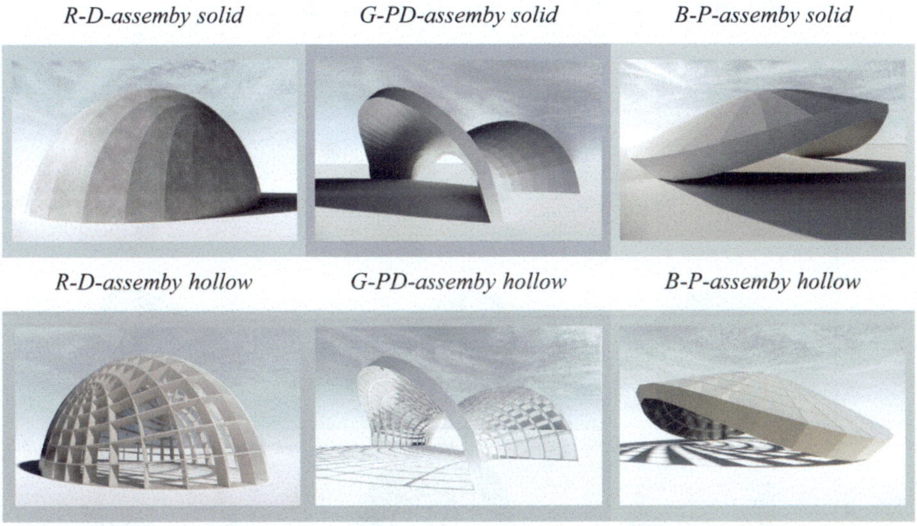

Fig. 14. Matrix of some of the prototypes based on the five degrees of design-freedom.

4 Realized Prototype

Next, we present a realized architectural prototype based on the design model above (Fig. 15).

Fig. 15. The final realized realized pavilion.

A. Form: We started by manipulating the five degrees of design-freedom. This has allowed us to explore different architectural interpretations. Next we narrowed them down to the following proposals seen in Fig. 16 (P-assembly Hollow).

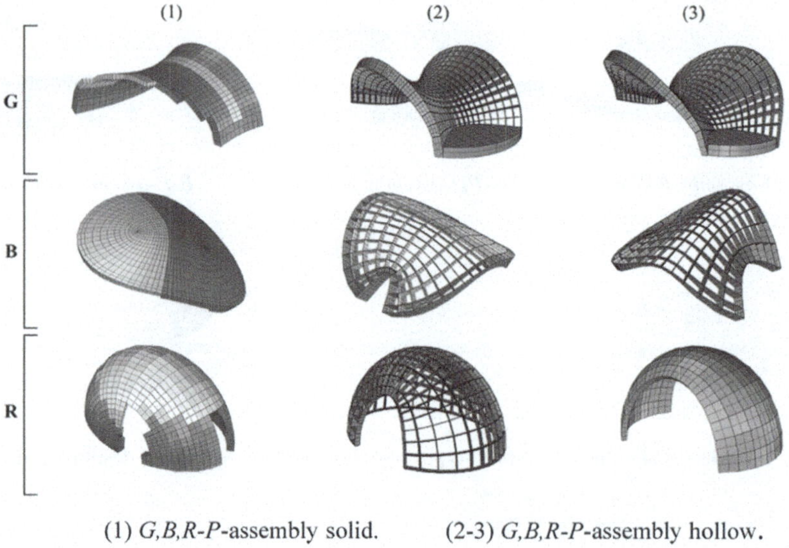

Fig. 16. (1) *G, B, R-P*-assembly solid. (2–3) *G, B, R-P*-assembly hollow.

From these proposals we decided on proposal (2B): the *B-P*-assembly hollow (version 1) as a base form upon which some adjustments will be made taking into account the size of the object, its spatial quality, and the needs of the project, as seen in Fig. 17.

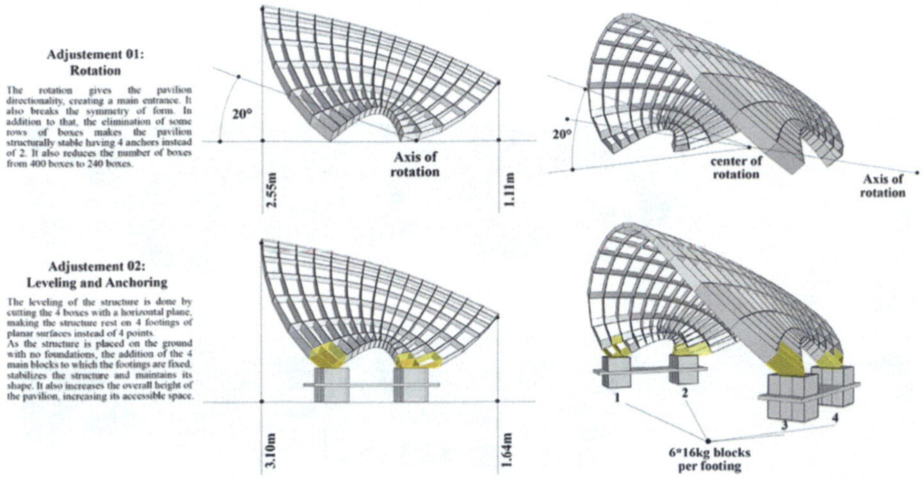

Fig. 17. Adjustments made to the B-P-assembly hollow version 01.

Once the exact final form was finalized, we made a 1:10 physical model in cardboard to test the form and the box-folding technique (Fig. 18).

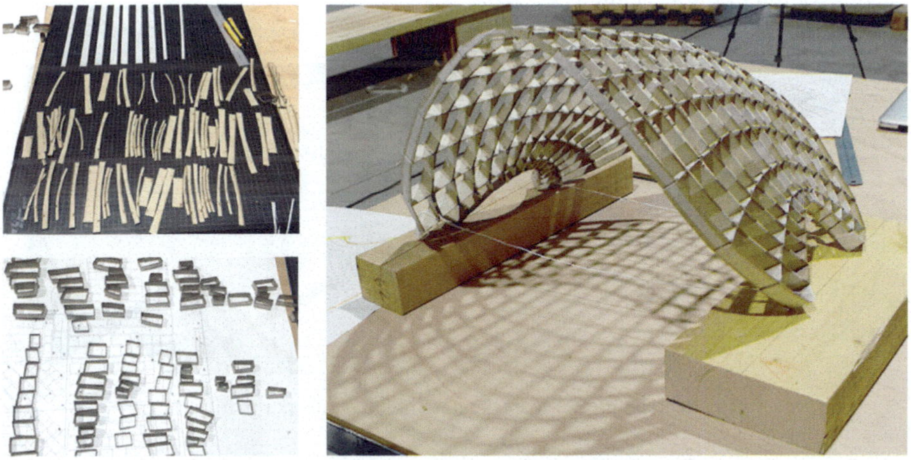

Fig. 18. 1:10 cardboard model of the final form.

B. Material: As we mentioned above, the choice of the P-assembly was fixed earlier on given the material we chose (27 sheets of plywood of 2.5 m length x 1.2 m width x

5 mm thickness) where we considered it to be mostly adapted to realizing planar surfaces given its limited bending capacity.

C. Fabrication and assembly: The advantage in working with P-assembly is that we were able to realize all the boxes (P-voxels) simply by folding flat pieces of plywood. These pieces were cut using the 3-axis milling and then folded as shown in Fig. 19.

Fig. 19. The folding technique employed for the assembly of the boxes.

Once the boxes were assembled, they were then grouped into partitions according to their position in the structure (Fig. 20).

Fig. 20. The grouping of the assembled boxes per partition.

The boxes of each partition were bolted together, then the partitions themselves were bolted together to form the final assembled structure as shown in Fig. 21. Notice the use formwork to maintains the levels of the partitions while they're being joined and the footings fixed to the blocks.

Fig. 21. The assembling of the partitions into the fully assembled model.

5 Conclusion

In this paper, we gave an exposition of a TOS showing its potential as a mathematical basis for generating design models. We showed that the geometric properties of the principal patches constituting the TOS can be used as tools to generate a large variety of architectural prototypes realized as assemblies of rational-voxels made from planar and developable surfaces. To end the paper, we would like to mention two ways for furthering this work; the first being, exploring the three-dimensional assemblies, in the sense of going into 'volumes' and not just shells. Next, it could be seen that the TOS presented here is to certain extent limiting in terms of architectural form, because of its symmetry and geometric logic. Thus, an interesting continuation for us, would be to explore certain deformations of TOS (and principal patches) that preserve the principal networks. Hence, allowing for more freeform shapes while maintaining the geometric properties presented here and their advantages for fabrication.

Acknowledgement. The realized prototype was the result of work done within the framework of a workshop for the master students at the national school of architecture and landscape in Lille. The supply of materials, production, and transportation of the pavilion was funded by the school. The pavilion was exhibited the 11th of June 2022 at Bazaar St-So during the 10th edition of the "*braderie de l'architecture*".

References

1. Bobenko, A., Matthes, D., Suris, Y.: Discrete and smooth orthogonal systems: C^{∞}-approximation. Available from arXiv:math/0303333 (2003)
2. Bobenko, A., Suris, Y.: Discrete differential geometry. Consistency as integrability. Available from arXiv:math/0504358 (2005)
3. Bobenko, A., Tsarev, S.: Curvature line parameterization from circle patterns. Available from arXiv: 0706.3221 (2007)
4. DoCarmo, M.: Differential Geometry of Curves and Surfaces. Prentice-Hall Inc, Englewood Cliffs, New Jersey (1976)
5. Eisenhart, L.: A Treatise on Differential Geometry of Curves and Surfaces. Ginn and Company, Boston (1909)
6. Gray, A., Abbena, E., Salamon, S.: Modern differential geometry of curves and surfaces with Mathematica, 3rd edn. Chapman & Hall/CRC (2006)
7. Liu, Y., Pottmann, H., Wallner, J., Yang, Y., Wang, W.: Geometric modeling with conical meshes and developable surfaces. ACM Tr., Proc. SIGGRAPH **25**(3), 681–689 (2006)
8. Pottmann, H., Liu, Y., Bobenko, A., Wallner, J., Wang, W.: Geometry of multi-layer freeform structures for architecture. ACM Tr., Proc. SIGGRAPH **26**(65), 1–11 (2007)
9. Pottmann, H., et al.: Freeform surfaces from single curved panels. ACM Tr. **27**(3), 1–10 (2008)
10. Filz, G., Schiefer, S.: Rapid assembly of planar quadrangular, self-interlocking modules to anticlastically curved forms. Available on researchgate.net/publication/ 271436191 (2014)
11. Kilian, A.: Fabrication of partially double-curved surfaces out of flat sheet material through a 3D puzzle approach. Available on researchgate.net/publication/30871581 (2003)

Towards DesignOps Design Development, Delivery and Operations for the AECO Industry

Marcin Kosicki(✉), Marios Tsiliakos, Khaled ElAshry, Oscar Borgstrom, Anders Rod, Sherif Tarabishy, Chau Nguyen, Adam Davis, and Martha Tsigkari

Foster + Partners, 22 Hester Road, London SW11 4AN, UK
mkosicki@fosterandpartners.com

Abstract. The overwhelming success of companies build on top of cloud computing technologies has been driven by their ability to create systems for processing big data at scale and designing high-quality digital products as well as being agile and capable of handling constant changes in the market. This runs somewhat contrary to the AECO industry, which generates an abundance of multidisciplinary data and faces numerous design challenges but is not as prone to agile management. The entire methodology for designing and delivering projects has historically been oriented toward getting all requirements defined and specified in advance. In that context, "change" of the workflow in AECO is often seen as an exception. Not only this is far from the paradigm or principles of today's business technologies, but today's enterprises are characterized by an opposing set of values. Latest software engineering methodologies, like DevOps and its design incarnation – DesignOps were created solely to tackle those issues in the IT industry. This paper will present how those methodologies could be successfully implemented in the AECO industry and increase the efficiency of existing design pipelines. We demonstrate a prototype of a software platform, an entire automated ecosystem where design operations are made in the cloud by a collection of automatic or semi-automatic microservices and where data flows seamlessly between various disciplines. The system leverages the potential of distributed computing, performance-driven design, evolutionary optimization, big data, and modern web design.

Keywords: DesignOps · DevOps · Cloud computing · Performance design · Optimization · High performance computing

1 Introduction

Design teams are frequently required to deliver iterations of their designs to the respective stakeholders, such as internal reviewers, clients, building authorities or consultants. This usually involves a time-consuming iterative process associated with the design cycle and its derivatives -such as documentation, costing or visualization. Some of those steps require intensive human input and analysis while some could be automated using various technologies. Since new design challenges constantly arise, these pipelines and the workflows build around them can never be static, they must be flexible to keep up. This

C. Gengnagel et al. (Eds.): DMS 2022, *Towards Radical Regeneration*, pp. 61–70, 2023.
https://doi.org/10.1007/978-3-031-13249-0_6

poses many challenges like the ones that the user experience and interface (UX/UI) community is facing and could be seen in a wider context of Design Operations – DesignOps [1]. It is an emerging movement advocating a much closer integration between design and technology escalated by the rise of agile development. It is understood as a practice of reducing operational ineffectiveness in the design workflow, as well as providing better quality design through technological advancement. DesignOps is about implementing design improvements and deploying them to users (designers) as quickly and as frictionless a way as possible. It is still in its formational stages and is an intentionally broad topic, because there are many elements to factor in when enabling consistently good quality design.

1.1 Road to DesignOps

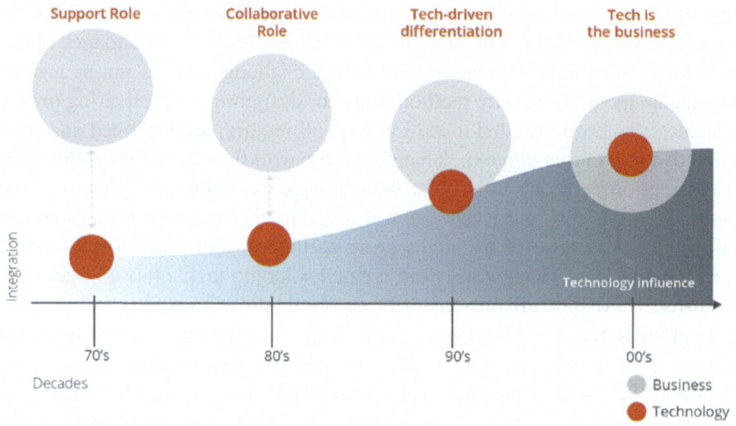

Fig. 1. Changing role of technology.

Understanding the principles of the DesignOps movement, requires a step back to look at how the role of IT in the business environment has evolved over the years (Fig. 1). From the 1960s and 1970s when it played primary a supportive role, where IT systems were used solely to make the existing processes faster and reduce cost [1]. Through the 1980s some forms of collaboration started to emerge, since the spread of PCs put machines on peoples' desks. In response, companies started asking for customized software. However, IT was still considered a tool which resulted in a very clear separation between development teams (people building IT systems) and operations (business users). The need for custom tools stared pulling IT teams closer to the operations side and by the mid 90's IT had emerged as an enabler which could provide competitive advantage through technology-driven differentiation. The trend continued through the dotcom explosion in the 00's as technology became the core of many businesses. Some companies did not simply differentiate themselves through technology but delivered technological products and solutions as their sole function. This required more robust processes and workflows which culminated in the idea of agile software development formed in 2001. Adopting agile methodologies pushed many organizations like Spotify [2] to product-centric rather than project-oriented development. It promoted fast delivery cycles [3] further reinforced by build-measure-learn feedback loops using Minimum Viable Product ideas (MVPs) from Lean Start-Up principles [4].

Agile also had its shortcomings since success was primarily measured on time and budget delivery and less on the quality of the product. This resulted in the operations (Ops) teams becoming increasingly skeptical about the quality of new releases from the development (Dev) teams [1]. Promoting closer collaboration resulted in DevOps, a methodology aimed at removing uncertainty by aligning development and system operations through automation. Increased demand for the delivery of technological products at scale and short design cycles started putting pressure on user experience (UX) and user interface design. Designers often worked in silos or as remote parts of dev teams and were sometimes individually responsible for a wide range of tasks from conducing UX research, wireframing to front-end coding. This loose structure was not scalable and could not cope with growing complexity of digital products or the demand for high-quality, consistent user customer experience. As a result, a separate set of processes and workflows attempting to operationalize design, inspired by DevOps had emerged – DesignOps [5].

DesignOps has been adopted by many organizations including Airbnb [6] and Salesforce [7]. However, each of them tailored it to their own specific business needs, DesignOps as an idea is centered around four main goals [8]: building efficient design workflows and processes, ensuring cross-functional team collaboration and alignment with different stakeholders ranging from engineering, UX researchers, UI designers, motion designer to marketing; standardising tools and systems as well as promoting consistent design culture. Although many of those activities had already existed within those companies, DesignOps as a separate role introduced structural changes and cultural shift focused mainly on scaling and amplifying design processes. Instead of segregating teams, it is seen as a methodology that enables highly integrated and effective design organizations.

1.2 AECO Perspective

There are many similarities between challenges addressed by DesignOps and the AECO industry. In architectural practices the design cycle timeline is a key factor in the success of any given project. The quicker a team can come up with design ideas and turn them into a viable design option, the better. Design is an iterative process; ideas are brainstormed then modelled and tested against others. As a design progress from the initial concept stage to final construction drawings, its concepts become more defined. This means that change comes with a constantly increasing overhead since design interventions can depend on decisions made earlier on in the process.

It is relatively easy to make changes during concept stages as they provide high flexibility, but the assumptions made during concept design can be critical to the future performance and operational cost of a buildings [9, 10]. Therefore, there is a constantly growing demand to guide the exploration of new design options not only based on their aesthetical criteria but also on hard metrics like environmental performance or user-experience. Such studies are usually conducted by consultants using specialized simulation software and the current process usually requires a design team to pass a digitized 3D massing model to a consultant who will run the required analysis and put together a report. This is a cumbersome workflow since massing options must be exported form the CAD software and sent over manually. At the same time, running performance analyses in a timely fashion comes with its own challenges. Models often need to be converted, the analysis can be computationally intensive and putting together a report which includes detailed feedback is usually a slow, manual process. All those steps add up to a considerable lag between the modeling of a design option and the understanding of its performance. The lag in the delivery cycle grows even more if the aspiration is to integrate data from many disciplines. It could take days or even weeks and by the time the performance of the design option is evaluated the actual design has progressed, rendering the returned feedback obsolete. This process is neither fast nor scalable and it is the result of workflows built around AECO software's that still follow a waterfall model [11]. By learning from the DevOps and DesignOps movements we can build more robust pipelines and deliver value at a much earlier stage in the project development cycle.

2 Methodology

To address these design cycle challenges, we have developed an approach that treats early-stage design like a modern software delivery cycle. In our approach an architectural design team would be analogue to the IT Ops team and a consultant team analogue to the IT Dev team. The product in this context is a set of highly optimized and performance-driven massing options tailored to a given architectural project and delivered at scale during early design stages.

The initial experiments started in late 2017 and were focused on accelerating existing performance analysis simulations using distributed computing. The initial goal was to build a pipeline inside a popular parametric CAD software which would execute a raytracing-based simulation with increased speed in the cloud, yet as seamlessly as the existing pipeline would on a single workstation. The case-study, which was a tower, provided a 3.8 times speedup by using 5 machines with 20 cores (40 threads) each in the cloud compared to a single 20-core Intel(R) Xeon(R) CPU E5–2660 v3 @ 2.60 GHz with 64 GB of RAM (Fig. 2).

The success of this case-study sparked many ideas on how this could be efficiently scaled up to run various complex analysis even on large-urban scale models. Such an approach would not only need to automatically coordinate jobs on hundreds of machines simultaneously but also communicate and store the generated data between them while visualizing the results. In this scenario, a rigorously tested and optimized massing option becomes a product of a software platform – an entire automated ecosystem where design operations are made by various automatic or semi-automatic microservices [12] and data flows seamlessly between various disciplines. Some of the components in such a system like classic parametric modeling [13] with performance simulations [9, 14] and multi-objective optimization [15] or interactive data dashboards [16] are not new to the AECO industry. However, combining them with distributed computing at scale and both modern web and cloud technology would require considerable new research

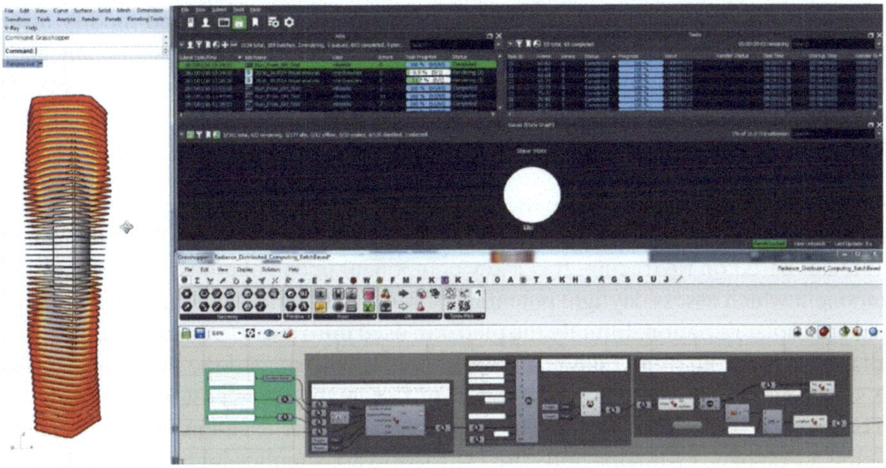

Fig. 2. Distributed daylight factor simulation. Each test floorplate was simultaneously calculated by 5 machines with 40 threads each using Rhino and Grasshopper software as an interface.

far exceeding traditional computational design. Since there is no architectural software that could take full advantage of such a technology stack, coordination of knowledge and expertise from different domains including full-stack software development, data science or parallel computing [17] would be required. Both DevOps and DesignOps workflows and practices would provide an efficient framework for how to drive this coordinated cross-domain software development effort while ensuring both consistency and high-quality design output.

Having that in mind, the first prototype of the platform was launched in early 2018 and tested on a large-scale urban project [18]. It had the core components in place namely a database with interoperability models, an evolutionary optimization solver capable of distributing performance simulations and a rudimentary webpage to display the results. The components were split into microservices ensuring that in the future small but quick and incremental changes could be implemented and immediately tested, using Continuous Integration and Deployment on live projects. The approach was successful and set out a roadmap for future development to improve the speed and reliability of the platform.

2.1 Producing Design Data at Scale

Interoperability. Having a common way of communicating coherent information between the various parts of the platform is essential for an uninterrupted workflow. In AECO different teams and disciplines use different tools and data exchange formats, which is a major obstacle for automation. For that reason, we chose to integrate our inhouse interoperability standard based on JavaScript Object Notation (JSON) [19] handling both geometry and analysis data. We developed shared data schemes which were made available to all disciplines and users in a format familiar to them through our bespoke messaging application. This eliminates the need to exchange static files in the traditional way. Interoperability allows us to streamline, compress and modify design data on demand through different components of the platform. It also enables any geometry of selected design options to be seamlessly imported directly into a CAD package and evaluated by an architectural team. The direct feedback was used to modify both parametric model and analytical objectives.

Simulation Engines. Essential to the success of a delivery cycle is the speed of which the relevant performance simulations can run. Some the most common analyses in architecture are based on raytracing. It is a key component in calculations of the vertical sky component which assess daylight potential, of sunlight hour analyses and when calculating annual solar radiation. Initially the platform used third party simulation software like Radiance [20] through the Ladybug plugin [21] and distributed the analysis on CPUs. But since raytracing also is a key component for creating renderings, which are used to visualise design options and assess quality of views for each option, we decided to develop a bespoke ray tracing software. This software supplies highly parallelized and performant analysis and rendering pipelines by utilizing Graphical Processing Units (GPUs). It implements the CUDA-centric [22] Nvidia Optix API [23], takes advantage of RTX-technology [24] and provides a comprehensive raytracing framework combined with lightweight scene representation.

With control over both geometry, ray generation, ray intersections and data transfer, the engine enables a flexible multi-GPU workflow which can easily be extended with new analysis pipelines to fit the varying needs of design projects. These pipelines can reach speeds of up to 15 giga rays/s on a single Nvidia A6000 GPU and run city scale models with hundreds of thousands of analysis points in seconds. The speedup offered by these bespoke GPU analyses does not only greatly improve the efficiency of the delivery cycle, but it also allows to run analyses at a higher resolution. Compared with previous studies [11], the analysis runs on average 6 times faster while being more accurate, with a 260x higher resolution.

2.2 Automated Reporting

The ability to quickly interpret and display results from tests and simulations is a key component in every feedback system. From extensive discussion with design teams, we developed a two-level reporting.

In the first stage a fast, triage-like, near real time feedback from the simulations was need. Since the platform could generate and test thousands of design options within hours, a tool for traversing their performances across key objectives and comparing them against each other was required. The tool had to be interactive and easily accessible to all stakeholders, so it was built as modern web application running in a web browser. For the front-end technologies like Bootstrap, developed initially at Twitter for responsive and consistent looking layouts, and D3.js for data-driven visualisation were used. Design data was pulled from options' database using JSON-based interop data formats.

When candidate options were selected, a second more in-depth feedback stage was required in the form of a detailed PDF report. To previously produce such reports required considerable manual work. A specialist software such as Adobe InDesign had to be opened to place images, adjust a layout, write descriptions and analysis results while ensuring consistent graphical quality, something that could take 2–3 h per report. A manual workflow would therefore be highly inefficient, given that the platform delivers thousands of analyses results. Thus, the process was automated by creating a reporting microservice driven purely from data using Node.js technology. It was developed using React [25], a front-end JavaScript library for building user interfaces created and maintained by Facebook (Meta). The service exposes different endpoints for document types which, based on JSON data, could automatically return a PDF document within a minute (Fig. 3).

Fig. 3. Sample payload sent from a client to the microservice defining the template to be used and the component types the template exposes. Sample template for a page containing a grid of captioned images in JSX.

3 Findings – The Process

The development of the platform has been highly successful. Since 2018 it has been used on 23 projects ranging from 50,000 to 2.17M sqm of GFA, most of whom were large-scale masterplans. It has generated over 300,000 design options and conducted 1.3M performance simulations. The decision to break down early design stages to sub-tasks such as project-specific massing creation, performance simulations, reporting and decision making and develop them as independent microservices fall under the DevOps model. As development progressed, the overall optimization design cycle time for a single project was significantly reduced, from 20 to 4 days. A development that could be implemented while the entire system was online, sometimes testing up to 140,000 options per project (Fig. 4). This was possible due to significant improvements in robustness and standardization of the analysis and reporting pipelines as well as to the progressive development of adaptive parametric models covering main building typologies. It also allowed further reduction in the time required to build custom parametric models for live projects from weeks to days. In recent projects the full cycle of model updates and simulation runs has decreased from the initial 1.5 months to 2–3 days per project while handling up to three large-scale optimizations projects simultaneously.

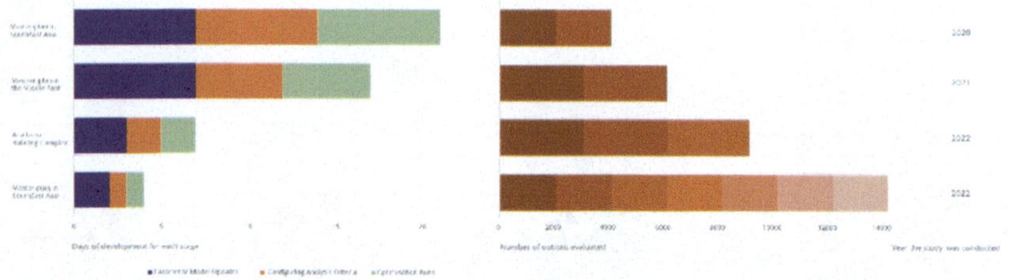

Fig. 4. Comparison between single design cycle time and a total number of options tested for an indicative project.

A good example of the capabilities in the DesignOps approach is the studies prepared for the Guangming Hub project. The project was a large-scale transport-oriented (TOD) design competition in China [11], that required four site-specific typologies covering a total GFA of 2,700,000 m2. The challenge was to produce a wide variety of massing options while maximizing environmental performance, cutting down average walking times and increasing views of greenery. For this study, three agile-like design optimizations sprints were conducted producing 18,000 options and running 80,000 analyses in total. Design solutions derived from this process and selected by the architectural team from each iteration were refined and used as basis for subsequent optimisations. The results from the last run were then manually finetuned and postprocessed by the architectural team to become the basis of the final massing distribution for the completion-wining proposal.

4 Conclusions

As demonstrated in the case study, the use of recent advancements in cloud computing supported by implementation of the latest software development standards, like DevOps and its design incarnation – DesignOps, can be transformative to architectural design. It has the capacity to reduce delivery cycles from months to hours by automating many complex design steps. This in turn can give architects an opportunity to rethink current design pipelines and workflows by making them more efficient and data oriented.

Now, this approach requires cross-domain teams of experts collaborating on a single software platform as well as a significant investment in dedicated hardware and high-performance computing infrastructure. Another obstacle is standardisation of both work stages and data formats which has traditionally been a major issue in the AECO industry and makes this methodology initially more suitable for large organisations which integrate many disciplines under a single roof. However, the wide spreading access to massive cloud computing power and data storage for a fraction of the previous cost is beginning to blur the boundaries and will in turn result in a wave of new cloud-based applications that expose elements of design-related functionality as a service. They could then be consumed by various stakeholders and integrated into their design workflows. Their adoption would be a direct reflection of both the design and data culture of the organisation that implements them. The implementation would require a comprehensive DesignOps approach including the combination of philosophy and tools of a given organization which in turn can facilitate an organization's capacity for delivering high-quality designs at scale and at a much faster pace than traditional practices.

References

1. Dornenburg, E.: The Path to DevOps. IEEE Software **35**(5), 71–75 (2018)
2. Kniberg, H., Ivarsson, A.: Scaling Agile at Spotify with Tribes, Squads, Chapters and Guilds (2012). [Online]. Available: https://blog.crisp.se/wp-content/uploads/2012/11/SpotifyScaling.pdf
3. Beck, K., et al.: Manifesto for Agile Software Development (2001). [Online]. Available: http://www.agilemanifesto.org/
4. Ries, E.: The Lean Startup: How Today's Entrepreneurs Use Continuous Innovation to Create Radically Successful Businesses. Crown Business, New York (2011)
5. Malouf, D.: DesignOps Handbook (2017). [Online]. Available: https://www.designbetter.co/designops-handbook/introducing-designops
6. Cleave, A.: DesignOps at Airbnb, How we manage effective design at scale (2018). [Online]. Available: https://airbnb.design/designops-airbnb/
7. Salesforce: (2020). [Online]. Available: https://medium.com/salesforce-ux/team-ops-and-product-ops-the-perfect-designops-pair-eb6508f1b435. Accessed 11 June 2022
8. Nick, B.: Shopify Partners. Shopify (13 Dec 2018). [Online]. Available: https://www.shopify.co.uk/partners/blog/designops. Accessed 11 June 2022
9. Li, S., Liu, L., Peng, C.: A review of performance-oriented architectural design and optimization in the context of sustainability: dividends and challenges. Sustainability **12**(4), 1427 (2020)
10. Stamatios, P., et al.: SandBOX - An Intuitive Conceptual. In: Impact: Design With All Senses. DMSB 2019. Cham (2020)
11. Kosicki, M., Tsiliakos, M., ElAshry, K., Tsigkari, M.: Big Data and Cloud Computing for the Built Environment. In: Bolpagni, M., Gavina, R., Ribeiro, D. (eds) Industry 4.0 for the Built Environment. Structural Integrity, vol 20. Springer, Cham (2022). https://doi.org/10.1007/978-3-030-82430-3_6
12. Newman, S.: Building Microservices. O'Reilly Media, Sebastopol (2015)
13. Martyn, D.: Rhino Grasshopper. AEC Magazine **42**(May/ June), 14–15 (2009)
14. Nguyen, A., Reiter, S., Rigo, P.: A review on simulation-based optimization methods applied to building performance analysis. Appl. Energy **113**, 1043–1058 (2014)

15. Rutten, D.: Galapagos: On the Logic and Limitations of Generic Solvers. Architectural Design, 132–135 (2013)
16. Chaszar, A., von Buelow, P., Turrin, M.: Multivariate Interactive Visualization of Data in Generative Design (2016)
17. Branke, J., Schmeck, H., Deb, K., Reddy, S.: Parallelizing multi-objective evolutionary. In: Proceedings of the 2004 Congress on Evolutionary, pp. 1952–1957 (2005)
18. Kosicki, M., Tsiliakos, M., Tsigkari, M.: HYDRA: Distributed Multi-Objective Optimization for Designers. In: Design With All Senses. DMSB 2019 (2019)
19. Crockford, D., Morningstar, C.: Standard ECMA-404 The JSON Data Interchange Syntax (2017). Patent https://doi.org/10.13140/RG.2.2.28181.14560
20. Ward, G.J.: The RADIANCE lighting simulation and rendering system. In: Computer Graphics (Proceedings of '94 SIGGRAPH conference) (1994)
21. Roudsari, M., Pak, M., Smith, A., Gill, G.: Ladybug: a parametric environmental plugin for grasshopper to help designers create an environmentally-conscious design. In: Proceedings of BS 2013: 13th Conference of the International Building Performance Simulation Association (2013)
22. Sanders, J.A.K.E.: CUDA by example: an introduction to general-purpose GPU programming. Addison-Wesley Professional (2010)
23. Parker, S.G., et al.: OptiX: a general purpose ray tracing engine. ACM Transactions on Graphics **29**(4) (2010)
24. Burgess, J.: Rtx on—the nvidia turing gpu. IEEE Micro **40**(2), 36–44 (2020)
25. React: (2022). [Online]. Available: https://reactjs.org/

Introducing Agent-Based Modeling Methods for Designing Architectural Structures with Multiple Mobile Robotic Systems

Samuel Leder[1,2(✉)] and Achim Menges[1,2]

[1] Institute for Computational Design and Construction, University of Stuttgart, Stuttgart, Germany
samuel.leder@icd.uni-stuttgart.de

[2] Cluster of Excellence IntCDC: Integrative Computational Design and Construction for Architecture, University of Stuttgart and Max Planck Institute for Intelligent Systems, Stuttgart, Germany

Abstract. As technology for multiple robotic systems (MRS) is becoming more and more robust, such systems are beginning to be introduced for various applications, such as within the Architecture, Engineering, and Construction (AEC) industry. Introducing MRS to construction requires a radical change to the current practices in the AEC industry as there currently exists little to no precedents for small, agile machines working on construction sites. Beyond the physical hardware, sensing communication and coordination strategies necessary to deploy MRS, the methods for designing structures assembled by MRS must be considered as they can influence what is possible with the novelties inherent to construction with such systems. In order to approach the question of design, this paper aims to break away from current standards of top-down design methods by introducing two agent-based modeling and simulation (ABMS) approaches for designing structures to be assembled with MRS that consider geometric and fabrication constraints in the process of design. Each approach is outlined at a conceptual level, further explained using an existing MRS at the operational level, and then analyzed based on its general workflow, interactivity, and adaptability. By providing the various approaches, we aim to understand how, not only the MRS themselves but, the method for designing structures with such systems can help to achieve the goal of inexpensive, adaptive, and sustainable construction promised by the application of MRS in the AEC industry.

Keywords: Multiple robotic systems (MRS) · Collective robotic construction (CRC) · Agent-based model and simulation (ABMS) · Robotic fabrication · Generative design

C. Gengnagel et al. (Eds.): DMS 2022, *Towards Radical Regeneration*, pp. 71–83, 2023.
https://doi.org/10.1007/978-3-031-13249-0_7

1 Introduction

From wheeled robots vacuuming our living spaces to drones delivering packages between retailers and consumers, mobile robots are currently revolutionizing our everyday lives including applications in both the public and private sectors. Current research on multiple robotic systems (MRS) is showcasing their potential application within the Architecture, Engineering, and Construction (AEC) industry [1]. This has included the co-design of mobile robots together with architectural systems that allow for the robots to build structures much larger than the machine themselves (Fig. 1). In contrast to existing practices in construction automation which generally deploy fixed position industrial robots, MRS are composed of teams of small, inexpensive mobile robots, which navigate the construction site. These teams of mobile robots can further inhabit the structure which they assemble, giving them the ability to rearrange, maintain and disassemble a building on the fly according to site, user, or environmental conditions. Therefore, the application of MRS in the life cycle of a building requires a transformation to current practices in the fields of AEC.

Fig. 1. MRS developed for construction processes. Heterogeneous team of bespoke robots for filament structures [2] (Left) and unmanned aerial vehicles (UAV) for the assembly and rearrangement of canopy structures [3] (Right).

The process of designing an architectural structure to be assembled or even further rearranged by a MRS is one of such practices that must be completely reenvisioned from existing architectural practice. The process of designing such a structure is a complex question which must not only address the novelties of the MRS and its highly unrestricted workspace but furthermore address issues of the material system, design intent or changes in the construction environment. However, the most common method for designing structures to be built with MRS takes a top-down approach by defining and then refining blueprints without specific relationship to the MRS. This highly traditional approach, which limits consideration of many of the potential benefits of deploying a MRS including the potential of using the intelligence, movement, or collaboration of the physical robots to inform or influence the design process, is mismatched to the technologically advanced construction automation systems in which it deals.

Recent research, on the other hand, is showcasing how agent-based modeling and simulation (ABMS) can integrate generation and materialization processes into a single workflow in order to break away from the rationalization based process of top-down design [4]. The characteristics, properties, and constraints of a physical building system as well as any of the accompanying fabrication or assembly system can be encoded in the ABM to allow for the complexity of the systems to be considered. ABMS is therefore one promising method to explore structures which can be built by MRS as it would allow for a bottom-up design process, in which the active drivers of design inform the exploration of emergent complexities.

The approach to how the agent-based model (ABM) is defined, however, can have direct implications on the overall design process, which can result in processes not dissimilar to top-down design processes where blueprints are algorithmically post-processed to derive assembly sequences for the robots to build the structure. To clarify the range of design processes for structures to be assembled by MRS and provide a framework for further discussion, this paper introduces and discusses two ABMS approaches. First, the conceptual model behind both approaches is explained and then sample operational models for each approach are elaborated on using an existing MRS.

2 Background: Agent-Based Models (ABMs) in Architecture

ABMS is a computational approach that aims at understanding complex systems through simulating autonomous agents and their interactions within an environment. Due to the ability of the approach to integrate varying constraints across the disciplines of AEC, which potentially even further change over time, ABMS is currently being utilized in architectural research as a design exploration modeling method [5]. This has largely led to the implementation of digital simulations in which the agents represent part of the physical building system and behaviors of the agents are based on the physical properties or constraints, which affect them. Specifically, ABMS for the exploration of fabrication-informed Zollinger lamella structures [6] or shells [7] and the generation of facade designs based on environment, structural and user preference information [8] have been systematically tested and evaluated for architectural design.

More recently with the growing robustness of MRS, ABMs have been created for architectural design exploration in which the agents represent the physical robots themselves. Therefore, the models are not focused on the negotiation of parts of a building, but rather on the movement of the robots in their environments and how they relate to design generation. Although some research was made on path correction for physical robots [9], ABMs that relate to physical hardware are generally used to iterate design options which are then afterwards sent to the physical robot [10, 11]. Furthermore, existing ABMs for MRS are generally derived for specific material-robot systems, therefore not addressing the overall implications of ABMS as applied to the designing of architectural structures with MRS.

3 ABM Approaches

The focal point of this paper is the definition and analysis of two different ABMS approaches aimed at designing structures with MRS. In order to define the two approaches, we used the instructions for defining a simulation as presented by Heath et al. [12]. First, we will outline a conceptual model for each and then further translate the conceptual model to an operational model using an existing MRS co-designed specifically for construction. To describe the conceptual model, we answer the first four questions for the initial development of an ABM from Macal and North [13]. The questions prompt the definition of (1.) the purpose, (2.) agent, (3.) environment and (4.) behaviors of the model.

The answer to the first question, the purpose of the model, is the same for both approaches: the models should derive architectural designs that can be assembled by a MRS. However, the further questions diverge between the two approaches and will be discussed further in following subsections (Fig. 2).

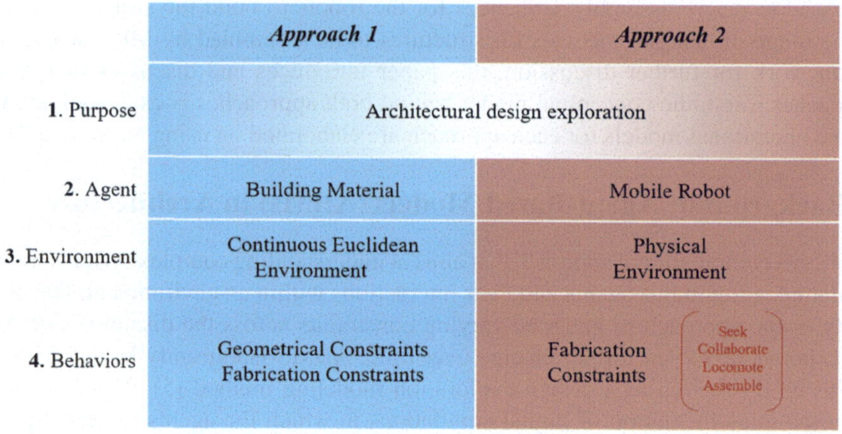

Fig. 2. Outline of conceptual models for ABM approaches based on questions developed in [13].

3.1 Approach 1: Agent Represents Building Material

Agent Representation. Approach 1 considers the agent to represent the building material that is assembled by the MRS into the design outcome of the model. In the case of continuous materials such a fibers, an approach for breaking down the material into discrete parts must be conceived. The attributes which define the agent are the geometric properties of the building material.

Environment. With this approach, the ABM negotiates the placement of the agents in digital space. The environment is thus a continuous digital Euclidean space.

Behaviors. Behaviors, which control the interaction of the agents, are based on the definition of the geometrical and fabrication constraints from the material system and the MRS that are being deployed. Geometric constraints inform the position of the agents relative to each other considering allowable orientations and connections between the material, while fabrication constraints maintain the ability of the design outcome to be assembled considering for example the reachability of each agent by a physical robot.

3.2 Approach 2: Agent Represents a Mobile Robot

Agent Representation. Approach 2 considers the physical mobile robots which are assembling the structure to be the agents. The attributes of the agents relate to the abilities of the robots themselves including such parameters as battery life and degrees of freedom (DOF).

Environment. In Approach 2, the environment transitions from an empty Euclidean space in which the agents interact to include information from the physical environment in which the robots are acting. This includes the status of the already assembled structure as well as other information known about the physical environment, for example the free volume in which the agents can build.

Behaviors. In this approach, the behaviors relate explicitly to how the robots would act in the real world. This can vary based on the instance of the mobile robot being used. Figure 2 expresses some examples of what these behaviors might be: seeking material to be assembled, locomoting in the environment, collaborating with another robot, or assembling material into the structure. In examples of MRS which work with continuous materials, the placement of material is a result of the movement of the robots, meaning that locomotion and assembly can be conflated into one behavior. Geometric constraints in this approach can be embedded in the assembly behavior of the agents, giving definition to how and where in the structure an agent can assemble something.

4 Case Study: Implementation of the ABM Approaches

To help clarify the two approaches and test their feasibility, two operational ABMs were created using the ICD ABM framework developed at the Institute for Computational Design and Construction (ICD) at the University of Stuttgart (Groenewolt et al. 2018). The goal of the framework is to provide an open code base written in both C# and Python to aid researchers and professionals in the AEC industry in the creation of ABMs. The core of the framework provides the basic functionalities for developing ABMs, while further application-specific and expert libraries have been developed since the creation of the framework, to aid with more targeted problems. The creation of the ABMs for this research have continued to lay the foundation for a new application-specific library for MRS (Fig. 3).

The two operational models are based on an MRS developed as a part of the Cluster of Excellence Integrative Computational Design and Construction for Architecture at the University Stuttgart [14]. The MRS within the research is composed of single-axis robotic actuators which must leverage linear timber struts for any form of locomotion or general movement (Fig. 4). Planar timber structures are assembled through the arrangement and rearrangement of multiple robotic actuators together with the timber struts.

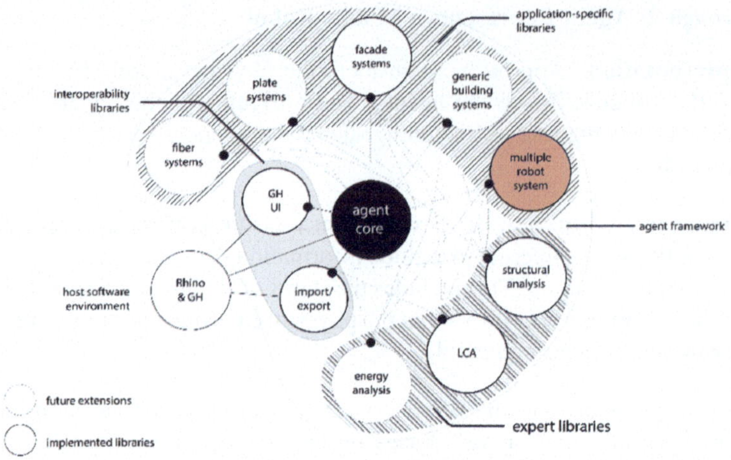

Fig. 3. Schematic overview of ICD ABM Framework, introducing MRS application specific library.

Fig. 4. Photograph of a MRS based on collaboration of single axis robotic actuator and timber struts for which the two operational models were developed.

For each approach as visualized in Fig. 5, the agent classes are derived from the CartesianAgent class of the ICD ABM Framework, inferring their ability to move in cartesian space at each iteration of the model. Figure 6 shows a Unified-Modeling-Language (UML) description of how the agent classes are derived from the framework. Using Approach 1, the agents represent timber struts, containing information on their cross section and length, and adjust their position with each iteration of the model to form planar assemblies (Fig. 7 Left). With Approach 2, the agent, rather than representing a single robotic actuator, represents a kinematic chain or combinations of timber struts and robotic actuators. As actuators cannot move on their own, defining the agent as a kinematic chain would give specification to the locomotion and manipulation abilities of each agent. At each iteration of the model, the agents change their position based on their current state with the aim of assembling a new strut into the environment (Fig. 7 Right).

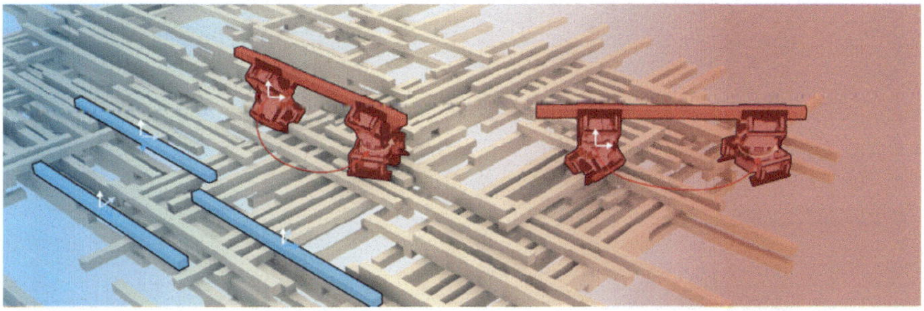

Fig. 5. Two ABM approaches graphically expressed using the MRS from [14]. Agent represents a timber struts, in blue, and Agent represents a kinematic chains composed of robotic actuators and timber struts, in red.

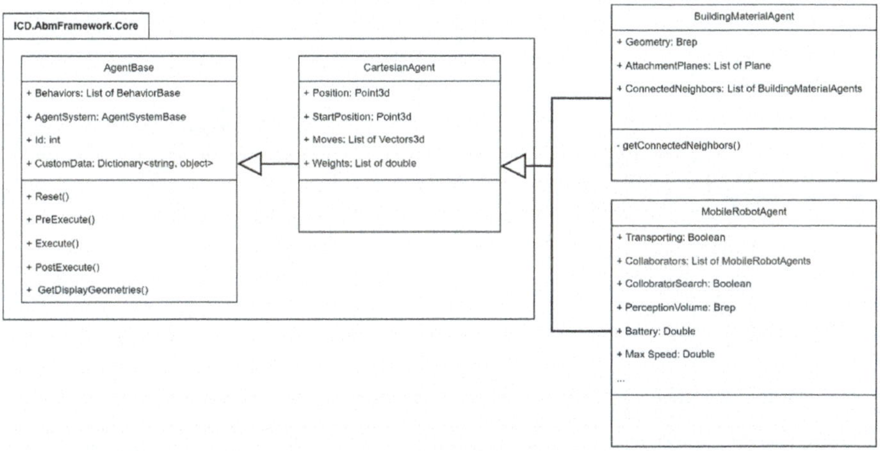

Fig. 6. UML description of agent classes derived for the two approaches

To give further definition to the behaviors of the agents in each approach, specifically how the agents move in Approach 1 and how the agents assemble material in Approach 2, mathematical fields were used to express design intent for the design outcome and stored in the environment of the agents (Fig. 7). The fields can give definition to the design outcome on a local or global level depending on its scale as related to the size of a single timber strut. Scalar fields are utilized to define the general density of timber struts and the priority of placing struts in specific areas, while vector fields define the general orientation of struts. The fields are an additional mechanism for a designer to interact with the ABMs, beyond adjusting the behaviors of the agents.

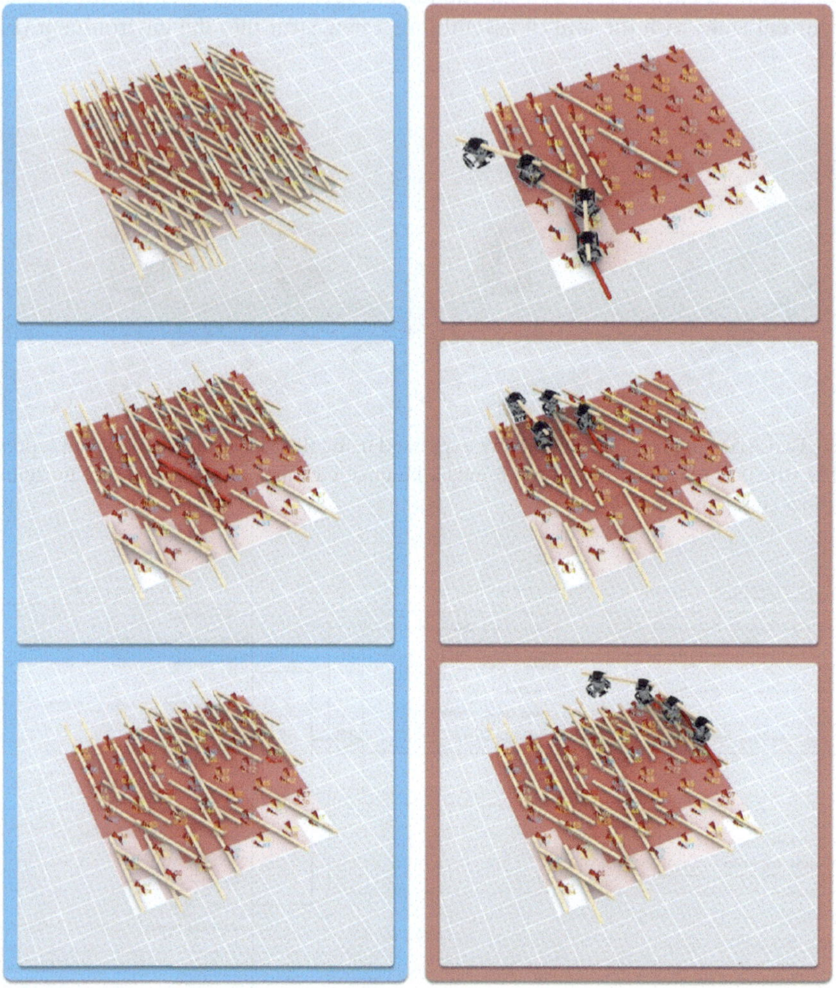

Fig. 7. Two ABM approaches shown as the model iterates with visualized design intent: Agent represents building material (Left) and agent represents a mobile robot (Right). The colors of the mesh faces represent a scalar value for density between struts, vectors represent orientation of struts and the numbers indicate placement priority.

5 Design Process Discussion

Although the two ABM approaches are viable options for designing structures using teams of mobile robots, each approach has a direct implication on how the (i) general workflow, (ii) interactivity, and (iii) adaptability of the overall design process functions. The observations in this section were made after conducting design explorations with each of the two operational ABMs described in the previous section (Fig. 8).

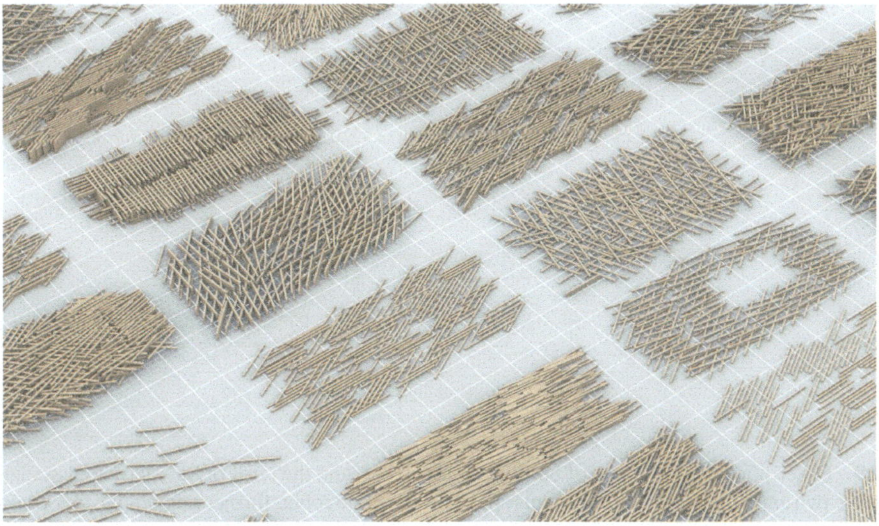

Fig. 8. Design explorations made using both approaches.

5.1 General Workflow

In the terms of general workflow, the use of each approach to ABM as it relates to the existing phases of building construction was analyzed (Fig. 9).

When implementing Approach 1 in which the agent represents building material, the ABM model converges on possible design outcomes. Each design outcome generated from the model must be then decomposed into an assembly sequence and then further into robotic motion plans, which can then be sent to the physical system for execution. As such, the design phase of the structure ends once the planning phase begins. Although the generation of the design outcome differs from the linear organization of current design practice in that it integrates geometric and fabrication constraints, the entire design process is similar in that the design, planning and construction phases of a building are explicitly delineated.

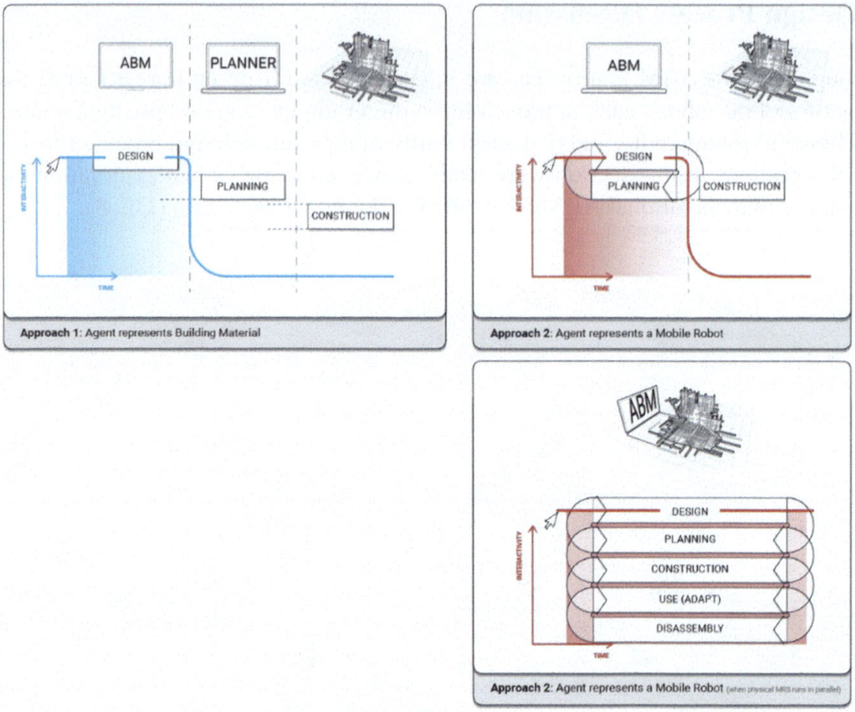

Fig. 9. Description of the phases of building construction and the level of interactivity associated with the two approaches. When the agent represents a mobile robot, further potential to incorporate construction, use, and disassembly phases into a single process is possible when the MRS runs at the same time as the ABM.

On the contrary, the motion of the robots is explicitly related to the design outcome in Approach 2. As the architectural design emerges, the planning required to achieve the design is determined in parallel. Syncing of the ABM to the physical MRS can allow for the construction phase of the structure to occur at the same time and even further accommodate any adjustments in the use or performance of the structure and its eventual disassembly.

5.2 Interactivity

In Approach 1, the position of the agents self-organize based on the defined behaviors in order to converge to a dynamic equilibrium. As such, the model begins with a visualization of the building material present in the system. As the model is running, the behaviors of the agents negotiate to reveal a final design outcome, which can be visualized in real time. A designer can therefore interact with the model with direct visual feedback on how the final design outcome would change. The various modes of interacting with the model developed for Approach 1 includes adding and removing agents, fixing or moving agents to specific positions, adjusting weights of behaviors on an individual agent or global level and adjusting the design intent fields (Fig. 10).

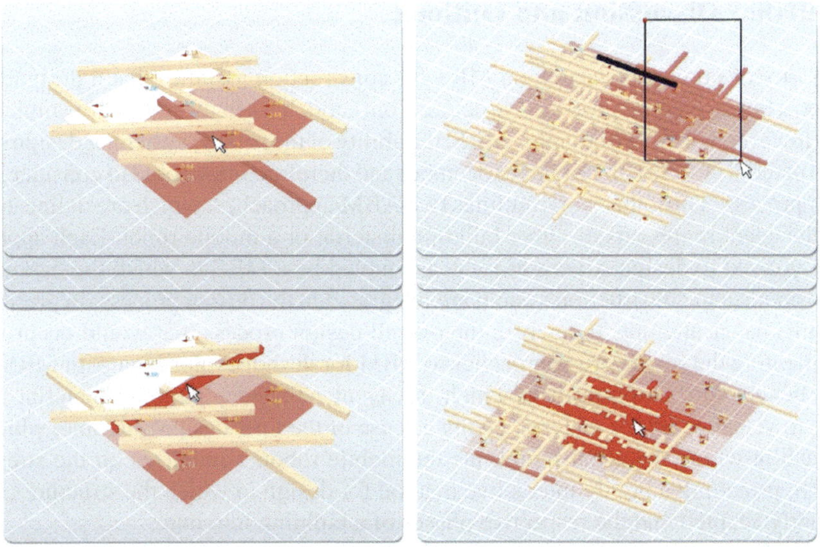

Fig. 10. Two modes of interactivity with ABM Approach 1: Single strut selection and rotation (Left) and multiple strut selection and movement to a fixed position (Right).

In Approach 2, the generation of final design outcomes is completed iteratively, considering the process of assembly. The agents assemble material to the structure as the model runs. In interacting with the model, direct feedback can be visualized on how the agents behave, however, not on how the final design outcome of the model will appear. As the agent represents the physical robots, the interaction of the designer becomes less about adjusting a final form but rather as a choreographer of the entire process, considering its general rhythm and flow. The various modes of interacting with the model developed for Approach 2 included adding, removing, or repositioning single struts, starting and stopping the design/assembly process, adjusting weights of behaviors on an individual agent and adjusting the design intent fields.

5.3 Adaptability

One opportunity provided by MRS is the potential of robots to rearrange the structure as they can inhabit it over its lifetime. The adaptability of the design outcome in the design process is such another evaluation criteria. In Approach 1, any changes to the design requires new plans to be generated, which can result in extended processes of exchange of information between the designing and planning sides of the overall design process. However in Approach 2, any desired adjustments to the design outcome as it is being assembled can be reflected in real time.

6 Further Discussion and Outlook

A majority of previous research on MRS for construction assumes that a designer can develop a blueprint for a proposed structure from scratch, as in the notable example from [15]. However, with more complex MRS, the ability of the robotic system or design space of the material system can be hard to decipher and therefore impossible to consider in the design process. Thus, this work outlines two ABM approaches, which are delineated by what the agent represents as either building material or a mobile robot. Each approach can be utilized to design structures to be assembled by a MRS in which the parameters of the robotic and construction system are integrated in the design process. In presenting each, this paper attempts to analyze the overall design process that would occur when implementing the respective approaches to ABM for design exploration with MRS. The major benefit of the approach in which the agent represents building material is the ability to visualize all the material giving a sense of the final design outcome, while the approach in which the agent represents the mobile robots capitalizes on the strengths of MRS, providing a highly interactive method for design in which the structure can be constantly adapted and the respective phases of a building into one.

Considering the goal of inexpensive, adaptive, and sustainable construction promised by the application of MRS in AEC, criteria of the automation system, construction system, and potentially even changing design intent should be considered in the design process. One promising design methodology to do this as outlined in this paper is ABMS. Further research will therefore be conducted to explore the potential of ABMS as it relates to the assembly of structures with MRS. Integration of global performance metrics beyond the more abstract mathematical fields into the ABMs will be considered for the generation of design outcomes with varied architectural and structural properties.

Acknowledgments. This research was funded by the Deutsche Forschungsgemeinschaft (DFG, German Research Foundation) under Germany's Excellence Strategy with grant number EXC 2120/1–390831618.

References

1. Petersen, K.H., Napp, N., Stuart-Smith, R., Rus, D., Kovac, M.: A review of collective robotic construction. Sci. Robot. **4**(28), 1–28 (2019)
2. Yablonina, M.., Menges, A.: Towards the development of fabrication machine species for filament materials. In: Robotic Fabrication in Architecture, Art and Design 2018 [Proceedings of the RobArch Conference 2018], pp. 152–166 (2018)
3. Wood, D., Yablonina, M., Aflalo, M., Chen, J., Tahanzadeh, B., Menges, A.: Cyber physical macro material as a UAV [re]configurable architectural system. In: Willmann, J., Block, P., Hutter, M., Byrne, K., Schork, T. (eds.) ROBARCH 2018, pp. 320–335. Springer, Cham (2019). https://doi.org/10.1007/978-3-319-92294-2_25
4. Baharlou, E., Menges, A.: Toward a behavioral design system: an agent-based approach for polygonal surfaces structures. In: ACADIA 2015: Computational Ecologies: Design in the Anthropocene [Proceedings of the 35th Annual Conference of the Association for Computer Aided Design in Architecture (ACADIA) ISBN 978-0-692-53726-8] Cincinnati, USA, pp. 161–172 (2015)

5. Groenewolt, A., Schwinn, T., Nguyen, L., Menges, A.: An interactive agent-based framework for materialization-informed architectural design. Swarm Intell. **12**(2), 155–186 (2017). https://doi.org/10.1007/s11721-017-0151-8
6. Tamke, M., Riiber, J., Jungjohann, H., Thomsen, M.R.: Lamella flock. In: Ceccato, C., Hesselgren, L., Pauly, M., Pottmann, H., Wallner, J. (eds.) Advances in Architectural Geometry 2010, pp. 37–48. Springer Vienna, Vienna (2010). https://doi.org/10.1007/978-3-7091-0309-8_3
7. Schwinn, T., Menges, A.: Fabrication agency: landesgartenschau exhibition hall. Archit. Des. **85**, 92–99 (2015)
8. Pantazis, E., Gerber, D.: A framework for generating and evaluating facade designs using a multi-agent system approach. Int. J. Archit. Comput. **16**(4), 248–270 (2018)
9. Vasey, L., et al.: Behavioral design and adaptive robotic fabrication of a fiber composite compression shell with pneumatic formwork. In: ACADIA 2015: Computational Ecologies: Design in the Anthropocene [Proceedings of the 35th Annual Conference of the Association for Computer Aided Design in Architecture (ACADIA) ISBN 978-0-692-53726-8] Cincinnati, USA, pp. 297–309 (2015)
10. Kayser, M., et al.: Fiberbots: Design and digital fabrication of tubular structures using robot swarms. In: Willmann, J., Block, P., Hutter, M., Byrne, K., Schork, T. (eds.) ROBARCH 2018, pp. 285–296. Springer, Cham (2019). https://doi.org/10.1007/978-3-319-92294-2_22
11. Pietri, S., Erioli, A.: Fibrous aerial robotics-study of spiderweb strategies for the sesign of architectural envelopes using swarms of drones and inflatable formworks. In: ShoCK! - Sharing Computational Knowledge! - Proceedings of the 35th eCAADe Conference - Volume 1, Sapienza University of Rome, Rome, Italy, pp. 689–698 (2017)
12. Heath, B.H., Ciarallo, F.W., Hill, R.R.: Validation in the agent-based modelling paradigm: problems and a solution. Int. J. Simul. Process Model. **7**(4), 229–239 (2012)
13. Macal, C.M., North, M.J.: Tutorial on agent-based modelling and simulation. J. Simul. **4**(3), 151–162 (2010)
14. Leder, S., et al.: Co-design in architecture: a modular material-robot kinematic construction system. In: 2020 IEEE/RSJ International Conference on Intelligent Robots and Systems (IROS) Proceedings of Workshop on Construction and Architecture Robotics (2020)
15. Petersen, K.H., Nagpal, R., Werfel, J.K.: Termes: An autonomous robotic system for three-dimensional collective construction. In: Robotics: Science and Systems VII, MIT Press, pp. 257-264 (2011)

Algorithmic Differentiation for Interactive CAD-Integrated Isogeometric Analysis

Thomas Oberbichler(✉) and Kai-Uwe Bletzinger

Chair of Structural Analysis, Technical University of Munich, Arcisstr. 21, 80333 Munich, Germany
thomas.oberbichler@tum.de

Abstract. Algorithmic differentiation is used to analyse the interaction between input and output parameters of arbitrary computational models. We discuss the application of algorithmic differentiation at the interface between architecture and civil engineering in the field of CAD-integrated structural analysis and form finding.

1 Introduction

The integration of analysis tools in computer aided design (CAD) enables structures to be generated and explored intuitively. To achieve a high degree of interactivity, the use of natural CAD geometric parametrisation – for example NURBS – is also desirable at the analysis stage (Cottrell et al. 2009). Beyond NURBS, modern CAD systems provide other descriptions of free-form geometries, such as discrete meshes or subdivision surfaces. To perform various types of analysis with different geometric descriptions, it is necessary to generalise the process of CAD-integrated isogeometric analysis (IGA) while also increasing the computational speed.

To address this issue, we present an efficient, and modular approach for implementing CAD-integrated analysis based on algorithmic differentiation (Griewank and Walther 2008). A feature-rich digital toolbox can be derived from a set of highly optimised mechanical and geometric building blocks (Oberbichler et al. 2021).

This paper consists of four sections. In Sect. 2 we present the fundamentals of algorithmic differentiation (AD) that are required in the context of the paper. We introduce the idea of algorithmic modelling based on a computational graph. Moreover, we present hyper-dual numbers as a possible implementation of AD. In Sect. 3 we apply these principles to isogeometric structural analysis and form finding. We highlight the important aspects for an interactive CAD-integrated implementation. Finally, we provide a conclusion and an outlook for further research.

C. Gengnagel et al. (Eds.): DMS 2022, *Towards Radical Regeneration*, pp. 84–95, 2023.
https://doi.org/10.1007/978-3-031-13249-0_8

2 Algorithmic Derivatives

Parametric models not only provide a concrete solution, but also a blueprint for constructing the solution based on certain input parameters. This blueprint can be represented as an algorithm on the computer. By executing it for different parameters, a wide variety of designs can be quickly generated (Fig. 1). However, it is often interesting to determine the parameters in such a way that a certain design is created. This is called an inverse problem because the underlying algorithm must be inverted to solve it. In structural mechanics, an equilibrium of forces for elastic problems can be considered as a minimum energy state. The energy state arising from elastic deformation can be directly calculated. The inverse question in this case is: how must the system be deformed in order that its energy state becomes minimal?

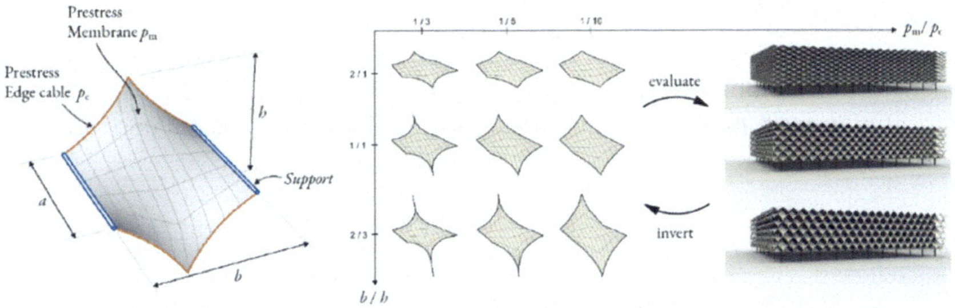

Fig. 1. Parametric model for a façade element. Numerical form-finding can be triggered for different configurations and the effect on the overall model can be considered (Oberbichler et al. 2019).

For nonlinear problems, a direct solution by inversion is usually too complex. The only remaining solution is to evaluate different parameter configurations and validate the output. To automate this process, classical numerical optimisation methods can be used. Evolutionary algorithms (e.g., Galapagos) generate a set of parameter configurations in each iteration and try to improve them based on the most promising candidates. The algorithm behind the problem does not have to be known. A black box is sufficient, which provides an output for certain input parameters. If the problem is a smooth function, the derivation can be used to iteratively improve an existing parameter configuration. First order algorithms which use the gradient of the problem (e.g., Kangaroo) usually deliver faster and more reliable results in such cases. For structural mechanics problems, the second derivative is often calculated (e.g., Karamba). Thus, a quadratic substitute problem can be solved for each iteration step, which increases the convergence speed. In this context, the first order derivative corresponds to the residual force vector and the second order derivative to the tangential stiffness matrix (Fig. 2).

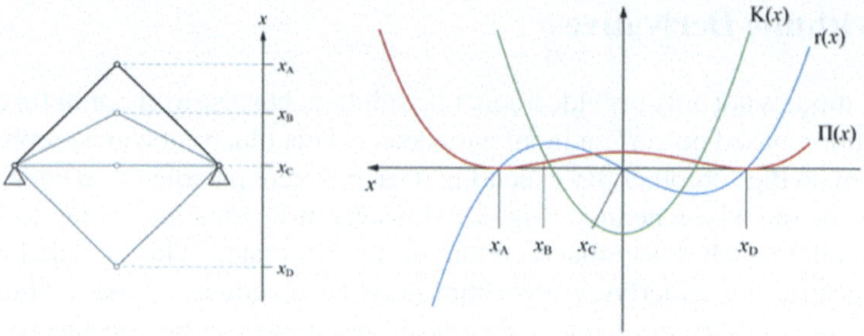

Fig. 2. Plot of the energy functional and its derivatives for a simple system (Oberbichler et al. 2021).

The implementation of this derivative for isogeomertric finite elements represents a significant effort. At the same time, the computation takes a lot of time. If the blueprint or algorithm for the computation of a model is available, the derivatives can be calculated automatically from it by using AD (McCarthy 1958). AD is a technique that transforms an algorithm describing a model into an algorithm that computes the derivatives of the model.

2.1 Computational Graph

In computer aided design (CAD), parametric modelling is supported by visual programming languages (VPL). A common example is Grasshopper for Rhinoceros. Algorithms are assembled from a set of nodes that represent individual basic functions. The newly created compositions can themselves be integrated as nodes (cluster) in more complex workflows. Node-based VPLs such as Grasshopper have the characteristics of functional programming languages. For example, each node represents a function that transforms input parameters into output parameters.

We take advantage of this functional notation and the representation as a flow in which given initial values pass through an algorithm while being transformed step by step to the solution. In principle, this way of thinking can also be transferred to common text-based programming languages.

As a simple example we visualise the expression $f_1 = (s_1 - s_2)/(s_2 \cdot s_3)$ in Fig. 3. First, the difference (blue) and product (green) are computed. Finally, the result (red) is obtained from the quotient of the intermediate results. In this graph we represent the interaction of parameters with edges. We weight the edges with the partial derivatives of the individual suboperations. Edges between two planes correspond to the first order partial derivatives. Edges within a plane correspond to the second order partial derivatives. If a partial derivative is zero, there is no interaction between the corresponding parameters. To improve clarity, we do not draw these edges.

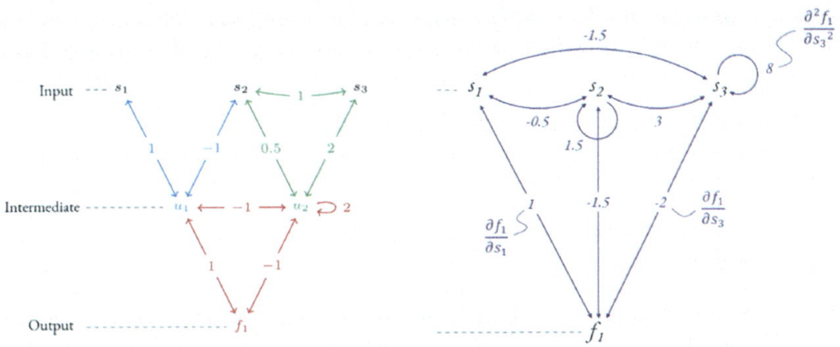

Fig. 3. Computational graph before (left) and after applying the transformation rules (right).

In this way, we obtain the connections between the levels or within the individual levels. However, we are interested in the relationship between the lowest and the highest level. Therefore, we apply three transformation rules to the graph to eliminate the layer of

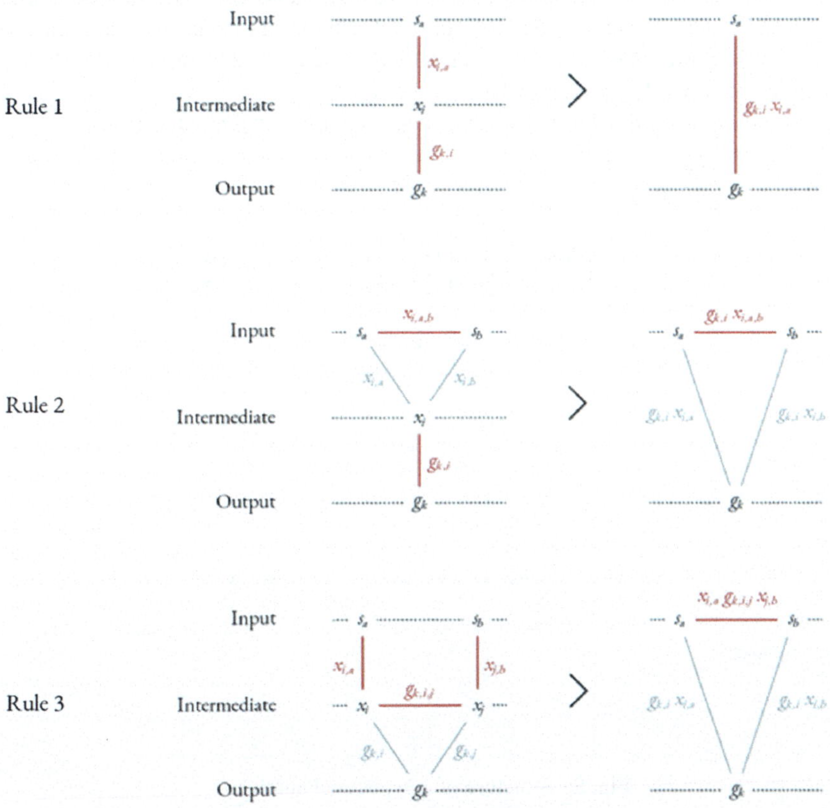

Fig. 4. Transformation rules for evaluating first and second order derivatives.

intermediate results (Fig. 4). The transformation of the graph is essentially a visualisation of the chain rule. The first transformation rule corresponds to the chain rule for the first order derivative. Transformation rules 2 and 3 are a consistent extension for the second order derivative and result from the repeated application of the chain rule. We end up with a new graph where the edge weights represent the direct influence of the input parameters on the output.

2.2 Hyper Dual Numbers

To apply the transformation rules, we make use of hyper dual numbers (Fike and Alonso 2011). A hyper dual number consists of one real and several dual components (Fig. 5). We use the real component to store the result of a function and the dual components to store the derivatives. To perform computations, we define the basic mathematical operations for these hyper dual numbers based on the transformation rules presented above. If an algorithm is evaluated using hyper dual numbers, the result obtained is also a hyper dual number. This not only contains the value of the function but also the values of the derivatives. The functional for different geometric parameterisations can now be evaluated to obtain the corresponding derivatives. To do this, it is only necessary to adjust the dual components of the input parameters. The implementation of the functional remains unchanged. This approach enables a declarative implementation of isogeometric finite elements. The user puts the focus on the implementation of the energy functional, while the derivatives are computed in the background, in a process that can be highly optimised e.g., by vectorisation using SIMD instructions.

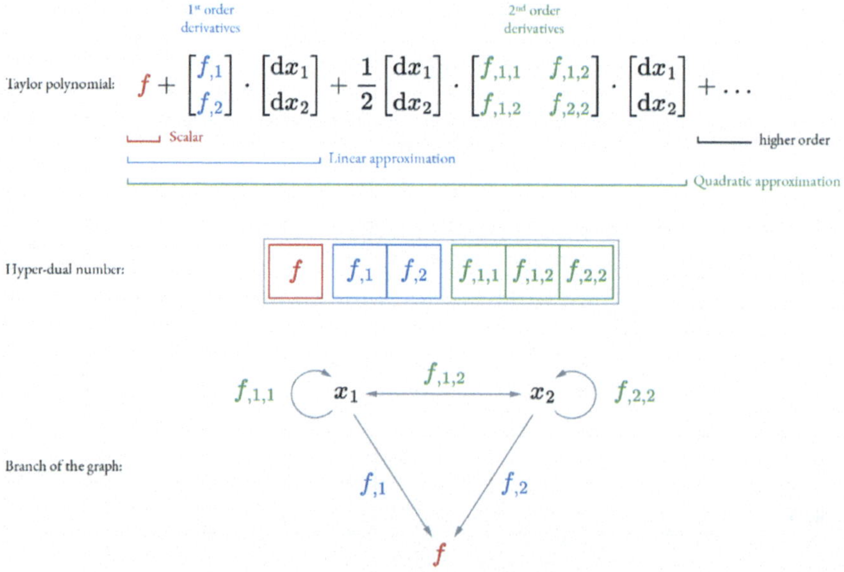

Fig. 5. Structure of a hyper dual number.

3 Application in Isogeometric Analysis

Isogeometric analysis (IGA) involves performing classical structural analysis using smooth geometries that are usually described by NURBS (Fig. 6). One advantage of this approach is that the geometric descriptions provided by a NURBS-based CAD environment can be used directly. Generally, however, geometric models consist not only of NURBS but of several different geometric parametrizations. For free form surface modelling meshes, NURBS and subdivision surfaces are commonly used. Each of these types requires modifications to the finite element formulation. For each combination, individual coupling conditions are implemented to model contacts and supports of the structure. To implement finite isogeometric elements, it is necessary to formulate complex expressions for the residual forces and the stiffness, which are then translated into source code. This is frequently time-consuming.

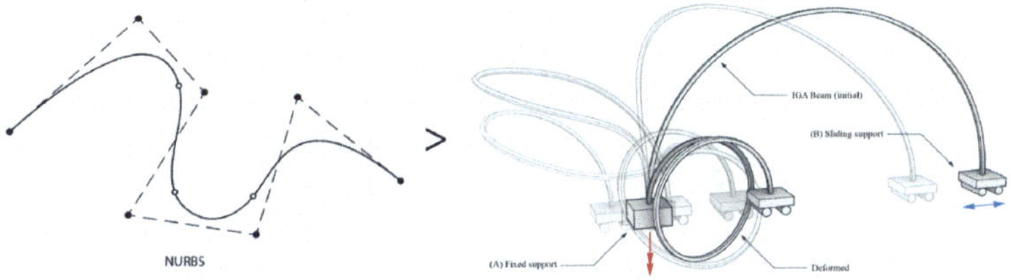

Fig. 6. Isogeometric analysis applies mechanical properties to smooth geometries.

Due to the complexity of the computation involved, an external finite element kernel is required. According to the workflow of a classic pre- and postprocessor, additional interfaces must be designed, implemented, and maintained to enable communication between CAD and the FEM kernel. As the sharing of functionalities and data structures through such interfaces is restricted, CAD functionalities are re-implemented within the FEM kernel and vice versa. Despite this, the computational speed of the approach remains rather poor. Even with a very simple example, the computation time precludes an interactive design.

To address these issues, we propose a modular and automatable approach that enables efficient implementation of IGA. It consists of four key elements: (i) an energy-based formulation of elastic elements from which the required residual forces and stiffnesses can be obtained by using AD to reduce the implementation effort; (ii) AD is applied in adjoint mode which dramatically increases the computational effort while reducing the memory requirements; (iii) a modular framework which allows the combination of different geometric representations with a set of mechanical core-elements, thus simplifying the development and maintenance of an IGA toolbox; (iv) a fully CAD-integrated IGA framework which can take advantage of the functionality provided by the CAD to support high-level features such as trimming and real-time interaction with the structure.

3.1 Automatic Energy-Based Implementation of Finite Elements

To compute the inner energy of a structure, we compare the current geometry with an initial stress-free configuration and weight each member according to its material stiffness which results in a quadratic term. Prestress can be modelled by an additional linear term which allows the implementation of form finding based on the Force Density Method (FDM) and Updated Reference Strategy (URS). We obtain a smooth function that depends on a set of degrees of freedom. A stable equilibrium is characterised by a local energy minimum. To find such a minimum, we use the derivatives with respect to the degrees of freedom (Fig. 7). The gradient indicates how the current configuration should be modified to decrease the energy. The second order derivative can be used to improve the direction and the amplitude of this modification. Moreover, it allows us to distinguish between stable and unstable equilibrium points.

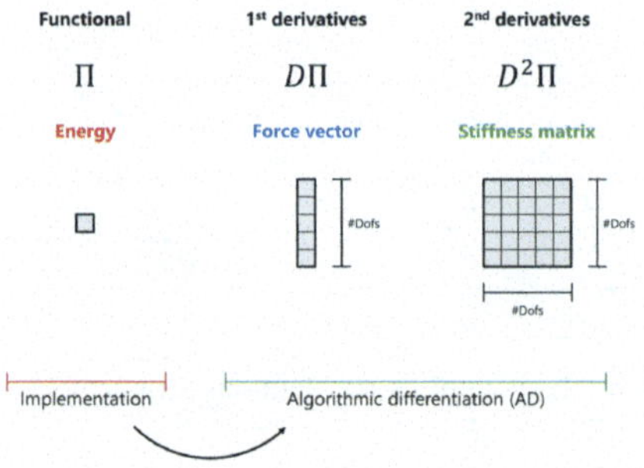

Fig. 7. Obtaining residual forces and stiffness matrix from the scalar energy functional by use of AD.

Existing packages can be used for applying AD to the algorithm. However, the implementations may differ significantly and may be better suited to different purposes. When used in the context of IGA, the following aspects should be considered: (i) the computational graph is identical for all IGA elements of the same type. Therefore, it should not be dynamically regenerated for each calculation. Instead, the graph should be implemented statically and ideally pre-built upon compiling. (ii) As supplied, the common packages support a set of basic mathematical operations. When generating the computational graph, the calculation is decomposed down to the level of these basic operations, making the graph more complex. To avoid this, the AD framework should be extended with IGA-typical functions. We propose a division into three components as shown in Fig. 1 evaluation of geometry (g), difference between reference and current configuration (d) and the evaluation of the energy (p). For the individual components, optimised algorithms can then be used to compute weights of the edges. E.g., the derivatives of the NURBS can be implemented according to (Piegl and Tiller 1995) and reused for all NURBS-based elements. This approach does not change the result but increases the computational speed. At the same time, it is still possible to use an automatically generated graph based on mathematical foundations for individual parts or the entire graph. This can be especially helpful in the development phase of new features.

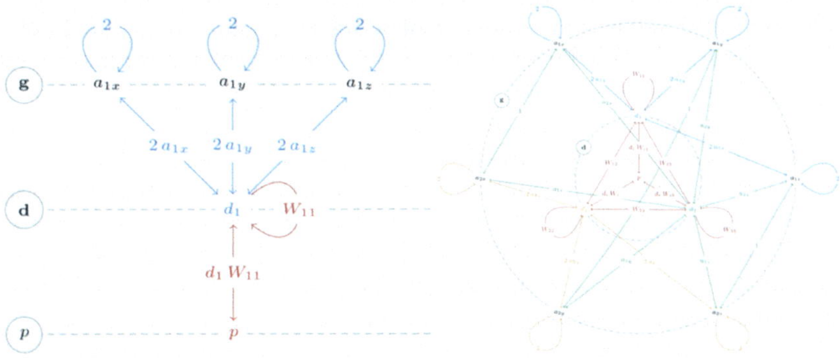

Fig. 8. Computational graph of a cable (left) and a membrane element (right) (Oberbichler et al. 2021).

3.2 Adjoint Formulation

To increase performance, we take advantage of the adjoint method to compute the derivatives. To compute the energy, we start with the current configuration of the degrees of freedom. We evaluate the geometry, the strains and finally the energy functional. The derivatives are usually evaluated in the same order. This constitutes a direct or forward approach of evaluating derivatives which provides the so-called B-Matrix as an intermediate result. It is also possible to evaluate the derivatives in the opposite direction, which corresponds to the adjoint or reversed approach. We simply apply our transformation rules in the reverse order. This does not affect the results but reduces the computational effort. A detailed study of this effect in IGA can be found in (Oberbichler et al. 2021).The benefit is even greater for higher polynomial degrees as well as for more complex elements, such as a nonlinear beam, for which manual optimisation is much more difficult. In addition, this approach avoids dynamic memory allocation and enables an efficient implementation with programming languages commonly used in CAD environments, such as C# or Python.

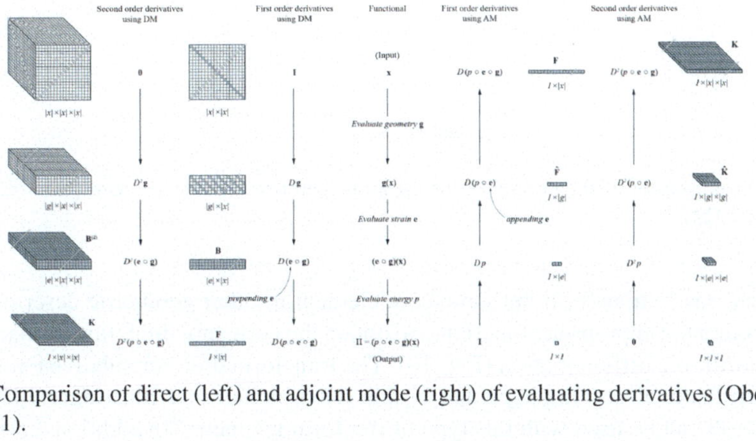

Fig. 9. Comparison of direct (left) and adjoint mode (right) of evaluating derivatives (Oberbichler et al. 2021).

The optimal evaluation of the adjoint method for IGA requires a careful implementation. We therefore propose three variants: (i) a complete computation in direct mode. This variant is the easiest to implement. However, the efficiency of the calculation is not optimal. It is therefore suitable for the early design phase of new components. (ii) A complete calculation in adjoint mode (Fig. 9). Here the computational performance is highest. However, the optimal implementation becomes more complex. (iii) A hybrid mode in which the calculation of a mechanical core element (Fig. 8) is performed in direct mode, while the transformation to the selected geometry description is performed in adjoint mode. In most cases, this variant represents a good compromise solution between performance and simplicity.

3.3 Modular Formulation

The adjoint method allows modularising IGA according to a core-congruential formulation (Crivelli and Felippa 1993; Felippa et al. 1994). This permits a clear separation between mechanics and geometry, which allows the creation of modules that can be combined for different types of analysis.

In the first step, the geometric description is integrated into the derivatives using the direct method. With the adjoint approach, the geometric transformation is moved to the end of the computation which simplifies the separation of geometry and mechanics. The mechanical component is referred to as the "core element". This is independent of the chosen geometric description. The core element clarifies the difference between classical FE and IGA. In classical FE, we approximate a cable described by a smooth curve with line segments. In IGA we do something similar. Each segment is tangential to the curve and represents a core element, however, the segments are not directly connected. The connection results from the geometric transformation, which translates forces and stiffnesses from the end points of the segments to the nodes of the control geometry.

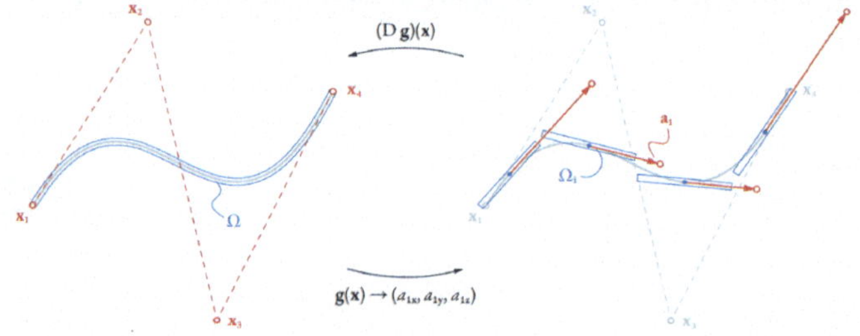

Fig. 10. Visualisation of the core-element for an isogeometric cable element (Oberbichler and Bletzinger 2022).

We can easily transform the same core element to other geometric descriptions by implementing the appropriate functional to obtain the corresponding transformation e.g., using algorithmic differentiation (Fig. 10). The transformation for subdivision surfaces with sharp features was implemented in this (Oberbichler and Bletzinger 2022). Any control mesh can be used with this type of freeform geometry. In addition, corners and edges can be "tagged" to create sharp features (Biermann et al. 2000; Catmull and Clark 1978). Different subdivision rules are applied dependent on the selected combination of mesh and tags. Therefore, the subdivision algorithm for calculating the smooth interface is much more complex than with NURBS.

3.4 CAD-Integrated Analysis

Due to the simplified formulation of IGA elements, an implementation of the analysis tool within a CAD environment becomes possible. This allows direct integration of the analysis tools in CAD, simplifies the data exchange between CAD and FEM, and enables direct interaction with the analysis process. Functionalities and data structures can be shared, for example, geometric operations from CAD can be used by the FEM kernel to compute trimmed geometries. At the same time, the computation is drastically accelerated, and the structural analysis of large models becomes faster (Fig. 11). For smaller models, real-time calculations are possible. The computation is performed in CAD, which allows direct interaction with the mechanical model.

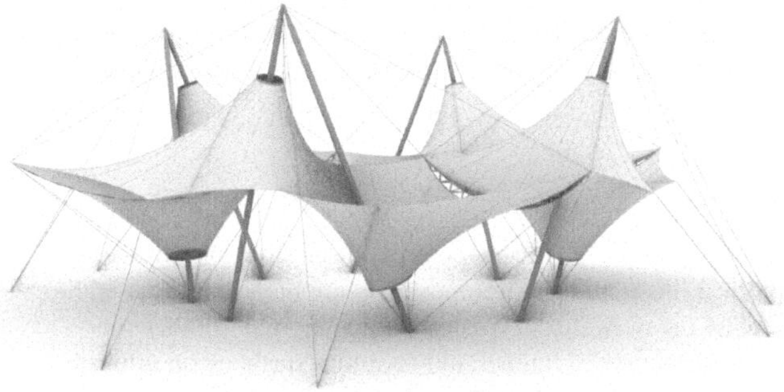

Fig. 11. Isogeometric form finding with URS inspired by Skysong ASU campus (Oberbichler and Bletzinger 2022).

4 Conclusion

We use an automatic implementation of finite elements based on energy functionals and algorithmic differentiation that simplifies the implementation and permits rapid prototyping of new elements. An adjoint approach for evaluating residual force and stiffness of the isogeometric elements does not affect the results but reduces the computational effort and memory consumption. This allows a direct integration of the analysis tools in CAD, simplifies the exchange of data and functionalities without interfaces and code duplication, and enables direct interaction with the analysis process.

We highlight the higher efficiency as an advantage that manifests itself in two ways. On the one hand, the computing speed can be significantly increased, leading to more efficient calculations. On the other hand, the modular approach leads to a more efficient implementation. This makes it easier to implement new features and functions.

The modular structure allows a clear division between mechanical and geometrical components. It permits numerous combinations, resulting in a versatile isogeometric analysis toolbox. As a result, we obtain a modular and automatable framework that enables efficient implementation of IGA including within CAD environments.

Visual programs like Grasshopper have an inherent similarity to computational graphs. The direct integration of AD into VPLs could be interesting approach in which it would be possible to combine the dynamic modelling of problem solutions with the advantages of gradient-based optimisation algorithms. AD would have to be applied directly to the node structure of visual programmes.

The application of AD was demonstrated for structural analysis and form finding. It can also apply to other analysis types e.g., graphical statics (Pastrana et al. 2021). Various analysis types can be clustered into building blocks which are then combined in a larger pipeline. For example, a structural analysis can be carried out based on a form finding. To model complex construction processes, several structural analyses must be run serially. Each run depends on the result of the previous run. When studying the influence of the input parameters on the final solution within such a pipeline, problems are quickly encountered due to the amount of data and the complex relationships. AD can provide a generalised and efficient solution. To do so, AD would have to be deeply integrated into an FE system. For such an approach, AD would have to be deeply integrated into FEM by becoming a key element in a "differentiable structural analysis".

References

Biermann, H., Levin, A., Zorin, D.: Piecewise smooth subdivision surfaces with normal control. In: Proceedings of the 27th Annual Conference on Computer Graphics and Interactive Techniques, pp. 113–120. ACM Press/Addison-Wesley Publishing Co. (2000)

Catmull, E., Clark, J.: Recursively generated B-spline surfaces on arbitrary topological meshes. Comput. Aided Des. **10**, 350–355 (1978). https://doi.org/10.1016/0010-4485(78)90110-0

Cottrell, J.A., Hughes, T.J.R., Bazilevs, Y.: Isogeometric Analysis. John Wiley & Sons Ltd, Chichester, UK (2009). https://doi.org/10.1002/9780470749081

Crivelli, L.A., Felippa, C.A.: A three-dimensional non-linear Timoshenko beam based on the core-congruential formulation. Int. J. Numer. Meth. Eng. **36**, 993 (1993). https://doi.org/10.1002/nme.1620362106

Felippa, C.A., Crivelli, L.A., Haugen, B.: A survey of the core-congruential formulation for geometrically nonlinear TL finite elements. Arch. Comput. Methods in Eng. **1**, 1–48 (1994). https://doi.org/10.1007/BF02736179

Fike, J.A., Alonso, J.J.: The development of hyper-dual numbers for exact second-derivative calculations. In: AIAA paper 2011-886, 49th AIAA Aerospace Sciences Meeting (2011). https://doi.org/10.2514/6.2011-886

Griewank, A., Walther, A.: Evaluating Derivatives. Soc. Ind. Appl. Math. (2008). https://doi.org/10.1137/1.9780898717761

McCarthy, J.: An algebraic language for the manipulation of symbolic expressions. AI Memo No. 1 (1958)

Oberbichler, T., Bletzinger, K.U.: CAD-integrated form-finding of structural membranes using extended catmull–clark subdivision surfaces. Comput.-Aided Des. **151**, 103360 (2022). https://doi.org/10.1016/j.cad.2022.103360

Oberbichler, T., Wüchner, R., Bletzinger, K.-U.: Efficient computation of nonlinear isogeometric elements using the adjoint method and algorithmic differentiation. Comput. Methods Appl. Mech. Eng. **381**, 113817 (2021). https://doi.org/10.1016/j.cma.2021.113817

Pastrana, R., Ohlbrock, P.O., Oberbichler, T., D'Acunto, P., Parascho, S., 2021. Constrained form-finding of tension-compression structures using automatic differentiation. https://doi.org/10.48550/arxiv.2111.02607

Piegl, L., Tiller, W.: The NURBS Book, Monographs in Visual Communications. Springer, Berlin Heidelberg (1995). https://doi.org/10.1007/978-3-642-97385-7

An Open Approach to Robotic Prototyping for Architectural Design and Construction

Andrea Rossi[1(✉)], Arjen Deetman[2], Alexander Stefas[3], Andreas Göbert[1], Carl Eppinger[1], Julian Ochs[1], Oliver Tessmann[4], and Philipp Eversmann[1]

[1] Universität Kassel, Universitätsplatz 9, 34127 Kassel, Germany
rossi@asl.uni-kassel.de
[2] Eindhoven University of Technology, P.O. Box 513, Eindhoven 5600 MB, Netherlands
[3] Kraenk Visuell GbR, Friedrich-Ebert-Platz 17, 64289 Darmstadt, Germany
[4] Technische Universität Darmstadt, El-Lissitzky-Straße 1, 64287 Darmstadt, Germany

Abstract. Emerging from research in computational design and digital fabrication, the use of robot arms in architecture is now making its way in the practice of construction. However, their increasing diffusion has not yet corresponded to the development of shared approaches covering both digital (programming and simulation) and physical (end-effector design, system integration, IO communication) elements of robotic prototyping suited to the unique needs of architectural research. While parallel research streams defined various approaches to robotic programming and simulation, they all either (A) rely on custom combinations of software packages, or (B) are built on top of advanced robotic programming environments requiring a higher skill level in robotics than conventionally available in an architectural context. This paper proposes an alternative open-source toolkit enabling an intuitive approach to the orchestration of various hardware and software components required for robotic fabrication, including robot programming and simulation, end-effector design and actuation, and communication interfaces. The pipeline relies on three components: Robot Components, a plug-in for intuitive robot programming; Funken, a serial protocol toolkit for interactive prototyping with Arduino; and a flexible approach to end-effector design. The paper describes these components and demonstrates their use in a series of case studies, showing how they can be adapted to a variety of project typologies and user skills, while keeping highly complex and specific functionality available as an option, yielding good practices for a more intuitive translation from design to production.

Keywords: Robotic fabrication · Hardware integration · Open-source

1 Introduction and Background

Emerging from research in computational design and digital fabrication, the use of robot arms in architecture is now making its way in construction (Kohler et al., 2014; Willmann et al. 2018). Due to the complexities of architectural fabrication, highly interactive processes, with integration of advanced hardware components are needed (Stefas et al.

C. Gengnagel et al. (Eds.): DMS 2022, *Towards Radical Regeneration*, pp. 96–107, 2023.
https://doi.org/10.1007/978-3-031-13249-0_9

2018), blurring the line between human and machine interfaces. To this date, there are still limited developments of shared approaches covering both digital (programming and simulation) and physical (end-effector design, system integration, IO communication) elements of robotic prototyping suited for architectural research. Most approaches limit themselves to simulation, while methods for development and management of hardware components are mostly lacking.

1.1 Software Platforms for Architectural Robotics

While parallel research streams defined approaches to robotic programming and simulation, they either rely on custom combinations of software, often for specific purposes and not open-source, or are built on top of advanced robotic frameworks, which require skills rarely available in an architectural context. For the first approach, various programming and simulation tools for the Grasshopper environment emerged, such as HAL (Schwartz 2013), KukaPRC (Braumann and Brell-Çokcan 2011), Robots (Soler et al 2017), Scorpion (Elashry and Glynn 2014), and Taco (Frank et al. 2016). With differences, all provide utilities for toolpath creation, code generation, and offline kinematic simulation within Grasshopper. Some, such as HAL, evolved into a standalone framework (Gobin et al. 2021). Machina (Garcia del Castilo Lopez 2019) focuses instead on intuitive interaction with robots through real-time programming. With some exceptions, several of these tools are closed-source, and hence they allow only limited adaptations and extension, when an API is available.

The second approach leverages state-of-the-art robotic frameworks, such as ROS (Robot Operative System) (Stanford Artificial Intelligence Laboratory et al. 2018) or V-REP (Rohmer et al. 2013), and integrate them in architectural interfaces. The most notable case is the CompasFab package (Rust et al. 2018) for the COMPAS framework (Mele et al. 2017), a Python framework for research in computational design. CompasFab enables to integrate the above-mentioned environments in CAD packages such as Rhino or Blender. While this simplifies their usage for architectural researchers, it still presents a significant entry barrier, caused by the need of sound knowledge in Python programming and understanding of complex robotics frameworks.

1.2 Hardware Platforms for Architectural Robotics

If the field of robot programming has seen significant development, hardware integration, another major element of robotic fabrication research, has lagged in the architectural domain. In most cases, the integration of sensors and end-effectors is either performed by specialized system integrators, or it is prototyped using custom approaches, leveraging low-cost computing devices such as Arduino microcontrollers (Braumann and Cokcan 2012), through Grasshopper plug-ins such as Firefly (Payne and Johnson 2013). While the first approach yields reliable fabrication processes at the cost of flexibility, the second often results in non-transferrable processes.

An attempt to address this is Robo.Op, a platform aimed at making robotic prototyping more accessible (Bard et al. 2014; Gannon et al. 2016). The toolkit provides a modular hardware platform for prototyping of end-effectors, and a software interface translating between ABB RAPID code and common creative coding tools. The hardware provides power supplies to mount various standard tools and an Arduino microcontroller connected to the robot IOs. While the project is open-source, at the time of writing development seems to have halted.

1.3 An Open Approach to Computational Tools Development

As mentioned, a key characteristics of architectural research in robotic fabrication is the importance of flexibility and adaptability of workflows, combining both low-cost prototyping components and industrial-standards systems. This is akin to what Eric Raymond described as a "bazaar" (Raymond 1999), where several specialized tools need to be integrated and orchestrated. This highlights the need to maintain openness, both in terms of hardware and software, as well as to allow the creation of shared communication interfaces between software and hardware elements (Stefa et al. 2018). This is why the tools and interfaces described here are provided as open-source packages, aiming at creating a flexible "toolkit" for robotic fabrication rather than a centralized closed "tool" (Mackey and Sadeghipour Roudsari 2018). While the reliance on open source tools might raise concerns in a safety-critical domain such as fabrication, it must be noted that its viability is being increasingly accepted in other domains where software safety has similar, if not higher, requirements than architectural robotics (Gary et al. 2011).

2 Methods

Given this context, this paper proposes an intuitive approach to the orchestration of hardware and software components required for robotic fabrication (Fig. 1). This is based on the definition of communication interfaces between human actors and robotic machines, as well as between various elements of a robotic system.

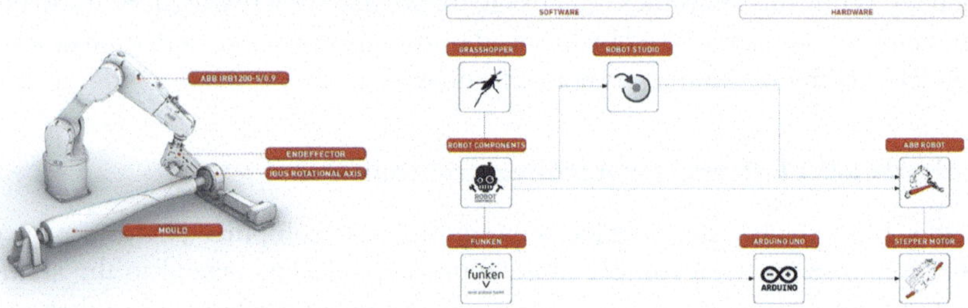

Fig. 1. Example of a process for robotic winding, entirely relying on the proposed pipeline.

Hence, we propose an open-source software and hardware pipeline, relying on three main components:

- Robot Components, an open-source plugin for intuitive robot programming;
- Funken, an open-source toolkit for serial communication;
- an open and modular approach to tool design.

2.1 Robot Components

The first element of the pipeline is Robot Components, an open-source robotic programming and simulation plug-in for Grasshopper (Deetman et al. 2022). This forms the basis to interface CAD geometries in Rhino with ABB robot arms.

Grasshopper Plug-in. The Robot Components plug-in provides an intuitive interface matching the logic of ABB's RAPID code for robotic programming. Given the difficulty of developing generic interfaces for different robot brands, while maintaining access to specific functionality, strictly following the RAPID code means that each RAPID code line is represented by one component. This is beneficial for teaching, since by visually programming with Robot Component students learn the basics of RAPID code without typing code lines. In parallel, to avoid an overwhelmingly complicated interface, Robot Components makes extensive use of casting methods and hidden parameters. This allows beginner users to quickly create robot programs, hiding unnecessary parameters and providing a simple introduction to robot programming, without removing the possibility of adding more advanced functionalities afterwards (Fig. 2).

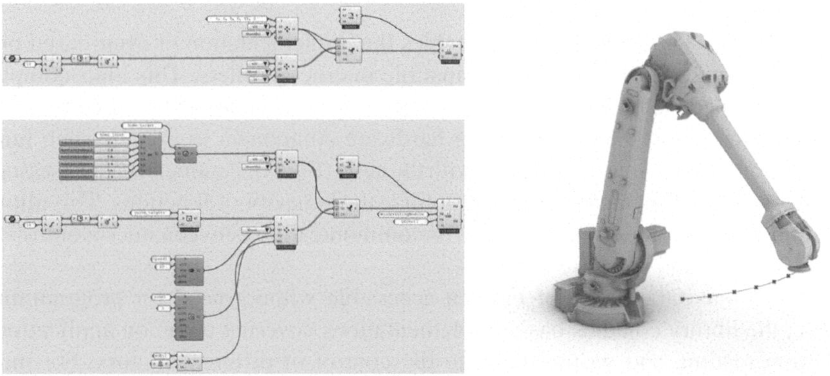

Fig. 2. Comparison between robot programming modes: beginner (top), where most parameters are hidden, and advanced (bottom), where all parameters are declared for more control.

Robot Components provides components for forward and inverse kinematics, allowing a quick visual check on the robot poses without leaving the Grasshopper environment. A controller utility category is available to send and get data from both physical (real robots) and virtual (simulated in ABB RobotStudio software) controllers. It relies on the ABB Controller API, allowing to set and get IO signals in real-time. This, in combination with other proposed tools, allows Grasshopper to become a central communication interface. The robot pose and tool position can also be read inside Grasshopper, simplifying calibration processes. This real-time connection also allows to check toolpaths within Grasshopper by remotely using the advanced kinematic simulation of RobotStudio.

API. For more advanced users and complex processes, an open-source API is available in Grasshopper via IronPython and/or C#. This allows to prototype processes requiring feedback loops that are harder to program relying exclusively on visual programming. These can be used in real-time during fabrication in combination with the controller utilities, but also during design, automatically validating fabrication feasibility through the embedded robot kinematics.

2.2 Funken

As already discussed, robotic programming and simulation is only one component of fabrication processes. In order to tackle the second necessary element, the control of sensors and end-effectors, we propose Funken (Stefas et al. 2018; Stefas 2020), an open-source toolkit for interactive prototyping with the Arduino framework (Mellis et al. 2007). Funken allows the definition of shared interfaces between hardware and software elements of a robotic system, simplifying the execution of complex tasks through a keyword-based callback method. It consists of an Arduino library and interfaces for different frameworks, such as Grasshopper, Python, Processing, and NodeJS.

Arduino Library. The Funken library enables the implementation of even-based programming (Faison 2006) on Arduino-compatible microcontrollers. This links complex functionality defined via Arduino code with simple keywords, that can be called from any serial communication-enabled software or hardware component to execute such functionality, without need to interact directly with the code (Fig. 3). Parameters can be associated with keywords, allowing to further customize the behavior of functions. This allows to create human-readable serial protocols for communication between microcontrollers and other hardware and software elements.

In order to make the usage of Funken accessible without need for programming experience, the library contains basic implementations covering common applications, ranging from reading and writing data, to the control of different motors. For more experienced users, Funken can be easily extended with custom functionality without editing the core library, by creating functions in Arduino code and linking them to keywords, making them accessible to any connected software or hardware.

Grasshopper Plug-in. Funken can connect to a variety of software frameworks, since it relies on serial communication, a basic and widely available protocol for hardware-software interfaces. In the context of robotic fabrication, the most relevant interface is the Grasshopper environment. The open-source GhFunken plug-in allows to remote-control any microcontroller running Funken, by either mirroring basic Arduino functionality, or by connecting to custom-defined Funken functions (Rossi 2020). It also allows to easily connect multiple microcontrollers to the same file, modularizing the control of an end-effector. The plug-in relies on the PySerial library (Liechti 2016), an open-source Python implementation for serial communication.

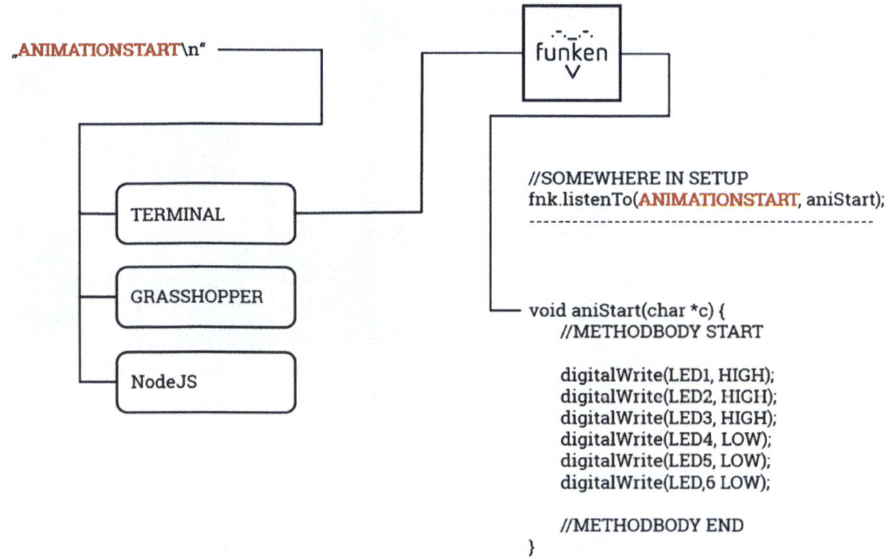

Fig. 3. Overview of Funken keyword-based call back method, linking the keyword "ANIMATIONSTART" to the execution of the code in the bottom left.

The plug-in enables users focusing on robotic programming to control custom end-effectors by simply knowing the defined interfaces, without needing complete knowledge of their internal functioning. It also allows to control the level of exposure of Arduino code, providing a platform for education in physical computing, gradually increasing the exposure of students to levels of programming complexity.

Virtual Prototyping of Hardware Interfaces. The combination of Funken and the controller utilities for Robot Components enables to quickly define and test communication interfaces between hardware components without need for hardware wiring. Indeed, by monitoring the robot IOs through Robot Components, and using them to trigger keyword messages through the GhFunken plug-in, it becomes possible to create virtual connections between specific IOs and Funken functions, remotely triggering different end-effectors behaviors (Fig. 4). While such communication determines relatively high levels of latency, potentially incompatible with fully automated production processes, it proved sufficient for prototyping research and testing of hardware systems.

Additionally, the combination of real-time IOs monitoring and Funken functions creates a parallel control system for end-effectors. Using routing functions, provided with Funken, it is possible to control specific end-effectors behaviors both through a wired connection to the robot IO system, as well as through a Grasshopper interface. This parallel control model is particularly valuable when prototyping processes, as it allows fully automated control of the process via wired IOs, but still allows over-writing of such behaviors from the Grasshopper interface.

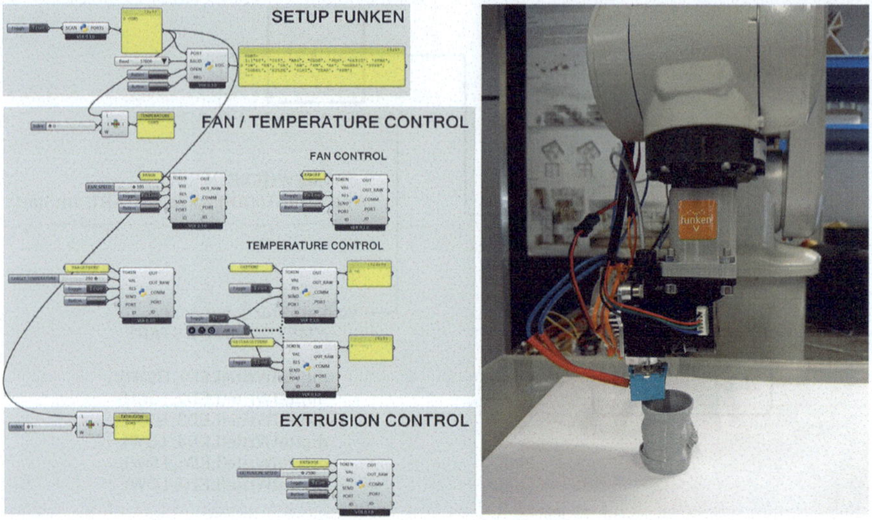

Fig. 4. Control dashboard for a robotic 3d printing process, showing the possibility of overwriting process parameters by changing Grasshopper sliders during execution on the robot.

2.3 Fast Physical Prototyping of End-Effectors

Beyond their control, end-effector prototypes can be easily designed in CAD and rapidly produced through low-cost FDM 3d printing. Using modular principles, such tools can integrate a variety of functions, such as cutting, heating, pumping, and distance measuring for real-time adjustments. Such functions can be controlled via Funken, and in parallel fitted with relay switches to work within the robotic system and outputs. For the bespoken example, low cost and vastly available electronics, such as, stepper motors, temperature sensors, distance sensors etc. can be combined into one prototype tool adapted for a specific application (Fig. 5).

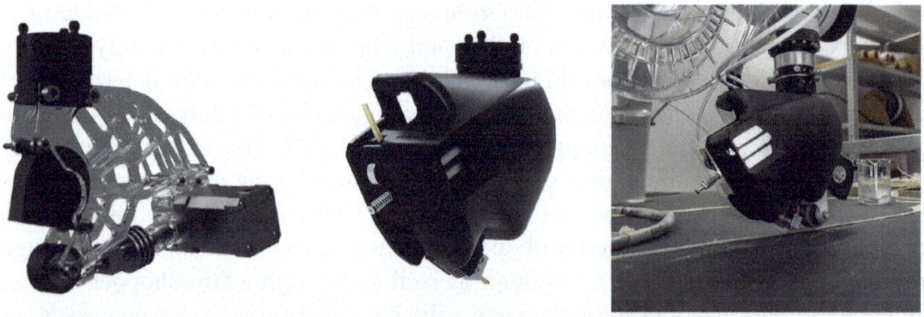

Fig. 5. Example of a 3d printed end effector for extrusion of continuous timber filament, from left to right: internal 3d structure and components, complete model, built extruder.

3 Results: Case Study Processes

The proposed pipeline and its components have been applied for various projects, highlighting their capabilities and application scenarios, as well as the flexibility provided by different soft- and hardware combinations.

3.1 Continuous Timber Filament Winding

As a first case study, the proposed pipeline has been applied for robotic winding of timber veneers (Göbert et al. 2022). The process has several parameters, which need to be synchronized: (a) extrusion speed of the filament, (b) axis rotation velocity and (c) robot Tool Center Point (TCP) speed. In addition, various I/O signals for the veneer extrusion, cutting, and automated adhesive application needed to be controlled during the process (Fig. 6).

The pipeline enabled fully-automated generation of winding processes for different setups. It also allowed fast shift from an initial prototyping setup for small profiles, featuring an Arduino-controlled external axis, to a full-scale production of architectural building components, integrating a robot-controlled external axis (Fig. 7). The entire code in Robot Components required minimal changes in the speed settings and IO-signals. This not only enabled a quick realization of larger demonstrators, but also allowed the use for both research and education, giving students with low programming knowledge the opportunity to generate their own winded elements.

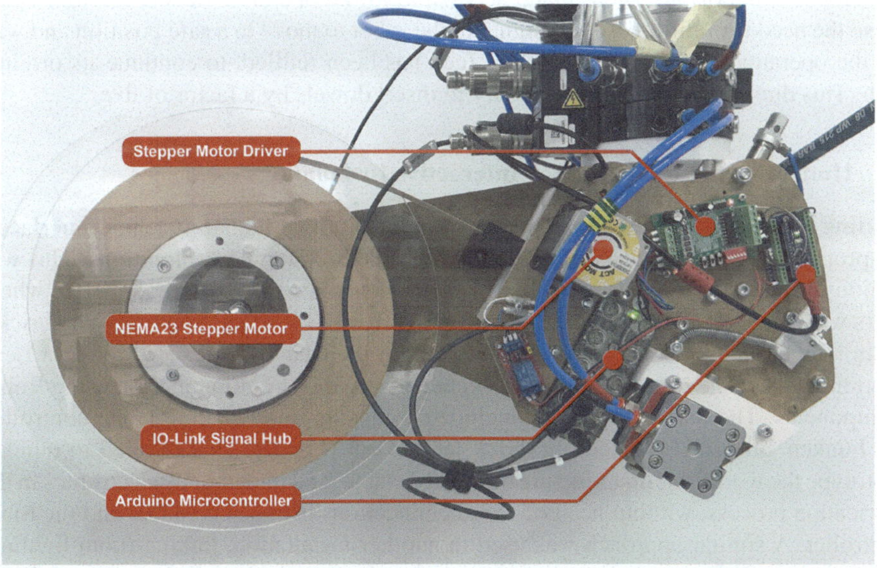

Fig. 6. Extruder for robotic winding with the various hardware components which are orchestrated in the proposed pipeline: a stepper driver and its relative motor, an Arduino microcontroller, and an IO-Link hub interfacing the signals with the robot ProfiNET network.

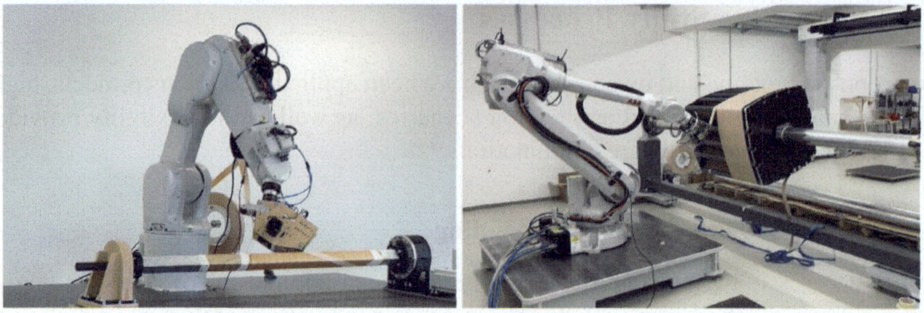

Fig. 7. Process scaling from initial prototyping studies (left) to full-scale production (right) of robotic winded components. As same control and hardware strategies were used, upscaling could be efficiently realized with our toolsets.

3.2 Automated Dowel Placement for Robotic Assembly

In a more focused and specific application, the proposed methods were used to integrate an automated dowel-placement tool for robotic assembly (Fig. 8). The tool recreates the manual placement process of wooden dowels, by using a pneumatic cylinder as hammer. Through rapid prototyping, we implemented an 80-dowels magazine, which required only limited refills during placement. A series of motors were synced to guide the dowels to the hammering position, and check for consistency in all steps. To guarantee a fluid fabrication process, the tool was equipped with several sensors, which, using Funken and Robot Components, were made accessible to the operator. Hence, the tool could sense the need for refilling, send a signal to the robot to move to a safe position and wait for the operator's confirmation that the feed has been refilled, to continue its original path. This diminished the time necessary to insert dowels by a factor of five.

3.3 Human-Robot Interfaces for Interactive Installations

Shifting the context to the development of more intuitive human-machine interfaces, the proposed tools could also be applied for flexible prototyping explorations. This was demonstrated in a series of installations, partly described in (Betti et al. 2018), where users were allowed to interact with a small ABB IRB120 robot arm cutting foam bricks. Visitors could control the shape of the cuts with their hands, tracked through a Leap Motion tracker. The resulting positions were converted into cutting motion using Robot Components. The hardware system included a custom-build rotating table, controlled via Funken, allowing to select the brick face to cut (Fig. 9). This enabled to quickly prototype the workflow, manufacture the custom table, and integrate its electronics in the fabrication process, without need of wired connections between the table and the robot controller. A similar approach was used in another installation, linking room lighting, controlled via DMX standards, with robot motion through Funken (Belousov et al. 2022; Wibranek 2021).

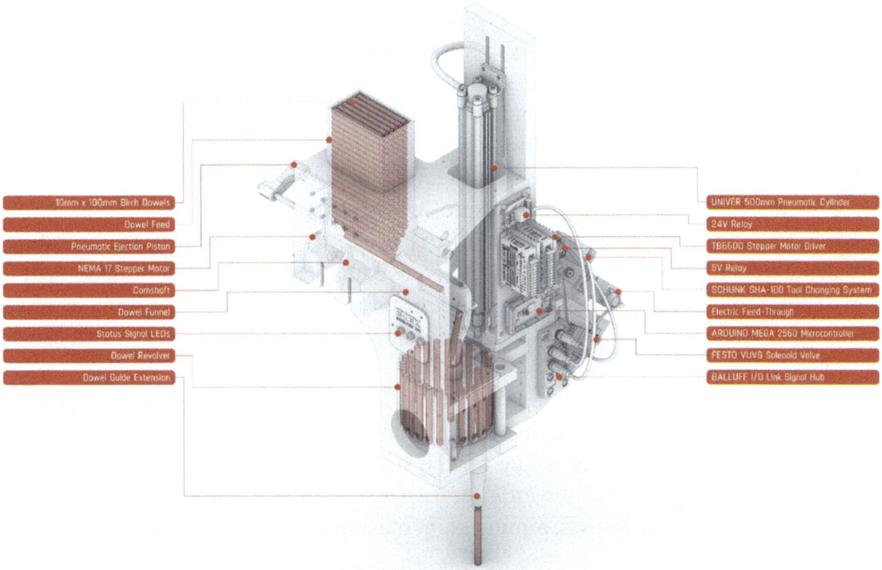

Fig. 8. Robotic doweling tool with the different components allowing for full automation of dowels loading, alignment and hammering through a pneumatic piston. A series of motors is used to guide the dowels from the magazine to the insertion position, and sensors track the process ensuring consistency.

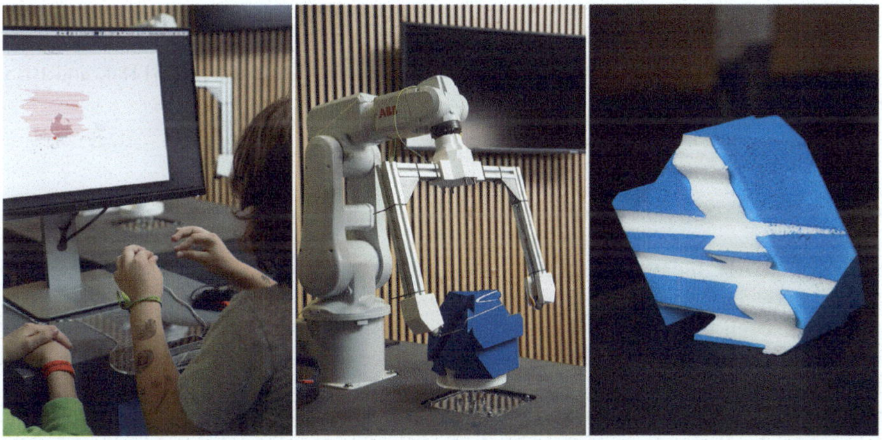

Fig. 9. Human-machine interface for translation of hand movement into robotic cutting instructions, from left to right: user hand tracking interface, robot cutting process, finished brick.

4 Conclusions

Results of case studies show that the workflow is adaptable to various projects and levels of user skills, while keeping complex and specific functionality available as an option.

Our framework can be applied in prototyping studies and interactive installations, as well as for structured fabrication workflows with large robots and a variety of actuators, sensors, and industrial protocols. The scale and variety of the applications demonstrates its suitability for research and prototypical production. The pipeline provides a framework for the definition of interfaces within complex human-machine fabrication processes, where roles are often redefined during research. As the structure of the different software elements relies on a layered structure, where only relevant information is made accessible to users with different levels of expertise, it allows to increase usability in safety-critical domains (Gary et al. 2011). Moreover, as all software components are built on top of common computational design and prototyping frameworks, relying on simple interfacing models, they have been designed maintaining the possibility of extension to other software environments.

At the current stage, the community revolving around the proposed tools is still small, and many users rely only on certain elements and not on the whole pipeline. As having an active community is key for the sustainability of an open-source project, future efforts will be directed towards the expansion of documentation and example files, covering various workflows, aiming at increasing the user-base and adoption of the proposed methods.

In conclusion, our research creates interfaces between the elements of fabrication systems through open-source tools and human-readable interfaces, rather than imposing specific and ultimately rigid workflows,. This yields good practices for a more intuitive translation from design to production and a flexible communication model between human and machinic actors. Overall, our research aims at providing shared models of human-machine interfaces, fostering open research and collaboration in robotic fabrication.

Acknowledgments. Case studies were partly funded by DFG, grant #436451184, and BBSR, grant #10.08.18.7-20.24.

References

Schwartz, T.: HAL. In: Brell-Çokcan, S., Braumann, J. (eds.) Rob | Arch 2012, pp. 92–101. Springer, Vienna (2013). https://doi.org/10.1007/978-3-7091-1465-0_8

Braumann, J., Brell-Çokcan, S.: Parametric robot control: integrated CAD/CAM for architectural design. In: Proceedings of the 31st Annual Conference of the Association for Computer Aided Design in Architecture, pp. 242–251. Banff, Alberta (2011)

Soler, V., Retsin, G., Jimenez Garcia, M.: A generalized approach to non-layered fused filament fabrication. In: Proceedings of the 37th Annual Conference of the Association for Computer Aided Design in Architecture, pp. 562–571. Cambridge, MA (2017)

Elashry, K., Glynn, R.: An approach to automated construction using adaptive programing. In: McGee, W., Leon, M.P. (eds.) Robotic Fabrication in Architecture, Art and Design 2014, pp. 51–66. Springer, Cham (2014). https://doi.org/10.1007/978-3-319-04663-1_4

Frank, F., Wang, S.-Y., Sheng, Y.-T.: Taco: ABB robot programming for Grasshopper (2016)

Gobin, T., Andraos, S., Schwartz, T., Vriet, R.: HAL robotics framework. In: Proceedings of the International Symposium on Automation and Robotics in Construction, vol. 38, pp. 733–740. IAARC Publications (2021)

Garcia del Castillo Lopez, J.L.: Enactive Robotics: An Action-State Model for Concurrent Machine Control. Ph.D. Dissertation, Harvard University (2019)
Stanford Artificial Intelligence Laboratory et al.: Robotic Operating System. https://www.ros.org. Last accessed 10 June 2022 (2018)
Rohmer, E., Singh, S.P.N., Freese, M.: V-REP: A versatile and scalable robot simulation framework. In: 2013 IEEE/RSJ International Conference on Intelligent Robots and Systems, pp. 1321–1326 (2013). https://doi.org/10.1109/IROS.2013.6696520
Rust, R. et al.: COMPAS FAB: Robotic fabrication package for the COMPAS Framework (2018). https://doi.org/10.5281/zenodo.3469478
Mele, T., et al.: COMPAS: A framework for computational research in architecture and structures (2017). https://doi.org/10.5281/zenodo.2594510
Braumann, J., Cokcan, S.-B.: Digital and physical tools for industrial robots in architecture: robotic interaction and interfaces. Int. J. Archit. Comput. **10**, 541–554 (2012)
Payne, A.O., Johnson, J.K.: Firefly: interactive prototypes for architectural design. Archit. Des. **83**, 144–147 (2013)
Bard, J. et al.: Seeing is doing: synthetic tools for robotically augmented fabrication in high-skill domains. In: Proceedings of the 34th Annual Conference of the Association for Computer Aided Design in Architecture, pp. 409–416. Los Angeles, California (2014)
Gannon, M., Jacobson-Weaver, Z., Contreras, M.: Robo.Op. https://github.com/peopleplusrobots/robo-op (2016). Last accessed 10 June 2022
Raymond, E.: The cathedral and the bazaar. Knowl. Technol. Policy **12**, 23–49 (1999)
Stefas, A., Rossi, A., Tessmann, O.: Funken - Serial Protocol Toolkit for Interactive Prototyping. In: Computing for a better tomorrow - Proceedings of the 36th eCAADe Conference, vol. 2, pp. 177–186. Lodz, Poland (2018)
Mackey, C., Roudsari, M.S.: The tool(s) versus the toolkit. In: De Rycke, K., Gengnagel, C., Baverel, O., Burry, J., Mueller, C., Nguyen, M.M., Rahm, P., Thomsen, M.R. (eds.) Humanizing Digital Reality, pp. 93–101. Springer Singapore, Singapore (2018). https://doi.org/10.1007/978-981-10-6611-5_9
Gary, K., et al.: Agile methods for open source safety-critical software. Softw. Pract. Exp. **41**, 945–962 (2011)
Deetman, A. et al.: Robot Components: Intuitive Robot Programming for ABB Robots inside of Rhinoceros Grasshopper (2022). https://doi.org/10.5281/zenodo.5773814
Stefas, A.: Funken - Serial Protocol Toolkit. https://github.com/astefas/Funken (2020). Last accessed 10 June 2022
Mellis, D., Banzi, M., Cuartielles, D., Igoe, T.: Arduino: an open electronic prototyping platform. In Proc. Chi **2007**, 1–11 (2007)
Faison, T.: Event-Based Programming. Springer (2006)
Rossi, A.: GhFunken. https://github.com/ar0551/GhFunken (2020). Last accessed 10 June 2022
Liechti, C.: PySerial documentation. https://pyserial.readthedocs.io/en/latest/ (2016). Last accessed 10 June 2022
Göbert, A., Deetman, A., Rossi, A., Weyhe, O., Eversmann, P.: 3DWoodWind: robotic winding processes for material-efficient lightweight veneer components. Const. Robot. **6**(1), 39–55 (2022). https://doi.org/10.1007/s41693-022-00067-2
Betti, G., Aziz, S., Rossi, A., Tessmann, O.: Communication landscapes. In: Willmann, J., Block, P., Hutter, M., Byrne, K., Schork, T. (eds.) ROBARCH 2018, pp. 74–84. Springer, Cham (2019). https://doi.org/10.1007/978-3-319-92294-2_6
Belousov, B., et al.: Robotic architectural assembly with tactile skills: Simulation and optimization. Autom. Constr. **133**, 104006 (2022)
Wibranek, B.: Robotic Digital Reassembly: Towards physical editing of dry joined architectural aggregations. Ph.D. Dissertation, Technische Universität Darmstadt (2021). https://doi.org/10.26083/tuprints-00018578

Augmented Intelligence for Architectural Design with Conditional Autoencoders: Semiramis Case Study

Luis Salamanca[1](✉), Aleksandra Anna Apolinarska[2], Fernando Pérez-Cruz[1], and Matthias Kohler[2]

[1] Swiss Data Science Center, ETH Zürich, Zürich, Switzerland
luis.salamanca@sdsc.ethz.ch
[2] Gramazio Kohler Research, ETH Zürich, Zürich, Switzerland

Abstract. We present a design approach that uses machine learning to enhance architect's design experience. Nowadays, architects and engineers use software for parametric design to generate, simulate, and evaluate multiple design instances. In this paper, we propose a conditional autoencoder that reverses the parametric modelling process and instead allows architects to define the desired properties in their designs and obtain multiple predictions of designs that fulfil them. The results found by the encoder can oftentimes go beyond what the user expected and thus augment human's understanding of the design task and stimulate design exploration. Our tool also allows the architect to under-define the desired properties to give additional flexibility to finding interesting solutions. We specifically illustrate this tool for architectural design of a multi-storey structure that has been built in 2022 in Zug, Switzerland.

1 Introduction

1.1 Motivation

Design tasks in architecture, engineering, and construction often entail many parameters, multiple constraints, and contradicting objectives. In a traditional design process, the architects rely on experience to craft a handful of candidate solutions or can use parametric modelling tools to easily create many variations of the design. However, to find designs that fulfil predefined performance requirements, the user needs to find the right combination of the parameters, which can be difficult and time-consuming.

We aim to invert the design paradigm by using machine learning, so that the user can specify the required attributes and in return be presented with different designs that fulfil them (Fig. 1, bottom). The purpose of our approach is not optimization but design exploration. Since the mapping from the geometric design parameters to the desired performance attributes is not injective, there may exist many designs with similar performance. This way, we expect the machine learning tool to augment the early design process, and enhance the architect's experience through a fast and intuitive exploration of the solution space.

L. Salamanca and A.A. Apolinarska—Contributed equally.

C. Gengnagel et al. (Eds.): DMS 2022, *Towards Radical Regeneration*, pp. 108–121, 2023.
https://doi.org/10.1007/978-3-031-13249-0_10

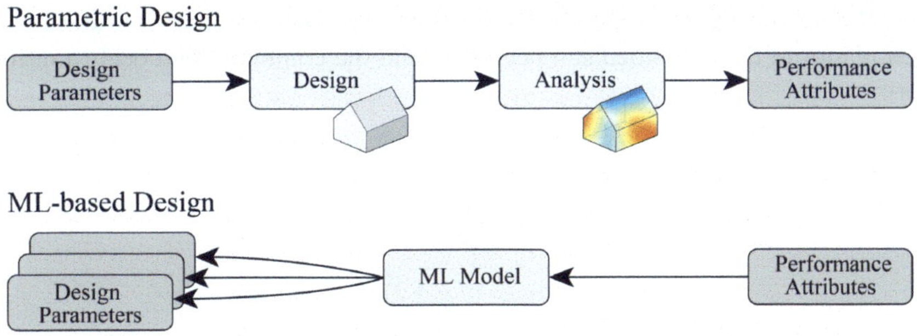

Fig. 1. In the parametric design process (top), the designer first chooses input parameters, creates a design instance, and simulates it (e.g. structure, solar gains) to evaluate its performance. Our proposed paradigm inverts this process (bottom).

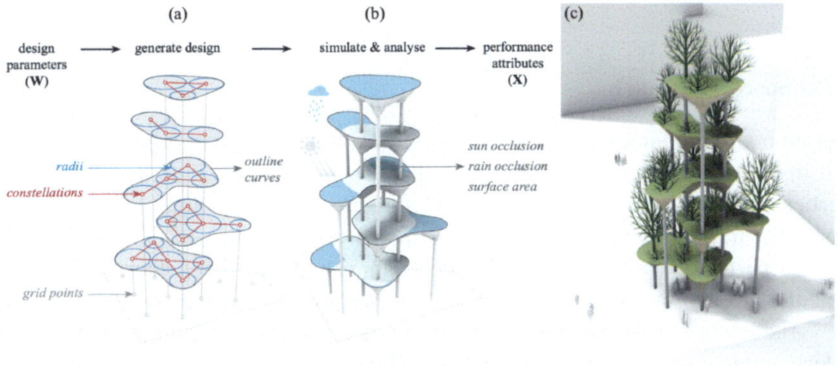

Fig. 2. The Semiramis project: a) setout and geometry constructed from design parameters **W**, b) analysis to obtain performance attributes **X**, c) the final design visualized in context and scale.

1.2 Case Study: Semiramis

Although the presented approach is generalizable to any design, to demonstrate this idea we apply it in a real architectural project. Our case study is a multi-storey structure of stacked plant platforms (an urban vertical garden project called Semiramis (Gramazio Kohler Research 2022), see Fig. 2c), that has been installed in 2022 in Zug, Switzerland. The particular design problem we address is to determine the outline curves of these platforms so that the total planting area and exposure to sun and rainfall satisfy requirements defined by the client and the landscape architect.

The five bowl-shaped platforms are placed at fixed heights and supported by columns on a triangular grid of 11 points. To design the platforms, we parametrize their outline shapes using a signed distance function that draws a smooth blend around the support points of each platform given the specified radii (see Fig. 2a). Each platform is supported by 3 to 5 columns in one of the ten hand-picked *base shapes* depicted in Fig. 3. All the possible positions, rotations and reflections of these base shapes within the given grid result in 115 *constellations*, i.e. subsets of the grid points that define support points for a platform. In summary, the *design parameters* are, for each platform: the choice of a constellation and the corresponding radii. The *performance attributes* are: the area of each platform, the combined sun occlusion and the combined rain occlusion for all platforms.

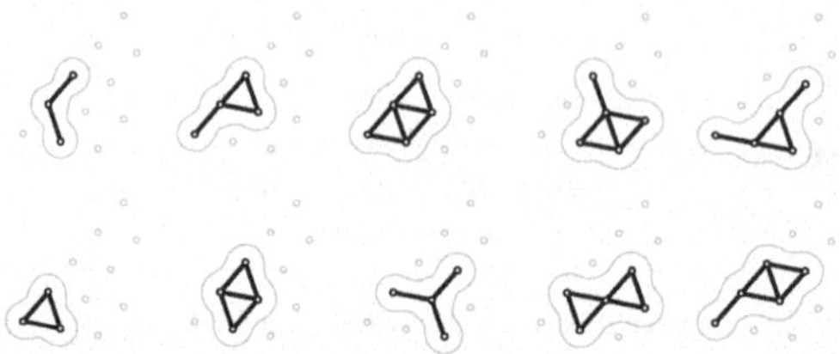

Fig. 3. Ten base constellation schemes. By including different positions and orientations of them within the 11-point grid, there are overall 115 possible constellations to choose from for each platform.

1.3 Proposed Approach

We set up a simple parametric *design+analysis* (DA) model that creates and analyses design instances from given input parameters. To generate designs that fulfill certain properties, we propose to use a *conditional autoencoder* (AE) (Sohn et al. 2015). This AE model is trained using a dataset consisting of tuples of design parameters and the associated performance attributes, obtained using the DA model. During training, in the latent space the performance measures are regressed together with the estimated latent variables to be able to propose multiple designs. Once the model is trained, the encoder emulates the parametric model, and the decoder maps from requested performance attributes to suitable designs, i.e. sets of design parameters. This tool enables the architect to explore different equally performing solutions to select their preferred design. When under-specifying the performance measures, the decoder can explore more freely the potential solutions.

2 Related Work

In architectural design and engineering, machine learning (ML) techniques are rapidly gaining interest in two main applications: for generative design and for design optimization. In the latter case, ML methods promise to solve many complex, black-box problems (Costa and Nannicini 2018) where gradient information is unavailable or impractical to obtain. Prominent examples employ evolutionary algorithms (Hornby et al. 2006), or convolutional neural networks (Takahashi et al. 2019), (Banga et al. 2018). In generative design, which we address in this paper, the role of ML methods is to help create many variations of designs. Models based on GAN or autoencoder architectures show considerable potential for this application, by encoding high-dimensional design variables in a low-dimensional design space. Notable work in this field includes generation of 2D floor plans (Hu et al. 2020; Chaillou 2020; Nauata et al. 2020), 3D shapes of family houses (Steinfeld et al. 2019) and of other building typologies (Miguel et al. 2019), or design of automotive parts (Oh et al. 2019). Other works provide general methodologies that allow encoding varied objects (Talton et al., 2012; Mo et al. 2019), using a hierarchy of the constituent parts. A common challenge in these applications is the parametrization/representation of the design, which impacts the flexibility of the design process and the quality of the solution (Brown and Mueller 2019).

Against this background, our method allows tailoring the trainable variables to the specific design question, and generating new designs with the same quality as those in the dataset. Hence, our approach can be applied to different design problems by adapting the type and dimensions of input parameters and of performance attributes. And finally, the user receives not one but multiple solutions that approximate the desired performance.

3 Methods

3.1 Implemented Solution

We choose a conditional autoencoder (AE) (Sohn et al. 2015) neural network architecture because, by having as inputs the design parameters **W**, we can enforce some performance attributes **X** in the hidden layer, and also the reconstruction at the output (Fig. 4a). After training, the decoder allows to, given some desired performance attributes $\mathbf{X^d}$, generate new sets of design parameters W^d (Fig. 4b).

The design parameters $\mathbf{W}_j$ for each platform j are: the constellation defining the support points, and a radius for each grid point (see 1.2). The constellations are encoded as a C-dimensional one-hot vector $\mathbf{W}_j^C$ (C = 115). The radii are represented in an R-dimensional vector $\mathbf{W}_j^R$ (R = 11) with non-negative real values for supporting grid points and otherwise zero. Thus, 630 variables as input W for P = 5 platforms:

$$\mathbf{W} = [\mathbf{W}_1^C, \ldots, \mathbf{W}_P^C, \mathbf{W}_1^R, \ldots, \mathbf{W}_P^R]$$

In the latent space we have the following. **X** is a D = 7 dimensional vector comprising the performance attributes (the surface area of each platform, and the sun and rain occlusions). **Z**, a DL-dimensional latent vector, encodes the input information, and allows during the generation phase to explore different solutions given some performance attributes.

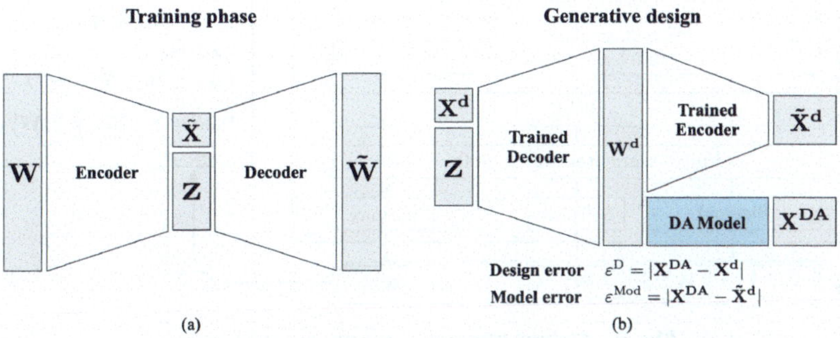

Fig. 4. In (a), a schematic of the architecture, with both **Z** and **X** in the latent space. In (b), the use of the trained encoder and decoder for the generation of new designs $\mathbf{W^d}$.

Further, $\tilde{\mathbf{W}}$ and $\tilde{\mathbf{X}}$ refer to the output during training of decoder and encoder respectively, that may differ from the fed training tuples $\mathbf{W}$ and $\mathbf{X}$. During generative design, $\mathbf{X^d}$ are the performance attributes values requested for the new designs $\mathbf{W^d}$.

3.2 Model Architecture and Training

The specific layers that comprise the neural network, dimensions used, and losses, are depicted in Fig. 5. For each one-hot vector $\mathbf{W}_j^{\mathrm{C}}$ we use different softmax blocks, and a sigmoid block for the output corresponding to the radii $\mathbf{W}^{\mathrm{R}}$. The losses on $\mathbf{Z}$ are regularization terms to enforce zero-mean and unit-variance. Each loss is weighted differently when computing the total loss.

Once the model is trained, all W in the training set are passed through the encoder. Given that $\tilde{\mathbf{X}}$ and $\mathbf{Z}$ in the hidden layer are approximately Gaussian, we model their joint distribution as a multivariate Gaussian with zero-mean and covariance matrix $\Sigma_{\tilde{\mathbf{X}}\mathbf{Z}}$, also beneficial for sampling purposes.

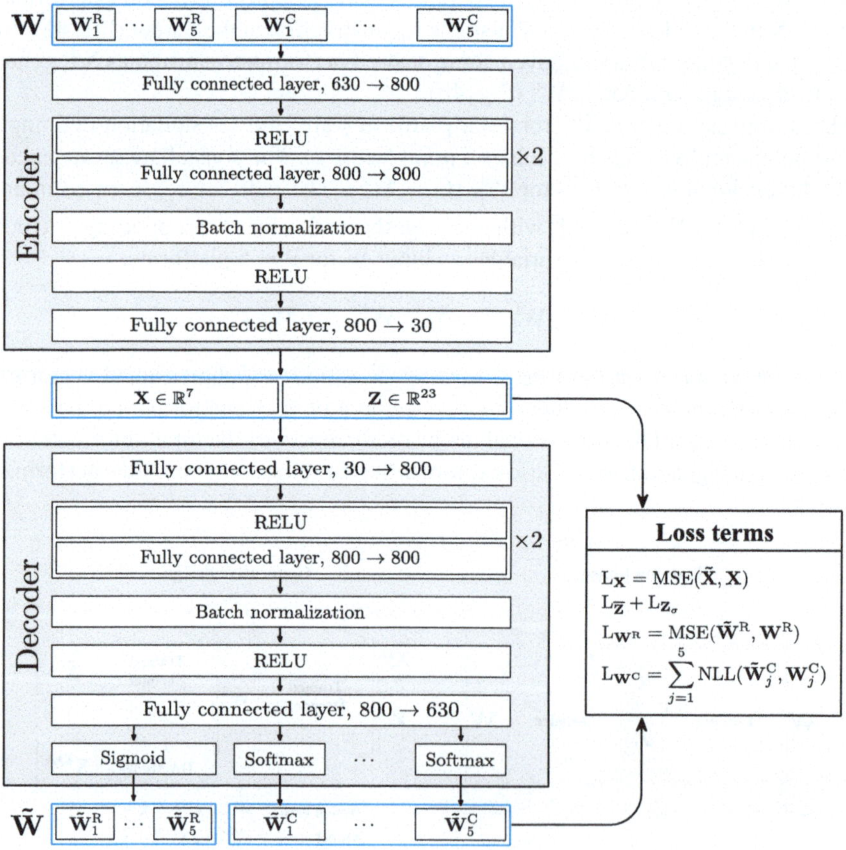

Fig. 5. Architecture of the AE model and losses.

3.3 Generative Design

To generate new designs, the designer specifies the desired performance attributes $\mathbf{X^d}$, or a subset $\mathbf{X^d_{sub}}$ (for a more flexible exploration of the solution space), and the number Nd of requested solutions. Now, the model can provide precise designs because:

1. It can sample $\mathbf{Z}$ (or the non-specified performance attributes and $\mathbf{Z}$), using the distribution of $\tilde{\mathbf{X}}$ and $\mathbf{Z}$ (Sect. 3.2). Sampling and evaluation can be run in parallel, hence are computationally light processes, enabling the quick generation of any number of designs (decoder in Fig. 4b).
2. The trained encoder acts as a surrogate model of the DA model and can be used to quickly estimate the performance attributes $\tilde{\mathbf{X}}^\mathbf{d}$ for all generated designs (encoder in Fig. 4b).

Hence, to further improve the performance of the generated designs, our strategy is to internally generate 100x more solutions and then use the encoder to choose the best Nd where $\tilde{\mathbf{X}}^\mathbf{d}$ is closest to $\mathbf{X^d}$ (or $\mathbf{X^d_{sub}}$). More details in Fig. 6.

Input: $\mathbf{X^d_{sub}}$, $\mathrm{N_d}$
Output: $\mathrm{N_d}$ sets $\mathbf{W^d}$ of generated design parameters
$\mathrm{N_{Tot}} = 100 \cdot \mathrm{N_d}$
$\varepsilon = [\,]$
for $n = 1$ **to** $\mathrm{N_{Tot}}$ **do**
  Different scenarios for the computation of $\mathbf{X^d}$ *and* $\mathbf{Z}$
  if No performance attributes requested **then**
    Sample $\mathbf{X^d}$ and $\mathbf{Z}$ from a Gaussian of mean-zero and covariance $\Sigma_{\tilde{\mathbf{X}}\mathbf{Z}}$
  else if All the performance attributes are indicated **then**
    $\mathbf{X^d} = \mathbf{X^d_{sub}}$, and sample $\mathbf{Z}$ conditioned on $\mathbf{X^d}$
  else
    if Total area is requested **then**
      Sample first from a Dirichlet to split the remaining area (total area minus other areas indicated)
    end if
    Sample all the remaining unspecified performance attributes to form vector $\mathbf{X^d}$,
    Sample $\mathbf{Z}$ conditioned on $\mathbf{X^d}$
  end if
  Computation of designs using the trained decoder
  $\tilde{\mathbf{W}}^\mathbf{d} = \mathbf{decoder}(\mathbf{X^d}, \mathbf{Z})$
  One-hot vectors from the logits of $\tilde{\mathbf{W}}^\mathbf{d}$:

$$\mathbf{W^d} = [\mathbf{argmax}\left(\tilde{\mathbf{W}}_1^\mathrm{C}\right), \ldots, \mathbf{argmax}\left(\tilde{\mathbf{W}}_\mathrm{P}^\mathrm{C}\right), \tilde{\mathbf{W}}_1^\mathrm{R}, \ldots, \tilde{\mathbf{W}}_\mathrm{P}^\mathrm{R}]$$

  Assessment of designs performance using the trained encoder
  $\tilde{\mathbf{X}}^\mathbf{d} = \mathbf{encoder}(\mathbf{W^d})$
  Computation of the performance attributes' errors:

$$\varepsilon_n = \sum_{i \in \mathbf{X^d_{sub}}} \frac{1}{\mathrm{std}(\mathbf{X}_i)} |\tilde{x}_i^\mathbf{d} - x_i^\mathbf{d}|$$

  Append ε_n to ε
end for
Return the $\mathrm{N_d}$ sets of parameters $\mathbf{W^d}$ with smallest values in ε

Fig. 6. Algorithm for prediction of new designs using the AE model

4 Evaluation

To evaluate the AE model we use the following sets of performance attributes (as observed in Fig. 4b): those requested by the user ($\mathbf{X^d}$), and the ones predicted by the encoder ($\tilde{\mathbf{X}}^\mathbf{d}$) and calculated through the DA model ($\mathbf{X}^{\mathrm{DA}}$) for some generated designs $\mathbf{W^d}$. For attribute i we then calculate the *design error* as $\varepsilon_i^{\mathrm{D}} = \left|\mathbf{X}_i^{\mathrm{DA}} - \mathbf{X}_i^{\mathrm{d}}\right|$, and the *model error* as $\varepsilon_i^{\mathrm{Mod}} = \left|\mathbf{X}_i^{\mathrm{DA}} - \tilde{\mathbf{X}}_i^{\mathrm{d}}\right|$.

4.1 Novelty of New Designs

To test the ability of the model to generate novel designs (never seen during training), we request 50 000 designs for a given $\mathbf{X^d}$, and select the best 100, i.e. with $\tilde{\mathbf{X}}^\mathbf{d}$ closest to $\mathbf{X^d}$. The values chosen for $\mathbf{X^d}$ are close to the mean of the distribution, which facilitates finding precise designs also in the training set. Then, we select the best 100 samples from the training set, i.e. closest to $\mathbf{X^d}$.

In Fig. 7 we depict the histograms of the average error distributions. For the designs generated through the AE model, we compute the design error (black) and the error $\left|\mathbf{X}_i^{\mathrm{d}} - \tilde{\mathbf{X}}_i^{\mathrm{d}}\right|$ (blue). Both are more shifted to the left, i.e. towards smaller errors, compared to the error for the designs drawn from the training set (red). This means the AE model delivers designs closer to the request than when selecting from the training set, which demonstrates the generation of novel, and more accurate, designs. This is exploited further thanks to the model ability to generate more designs and return the N best ones.

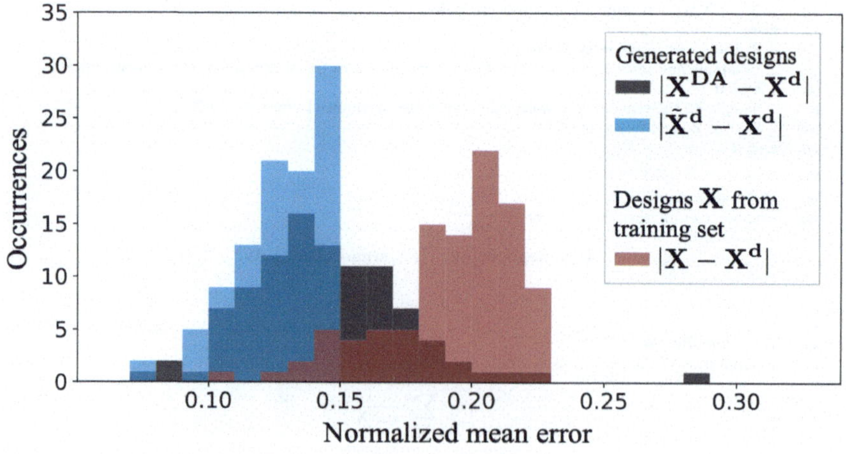

Fig. 7. Errors' distribution for generated designs and for samples from the training set.

4.2 Exploration of Design Space

By interrogating the AE model, the designer can understand the space of solutions: constraints, correlations, feasibility, etc. We illustrate this through seven different requests

$\mathbf{X^d}$ of rain occlusion and total area (red dots in Fig. 8). For each, we randomly pick three designs from the best 1% out of 1 000, and plot $\mathbf{X}^{\mathrm{DA}}$ (black diamonds) and the corresponding design errors (black line).

The designs generated for requests B, C and E approximate accurately the intended values, as they fall in an area where the model has observed more training samples. Conversely, the large error for designs associated to G suggests that such request is unachievable: low rain occlusion (scattered platforms) and large total area (big platforms stacked vertically) are clearly contradictory requirements.

For requests A, D and F, in areas with few training samples (<5), the AE model was still able to discover new feasible and accurate solutions. In Fig. 9 we present the richness of proposed solutions by showing 10 exemplary designs for each request.

4.3 Assessment of Model Performance

The validation is performed by assessing how well generated designs match the requested performance attributes, for their full distribution. For each performance attribute, its distribution is split into bins, and for each we request 1 000 designs.

This allows calculating $\mathbf{X}_i^{\mathrm{DA}}$ and $\varepsilon_i^{\mathrm{D}}$ for each sample, leading to a mean design error per bin. Figure 10 shows these errors when requesting only one performance attribute (a, b) and two simultaneously (c, d).

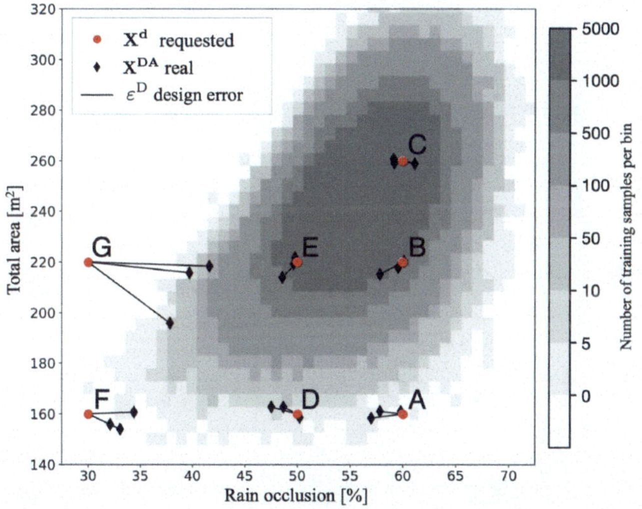

Fig. 8. Design errors for rain occlusion and total area, points A–G, and the distribution of training samples values **X** (grey heat map). Out of the best 1% designs provided for each request, we depict 3, randomly selected.

For the rain occlusion, the design error is smaller for performance values for which the model observed more data (Fig. 10a). For the total area attribute (Fig. 10b), however, the error increases as the attribute values grow – the errors per platform accumulate. Nevertheless, both cases lead to accurate designs for the best 10% of the samples per bin: <2 pp for the rain occlusion and <5 m^2 for total area, for almost all the distribution. Studying multiple performance values allows to detect areas where the attributes' correlations lead to unfeasible requests, e.g. higher error in bottom-right corner in Fig. 10c–d. Besides, it also helps discovering areas that intuitively seem unfeasible, but where the AE model can still generate accurate designs (e.g. <5 pp for a high rain occlusion of 64% and much lower sun occlusion of 21.7%, as observed in Fig. 10c–d).

The trained encoder has an important role as surrogate model: to perform the selection of the best designs, and to allow an agile and time-efficient exploration of the solution space. In Fig. 10a–b we additionally plot the model error (red dashed-line). In both cases its value is small (∼1 pp and 1 m^2 respectively). This helps concluding that the

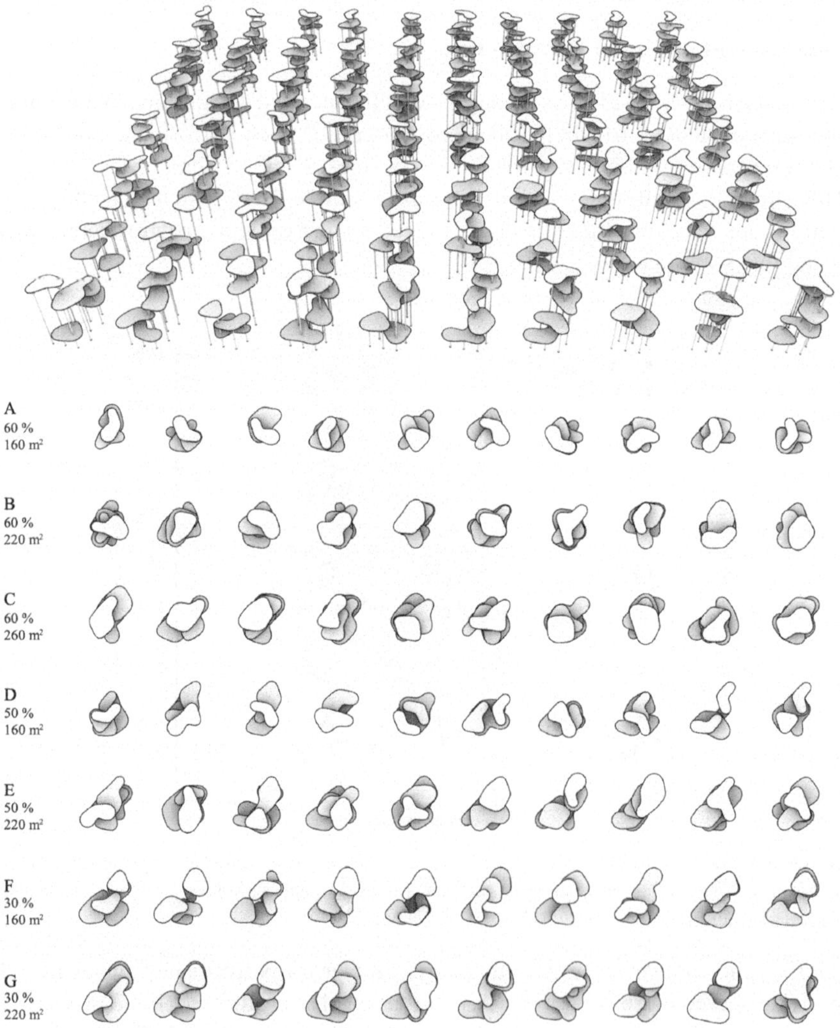

Fig. 9. Exemplary designs for seven different requests of rain occlusion and total area in perspective and top view.

encoder provides on average predictions $\tilde{\mathbf{X}}^{\mathbf{d}}$ similar to those calculated in the DA model, verifying its validity for the selection the best designs.

5 Application in Architectural Design

5.1 Design Workflow

The overall design workflow using the proposed approach consists of the following steps: 1) set up the project- specific parametric design+analysis (DA) model, 2) generate the dataset, 3) train the AE model, and finally 4) deploy the trained model to explore similarly performing design options, as explained in 3.3. The DA model can be used three-fold: a) to create an instance of the design for input parameters $\mathbf{W}$ given explicitly by the user (optional), b) to generate the dataset using input parameters $\mathbf{W}$ sampled randomly in step 2, and c) to construct and analyse design instances from design parameters $\mathbf{W}^{\mathbf{d}}$ predicted by the AE model for the aforementioned design exploration in step 4 (see Fig. 11).

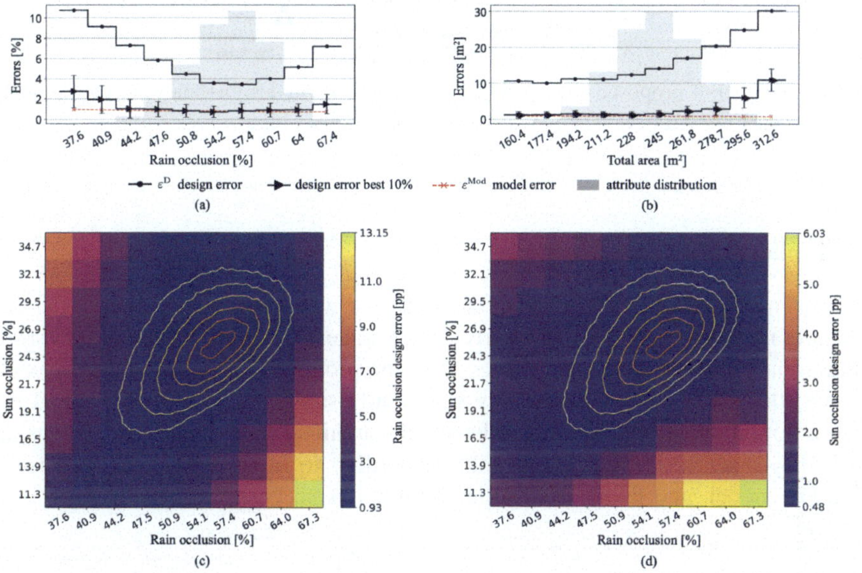

Fig. 10. In (a) and (b), average design and model errors. For a simultaneous request of rain and sun occlusion, rain (c) and sun (d) design errors for the best 10% designs. The isocurves depict the distribution in $\mathbf{X}$ of rain and sun occlusion.

5.2 Implementation

In the presented case study, the project-specific DA model was built using Grasshopper (McNeel and Rutten, 2010) and GhPython. The dataset $D = \{\mathbf{W},\mathbf{X}\}$ consisting of 470 000 random designs was split into a training (90% of samples) and a validation set. The latter is used for hyperparameters' tuning of the weighting terms and the learning rate using GPyOpt (The GPyOpt authors 2016), process that took approximately 72 h on a machine with a single GPU (Tesla P100). The AE model, implemented using PyTorch (Paszke et al. 2019), is finally trained for 200 epochs with a batch size of 1024, using Adam optimizer with default values. The script for the rollout of the trained model runs as a server outside of the Rhino environment and communicates with the DA model in Grasshopper over a http protocol[1]. The requests are processed quickly, e.g. returning 100 design within <5 s.

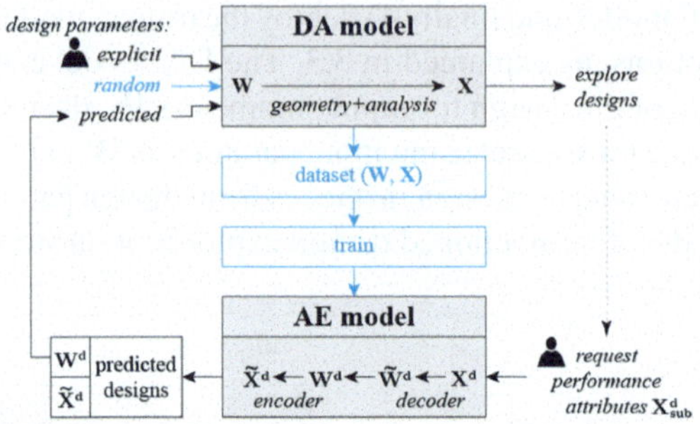

Fig. 11. The overall proposed workflow using the parametric design+analysis (DA) model and the autoencoder (AE) model. The pawn symbol indicates user input.

5.3 Application in the Case Study

The implemented model was used by an architect, who was not involved in the development of the presented tool, in an early design stage of the project. By interacting with it, the initial target performance values from the project brief were revised, as well as observed that sun occlusion is usually sufficiently satisfied and can be omitted in the request. For the selection of the final design, the architect requested 100 solutions, from which they discarded designs unsuitable according to further criteria not considered in the DA model, and finally chose the winning design based on other values such as aesthetics.

5.4 Observations

We observed that the solutions proposed by the autoencoder are very versatile, both geometrically (as exemplified by a selection in Fig. 12), and in terms of strategies to accomplish the requested performance, some of which the architect did not intuitively anticipate. For example, the autoencoder "discovered" that a needed percentage of rain-exposed areas can be achieved by having large platforms at the top and very small ones at the bottom (Fig. 13). Another observation was that requesting a small total area with low occlusions will result in small platforms spread far apart (compare designs A and F

[1] Video demonstrating the design exploration process here.

in Fig. 9). In both cases such solutions might be unfavourable, e.g. for structural reasons, and to curb the first effect, the user can opt to explicitly request a small area for the top platform. Overall, such findings help the designer understand the solution space and how the performance values translate into geometries.

6 Discussion and Outlook

We observed that the AE model undoubtedly can present many versatile solutions, and that this variety can give valuable insights that enhance the architect's understanding of the design task. Additionally, the analysis of the dataset can assist the architect in setting up and revising the DA model, e.g. by revealing hidden correlations.

Fig. 12. Different geometries generated from design parameters produced by the AE model, all for the same requested performance attributes.

Fig. 13. Example of a strategy found by the AE model to easily achieve a large area exposed to rain though a very large top platform (a). The user can counteract it by requesting explicitly a small top platform (b). Both designs have the same total area and rain occlusion.

Throughout the pipeline, from the way the DA model is implemented to the human exploration and the AE methodology, there are biases. To understand them better, the future work could conduct studies on the data diversity during generation, biases on the encoding process, etc. Nonetheless, we propose a human-in-the-loop interactive approach as we believe is the best way to tackle these biases. By exploring the solutions, the designer can learn from them, tune the parameters accordingly, and finally understand the implicit biases and adjust to them.

For a widespread adoption, the methodology should allow a seamless and intuitive interaction. In our proof-of- concept, the trained model returns up to 100 designs within a few seconds, making it fairly interactive, and enabling a deep and exhaustive exploration of possible solutions.

In future work, we aim to generalize the tool to not only accept vectorial inputs, but all type of data presentations (graphs, images, etc.). This will require using alternative deep learning architectures, such as convolutional neural networks. Besides, other generative models, based on GANs or variational AE, will be implemented and tested on similar use cases.

Acknowledgements. We would like to thank Sarah Schneider of Gramazio Kohler Research, ETH Zürich, and the planning team of the Semiramis project for the valuable user feedback from the architect's perspective. This project was supported by the Swiss Data Science Center (project C20-09).

Contributions. Matthias Kohler developed the basic concept and the initial software prototype for the case study. Luis Salamanca and Fernando Pérez-Cruz improved the methodology beyond this initial idea, implementing the final AE model, and carrying out the analyses for performance, creativity, etc. Aleksandra Anna Apolinarska developed the final DA model, its parametrization and analysis, the interface to the AE model, and contributed to assessment of its performance.

References

Chaillou, S.: Space layouts & GANs Jan (2020). https://towardsdatascience.com/space-layouts-gans-19861519a5e9

Banga, S., Gehani, H., Bhilare, S., Patel, S., Kara, L.: 3D topology optimization using convolutional neural networks (2018). https://arxiv.org/pdf/1808.07440.pdf

Miguel, J.de, Villafañe, M.E., Piškorec, L., Sancho-Caparrini, F.: Deep form finding – using variational autoen- coders for deep form finding of structural typologies. In: Sousa, J.P., Xavier, J.P., Castro Henriques, G. (eds.) Architecture in the Age of the 4th Industrial Revolution – Proceedings of the 37th eCAADe and 23rd SIGraDi Conference, 11–13 Sep, vol. 1. University of Porto, Porto, Portugal (2019). http://papers.cumincad.org/data/works/att/ecaadesigradi2019_514.pdf

Costa, A., Nannicini, G.: RBFOpt: an open-source library for black-box optimization with costly function evaluations. Math. Program. Comput. **10**(4), 597–629 (2018). https://doi.org/10.1007/s12532-018-0144-7

Gramazio Kohler Research: Semiramis (2022). https://gramaziokohler.arch.ethz.ch/web/e/projekte/409.html

Hornby, G., Globus, A., Linden, D., Lohn, J.: Automated Antenna Design with Evolutionary Algorithms. Space AIAA SPACE Forum. American Institute of Aeronautics and Astronautics (2006). https://arc.aiaa.org/doi/ https://doi.org/10.2514/6.2006-7242

Hu, R., Huang, Z., Tang, Y., Van Kaick, O., Zhang, H., Huang, H.: Graph2plan: Learning floorplan generation from layout graphs. In: CM Transactions on Graphics 118-1 (2020)

McNeel, B., Rutten, D.: Grasshopper (2010). www.grasshopper3d.com

Mo, K., Guerrero, P., Yi, L., Su, H., Wonka, P., Mitra, N., Guibas, L.J.: Structurenet: Hierarchical graph networks for 3d shape generation (2019)

Nauata, N., Chang, K.-H., Cheng, C.-Y., Mori, G., Furukawa, Y.: House-GAN: relational generative adversarial networks for graph-constrained house layout generation. In: European Conference on Computer Vision, pp. 162–177. Springer (2020). https://arxiv.org/pdf/2003.06988.pdf

Oh, S., Jung, Y., Kim, S., Lee, I., Kang, N.: Deep generative design: integration of topology optimization and generative models. J. Mech. Des. **141**(11) (2019). https://asmedigitalcollection.asme.org/mechanicaldesign/article/141/11/111405/955342/Deep-Generative-Design-Integration-of-Topology

Paszke, A., et al.: PyTorch: an imperative style, high-performance deep learning library. In: Advances in Neural Information Processing Systems, vol. 32, pp. 8024–8035. Curran Associates (2019)

Sohn, K., Lee, H., Yan, X.: Learning structured output representation using deep conditional generative models. In: Advances in Neural Information Processing Systems, pp. 3483–3491. Curran Associate (2015)

Steinfeld, K., Park, K., Menges, A., Walker, S.: Fresh eyes: a framework for the application of machine learning to generative architectural design, and a report of activities at smartgeometry 2018. In: Computer- Aided Architectural Design."Hello, Culture", Communications in Computer and Information Science, pp. 32–46. Springer, Singapore (2019)

Takahashi, Y., Suzuki, Y., Todoroki, A.: Convolutional neural network-based topology optimization (CNN-TO) by estimating sensitivity of compliance from material distribution (2019). http://arxiv.org/abs/2001.00635

Talton, J., Yang, L., Kumar, R., Lim, M., Goodman, N., Mech, R.: Learning design patterns with bayesian grammar induction. In: Proceedings of the 25th annual ACM symposium on User interface software and technology, pp. 63–74 (2012)

The GPyOpt authors: GPyOpt: a Bayesian optimization framework in python. (2016). http://github.com/SheffieldML/GPyOpt

Brown, C.T., Mueller, N.C.: Design variable analysis and generation for performance-based parametric modeling in architecture. Int. J. Arch. Comput. **17**(1), 36–52 (2019). https://doi.org/https://doi.org/10.1177/1478077118799491

Reducing Bias for Evidence-Based Decision Making in Design

Matthias Standfest(✉)

Archilyse AG, Flüelastrasse 31B, 8047 Zürich, Switzerland
standfest@archilyse.com

Abstract. This paper presents a strategy to help designers confronted with a massive flow of new technologies, exhaustive regulations, and design requirements. It summarises the existing proposals for AI-driven design development strategies, and lists frequent pitfalls like the focus on local optima or the lack of backpropagation. It identifies the main source of bias in generative design as a lack of detail and contextual complexity. The paper introduces an augmented AI process for preparing real-world data and its meta information to be used in design processes (BIM, GIS, external statistics including information such as rental price or spatial cognition). This evidence-based approach for deriving verifiable fitness functions presents a way to create holistic designs that reflect the complexity of today's built environment by using a posteriori and unbiased statistical models to substitute existing speculative categorisations. Hence, it allows avoiding naïve and overfitted solutions, and it inverts the dominant paradigm of automated generation and manual curation.

1 Old Wine in New Bottles: Design Development in the AI Era

Thomas Kuhn[1] is frequently cited on the effects of a Human-in-the-loop (HIL) approach to cover the complexity of true data-driven design creation (Carrier 2006). His research highlights the importance of subjective elements in the comparative evaluation of theories as well as its negative effects. According to him, the accumulation of failures in processes with significant subjective contributions can be attributed to the lack of empirical evidence and the resulting difficulty of clear decision-making in such matters.

This unparalleled level of complexity (Vrachliotis 2012, p. 163) comes with a high degree of responsibility and unavoidable cognitive overload (Matthews et al. 2020), while existing problem-solving strategies reveal themselves as inherently flawed: they are mainly based on human experience which is gathered over time. This process is

[1] According to Thomas Kuhn, competing paradigms are frequently incommensurable. That is, they are competing and irreconcilable accounts of reality. Thus, our comprehension of science can never rely wholly upon "objectivity" alone. Science must account for subjective perspectives as well, since all objective conclusions are ultimately founded upon the subjective conditioning/worldview of its researchers and participants.

C. Gengnagel et al. (Eds.): DMS 2022, *Towards Radical Regeneration*, pp. 122–132, 2023.
https://doi.org/10.1007/978-3-031-13249-0_11

based on the assumption that environmental requirements are not changing too fast – an assumption which obviously outlived its validity (Alexander 1978; Joedicke 1976; Stiny and Mitchell 1978; Vitruv 2015). As a result, designs struggle with an increasing imbalance between workload and creative freedom while still needing to adapt to technological requirements rather than the other way around. This imbalance is further nourished by the fact that the real estate sector is one of the least digital sectors in western economy (Kane et al. 2015). Therefore, coordination of projects always involves considerable additional efforts in communication. So, how can one reduce the cognitive overload for designers while empowering them with evidence-based knowledge to help them design better?

1.1 Current AI-Driven Design Development Strategies

Currently, the dominant paradigm in computer-aided architectural design (CAAD) is to improve the design space exploration of generative designs (Hester et al. 2018; Reisinger et al. 2021). To reduce the cognitive load many researchers try to arrange the design space in a two-dimensional grid. Unsupervised clustering or self-organising maps are used frequently, as are the concepts of Pareto fronts. Because this process is complex and difficult, ready-made "Volume Solvers" are increasingly popular. Dominant players (like spacemaker, skyline, testfit, architechtures [sic!], proving ground, engrain, archistar, sitesolve, metabuild, digital blue foam, etc.) provide Pareto fronts designers can pick from. This comparative methodology is a common tool to iteratively improve competing designs.

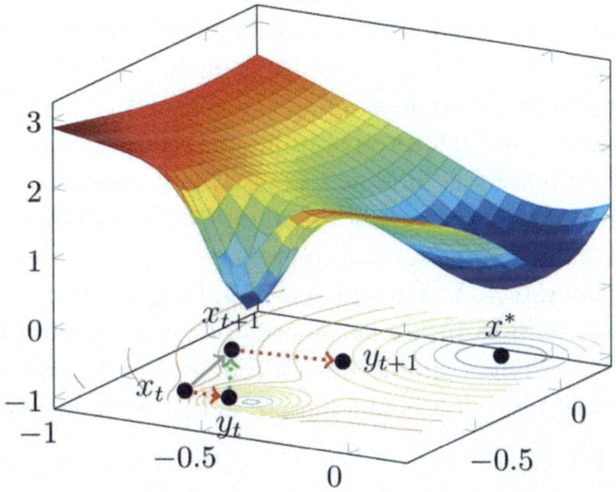

Fig. 1. Stochastic gradient descent (SDG). Image courtesy by Kleinberg et al. for an alternative view: When does SGD escape local minima? ICML 2018/02/17.

While architectural design is often represented as a linear optimisation process (Bre et al. 2016), it is not. Architectural design is more related to a stochastic gradient descent process [Fig. 1] in a hyper-dimensional solution space (De et al. 2020). Every design is a sample in this solution space, and the level of detail is the amount of dimensions to probe in this space. Of course both, the amount of samples and of dimensions, correlate with the probability of finding the global optimum. The hypothesis is that algorithms can be used to generate an abundance of geometric variants which are derived from gradually changing the underlying numeric input parameters. While the number of versions generated this way indeed might be infinite, they do not cover the whole design space but merely a subset which is reflecting the specified input-output mapping. So rather than approaching a local or global optimum of the solution space by iterative re-design, as has been done so far, now the generative design methods predefine a solution space which can be searched instead. Such searches are mostly implemented as fitness functions for multi-criteria optimisations (MCO) and reproduce the best set of input parameters for the given parametric design. Often these results get falsely interpreted as global optima for the given design task but given the limitations of the investigated solution space (which can be read as bias) these claims are clearly exaggerated. If one only analyses a single subsection of the solution space, albeit very thoroughly (which is exactly what is done when using a simplified shape grammar approach), one will most likely not find a global optimum. To mitigate the pitfalls of the local optima, domain experts have already come up with their own "stochastic" process: the design competitions where topologically very different setups can be compared.

Additionally, the synthesis-analysis loop of design is strongly affected by the analysis tools where errors in the analysis are typically propagated into the design. And prominent MCO strategies for building layouts (Sangani 2021) lack both scope of analysis and scope of design space. The oversimplifications they are built upon introduce additional bias and reduce the necessary design diversity even further. Most current MCO tools used in design development are not leveraging true alternative concepts. Also, they strip designers of both creative and interpretive influence. MCO are trojan horses claiming to reduce the HIL issue just by adding new bias. This is caused by insufficient linking of complex subjective parameters to the geometry of the design studies at the very beginning. Current volume optimisers are merely overfitting without taking necessary information into consideration. But designers should be able to comparatively evaluate the alternative theories with as little bias as possible. Simply put, the machine should rank or select, and the architect should design rather than the other way around. This challenges current strategies, but it is much more in line with the historic developments of other engineering disciplines.

1.2 Issues with False Design Automation

This false understanding of better design development through automation causes many issues:

a) **Focus on the early stage**. Currently, most effort is put on early-stage design and analysis (e.g., spacemaker).

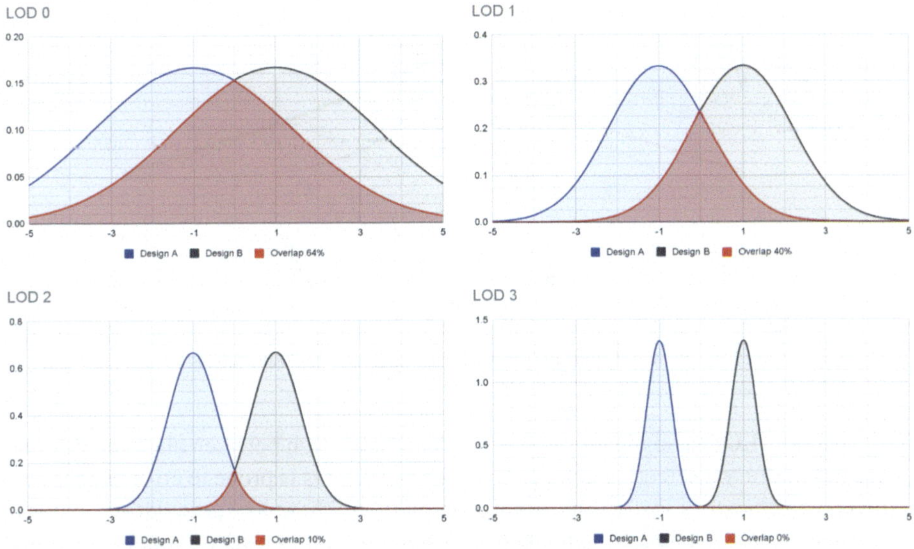

Fig. 2. Level of development in decision making. This conceptual illustration highlights the effect of comparing probability distributions of different standard deviations. This is due to the fact that in low levels of detail the final performance for each simulation dimension can only be predicted with low accuracy, while with higher levels of detail the accuracy of the performance prediction with simulation increases. The overlapping areas are those, where it cannot be decided which of the designs is finally going to perform "better". This uncertainty area (the overlap) is smaller, the higher the level of detail of the underlying model, which of course is limited according to the MacLeamy Curve. Image courtesy by the author.

The MacLeamy Curve [Fig. 3] tells us that changes applied soon cost less money. The lower the level of development (LOD) of a design, the cheaper it is to apply changes. [Fig. 2].

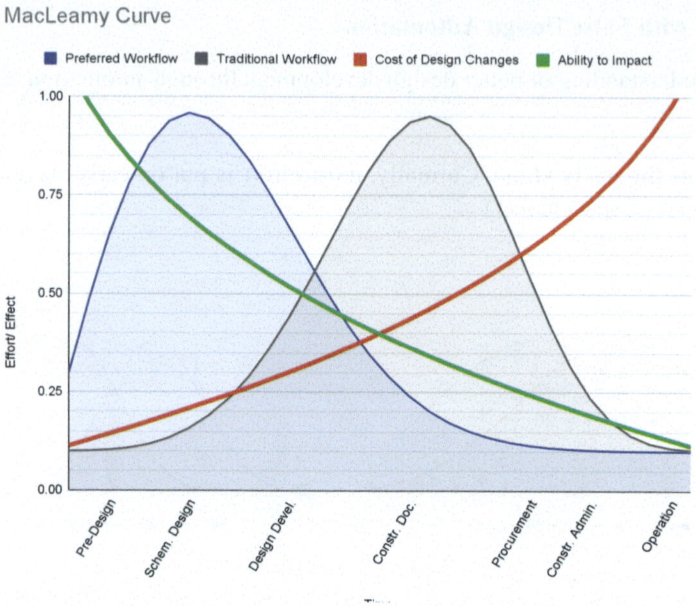

Fig. 3. MacLeamy Curve. Image courtesy by the author.

But low LOD also means there is no clear distinction between alternate designs. On the contrary – selecting designs based on low LODs is prone to error because too much can change [Fig. 2]. Any simulation based on these low granularities tends to be misleading. Additionally, many simulations do not consider enough contextual information like the built environment.

b) **Lack of holistic analysis**. The majority of simulations in architecture come with interfaces to the frequently used design environments, middleware like Speckle or plugin ecosystems. This way, simulating many physical or geometrical aspects of designs becomes a commodity. And all these deterministic simulations share a materialistic origin – their scope hardly contains social or psychological evaluations (complex relationships between architecture and humans like rental prices or social and environmental impacts). The disregard of such factors leads to paradoxic results: even the most energy-efficient building is hardly sustainable, if it is vacant for years because the local market needs were not met properly. This perfectly illustrates Kuhn's initially mentioned critique when subjective, difficult-to-measure parameters are ignored (Carrier 2006). Clearly this drift towards materialistic simulations has a reason: psychological or social implications are much harder to measure. At least with *a priori* models (Halevy et al. 2009).

Only data-intensive statistical modelling of contextual features has the potential to reduce bias with *a posteriori* models to predict such psychological and social effects. This mitigates the negative effects of incomplete evaluation scopes the domain still struggles with. On the other hand, ignoring complex data results in misleading optimisations. So the lack of a large data set for holistic modelling is evident.

c) **Feature weights in multi-dimensional space.** Not all features are scaled and distributed equally. This can cause irregularities in analysis and design. To identify strengths and weaknesses, it is required to normalise the parameters of a dataset. Otherwise, features with outliers are overrepresented. Such a normalisation can only be done if a sufficient amount of real-world data is provided for benchmarking.

It can thus be stated that the Pareto fronts generated by current generative design tools are far from optimal, even misleading at worst. To mitigate this, focusing on more details, a larger scope, and normalised scales is recommended. The method we propose allows for improving these parameters at a novel level of cost-effectiveness.

2 Method

Two extremes of the application at hand can be illustrated: either one creates evidence-based design processes which empower designers and strengthen the creativity of the designers or one is reducing their roles to become mere curators of deterministic solution sets. The resulting hypothesis is that offering an HIL-friendly AI solution could help designers to regain decision power. Also in the context of explainable and collaborative processes AI augmentation shows potential for efficiency and transparency increase in data homogenisation compared to black box ML projects. In these regards, the data pipeline developed is able to support real estate decision makers with validated and reliable benchmarks. Methodologically, these steps are the basis for better data in an automated design development:

a) **Homogenisation process via augmented AI process.** The most important step is to convert any plan from any file (including 2D raster and vector files) into structured BIM data (IFC 4.0 LOD 200) with an economically efficient augmented AI process. A HIL-strategy is commonly used in industrial applications to improve the accuracy of object recognition when confidence levels of ML predictors are low. At its core, this process is focusing on the validation of the geometrical information with multiple sources of truth, like provided room lists or governmental GIS data regarding the building hull geometry. The resulting IFC files of the reconstructed Digital Twins serve as a validated single source of truth (SOT) for the subsequent analysis. [Fig. 4].

Objects and areas of these IFC files are not only labelled and attributed according to widespread standards, they are modelled with similar generative processes. This homogeneous data structure is important to avoid geometric/semantic bias in subsequent statistical analysis.

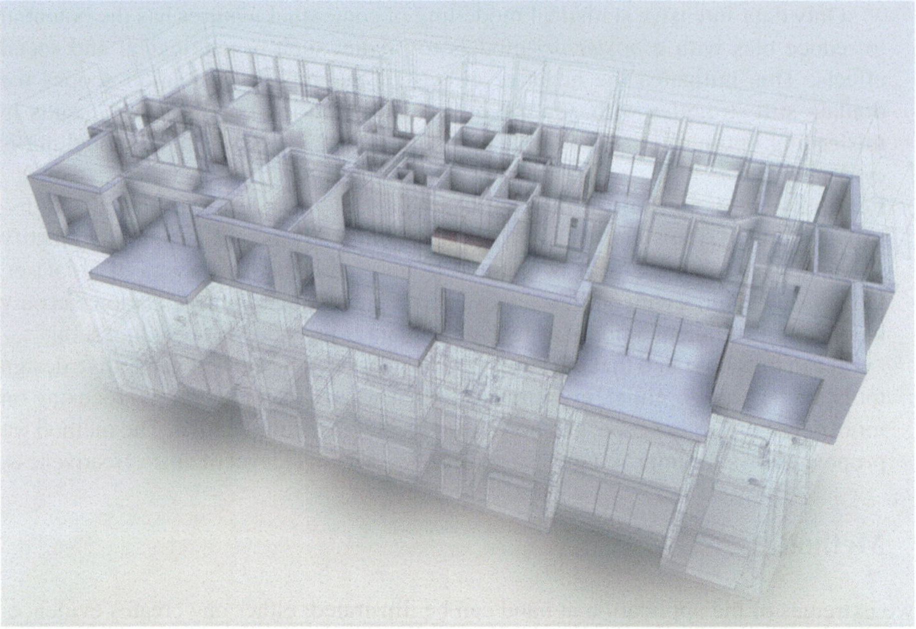

Fig. 4. IFC file rendered in autodesk online viewer. Image courtesy by Archilyse.

b) **Data enrichment.** Once the validated digital twin of a building is created, it has to be positioned within a GIS model of the environmental geometry. [Fig. 5]. This way, semantically annotated geometry containing building hulls, topographic information, and additional map layers of a 20 km x 20 km area is fused with the annotated geometry of the digital twin located at its centroid. This contextual information of the built environment is gathered from commercially available data sources, open government data sources and open-source repositories. The resulting data has a high level of detail regarding interior and exterior geometry. This is a prerequisite for detailed spatial analysis – and mandatory if bias known from purely building-hull driven models needs to be avoided.

c) **Data densification.** After generating a comprehensive semantically annotated geometric representation of the digital twin and its environment, different simulations to calculate spatial qualities are applied. [Fig. 6]. Using a hex-grid in a 25 cm resolution to define the location of the observation points has shown a high number of benefits. For every point the following simulations are calculated: semantic spherical viewshed (how much of which label can be seen?), 3D Isovist, traffic noise (at different times of the day), natural light (including atmospheric illumination and direct sunlight every 2 h throughout the year), accessibility (multiple centrality metrics like betweenness or closeness). In addition to these 50 simulations per observation point, 25 discrete features for every room are computed (like area type, net area, perimeter length, largest inscribed rectangle, furnishability, etc.) and additional 68 features per

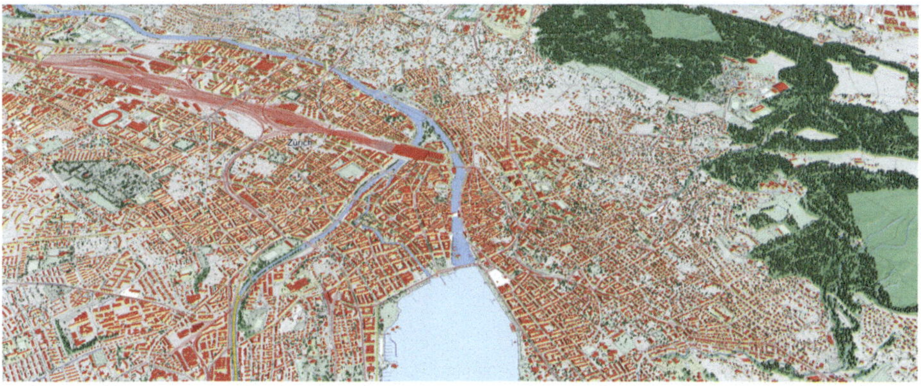

Fig. 5. Screenshot showing the topography of Switzerland, using open-gov geo data. Image courtesy of SwissTopo.

floor level (projected wall areas, number of doors, surface of bathroom walls, etc.). For individual rooms it has proven to be useful to also aggregate the values of the observation points into seven figure summaries (STD, Min., Mean, Max., P20, P50, P80). Covering a multitude of simulations allows for reducing the bias in the data set drastically.

Fig. 6. High-resolution heatmap. Image courtesy by Archilyse.

d) **Data normalisation (benchmarking).** Having access to a large data set of built apartments (more than 7 million m^2), this information is used to normalise the results of the simulations. This way, all the values aggregated per room have been converted from raw values into percentiles. Resulting in values between 0 and 1, indicating how many percent of the data set have lower or higher values regarding the respective figure. This allows for intuitive outlier detection (everything smaller than 0.1 or larger 0.9), interpretation of significant strengths and weaknesses and, of course, identification of average values close to the expectation (median). [Fig. 7].

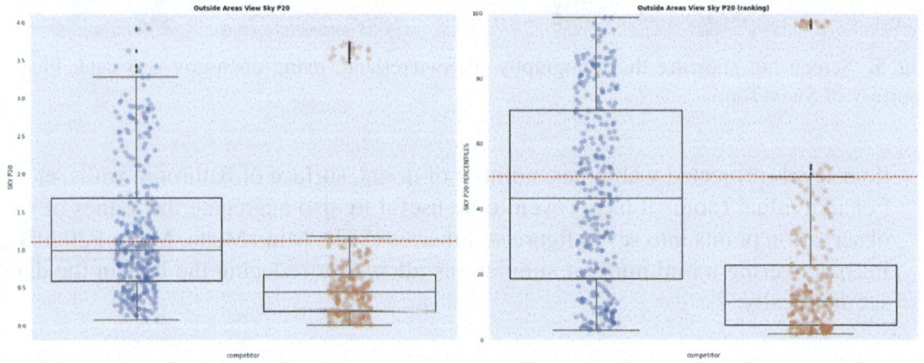

Fig. 7. The minimal skyview of all private outside areas of two competing designs (blue and orange). On the left in steradian, and on the right as their respective percentile rank. Image courtesy by Archilyse.

3 Interpolation/Application

The data generated can be used in two ways: Either directly as features for established processes (like building masses for cost estimation), or via supervised learning to enhance statistical models. In both cases, the processes benefit from a standardised data input and normalised feature vectors.

a) **Estimation of costs.** Applied to building cost estimations (CE), the provided data can be used to increase speed and precision of the process. Building masses are labelled and constructed in a homogeneous way, so comparing different architectural designs is not impacted by the different drawing styles of the different architects. Usually, BIM-based CE suffers from labelling mistakes caused by the complex UX of off-the-shelf CAAD environments. Consequently, the costly and tedious process of quality control for building mass extractions can be replaced by the presented method. Additionally, the level of detail and feature aggregation (e.g., m^2 of bathroom surfaces for the cost of tilings) is a magnitude higher than the usually used purely m^3-based CE approaches.

b) **Estimation of revenue.** State-of-the-art methods for automated valuation models (AVM) are based on some flavour of hedonic regression. Traditionally, categorical input related to certain architectural qualities is provided on a 5-point scale (low, below average, average, above average, high). Both the number of features as well as their resolution are limited to human perception. The presented method ensures a holistic scope of highly accurate features regarding spatial qualities that are outperforming both precision and accuracy of existing manual processes by far. This drastically improved dataspace has reduced noise and higher density levels. AVMs based on these provided features have shown a reduction in the respective prediction error by more than 50%.
c) **Estimation of sustainability.** For questions related to sustainability, it is important to have accurate building masses and areas, but it is even more important to have data about physical exposure of the individual areas. Calculating the grey energy needed for a design is strongly correlated with the amount of materials used. And using digital twins for thermal analysis is superior to a building-hull-only approach since the internal arrangement is impacting the overall thermal behaviour. Additionally, having information about solar exposure of the different rooms can provide further insight on ways to optimise heating and artificial light. And the generated solar profile of the rooms can be used for predictive heating control too. Furthermore, vacancy rates are a waste of grey energy. So, reducing vacancy is improving the energy footprint of buildings drastically. By using real-life energy consumption of buildings and the digital twin strategy presented above, some stakeholders have even been able to successfully deploy statistical models to derive the thermal resistance coefficients of the different building materials – and hence the proper thermal model of the building.
d) **Analysis for judging competitions.** For choosing from a set of competing designs, especially in real-world scenarios with high investment risks, stakeholders involved tend to hire a high number of experts for design quality assessment. The provided features, including meta-models like cost and revenue estimations, help to drastically reduce the communication overhead and to increase both speed and accuracy of this process. Detailed comparative analysis is significantly more objective when compared to its conventional counterpart. Reducing the bias in this step is directly reducing the planning and decision-making risk involved otherwise.
e) **Aggregating into fitness functions for better comparative analysis.** When applied in generative design loops, the generated benchmarks and simulation results provide a much more accurate indication of architectural qualities for fitness functions. Multi-criteria optimization (MCO) using the provided benchmarks and meta-models can generate significantly improved design proposals.

4 Conclusion

A process for built and to-be-built architecture was presented that allows for drastically increased accuracy in related decision-making. A novel approach was presented to include features previously subjectively decided upon. The dominant paradigm of partial optimisation was challenged. The dominant paradigm of automated generation and manual curation was inverted.

Acknowledgements. We thank Archilyse AG Zürich who provided insight and expertise that greatly assisted the research. We thank Anja Mutschler for her comments that greatly improved the manuscript.

References

Alexander, C.: A Pattern Language. Towns, Buildings, Construction. Oxford University Press, Oxford (1978)

Bre, F., et al.: Residential building design optimisation using sensitivity analysis and genetic algorithm. Energy and Buildings **133**, 853–866 (2016)

Carrier, M.: Ziel und struktur der methodologischen theorien. Deut Z Philos **54**(3), 387–400 (2006)

De, S., Maute, K., Doostan, A.: Bi-fidelity stochastic gradient descent for structural optimization under uncertainty. Comput. Mech. **66**(4), 745–771 (2020)

Halevy, A., et al.: The unreasonable effectiveness of data. IEEE Intell. Syst. **24**, 8–12 (2009)

Hester, J., et al.: Building design-space exploration through quasi-optimization of life cycle impacts and costs. Build. Environ. **144**, 34–44 (2018)

Joedicke, J.: Angewandte Entwurfsmethodik für Architekten. Karl Krämer, Stuttgart (1976)

Kane, G.C., et al.: Strategy, not Technology, drives Digital Transformation. MIT Sloan Management Review and Deloitte University Press, Becoming a Digitally Mature Enterprise (2015)

Matthews, G., et al.: Stress, skilled performance, and expertise: overload and beyond. In: Ward, P., et al. (eds.) The Oxford handbook of expertise, pp. 490–524. Oxford University Press (2020)

Reisinger, J., Knoll, M., Kovacic, I.: Design space exploration for flexibility assessment and decision making support in integrated industrial building design. Optim. Eng. **22**(3), 1693–1725 (2021)

Sangani, P.: Know your architects. 10 Firms That Use Artificial Intelligence Technologies for AEC. Re-Thinking the Future (2021). https://www.re-thinkingthefuture.com/know-your-architects/a3670-10-firms-that-use-artificial-intelligence-technologies-for-aec/. Accessed 01 Apr 22

Stiny, G., Mitchell, W.J.: The palladian grammar. Environ. Planning B **5**, 5–18 (1978)

Vitruv, 2015: Zehn Bücher über Architektur. De Architectura Libri Decem. Wiesbaden: Marixverlag

Vrachliotis, G.: Geregelte Verhältnisse: Architektur und technisches Denken in der Epoche der Kybernetik. Springer, Vienna (2012)

Artificiale Rilievo GAN-Generated Architectural Sculptural Relief

Kyle Steinfeld(✉), Titus Tebbecke, Georgieos Grigoriadis, and David Zhou

College of Environmental Design, University of California, Berkeley, 345 Bauer Wurster Hall, Berkeley, MC 1800, USA
ksteinfe@berkeley.edu

Abstract. This paper describes "Artificiale Rilievo", the first work of architectural sculptural relief produced by a generative adversarial network (GAN). Technically, the authors present novel methods developed for the generation of three-dimensional sculptural designs using a pseudo-3d description of form based on vector displacement maps (VDMs). Our approach improves on existing methods by expanding the range of possible forms, and suggests broad application in ornamental architectural design. Conceptually, the artistic work described here brings tiling geometries found in contemporary architectural ornament into dialog with forms drawn from the Western architectural canon, and reflects on the dataset as a retrograde influence in the otherwise avant-garde field of creative AI. In contrast with other AI-driven tools that center efficiency at the expense of expressiveness and authorial jurisdiction, the methods described here stand as an alternative approach to the application of machine learning in architectural design. Negotiating the "uncanny" boundary of individually-recognizable forms within a differentiated field, the piece materializes an animated walk through the latent space of a GAN in the solidity of cast bronze.

1 Background and Motivation

Sculptural relief applied as a decorative treatment on buildings is perhaps as old as building itself. The design of decorative relief has been practiced by a dynamic set of design traditions across history and by a wide diversity of cultures. As architectural styles and building practices change, the design methods and craft knowledge related to certain forms of ornamental relief are lost, while new methods emerge and develop. In the context of Western architecture, industrialization marks a moment of pronounced, perhaps exaggerated [9], change in architecture's relationship to ornament. More recently, the advent of digital design has brought about another reconfiguration of the relationship between architecture and ornament [10]. This change corresponds with a new interest in geometric pattern, in computational design methods (e.g. generative and procedural systems [6], fractals [8])) and in certain fabrication practices (e.g. 3d printing [13]) that hold relevance to ornamental design. It is in this context that this project brings to bear a nascent design technology based on machine learning - the GAN - as an additional tool in the digital designer's toolbox (Fig. 1).

C. Gengnagel et al. (Eds.): DMS 2022, *Towards Radical Regeneration*, pp. 133–148, 2023.
https://doi.org/10.1007/978-3-031-13249-0_12

Fig. 1. The Artificiale Rilievo piece, as installed. Photo by authors.

Across the AEC industry, technologists, developers, and designers have begun to explore the integration of machine learning processes into a new generation of digital tools [1]. While the promise of machine-augmentation in design is expansive [17], most research has focused on tools and processes valued for their greater efficiencies; this includes automated construction processes [4], more accurate predictive urban models [18], and more flexible building controls [15]. While such efforts hold quantifiable and immediate value to conventional design practice, less attention has been focused on novel approaches valued for qualities that threaten greater disruption to the socio-technical landscape of architectural design; this includes tools that offer radically greater accessibility [14], that complicate narratives of singular authorship, or that leverage the unique capacity of machine learning to embed and address historical or cultural issues. The processes prototyped here seek just this, and envision this new generation of tools as data-driven instruments that openly announce an easily-overlooked fact: that all software, even the most neutral-seeming of our CAD tools, is intertwined with culture and enmeshed in history.

This paper documents "Artificiale Rilievo", the first work of architectural sculptural relief produced by a GAN. The first section below describes the technical processes and methodological approaches developed over the course of the project. Here we articulate contributions to certain artistic domains - including methods that offer new possibilities for sculpture practice in general, and for creative practice in generative architectural ornament in particular. The following section documents the realization of the project as an artistic installation, and as a case study for the application of these methods.

1.1 Creative Motivation

The inspiration for the project was borne from the ongoing pandemic. Confined to a single neighborhood in Oakland, California in the lockdown of 2020, the authors began to take notice of the modest architecture of everyday buildings. Of particular interest were small ornamental pieces, such as the ones shown in the nearby figure, expressed as deformations of stucco that hold imagistic qualities. They may manifest as flowers, faces, or soft-serve ice-cream, and are applied to recall some vague Western tradition: Greek, Roman, Italian, French - it's difficult to tell, and hardly matters. In their historicism, these pieces play on our capacity for recognition and recall. In their constructed illusion of high relief, they play on our tendency to perceive three-dimensional form (Fig. 2).

Fig. 2. A single-family home in Oakland, CA. Photo by authors.

Recognizing the centrality of historical data in creative AI work, we observed that a project operating in this space requires a historical reference as a starting point, and as an object to reconsider. For this, we identify a prominent piece of architectural sculptural relief in the Western canon, and arguably the ancestor of the kitschy sculptural stucco details found in Oakland. The Pergamon Altar is a Greek construction originating in modern-day Turkey, which was disassembled in the late 19th century, and later re-assembled in the early 20th century in a Berlin museum [5]. Thus, the origin of the project operates in a manner that mimics the fate of the Pergamon: It begins with a disassembly of selected sculptural forms into fragments. These extracts serve to train a neural network to understand and reproduce a specific form-language, and are the basis of the realized "Artificiale Rilievo" piece.

2 Technical Contributions

This section details technical processes for representing a constrained subset of three-dimensional polygon mesh forms as vector displacement maps (VDMs) [3], for processing and manipulating these maps, and for translating them "back" into pseudo-3d forms. This pipeline is developed in preparation for, but independent from, GAN training and image synthesis. Apart from considering GAN-generated synthetic VDMs, the "round-trip" from 3d mesh to VDM to pseudo-3d mesh is a lossy process, and imposes certain creative constraints that are detailed below. In a parallel thread of work, the authors confirm experimentally that a generative adversarial network can be trained to convincingly reproduce certain classes of pseudo-3d forms described as VDMs.

We begin with a description of early experiments that employ a simplified framework using raster heightmaps, and that provide motivation for the development of a more ambitious framework that includes VDMs.

The capacity of a GAN to usefully synthesize pseudo-3d forms has been established [16] using the well-known "heightmap" format. Initial experiments in this project modestly extend this format as a "two-sided" heightmap, by encoding depth information from different directions into the separate channels of an RGB image (Fig. 3).

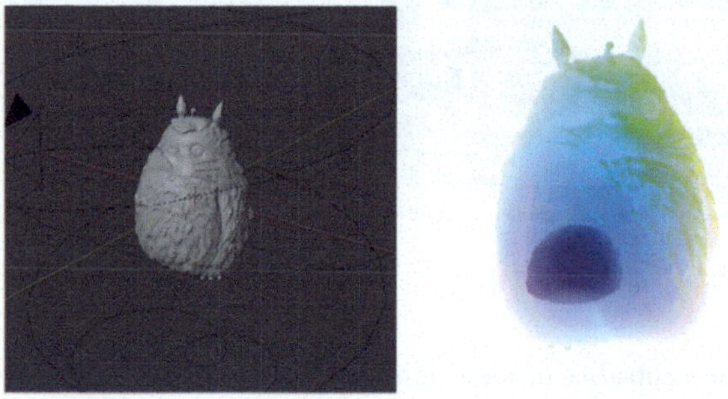

Fig. 3. A "two-sided" depthmap is extracted from a 3d model. Here, depth information from different directions is encoded into the channels of an image.

The results shown in the nearby figure demonstrate that, although our "two-sided" heightmap representation is somewhat abstracted away from the source 3d geometry, our chosen GAN architecture, SyleGAN2-ADA [7], is able to capture the "form language" of the dataset (Fig. 4).

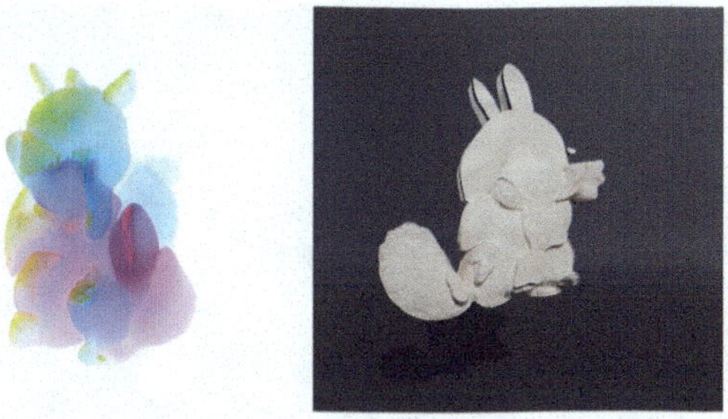

Fig. 4. A synthetic heightmap on the left, and a form derived from this map on the right.

2.1 Why Vector Displacement?

The experiments above show that our selected GAN architecture can learn a distinct language of pseudo-3d forms, and thereby establish a rationale for the development of a more ambitious framework. We observe that additional creative freedom would be offered by a format which describes displacement more robustly. To explore this possibility, we invented a new workflow for encoding pseudo-3d forms as vector displacement maps (VDMs).

Vector displacement [2] is an approach to digital modeling developed to address the problem of depicting geometrically detailed sculptural forms in a lightweight way. Here, displacements are stored not along a single direction, but rather along a vector of arbitrary length and direction and stored as the RGB channels of a raster image (Fig. 5, 6).

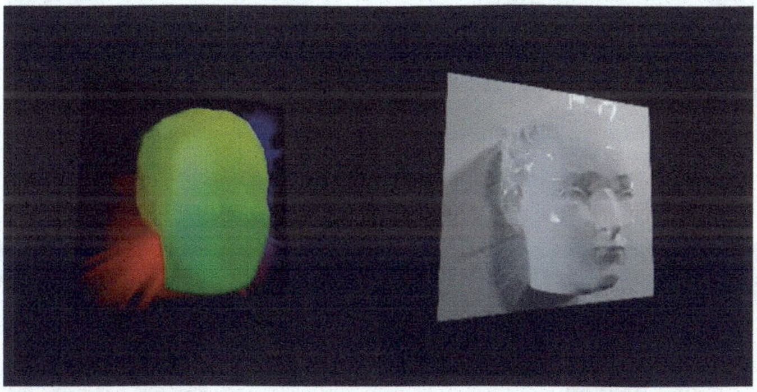

Fig. 5. In a vector displacement map, displacements are described as vectors, and stored in the RGB channels of an image.

A process for representing 3d polygon meshes as 2d vector displacement maps is summarized as: given a sample polygon mesh (1), a selected fragment (2) is "squashed" onto a plane, with displacements stored as vectors (3). This information is stored as the RGB channels of a raster image (4), a format that is both amenable to a GAN, and is able to be re-interpreted as displacements from a base raster plane (5) to reproduce similar sculptural forms (6) (Fig. 7).

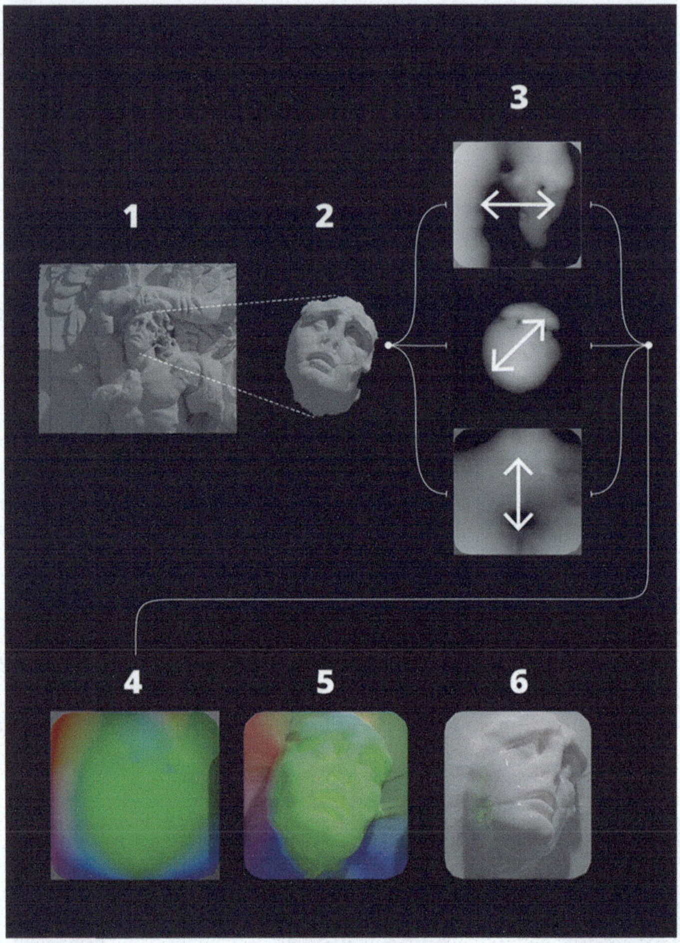

Fig. 6. A pipeline for representing 3d sculptural relief as raster data

Fig. 7. A "round trip" from a 3d polygon mesh (left) to a VDM (center) and "back" to a pseudo-3d form.

2.2 3d- > VDM

We begin with a large and densely-detailed polygon mesh, and some meticulous hand-work in CAD. For the purposes of the project, this is a 3d scan of a physical object. A smaller fragment of this large mesh is then identified and isolated. The resulting fragment must be a manifold surface with one open edge, must contain no self-intersections, and must display a reasonable ratio between total surface area and open edge length. In the final step completed "by hand", the fragment is oriented, scaled and translated to a unit cube, then exported as an STL file.

The remaining steps in converting a polygon mesh to a VDM are automated. First, the mesh fragment is "squashed" from a 3d unit cube to a 2d unit circle. Using the Kangaroo [12] particle-spring simulation tool, we ensure that the open edge of the 3d mesh is coincident with the edge of the unit circle, and that each 2d mesh vertex is close to its original 3d position without producing overlapping faces. These two related meshes are together stored as XYZ and corresponding UV information in a OBJ file, which now holds all the information required for a VDM. In a final step, we produce VDMs as 32-bit TIFFs by walking the unit square of the UV information stored in our OBJ, and tracing a vector from the face-wise position on the 2d "squashed" mesh back to its corresponding 3d location (Fig. 8).

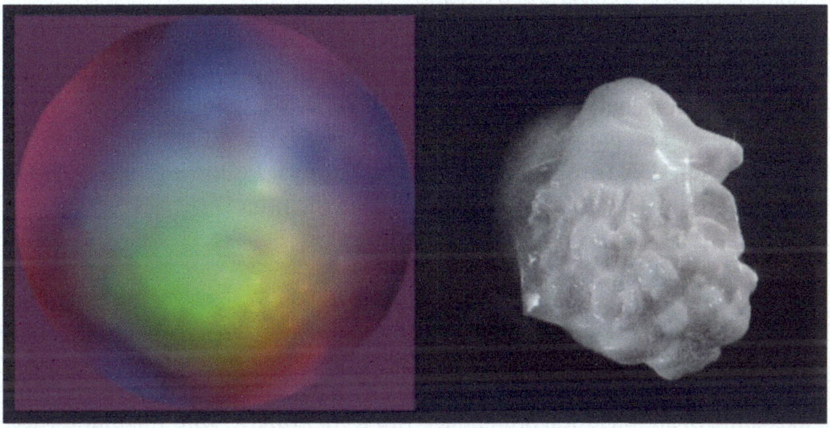

Fig. 8. A sample described as a VDM (left) and a pseudo-3d reconstruction of the original form (right).

2.3 Training and Synthetic VDM Generation

Here we survey the methods used for training a generative adversarial network on VDM data, and for generating synthetic VDMs. Since we employ the well-known technology of StyleGAN2 [7], we describe these procedures only in overview. Since the process described above for producing VDMs from 3d models holds practical limits on the number of samples produced, steps are taken to first confirm that our VDM format is amenable to the GAN architecture, and then to mitigate over-fitting by pre-training on a larger synthetic dataset of VDM samples. We first validate that StyleGAN2-ADA can be usefully trained on our format by using a synthetic dataset of approximately 3000 VDMs extracted from simple Perlin noise forms (Fig. 9).

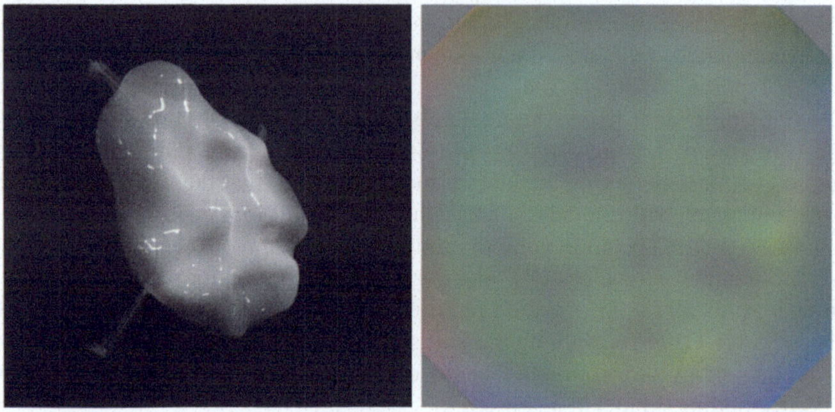

Fig. 9 A GAN-generated VDM displaying formal qualities similar to Perlin noise (right), the corresponding pseudo-3d reconstruction (left).

Given the small size of our primary dataset, we mitigate the risk of over-fitting by pre-training a model on a second synthetic dataset of 1500 VDM derived from a 3d moprhable model [11]. Training on the primary dataset then proceeds with minimal modifications of the StyleGAN2-ADA recommended training settings. Following training, standard latent walks form the basis of all synthetic VDM in the project.

2.4 VDM Processing and Manipulation

The manipulation of VDMs required by the project includes steps completed in anticipation of training, as well as processes in post-processing in preparation for various artistic applications as they are translated "back" into pseudo-3d forms. As discussed below, the nature of architectural ornament is often "field-like", with singular forms often embedded in a context of larger arrangements (such as those found on an ornamental frieze) or patterns (such as those found on a contemporary rainscreen design). To support these strategies, the project develops ways to set individual VDM samples in different combinations to produce larger compositions. For example, one of the processing steps includes the application of a falloff vignette to a given sample, with the magnitude of the Z-component of the vector displacement reduced as the boundary is approached - this allows for a smoother transition between adjacent samples. Similarly, a manipulation operation is developed for performing arbitrary rotations and reflections of a given sample - applied as a pre-training step this supports basic data augmentation, while applied to a generated VDM it allows for compositional maneuvers such as tiling along a glide reflection pattern. Since the VDM is encoded as a standard 32-bit raster image, many of these manipulations are well-supported by raster editing techniques. The spatial nature of the vector information, however, introduces some nuance that requires bespoke procedures for even these basic manipulations. For example, simple reflection operations performed on the image must correspondingly manipulate the X- and Y-components of each vector by directly altering the underlying color channels, such that the image reflection also alters the displaced form.

2.5 VDM- > pseudo-3d

Two parallel processes are developed for translate VDMs "back" into pseudo-3d forms. One process is developed for visualizing aggregations and compositions as renderings. For this purpose, we rely heavily on the vector displacement support offered by the Blender software application, and focus our efforts on developing aggregation techniques that translate latent walks into composite forms. A nearby figure showcases a sampling of these techniques, each of which combine a related series of VDM images into a single composite VDM prior to the render-time displacement process (Fig. 10).

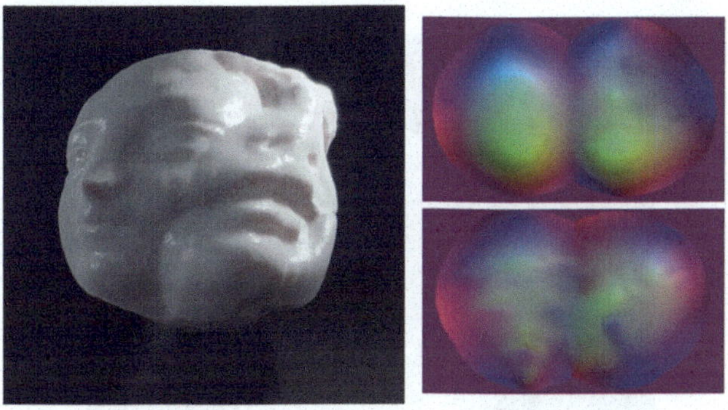

Fig. 10. Selections of VDMs from the Pergamon dataset are composited to create a "merged" pseudo-3d form.

A separate process is developed for preparing a given composition for fabrication. Here, a bespoke process of displacement is developed as a stand-alone Python application. This application handles a number of steps required for fabrication, as illustrated in a section below. In summary, these steps include: the pixel-wise displacement of a plane to construct a 3d polygon mesh, the trimming and cleaning of this mesh to ensure geometric integrity, and a set of Boolean operations to create solid 3d tiles suitable for 3d printing.

3 Application and Installation

This section documents the realization of the "Artificiale Rilievo" project as an artistic installation. As the first GAN-generated work of architectural sculptural relief, the project is novel on a technical level, and demonstrates techniques that suggest broad application in machine-augmented architectural design. The project also operates on a critical level, and reflects on the influence of the Western architectural canon in ornamental design, while simultaneously revealing the dataset as a central influence. Here, a dataset of decontextualized historical sculptural relief underlies the generation of uncanny forms that straddle the unrecognizable and the familiar. Individual ornamental tiles recall forms familiar to the Western canon, while the global composition introduces fragmentations that set up a resistance to any clear singular recognition.

Following the rationale described in a section above, the project begins with VDMs extracted from a 3d scan of the Pergamon Altar[1], which serve to train a neural network to understand and reproduce a specific form-language.

Since VDMs can be manipulated and combined using modified versions of standard image editing methods, the project embraces opportunities to combine, aggregate, and merge GAN-generated forms. This allows us to apply some of the combinatorial strategies described above. In the nearby figure, two distinct forms are shown, each related to a distinct latent walk. Recognizable features and symmetries appear, only to be disrupted as these twins slide apart, and merge into one another. A number of similar animations are thereby produced (Fig. 11).

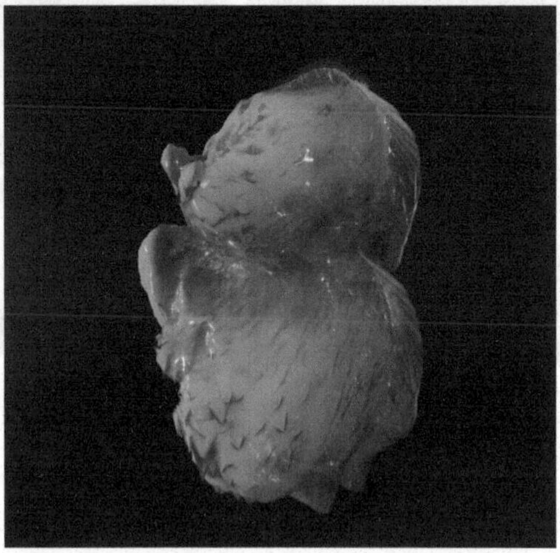

Fig. 11. A still-frame of a composite latent walk.

While a variety of options for physical fabrication may be employed to bring these synthetic forms into the world, a casting in bronze is a particularly appropriate selection given the classical subject matter. Accordingly, applying methods detailed in the previous section, a procedure is developed to produce geometries required by a 3d-printed investment casting fabrication processes. In summary, this process is as follows:

Synthetic forms are directly derived from the GAN-generated vector displacement map - pseudo-3d polygon meshes. As shown by the nearby figure, a single well-bounded form is represented fairly clearly, and the base plane is clearly discernible (Fig. 12).

[1] Scans of similar samples of architectural ornament were also included, and acquired under a creative commons license from "Scan the World" at www.myminifactory.com/scantheworld

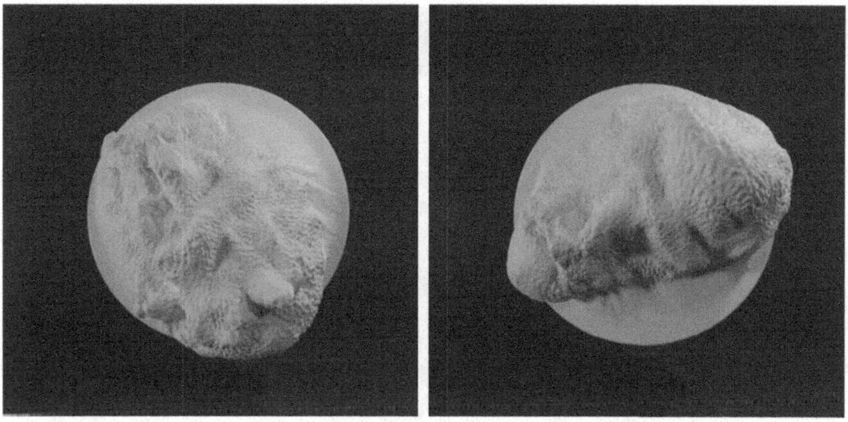

Fig. 12. Two synthetic forms directly derived from a GAN-generated VDM.

In order to break up the reading of the single latent space sample, an individual sample is broken into a simple geometric tiling. This also allows for easier aggregation when installed. Also at this stage, practicalities such as material thicknesses and labels are accounted for (Fig. 13).

Fig. 13. A synthetic form broken into simple tiles.

Because an approach to arranging the pieces in aggregate is required, another of our compositional strategies is applied here. The pieces are designed to be modular, and while the position of each within the aggregate is fixed, individual orientations are adjusted in order to break down the singularity of the samples, and to encourage multiple simultaneous readings of form (Fig. 14).

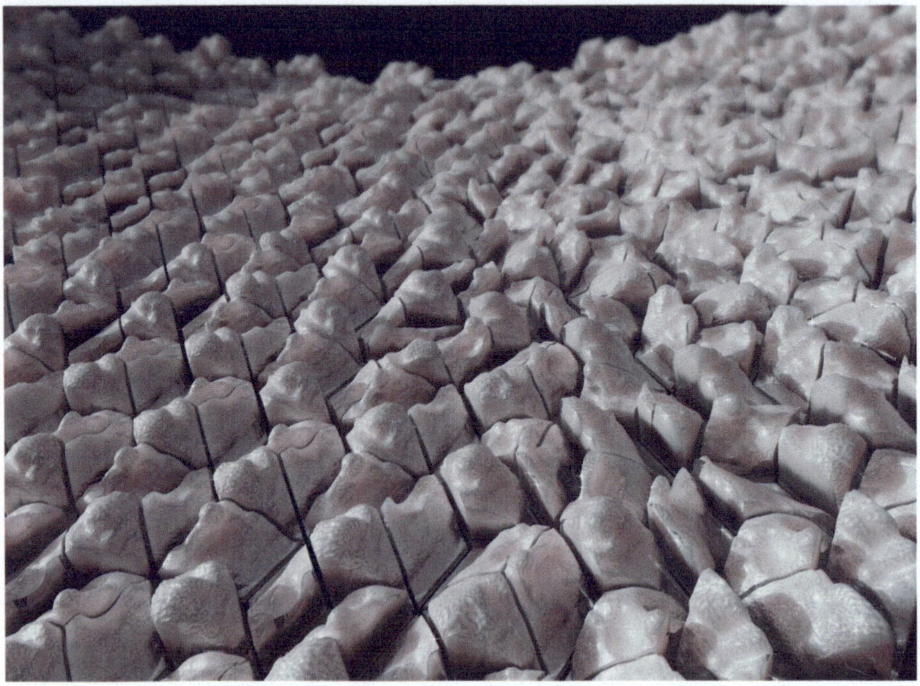

Fig. 14. Recalling the rhythmic symmetry of frieze, a "walk" through the latent space of a GAN is aggregated across a surface in high relief.

With the units defined locally and in aggregate, fabrication begins by printing the GAN-generated forms on a standard SLA printer using a filament designed for investment casting (Fig. 15).

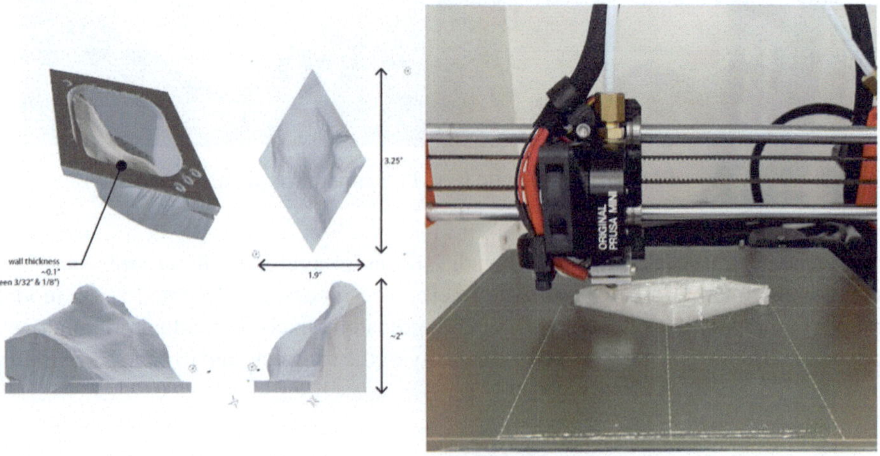

Fig. 15. The process of 3d printing.

These prints are then cast in bronze. This scope of work requires 48 samples to be cast using a lost-wax process (Fig. 16).

Fig. 16. The process of investment casting.

The pieces as they arrive from the foundry are in an unfinished state, and appear in the nearby figure prior to finishing, mounting, and patina (Fig. 17).

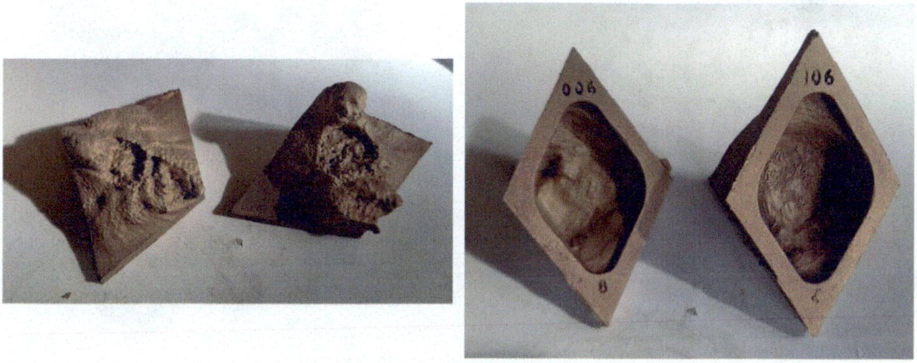

Fig. 17. Unfinished bronze pieces.

Finishing steps include the typical clipping, sanding, and de-burring, as well as the welding of brass mounting plates, and the application of a cold patina (Fig. 18).

Fig. 18. Details of the finishing process.

As installed, the final piece evokes a suggestion of the historical material from which it is formed - a vestige of the "form language" inherited from the Western canon of architectural ornament generally, and the Pergamon Alter in particular - while maintaining

Fig. 19. Details of the "Artificiale Rilievo" project as installed.

a dialog with geometries found in contemporary architectural ornament. Individually-recognizable historical forms flicker within a differentiated field, and the overall composition recalls the now-ubiqtious animated walks through the latent space of a GAN, but materialized in solid bronze (Fig. 19).

4 Reflection and Conclusion

The "Artificiale Rilievo" project offers methodological contributions that support new pathways in practice, and also documents a realized case study that seeks a critical positioning of these methods. The vector displacement pipeline described in the first section above is valuable insofar as it opens up new subjectivities in design practice, and establishes new opportunities for creative expression. As such, the project detailed in the second section not only stands as a critical creative work, but also serves to substantiate that the techniques developed here hold value in contributing to new practices in machine-augmented architectural design.

References

1. As, I., Basu, P. (Eds.): The Routledge Companion to Artificial Intelligence in Architecture. Routledge (2021)
2. Barr, C.: Vector Displacement in the Sculpting Workflow. In: Game Development Tools, Ansari, M. (ed.): CRC Press, pp. 219–230 (2016)
3. Cook, R.L.: Shade trees. In: Proceedings of the 11th Annual Conference on Computer Graphics and Interactive Techniques (SIGGRAPH '84), Association for Computing Machinery, New York, NY, USA, pp. 223–231 (1984). https://doi.org/10.1145/800031.808602
4. Fang, Z., Wu, Y., Hassonjee, A., Bidgoli, A., Llach, D.C.: Towards a distributed, robotically assisted construction framework. In: Proceedings of the 40th Annual Conference of the Association of Computer Aided Design in Architecture (ACADIA), pp. 320–329 (2020)
5. Gossman, L.: Imperial Icon: The Pergamon Altar in Wilhelminian Germany. J. Modern History **78**, 3 (September 2006), pp. 551–587 (2006). https://doi.org/10.1086/509148
6. Hansmeyer, M., Dillenburger, B.: Mesh grammars: procedural articulation of form. In: Open Systems: Proceedings of the 18th International Conference on Computer-Aided Architectural Design Research in Asia, Singapore, pp. 821–829 (2013)
7. Karras, T., Aittala, M., Hellsten, J., Laine, S., Lehtinen, J., Aila, T.: Training generative adversarial networks with limited data. In: Proc. NeurIPS (2020)
8. Kolarevic, B., Klinger, K.: Manufacturing Material Effects: Rethinking Design and Making in Architecture. Routledge (2013)
9. Long, C.: Ornament, crime, myth, and meaning. In: Architecture, Material and Imagined: Proceedings of the 85th ACSA Annual Meeting and Technology Conference, pp. 440–445 (1997)
10. Moussavi, F., Kubo, M. (eds.): The function of ornament. ACTAR, Harvard Graduate School of Design, Barcelona (2008)
11. Paysan, P., Knothe, R., Amberg, B., Romdhani, S., Vetter, T.: A 3D face model for pose and illumination invariant face recognition. IEEE, Genova, Italy (2009)
12. Daniel Piker. 2013. Kangaroo: Form Finding with Computational Physics. Architectural Design 83, 2 (2013), 136–137

13. Rael, R., San Fratello, V.: Printing Architecture: Innovative Recipes for 3D Printing. Chronicle Books (2018)
14. Rafner, J., et al.: Utopian or Dystopian?: using a ML-assisted image generation game to empower the general public to envision the future. In: Creativity and Cognition (C&C '21), Association for Computing Machinery, New York, NY, USA, pp. 1–5 (2021). https://doi.org/10.1145/3450741.3466815
15. Ida, S., Smith, L.: Machine learning integration for adaptive building envelopes: an experimental framework for intelligent adaptive control. In: Proceedings of the 36th Annual Conference of the Association for Computer Aided Design in Architecture (ACADIA), CUMINCAD, Ann Arbor, pp. 98–105 (2016)
16. Spick, R.J., Cowling, P., Walker, J.A.: Procedural generation using spatial GANs for region-specific learning of elevation data. In: 2019 IEEE Conference on Games (CoG), pp. 1–8 (2019). https://doi.org/10.1109/CIG.2019.8848120
17. Steinfeld, K.: Dreams may come. In: Proceedings of the 37th Annual Conference of the Association for Computer Aided Design in Architecture (ACADIA), Association for Computer Aided Design in Architecture, Cambridge, MA (2017)
18. Grignard, Y., Zhang, A.: Machine learning for real-time urban metrics and design recommendations. In: Proceedings of the 38th Annual Conference of the Association for Computer Aided Design in Architecture (ACADIA), CUMINCAD, Mexico City, Mexico, pp. 196–205 (2018)

Harnessing Game-Inspired Content Creation for Intuitive Generative Design and Optimization

Lorenzo Villaggi[1](✉), James Stoddart[1], and Adam Gaier[2]

[1] Autodesk Research – AEC Industry Futures, New York, US
lorenzo.villaggi@autodesk.com
[2] Autodesk Research – AI Lab, New York, US

Abstract. A generalizable and example-based model for multi-scale generative design is presented. The model adapts the Wave Function Collapse (WFC) algorithm, a procedural approach popularized in game development, to a quality-diversity (QD) framework, a state-of-the-art multi-solution optimization approach. QD enables the search of high-performing solutions not only against objectives, but along a set of qualitative features -- explicitly ensuring diversity within the solutions. We demonstrate the challenges and opportunities in applying these novel methodologies to AEC-focused problems through a real-world residential complex case study.

Keywords: Procedural modeling · Quality diversity · Generative design

1 Background

1.1 Quality-Diversity (QD)

In AEC, optimization is often used to understand the constraints and possibilities of a design space and discover novel solutions by exploring a diverse set of high-performing alternatives [1]. The most common methodology is multi-objective optimization (MOO) [2, 3]. Though MOO is often used for exploratory analysis, in many ways it is poorly suited for it. MOO requires the definition of objectives to be minimized—design exploration calls for features to be explored.

Quality-Diversity [4] algorithms move beyond MOO to produce sets of high performing designs organized by high-level features better suited to the judgement of domain experts. QD searches explicitly for high performing solutions with varied qualities, such as the perimeter of a building or the number of bedrooms in a unit. In contrast to a pareto curve of non-dominated solutions, the most widely used QD algorithm MAP-Elites [5] produces a grid or 'map' of the solutions – with each axis corresponding to a feature. This map provides an intuitive overview of the performance potential for each region of the feature space. Though originally designed for applications in robotics [6] and artificial life [7], QD techniques have begun to be applied in design applications such as engineering optimization [8, 9] and procedural content generation [10–12].

C. Gengnagel et al. (Eds.): DMS 2022, *Towards Radical Regeneration*, pp. 149–160, 2023.
https://doi.org/10.1007/978-3-031-13249-0_13

1.2 Wave Function Collapse (WFC)

Though generative design is gaining broader adoption in the AEC industry, its impact is limited by the level of technical skill required to operate computational design tools and the challenge of building generalizable applications[13]. We address both issues with a versatile design space model for semi-constrained designed systems, like modular or prefab, compatible with traditional design methods.

Our design space model adapts WFC[14], a texture synthesis approach popular in the game development community. WFC is a constraint-based procedural content generation method which extracts local patterns from a sparse set of examples and transforms them into a set of local constraints. These constraints drive generation and ensure that every local patch of the output also exists in the set of input examples. The inner workings of the algorithm have been extensively described [14, 15].

We extend the WFC algorithm to architectural applications (Fig. 1) where discrete architectural tiles are manually composed into larger assemblies and supplied to the algorithm as design examples (Fig. 2). This example-based approach makes this methodology compatible with traditional architectural design workflows where experienced designers can *show and teach* what good designs look like and have the computer replicate virtually infinite variations of the provided examples. As discussed by Karth and Smith, the WFC "is particularly suited to non-programmers" [15], an uncommon feature among many advanced computational and generative design frameworks.

While in the original implementation the probability of certain patterns to appear in the final output is determined by pixel frequencies in the design samples [14], in our work that probability is guided through exposed normalized *weights* assigned to each individual tile.

Related to texture synthesis, model synthesis is one of the earliest applications of procedural constraint solving for 3D environments [16]. Recently, a renewed interest in such methods has attracted designers beyond game design applications including urban and building scale applications in conjunction with machine learning methodologies [17–19]. Our work further extends these by allowing a search algorithm to manipulate the probability-weights and placement of fixed tiles to control diversity of output and optimization along a set of objectives and features. Despite the observed growing interest in procedural constraint solving methods, viable applications for architectural design are still unexplored.

Observed Limitations. WFC is remarkable for its simplicity but, despite some work on extending its functionality [20] has several limitations:

- WFC does not offer control over global constraints.
- Constraints are purely spatial (adjacency).
- Lacks input controls for a search algorithm.
- Lacks domain specific constraints.

Our approach addresses these limitations via:

- Control of formal massing via global performance metrics (e.g., natural ventilation and noise) and global geometric features (e.g., building façade area) via integration with a QD optimization framework.
- Dynamic weighting for tile unit selection as optimization controls.
- Dynamic pre-constraining of tiles for improved searchability.
- Fixed pre-constraining with boundary solution tiles for design-domain ease of use.

The manual nature of crafting design examples and building a catalog of units makes this design space model highly versatile, accessible, and compatible with traditional modeling techniques and design approaches.

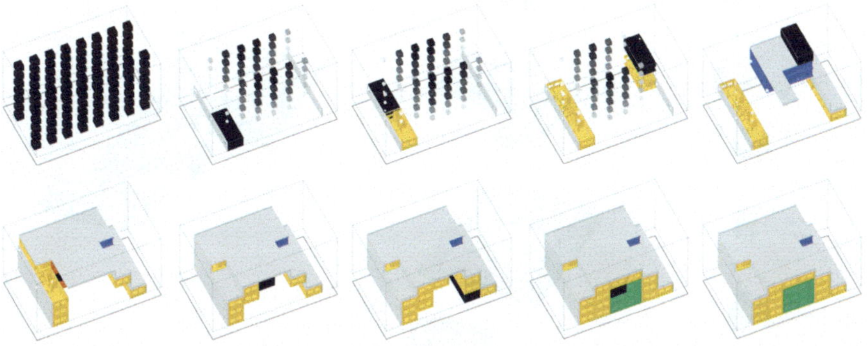

Fig. 1. Collapsing process of the WFC. Refer to this video featuring this process in action: https://vimeo.com/668784164

2 Methods and Data

2.1 Geometry System

The tiles catalog supplied to our model includes basic building components: façade, apartment, and stair core tiles. We also use empty tiles to govern building boundaries and open spaces (Fig. 2). This tile set is then used to manually create a defined set of example designs that can represent the kinds of desired variations and provide the WFC algorithm with tile-to-tile adjacency rules. Using this set of design examples in conjunction with the WFC algorithm we automatically generate a wide variety of site building layouts (Fig. 3).

To improve control over the WFC output we extend the algorithm's basic functionality with tile probability weights and variable tile pre-constraints. The set of weights, one for each tile type, can be varied to control the probability of the associated tile to appear in the WFC output (Fig. 4 top). Variable pre-constraints – tiles which are fixed at the start of WFC – 'lock-in' parts of the design while leaving the rest to the WFC generation process. These fixed tiles are added and removed from solutions as part of the search process (Fig. 4 bottom).

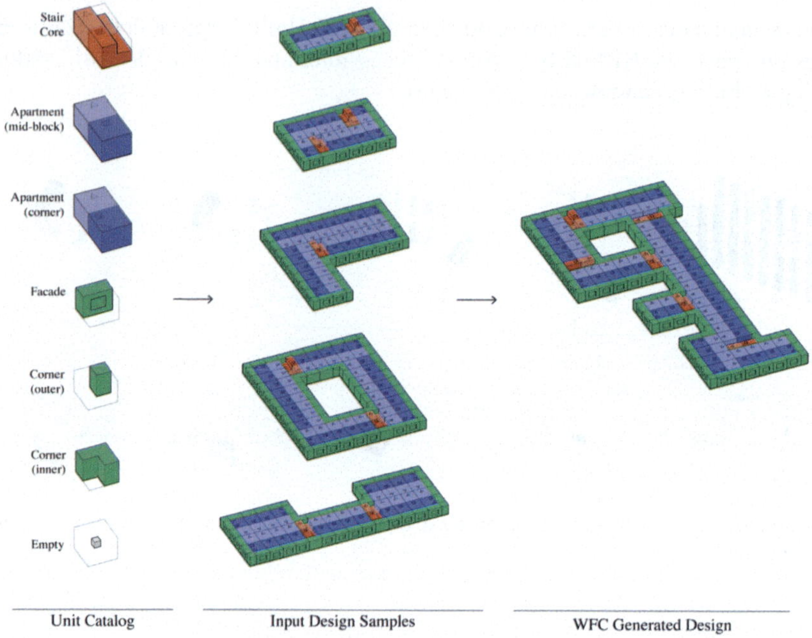

Fig. 2. WFC steps: definition of a catalog of units (left), design of examples (middle), and example WFC solution based on provided examples (right).

2.2 Features and Objectives

The QD algorithm MAP-Elites produces designs that are high performing along a set of objectives and diverse along a set of features (Fig. 5). While objectives are functions to be minimized or maximized (such as ventilation or site noise), features are quantifiable design characteristics to be fully explored (such as perimeter length or number of buildings). In our case study project, we included sustainability, livability, and penalty types of objectives and features that define geometric attributes.

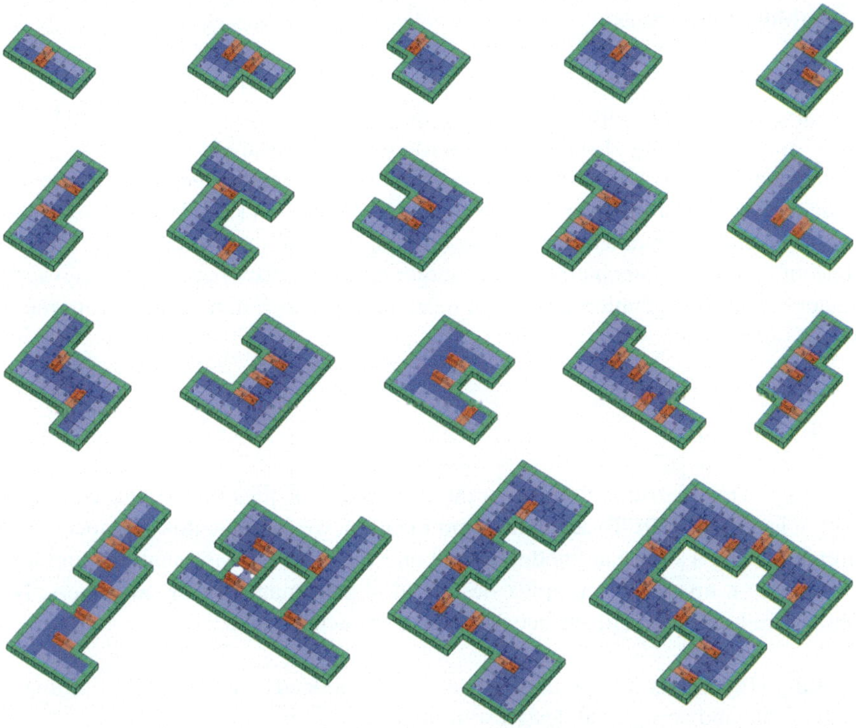

Fig. 3. Sample WFC outputs. Varied and diverse results that comply with the rules encoded by the example designs. This represents a new way of generating industry-specific design solutions beyond typical parametric approaches.

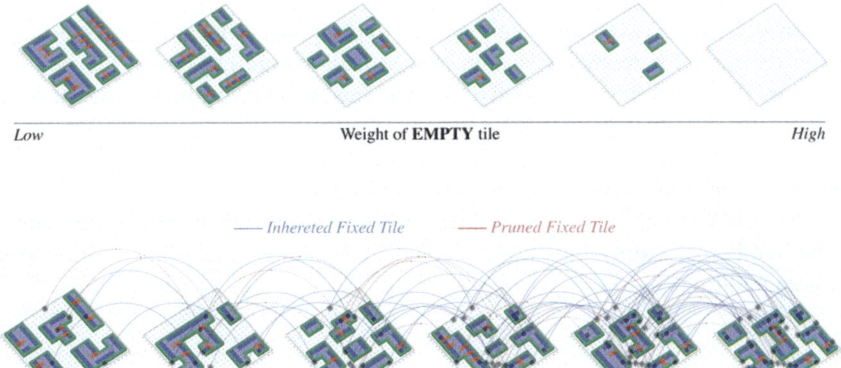

Fig. 4. Model parameters: (above) weights are assigned per tile type and drive the probability of it appearing in the final output. (below) Variable pre-constraints: location, addition and removal of pre-determined tiles are additional variables to influence collapsing process.

Sustainability Objectives

- *Indoor Ventilation.* This metric defines the natural indoor ventilation potential for each apartment. A simplified version of the air flow network (AFN) methodology is used—the connectivity distance of each room to the apartment's windows.
- *Landscape Capacity for Carbon Sequestration.* This metric measures the potential capacity for outdoor green areas to store and avoid carbon [21]. We approximate this capacity from a 'clearance' metric, the amount of clear space green areas have form adjacent buildings. This metric values larger areas of clear space, which can support greater levels of vegetation and trees, more highly – differentiating it from the total area of open space.

Livability Objectives

- *Site noise.* This metric is defined as the percentage of tiles on the site with a noise level of less than 50db. To accelerate optimization, we estimate this site-specific measurement using a surrogate model trained on a large set of noise analysis simulations performed on apartment complex designs designed manually by customers. Noise sources are highways and surface roads near the actual site.

Penalty Objectives. The penalty metrics are introduced to help steer the optimization towards viable and acceptable design solutions.

- *Number of Apartment Units.* Count of apartment units.
- *Proximity of Units to Building Cores.* Tile distance of each apartment unit to the building's cores. The distance to a core must be less than 5 tiles.

Features. This set was deliberately chosen to promote as much diversity as possible based on the model's geometry system. Three features are chosen as the number of feature pairs (three pairs, versus six for four features) can be succinctly communicated.

- *Façade Length.* Ratio of number of façade tiles to number of units
- *Number of Buildings.* Count of complete buildings.
- *Total Size of Open Spaces.* Surface area of open spaces. Open spaces are computed as the number of empty tiles not occupied by buildings excluding 1 tile corridors between buildings.

2.3 Encoding and Optimization

MAP-Elites. The QD algorithm MAP-Elites [5] is used to optimize an encoding composed of two parts: a vector of tile weights and a set of fixed tiles. MAP-Elites first divides the feature space into a set of discrete bins, or map. The map houses the population, with each bin in the map holding a single individual. When a new solution is evaluated, it is assigned a bin based on its features and, if that bin is empty, it is added

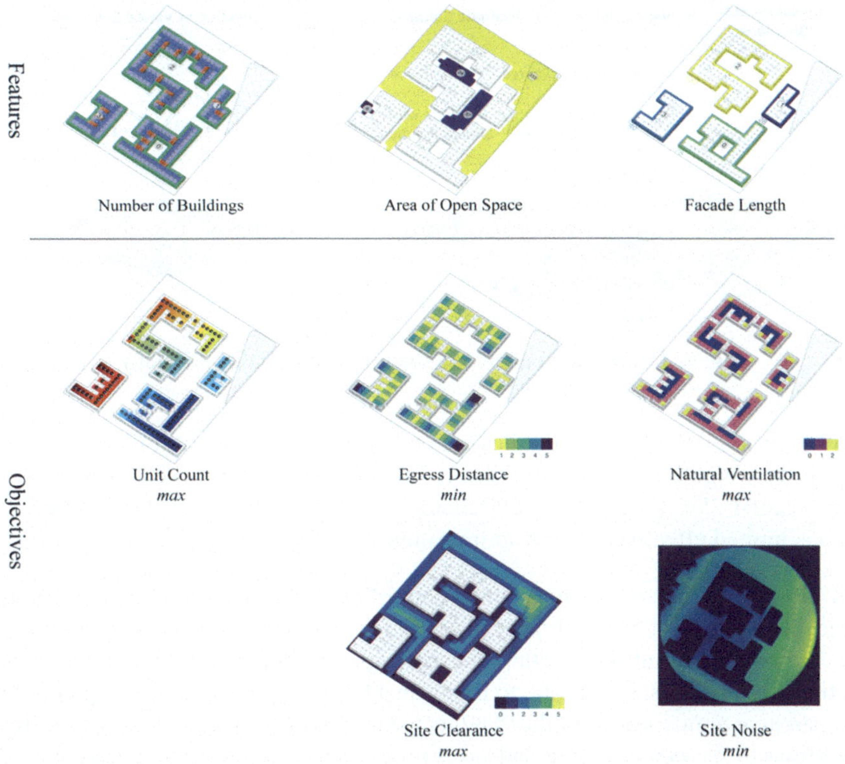

Fig. 5. Features (above) and objectives (below). Greyed out region indicates area unavailable for development.

to the map. If the bin is occupied by another solution the solution with a higher fitness is kept in the bin and the other discarded. In this way each bin contains the best solution ever found for that combination of features. These best solutions are known as elites.

To produce new solutions parents are chosen randomly from the elites, mutated, evaluated, and assigned a bin based on their features. Child solutions have two ways of joining the map: discovering an unoccupied bin, or out-competing an existing solution for its bin. Repetition of this process produces an increasingly explored feature space and an increasingly optimal collection of solutions. The optimization process is illustrated in Fig. 6.

Multiple objectives are optimized using the T-Domino [22] variant of MAP-elites. T-DominO ranks solutions according to the number of other solutions in the map that are dominated on each objective – rewarding solutions with balanced performance over those which excel at only a single objective. Solutions which follow constraints are always preferred over those which do not.

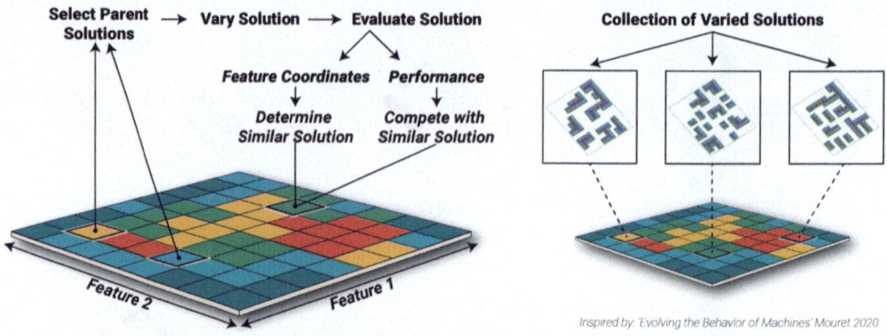

Fig. 6. Optimization of diverse solutions with Quality-Diversity.

Tile Weights. In WFC generated tiles are chosen from a set of valid tiles probabilistically – if the weighting of two valid tiles is 3:1, the first will be chosen 75% of the time and the second 25% (Fig. 4 top). Differences in this weighting has broad effects, but alone has limited effectiveness for optimization (Fig. 7).

Fixed Tiles. To achieve the fine-grained control necessary for optimization, solutions are encoded with a set of fixed tiles. MAP-Elites is an evolutionary algorithm, which produces new child solutions by altering existing parent solutions. Children inherit these fixed tiles from parents, in addition to fixing an additional tile from the design produced by the parent or removing one of the tiles that were fixed by the parent. Fixing tiles freezes key portions of the parent design and saves progress toward interesting designs – while still allowing substantial deviation from the parent, as the rest of the tiles are generated stochastically with WFC (Fig. 4 bottom).

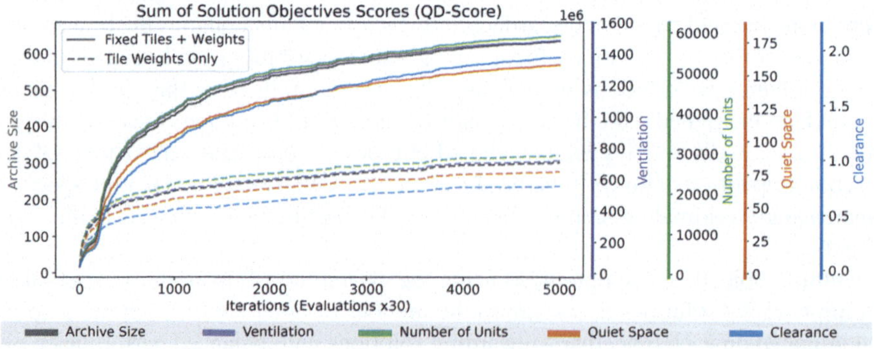

Fig. 7. Sum of objective values of best designs found in every feature region by MAP-Elites. Fixing tiles (solid line) as part of the optimization process dramatically improves the diversity of solutions found (Archive Size) as well as the performance of the solutions on the target objectives vs. using only global weights (dotted line).

Optimization Settings. At each generation 30 new individuals were created by mutating parent individuals with at 50% probability of adding a tile and a 50% probability of removing a tile. One run consists of 5000 generations. One full run takes approximately 8h on a 32 core workstation. The feature space was divided into a 10x10x10 grid, to create a collection of solutions, or archive, of up to 1000 solutions. Features were explored between the ranges of [1.8–3.6], [300–500], [4–14] for the Façade Per Unit, Open Area, and Number of Buildings respectively.

3 Findings and Discussion

Our approach generates a high performing set of apartment layouts which vary along the provided features -- illuminating the relationship between these features and performance. Viewing the performance of designs organized by these features we see that layouts with fewer large buildings tend towards poor natural ventilation —- an effect

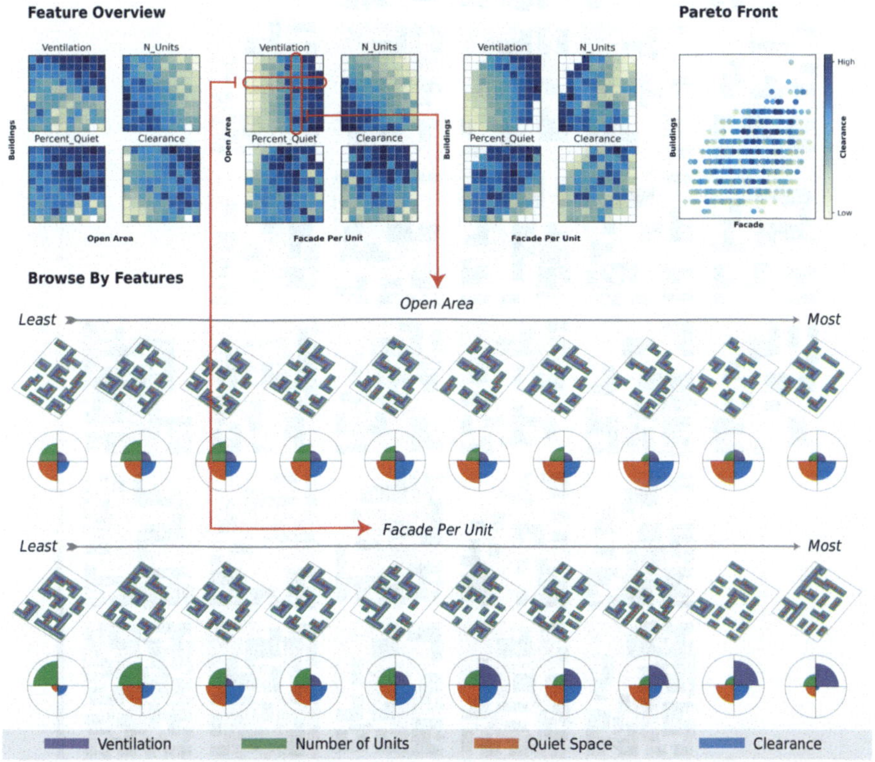

Fig. 8. Example visualization approaches made possible by the feature-centered Quality-Diversity approach. Top Left: Top solutions are organized by feature combinations, with each feature region represented as a bin colored by performance on each objective. Relationships are visible at a glance, allowing rapid identification of promising regions. Top Right: The same overview presented with a single objective as a Pareto Front. Bottom: Browsing designs by features. The highlighted feature regions are explored as a walk-through feature space, allowing designers to browse designs in an intuitively structured way.

that can be remedied with longer, more convoluted facades. Conversely, layouts with few large buildings and larger open areas interspersed across the site tend to have less noise. These insights are easily identifiable with the map-based visualization approaches (Fig. 7).

Our method of encoding and optimizing WFC-based solutions makes this type of exploration possible. Optimization of tile weights guides the direction of WFC and iteratively fixing tiles provides further control -- resulting in improvements in the quality of solutions produced while accelerating optimization by an order of magnitude. The fixed tile approach adds an intuitive method of steering optimization. Fixed tiles can be manually or parametrically placed to guide design outcomes around constraints like stair locations, courtyards, or existing structures (Fig. 8).

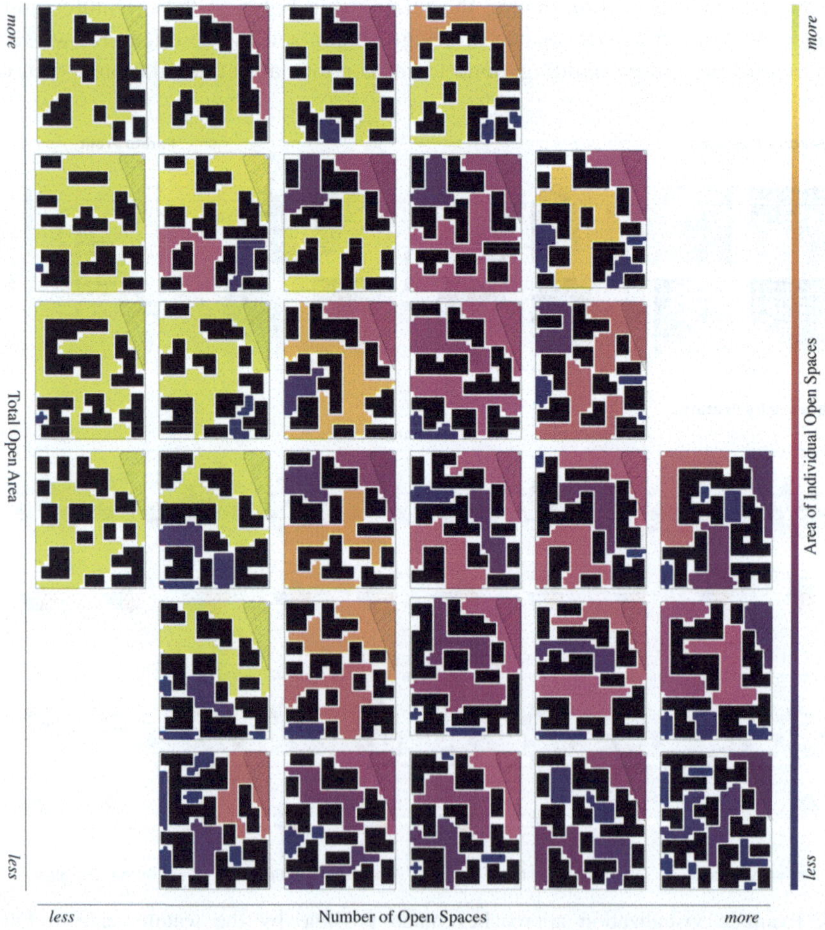

Fig. 9. An example of 3D feature mapping for intuitive navigation of design spaces. Here we show how additional attributes can be extracted (such as number of open spaces and area of individual open spaces) to further aid the design space navigation.

High-level features are valuable to designers for decision-making, but difficult to integrate into multi-objective frameworks -- QD allows the design space to be viewed through the lens of these features (Fig. 9). WFC allows design intent to be communicated through concrete, visual examples as well as promoting a multi-scalar approach to design where internal layouts, building footprint and site organization happen all simultaneously. Both advances flow from the same principle: for generative design to be useful, it must be intuitive.

Acknowledgments. The authors would like to thank Renaud Danhaive, Jeffrey Landes, and the entire Spacemaker team for their invaluable site analysis tool and expertise as well as Mark Davis and David Benjamin for their guidance and support.

References

1. Bradner, E., Iorio, F., Davis, M.: Parameters Tell the Design Story: Ideation and Abstraction in Design Optimization. In: Proceedings of the Symposium on Simulation for Architecture & Urban Design, vol. 26 Society for Computer Simulation International (2014)
2. Gerber, D.J., et al.: Design optioneering: multi-disciplinary design optimization through parameterization, domain integration and automation of a genetic algorithm. In: Proceedings of the 2012 Symposium on Simulation for Architecture and Urban Design, pp. 1–8 (2012)
3. Druot, T., et al.: Multi-objective optimization of aircrafts family at conceptual design stage. In: Inverse Problems, Design and Optimization Symposium. Albi, France, vol. 29, pp. 773–79 (2013)
4. Cully, A., Demiris, Y.: Quality and diversity optimization: a unifying modular framework. IEEE Trans. Evol. Comput. **22**(2), 245–259 (2017)
5. Mouret, J.B., Clune, J.: Illuminating Search Spaces by Mapping Elite. ArXiv Preprint ArXiv:1504.04909 (2015)
6. Cully, A., Clune, J., Tarapore, D., Mouret, J.B.: Robots that can adapt like animals. Nature **521**(7553), 503–507 (2015)
7. Lehman, J., Stanley, K.O.: Evolving a diversity of virtual creatures through novelty search and local competition. In: Proceedings of the 13th annual conference on Genetic and evolutionary computation, pp. 211–218 (2011)
8. Gaier, A., Asteroth, A., Mouret, J.B.: Data-efficient exploration, optimization, and modeling of diverse designs through surrogate-assisted illumination. In: Proceedings of the Genetic and Evolutionary Computation Conference, pp. 99–106 (2017)
9. Gaier, A., Asteroth, A., Mouret, J.B.: Data-efficient design exploration through surrogate-assisted illumination. Evol. Comput. **26**(3), 381–410 (2018)
10. Khalifa, A., Lee, S., Nealen, A., Togelius, J.: Talakat: Bullet hell generation through constrained map-elites. In: Proceedings of The Genetic and Evolutionary Computation Conference, pp. 1047–1054 (2018)
11. Alvarez, A., Dahlskog, S., Font, J., Togelius, J.: Empowering quality diversity in dungeon design with interactive constrained map-elites. In: 2019 IEEE Conference on Games (CoG) IEEE, pp. 1–8 (2019)
12. Fontaine, M.C., et al.: Illuminating mario scenes in the latent space of a generative adversarial network. arXiv preprint arXiv:2007.05674 (2020)
13. Nagy, D., Villaggi, L., Zhao, D., Benjamin, D.: Beyond heuristics: a novel design space model for generative space planning in architecture (2017)

14. Gumin, M.: Wave Function Collapse. https://github.com/mxgmn/WaveFunctionCollapse (2016)
15. Karth, I., Smith, A.M.: WaveFunctionCollapse is constraint solving in the wild. In: Proceedings of the 12th International Conference on the Foundations of Digital Games, pp. 1–10 (2017)
16. Merrell, P.C.: Model synthesis. Ph.D. Dissertation. University of North Carolina at Chapel Hill (2009)
17. Lin, B., Jabi, W., Diao, R.: Urban space simulation based on wave function collapse and convolutional neural network. In: Proceedings of the 11th Annual Symposium on Simulation for Architecture and Urban Design, pp. 1–8 (2020)
18. Mintrone, A., Erioli, A.: Training spaces – fostering machine sensibility for spatial assemblages through wave function collapse and reinforcement learning. In: proceedings of the 39th eCAADe Conference, Volume 1, University of Novi Sad, Novi Sad, Serbia, pp. 8-10 (2021)
19. Chasioti, E.: Gameplay with encoded architectural tilesets: A computational framework for building massing design using the Wave Function Collapse algorithm. BARC0141: Built Environment Dissertation (2020)
20. Sandhu, A., Chen, Z., McCoy, J.: Enhancing wave function collapse with design-level constraints. In: Proceedings of the 14th International Conference on the Foundations of Digital Games (FDG '19: The Fourteenth International Conference on the Foundations of Digital Games, San Luis Obispo California USA: ACM), pp. 1–9 (2019)
21. i-Tree, Califronia Urban Forests, Urban Ecos and Cal Fire (2020). https://www.itreetools.org/support/resources-overview/i-tree-methods-and-files/new-carbon-equations-and-methods-2020
22. Gaier, A. Stoddart J., Villaggi L., Bentley, P.J.: T-Domino: Exploring Multiple Criteria with Quality Diversity and Tournament Dominance Objective. In: Parallel Problem Solving From Nature. Springer (2022). https://doi.org/10.1007/978-3-031-14721-0_19

Design with Digital and Physical Realities

Digitization and Energy Transition of the Built Environment – Towards a Redefinition of Models of Use in Energy Management of Real Estate Assets

Daniele Accardo[1](✉), Silvia Meschini[2], Lavinia Chiara Tagliabue[3], and Giuseppe Martino Di Giuda[1]

[1] Department of Management, University of Turin, 10134 Turin, Italy
daniele.accardo@unito.it
[2] Department of Architecture, Built Environment and Construction Engineering, Politecnico di Milano, 20133 Milan, Italy
[3] Department of Computer Science, University of Turin, 10149 Turin, Italy

Abstract. Thousands of years of progress in urban development led to complex urban environments that can be considered as complex systems. The management of complex systems must account for uncertainty, unpredictable future states and nonlinear behaviour. In this context, digitalisation can play a key role in the development of innovative tools to support strategic decisions and emergency management. In particular, the management of university building stocks can be facilitated with the creation of digital environments.

In the Italian case, university campuses are often complex assets composed of widespread buildings and the management process is still based on fragmented databases handled by different administrative divisions resulting in a lack of information among stakeholders.

The integration between Building Information Modeling (BIM) and Geographic Information System (GIS) is already making some steps towards the creation of a digital model of the city. The combination of BIM-GIS with a platform for the data management is the base to develop an Asset Management System to exploit Business Intelligence (BI) tools for Operation and Maintenance (O&M) in smart campuses.

This research aims to integrate BIM, GIS and BI tools in a digital framework for the development of an AMS and web-based application for the improvement of the experience among users and the optimal use of resources. A real case study is proposed for the development of the research project, namely the University of Turin building stock, in Italy.

Keywords: BIM-GIS · Asset management · Energy transition · Information value · Sustainability

C. Gengnagel et al. (Eds.): DMS 2022, *Towards Radical Regeneration*, pp. 163–174, 2023.
https://doi.org/10.1007/978-3-031-13249-0_14

1 Background

University campuses are difficult to maintain since they are made up of many diverse buildings, many of which were erected at various times, especially in Italy. Their administration is frequently structured on fragmented databases that are difficult to access and are still document-based, resulting in insufficient and asymmetrical information that leads to unproductive choices and resource utilization, particularly during the operation and maintenance (O&M) phase. In terms of the total cost of the asset life cycle, this period proved to be the costliest [1]. Building usability and energy consumption are strongly influenced by the effective use of spaces (i.e., space management connected to occupancy flows), users' behaviour, and the demands for supply and services. As a result, if they are not properly managed, they can waste resources and raise management, operating, and maintenance expenses. The digital transformation of process management has become critical in the development of digital Asset Management Systems (AMS) as decision-support tools for managing and optimizing university spaces and activities [2, 3]. To make the transition to full digital management easier, a method for Information Management (IM) and information protocols for tailored data modelling are required. These methods and protocols must ensure the availability of accurate information at the right time and in the required format for the appropriate subject [4]. To construct digital AMS solutions capable of managing all the data required over the whole university asset lifespan, information regarding how and when data exchanges should occur, as well as among which stakeholders, must be established.

The combined use of BIM (Building Information Modelling) and GIS (Geographic Information System) has shown promise in the development of Smart Cities and Digital Twins (DTs) in recent decades [5, 6]. BIM is essential for creating extremely complex architectural models, whereas GIS allows them to be managed and analysed using a worldwide geographic reference system [7–9]. Furthermore, the BIM-GIS merger, also known as GeoBIM received little attention in the field of Asset Management (AM) [10]. The current availability of both new technologies and large data quantities, along with BIM-GIS compatibility, can optimize management operations, particularly during the O&M phase. It can also help to design successful AMS [11]. The capacity to digitally manage the asset from the geographical macroscale to the microscale of the single asset component is one of the major capabilities of BIM-GIS integration [12].

Based on these assumptions, the research project intends to design a digital and repeatable approach for developing an AMS based on BIM and GIS integration via a web platform (i.e., AMS-app) targeted at improving information management and decision-making processes in large and distributed assets. The AMS app should collect all the data that is presently managed independently by multiple administrations, allowing for independent yet collaborative management. As a result, rather than managing each building, the building asset may be managed at the system level. Another significant result of the study is the creation of an information protocol to improve distributed university asset modelling using BIM-GIS platforms. The study focuses on the tremendous effort to aggregate data from multiple separate databases into a consolidated and easily accessible one, and it describes the reproducible methodological approach used to construct the AMS app. It's also shown how the AMS app was created using a centralized database to enable real-time representation of the whole distributed asset in an interactive 3D map. The product is a "GeoBim" system that allows users to store, examine, and exchange continually updated geometric, spatial, and functional data, which will help the institution manage its assets more efficiently and sustainably. Finally, the paper demonstrates how to put the theory into practice. Furthermore, the study describes how to apply the stated technique using two demonstrators, highlighting preliminary findings, potentials, and limits. The two demonstrators are part of the University of Turin building stock, which is a pilot use case. It is one of Italy's most dispersed campuses, with a large catchment area and unstructured management. Due to these characteristics, there is a significant information asymmetry, which prevents knowledge of its consistency and usage, resulting in resource waste and efficiency losses. Furthermore, one of the targeted aims in the University's strategic plan, which stands out as a novelty in the Italian panorama, is the digitalization of building stock and the employment of digital technologies for AM.

2 Methods and Data

Starting with a state-of-the-art examination of methodologies and tools for producing AMS of large building stocks using BIM-GIS integration, the research approach was designed. In the second phase, which focused on the examination of the pilot use case, namely the University of Turin building stock and its present management practices, the real definition of the methodological approach began. Following the analysis, processing, and structuring of the collected data, a centralized relational database was created to collect all the disparate data about the asset's geographical and functional features. Finally, a unique BIM-GIS online platform was created to allow the university's building assets and characteristics to be seen in an interactive 3D digital environment. Figure 1 depicts the study methodology, which might be applied to various scenarios.

The examination of the asset's consistency and the identification of the administrations engaged in its management were the initial phases of the analytical approach. The major goal was to determine which data were required, from which administrations, which were previously available or not, how they were maintained and shared, and where the missing data might be gathered. To achieve this, the institution's communication channels (i.e., the website and its Transparent Administration section, which contains official documentation related to the institution's core activities and people working there) were investigated to collect data and identify administrations involved in the management processes (Table 1). Subsequently, it was investigated how a web platform called Opensipi was being used to handle data on the building stock (e.g., dimensions, occupancy, mechanical assets, and so on). The "Information systems and e-learning portal Directorate" was in charge of this platform, which may be updated by the "Building, logistics, and sustainability Directorate". Academic activity data was also discovered to be held in heterogeneous excel sheet files handled by administrative workers from the "Educational Services Directorate." Finally, an external database, namely Cineca, was recognized as having students' data handled through specific interfaces but not available for consultation.

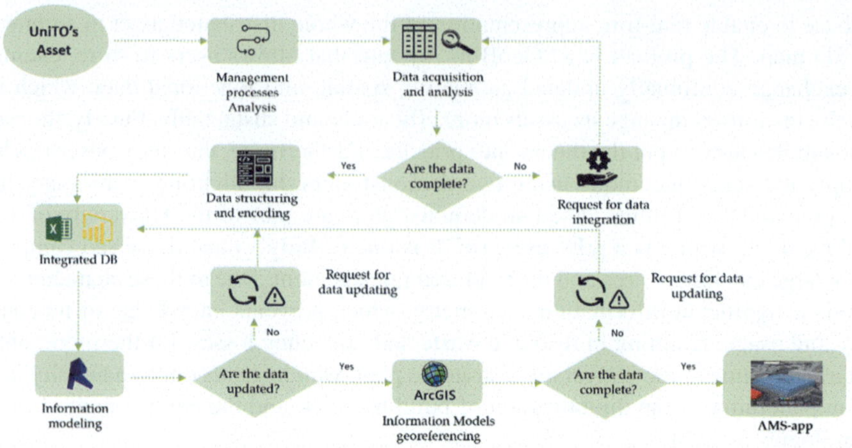

Fig. 1. Methodological approach

The interviews with administrative and managerial professionals from numerous departments verified this fragmentation and underlined the difficulty in acquiring the necessary data because extrapolation must be done on purpose, resulting in extended wait periods. Then, to identify which data should be linked and connected for the effective administration of the University's assets and operations, an in-depth investigation of the databases was done.

The Opensipi platform was created to make a GIS-based solution for space and asset management. Using Google Maps, Opensipi allows to search and see the whole University building stock. The software also displays spatial features and specific premises of each building via digital interactive plans (Table 1). It's worth noting that the functional characteristics of places (such as occupancy, furniture, and mechanical systems) that are important during the O&M phase aren't present. Another issue is that the platform is not updated regularly, therefore its data may not reflect the current state of the spaces. Subsequently, current building stock data and plans were acquired through short interviews and documentation requests to the technical office.

The "Educational Services Directorate" employed heterogeneous excel sheet files to manage schedules and course data, as shown in Table 1. On the other hand, Cineca, which is also responsible for data storage, provides a specific application for managing data linked to students' careers, fees, and course catalogues. Short interviews with workers were also done in this case to better understand which data they manage and how various administrations interact. It was noted that the "Educational Services Directorate's" excel sheet files were not linked to the data given by the database supplier Cineca.

As a result, there was a lack of communication across administrations, resulting in highly fragmented data, which added to staff workload and inhibited job automation. Cineca, as previously stated, is a third-party software and service provider in charge of data administration. The administrative staff cannot directly alter data contained in the database; instead, a specific program is required to see information, and Cineca must be queried using bespoke queries, resulting in high processing times and inefficient administration. Direct access to a centralized database and the necessity for efficient

Table 1. Administrations involved in management processes and dedicated databases

Database	Data	Administrative directorate
OpenSIPI	Building name Geometric data Building location State of use	Information systems and e-learning portal Building, logistics and sustainability
Excel sheet files	Timetables Courses	Educational services
Cineca	Student career Course catalogue Fees	Not directly accessible

functionality is critical for developing AMS for scattered and heterogeneous building stock [13]. Therefore, an integrated database was required, in which data from Cineca and other databases converged, resulting in a data source that was easily accessible and queryable. Such an AMS may be a game-changer for the "Educational Services Directorate," which would be able to rely on a single source of data on the number of students, available spaces, and courses.

The structure of the centralized database was established considering the existing situation while keeping in mind the possibility of future deployments. Following the acquisition of the data, a state-of-the-art study was undertaken to determine the optimum approach to preserve them to give flexibility, easy updating, and user accessibility. Relational databases (RDBs), which enable data to be structured according to a specific hierarchy, resulted to be a well-established method of storing data [14]. RDBs allow numerous sorts of connections between data to be established, allowing for customised queries as well as the ability to filter and aggregate data based on the intended purpose. The database structure was separated into 9 tables that were supplied with the acquired data and were also based on future database usage and queries related to asset management operations (Fig. 2).

This allowed us to see how the various levels and types of data should be linked. The initial stage was to select the primary data to which additional data would be linked. Because spatial data are required for most management operations, it was chosen to start there. As a result, the database structure was built on spatial data in a core table called "Building Stock". Then, on the left side, there were connections to more information regarding "Spaces", "Occupancy" and "Timetables". Data regarding "Real property titles", "Rental revenues", "Rental costs," and "Degree programs" were branched out on the right side.

One of the most important steps in identifying the building's areas was to create a custom encoding system that could also connect data through the database. It is arranged as follows, according to the various building stock levels of definition:

$$PR^{1}\_000^{2}\_000^{3}\_A^{4}\_P00^{5}\_0000^{6}$$

Province 1; venue 2; settlement 3; building code 4; floor 5; rooms 6.

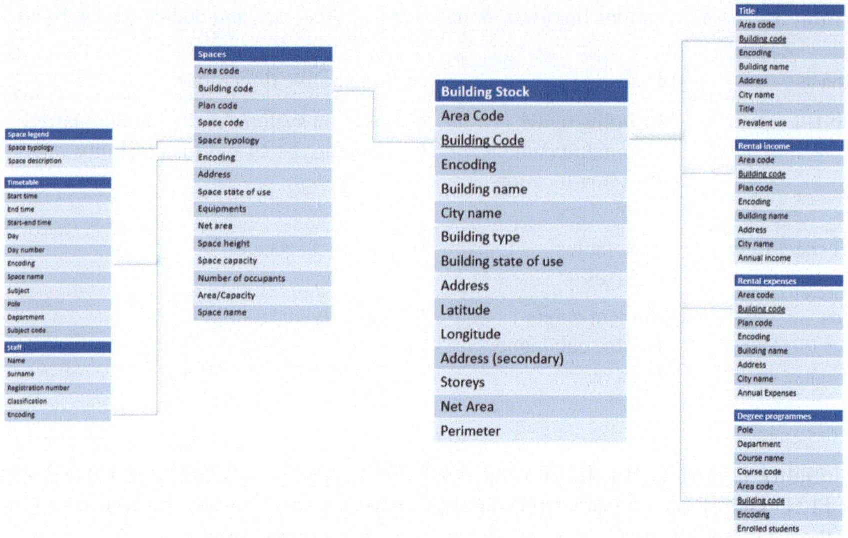

Fig. 2. Database structure

This technology allowed data to be connected following the building's level of detail. Both the encoding strategy and the centralized database structure were important in the AMS app since they allowed users to filter shown information using the single data set described above. The assignment of the semantic data recorded in the centralized database to the building stock was the fourth phase. The initial step was to create BIM models that were to be enhanced with building characteristics, beginning with two pilot cases: the Faculty of Computer Science's headquarters and the Faculty of Humanities' headquarters. The major goal of this early phase was to get a level of data that was appropriate for representing the entire university asset without being burdened by an excessive volume of data. As a result, a significant decision was to model the building stock using masses, floors, and rooms rather than a high level of detail. The Autodesk Revit® authoring tool was chosen, which is one of the most widely used BIM tools in the AEC sector. This decision was based on three factors: the ability to represent only the main volumes of the university building stock as masses, floors, and rooms (which will be implemented in the future); the availability of the Dynamo plugin for Revit®, which automates both modelling and parameter assignment processes; the high level of interoperability between Revit® and GIS platforms (such as Esri's ArcGIS Pro), which allows for the import of BIM Models [15].

The individual categories of created elements (i.e., masses, plans, rooms) were assigned to the spatial and functional attributes previously stored in the centralized digital database using VPL (Visual Programming Language) [16], which is widely used in the AEC industry with significant improvements both in modelling and data management [17]. VPL is a programming language that substitutes ordinary computer programming with special objects (i.e., nodes) that have unique functions. Dynamo is an open-source

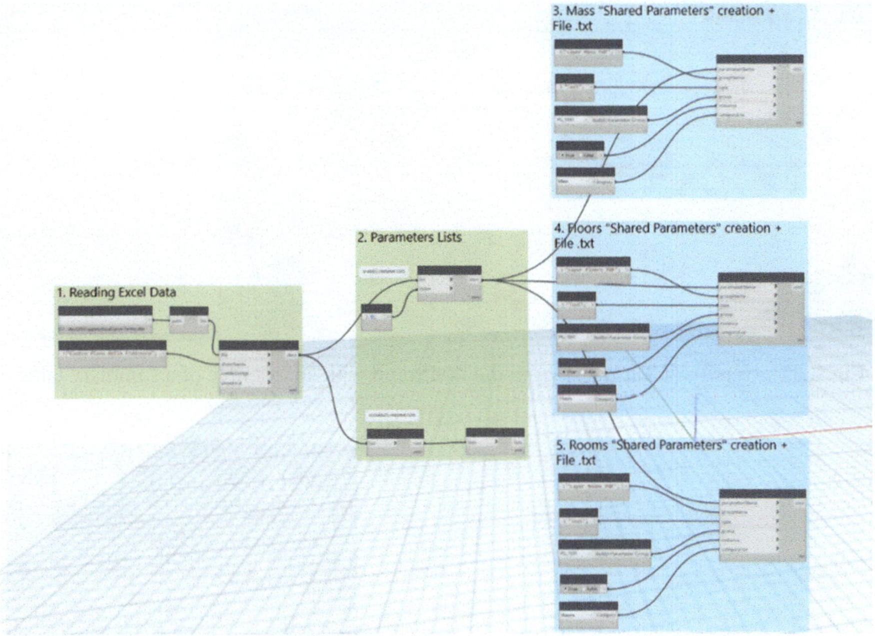

Fig. 3. Dynamo nodes

interface that connects VPL and Revit API nodes to construct highly customizable algorithms [18]. The geographical and functional data were first organised separately before being coupled to the parametric models. Extrapolating data from spreadsheets, creating new-shared parameters in.txt format, and assigning them to three kinds of parametric components (masses, planes, and rooms) in BIM models were all done with customized nodes (Fig. 3).

The fifth phase of this research project focused on the creation of a web-based BIM-GIS platform (i.e., AMS-app) that allows for real-time display of assets and their properties in a 3D digital map (Fig. 4). This platform may be thought of as a GeoBIM system since it allows users to link information models, geographical data, and functional data all at once. Therefore, georeferenced masses with geometric and semantic information were imported into the GIS platform. The building stock may be explored and filtered at several information levels (asset, single building, floor, and local), allowing for asset analysis and strategic decision-making by gathering and visualizing only the information required.

Finally, through business intelligence (BI) technology, it was feasible to turn data into meaningful information and give important instruments to assist strategic choices. Microsoft Power BI® was chosen to provide the best interoperability with excel sheet files used by administrations and minimize data loss. It is often regarded as one of the best ideal applications for managing massive amounts of data [19], and it supports the presentation of Arcgis maps via a custom plugin.

Fig. 4. University of Turin's asset in the AMS-app with an example of a summary table

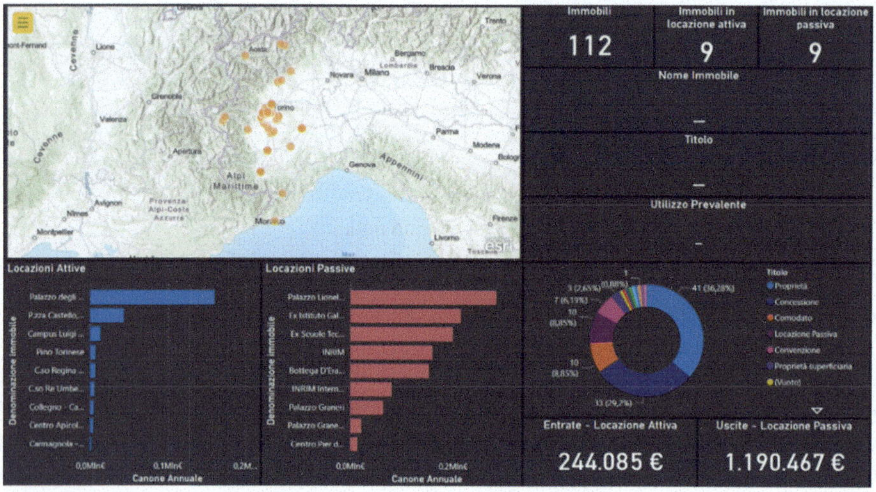

Fig. 5. Building stock dashboard 1

The capacity to work on datasets without altering the data source is an important advantage. Using this functionality and the previously mentioned centralized database, various interactive and detailed dashboards are offered to let university building users visualize and interpret relevant information (Figs. 5, 6, 7 and 8).

A first dashboard was constructed containing the asset's general analytic features (Fig. 5). It shows an interactive map with connected data (e.g., number of buildings, building name, building title, prevalent usage, rental revenue, rental expenditures) and certain user-interactive components. Data may be filtered or aggregated by clicking them, and key performance indicators (KPI) can be changed interactively via maps, bar graphs, and ring graphs. As a result, without having to wait for reports, the online application provides a comprehensive picture of the building stock consistency, allowing administrators to better allocate resources.

Fig. 6. Building stock dashboard 2

In a second dashboard (Fig. 6) assets and associated information might be sorted by educational pole or department for the depiction of the whole building stock. These data, when combined with information on classroom occupancy and course schedules, allow for better course allocation, facility use, and user mobility around the campus, resulting in a more sustainable campus. Then, for the examination of the first demonstrator, two further types of dashboards were created (Fig. 7 and 8).

The first dashboard provides for a more detailed study of the various activities at the building space level (Fig. 7). Some classes were underutilized, while others were

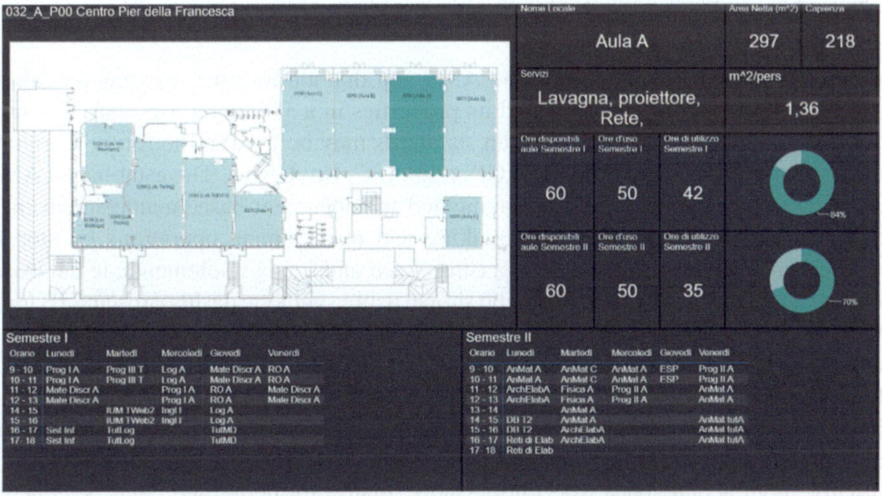

Fig. 7. Piero della Francesca centre's dashboard 1

overutilized, resulting in user discomfort and leasing costs for exterior areas. The Educational Services Directorate and other authorities may now readily examine and reference this data to allocate study courses optimally based on real space availability.

The last dashboard (Fig. 8) was dedicated to the offices' floor in which a detailed analysis regarding the utilization of the space could be done, visualising briefly the distribution of persons and the percentage of spaces used by role.

Fig. 8. Piero della Francesca centre's dashboard 2

3 Results and Conclusions

The main output of this research is the AMS app that allows users to view the whole University of Turin building stock and its properties in a 3D interactive map. It is now confined to authoring formats, but open standards may be used in the future to enable more compatibility between BIM and GIS (e.g., IFC, CityGML). The established AMS allowed for the resolution of present fragmented and document-based management challenges. It enabled effective decision-making and management operations by storing comprehensive and up-to-date data in a centralized and easily implementable database. Useful strategic data and graphs might be shown using synthetic dashboards and BI tools linked to the AMS app. These characteristics allowed for improved financial and geographical resource management at the institution, as well as waste reduction and cost reductions. It is possible to: rationalize space use based on actual availability, course schedules, and occupancy; optimize real estate investments by visualizing over or under exploitation of both buildings and spaces; manage maintenance and cleaning operations optimally about actual space use. Furthermore, thanks to the unified database, energy bills are also linked to each building, making it possible to highlight anomalies in energy consumption compared to the average per square metre. The AMS-App allows users to access information through a webapp without any installation requirement. With different levels of permissions every stakeholder can access to valuable information and participate in its improvement through dedicated text boxes. Sensor networks to detect thermal comfort, energy consumption, people presence, and any other valuable data for effective management of the O&M phase are scheduled to be installed in most major buildings in terms of dimension, occupancy and complexity. These data will be collected using the database that has been created and displayed using custom dashboards. As a result, BIM models might be used to create important DTs that allow for real-time optimization of internal comfort conditions, energy consumption, emergency evacuation, supported maintenance operations, and optimal occupancy using VR/AR tools. As a result, real-time or predictive building management might be given, along with improved user experience and lower administration costs and resource usage.

References

1. Seghezzi, E., Locatelli, M., Pellegrini, L., et al.: Towards an occupancy-oriented digital twin for facility management: test campaign and sensors assessment (2021). https://doi.org/10.3390/app11073108
2. Lu, Q., Xie, X., Parlikad, A.K., Schooling, J.M.: Digital twin-enabled anomaly detection for built asset monitoring in operation and maintenance. Autom. Constr. **118**, 103277 (2020). https://doi.org/10.1016/j.autcon.2020.103277
3. Ward, Y., Morsy, S., El-Shazly, A.: GIS-BIM data integration towards a smart campus. In: El Dimeery, I., et al. (eds.) JIC Smart Cities 2019, pp. 132–139. Springer, Cham (2021). https://doi.org/10.1007/978-3-030-64217-4_16
4. Chen, K., Lu, W., Peng, Y., et al.: Bridging BIM and building: from a literature review to an integrated conceptual framework. Int. J. Proj. Manag. **33**, 1405–1416 (2015). https://doi.org/10.1016/J.IJPROMAN.2015.03.006
5. Zaballos, A., Briones, A., Massa, A., et al.: A smart campus' digital twin for sustainable comfort monitoring (2020). https://doi.org/10.3390/su12219196
6. Bryde, D.B.M.V.J.M.: The project benefits of Building Information Modelling (BIM). Int. J. Proj. Manag. **31**, 971–980 (2013)
7. Zhu, J., Wu, P.: Towards effective BIM/GIS data integration for smart city by integrating computer graphics technique. Remote Sens. **13** (2021). https://doi.org/10.3390/rs13101889
8. Burrough, P.: Principles of geographical information systems for land resources assessment (1986)
9. Liu, X., Wang, X., Wright, G., et al.: A state-of-the-art review on the integration of Building Information Modeling (BIM) and Geographic Information System (GIS). ISPRS Int. J. Geo-Inf. **6**, 53 (2017)
10. Moretti, N., Ellul, C., Re Cecconi, F., et al.: GeoBIM for built environment condition assessment supporting asset management decision making. Autom. Constr. **130**, 103859 (2021). https://doi.org/10.1016/j.autcon.2021.103859
11. Pärn, E.A., Edwards, D.J., Sing, M.C.P.: The building information modelling trajectory in facilities management: a review. Autom. Constr. **75**, 45–55 (2017). https://doi.org/10.1016/J.AUTCON.2016.12.003
12. Yamamura, S., Fan, L., Suzuki, Y.: Assessment of urban energy performance through integration of BIM and GIS for smart city planning. Procedia Eng. **180**, 1462–1472 (2017)
13. Kensek, K.: BIM guidelines inform facilities management databases: a case study over time. Buildings **5**, 899–916 (2015). https://doi.org/10.3390/buildings5030899

14. Atzeni, P., De Antonellis, V.: Relational database theory (1993)
15. Song, Y., Wang, X., Tan, Y., et al.: Trends and opportunities of BIM-GIS integration in the architecture, engineering and construction industry: a review from a spatio-temporal statistical perspective. ISPRS Int. J. Geo-Inf. **6**, 397 (2017)
16. The Dynamo Primer (2021). https://primer.dynamobim.org/. Accessed 28 Jan 2022
17. Boshernitsan, M., Downes, M.: Visual programming languages: a survey (2004)
18. Salamak, M., Jasinski, M., Plaszczyk, T., Zarski, M.: Analytical modelling in dynamo. Trans VŠB Tech. Univ. Ostrava Civ. Eng. Ser. **18** (2019). https://doi.org/10.31490/tces-2018-0014
19. Shaulska, L., Yurchyshena, L., Popovskyi, Y.: Using MS power BI tools in the university management system to deepen the value proposition. In: 2021 11th International Conference on Advanced Computer Information Technologies (2021). https://doi.org/10.1109/ACIT52158.2021.9548447

Collective AR-Assisted Assembly of Interlocking Structures

Lidia Atanasova[1(✉)], Begüm Saral[1], Ema Krakovská[1], Joel Schmuck[1], Sebastian Dietrich[1], Fadri Furrer[2], Timothy Sandy[2], Pierluigi D'Acunto[1], and Kathrin Dörfler[1]

[1] Technical University of Munich, 80333 Munich, Germany
lidia.atanasova@tum.de
[2] ETH Zurich, Zurich, Switzerland

Abstract. Research on mobile Augmented Reality (AR) technologies has proven many potentials and benefits for assisting craftspeople in various building applications within the AEC domain. However, little research has been done on the use of multi-user mobile AR systems coordinating several people at the same time. This paper examines the potentials of a collective construction process enabled by AR technology that distributes and guides manual assembly tasks for multi-user participation. For this purpose, a custom mobile AR app is developed that uses cloud services and allows multiple people to participate in the construction and be coordinated with each other at the same time. In the proposed setup, digital building instructions for the stepwise assembly of a physical building structure can be retrieved by multiple users via the app. The app positions these building instructions in 3D space, visually superimposed on the building site, where the building structure is being assembled. Methods are proposed for synchronizing the construction progress via user-specific AR content visualization over the app's user interface (UI). Based on the principle of topologically interlocking structures, the material system developed for this research offers several form-fitting connections with one modular wooden component without the need for mechanical fasteners for their assembly. This principle enables the manual implementation of various complex building structures at full architectural scale, which are reconfigurable and fully disassemblable. The proposed methods were experimentally validated in a 1:1 scale demonstrator. A pavilion was assembled collectively by students and researchers over two days, and the UI was evaluated through a qualitative user study. As an outlook, the paper discusses the potential of such AR systems to make digitally-driven construction processes more tangible and accessible to laypersons and unskilled people and thus encourage community participation.

Keywords: Augmented reality · Multi-user mobile application · Topological interlocking · Cloud data · Participatory digital fabrication

C. Gengnagel et al. (Eds.): DMS 2022, *Towards Radical Regeneration*, pp. 175–187, 2023.
https://doi.org/10.1007/978-3-031-13249-0_15

1 Introduction and Context

In the past decade, the availability of mobile handheld devices and their increasing use for AR applications have led to a vast rise in the adoption of AR technologies in the AEC domain. Although these applications have already proven to be useful for managing construction processes, the potentials of the AR technology are not yet fully exploited and appropriate application areas will continue to expand [1]. The rapidly growing capabilities and features of mobile devices combined with the ubiquity of Internet access as well as the multitude of embedded sensors make the development of mobile AR applications utilizing remote servers possible [2]. In combination with existing technologies such as cloud technology and wireless communication, AR shows a large potential for collaborative and networked applications. In this regard, recent research developments have investigated AR approaches for computer-supported collaborative work [3], the possibility to use handheld devices for massively multi-user AR [4] and mobile collaborative augmented reality [5].

The possibilities of using AR to support craftspeople in manual production processes in architecture have recently been highlighted by a growing number of research efforts. In early projects, AR technologies were used to guide human actions via projection-based mapping [6, 7]. More recently, AR headsets were used by people to interface with robots enabling collaborative fabrication processes [8], as well as to instruct craftspeople for various manual tasks via holographic 3D models in space [9, 10]. Custom-built AR systems providing extended features such as the 3D registration of discrete objects together with highly accurate pose estimation [11] have shown the real-time guidance of construction tasks using craft-specific user interfaces, as well as the ability to register and measure the as-built structure with sub-centimeter accuracy [12]. Recently developed mobile AR apps for the visual guidance of manual construction processes provide virtual information based on exported building plans [12] or a real-time stream of digital models over Wi-Fi from the design environment to connected mobile devices [9].

Available AR methods generally constrain fabrication to prescribed sequences, and streamed models provide users with uniform visual content, thus limiting the deployment of these systems for multi-user task coordination, in which user-specific information must be supplied. Instead, we propose an AR app that uses a centralized design model to synchronize solely fabrication-related data between multiple users employing cloud services. Asynchronous multi-user assembly tasks are coordinated in real-time via the cloud by providing user-specific AR content over a custom-developed user interface (UI). The proposed workflow has been explored and evaluated via the AR-assisted collective assembly and disassembly of a complex wooden structure consisting of identical interlocking modules.

2 Methods

2.1 Material and Structural System

Material System. As part of the overall concept, a material system consisting of discrete wooden units, offering a variety of form-fitting connections was developed. The units can be assembled and reassembled into various configurations, and eventually disassembled. One unit is composed of five wooden pieces, cut from a square timber member with a 45 × 45 mm cross-section, and mechanically fastened with screws. The length and arrangement of the pieces in the unit follow a square grid defined by the timber member's cross-section. Three of such units are then organized into a module. By introducing an asymmetric module shape and following a defined assembly logic, the modules are topologically interlocking, allowing building structures to be assembled without the need for mechanical fasteners, even for complex structural configurations such as openings or overhangs. During the assembly, these modules are not only used as structural components of the building structure but also as temporary support – for configurations such as overhangs and openings – that can be removed subsequently and reused in another part of the structure. Several module iterations and their interlocking properties were examined to satisfy the conditions for both a structural system and ease of manual assembly, to finally select an asymmetric module composed of three units (see Fig. 1).

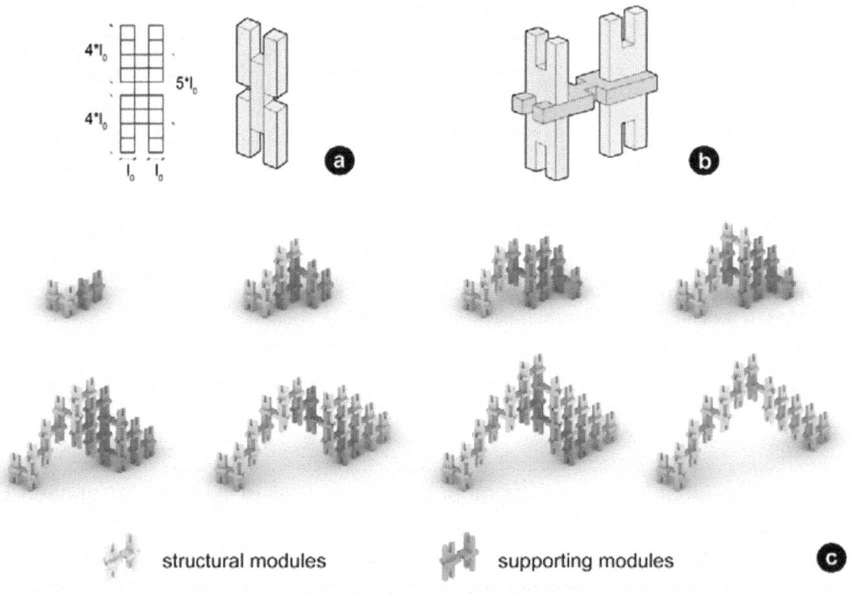

Fig. 1. a) The length and the arrangement of the timber members in the unit follow a square grid defined by the cross-section dimension of the square timber; b) each discrete wooden module consists of three units. c) Due to their asymmetric geometry the modules interlock with each other and eliminate the necessity for additional mechanical connectors. Modules can also be used as a temporary support during assembly and removed after.

Structural System. The units are designed to generate horizontal and vertical interlocking mechanisms by connecting the individual units in various directions in space. Due to this topologically interlocking configuration, compression force transmission between units is made possible. A distinctive feature of the interlocking mechanism is that it allows some tolerance in the connection between the various units, thus facilitating the assembly and disassembly process. In terms of the static system, this means that slight movements between the units are allowed without affecting the overall stability of the structure. An FEM model allows for the structural analysis of a cluster of modules with respect to its overall stability and deformation behavior across different design iterations. This model reproduces the interlocking mechanism by taking into consideration the stacking configuration and the tolerance between connected modules (see Fig. 2).

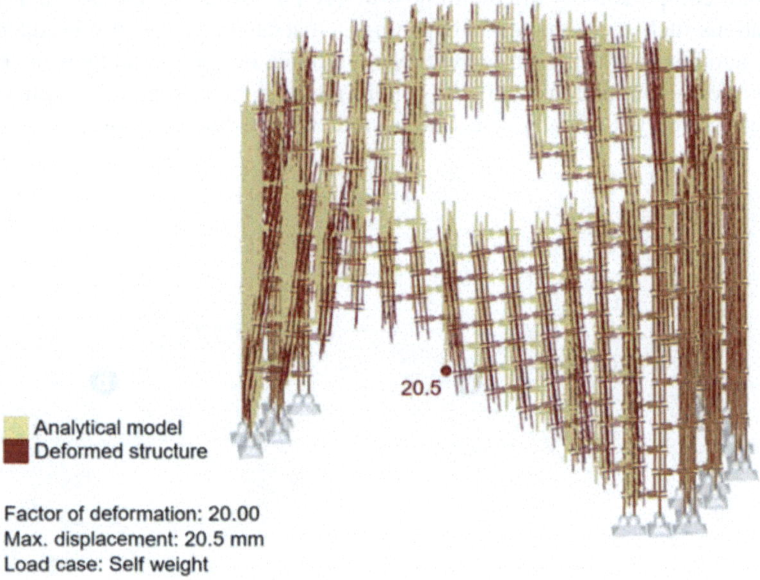

Fig. 2. FEM structural model showing the predicted deformation behavior of an assembly of modules.

In the FEM model, each unit is modeled as a cluster of beam elements and fully rigid connection elements. While the beam elements reproduce the timber members of a unit, the rigid connection elements are used to model the eccentricity between the individual timber members and thus connect the timber members together. The interlocking mechanism between the individual modules of three units is modeled via customized joints. Based on the position of the joints within the unit, the six degrees of freedom of each joint are individually defined either as fixed, released or with a non-linear mechanical behavior, such as compression-only or with localized slippage (see Fig. 3). The parameters for local slippage are estimated from material tests on 1:1-scale physical models.

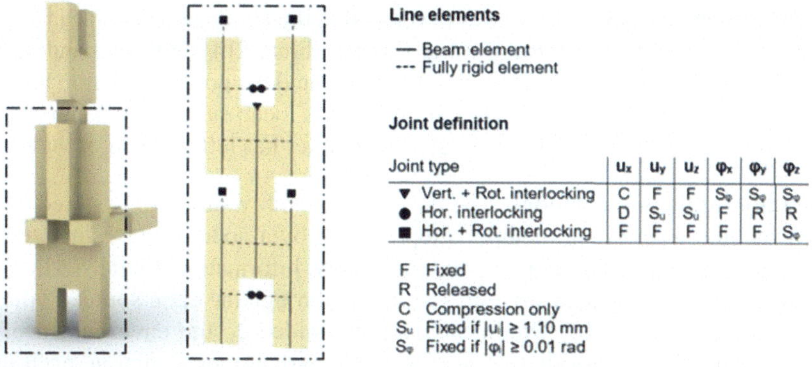

Joint type	u_x	u_y	u_z	φ_x	φ_y	φ_z
▼ Vert. + Rot. interlocking	C	F	F	S_φ	S_φ	S_φ
● Hor. interlocking	D	S_u	S_u	F	R	R
■ Hor. + Rot. interlocking	F	F	F	F	F	S_φ

Fig. 3. Detailed model view of a single unit illustrating the implementation of the local interlocking mechanism in the FEM structural model.

2.2 Fabrication Setup and Workflow

Experimental System Setup. In the experimental workflow, user-dependent inputs and a digital model hosted on a cloud can be synchronized across multiple devices via a custom-developed mobile AR app (Fig. 4). The building actions of individual participants are updated and continuously synchronized via the app in real-time. This synchronization makes it possible to simultaneously coordinate and visually guide several people involved to complete assembly tasks, thus enabling the proposed collective assembly process. As an extension to existing AR-guided fabrication systems, the task sequence is not entirely specified a priori but results from user-based decision-making processes during assembly.

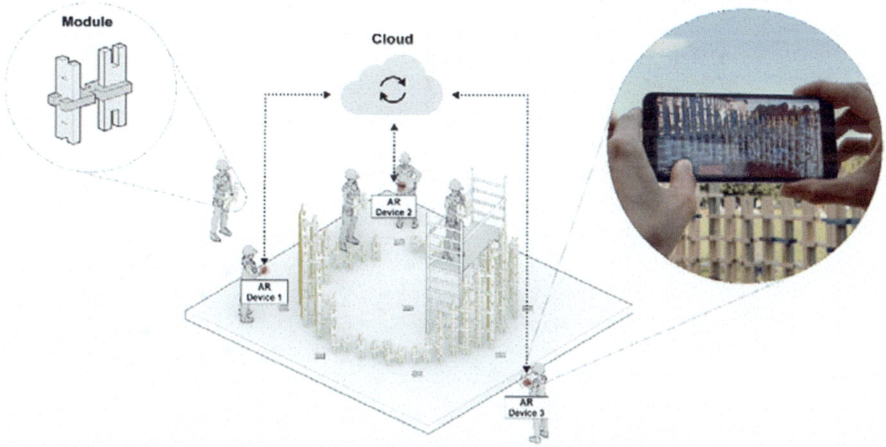

Fig. 4. In the experiment, the proposed multi-user AR-System setup features three mobile devices communicating with a central cloud database over the mobile AR app.

The prototype of the AR app was developed in the Unity game engine [13] using 2D marker-based tracking with the Vuforia Engine library [14]. Vuforia handles both the spatial positioning of the digital model projection and the pose calculation based on the markers including extended SLAM (Simultaneous localization and mapping) tracking for continuous content visualization after the marker has left the field of view.

Assembly Information Model. The designed assembly structure is stored as a graph data structure, where each module is represented by a node of the graph (see Fig. 5). The digital model, referred to as the Assembly Information Model (AIM) [15], is implemented using the data structures available through the open-source Python-based COMPAS framework [16] within the computer-aided design environment of Rhino-Grasshopper [17]. Alongside the module's geometrical representation, each module's states, i.e., *built*, *buildable*, is_*support*, *removable*, and *selected*, are stored as Boolean variables inside the node attributes. The edges of the graph store topological data about the connectivity of the individual modules, where physically connected modules are referred to as neighbors and are interfaced via an edge in the digital model (see Fig. 5). The topology and individual module's states allow to continuously compute the states of other modules according to their assigned states: a module is assigned the state *built* if it is already built in the structure; a module has the state *buildable* when both lower neighbors's state is set to *built*; the state of a module that temporally serves as supporting structure is defined as *is_support*. As the structure gradually gains sufficient structural stability during construction, the removable state of the supporting modules, which are not required anymore, changes such that these modules can then be removed and used further in the building process.

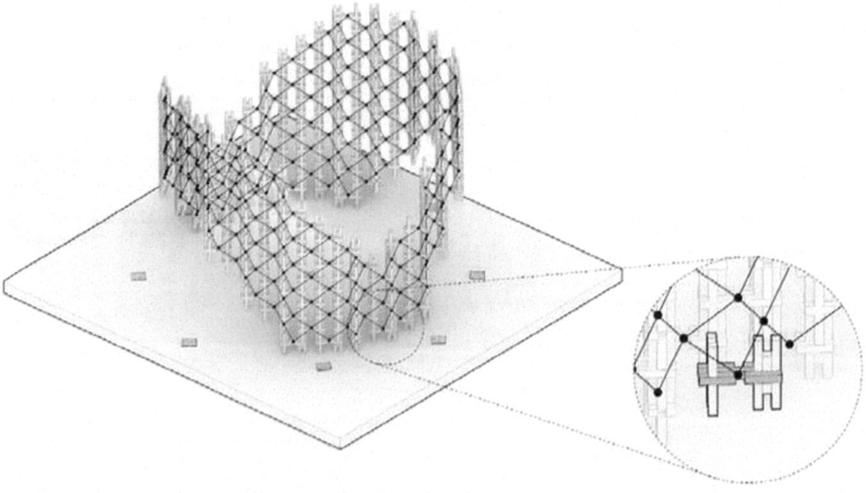

Fig. 5. The digital model includes a graph data structure that stores topological data (edges) about the connectivity and state of the modules (nodes).

A set of geometrical rules based on the module's shape is developed and organized into design patterns, allowing for larger configurations and various design possibilities emerging from the discrete aggregation of a single module.

Data Exchange and Synchronization. To transfer the AIM data structure from the architectural design and planning environment to the mobile app, the assembly model is serialized to a JSON file format and later deserialized to reconstruct the data structure and the geometry in the app's environment. For this purpose, a custom implementation of the Pyrebase Python wrapper for the Firebase API [18] is utilized to enable the transfer of the assembly model directly from the design environment in Rhino-Grasshopper to the Google Firebase Cloud Storage. To then synchronize the model's parameters – i.e., a continuous update of the elements' states and the indication of already selected buildable elements – across multiple connected devices, a cloud-hosted real-time database – Firebase Realtime Database – is used (see Fig. 6).

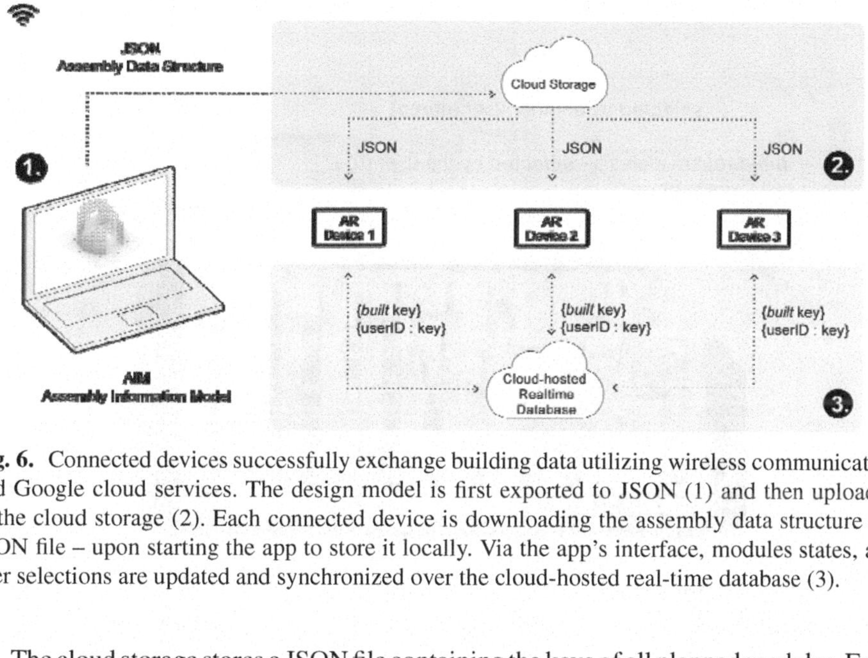

Fig. 6. Connected devices successfully exchange building data utilizing wireless communication and Google cloud services. The design model is first exported to JSON (1) and then uploaded to the cloud storage (2). Each connected device is downloading the assembly data structure – a JSON file – upon starting the app to store it locally. Via the app's interface, modules states, and user selections are updated and synchronized over the cloud-hosted real-time database (3).

The cloud storage stores a JSON file containing the keys of all planned modules. Each time a newly built module is confirmed by a user over the app's interface by pressing the "Build"-button, the state of the respective module's key is updated in the real-time database. In addition to modules' states, the activity of the user is continuously tracked by storing the key of the currently selected module as part of the stored user data. Each client, i.e., connected mobile device, listens to changes on the real-time database. As soon as a change is detected, it is instantly forwarded to all clients triggering UI actions such as the display of a text notification or updating the visualization of the elements according to their changed states and users' current selections.

User Interface. The UI was designed to promote both personalized user experience via user input and inclusiveness in a collective endeavor. The custom UI elements include three input controls and one informational component (see Fig. 7). The input controls are shown as a drop-down list to choose the desired display mode, a slider to select a desired buildable module and a "Build"-button to confirm a newly built module. Users can choose between four pre-programmed display modes – *show all, show built + current phase, show all + buildable, show only built* – to display the assembly structure based on the modules' states and to keep track of the overall building progress. Since the digital model is stored locally on the phone and the app does not rely on the streaming of a single digital model, the app interface supports different display modes on each connected device at a time.

As module states are continuously computed, users are informed about which modules can be assembled at the current building stage. By allowing users to select modules to assemble via the slider, the proposed workflow aims to liberate participants from prescribed assembly sequences and procedural instructions and promote their decision-making and participatory engagement instead [19] (see Fig. 8).

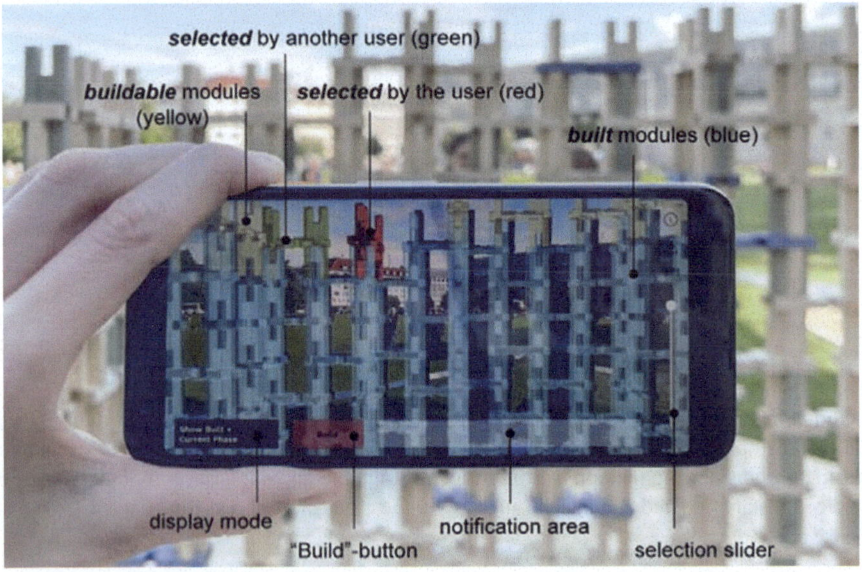

Fig. 7. The elements of the custom UI include three input controls – a display mode drop-down list, a selection slider for selecting a buildable module, and a "Build" – button to confirm a just built module – and a notification area. A color-coding according to the module's state guides the users during assembly: *built* = blue, *buildable* = yellow, *selected* by the user = red, *selected* by another user = green.

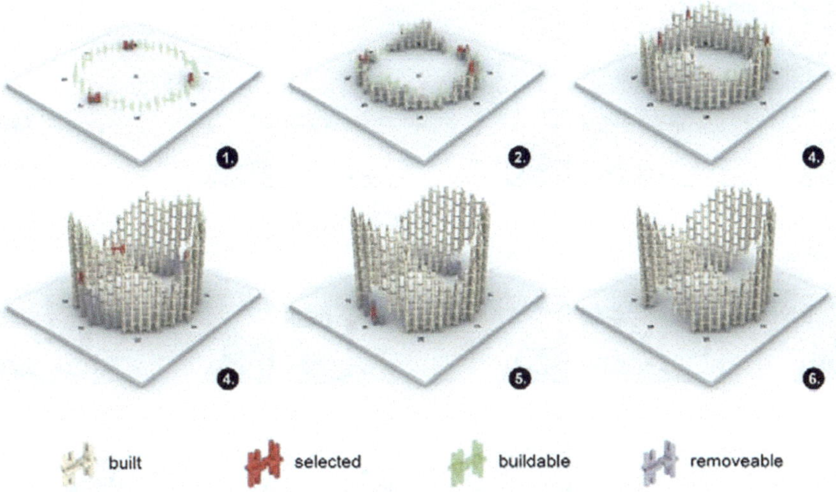

Fig. 8. The assembly order is not specified. Instead, the participating users select and place any module from a subset of *buildable* modules or remove from *removable* modules, defining the order of assembly on the fly.

3 Case Study

Assembly Process. The proposed approach was applied to construct a pavilion collectively. For this purpose, 945 timber units – a total of 315 modules – were prefabricated to be later assembled on-site over two days. Sixty modules were built to support the structure temporally and were gradually removed and reused as a part of the final design. The final design had a circular shape with an outer diameter of 4.26 m and a total height of 3.0 m.

For the marker-based tracking, nine fiducial markers were distributed around the base of the planned structure to ensure that at least one will remain in the field of view during assembly. To prevent positioning errors and discrete jumps of the AR content due to misplacement of the markers, a precise measurement of them using a total station was performed before construction. The measured location coordinates and the orientation of the individual fiducial markers were then embedded into the app's final version.

The app was used and explored by eight different participants using up to three Android smartphones connected to a public Wi-Fi network to enable data transfer to and from the cloud simultaneously. The participants assembled modules of the entire structure either independently or in teams of two – i.e., one person was assembling modules following the instructions displayed on the mobile device, which was held, and operated by a second person (see Fig. 9).

Fig. 9. To test the proposed methods and evaluate the multi-user AR app, a pavilion was assembled collectively with students over the course of two days.

User study. A qualitative survey in the form of a user questionnaire with five of the eight participants of the case study was performed to evaluate the AR app's usability and the custom UI. The answer given to the questionnaire confirmed the potential of the custom UI to improve task coordination during construction by providing individual guidance. According to the participants' feedback, this resulted in a spontaneous and playful experience enabling them to start or stop building or take turns at any time without obstructing the digitally coordinated workflow. The integration of communicative features, namely being informed about the actions of other participants via the UI, such as being visually informed about other participants selecting a module to build or completing a task, led to an indirect interaction between all participants via the app and was perceived as motivating.

4 Conclusion and Future Outlook

This research has explored and developed methods for a mobile AR app enabling multi-user participation in a digitally-driven construction process. The collective construction of a pavilion structure enabled the experimental validation of the material and structural system using the AR app and its custom-developed functionalities (see Fig. 10).

Concerning the applied AR technology, the proposed system could greatly benefit from an extended context-awareness, featuring direct object-tracking instead of marker tracking as introduced by [11], as well as the automated recognition of user actions instead of active user interaction via the app, e.g., automatically registering added modules instead of pressing the "Build" button once a module has been added.

With respect to community involvement and participation, well-conceived, people-centered public spaces have vast potential to become assets that cities can leverage to transform the quality of urban life and improve city functioning. The presented research could allow people in the future to design and build their co-created multi-purpose structures, engaging the public in establishing a profound sense of belonging and responsibility for our built urban environment.

Fig. 10. The final pavilion consisting of 315 modules was realized as part of the Kunstareal-Fest 2021 in Munich.

Acknowledgements. This research was supported by the Technical University of Munich, School of Engineering and Design, Department of Architecture. We thank Design Factory for their expertise and support in prefabricating the timber modules. Wolfgang Wiedemann (Chair of Engineering

Geodesy, TUM Department of Aerospace and Geodesy) conducted the measurements of the fiducial markers. Empfangshalle conceived and realized the wooden platform on which the pavilion was assembled. The authors would like to thank the students Zirui Huang and Andre Nikolai Berlin, who helped build the pavilion and participated in the survey. Additionally, we thank the students who participated in the design studio, "Participative Digital Fabrication": Emmanuel Appiah Acheampong, Badr Ghammad, Veronica Giancola, Iuliia Larikova, Egzon Musa, Chiara Nespoli, Zhan Shi, Abdulhakeem Folorunsho Yusuff and Mohamed Elyes Zahrouni. The pavilion was realized as part of the program of the Kunstareal-Fest 2021 in Munich, Germany. Special thanks go to Laura Schieferle (Kunstareal München), who supported the project by providing the venue.

References

1. Rankohi, S., Waugh, L.: Review and analysis of augmented reality literature for construction industry. Visual. Eng. **1**(1), 1–18 (2013). https://doi.org/10.1186/2213-7459-1-9
2. Chatzopoulos, Di., Bermejo, C., Huang, Z., Hui, P.: Mobile augmented reality survey: from where we are to where we go. IEEE Access **5**, 6917–6950 (2017). https://doi.org/10.1109/ACCESS.2017.2698164
3. Billinghurst, M., Weghorst, S., Furness, T.: Shared space: an augmented reality approach for computer supported collaborative work. Virtual Real. **3**(1), 25–36 (1998). https://doi.org/10.1007/BF01409795
4. Wagner, D., Pintaric, T., Ledermann, F., Schmalstieg, D.: Towards massively multi-user augmented reality on handheld devices. In: Gellersen, H.-W., Want, R., Schmidt, A. (eds.) Pervasive 2005. LNCS, vol. 3468, pp. 208–219. Springer, Heidelberg (2005). https://doi.org/10.1007/11428572_13
5. Reitmayr, G., Schmalstieg, D.: Mobile collaborative augmented reality. In: Proceedings of IEEE and ACM International Symposium on Augmented Reality, ISAR 2001, pp. 114–123 (2001). https://doi.org/10.1109/ISAR.2001.970521
6. Yoshida, H., et al.: Architecture-scale human-assisted additive manufacturing (2015)
7. Johns, R.L.: Augmented materiality: modelling with material indeterminancy. In: Fabricate, pp. 216–223 (2017)
8. Kyjanek, O., Al Bahar, B., Vasey, L., Wannemacher, B., Menges, A.: Implementation of an augmented reality AR workflow for human robot collaboration in timber prefabrication. In: Proceedings of the 36th International Symposium on Automation and Robotics in Construction (ISARC) (2019). https://doi.org/10.22260/isarc2019/0164
9. Jahn, G., Wit, A.J., Pazzi, J.: [BENT] Holographic handcraft in large-scale steam-bent timber structures. In: Ubiquity and Autonomy: Paper Proceedings of the 39th Annual Conference of the Association for Computer Aided Design in Architecture, ACADIA 2019, pp. 438–447, October 2019
10. Jahn, G.: Holographic Construction, December 2019
11. Sandy, T., Buchli, J.: Object-based visual-inertial tracking for additive fabrication. IEEE Robot. Autom. Lett. **3**(3), 1370–1377 (2018). https://doi.org/10.1109/LRA.2018.2798700
12. Mitterberger, D., et al.: Augmented bricklaying. Constr. Robot. **4**(3–4), 151–161 (2020). https://doi.org/10.1007/s41693-020-00035-8
13. Unity Real-Time Development Platform | 3D, 2D VR & AR Engine
14. Vuforia Library
15. Lharchi, A., Thomsen, M.R., Tamke, M.: Towards assembly information modeling (AIM). Simul. Ser. **51**(8), 51–56 (2019)

16. Van Mele, T., et al.: Compas-dev/compas: COMPAS 1.14.1. Zenodo, Febraury 2022. https://doi.org/10.5281/zenodo.6108431
17. McNeel, R., et al.: Rhinoceros 3D, Version 7.0. Robert McNeel Assoc., Seattle (2022)
18. Pyrebase. https://github.com/thisbejim/Pyrebase
19. Parmentier, D.D., Van Acker, B.B., Detand, J., Saldien, J.: Design for assembly meaning: a framework for designers to design products that support operator cognition during the assembly process. Cogn. Technol. Work **22**(3), 615–632 (2019). https://doi.org/10.1007/s10111-019-00588-x

Print-Path Design for Inclined-Plane Robotic 3D Printing of Unreinforced Concrete

Shajay Bhooshan[1,2,3(✉)], Vishu Bhooshan[2,3], Johannes Megens[2,3], Tommaso Casucci[2,3], Tom Van Mele[1,3], and Philippe Block[1,3]

[1] Block Research Group (BRG), Institute of Technology in Architecture, ETH Zurich, Zurich, Switzerland
bhooshan@arch.ethz.ch
[2] Computation and Design Group (CODE), Zaha Hadid Architects, London, UK
[3] Incremental3D (In3d), Innsbruck, Austria

Abstract. The paper details the computational toolkit for print-path synthesis and execution that was used in the physical realisation of an arched, bifurcating, unreinforced masonry footbridge spanning 16 m, composed of 53 3D-printed concrete blocks. The printed concrete filaments of every block were placed in layers that are orthogonal to the expected, compressive force flow, resulting in the need for non-parallel, inclined print-path planes, thus also resulting in non-uniform print-layer heights. In addition, the bridge's global structural logic of stereotomic masonry necessitated the precise coordination of the interface planes be- tween blocks. Approximately 58 km of print path, distributed over 7800 inclined layers, were generated and coordinated such that the resulting print paths meet printing-related criteria such as good spatial coherence, minimum and maximum layer thickness, infill patterns etc. We describe a schema based on Function Representation (FRep) for inclined-plane print-path generation, and its full implementation for practical and large-batch production. We also implement specific extensions to generate the infill print paths typically needed in 3D concrete printing.

Keywords: 3D concrete printing · Digital fabrication · Shape design · Automation pipeline · Bridge design · Computer aided design · Mesh-based geometry processing · Structure and fabrication-aware shape design · Software encapsulation

1 Unreinforced Masonry, 3D Concrete Printing and Print-Path Synthesis

The unreinforced masonry design paradigm and techniques are highly compatible with the compression-dominant, orthotropic material properties of layered 3D Concrete Printing (3DCP) [1–3]. The compatibility is at two scales: the overall funicular form of the structure and the force flow through each 3DCP block. Globally, the structure's funicular form engages the compression-dominant properties of 3DCP, offers a clear discretisation strategy orthogonal to the designed force flow, enables the separation of compressive

C. Gengnagel et al. (Eds.): DMS 2022, *Towards Radical Regeneration*, pp. 188–197, 2023.
https://doi.org/10.1007/978-3-031-13249-0_16

and tensile structural elements, makes dry assembly of the prefabricated components, results in significant lower stresses overall in comparison with bending solutions, etc. Locally, the alignment of the printed layers orthogonal to expected compressive force flows engages the compressive strength of 3DCP whilst eliminating the need for tensile reinforcement or post-tensioning to prevent shear failure along the print planes in the 3DCP blocks [4]. Furthermore, the wider benefits of structural geometry and the masonry paradigm to improve recyclability, maintenance and repair and reuse of material due to dry-assembly and clean separation of tensile and compressive materials has also been recently highlighted [5, 6]. Aligning the print layers of 3DCP orthogonal to the expected, compressive force flow results in the need for non-parallel, inclined print-path planes, thus also in non-uniform print-layer heights. In addition, the global structural logic of stereotomic masonry necessitates precise coordination of the interface planes between blocks.

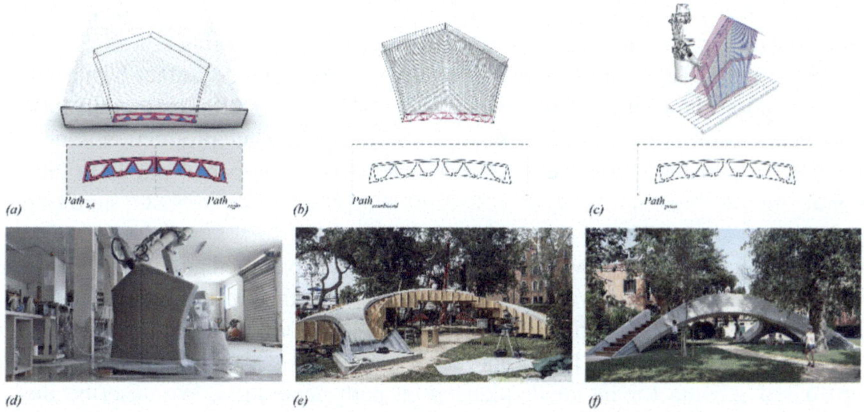

Fig. 1. (a) Print-path generation using function representation (FRep), (b) post processing of print paths, (c) evaluation of printability & generation of machine code (d) robotic 3d concrete printing of blocks, (e) assembly process, (f) finished bridge

1.1 Key Contributions

Striatus is an arched, bifurcating, unreinforced masonry footbridge spanning 16 m, composed of 53 3D-printed concrete blocks. Approximately 58 km of print path, distributed over 7800 inclined layers, were generated and coordinated such that the resulting print paths meet printing-related criteria such as good spatial coherence, minimum and maximum layer thickness, infill pattern etc. (Fig. 1). The print files generated were used to robotically print the 53 blocks in 85 h. The design of the global compressive surface shape and the decomposition of the offset surface into blocks (stereotomy) is beyond the scope of this paper. For more, we refer the reader to Bhooshan et al. [7].

This paper describes the workflow used to synthesize printing paths and to batch manufacture the blocks. The key contributions of the paper are the print-path- synthesis-to-manufacture workflow. The bespoke workflow and tools developed enabled: (i) integration of print-path synthesis within the interactive and iterative design cycles; (ii) synthesis of print paths aligned orthogonal to expected, compressive force flow; and (iii) coordinated, large-batch production of 53 blocks with negligible tolerance errors at the interface planes, which meant that the blocks were easily assembled on site into the global form.

2 State of the Art

The layered deposition process of 3DCP implies that the synthesis of print paths has a significant bearing on the physical outcome [8–10]. The topic of process-appropriate print-path synthesis is beginning to receive research interest, particularly to enable support-free, in-place 3DCP of large compression shells [11]; overcome the limitations of horizontal print paths, including the so-called staircase effect using non-planar, spatial curved print paths [12]; achieve surface textures in the printed artefact by visualising and manipulating the printing paths and/or the associated robotic instructions [10, 13, 14]; synthesise print-critical infill patterns [15, 16]; and alleviate production issues such as jittery motion of the robotic arm caused by excessive number of points in the print path, non-smooth transition between consecutive layers, feature-agnostic sampling etc. [17, 18].

Our work shares interests with all of the above. We developed an integrated, design-to-production toolkit to print-and-assemble 3DCP blocks into masonry structures as demonstrated in the Striatus bridge prototype (Fig. 1). In particular, Bhooshan, Ladinig, Van Mele and Block [2] and Bhooshan et al. [4] describe a Function Representation (FRep) based schema for inclined- plane print-path generation. We describe the full implementation and extension of the schema for practical and large-batch production. We also implemented specific extensions to visualise the toolpaths at design-time, achieve expressive textures in the printed blocks, and generate the infill print paths and post-processing routines to alleviate print-time issues.

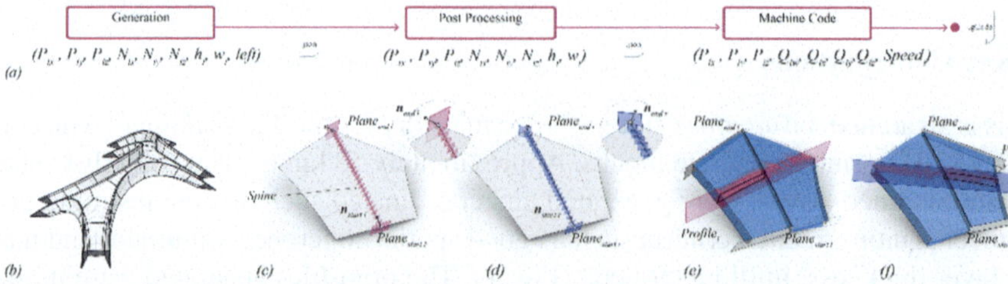

Fig. 2. (a) Workflow diagram for print-path synthesis to production, (b) global shape with stereotomy, (c, d) left & right planes interpolated between start (P_{sl}, P_{sr}) & end planes (P_{el}, P_{er}). (e, f) base cross section profile.

3 Print-Path Synthesis to Production

Given a global compressive shape and the associated stereotomy (Fig. 2b), three main steps follow: synthesis of the print paths, their preparation for print readiness and the generation of robotic instructions or so-called GCode (Fig. 2a and Fig. 1a–c). Information is passed from encapsulated step into the next using a simple text-based file format that encodes information using the JavaScript Object Notation (JSON).

4 Print-Path Generation

The interface planes and spines of the deck and balustrade 3D blocks (Fig. 2a, b) are the input information needed to create the print paths per block. Each block has a start and an end plane.

The spine of edges associated with each block is subdivided at equal input distances set as the mean of minimum and maximum print layer height permissible. New planes, going through each of the subdivided points, are then generated (Fig. 2a, b). The normals of the planes (n_i) are computed by interpolating between the start (n_{start}) and end (n_{end}) planes, using a weighting factor (w_i).

$$n_i = (1 - w_i) * n_{start} + w_i * n_{end}; \, where w_i = (0.0, 1.0)$$

A weighted nonlinear interpolation scheme was implemented to ensure that the print height between subsequent planes lies within the domain specified by the robotic printing constraint (see Sect. 7). It can be noted that, because of the bifurcating topology of the bridge, all the deck blocks have two sets of start and end planes – a left and right set. Consequently, there are two sets of interpolated planes (Fig. 2c, d). The balustrade blocks consist of only one set.

Next, a base cross-sectional profile was computed, for each of the planes, as the zero contour of a so-called signed distance field (SDF) (Fig. 2e, f). This profile is then used to create additional SDFs on the interpolated planes to update the cross-sectional profile.

For both the deck & balustrade blocks, the main SDFs are as follows (Fig. 3):

- base profile polygon f_0;
- offset polygon $f_1 = f_0 + 0.5$ * print width;
- offset polygon $f_2 = f_0 + 1.5$ * print width; infill f_3;
- trim f_4 = line SDF at pattern/brace point;
- resultant fresult $= (f_1 - (f_2 - f_3)) - f_4$.

The resultant SDF (f_{result}) is constituted as the Boolean of five main individual SDFs. Each of the five SDFs serve a specific purpose:

- base profile SDF f_0;
- two boundary SDFs (f_1, f_2) to control the cross-sectional thicknesses based on specified print width & f_0;
- infill SDF (f_3) to provide local stiffeners in each cross section; and
- trim SDF (f4), at pattern points, aiding in the creation of one continuous print profile

Together, this step creates two sets of profile curves for the left and right planes respectively for the deck blocks and a single set for the balustrade blocks. All the print paths on the left and right set of planes of the deck blocks are combined in the next post-processing step.

The combination of individual SDFs to compose the final print-path allowed the accommodation and fine control of discoveries made during the material and prototyping phase (Sect. 8). For example, the visible ribs in the blocks (Fig. 4d, e) were developed as a separate SDF after it was discovered that the turning of the robot-head necessitates a reduction of angular velocity which in turn produces a naturally occurring grove (Fig. 8c, d).

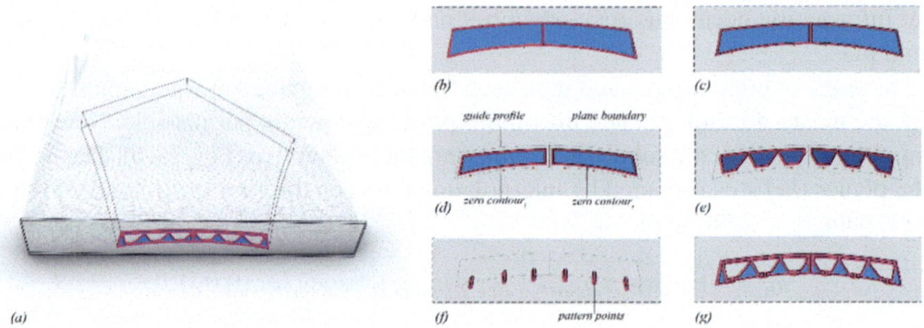

Fig. 3. (a, g) The resultant SDF (f_{result}) is constituted as the Boolean of five individual SDFs - f_0 (b), f_1 (c), f_2 (d), f_3 (e) & f_4 (f).

5 Post Processing

The resulting SDF contours generated from the previous step are polylines. The number of points and segments in each of the curves (sampling) depends on the cell size used in computing the SDF (Fig. 4a). The sampling of the curve is a critical parameter in 3D printing, as it affects the speed and interpolated trajectory of the robot head and thus the deposition of concrete. As such, the contour curves need to be resampled.

Two resampling methods were tested: uniform resampling and feature-based adaptive resampling. Uniform resampling rebuilds the input curve with even spacing of the points (Fig. 4b). Whilst this method successfully reduces the number of control points to be lower than the robot constraint (60,000 target points), it can cause misalignment of the points of the paths in consecutive layers, and thus artefacts to appear in the printed blocks (Fig. 4d). Feature-based adaptive resampling splits the input curves into a set of individual segments based on feature points (Fig. 4c). The segments are then uniformly resampled. The position of the feature point is spatially coherent across consecutive printing layers. Consequently, the occurrence of artefacts in the printed blocks is significantly reduced (Fig. 4f). Additionally, this method allows adaptive sampling, the sampling can be lower in the areas that do not necessitate high-quality finish, such as internal infills, and higher in areas which need a higher finish, such as visible faces of the blocks.

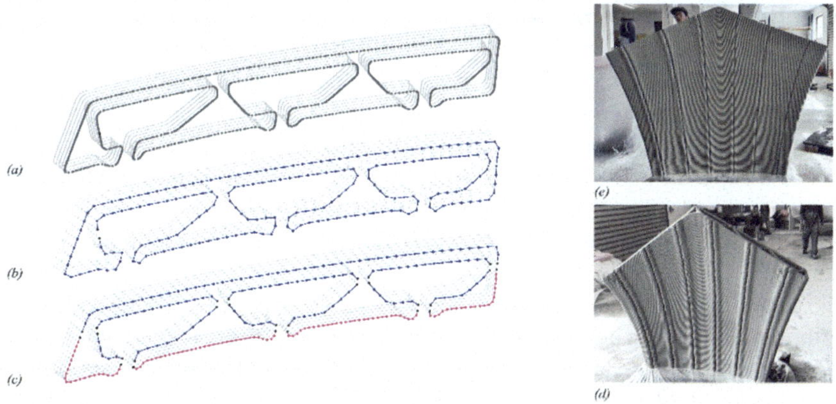

Fig. 4. Printing paths sampling: (a) print-path synthesis from SDFs, (b) uniform spacing, (c) adaptive feature-based resampling, (d, e) result of the printing for uniform and adaptive feature-based resampling, respectively.

6 Contiguous Print Paths

Resampled sets of contours are further processed in order to generate a single continuous printing path for each of the blocks. This is achieved in two steps: First, the left and right curves of each layer are connected together by splitting the curves based on an input guideline curve (g_L and g_R for left and right path respectively) and then reconnecting the left and right paths with two straight segments (pt_1-pt_2 and pt_3-pt_4). The guideline curves are generated for every layer to have a consistent reference position at the inner upper corner of each path (L_1 and R_1) (Fig. 5). Next, consecutive layers are connected into a continuous spiraling path connecting the end of each layer path to the start of the next.

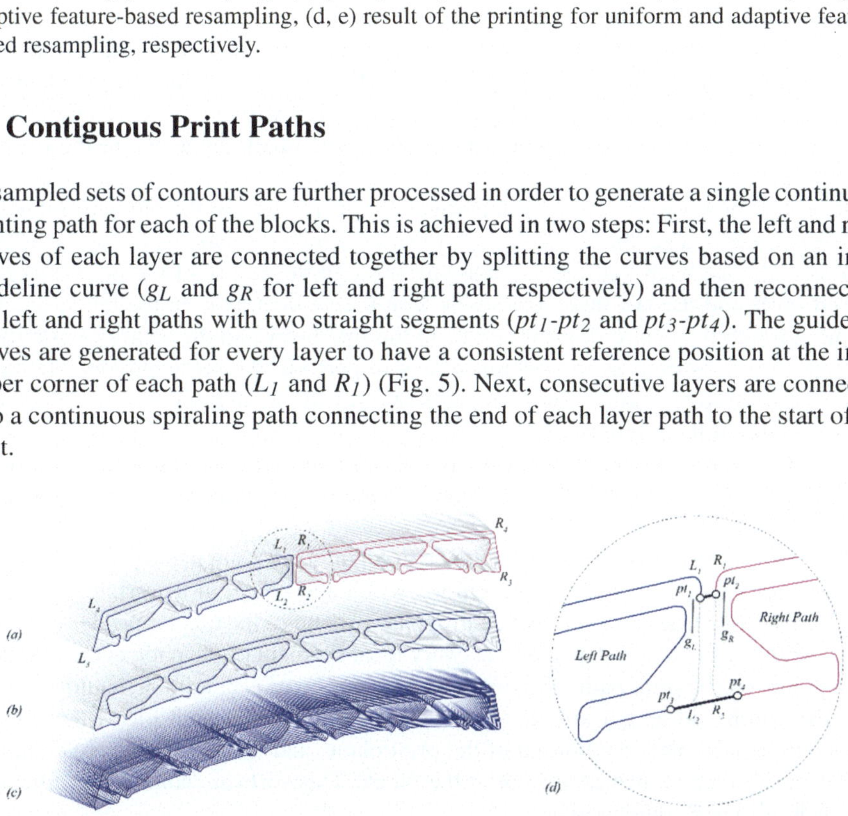

Fig. 5. Generation of continuous printing path steps: (a) left and right contours; (b) connected layer, and (c) spiralling path; (d) Sequence for the connection of the left and right paths.

7 Machine Code Generation

Each block to be 3D printed is represented by one contiguous, spiralling curve as generated by the previous step (input path file). This file is further processed to generate the robotic instructions or so-called GCode, to print the block. As a quality check, a validation was performed by simulating and tracing the robot path as produced by the GCode (output path file). Both the input path file and output path file were meshed to visualise the tubular print layers. The two meshes were aligned and visually checked for deviation.

The print sequence of a single block consists of two steps: establishing a planar basis and printing the block. Thus, the final path includes base layers that are not part of the block. The raft levels out any unevenness in the print bed and ensures precise interfaces between blocks. A plastic sheet was inserted mid-print, to separate the raft from the actual block.

We used the raft additionally to establish stable print orientations for all of the blocks. In other words, we use the raft layers to print a plinth that aligns the blocks in an appropriate orientation that improves printability (Fig. 6a).In particular, thin deck blocks and balustrade keystones would collapse during the print without this correction, as the centre of gravity of the blocks would fall outside the block geometry in the original orientation.

Two approaches were tested for the orientation correction of the block (print paths) (Fig. 6b) – so-called geometric and centre-of-gravity transformations. In the former, the block is realigned such that the axis passing through the centre of the print-curves in the first layer and the last layer, coincides with the vertical axis. In the latter, the axis passing through the centre of print-curves in the first layer and the centre of gravity of the entire block, is used as the reference axis to perform the reorientation. The latter solution proved to be a good compromise between vertical alignment of the block and the inclination of the plinth itself.

The 3D concrete printing was executed with a 6-axis industrial robot, which allowed printing with inclined planes. The kinematic solution was checked primarily for robot reach and access.

A speed variable (*robot target speed*) was computed per point in the print-path, using the parameters l_w and l_h which represent the width and height of the printed extrusion (Fig. 7). The speed variable in effect controls the volume of the tubular print layers, as it affects volume of material deposition at every local segment of the path, centred around target points on the print path. The depth of the print layer is determined automatically since the width and height are al- ready defined in the input path file. Together, the points on the print path, the normal of the print plane, and speed variable are compiled into GCode, which is then used to print the block. This software solution significantly reduces hard- ware complexity.

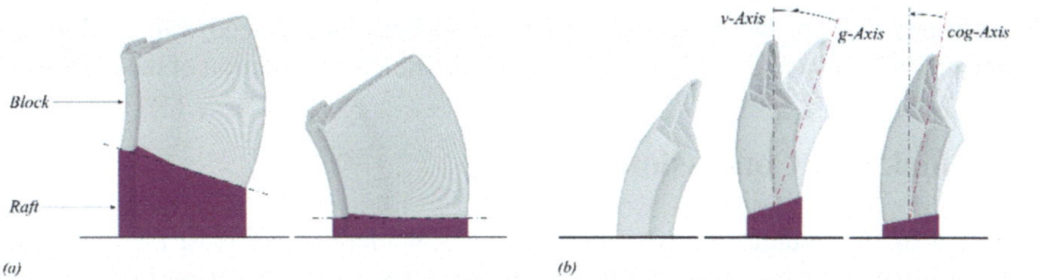

Fig. 6. Block reorientation for printing: (a) horizontally aligned unstable axis of base layer to correct reorientation axis and minimise raft print (magenta). (b) rotation around unstable axis; alignment along centre of gravity axis avoids slipping between block and raft as well as providing a stable print position.

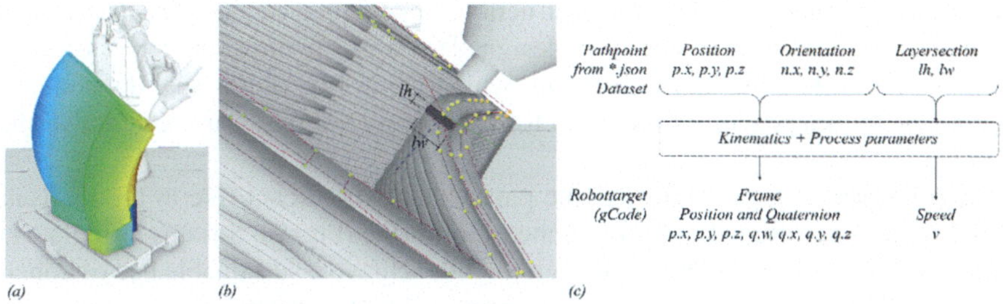

Fig. 7. Kinematic simulation with colour indication of print speeds (a) via Mesh and (b) via Robot Targets. (c) Per print point layer, dimensions l_w and l_h are transformed to speed settings for each robot target along with common kinematic parameters.

8 Material Tests and Prototyping

Preliminary print tests addressed material processing and quality, curing time and, essential for the path generation, the definition of minimum and maximum limits for the layer heights (Fig. 8). These parameters affect the actual print speed in relation to the material flow rate. These details were defined by multiple print-tests that mostly consisted of block prototypes, which allowed developing an efficient workflow between the collaborating parties. As a result, two basic settings for deck and balustrade blocks were defined:

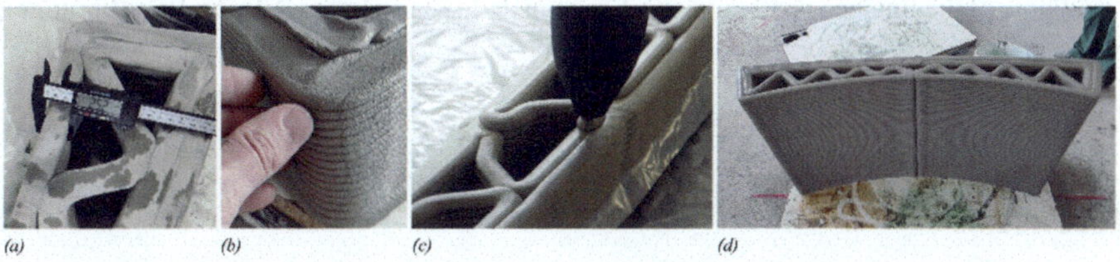

Fig. 8. (a) path and infill tests, (b) defining material and processing parameters for consistency, and (c, d) Learnings had been reintroduced for the final design, most relevant being the seam details for the deck blocks. (See also Fig. 4)

a lower pump rate for the balustrade blocks to address the reduced wall thickness and delicate details, and a higher flow rate to allow an efficient production of the deck blocks.

9 Outlook and Conclusions

Two areas of immediate improvement can be considered. Firstly, a fast constrained optimisation routine to guarantee that interpolation (Sect. 2) will meet the layer-height limits would alleviate the need for manual intervention. Additionally, a robust, automated calibration procedure to check and minimise the deviation between as-printed and as-designed blocks would improve construction tolerance and production time. In the current project, we relied on spot-checking and experience.

In summary, the custom toolchain that was developed enables print-path synthesis, verification and generation of robotic instructions or so-called GCode. The toolchain and the constituent, standalone applets were designed to be easy to implement, fast to execute, and enable rapid iteration and refinement, whilst being free of external dependencies. Together, the toolkit provides a blueprint for real-time, printing-aware, interactive shape design.

The print path design detailed in the paper had a significant role in the realisation of Striatus, an unreinforced masonry arched footbridge (Fig. 9). Striatus articulates the relevance of unreinforced masonry design to both achieve a sustainable use of concrete and realise the benefits of 3D concrete printing (3DCP). Striatus thus demonstrates a practical pathway to design and construct bridge structures that are repair, reuse and recycling friendly.

(a) (b)

Fig. 9. Photo of the finished project on site, Giardini della Marinaressa, Venice, Italy. Photographs by Naaro

References

1. Khoshnevis, B.: Automated construction by contour crafting-related robotics and information technologies. Autom. Constr. **13**(1), 5–19 (2004)

2. Bhooshan, S., Ladinig, J., Van Mele, T., Block, P.: Function representation for robotic 3D printed concrete. In: Willmann, J., Block, P., Hutter, M., Byrne, K., Schork, T. (eds.) Robotic Fabrication in Architecture, Art and Design, pp. 98–109. Springer, Cham (2018). https://doi.org/10.1007/978-3-319-92294-2_8
3. Carneau, P., Mesnil, R., Roussel, N., Baverel, O.: An exploration of 3D printing design space inspired by masonry. In: Proceedings of IASS Annual Symposia, no. 6, pp. 1–9. International Association for Shell and Spatial Structures (IASS) (2019)
4. Bhooshan, S., Van Mele, T., Block, P.: Morph & Slerp: shape description for 3D printing of concrete. In: Symposium on Computational Fabrication, pp. 1–10 (2020)
5. Block, P., Van Mele, T., Rippmann, M., Ranaudo, F., Calvo Barentin, C.J., Paulson, N.: Redefining structural art: Strategies, necessities and opportunities. Struct. Eng. **98**(1), 66–72 (2020)
6. Ranaudo, F., Van Mele, T., Block, P.: A low-carbon, funicular concrete floor system: design and engineering of the HiLo floors. In: Proceedings of IABSE Congress 2021 (2021)
7. Bhooshan, S., et al: The Striatus arched bridge Computational design and robotic fabrication of an unreinforced, 3D-concrete-printed, masonry bridge. Archit. Struct. Constr. (2022, in press)
8. Gosselin, C., Duballet, R., Roux, P., Gaudillière, N., Dirrenberger, J., Morel, P.: Large-scale 3D printing of ultra-high performance concrete–a new processing route for architects and builders. Mater. Des. **100**, 102–109 (2016)
9. Bhooshan, S., Van Mele, T., Block, P.: Equilibrium-aware shape design for concrete printing. In: De Rycke, K., et al. (eds.) Humanizing Digital Reality, pp. 493–508. Springer, Singapore (2018). https://doi.org/10.1007/978-981-10-6611-5_42
10. Breseghello, L., Sanin, S., Naboni, R.: Toolpath simulation, design and manipulation in robotic 3D concrete printing (2021)
11. Motamedi, M., Oval, R., Carneau, P., Baverel, O.: Supportless 3D printing of shells: adaptation of ancient vaulting techniques to digital fabrication. In: Gengnagel, C., Baverel, O., Burry, J., Ramsgaard Thomsen, M., Weinzierl, S. (eds.) DMSB 2019, pp. 714–726. Springer, Cham (2020). https://doi.org/10.1007/978-3-030-29829-6_55
12. Lim, S., Buswell, R.A., Valentine, P.J., Piker, D., Austin, S.A., De Kestelier, X.: Modelling curved-layered printing paths for fabricating largescale construction components. Addit. Manuf. **12**, 216–230 (2016)
13. Anton, A., Yoo, A., Bedarf, P., Reiter, L., Wangler, T., Dillenburger, B.: Vertical modulations. In: Proceeding of ACADIA 2019 (2019)
14. Westerlind, H., Hernández, J.: Knitting concrete. In: Bos, F., Lucas, S., Wolfs, R., Salet, T. (eds.) DC 2020, pp. 988–997. Springer, Cham (2020). https://doi.org/10.1007/978-3-030-49916-7_96
15. Zhao, H., et al.: Connected fermat spirals for layered fabrication. ACM Trans. Graph. (TOG) **35**(4), 1–10 (2016)
16. Bi, M., et al.: Continuous contour-zigzag hybrid toolpath for large format additive manufacturing. Addit. Manuf. **55**, 102822 (2022)
17. Anton, A., Reiter, L., Wangler, T., Frangez, V., Flatt, R.J., Dillenburger, B.: A 3D concrete printing prefabrication platform for bespoke columns. Autom. Constr. **122**, 103467 (2021)
18. SlicerXL: SLicerXL (2021)

Reusable Inflatable Formwork for Complex Shape Concrete Shells

Camille Boutemy[1(✉)], Arthur Lebée[1], Mélina Skouras[2], Marc Mimram[3], and Olivier Baverel[1,4]

[1] Laboratoire Navier UMR8205, Ecole des Ponts Paristech, Université Gustave Eiffel, CNRS, Champs-sur-Marne, 77455 Marne-la-Vallée Cedex 2, France
camille.boutemy@enpc.fr

[2] Université Grenoble Alpes, Inria, CNRS, Grenoble INP, LJK, Grenoble, France

[3] Laboratoire OCS/UMR AUSser 3329, ENSA Paris-Est/Université Gustave Eiffel, Marne-la-Vallée, France

[4] GSA/Ecole Nationale Supérieure d'Architecture de Grenoble, Grenoble, France

Abstract. Construction of concrete shells is expensive and generates wastes from the fabrication of formworks. Being non-reusable, these elements have a negative impact on the life-cycle assessment of the construction. The purpose of this research is to design and build a new inexpensive formwork system made of inflatable structures for precast and thin concrete shells construction.

By sealing two membranes according to a pattern, this system allows the construction of complex inflated shapes. The sealing pattern is designed such that, once inflated, the planar metric becomes not uniform and generates a 3D surface following Gauss's Theorema Egregium, a classical result of differential geometry. This design of the seal pattern is guided by a numerical tool capable of accurately predicting the inflated shape. The simulations are compared to physical models made of fabrics, before manufacturing inflatable formwork prototypes in composite membranes from about 1 to three metres wide. Support is set up to pour concrete on the inflatable formwork without damaging it for reuse. The resulting thin concrete shell and its fabrication method are eligible for wider-scale application in the AEC industry.

Keywords: Inflatable structure · Pneumatic structure · Formwork · Concrete · Membrane

1 Interest of Creating Concrete Elements from Inflatable Structures

Custom-made concrete formworks are expensive and represent a large amount of waste in the Architecture, Engineering, and Construction (AEC) industry. Regarding standard construction with planar concrete panels, 53% (formwork materials and installation) of the concrete structure costs are spent on formworks. When complex and unique formwork must be created, they represent 88% of the total concrete construction costs (see

C. Gengnagel et al. (Eds.): DMS 2022, *Towards Radical Regeneration*, pp. 198–210, 2023.
https://doi.org/10.1007/978-3-031-13249-0_17

Fig. 1). These formworks are themselves complex and expensive elements without being reusable: they have a negative impact on the life-cycle assessment of the construction project.

These difficulties partly explain why concrete shells disappeared at the end of the 20th century, despite the undeniable architectural quality they provided to the spaces created [1].

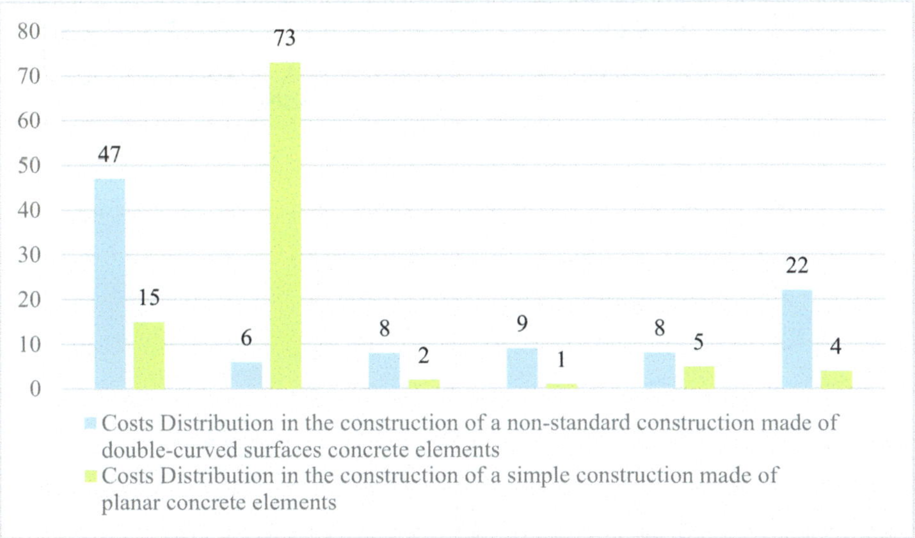

Fig. 1. Costs distribution in the construction of a simple and a complex concrete structure [2, 3].

This research aims to create a new economic formwork system for precast concrete shells. Inflatable structures are attractive for their lightweight properties and diverse realisable shapes, even complex. Moreover, they are inexpensive because they are composed of standard sheet materials and air.

Inflatable structures have already been used as formwork elements to fabricate concrete shells, for example, Domecrete by Heifetz, Bini-Shells by Dante Bini [4], the two-chamber system formwork for Bubble Housing System by Heinz Isler [5] and the Pneumatic Formwork Systems in Structural Engineering by [6]. Both these examples consist of a unique patterned membrane such that once inflated, it becomes a constant mean curvature surface (see Fig. 2 1a, 1b, 1c). This inflated volume is used to lift fresh concrete or set precast concrete elements to create domed shells.

In the present work, we propose to create an inflated surface made of two flat membranes without patterning, sealed according to a 2D pattern (see Fig. 2 2a, 2b, 2c). Once inflated, the seals become valleys delimiting mountains, inflated cavities. This technique was introduced by [7], who created models made of sealed paper that bends once inflated.

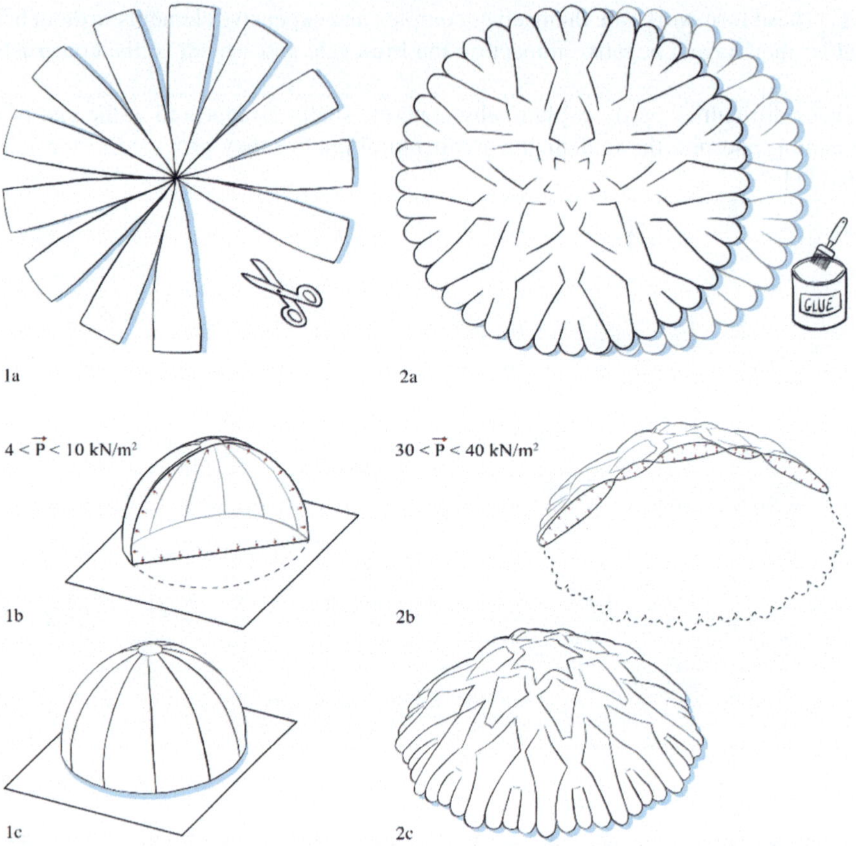

Fig. 2. Patterning (1a, 1b, 1c) and 2 layers glued (2a, 2b, 2c) technique for fabricate an inflatable structure.

We first describe the geometry principles and simulation tools. Then, the design of the 2D pattern to reach a chosen shape is explained. Finally, we present the fabrication of the inflated formwork and the construction of the concrete shell.

2 Interest of Creating Concrete Elements from Inflatable Structures

As seen in Sect. 1, we choose to explore a method without patterning. An in-house tool, InflatableSheetSimulator [7], simulates the inflated sheets numerically using the finite element method, starting from the 2D pattern made of an external contour and internal lines. The tool generates two meshes connected by the 2D pattern. Custom parameters are the resolution of the mesh, the thickness and Young's Modulus of the fabric, and the internal pressure. The sheets are modelled with tension field theory [8], a very efficient convexified membrane model: when membrane stress is under compression, its stiffness vanishes so that wrinkling instabilities are regularised. A linear triangle interpolation is sufficient for this modelling.

Fig. 3. Manufactured models with 2 sheets of fabric sealed according a 2D pattern.

We also use physical models to observe the 3D shapes created by the sealed patterns on the two layers sheets. Two TPU coated Nylon fabric layers are sealed according to a 2D pattern with an air supply to inflate the created model (see Fig. 3).

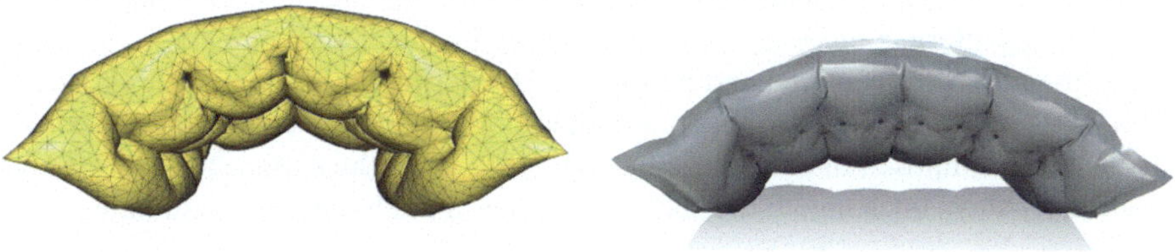

1a 1b

Fig. 4. Simulation of inflated sheets (1a) and fabricated prototype (1b) with the same 2D pattern.

Prototypes in composite membranes, a full-scale construction material, are also manufactured and compared to the numerical simulation in which the shape is very close. (see Fig. 4). These simulations are very realistic and help the design of the sealing patterns by quickly verifying their effect on the inflated geometry.

Let us illustrate how inflation may change the metric of the surface in a simple case (see Fig. 5). We seal two sheets according to parallel lines. When inflating, tubes appear, and the metric changes only in the transverse direction. The distance between two seals d_1 tends to be a half-circle with air pressure in two dimensions. With d_2 as the orthogonal projection in the plane of the pattern between two seals, for a maximal contraction, we have $d_2 = \frac{2}{\pi}d_1$. The contraction factor λ, induced by air pressure, can vary from $\frac{2}{\pi}$ to 1. In our case, the contraction is maximal in the tubes transverse direction, $\lambda_1 = \frac{2}{\pi}$ and non-existent along with the seals, $\lambda_2 = 1$. This lack of contraction makes assembly lines behave as soft hinges. When varying the orientation of seals, the sheets take on a 3D shape.

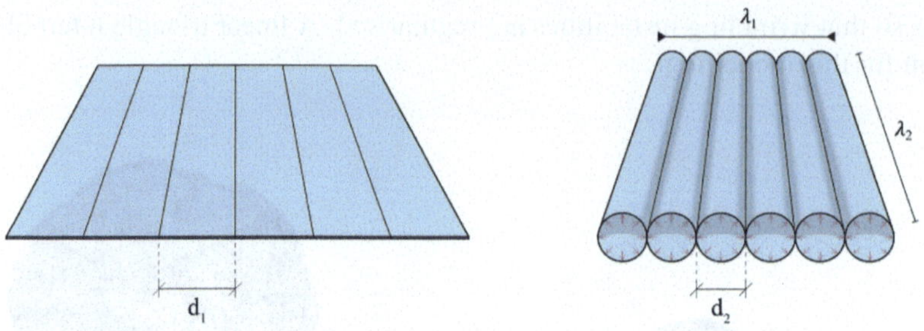

Fig. 5. Two layers sealed according to parallel lines and inflated.

3 Design of a Pattern for a Concrete Dome

For the first project of inflatable formwork conception and fabrication, the target shape is a dome. This positive double curvature form is chosen to simplify the boundary conditions.

From 2 discs of membrane, we can obtain positive or negative double curvature surfaces by changing the plane metric thanks to air pressure and non-parallel seals. The patterns are composed of long seals (curves or straight lines) and smaller ones as subdivisions (see Fig. 8 1b) to harmonise the cavity sizes and discretisation that refine the shape. With a pattern composed of radial lines (see Fig. 6 1), the conical tube retraction changes the ratio between area and perimeter. The perimeter becomes shorter, and the surface goes out of its plane to reach a double positive curvature. On the other hand, with a pattern composed of concentric lines (see Fig. 6 2), the surface area shrinks, and the perimeter stays unchanged: the surface becomes a double negative curvature.

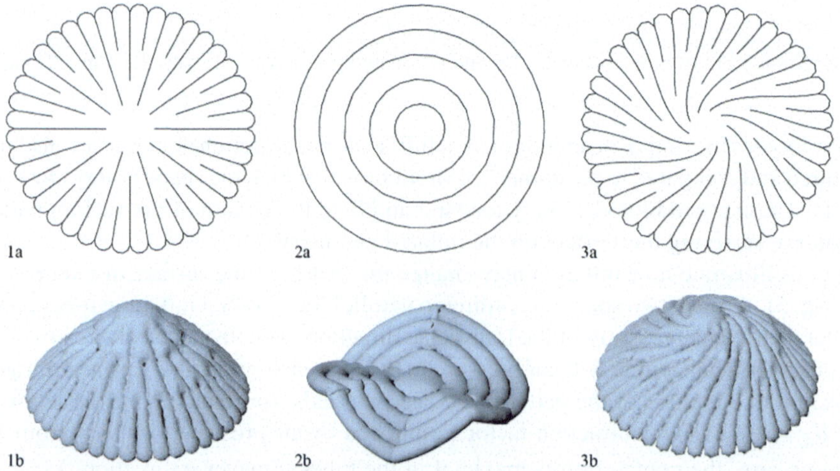

Fig. 6. 2D patterns and simulation of 2 sheets sealed according to the pattern. Radial lines 2D pattern (1a) and simulation (1b). Concentric curves 2D pattern (2a) and simulation (2b). Curves 2D pattern (3a) and simulation (3b).

When we pour concrete on the inflated structure, these curves become the ribs of the concrete element. A parallel numerical structural study has been made on concrete shell structures cast by our inflatable formwork system [9]. Two concrete shells, one smooth (Shell 1) and the other one ribbed (Shell 2) (see Fig. 7), have been structurally analyzed with a karamba solver.

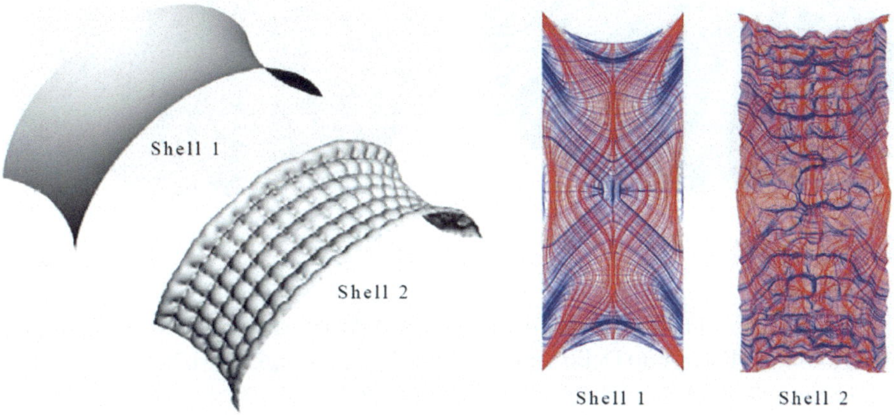

Fig. 7. Shell numerical models and principal stresses simulation (compression in red, tension in blue). [9].

The two shells are 5 m long and 2.5 m large. Shell 1 is 34 mm thick, determined by the maximal admitted compression and tension stresses under a classical load case with a bending displacement less than L/300 (With L as the length between the two supports); his weight is 1537 kg. Shell 2 has the same weight but is only 30 mm thick despite a larger surface with the ribs. The ribs allow reducing the quantity of concrete.

The structural analysis shows the principal compression and tension stresses; if we compare the two shells, the behaviour of Shell 2 is close to Shell 1. The buckling load factor appears higher on the ribbed shell (144) than on the smooth shell (57), thus the ribs make Shell 2 more resistant to buckling. To conclude, if the ribs are well placed, they benefit the shell's structural behaviour. They permit to slim down the thickness of the shell and be more resistant to buckling.

Siefert et al. [10] propose an array of curves as a sealed pattern to reach parabola shape (see Fig. 6 3), which is ideal for the compression behaviour of the concrete shell. Unfortunately, the ribs generated by this pattern are not efficient for the concrete shell stability. The inflated sheet numerical simulation form is as satisfying as the fabricated prototype but is very unstable. During inflation, a rotation appears to shape the parabola. This rotation repeats itself when we apply a load on the upper part of the inflated dome (see Fig. 8 1). In contrast, the prototype made with the conic pattern (see Fig. 8 2) shows no rotation during inflation. The radial tubes make the inflatable structure very stiff.

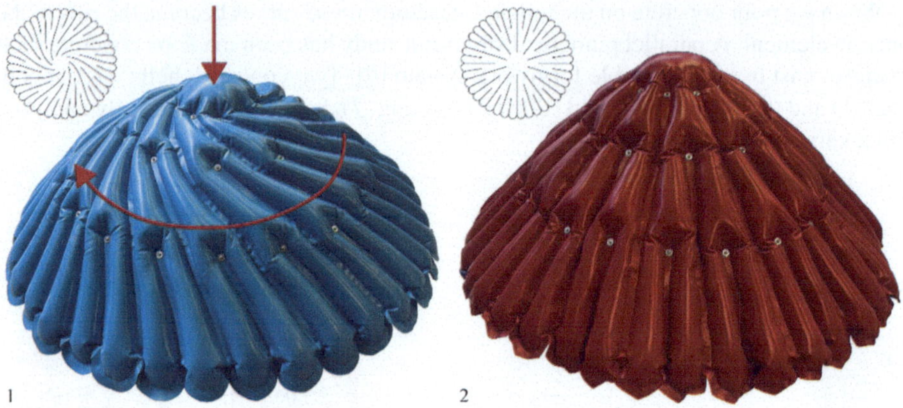

Fig. 8. Prototypes with spirals pattern (1) and cone pattern (2).

We propose to add a second mirror array of spirals to the first (see Fig. 9 1c), inspired by the ribs of the Palazzetto Dello Sport in Rome by Pier Luigi Nervi, built in 1957 [11]. From this array, we propose three ways to design the pattern (see Fig. 9 2):

2a. To keep the curves of the array to achieve effective cross-concrete ribs. In this configuration, the inflatable structure is quite flexible since there are no air continuities from one limit to another of the pattern and the seals behave like soft hinges.

2b. To draw rhombus by keeping inflated spiral tubes. The spiral tubes bring stiffness to the inflated dome and favour the inflatable deployment into parabola shape, but the discontinuous ribs are not efficient.

2c. This third model is an arrangement between air continuities ensuring the stability of the inflatable and continuous joints generating efficient ribs for the concrete shell. Radial air continuities reinforce the inflatable formwork as observed on the inflatable cone (see Fig. 8 2), which is very stiff. The peeling force at seals is higher when the cavity is more extensive, and there is stress concentration at the ends and open angles of seals. The layout of this pattern is very convenient since the opened angles turn toward the small cavities.

It is necessary to tweak these patterns through different parameters: the minimal distance between 2 seals to ensure a correct air supply, the number of units (set of repeating lines) in the polar array and the number and hierarchy of the intermediate curves to have a good definition and discretisation of the shape. These parameters settings depend on the size of the manufactured inflatable, as was done with pattern 2c (see Fig. 9). Indeed, the difference in the size of the cavities of the pattern c is reduced to uniformise the tension in the membrane and thus correctly inflate each cavity at a certain pressure, avoiding deflated cavities or seal peeling off.

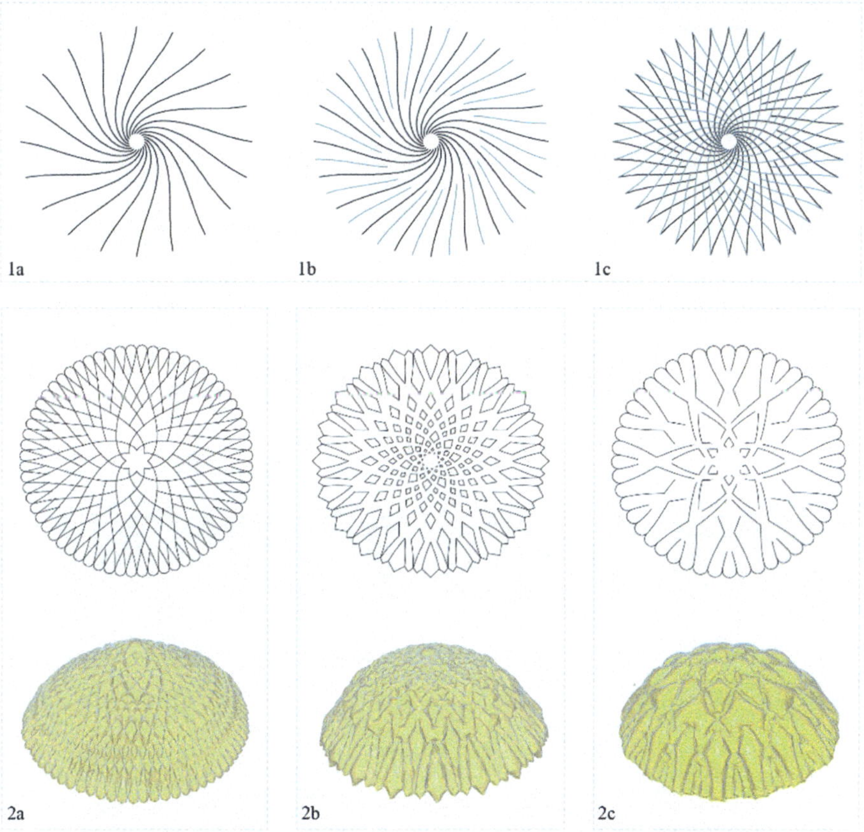

Fig. 9. From an array (1), 3 patterns are drawn (2a, 2b, 2c).

The relation between the peeling force per unit length ρ (N/m) along the seam, the pressure p (N/m^2), the area A (m^2) and the perimeter l (m) of the cavity writes as follows:

$$\rho = \frac{pA}{l} \tag{1}$$

The number of units in the polar array stays unchanged, but the unit is modified by adding a third degree of discretisation. These new curves divide the perimeter of the inflatable twice as much, from 40 parts in the first version to 80 in the second, which refines the shape.

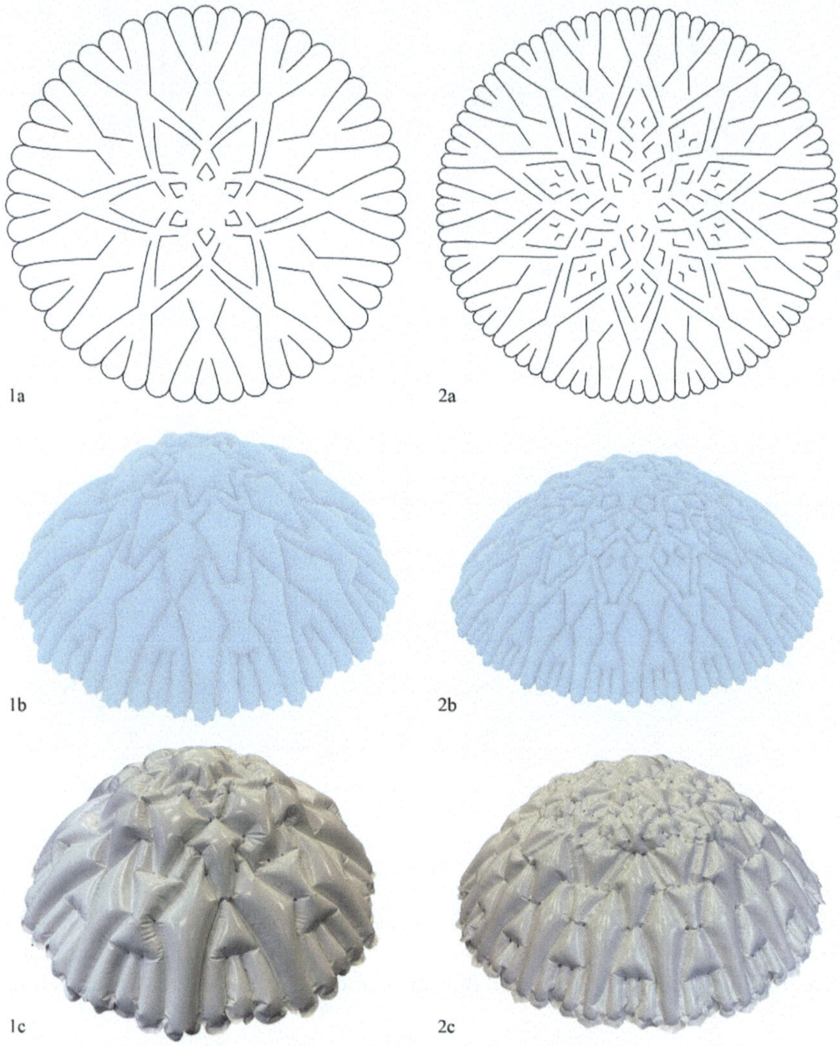

Fig. 10. Modifications of the pattern 2c (Fig. 9).

4 Fabrication of the Inflatable Formwork and Concrete Shell

The used membranes are inextensible to control the deformation and construct stiff inflatable structures in order the formwork carries the concrete. The tension in the membrane has to be sufficient to resist the applied load, thanks to the air pressure between the two membranes.

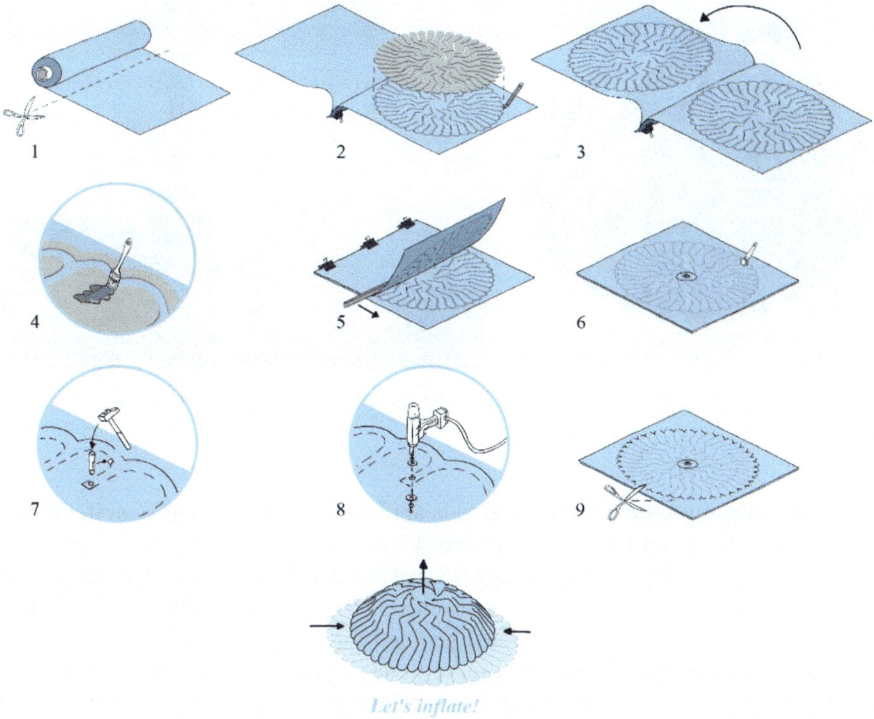

Fig. 11. Protocol of an inflatable fabrication, from step 1 to step 9.

Since the joints are subjected to peeling stresses, their strength is much lower than traditional membrane assembly exposed to shear stresses, so they must be reinforced for large scale models. We use a composite membrane composed of a prestressed polyester fibres mesh; with two sides coated with PVC. The PVC coat is thicker than the tiny model's fabric, allowing us to create more robust assemblies.

These membranes are generally assembled by high-frequency welding, seaming or gluing. The high-frequency welding works with electrodes, and the standard ones are straight with little diversity. The patterns are curved; it would be necessary to discretize the curves, as made on prototype Fig. 8.1, and in certain cases fabricate specific and expensive electrodes. Hence, we are using glue which is cheaper and allows the creation of assemblies of whatever shape. We measured the ultimate tensile strengths of these two assemblies through tensile testing with several test specimens: membrane strips assembled by high-frequency welding or gluing. For each test, the membrane failed and not the assemblies, showing that the assemblies are stronger than the membrane. The glue seems to be the best option for its ease of application and resistance capacities.

After several manufacturing trials, we established a protocol to ensure the accuracy and robustness of the creation of the inflatable and to gain in rapidity (see Fig. 11). This method has been used to produce prototypes with diameters of 1.35 m and 2.70 m (deflated).

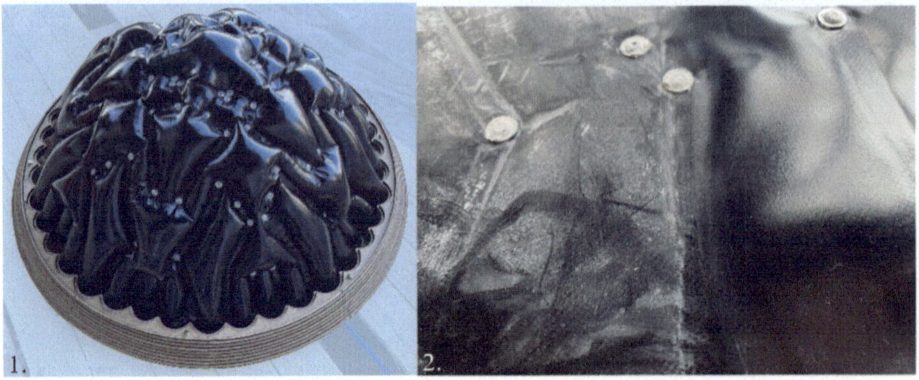

Fig. 12. Inflatable Formwork with support (1). Reusable inflatable Formwork after concrete shell fabrication (2).

The pattern is reproduced and the glue is applied using a laser-cut stencil. Once the glue has dried, the glued pattern is reinforced with rivets at the ends and corners. The rivets are positioned with washers on both sides of the membrane. They support the tensile strength locally by applying pressure. A dozen 1.35 m prototypes with different patterns have been made to observe the generated geometry using the full-scale membrane and this fabrication technique. These tests serve to determine the width of the glued seals and the minimal space between them. Also, the air distribution has been checked and adapted. Three inflatables of 2.70 m diameters (deflated) are made with this technique: one with the spiral pattern (see Fig. 6 3) and two with the pattern 2c (see Fig. 10). The last inflatable is the most accomplished, stiff and with a well-defined shape.

Some concrete shells have been made with a formwork of 1,35 m diameter deflated. Wooden support is a traction ring at the basement of the inflatable structure (see Fig. 12 1), designed from the numerical simulation to fit the edges of the inflatable. The air pressure inside the glued membranes is up to 40 kN/m^2 to construct the shell. The support is attached to the inflatable with adhesive tape, a reversible fastener to avoid damage to the membranes. A silicone seal ensures continuity between the inflatable and the support. Silicone covers the rivets to prevent them from being sealed in the concrete.

The concrete mix is composed of 45% sand, 45% cement and 10% of water. Some glass fibres (1% of the total volume) are added to prevent eventual cracking during shrinkage when the concrete hardens. First, a thin slip (a fluid mix of cement and water only) is poured on the inflatable to fill the valleys. After this first layer dries, a thicker layer of 1,5 cm of concrete mix is added. The concrete sets in about 2 h, and then the inflatable formwork is deflated to remove the membranes. The inflatable formwork is intact and perfectly reusable (see Fig. 12 2) (Fig. 13).

Fig. 13. A concrete shell made with an inflatable formwork.

5 Conclusion and Outlooks

This new concrete casting technique attempts to answer the costs and environmental problems of the formworks in the AEC industry. Pneumatic solutions exist and are used on construction sites such as Resair®, for instance, to make reservations in concrete structures using inflatable elements.

Unlike other pneumatic formwork solutions, our proposition can achieve a wider variety of shapes and can be automated. The next step of this research is to cast a wider shell with the bigger inflatable formworks (see Fig. 10) by using concrete spraying.

Other shapes, such as a double negative curvature surface, are the next goal for fabrication and casting. For the moment, the limiting size of the inflatable formwork is determined by the width of the membrane, but we can make a seal between two widths to create larger inflatables. We can use standard scaffolding elements to help the inflated skin to cross and reinforce it locally with future strategies to create an extensive inflatable formwork. It will undoubtedly be necessary to reinforce the concrete with steel for such a wide shell.

Acknowledgement. We thank *Serge Ferrari* company for their support, the supply of composite membrane for the realization of prototypes. We thank the students of the Ecole Nationale Supérieure d'Architecture de la Ville et des Territoires from the Master Matière à Penser: Doriane Antoine, Guillaume Brezak, Pierre Kraft, Nathan Laglace, Emelyne Matilde and Benjamin Ortega for their great work on this subject the last semester. We also thank the supervisor team of the department: Marc Mimram, Jean-Aimé Shu, Margaux Gillet and Julien Glath.

References

1. Tang, G.: An overview of historical and contemporary concrete shells, their construction and factors in their general disappearance. Int. J. Space Struct. **30**, 1–12 (2015)
2. Lab, R.: Think formwork - reduce costs. Struct. Mag., 14–16 (2007)
3. De Soto, B.G., Agust-juan, I., Hunhevicz, J., Habert, G., Adey, B.: The potential of digital fabrication to improve productivity in construction: cost and time analysis of a robotically fabricated concrete wall. Autom. Constr. **2018** (2018)

4. Pugnale, A., Bologna, A.: Dante Bini's form-resistant Binishells. Nexus Netw. J. **19**(3), 681–699 (2016). https://doi.org/10.1007/s00004-016-0323-7
5. Beckh, M., Boller, G.: Building with air: Heinz Isler's bubble houses and the use of pneumatic construction techniques. In: Water, Doors, and Buildings: Studies in the History of Construction. The Proceedings of the Sixth Conference of the Construction History Society, pp. 494–506 (2019)
6. Kromoser, B., Huber, P.: Pneumatic formwork systems in structural engineering. Adv. Mater. Sci. Eng. **2016**, 13 pages (2016). Article ID 4724036
7. Ou, J., et al.: aeroMorph - heat-sealing inflatable shape-change materials for interaction design. In: UIST 2016, Tokyo (2016)
8. Skouras, M., et al.: Designing inflatable structures. ACM Trans. Graph. **33**(4), 1–10 (2014)
9. Clédat, A., Lefeu, C.: Calcul de coques nervurées aux éléments finis: influence des nervures et sensibilité du maillage, COAST 21, ENPC (2021)
10. Siefert, E., Reyssat, E., Bico, J., Roman, B.: Programming stiff inflatable shells from planar patterned fabrics. Soft Matter **16**, 7898–7903 (2020)
11. Iori, T., Poretti, S.: Pier Luigi Nervi: his construction system for shell and spatial structures, vol. 54. International Association for Shell and Spatial Structures (IASS) (2013)

Upcycling Shell: From Scrap to Structure

Timo Carl[1(✉)], Sandro Siefert[1], and Andrea Rossi[2]

[1] Frankfurt University of Applies Sciences, Frankfurt, Germany
timo_carl@fb1.fra-uas.de
[2] University of Kassel, Kassel, Germany
rossi@asl.uni-kassel.de

Abstract. The reuse of building materials has a long tradition in construction. More recently, the concept of digital materials envisions universal building blocks, that can be reused and reconfigured multiple times, like Lego© bricks, and placed freely in any structurally sound assembly. We continue our investigations along this trajectory, by exploring mass-produced everyday objects, mined from the waste of our consumer society, to produce those reusable and universal building blocks. We investigated the potential of such an approach through a computational design & build case study realized in an educational context: a *summer igloo* comprised of an aggregated shell structure with 10.000 reused wire hangers. We present design methods, materials selection processes, as well as fabrication and assembly strategies. Lastly, we discuss the impact of a creative and multiple reuse of standardized construction components for the construction industry.

Keywords: Circular design · Discrete assemblies · Digital material · Form-finding · Structural simulation · Parametric modelling

1 Introduction and Related Work

It is common knowledge that we need to reduce the energy footprint of our built environment. Evaluating building materials not solely based on their embodied primary energy, but also for their potential to be reused in circular design seems appropriate – given the enormous resource consumption of the building sector, amounting in Germany alone to 700 million metric tons per year (Mettke 2022) (Fig. 1).

Circular design strategies tap into the potential for multiple use and reuse of existing components. Projects range from research driven investigations that focus on large-scale prefabricated concrete slab elements, harvested from demolished residential housing blocks (Mettke 2022), to educational design & build work of the Rural Studio at Auburn University, reusing unconventional materials to construct community buildings, e.g. Yancy Tire Chapel (Durden and Tretheway 1995). In both cases, waste materials (i.e. old tires) are reused to substitute specific parts (i.e. walls) of new buildings.

The *Mine the scrap project* (Nolte and Witt 2016) takes a step further by considering for reuse waste material geometries with similar topologies, but with different dimensions and orientations. Discrete 2-dimensional elements, like wooden demolition-boards, are photogrammetrically scanned and stored in a database, that in turn informs

C. Gengnagel et al. (Eds.): DMS 2022, *Towards Radical Regeneration*, pp. 211–222, 2023.
https://doi.org/10.1007/978-3-031-13249-0_18

Fig. 1. Summer Igloo constructed from approx. 10.000 reused wire hangers

geometric design possibilities. The inherent geometric parameters of the existing building parts constrain the emergent formation of new structures. While the workflow shows potential, such an inventory-constrained design strategy allows for limited customization - the outcome being highly dependent on the geometry of the input material.

Gilles Retsin on the other hand, focuses on the combination of a stock of identical building blocks for architectural structures, hence enabling disassembly and reuse. Similar to a set of LEGO© elements, or the traditional brick, a variety of objects can be constructed - limited only by the scale of the base elements and their respective resolution. In this context, the prototypical architecture of the *Diamond House* hints at a reversible construction method allowing the multiple reuse of its material - like digital data that "[…] can be recombined, are universal and versatile" (Retsin 2018). Retsin's work builds on research from the *Center for Bits and Atoms* at MIT, led by Neil Gershenfeld, who describes a "digital material", in contrast to "analogue materials", as not continuous and homogeneous, but consisting of discrete repetitive elements, joined together with reversible connections (Ward 2010; Gershenfeld 2012). This highlights a radical difference from a Modernist understanding of modularity and discreteness, where each building element still had a unique functional role in the overall assembly (column, wall, floor). In a computational translation, discrete elements do not have any fixed functional connotation, but are freely placed in an assembly, losing the meaning inherent in architectural types, becoming generic physical bytes in a digital assembly (Tessmann and Rossi 2019). Through this, digital materials link the physical and digital realm through a programmability of both design and assembly processes (Rossi and Tessmann 2017).

In the following paper, we speculate about the material DNA of bespoke digital materials, by proposing building blocks that exploit mass-produced waste materials for their production. We aim for substituting traditional building resources with serially produced industrial waste objects of our consumer society, which are no longer needed elsewhere.

Our goal is to extend previous investigations of materials beyond the end of their first life cycle. Most studies focus either on the reuse of second hand building parts for only one additional cycle, or on the production of new building materials, which can be easily disassembled and reused in multiple cycles. By eliminating the primary energy demand for the production of additional building materials in the first place, we reduce the energy footprint further, only requiring energy for processing and transport. We will discuss the scalability of our approach and the usefulness of the applied computational techniques at the end of this paper.

1.1 Contribution

We describe a series of computational methods that facilitate the creative and efficient reuse of standardized components with unique geometric and topologic characteristics. Moreover, we present a proof-of-concept design & build case study with the proposed workflow, demonstrating the sourcing of building materials from industrial consumer waste. The unique aesthetic qualities of our *summer igloo* sparks further discussion and might inspire future research. The presented demonstrator is intended as a playful provocation that links construction material flows with our private and industrial consumption.

2 Methods

The design and materialization of our demonstrator involves a multi-step computational process that is deeply interconnected with aspects of waste material selection and assembly strategies:

2.1 Aggregation of Base Module(s) and Geometric Constraints

- Selection of suitable readymade waste materials
- Structural evaluation and material optimization
- Connection between digital simulation, serial fabrication and assembly

2.2 Aggregation Logics and Geometric Constraints

To test both the limits of digital materials (comprised of industrial readymades) and to achieve the widest possible implementation, we decided to translate the continuity of smooth and double curved shell geometry into a tessellated load-bearing construction (Fig. 2). This test case is structurally elegant - in the spirit of saving material - and geometrically complex enough to probe the threshold of different aggregation strategies.

The computational workflow is based on a parametric design process that begins with a globally form-found shape. In hindsight, pursuing a 6 × 6 m compression-optimized shell was highly useful in an educational design & build context. The output of basic form-finding strategies transfers seamlessly to both aggregation and structural simulations.

Fig. 2. Form-found shell (l), aggregation with two base modules (m) and instantiated detail geometry (r)

The Grasshopper plug-in Wasp (Rossi 2022) provides the framework for generating discrete aggregations for a large number of repetitive units that were combined with structural optimization using Karamba3D (Preisinger and Heimrath 2014).

Wasp provides an ideal toolset to describe both geometry and topology for each unit, as well as defining connectivity rules. The main challenge lies in designing a seamless interlocking system that - despite being composed of a discrete number of units - can still achieve an articulated morphological variation and a sufficient level of design flexibility. One key finding is, that platonic solids like cubes and tetrahedrons can be assembled in a mutually interlock way without any gaps. We used a thickened shell geometry that acts as container for seamlessly aggregating two interlocking solids: tetrahedron and octahedron. An additional constraint states that only units with their center point inside this container are considered for aggregation (Fig. 3) Other geometries, like L-shaped elements (Fig. 4) have been considered, but led to branching systems that are unfit for generating a continuous and seamless shapes with adequate structural performance.

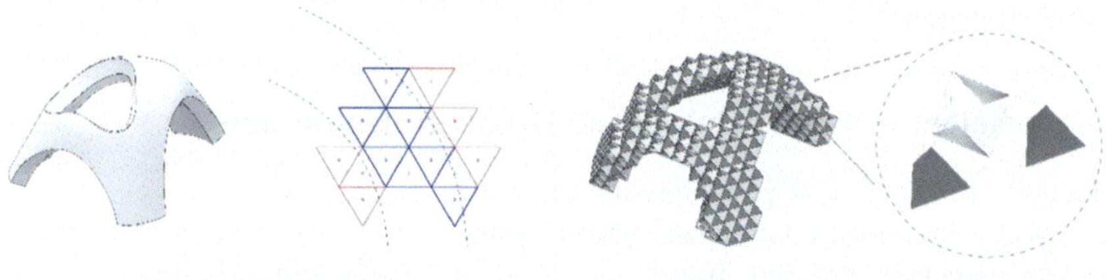

Fig. 3. Form-found shell (l), aggregation with two base modules (m) and instantiated detail geometry (r)

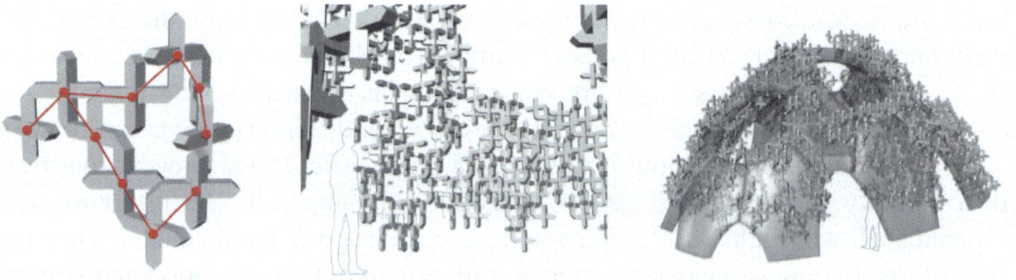

Fig. 4. Testing potential readymades and their geometric properties for various topologies

2.3 Finding Suitable Readymades

The selection process of a weatherproof, structurally and geometrically suitable mass produced object is at the center of any assembly relying on the proposed workflow. Moreover, creating reusable components that are fit for circular design requires the implementation of reversible connection details. Otherwise, the reused materials will be limited to a single lifecycle. Another constraint is the geometry of the parts constituting the building blocks - as not every waste material is fit to generate platonic solid geometries. Lastly, the size of any reused object is a fundamental constraint, as it determines the resolution of any newly designed structure.

We narrowed the initial (re)search for readymades to everyday consumer products with a high level of availability. Overall, five object typologies (milk cartons, cardboard rolls, large nail files, thermoformed plastic packaging, and wire hangers) were found and rated according to the following criteria:

- load-bearing performance
- weather resistance
- geometric suitability for the construction of interlocking building blocks
- ease of fabrication and assembly.

Fig. 5. Testing potential Readymade connectors and reversible 3D printed connection details

Simultaneously, we investigated connections details for the multiple reuse of all materials - either in form of adapted waste materials (i.e. bicycle tires) or in form of secondary custom connectors (i.e. 3D-printed or CNC machined parts).

We connected the wire hanger elements along the side edges (Fig. 5). In this way, only two elements had to be considered for the design of the connection detail. In contrast, up to six building blocks can meet at the corner nodes. Introducing reversible connection elements is in hindsight an essential parameter for disassembly and multiple reuse. However, this was not considered at the time of construction, because of inaccuracies that resulted from the manual assembly of wire hanger modules. We used tying wires instead - an inexpensive and forgiving joint detail, which helped compensating for construction tolerances. We used the tying wires throughout all component levels: To build the individual blocks, to connect larger segments for pre-assembly and lastly to merge the latter into the final shape. The tie wires are an adequate expression for the pavilion's architectural ambition, but too labor intensive for disassembly.

Nevertheless, the reuse of obsolete wire hangers proofed very successful for the construction of building blocks. A ubiquitous material, often unwillingly provided when taking your clothes to the laundry, the wire hangers are thrown away eventually, often immediately after purchase. Students sourced the initial set of material form local laundry stores, which normally reuse those the B-stock hangers internally before they purge them out. Given the short time span of the project, approximately half of the hangers were bought in bulk. Further research should look into alternative models of sourcing, also involving local communities (Fig. 6).

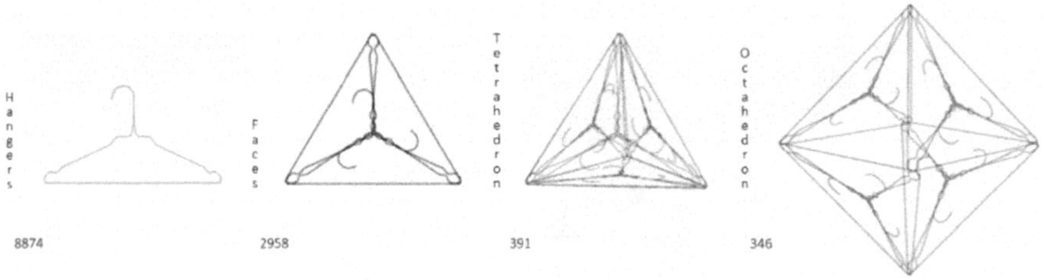

Fig. 6. Building blocks constructed from tri-equilateral faces

A two-sided isosceles wire hanger functions as the lowest common denominator and permits the generation of a recursive modularity strategy. Three-sided isosceles triangle faces constitute Tetrahedron and Octahedron solids (Fig. 3). The material itself is resilient enough to build light and temporary structural elements.

Our *summer igloo* consists of roughly 10.000 individual wire hangers, preassembled into 391 Tetrahedron and 346 Octahedron solids units that aggregate seamlessly (Fig. 7) along a predefined shell geometry - the overall weight is roughly 294 kg.

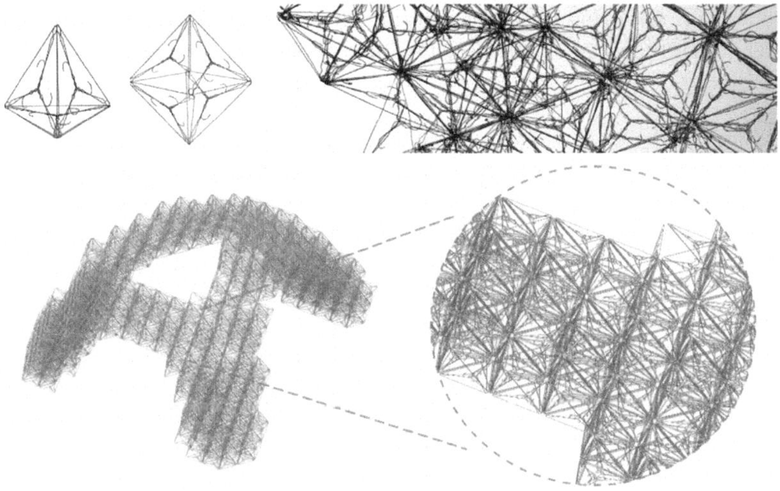

Fig. 7. Linear and curved shape aggregation tests with two base units

2.4 Structural Assumptions

The aggregated solids are the basis for the structural simulation set-up. Instead of considering the units as a rigid and non-deformable solid elements, only the edges of the building blocks are considered, approximating the its topology.

Inevitably, the aggregation of solid units generates double faces. While these additional strands of double edges add stability, they also result in extra weight. Therefore, we decided to reduce the amount of wire hangers by deleting the four transverse double faces of the Octahedrons (Fig. 8), hence saving 2076 wire hangers.

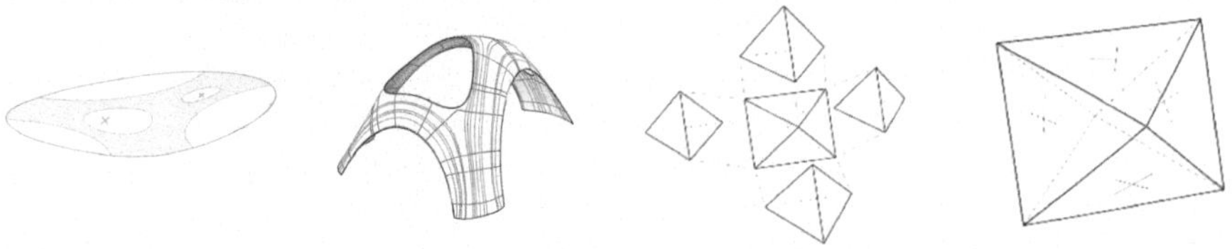

Fig. 8. Form-finding and optimization of shell geometry (l) and eliminating congruent double faces (r)

Thereafter, the edges of the units are converted into a network of lines that function as beams. The resulting wireframes are split at the location of the tie wire connections, creating nodes. These are interconnected with lines, which act as springs for the simulation (Fig. 9).

The structural nodes are not assumed fully rigid, but stiff enough to prevent the rotation of the beam elements. Therefore, the connecting springs are clamped on the one end and hinged with minimal stiffness at the other end. The normalizing of the

Z-orientations for all spring lines was determined with help of the associative geometry model - by vector matching of the projected line geometry.

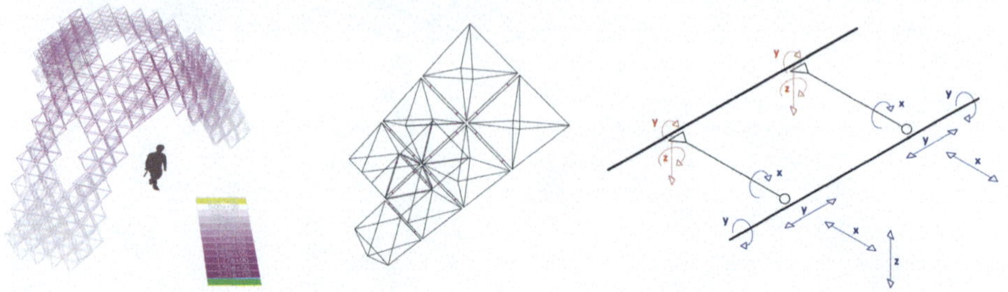

Fig. 9. Detailed structural model, based on the aggregated building block edges (l) and parametric FEM analysis setup (r)

The structural simulation provided qualitative results for deflection under self-weight and quantitative load values for the three supports. For additional safety, we added one impact load of 70 kg at the lower base of the shell. We used aluminum as material input and disregarded wind, due to the open nature of the wire structure. Nevertheless, the simulation results are accurate enough for geometric optimization of the shell shape.

The crystalline matrix of the modules distributed loads equally throughout the structure, avoiding any deformation of the lower units, which carry all the weight of the above elements (Fig. 10).

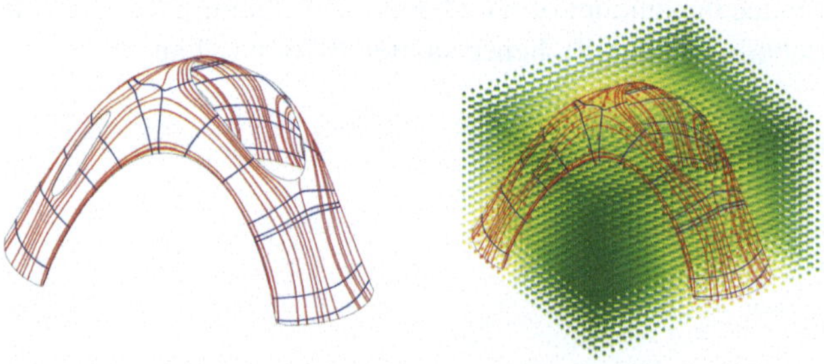

Fig. 10. Structural optimization of shell geometry in relation to stress lines

Lastly, we tested scalar fields to inform further structural optimization, based on the flow of forces. The assumption was to use stress lines to control the material density of the field and lead to a local density differentiation in the aggregation. However, the resolution of the field was too small, compared to the size of the units, to yield any meaningful results. This means that flow of forces found in compression-active vernacular arches, vaults, shells are only partially relevant for aggregated structures.

2.5 Fabrication and (Dis)-Assembly

Thanks to the serial characteristics of the repetitive units, a team of 12 students produced all building blocks - with the help of a jig - within seven days (Fig. 11). This hints at a potential relevance of digital materials for the informal building sector, which relies mainly on manual labor. Moreover, the reuse of industrial waste supports cost effective structures - all expenses amounted to less than 1000 Euros (Fig. 12).

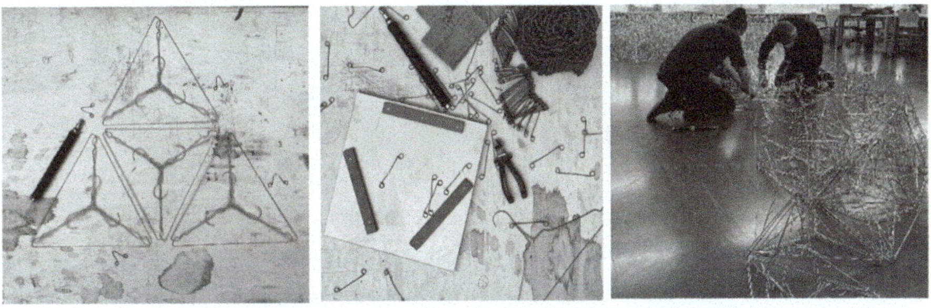

Fig. 11. Assembly process and first connection tests

Fig. 12. Assembly and finished demonstrator displaying the aesthetic qualities of the approach

We divided the structure into medium sized horizontal segments for assembly.

Two people could easily carry those and tagging and logistics were managed with the help of a digital model. This underscores the potential of identical building blocks for reconfiguration, creating constantly new structural through assembly and disassembly (Jenett and Cheung 2017).

In theory, compression-only structures generate horizontal thrust in their supports and performs only after the last element (i.e. keystone) is placed. This is often a geometric complex endeavor. Due to a slightly sloped terrain, the *summer igloo* required a scaffold for assembly and was built form the top down. However, the units of aggregate structures could be also unequivocally positioned, in relation to each other, and assembled with the help of HD cameras or IR distance sensors, thanks to the repetitive geometry of the units as communication interface between designer and fabrication equipment (Rossi 2022).

Disassembly follows the same sequence in reverse order. However, using tie wires for the connectors proofed disadvantageous, as the disconnection process required too much time.

Another finding suggests the potential for differentiation of the connectors itself - by either color or material. While a single connection detail underlines the architectural quality of the wire hanger structure, it makes it much harder to distinguish the various connector hierarchies (i.e. building block vs. segments shown in Fig. 13).

Fig. 13. Subdivision of the structure to generate preassembled segments with limited weight

3 Results and Discussion

The project is clearly a speculation in an educational context, negating necessary architectural requirements, such as climate regulation or durability. Nevertheless, it demonstrates that mass-produced everyday objects can be reused as construction materials. Moreover, it brings forward a series of considerations that could apply to the wider architectural context and construction sector.

While concrete is still the most common used material in construction (resulting in irregular, hard to reuse waste elements), the majority of waste of the build environment actually consists of standardized, industrially produced elements (e.g. the traditional brick stone, from which the load bearing walls of most postwar European cities were rebuild). This also holds true for various elements with shorter lifespans than structural components, such as interior fittings, windows and doors, façade panels, as well as most structural and non-structural timber elements. In this sense, developing methods for the creative and efficient reuse of standardized construction components with their unique geometric and topological characteristics could have a positive impact. Increasing the probability for reintroducing them into circular usage cycles could improve the energy footprint of the construction sector. Looking into possible material DNA combinations that constitute universal digital materials on different scales seems therefore worth pursuing also for more permanent structures.

Common to both form-finding methods, which forefront geometric constraints and specific material properties (Carl and Schein 2019) and the emerging field of digital materials in architecture, which assume the aggregation of many discrete units, is the reliance on an adequate computational framework. Building a shell structure that consists of 10.000 wire hangers with a team of students in one semester is proof of concept for the robustness of the computational methods.

The outlined workflow also hints towards an alternative path for an otherwise fully automated high-tech construction industry. While such vision might be the ultimate goal, the process to achieve it will proceed through intermediate steps where manual and automated processes will need to mix.

The repetitive nature of the units enables both a high precision assembly (i.e. with industrial robots) and low precision ones (i.e. students or construction workers – with all the discussed drawbacks). This is one of the most relevant features of an approach based on digital materials (Jenett and Cheung 2017) - enabling the implementation in contexts with a variety of levels of technological adoption. Working with standardized digital materials seems key to provide a platform for integration and collaboration at larger scales. Further research will have to address the already discussed connection and assembly issues.

Lastly, we hint at the fuzzy nature of the demonstrator and it´s novel aesthetic quality that entails both a constructive and an artistic narrative within a unified architectural process. Looking for reusable materials that facilitate this synergy of aesthetic and ecological parameters seems of real importance to promote the latter. We agree with Sambo Mockbee from Rural Studio, that it is the right approach "[…] to be over the edge, environmentally, aesthetically, and technically".

Acknowledgements. Agnes Weilandt contributed generously to the development of the presented structural analysis concept. A group of highly dedicated students at the Frankfurt University of Applied Sciences built the *summer igloo* during a 5th year BA Digital Studio in one semester.

References

Retsin, G.: Discrete assembly and digital materials in architecture. In: Herneoja, A., Östlund, T., Markannen, P. (eds.) eCAADe 34 – Complexity & Simplicity, vol. 1, pp. 143–151. eCAADe, Brussels (2016)

Retsin, G.: MAJA, digital material. In: MAJA, Material and Novelty, pp. 61–67 (2018)

Preisinger, C., Heimrath, M.: Karamba – a toolkit for parametric structural design. Struct. Eng. Int. **24**(2), 217–221 (2014)

Jenett, B., Cheung, K.: BILL-E: robotic platform for locomotion and manipulation of lightweight space structures. In AIAA 2017, p. 1876 (2017). https://doi.org/10.2514/6.2017-1876. Accessed 23 Mar 2022

Mettke, A.: Aufbau eines Netzwerkes zur Wiederverwendung von gebrauchten Bauteilen in Deutschland, DBU-Projekt (2022). https://www.dbu.de/projekt_23023/_db_1036.html. Accessed 23 Mar 2022

LEGO® is a trademark of the LEGO Group of companies that does not sponsor authorize, or endorse this material

Gershenfeld, N.: How to make almost anything: the digital fabrication revolution. Foreign Aff. **91**, 43 (2012)

Ward, J.: Additive Assembly of digital materials. Ph.D. dissertation, Massachusetts Institute of Technology, Cambridge (2010)

Carl, T., Schein, M.: A parametric framework for solar tensile structures – bridging the gap between conception and construction. In: Thomsen, M.R., Tamke, M., Gengnagel, C., Faircloth, B., Scheurer, F. (eds.) Impact: Design with all senses, pp. 82–92. Springer, Cham (2019). https://doi.org/10.1007/978-3-030-29829-6_7

Tessmann, O., Rossi, A.: Geometry as interface: parametric and combinatorial topological interlocking assemblies. J. Appl. Mech. **86**(11), 111002 (2019)

Rossi, A., Tessmann, O.: Geometry as assembly-integrating design and fabrication with discrete modular units. In: Fioravanti, A., et al. (eds.) ShoCK! - Sharing Computational Knowledge! - Proceedings of the 35th eCAADe Conference, vol. 2, pp. 201–210 (2017)

Rossi, A.: Wasp—Discrete Design for Grasshopper (2022). https://github.com/ar0551/Wasp

Modelling and Simulation of Acoustic Metamaterials for Architectural Application

Philipp Cop, John Nguyen, and Brady Peters(✉)

University of Toronto, Toronto, ON M5S 2J5, Canada
{philipp.cop,johnnie.nguyen}@mail.utoronto.ca,
brady.peters@daniels.utoronto.ca

Abstract. Acoustic metamaterials are novel engineered materials with geometric features of subwavelength size that create highly exotic acoustic behaviors such as negative refractive indexes, perfect sound absorption and sound waveguiding. While these new materials hold much promise to be useful for the architectural engineering and construction sector, there has been little research done on the application of acoustic metamaterials for architectural application. The research presented in this paper investigates acoustic metamaterials for architectural acoustics and demonstrates how architects can leverage parametric design, digital fabrication, and computer performance simulation to develop new metamaterial designs tuned for customized acoustic performance. This paper proposes a new definition of 'metamaterials for architectural acoustic application' and provides a brief overview of the history and theory of acoustic metamaterials alongside a discussion of the relevance of such materials for implementation in architectural acoustic applications. A design and simulation workflow is presented that demonstrates the parametric design and iteration of metamaterial geometry, performance evaluation through computer simulation, and digital fabrication of functional prototypes. A set of well-performing metamaterial geometries are presented that can be used to design architectural acoustic surfaces.

Keywords: Acoustic metamaterials · Acoustic surfaces · Performance simulation · Computational design

1 Introduction

Since the early 2000s, metamaterials gained interest from academic researchers and led to the discovery of many exotic properties and the proposal of many new electromagnetic and acoustic metamaterials (Liu et al. 2000; Pendry 2000; Shelby et al. 2001). The diverse range of metamaterial properties, material composites, and geometries has made it difficult to formulate one distinct definition for acoustic metamaterials (Shamonina and Solymar 2007). Discussion around the term "metamaterials" itself was part of this turn of the century emerging field (Welser 1999). Especially in comparison to other composite materials, metamaterials were distinguished to go beyond any composite

C. Gengnagel et al. (Eds.): DMS 2022, *Towards Radical Regeneration*, pp. 223–236, 2023.
https://doi.org/10.1007/978-3-031-13249-0_19

properties that solely rely on the addition of their constituent properties. Walser (2003) points out that metamaterial geometry "is aimed at achieving performance beyond that of conventional macroscopic composites", hence the prefix "meta". These exotic new properties are possible by carefully designing the composite unit cell and achieving homogenized overall properties based on individual unit cell effects. More recently Cummer et al. (2016) define metamaterials as "a material with 'on-demand' effective properties" where the "internal structure is used to induce effective properties in the artificial material that are substantially different from those found in its components".

2 Background

2.1 Metamaterials History

In acoustic metamaterials, exotic behaviors such as acoustic cloaking (Popa et al. 2011) or perfect sound absorption (Jiménez et al. 2016) and the creation of complex sound fields from simple tones (Xie et al. 2016) become possible. Manmade geometry has been used to modulate the sound in rooms for a long time. Vitruvius claimed that the Greeks used resonating vases, to control sound and improve the room acoustics (Vitruvius 1960). Today, Helmholtz resonators and perforated panels are used to tune the acoustic properties in large rooms to absorb sound at specific frequencies. Acoustic scattering geometry is used in theaters and music venues to improve the quality of reflected sound. However, these structures must be in the same spatial order as the operating wavelength to function. Audible low frequency wavelengths (20–800 Hz) are in the range of 0.4–17 m, therefore rendering application of traditional geometries unpractical. Porous materials such as mineral wool or open-cell foam, are commonly used in architectural acoustic absorber products, but these do not perform well at low sound frequencies (Maekawa et al. 2011). Low frequency sounds are not well incorporated in current acoustic design strategies although they penetrate walls readily and contribute to room modes due to their large wavelengths. One of the potential benefits of acoustic metamaterials is to break the effective size limits as the individual unit cell can be tuned to function for lower frequencies.

Metamaterial properties were first proposed for the electromagnetic wave spectrum. Veselago (1968) speculated that artificially modifying material properties into the negative would result in negative material parameters and unusual wave characteristics. Electromagnetic emission was observed to not propagate in periodically changing dielectric constant media (Yablonovitch 1987). In 1993 these electromagnetic waves stopping effects were translated to pressure acoustic wave propagation, and when this was done, the "acoustic bandgap" was observed (Kushwaha et al. 1993). The "acoustic bandgap" or sonic or phononic band gap is a frequency range where no sound propagation is possible, this is analogous to the electromagnetic band gap. The sonic bandgap occurs due to destructive interference patterns caused by the periodic scattering of sound waves. This phenomenon is called "Bragg scattering" and is named after W. H. Bragg and W. L. Bragg who in 1934 observed destructive and constructive X-ray wave patterns when the wavelength was in the same order as the atom structure in a crystal (Bragg and Bragg 1934).

The periodically changing acoustic impedance of two media produces strong scattering and interference when the distance between the scatterers is a multiple of the operating wavelength (Hussein et al. 2014). In 2000, this acoustic "stop band" was produced by periodic local resonance (Liu et al. 2000). The locally resonant unit cell produces scattering similar to the impedance difference induced scattering. However, instead of scattering at the interface of a media change, the wave scattering was produced by a mass spring system, or resonance. The resonance mechanism functions for far lower frequencies and local scattering is therefore no longer linked to a spatial period. This meant that geometry could remain effective for much lower frequencies at a much smaller scale. Since the application of local resonance in acoustic metamaterials, a diverse range of sonic metamaterials has emerged following suite to their electromagnetic counterparts.

2.2 Defining Architectural Acoustic Metamaterials

Although metamaterial properties are mentioned to be very applicable for sound control in the build environment, not much research has emerged on the specific integration and application. A literature review discussed various metamaterial geometries for their application to the architectural and urban context (Setaki et al. 2014) and (Kumar and Lee 2019). A few projects using acoustic metamaterials have been proposed: a soundproof, air and light transparent window was constructed (Kim and Lee 2013), an air flow permitting metamaterial is used to suppress fan noises (Ghaffarivardavagh et al. 2019), a omnidirectional sound mitigating and air transparent structures is investigated (Shen et al. 2018), and resonant U-shaped scatterers are proposed as traffic noise barriers (Romero-García et al. 2011). A few metamaterials have become commercially available for sound mitigation in the built environment (MetaAcoustic and Acoustic Metamaterial Group - AMG).

In this paper we define acoustic metamaterials as engineered structures that are designed to interact with sound waves in a specific and desired way. They exhibit acoustic properties that are not attainable by existing materials found in nature. The structure is usually composed of many elements smaller than the wavelength that they operate on. Due to their micro, subwavelength unit cell structure, they can obtain new homogenized material properties for the macro, overall structure and appear as a cohesive continuous material to the propagating pressure wave. Since the performance is inherent to the geometry of the structure, the acoustic absorption, diffusion, and transmission characteristics can be modified as needed, making it enticing for designers to explore geometries that fit their performance needs.

2.3 Selected Metamaterials

Based on the outlined benefits, the following will select acoustic metamaterials best suited for architectural deploy. Metamaterials will be differentiated by whether they produce acoustic effects solely due to their geometry or if they rely on a specific material configuration to achieve overall acoustic effects. Different media engineered metamaterials such as rubber coated lead scatterers (Liu et al. 2000) will not be discussed. This research also does not address active acoustic metamaterials such as the sensing and electronically tunable membranes to control wave propagation (Popa et al. 2015). Our studies will focus on geometry-based metamaterials due to ease of fabrication and practical implementation in room acoustic surfaces. Three different acoustic metamaterial types and their applications will be discussed in greater detail.

Small Subwavelength Coiled Channel Metamaterial
Coiled structures have been inserted in the body of a Helmholtz resonator to allow smaller resonators to maintaining sound absorption performance (Li and Assouar 2016) see Fig. 1(1). Coiled spaces with different coupled apertures have been used to produce resonators with a broader absorption peak (Chen et al. 2017) see Fig. 1(6). The sound propagation direction or refraction can be controlled by coiled space unit cell assemblies (Liang and Li 2012) see Fig. 1(9). The coiling of space has also been used to produce destructive interference for sound absorption (Godbold et al. 2007) see Fig. 1(11).

Helmholtz-Like Resonance Metamaterials
A Helmholtz acoustic absorber is an air-spring and mass-based system that takes acoustic energy out of a room (Maekawa et al. 2011). Helmholtz-like resonators have been used in metamaterials. Jimenez et al. (2016) have demonstrated perfect sound absorption by inserting Helmholtz resonators into quarter wavelengths resonant slits; by inserting the Helmholtz resonators, the slit resonant frequency can effectively operate for subwavelength overall dimensions, and the overall assembly remains air transparent see Fig. 1(7). Furthers studies have shown that nested resonators with multiple neck apertures can be used to achieve broadband absorption. (Wu et al. 2019) see Fig. 1(5).

Periodic Scatterers
Periodically arranged scatterers produce a "stop band" where no sound is permitted to propagate. This happens due to the interference patters that emerge when the period of scattering matches the wavelength of the operating frequency. The scattering unit cell can either be an impedance different medium or a locally resonant unit cell. The impedance different scatterer or also known as sonic crystal produces the previously discussed Bragg scattering. The locally resonant unit cell scatters sound as an effect of resonance in the unit cell. Both phenomena produce periodic scattering and produce a "stop band". In the early 90s a sonic crystal was adopted to an outdoor sculpture for noise mitigation (Martínez-Sala et al. 1995) and sonic crystals have been explored as road noise control barriers (Peiró-Torres et al. 2016) see Fig. 1(3). Sound stopping was achieved by locally resonance of lead spheres in soft matrix see Fig. 1(4). Other assemblies, such as lattice systems, with a solid inclusion and a soft matrix material have been shown to produce sound stopping effects due to local resonance in combination with simultaneous Bragg scattering (Chen and Wang 2014) see Fig. 1(8). Local resonance is also produced by single-material unit cells to dissipate sound energy mechanically (Krushynska et al. 2017) see Fig. 1(12). C-shaped Helmholtz like unit cells have been tuned to perform as locally resonant scatterers for attenuation of sound energy, these assemblies have also been nested to achieve more broadband results (Elford et al. 2011.) see Fig. 1(10).

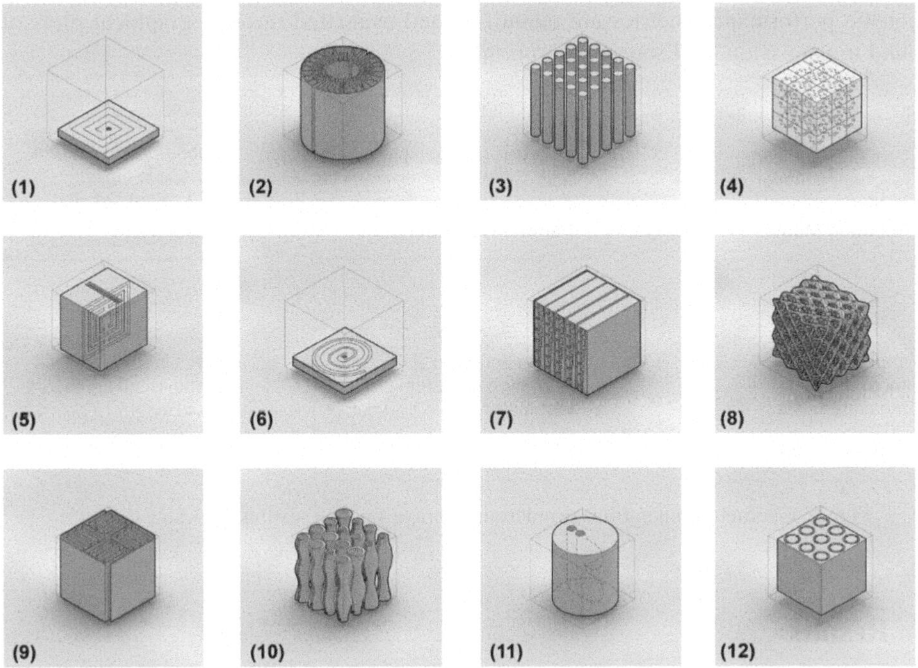

Fig. 1. (1) Li and Assouar 2016 (2) Shao et al. 2019 (3) Peiró-Torres et al. 2016 (4) Liu et al. 2000 (5) Wu et al. 2019 (6) Chen et al. 2017 (7) Jimenez et al. 2016 (8) Chen and Wang 2014 (9) Liang and Li 2012 (10) Elford et al. 2011 (11) Godbold et al. 2007 (12) Krushynska et al. 2017

3 Methods

In architectural acoustics, computational tools for performance evaluation are focused on fast computation times to facilitate rapid design evolution, as well as good integration with the design CAD environment. As most acoustic simulation tools are created to solve room acoustic problems, these tools rely on geometric methods and their algorithms approximate sound as rays. The results from these simulation tools can describe room scale acoustics in sufficient resolution. Tools such as Pachyderm, a plugin to Rhino3D CAD software, integrate well into the architectural design environment (Peters and Nguyen 2021). However, tools that integrate well into an architectural design workflow for acoustic metamaterials do currently not exist. Metamaterial geometry is frequency dependent and operates based on specific wave characteristic behavior that require the numerical modelling of the propagating wave. In our experiments, geometries were designed using Grasshopper in the Rhino3D CAD software. To evaluate the performance of metamaterial geometries, the Finite Element Modelling software COMSOL Multiphysics version 6 was employed. Metamaterials from the literature were used to establish control simulations to verify the simulation set-up, and the geometry exchange pipeline between the modelling environment and the simulation environment. The intention is to propose a proof-of-concept strategy to integrate these novel acoustic performance possibilities into an architectural parametric design workflow (see Fig. 2). Relevant acoustic performance metrics are quantified and evaluated through graphical plots of sound transmission, and sound absorption.

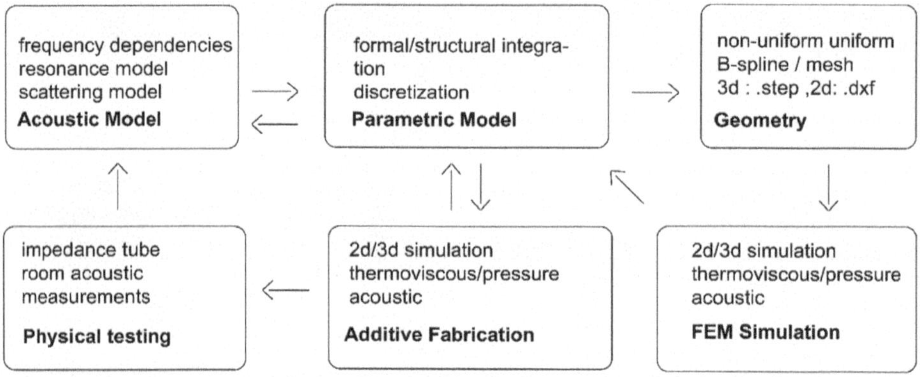

Fig. 2. Geometry generation, simulation, fabrication, and testing workflow diagram

4 Results

Of the selected geometry in Sect. 2.3 we have identified five geometries for further investigation: resonators with coiled backed space (Li and Assouar 2016); resonators with inserted channels (Shao et al. 2019); periodic spatial scatterers (Martínez-Sala et al. 1995); locally resonant unit cells based on Helmholtz resonance (Elford et al. 2011); and resonators inserted in slits (Jiménez et al. 2016). The selection criteria were based on our ability to model, fabricate, and simulate the given geometries. Single material geometries were favored over multi-material assemblies for architectural integration and digital fabrication processes. For each type, a control geometry from the literature was modelled and simulated and evaluated for their potential for architectural application (Fig. 3).

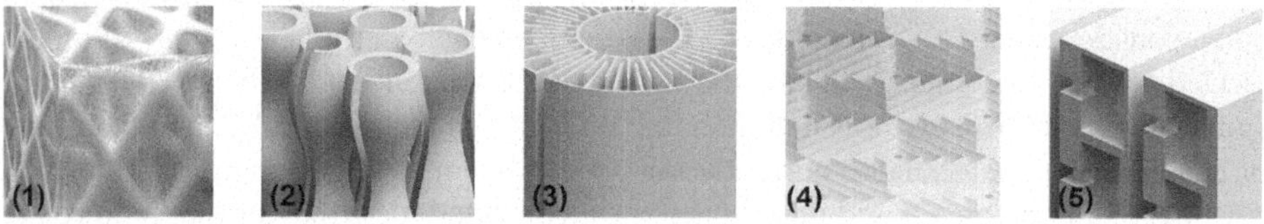

Fig. 3. Selection of 5 geometries: (1) Krushynska et al. 2017 (2) Elford et al. 2011 (3) Shao et al. 2019 (4), Li and Assouar 2016 (5) Jimenez et al. 2016

4.1 Periodic Scatterer

The periodic scatterer is not a metamaterial due to its large size, however it is an example of geometry producing exotic wave stopping effects. A transmission reduction can be observed in the frequency range that corresponds to the period of the scatterer. The lattice period was adjusted in the geometry model to perform at a design frequency, simulation results are in good agreement with the geometry model and produce sound stopping effects at the design frequency, see the graphical plots in Fig. 5.

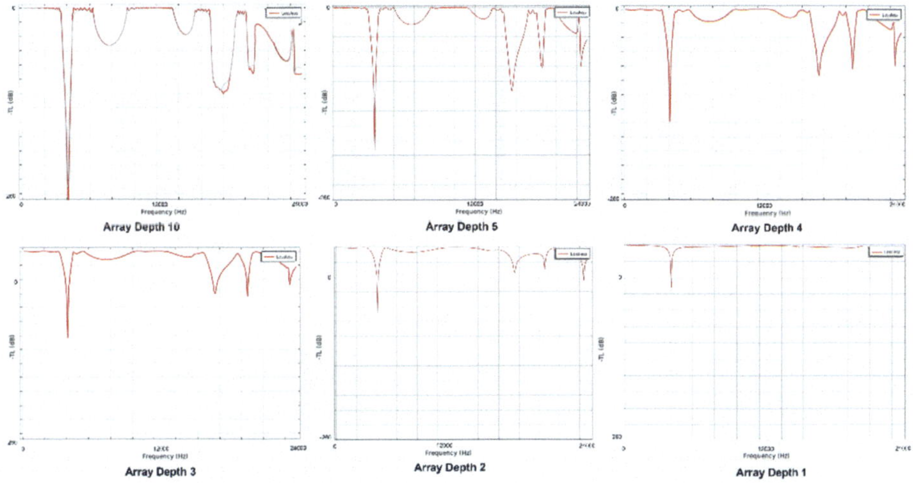

Fig. 4. C-shaped resonators sound transmission loss plot with different unit array depth

To shrink the size of a sonic crystal scattering system, Elford et al. 2011 replaced the cylindrical unit cell scatterer with a Helmholtz-like resonator. An array of periodic resonators produces the same periodic scattering and therefore a sound stop band. In our simulations we have replicated the wave stopping effect, the resulting transmission reduction can be seen in Fig. 4. The depth of required resonator count was tested for multiple different array depths.

4.2 Coiled Channels in Helmholtz Resonator

Coiled channel metamaterial resonators are investigated for sound absorption. Li and Assouar (2016) have shown that it is possible to reduce the overall size of a resonators by introducing coiled structures in the resonator body. The volume of the resonator must be sufficiently large to effectively function as an air spring for a given frequency. To decrease the overall thickness, Li and Assouar have proposed to introduce coiled structure in the resonator body as they have found that the length of the channel is important in providing a compensation for the viscosity and reactance in the neck of the resonator and providing a sufficient air spring to move air in a resonant state. By introducing coiled channels, the effective length of the channel can be increased while maintaining a thin profile. Then, resonance and sound absorption can be produced with a geometry well below the operating wavelength scale. In our simulation we have found that a normal Helmholtz resonator of the same frequency tuning must be 5 times as deep (see Fig. 6) to produce the same absorption coefficient as the coiled counterpart.

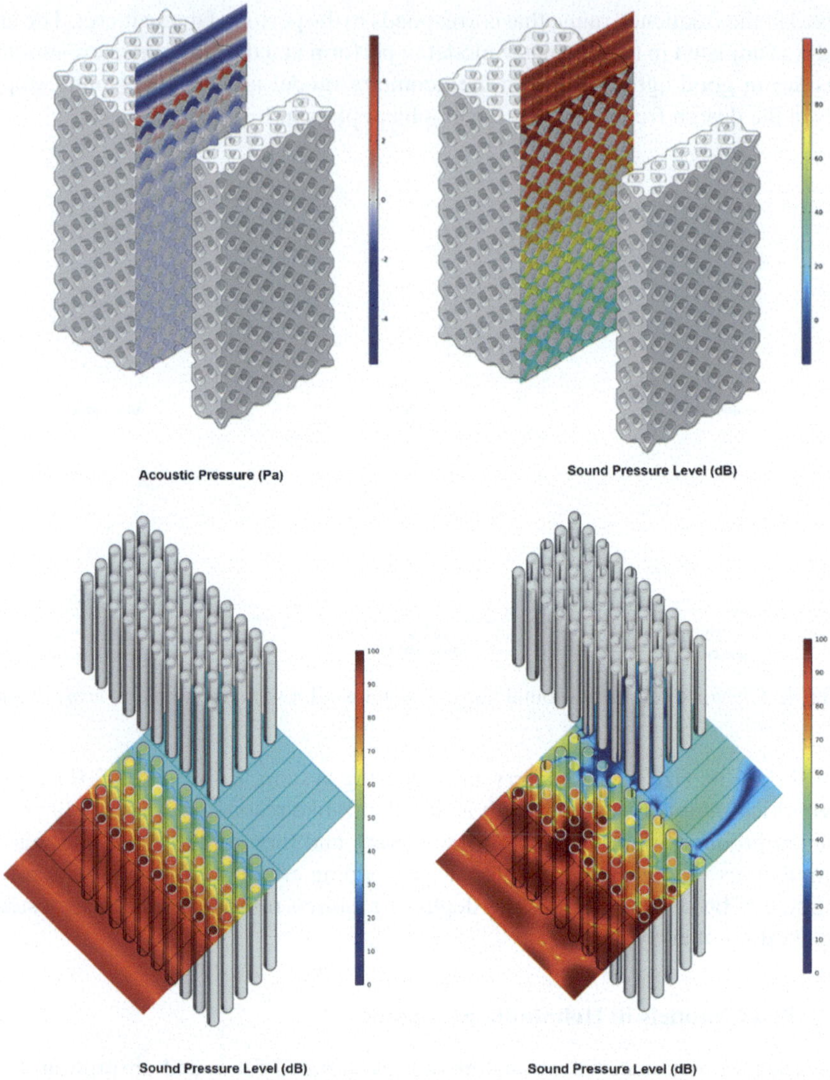

Fig. 5. (1) Sound pressure level across octet lattice, (2) Sound pressure level across locally resonant c-shaped scatterers

Shao et al. (2019) have demonstrated low frequency subwavelength by rigidly backed resonators with an inserted coiled channel for resonance absorption. The coiled space amplifies the energy concentration in the resonator and sound energy is lost. Our simulation results show multiple absorption peaks for the coiled resonator. The absorption coefficient in our tests peak at 0.75 at 130 Hz, the location of the resonant peak is in good agreement with the literature.

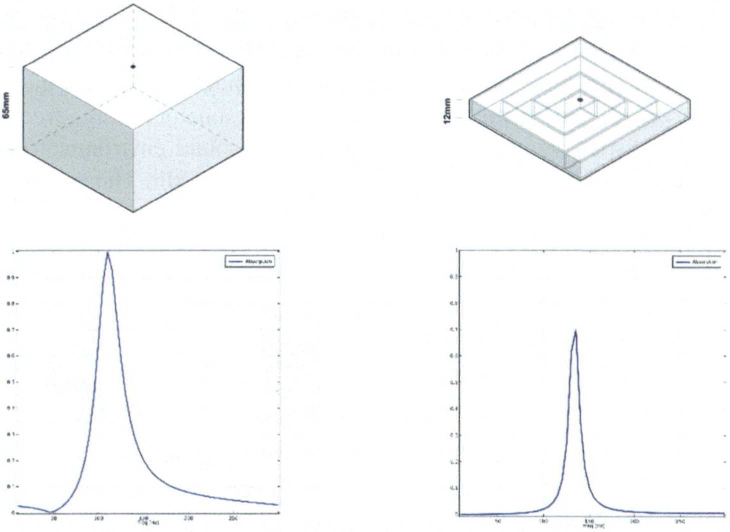

Fig. 6. Sound absorption plots and renders of regular Helmholtz resonator and coil inserted in Helmholtz resonator

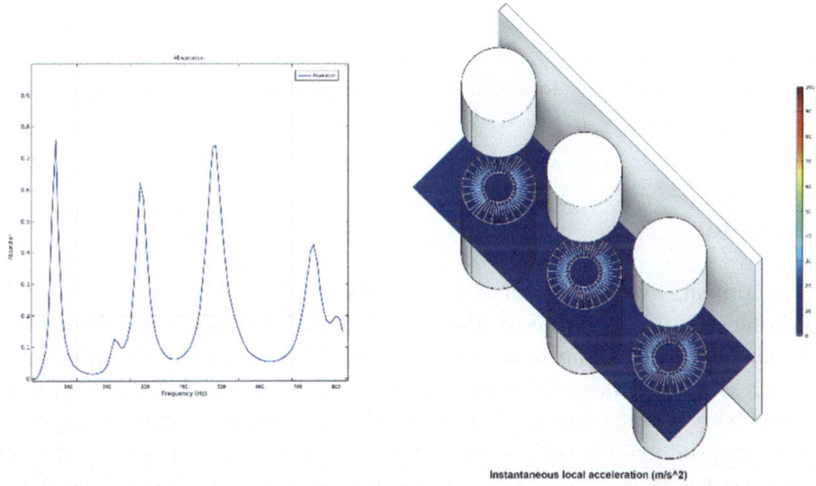

Fig. 7. (1) Sound absorption plots, (2) Instantaneous local acceleration plot

Jimenez et al. (2016) have demonstrated perfect sound absorption through inserting Helmholtz resonators into quarter wavelengths resonant slits. By inserting the Helmholtz resonators, they are able to reduce the resonant frequency of the slit to operate in subwavelength dimensions. We have verified their geometry performance for a design frequency. The absorption peak occurs at the same time as increased transmission loss is apparent (see Fig. 8). When these two conditions occur at the same time, perfect absorption is apparent with no propagating sound energy and no reflected sound energy for the frequency range in which these two conditions are true. The slit however also remains air transparent which could have promising applications. They describe that the dispersion of the incident wave occurs at the inserted Helmholtz resonators design frequency, and that the strongly dispersive environment creates slow sound environment in the slits which effectively decreases the operational frequency of the slit. The geometry parameters from the literature were used to replicate the results both in two-dimensional and three-dimensional simulations.

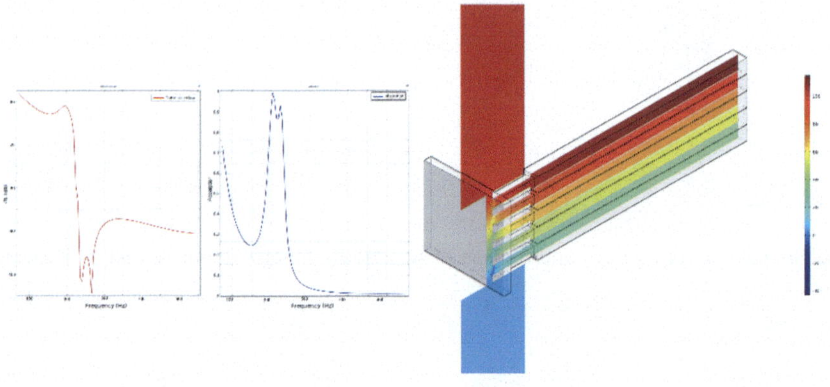

Fig. 8. Transmission loss plot, absorption plot, sound pressure level plot

5 Discussion

In this experiment, five metamaterial geometries were modelled and simulated. The simulation data was to be validated against several 3D printed starch models (see Fig. 10) to be tested in a physical impedance tube. Covid-19 lockdowns at the time, hindered us from accessing facilities to produce and test physical model, future research will address these shortcomings. Instead, we verified our simulation and modeling methods against the acoustic models within our reviewed papers. The results show that spatial and locally resonant scattering in acoustic metamaterials as well as resonance behavior in intricate bulk resonator assemblies is possible using metamaterials. The tuning of acoustic performance based on geometry parameters was demonstrated. We were most successful in tuning the scatterers both for sonic crystals and Helmholtz based locally scatterers. These rely on simple geometry models such as the simplified Helmholtz equation to retrieve the resonance frequency and thereby tune the scattering and sound stopping frequencies for the periodic resonant scatterer.

The first geometry investigated was the octet lattice as a sonic scatterer (see Fig. 4). The simulation demonstrated that the lattice period could be tuned to produce sound transmission loss through the sample for a desired frequency due to Bragg scattering. The performance is however locked to the lattice period and therefore is not ideal for low frequencies due to the large scale required. The c-shaped locally resonant periodic scatterers (see Fig. 5) proved to reduce the size required for sound transmission loss to occur as outlined by the background. This Helmholtz like resonance is a simple and geometric way of producing local resonance and would therefore be very beneficial for fabrication and implementation. The next geometry simulated was the coiled channels inserted in a resonator (see Fig. 6). This geometry produced similar sound absorption levels comparable to its non-coiled space counterpart however with a 5 times thinner profile. The other coiled structure sample also produced high absorption values see Fig. 7. In an architectural context coiled space inserted in resonators might help in tuning perforated panels for much lower frequencies while retaining at thin profile for practical fabrication and application. Lastly, we simulated the Helmholtz resonators inserted in a quarter wavelength slit (see Fig. 8). Both strong transmission loss (−105 dB) and high absorption values (0.9) occur simultaneously. This geometry could be applicable to the architectural context as it is single materials and produces very good results that are applicable both for sound blocking and sound absorption.

It is clear that the promise of acoustic metamaterials could be very beneficial to architectural acoustic and their application in our surrounding environment is an inherently architectural problem. A successful integration is however reliant on several factors. They must be easily tunable and therefore parametrically defined through a simple rule set, they must be fabricable by single material means and further a simulation and testing workflow must be feasible for a rapid design environment to be able to adapt to performance and design goals (Fig. 9).

Fig. 9. Octet lattice, locally resonant sonic crystal

Fig. 10. Starch 3D-printed octet lattice and gyroid minimal surface

6 Conclusion/Future Outlook

Acoustic metamaterials exhibit great potential for architectural acoustic application. Industry today relies on an exclusive selection of sound-mitigating materials that have performance constraints, especially in the low frequency regime, in addition these materials are difficult to custom-tune for specific spaces or desired performance and might be impractical to implement. With acoustic metamaterials, performance can be embedded in the geometry. The acoustic effects can be custom designed for a desired outcome by adjusting simple parameters in the geometry model. This proves highly valuable for an architectural workflow as there is no need to rely on specific materials and could enable an integrated strategy to embed performance tuning through geometry to ultimately design for sound more effectively. Current advances in digital fabrication and computational simulation enable us to design, verify and fabricate these novel materials for use in architectural acoustics.

References

Author, F.: Article Title. Journal **2**(5), 99–110 (2016)

Acoustic Metamaterials Group. https://acousticmetamaterials.org. Accessed 2 Apr 2022

Chen, C., Du, Z., Hu, G., Yang, J.: A low-frequency sound absorbing material with subwavelength thickness. Appl. Phys. Lett. **110**(22), 221903 (2017). https://doi.org/10.1063/1.4984095

Chen, Y., Wang, L.: Periodic co-continuous acoustic metamaterials with overlapping locally resonant and Bragg band gaps. Appl. Phys. Lett. **105**(19) (2014). https://doi.org/10.1063/1.4902129

Cox, T.J., D'Antonio, P.: Acoustic Absorbers and Diffusers: Theory, Design and Application. Taylor & Francis, London (2009)

Cummer, S.A., Christensen, J., Alù, A.: Controlling sound with acoustic metamaterials. Nat. Rev. Mater. **1**(3) (2016). https://doi.org/10.1038/natrevmats.2016.1

Bridgman, P.W.: Crystal structure: the crystalline state. Edited by Sir W. H. Bragg and W. L. Bragg. Vol. I. A general survey, by W. L. Bragg, xiv + 352 pages, 23 × 14.5 cm, with 186 figures and 6 appendices. Published by Macmillan and Company, 60 Fifth Ave., New York City, 1934, $5.50. Science (Am. Assoc. Adv. Sci.) **80**(2074), 290–291 (1934). https://doi.org/10.1126/science.80.2074.290.b

Elford, D.P., Chalmers, L., Kusmartsev, F.V., Swallowe, G.M.: Matryoshka locally resonant sonic crystal. J. Acoust. Soc. Am. **130**(5), 2746–2755 (2011). https://doi.org/10.1121/1.3643818

Godbold, O.B., Soar, R.C., Buswell, R.A.: Implications of solid freeform fabrication on acoustic absorbers. Rapid Prototyp. J. **13**(5), 298–303 (2007). https://doi.org/10.1108/13552540710824805

Godbold, O.: Investigating broadband acoustic adsorption using rapid manufacturing. Loughborough University (2008)

Huang, H.H., Sun, C.T.: Wave attenuation mechanism in an acoustic metamaterial with negative effective mass density. New J. Phys. **11**(1), 013003 (2009). https://doi.org/10.1088/1367-2630/11/1/013003

Hussein, M.I., Leamy, M.J., Ruzzene, M.: Dynamics of phononic materials and structures: historical origins, recent progress, and future outlook. Appl. Mech. Rev. **66**(4) (2014). https://doi.org/10.1115/1.4026911

Jiménez, N., Huang, W., Romero-García, V., Pagneux, V., Groby, J.P.: Ultra-thin metamaterial for perfect and quasi-omnidirectional sound absorption. Appl. Phys. Lett. **109**(12) (2016). https://doi.org/10.1063/1.4962328

Joannopoulos, J.D., Johnson, S.G., Winn, J.N., Meade, R.D.: Photonic Crystals: Molding the Flow of Light. Princeton University Press, Princeton (1995)

Kladeftira, M., Pachi, M., Bernhard, M., Shammas, D., Dillenburger, B.: Design strategies for a 3D printed acoustic mirror. In: Haeusler, M., Schnabel, M.A., Fukuda, T. (eds.) Intelligent & Informed – Proceedings of the 24th CAADRIA Conference – Volume 1, Victoria University of Wellington, Wellington, New Zealand, 15–18 April 2019, pp. 123–132 (2019)

Krushynska, A.O., Miniaci, M., Bosia, F., Pugno, N.M.: Coupling local resonance with Bragg band gaps in single-phase mechanical metamaterials. Extreme Mech. Lett. **12**, 30–36 (2017). https://doi.org/10.1016/j.eml.2016.10.004

Kumar, S., Lee, H.P.: The present and future role of acoustic metamaterials for architectural and urban noise mitigations. Acoustics **1**(3), 590–607 (2019). https://doi.org/10.3390/acoustics1030035

Kushwaha, M.S., Halevi, P., Dobrzynski, L., Djafari-Rouhani, B.: Acoustic band structure of periodic elastic composites. Phys. Rev. Lett. **71**(13), 2022–2025 (1993). https://doi.org/10.1103/PhysRevLett.71.2022

Lu, S., Yan, X., Xu, W., Chen, Y., Liu, J.: Improving auditorium designs with rapid feedback by integrating parametric models and acoustic simulation. Build. Simul. **9**(3), 235–250 (2016). https://doi.org/10.1007/s12273-015-0268-x

Li, Y., Assouar, B.M.: Acoustic metasurface-based perfect absorber with deep subwavelength thickness. Appl. Phys. Lett. **108**(6), 63502 (2016). https://doi.org/10.1063/1.4941338

Liu, Z., et al.: Locally resonant sonic materials. Science (2000). https://doi.org/10.1126/science.289.5485.1734

Liang, Z., Li, J.: Extreme acoustic metamaterial by coiling up space. Phys. Rev. Lett. **108**(11), 114301 (2012). https://doi.org/10.1103/PhysRevLett.108.114301

Maekawa, Z., Rindel, J., Lord, P., Takahashi, T.: Environmental and Architectural Acoustics. Taylor & Francis Group, Florence (2011)

Martínez-Sala, R., Sancho, J., Sánchez, J.V., Gómez, V., Llinares, J., Meseguer, F.: Sound attenuation by sculpture. Nat. (lond.) **378**(6554), 241 (1995). https://doi.org/10.1038/378241a0

Metacoustic. https://metacoustic.com/. Accessed 2 Apr 2022

Peters, B., Nguyen, J.: Parametric acoustics: design techniques that integrate modelling and simulation. In: Proceedings on Euronoise Congress (2021)

Popa, B.-I., Zigoneanu, L., Cummer, S.A.: Experimental acoustic ground cloak in air. Phys. Rev. Lett. **106**(25), 253901 (2011). https://doi.org/10.1103/PhysRevLett.106.253901

Popa, B.I., Shinde, D., Konneker, A., Cummer, S.A.: Active acoustic metamaterials reconfigurable in real time. Phys. Rev. B Condens. Matter Mater. Phys. 91(22) (2015). https://doi.org/10.1103/PhysRevB.91.220303

Scelo, T.: Integration of acoustics in parametric architectural design. Acoust. Aust. **43**(1), 59–67 (2015). https://doi.org/10.1007/s40857-015-0014-7

Shao, C., Long, H., Cheng, Y., Liu, X.: Low-frequency perfect sound absorption achieved by a modulus-near-zero metamaterial. Sci. Rep. **9**(1), 13482–13488 (2019). https://doi.org/10.1038/s41598-019-49982-5

Vitruvius: The Ten Books on Architecture. Dover, New York (1960). Translated by: Morgan, M.H.

Yablonovitch, E.: Inhibited spontaneous emission in solid-state physics and electronics. Phys. Rev. Lett. **58**(20), 2059–2062 (1987). https://doi.org/10.1049/cp.2012.0663

Walser, RM.: Metamaterials: an introduction. In: Introduction to Complex Mediums for Optics and Electromagnetics, pp. 79–102. SPIE Press (2003). https://doi.org/10.1117/3.504610.ch13

Walser, R.M.: Metamaterials: What are they? What are they good for? Bull. Am. Phys. Soc. **45**, 1005 (2000)

Wu, P., et al.: Acoustic absorbers at low frequency based on split-tube metamaterials. Phys. Lett. A **383**(20), 2361–2366 (2019). https://doi.org/10.1016/j.physleta.2019.04.056

Zigoneanu, L., Popa, B.I., Cummer, S.A.: Three-dimensional broadband omnidirectional acoustic ground cloak. Nat. Mater. **13**(4), 352–355 (2014). https://doi.org/10.1038/nmat3901

Kim, S.-H., Lee, S.-H.: Air transparent soundproof window (2013). https://doi.org/10.48550/arXiv.1307.030

Ghaffarivardavagh, R., Nikolajczyk, J., Anderson, S., Zhang, X.: Ultra-open acoustic metamaterial silencer based on Fano-like interference. Phys. Rev. B **99**, 024302 (2019). https://doi.org/10.1103/PhysRevB.99.024302

Peiró-Torres, M.D.P., Redondo, J., Bravo, J.M., Pérez, J.S. Open noise barriers based on sonic crystals. Advances in noise control in transport infrastructures. Transp. Res. Procedia **18**, 392–398 (2016). https://doi.org/10.1016/j.trpro.2016.12.05

Pendry, J.B.: Negative refraction makes a perfect lens. Phys. Rev. Lett. **85**(18), 3966–3969 (2000). https://doi.org/10.1103/PhysRevLett.85.3966

Romero-García, V., Sánchez-Pérez, J.V., Garcia-Raffi, L.M.: Tunable wideband bandstop acoustic filter based on two-dimensional multiphysical phenomena periodic systems. J. Appl. Phys. **110**(1), 014904–1–014904–9 (2011). https://doi.org/10.1063/1.3599886

Xie, Y., et al.: Acoustic holographic rendering with two-dimensional metamaterial-based passive phased Array. Sci. Rep. **6**, 35437 (2016). https://doi.org/10.1038/srep35437

Setaki, F., Tenpierik, M., Turrin, M., van Timmeren, A.: Acoustic absorbers by additive manufacturing. Build. Environ. **72**, 188–200 (2014). https://doi.org/10.1016/j.buildenv.2013.10.010

Shamonina, E., Solymar, L.: Metamaterials: how the subject started. Metamaterials **1**(1), 12–18 (2007). https://doi.org/10.1016/j.metmat.2007.02.001

Shelby, R.A., Smith, D.R., Schultz, S.: Experimental verification of a negative index of refraction. Sci. (Am. Assoc. Adv. Sci.) **292**(5514), 77–79 (2001). https://doi.org/10.1126/science.1058847

Shen, C., Xie, Y., Li, J., Cummer, S.A., Jing, Y.: Acoustic metacages for omnidirectional sound shielding. J. Appl. Phys. **123**, 124501 (2018). https://doi.org/10.1063/1.5009441

Sihvola, A.: Metamaterials in electromagnetics. Metamaterials **1**(1), 2–11 (2007). https://doi.org/10.1016/j.metmat.2007.02.003

Veselago, V.G.: The electrodynamics of substances with simultaneously negative values of and μ. Sov. Phys. Uspekhi, **10**, 509 (1968). https://doi.org/10.1070/PU1968v010n04ABEH003699

ADAPTEX

Physical and Digital Prototyping of Smart Textile Sun Shading

Paul-Rouven Denz[1(✉)], Natchai Suwannapruk[1], Puttakhun Vongsingha[1], Ebba Fransén Waldhör[2], Maxie Schneider[2], and Christiane Sauer[2]

[1] Priedemann Fassadenberatung GmbH, Am Wall 17, 14979 Großbeeren/Berlin, Germany
paul.denz@priedemann.net

[2] Weißensee Academy of Art Berlin, Bühringstraße 20, 13086 Berlin, Germany

Abstract. The objective of the R&D project ADAPTEX aims at developing a novel sun-shading system that contributes to a building's performance efficiency without taxing the economy of its subsystems. By utilizing the potentials of textile construction and the integrity of Shape Memory Alloy (SMA), which reduces material weight and operation energy for a dynamic shading system. SMAs are generally suitable for construction applications due to their maintenance free, function through multiple cycles without showing wear, and replacement of complex motor and driving mechanism. The design system is developed to be driven by changes in the environment, that allows autonomous and adaptive reactions to external stimuli like ambient heat or solar radiation. Because SMAs are designed for the accuracy and standards of mechanical engineering rather than for architectural facade applications, ADAPTEX closes the inherent gaps established in systems engineering by integrating SMA wire into large scale architectural surfaces by integrated SMA into light weight material as textile. The potential of a larger scale of ADAPTEX will be explored in the continuation project ADAPTEX KLIMA+ in Muscat, Oman, which also allows the autarkic operation of ADAPTEX to be tested under ambient condition. This article discusses the development process of implementing and exchanging between digital analysis and physical prototype. This includes the process of analyzing the local climate, deeper understanding of SMA performance for a specific project site, and finally the execution of these data into full scale façade prototype and its monitoring plan for cross validations.

Keywords: Adaptive façade · Textile sun shading · Self-sufficient climate responsive operation · Shape Memory Alloy (SMA) · Physical prototyping · Digital prototyping

1 Introduction

1.1 Background

To combat the effects of climate change, the European Green Deal calls for a resilient climate adaptation of future building systems (European Commission 2021). The façade plays a key role in regulating the building's energy performance and one of its important

C. Gengnagel et al. (Eds.): DMS 2022, *Towards Radical Regeneration*, pp. 237–248, 2023.
https://doi.org/10.1007/978-3-031-13249-0_20

functions is the prevention of overheating of interior spaces (Hoces 2018). In this sense, the façade significantly contributes to the energy efficiency of buildings by reducing cooling loads and respective expenses for building services (Klein 2013). Overheating usually occurs as a result of direct solar radiation. Shading strategies are therefore considered as the best method to exclude solar impacts from penetrating the interior space and therefore from causing high indoor temperatures (Hoces 2018).

Adaptability contributes to the efficiency of sun shading, as it allows for altering behaviors in response to changing environmental conditions, like the constantly changing sun position or different intensities of solar radiation. The implementation of such adaptive system has been widely explored in recent years by various researchers and designers, for instance, the kinetic solar shading device in ABI's Tessellate™ (Drozdowski 2011) and the application of exoskeletal, dynamic building envelope through the means of actuated tensegrity by Tristan d'Estree Sterk (Sterk 2006). The multifunctionality of the façade leads, following a polyvalent approach of an adaptive implementation, to sophisticated and potentially more complex systems. The application of "smart materials" that provide an inherent control or self-regulating capability can help to reduce over-complexity, maintenance requirements and also recycling issues (Barozzi et al. 2016).

Textiles offer a number of advantages for solar shading applications, for example their configurability, their light-weight nature and flexibility. The R&D project ADAPTEX showcases a material-driven and computationally informed design approach to adaptive textile façade through the integration of Shape Memory Alloy (SMA) as an actuator creating textile sun shading solutions. SMA respond to temperature changes in their surrounding environment by geometrically expanding or shrinking. This effect can be used to activate a textile surface and control its permeability: parameters such as light transmission and reflection are adjusted. In ADAPTEX, the textile, the SMA and the environment are understood as a coherent system in which the elements interact to create a cyclic movement between the closing and opening of textile sun-shading surfaces (Denz 2015).

The objective of the project is on an autarkic operation of the adaptive textile sun shading, which is passively reacting without additionally introduced energy to temperature changes of the environment. Therefore, the knowledge and understanding from the previous tests, simulations and prototypes laid a fundamental understanding for the implementation of ADAPTEX on an actual façade planned in the follow up project - ADAPTEX KLIMA+ which is developed for monitoring and validating the autarkic operation of the ADAPTEX concepts under the ambient condition in Muscat, Oman.

1.2 ADAPTEX Concepts Development

The research is a collaboration of an interdisciplinary team from the fields of architecture, engineering, textile design, textile fabrication and smart materials research. Through a research-by-design approach, various application scenarios were developed to integrate the SMA as actuator in textile shading solutions (Denz et al. 2021). In further elaboration of the project the possibilities were narrowed down to two most promising concepts: ADAPTEX Wave and ADAPTEX Mesh. As a part of the research process, both concepts were digitally simulated and realized as demonstrators and prototypes in various scales.

The prototypes were utilized in a series of tests to establish further understanding on their operational behavior under electrical input as well as from thermal influences (Fig. 1).

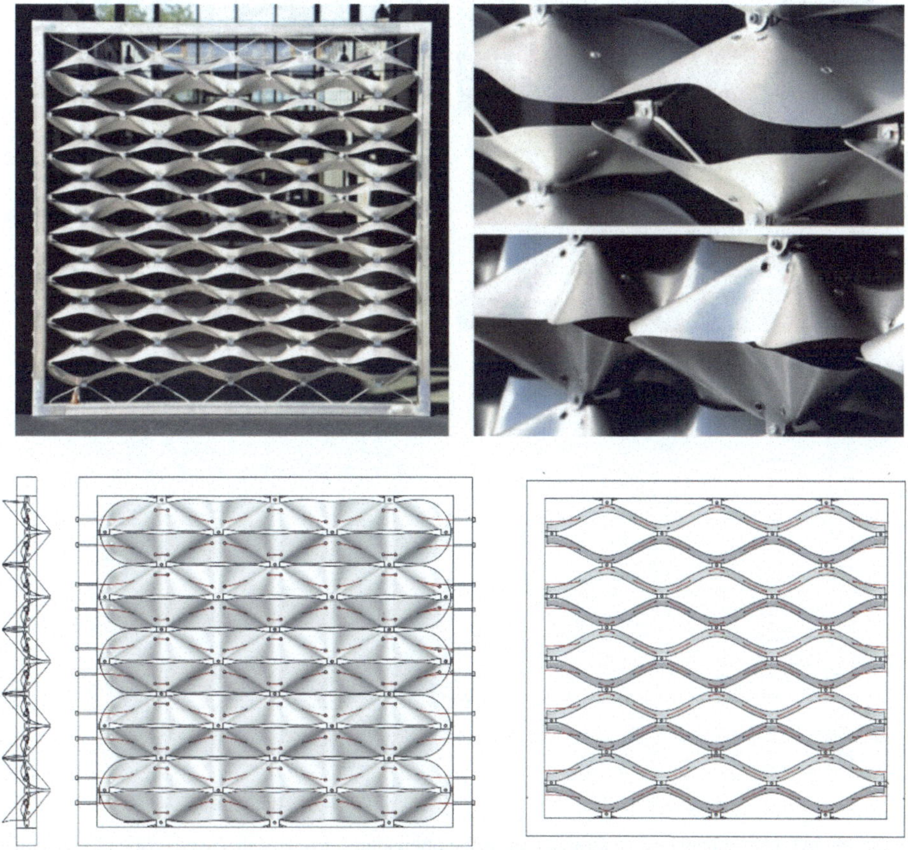

Fig. 1. ADAPTEX Wave demonstrator (top/left); open state/closed state (top/right); CAD details of the textile transformation between inactivated and activated stage (bottom)

ADAPTEX Wave, cyclic actuation is achieved by the interaction between the material tension of a textile band and a SMA wire. An elongation is created in the material by the selective attachment of wave-shaped, semi-rigid textile tapes to a cable mesh. Each tape is interwoven along its entire length with a linear SMA wire that, when activated, contracts 3–5% of its initial length and forces the tape to buckle. (Denz et al. 2021) This results in a closure of surface, configurating the openness factor which ranges between 70% during inactivation to 5% when activated (Schneider et al. 2020). Due to the material elasticity, the textile tape can move back to its original position after the SMA is deactivated. In the first full-scale prototype, the textile bands are made of an opaque, durable, flexural-resistant, glass fiber reinforced fabric, which protects from solar radiation and glare (Fig. 2).

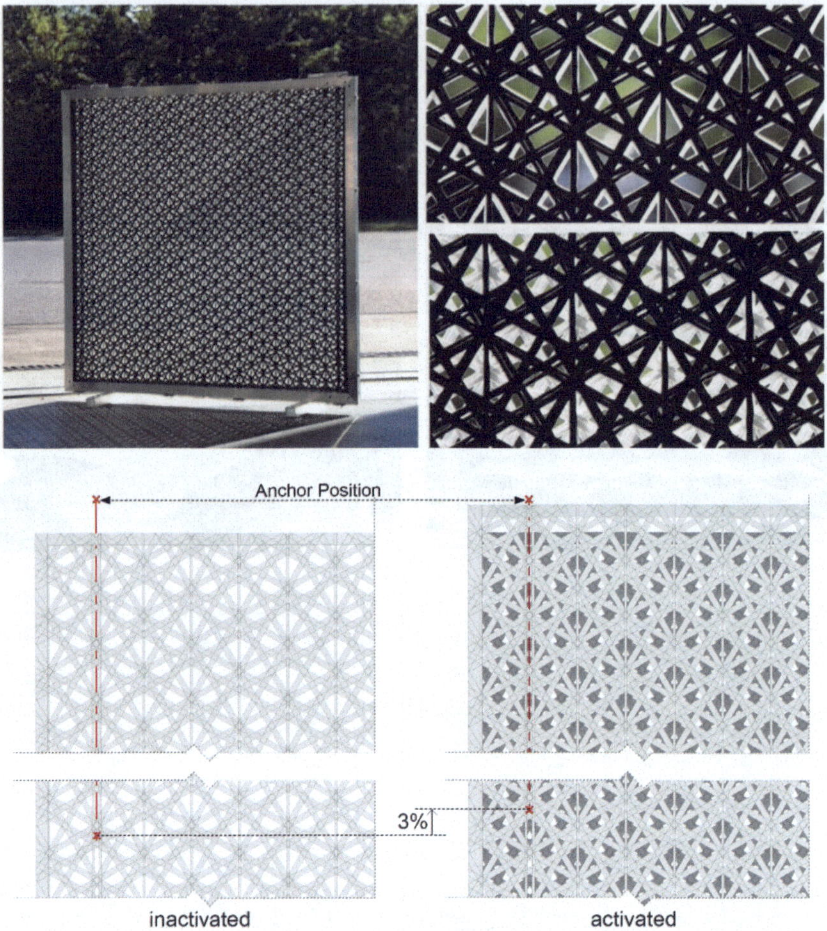

Fig. 2. ADAPTEX Mesh demonstrator (top/left); open state/closed state (top/right); CAD details of the textile transformation between inactivated and activated stage (bottom)

ADAPTEX Mesh, aims to develop a resource-efficient yet simple system. Two membranes with identical perforations are placed in front of each other to achieve maximum permeability. When one of the membranes slides up, the overlapping of the patterns provides a reduction in overall permeability across the surface. When inactivated, the openness factor is at 63% and when it is activated the openness factor is reduced to 39% (Schneider et al. 2020). In principle the operation required low vertical movement, to achieve shading efficiency. (Denz et al. 2021).

2 Implementation of ADAPTEX KLIMA+ Under Ambient Condition in Muscat, Oman

To facilitate the design and planning process, design criteria such as filament sizes, angle and transition behavior were translated into algorithmic parameters which were used in investigating and optimizing the openness factor of both fabric and system, the permeability, as well as material interaction of ADAPTEX. Furthermore, the environmental context was explored by a comprehensive analysis of weather data in various climatic settings. Relevant weather data such as dry-bulb temperature, day/night temperature differences, solar radiation, sunshine hours, and solar hours were mapped against the SMA activation temperature (Dilibal et al. 2013). The evaluation enhances the selection process of the optimal SMA for the context of a specific project site.

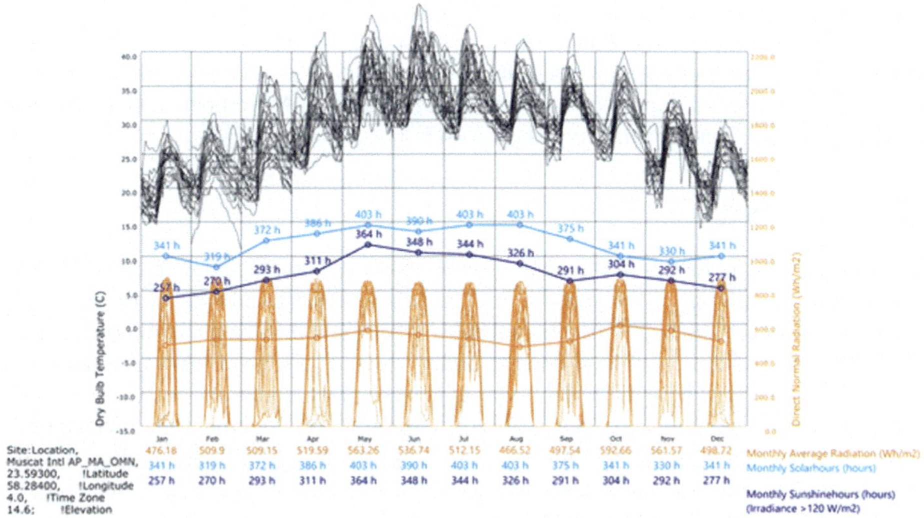

Fig. 3. Weather data analysis of Muscat, Oman

According to (Schneider et al. 2020), Muscat is a suitable location for ADAPTEX prototype. A detailed climate analysis of Muscat is shown in Fig. 3. Despite the fluctuating temperature between season, as a subtropical area, it presented high differences in day and night temperature. Due to the daily temperature variation, the operation of the prototype can be comprehensively studied. Based on the air temperature data, the system will be activated during daytime in summer, due to high temperature. While due to the lower temperature, the system is predicted to remains inactivated in most part of winter. However, the high level of solar radiation throughout the year would also act as an ancillary driver to activate the SMA. The climate condition allows further investigation on the influence of solar exposure towards ADAPTEX operation to be monitored.

2.1 SMA Selection Criteria

With the changes in temperature, SMA undergoes solid-state phase transformation between martensite and austenite states. The SMA is activated with high strength or austenite state at a higher temperature and when cooled down, the material becomes inactivated to the martensite state with lower strength (Josephine et al. 2020).

When SMA temperature increases, the austenite start (As) phase is initiated, gradually changing from the inactivated martensite phase to the fully activated state or the austenite finish (Af) state. On the other hand, from austenite phase, when the temperature decreases, at a certain temperature, the material starts to change to martensite phase, which is martensite start (Ms) and reaches the complete inactivate stage at martensite finish (Mf).

To ensure the cycle of activation and deactivation of the sun shading for a specific site, the selected SMA needs an Af which is lower than the maximum reachable temperature, while Mf should be higher than the lowest reachable temperature. According to the weather data analysis, Muscat has sufficient temperature difference between seasons, and high difference between day and night. (Schneider et al. 2020). In this project, a broad range of operation period is selected from March to November, with high solar radiation and temperature range between 20–45 °C.

2.2 Testing of SMAs

To understand the behavior of the SMAs, three tests were conducted both before and in parallel to the planning of the prototypes. The first test was conducted with electricity activation under constant operation in multiple cycles to establish a prove of concept that the selected SMAs are capable to operate under loads. The second test was a temperature correlation test, where the SMA and textile were place under direct solar radiation. Arduino and temperature sensors were used to monitor and document the surface temperature of the materials, to find the correlation between dry bulb temperature, textile surface temperature, and SMA temperature (Denz et al. in press). Lastly, the autarkic test was conducted. The SMAs with attached loads were activated with a similar air temperature level of Muscat. This will be further elaborated in the following section.

2.3 Autarkic Test for SMA Selection

Six SMAs were tested under a simulated temperature condition with a range between 25 °C to 120 °C. The test setup is shown in Fig. 4. The SMA is hanged from the top of a transparent tube with an attached load. To replicate the actual scenario, the SMAs carry 2.77 kg (27.2 N) resembling the textile weight for ADAPTEX Mesh and 15 N as textile reaction force for ADAPTEX Wave. The test was documented through the means of Arduino, a microcontroller which controls two - MAX31855 with K-Type Thermocouple sensor, one on the top of the tube and one at the base of the tube, to measure air temperature, along with a distance ranging sensor, and VL6180x sensor to measure the displacement distance as the weight is being lifted.

As the air in the tube is heated up via a hair dryer, the solid-state transformation of the SMA begins to occur at their Austenite Start temperature. The contraction then allows the weight to be lifted. The measurement of the contracted range was documented as shrinkage percentage. The shrinkage was then compared to the average air temperature between the top and the bottom of the tube. The results showed the association between the temperature and shrinkage of SMA under load.

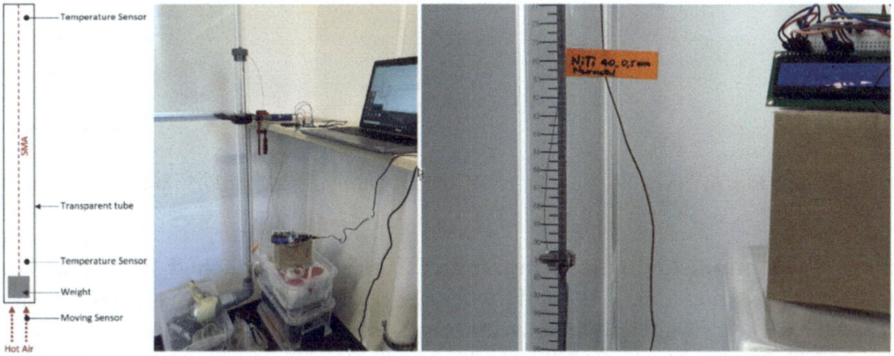

Fig. 4. SMA temperature test setup

From the results, four suitable SMAs were selected, including: NiTiCu30/0.5 mm and NiTi40/0.5 mm for ADAPTEX Mesh and NiTi30/0.3 mm and NiTi40/0.3 mm for ADAPTEX Wave. The results have shown that with an attached weight, all selected SMA can lift the weight to the shrinkage percentage required with relatively high temperature (ADAPTEX Mesh > 3.25%, ADAPTEX Wave > 3%) (Fraunhofer IWU 2016). However, NiTiCu30/0.5 mm can achieve the required shrinkage at ~52 °C, which is the lowest in comparison to the other SMAs. The result show that none of available SMA can be activated solely by air temperature in Muscat. However, according to (Denz et al. in press), under direct solar radiation, the temperature of SMA is significantly higher than dry-bulb temperature and lower than the textile surface temperature. Therefore, due to increased heat from high solar radiation, it is expected that the autarkic operation can still be managed. This hypothesis shows that the sun shading would remain in deactivated position, even under environment with high air temperature, and would only be activated under direct solar radiation as additional heat source (Figs. 5 and 6).

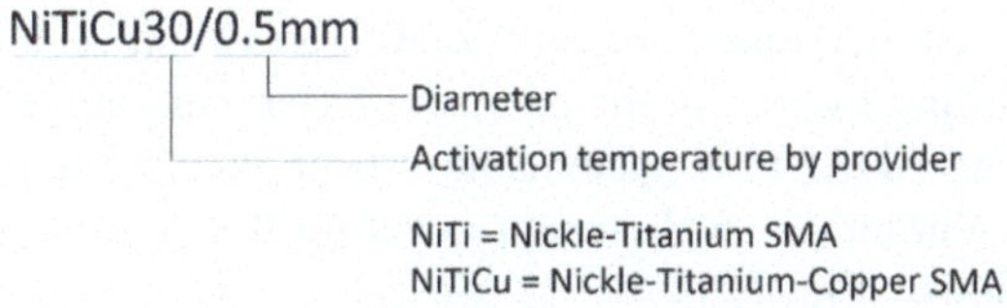

Fig. 5. Definition of SMA name

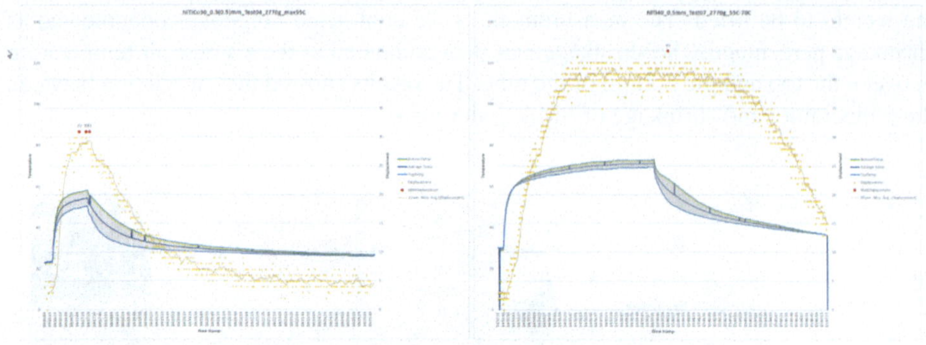

Fig. 6. SMA test results: NiTiCu30/0.5 mm and NiTi40/0.5 mm

NiTiCu30/0.5 mm: Maximum Temperature Test 55 °C, Attached Weight 2770 g. The shrinkage started almost instantly when the temperature increases from 25 °C. It reached the require shrinkage at approximately 52 °C. When the heat source was turned off, the SMA is gradually extended to its initial length.

NiTi40/0.5 mm: Maximum Temperature Test 70 °C, Attached Weight 2770 g. The shrinkage started almost instantly when the temperature increases from 25 °C but at a slower speed in comparison to NiTiCu30/0.5 mm. It reached the require shrinkage at approximately 66 °C. When the heating was turned off, the SMA maintain the activation for a few minutes, then gradually extend to its initial length (not in the chart).

This test presents a general understanding which serves as a design parameter for the prototype in Muscat. In order to validate the results, the selected SMAs will be tested in Fraunhofer IWU to determine the conversion temperature in load dependent SMA. Furthermore, they will be implemented under ambient condition in Oman to validate their operational behavior.

3 Implementation of ADAPTEX KLIMA+ Under Ambient Condition in Muscat, Oman

The knowledge and understanding from the previous tests, simulations and prototypes laid a fundamental understanding for the implementation of ADAPTEX KLIMA+. The project is planned to be tested under ambient condition as a full-scale façade application at the EcoHouse building located in the campus of German University of Technology in Muscat, Oman. The ADAPTEX systems have been planned to replace the existing 1,500 × 7,500 mm aluminum mesh screen covering the entire south and south-east window of the EcoHouse (Fig. 7).

With considerable influences on the thermal condition of the SMA and fabric, the ambient condition plays a crucial role in the autarkic operation of ADAPTEX. Therefore, the influence of the external stimuli as well as the intercorrelation between the build-up material such as the SMA and fabric must be thoroughly investigated. To allow various configurations of both the SMA and textile to be examined, the screens are subdivided.

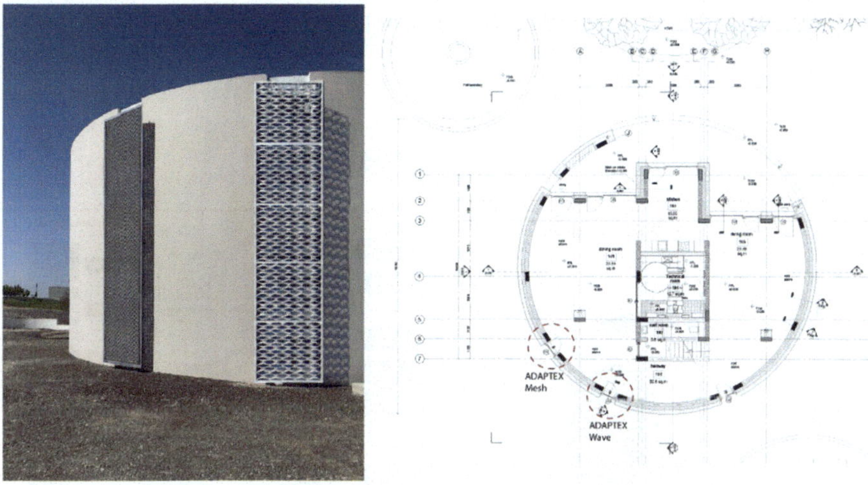

Fig. 7. Visualization of ADAPTEX Mesh and Wave at EcoHouse.

ADAPTEX Wave is designed as three separate modules with 1175 × 2539 mm steel frame and integrated steel net bracing. With a similar approach, ADAPTEX Mesh is subdivided into five smaller modules. Each module consists of 1236 × 1474 mm aluminum frame with capping. The planned scheme is depicted in Fig. 8.

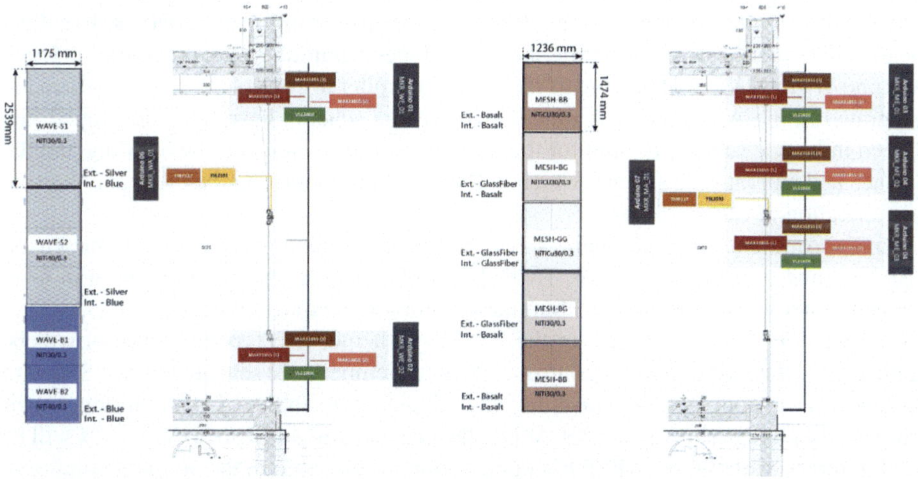

Fig. 8. Design and sensor setup for ADAPTEX Mesh and Wave for Muscat, Oman.

To establish a comprehensive study on the influence of the external stimuli and behavior of the different configurations of ADAPTEX Mesh and Wave, various sensors have been planned as an integrative part of the modules. Through the means of a weather station, microcontrollers, and sensors; five main parameters including the weather data, surface temperature, operational behavior, day light penetration and temperature behind the screen will be monitored (Fig. 9).

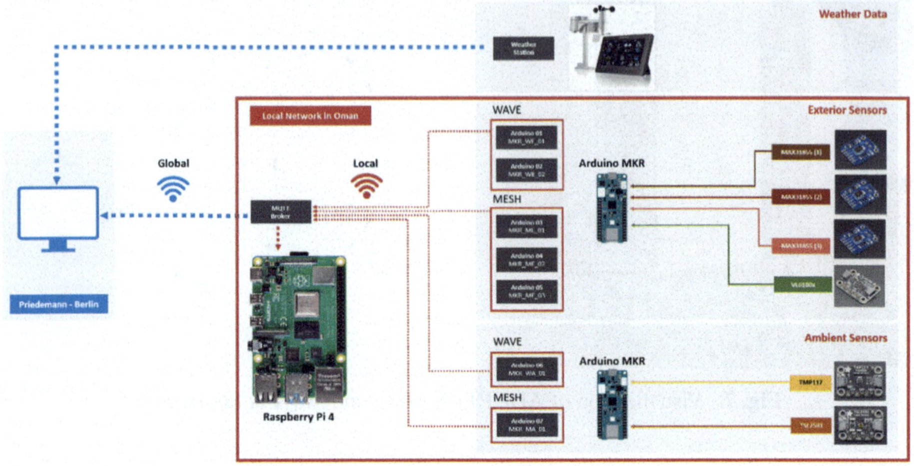

Fig. 9. Sensor network scheme for ADAPTEX KLIMA+.

With a Raspberry pi microcomputer acting as a server, a local network set-up will be established to collect and document relevance data. These data will be transmitted via the internet network for further interpretation in Berlin. The envisioned scheme is presented in Fig. 9. An on-site weather station will be installed on the rooftop of EcoHouse to collect weather data such as dry-bulb temperature, wind speed, humidity, and precipitation level. Arduino Wi-Fi MKR will be used as the microcontroller to control the sensors which are integrated in the ADAPTEX modules. Two sensors scheme have been developed, the Screen influence scheme to monitor the influences of the screen on the building and the Surface and Operational scheme which will be used in monitoring the thermal condition and operation of the modules.

In the Ambient sensors scheme shown in Fig. 9, a high dynamic range light sensor, TSL2591 and temperature sensor, TMP117 will be installed behind the screen in front of each window to examine the illumination and temperature influence of ADAPTEX on the window opening. While in the Exterior scheme, two types of sensors will be connected to the Arduino microcontroller. Surface temperature sensor, MAX31855 with K-Type Thermocouple will be installed on both the fabric and the SMA to monitor their temperature changes during the day. While the distance ranging sensor, VL6180x will be used in monitoring the ADAPTEX's operation in relationship to the ambient condition.

In ADAPTEX Wave, two sets of surface and operational sensors are planned to be incorporated at the top panel and the bottom panel to investigate the thermal behavior of Silver and Black fabric as well as their influence on the SMA. Moreover, the operation routine will also be documented. With a similar approach, three sets of surface and operational sensors will be integrated into ADAPTEX Mesh modules to investigate on the thermal surface condition of three fabric configurations (Basalt/Basalt, Glass/Basalt, and Glass/Glass) as well as the behavior of the SMA.

With the data from the sensors network, the correlation between on-site weather condition and the operational behavior of ADAPTEX as well as the thermal relationship between the SMA and the fabric can be examined and interpreted. Moreover, the operational routine of the ADAPTEX modules will allow further investigation on glare protection and reduction of solar radiation exposure of building interiors to be conducted. By interpreting the monitored data, the design process regarding SMA selection and fabric materials could be validated and further optimized.

4 Conclusion

The development of adaptive textile façade actuators based on Shape Memory Alloys requires highly interdisciplinary expertise as well as multi-tool-based design, material, construction and function development. ADAPTEX Wave and ADAPTEX Mesh, with their distinct properties arising from their materials and manufacturing processes, both illustrate a material system where the parameters of the SMA, the textile and the surrounding environment work together to constitute a resilient cyclic actuation. By avoiding mechanical construction and relying on a soft textile operation logic with few components, the aim is to achieve a durable, simple and flexible design solution. The two concepts propose a great potential of lightweight textile structures through the reduced amount of deployed materials and mechanical devices in comparison to conventional products. This underlines their potential to minimize the energy and material consumption required for construction, transportation, and operation. The alliterative research approach, by constructing a 1:1 prototype and installing it in a location with suitable climate such as Muscat, allows the concept to be monitored and validated. Furthermore, this allows a thorough investigation on the correlation on how air temperature and solar radiation has on the operation of ADAPTEX. It enhances the optimization process in creating a dynamic and demand-oriented sun shading solution for buildings, as the autarkic textile façade controls solar transmission and thereby regulates the energy-efficiency as well as user comfort of the building.

References

Barozzi, M., Lienhard, J., Zanelli, A., Monticelli, C.: The sustainability of adaptive envelopes: developments of kinetic architecture. Procedia Eng. **155**, 275–284 (2016)

Denz, P.-R.: Intelligente Gebäudehüllen und Smart Textiles - eine ideale Kombination (2015)

Denz, P.-R., Sauer, C., Waldhör, E.F., Vongsingha, P.: Smart textile sun shading development of functional ADAPTEX prototypes. J. Facade Des. Eng. **9**(1), 101–116 (2021)

Denz, P.-R., et al.: Development and testing of real size smart material sun-shading R&D ADAPTEX (in press)

Dilibal, A., Adanir, H., Cansever, N. Saleeb, A.F.: Comparison and characterization of NiTi and NiTiCu shape memory alloys (2013)

Drozdowski, Z.: The adaptive building initiative: the functional aesthetic of adaptivity. Archit. Des. **81**(6), 118–123 (2011)

European Commission, D.-G. f. R. a. I.: Europe's 2030 climate and energy targets: research & innovation actions. Publications Office (2021)

Fraunhofer IWU: Smart Tools for Smart Design (2016). http://st4sd.de/

Hoces, A.P.: COOLFACADE - Architectural Integration of Solar Cooling Technologies in the Building Envelope (2018)

Josephine, S., Ruth, D., Rebekah, S.D.: Shape memory alloys. In: Alloy Materials and Their Allied Applications, pp. 213–223 (2020)

Klein, T.: Integral Facade Construction: towards a new product architecture for curtain walls. Doctoral thesis (2013)

Schneider, M., et al.: Adaptive textile façades through the integration of Shape Memory Alloy (2020)

d'Estree Sterk, T.: Shape Change in Responsive Architectural Structures. ACADIA (2006)

Thinking and Designing Reversible Structures with Non-sequential Assemblies

Julien Glath[1](✉), Tristan Gobin[2], Romain Mesnil[1], Marc Mimram[3], and Olivier Baverel[1,2,4]

[1] Laboratoire Navier, UMR 8205, Ecole des ponts Paristech, Université Gustave Eiffel, CNRS, 77420 Champs-sur-Marne, France
julien.glath@enpc.fr
[2] Laboratoire GSA, Ecole Natioale Supérieure d'Architecture Paris-Malaquais, Paris, France
[3] Laboratoire OCS, UMR AUSser 3329, Ecole d'Architecture de la Ville et des Territoires Paris-Est/Université Gustave Eiffel, Champs-sur-Marne, France
[4] Ecole Nationale Supérieure d'Architecture de Grenoble, Grenoble, France

Abstract. In the face of dwindling resources and to reduce the construction sector's carbon footprint, it is important to envisage the building's end of life. It is necessary to consider buildings and their structures as elements that can be dismantled/disassembled for recycling/reuse. To address these issues, this work focuses on non-sequential assemblies that allow reversibility. A full-scale Nexorade prototype has been produced and is used to illustrate the entire article demonstrating the relevance of these assemblies. The first part defines non-sequential assemblies and the multiple possible kinematics for an assembly. The second part focuses on the pavilion's geometry: the choice of the tessellation and its transposition to a nexorade. The article continues with the mechanical analysis, showing that tilting the beams by 30° reduces the deflection by 35%. The assembly sequence is then chosen to ensure that each node closes non-sequentially. Finally, the pavilion's construction is detailed, showing the transition from the digital to the physical model and the adjustment needed to switch between these two models. To conclude, the authors discuss the results obtained on non-sequential connections and the inclination of the beams of a nexorade. Possible extensions and improvements for future research are suggested.

Keywords: Non-sequential assembly · Kinematic · Nexorade · Joinery · Digital fabrication

1 Introduction

Nowadays, in the face of climate change, it is necessary to think about the impact of our design or build decisions in the construction sector. [1] shows that in 2017, the construction sector produced 224 million tons of waste in France, corresponding to 70% of the national waste production. It is necessary to consider buildings and their structures as elements that can be dismantled or disassembled for recycling or reuse. This work focuses on non-sequential assemblies to address these issues. These assemblies require

C. Gengnagel et al. (Eds.): DMS 2022, *Towards Radical Regeneration*, pp. 249–259, 2023.
https://doi.org/10.1007/978-3-031-13249-0_21

a perfectly simultaneous movement of at least three different parts to be assembled. It is locked only by its kinematics, which allows reversibility, unlike glue or nails. To demonstrate the relevance of these assemblies, a full-scale Nexorade prototype has been produced and is used to illustrate the entire article. The first part defines non-sequential assemblies and the possible kinematics. The second part focuses on the determination of the form of the prototype. Then a mechanical model is used to verify the behaviour of the structure. The assembly sequence is discussed in Sect. 5, and the construction process is developed in Sect. 6. To conclude, the authors discuss the results obtained on non-sequential connections and the inclination of the beams of a nexorade. Possible extensions and improvements for future research are suggested.

2 Non-sequential Assembly

A non-sequential assembly can be defined by its valence and its degree of connectivity: the number of parts in contact with another. In Fig. 1, we can see that the node valence is four, and each part has two degrees of connectivity. Non-sequential assembly is composed of multiple parts $\mathbf{P}_i$ connected by unidirectional slides. One can analyse the contacts between each $\mathbf{P}_i$ to obtain the connectivity graph, Fig. 1b. In Fig. 1c, we can see in green the slide directions. The combination of the connectivity graph and the slide directions can give the oriented connectivity graph (OCG) in Fig. 1d. As a connectivity graph, each node is a part, and each line is a link. The difference is that lines are oriented according to slide vectors $\boldsymbol{\alpha}_i$ between two parts.

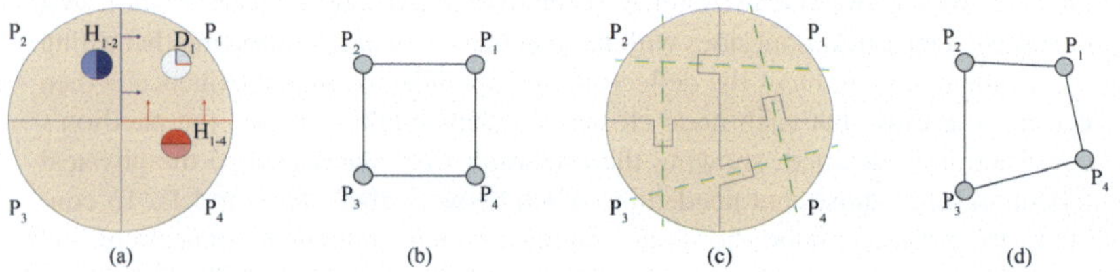

Fig. 1. Graph for an assembly; (a) contacts analyse, (b) connectivity graph, (c) slides directions, (d) oriented connectivity graph (OCG).

There are two ways to obtain a non-sequential assembly: define the displacement vectors and find the slides or define the slides and determine the displacement vectors. Both methods are developed in more detail in [2].

The first method analyses the contact surfaces between each part, Fig. 1a. This analyse gives displacement cone $\mathbf{D}_i$, where any displacement vector $\mathbf{V}_i$ can be used without collision with other parts. Then $\mathbf{V}_i$ can be defined in Fig. 2a. To obtain the slides, Fig. 2b, it is necessary to subtract:

$$V_j - V_i = \alpha_i \tag{1}$$

The second method is to define the slides for each part in Fig. 2c. We can extend the slide directions to obtain an intersection corresponding to the OCG. In Fig. 2d, we can define an origin point **O**, anywhere in the plane, to create displacement vector $\mathbf{V}_i$ with the intersection of the slide directions. The oriented connectivity graph can be interpreted as a velocity graph, where vertex positions correspond to the velocity of the components. The length between the vector's origin **O** and a node of the OCG gives the relative speed of a part $\mathbf{P}_i$.

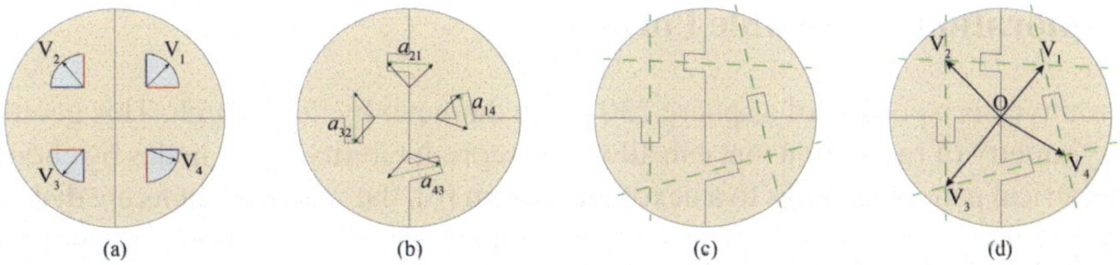

Fig. 2. Generation of non-sequential assembly; (a) choice of displacement vectors, (b) obtaining the slides, (c) choice of slides directions, (d) obtaining the displacement vectors.

Note that for both methods, it is possible to define the length and thickness of the tenons (by multiplying α_i by a coefficient λ for the length and with an offset of αi for the thickness).

The second method gives the designer more freedom to choose the kinematics. When **O** moves, the global kinematics change, but the relative speed between parts is maintained. Moreover, thanks to Combescure transformations in [3] and [4], also known as parallel transformations, we can deform the OCG to obtain the desired kinematics. This transformation preserves angles and gives many geometries for the OCG: it is possible to deform the OCG from Fig. 3a to Fig. 3b, where the quadrangle is flattened to be thinner.

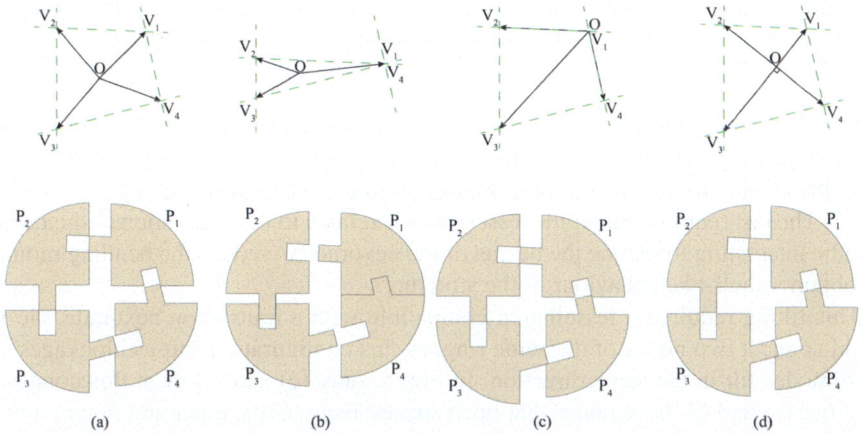

Fig. 3. Different kinematics for a non-sequential assembly; (a) iso-velocity, (b) two parts that do not separate from each other, (c) O coincident with a vertex, (d) same angle between displacement vectors.

Depending on the position of O and the Combescure transformation, the same assembly can have many kinematics. Figure 3 shows different kinematics: iso-velocity, two parts that remain joined, a part that remains stationary and the last kinematic where the angles are equal between each displacement vector. [5] detailed other rules that may apply to non-sequential assemblies to help the designer design these assemblies.

3 Exploration of Possible Forms

The prototype is a portion of a sphere 230 cm in diameter by 70 cm high. This geometry has a symmetry of revolution that can allow the reproducibility of the nodes by choosing a symmetrical pattern. Thanks to a nexorade system [6], the wooden beams are distanced to simplify the assembly node. This distancing does not affect the non-sequentiality of the nodes.

The first step is to choose the tessellation according to the different requirements to determine the form. Generally, the dome has a single singularity at the top. For a nexorade configuration, it is easier to have similar-sized cells, e.g. by increasing the number of singularities from the Euler polyhedron formula. In Figs. 4b–4d, one can see a sphere decomposed by four, eight and n singularities and their influence on the size of the cells.

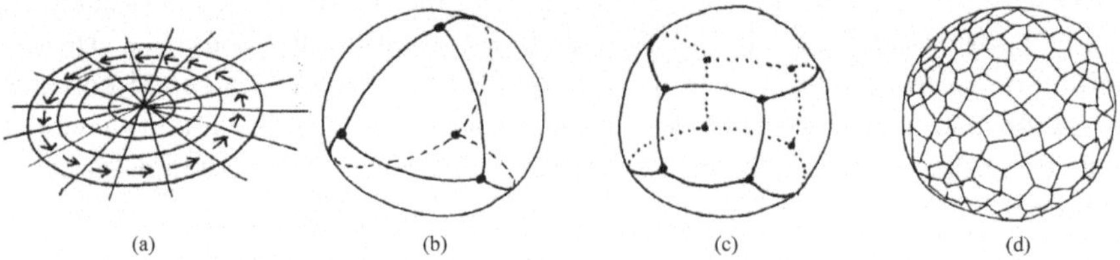

Fig. 4. Images from [7]; (a) tessellation with 1 singularity, (b) decomposition of the sphere with 4 vertices, (c) decomposition with 8 vertices, (d) decomposition with many vertices.

It is also required to choose a pattern where beams can be connected as a succession of non-sequential assemblies. It is interesting to consider a ring assembly starting from the centre of the dome, closing one ring to continue to the next radially, as shown in Sect. 5. The last requirement of the tessellation is related to the inclination of the beams: it can be interesting to incline the beams of the nexorade to reduce the bending moment and obtain a shell-like behaviour of the structure.

This tilting required a tessellation compatible with a translation nexorade. No bar should connect two nodes of the same ring, as this configuration causes blockages and not all nodes tilt in the same direction. In Fig. 5, only (c) and (e) meet this condition, while (a), (b) and (d) have nodes that open successively on the outer and inner parts.

Finally, the final pattern in Fig. 5e has a rotational symmetry of order three that satisfies all the previous requirements. A ring system develops from the centre of the pattern, and the addition of nodes to each ring avoids cell enlargement. In Fig. 6a, the pattern allows generating a nexorade by translation or rotation. This pattern can be

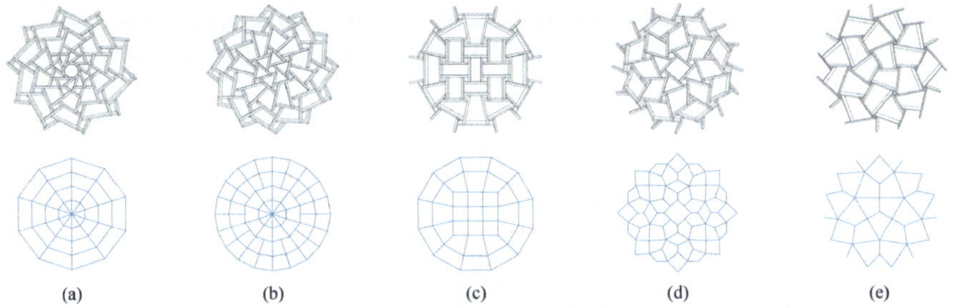

Fig. 5. Pattern research; (a) radial with 1 singularity, (b) fractal: subdivision of cells, (c) 4 singularities, (d) quarter circle duplicated by rotation, (e) third of a circle duplicated by rotation.

extended by adding singularities; the development of four rings of 230 cm diameter was chosen for this pavilion.

Once the pattern is defined, it is applied to the portion of the sphere using a radial projection from the sphere's centre to the target surface. This projection allows keeping a homogeneity of cell size, unlike a vertical projection.

To generate the nexorade, a Grasshopper plug-in, developed in [8], is used. The inputs are the mesh, the engagement length, a lambda (a weight factor for engagement length and eccentricity), and whether the result should be an alternating nexorade. This plug-in outputs a wireframe model to which the beam dimensions must be added to give the final geometry Figs. 6b and 6c.

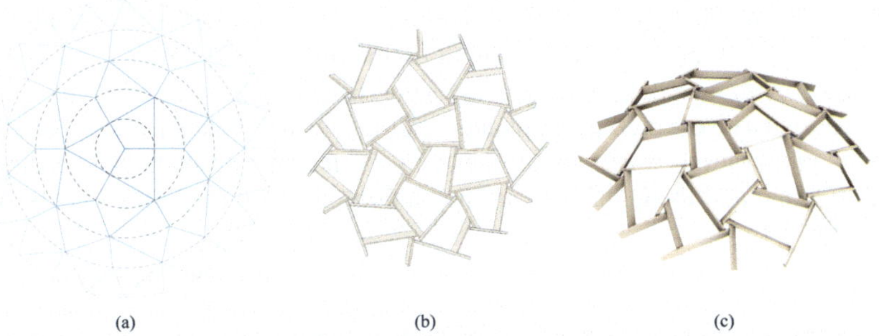

Fig. 6. Selected pattern; (a) pattern expansion, (b) top view, (c) perspective view.

4 Mechanical Model

To verify the mechanical behaviour of the structure, a linear elastic analysis in Karamba3D was performed. The nexorade structure has an eccentricity of the bars at the nodes, so it is necessary to connect the extremities to create the model in [9]. Springs that are very stiff relative to the beams are used to model this eccentricity. The vector of translational spring stiffness Ct is 1010 kNm/rad, for x, y and z about the local direction, and the vector of rotational spring stiffness Cr is 104 kNm/rad, for x, and 1010 kNm/rad, for y and z. In Fig. 7, one can see in blue the springs that connect the ends of the beams to the closest point on the adjacent beam. The beams have the correct neutral axis, and the springs allow the eccentricity to be managed.

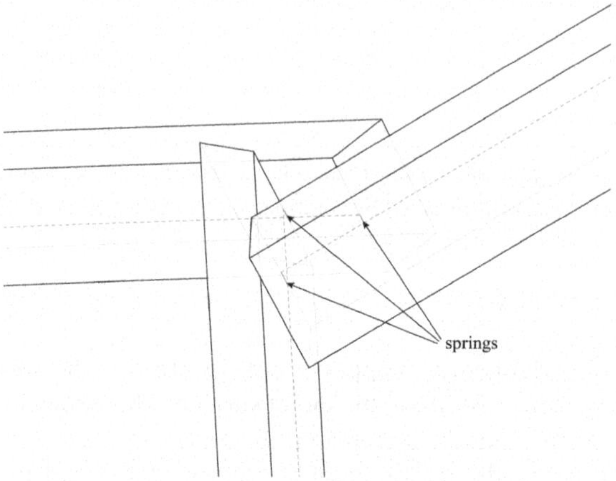

Fig. 7. Modelling eccentricity with spring between beams.

In [10], unbraced nexorades suffer from poor structural behaviour, and the scalability of these structures is very limited: the area density increases considerably with the span and having a span greater than 25 m is unviable. To deal with this problem, this research investigates tilting the beams from their neutral axis. Tilting the beams can also be used to have a node with no engagement length on the outside and with engagement length on the inside. Although this work focuses on the structure, the node without engagement length can easily be covered using planar quadrangles to create a roofed space.

Two configurations are studied: one with the beams perpendicular to the surface normal, Fig. 8a, and the other with an inclination of 30° to the beam's neutral axis, in Fig. 8b (Table 1).

Table 1. Comparison of the pavilion 0° or 30° under uniform load of -1 kN/m^2, gains and losses are calculated considering 0° as 100%.

	Deflection (cm)	My (Nm)	Mz (Nm)	Mt (Nm)	Vy (N)	Vz (N)	N (N)
0°	0.030	22.5	12.0	3.34	574.5	139.7	791.9
30°	0.019	14.6	9.1	2.40	476.1	337.1	786.3
Gains/ Losses	-35.0%	-34.9%	-24.2%	-28.0%	-17.1%	+141.2%	-0.7%

Both geometries are analysed through linear analysis in *Karamba3D*, with the same support conditions and sections (7 × 2,5 cm). The bending moments in the structure change between 0° and 30°. Nevertheless, they remain in the same order of magnitude; for 30°, the bending moment y decreases by 35%, while the bending moment z decreases by 24%. The gain of the inclination is also on the deflection, which decreases by 35%. In Fig. 8, one can see the displacement of both structures under a uniform load of $-10\,\text{kN/m}^2$.

There is a gain in shear forces $\mathbf{V}_y$ of 17%, which is at the detriment of $\mathbf{V}_z$ because the compression is no more transmitted in the thinner part of the beam. For this geometry, tilting the beams seems favourable for the structure, but the model proposes a 50° tilt for optimal deflection. Beyond 45°, the structure approaches a gridshell structure braced with panels, hence the decrease in maximum deflection.

Note that the gain due to the inclination of the beams depends on the pavilion's geometry. For some meshes, 0° is more favourable than 30° inclination.

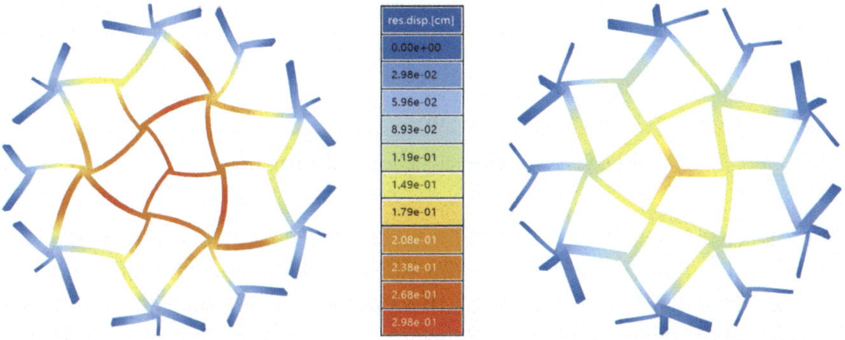

(a) (b)

Fig. 8. Displacement under a uniform load of -10 kN/m^2, seen from the top (displacements are magnified 25 times); (a) beams normal to the surface, (b) beams with a rotation of 30° to the normal.

5 Sequence of Construction

For this prototype, each node of the nexorade must be non-sequential. The edge of the mesh is an exception to this rule as it contains nodes of valence two, which are sequential by definition. There are many possible sequences; one is proposed in Fig. 9. The minimum valence to make a non-sequential assembly is three. Some nodes are sometimes first built partly with a sequential assembly, circled in Fig. 9e and finished by a non-sequential assembly in Fig. 9f. The successive locking of nodes has been privileged in this method, which implies the manipulation of several bars at once. Another method may be to add only two bars per step, thus reducing the number of robotic handlers, but this is more restrictive and does not work for edges.

This sequence consists of nineteen steps, but the geometry is a radial symmetry of three, so the number of steps can be reduced to seven. The first step is to assemble the central node of the dome, which has a valence of three (a). Three bars are added to close the valence node four of the second ring (b). It is necessary to privilege the closing of the

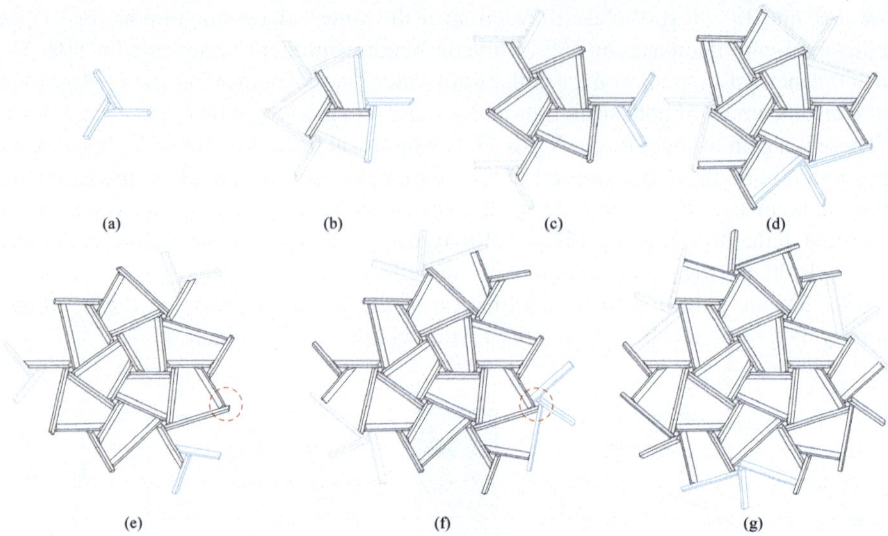

Fig. 9. Assembly sequence of the pavilion.

smallest valences to avoid blocking the kinematics. The third step is to close the valence node three of the third ring rather than the valence node five of the second (c). After that, one can close the valence node five (d). As before, valence node three of the fourth ring is preferred (e). Then, one of the valence five nodes of the last ring is assembled (f). Finally, the last node is assembled (g).

6 Construction Process

The final pavilion is a wooden nexorade dome made from non-sequential assemblies. The wood used for this project is reclaimed Douglas from the offcuts of a landscaping project.

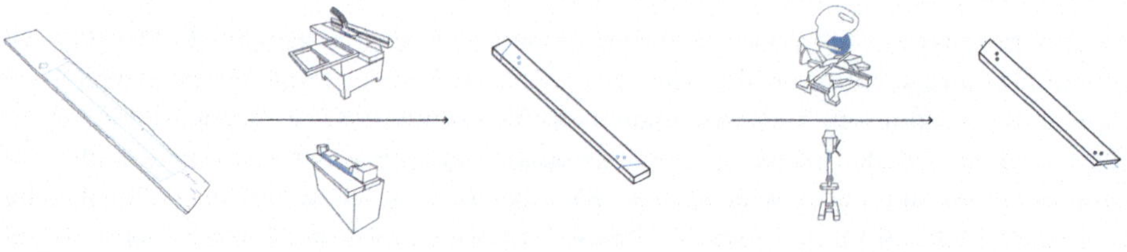

Fig. 10. Machining of beams

The recovered boards have been deformed due to outdoor storage. They are therefore machined to obtain beams with regular faces. Figure 10 shows the machining process.

First, the beams are sanded and cut to the correct dimensions (7 × 2,5 cm). The various drilling points and mitre cuts must then be marked out. Laser-cut guides are used, in Fig. 11, to drill the beams with the required orientation.

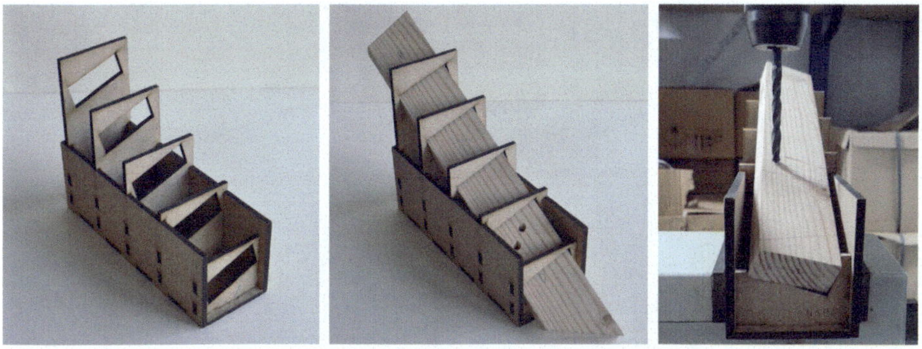

Fig. 11. Guides used with a drill press

As mentioned in Sect. 2, it is required to have a unidirectional slide between each part of a node. They are made with hardwood tenons inserted into the beams in Fig. 12. In order to have tenons with sufficient length in the beam, the outer node has an engagement length. Reducing this engagement length to zero may be possible, but perfect management of the tenon's location is required. Once all the parts have been prepared, the pavilion should be assembled using the sequence selected in Sect. 5. As the pavilions are assembled by humans and not by robots, it was necessary to introduce gaps in the assemblies. Thus the tenons are 0.8 cm and the mortises 0.85 cm in diameter.

Fig. 12. Final pavilion

Once the pavilion is assembled, a wooden piece is added to each extremity of the beam for the supports. There are 15 supports, and a steel cable runs between each to create a compression ring. When this ring is in place, the structure becomes rigid, functioning like a shell. The pavilion can be assembled or disassembled in twenty minutes by three people. Once dismantled, the 51 beams, the 15 supports and the steel cable do not occupy an ample space (Fig. 13) and can be easily moved to build the structure in another spot.

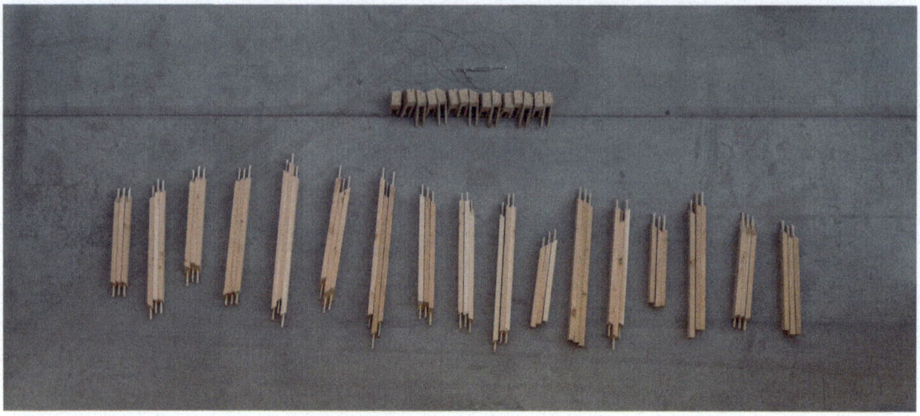

Fig. 13. Dismantled final pavilion

7 Conclusion and Future Works

In this paper, the non-sequential nexorade pavilion highlights some interesting structural aspects. First, non-sequential assemblies may be the solution for designing reversible structures. As seen in Sect. 2, these complex assemblies follow several rules that can help the architect design these assemblies. Although it is made by humans and not robots, the prototype shows that the assemblies can be used to create a structure that can be easily assembled and dismantled.

Secondly, in some cases, it is interesting to incline the beams of a nexorade to optimise the section used and facilitate the structure's cover: quadrangles are flat, and the absence of an engagement length on the outside facilitates the connection of the panels.

Nevertheless, a theorem remains to be determined about the assembly sequence and the number of simultaneous operations, which will be the focus of a forthcoming paper. Further research is necessary to quantify the potential gains from tilting the beams. Finally, it may be important to consider how insulation and cladding could be integrated into this assemble/disassemble process.

Acknowledgement. The authors would like to thank the EAVT Paris-Est. In particular, the STEFANS student group: Vincent Barazzutti, Antoine Bayard, Gaspard Leveque, Sosava Peka, Lancelot Senlis and Paul Thieffry. Thanks also to the Master Matière à Penser and the seminar teachers: Jean-Aimé Shu, Margaux Gillet and Camille Boutemy. This project is supported by Future I-Site.

References

1. ADEME: Déchets chiffres-clé - édition 2020. ADEME, Angers (2020)
2. Glath, J., Gobin, T., Mesnil, R., Mimram, M., Baverel, O.: Design method for non-sequential assembly. In: Proceedings of the IASS Annual Symposium 2020/21 and the 7th International Conference on Spatial Structures, pp. 2367–2376 (2021)

3. Pottmann, H., Liu, Y., Wallner, J., Bobenko, A., Wang, W.: Geometry of multi-layer freeform structures for architecture. In: ACM SIGGRAPH 2007 Papers, pp. 65–es (2007)
4. Mesnil, R., Douthe, C., Baverel, O., Léger, B.: Morphogenesis of surfaces with planar lines of curvature and application to architectural design. Autom. Constr. **95**, 129–141 (2018)
5. Glath, J., Mesnil, R., Mimram, M., Baverel, O.: Theory and design method for non-sequential assembly. Autom. Constr. (2022, under review)
6. Baverel, O., Nooshin, H., Kuroiwa, Y., Parke, G.A.R.: Nexorades. Int. J. Space Struct. **15**(2), 155–159 (2000)
7. Petit, J.-P.: Le Topologicon. Belin, Paris (1992)
8. Mesnil, R., Douthe, C., Gobin, T., Baverel, O.: Form finding of nexorades using the translations method. Autom. Constr. **95**, 142–154 (2018)
9. Sénéchal, B., Douthe, C., Baverel, O.: Analytical investigations on elementary nexorades. Int. J. Space Struct. **26**(4), 313–320 (2012)
10. Mesnil, R., Douthe, C., Gobin, T., Baverel, O.: Form finding and design of a timber shell-nexorade hybrid. In: Advances in Architectural Geometry 2018 (AAG 2018), Göteborg, Sweden (2018)

Data Based Decisions in Early Design Stages

Niklas Haschke(✉), Alexander Hofbeck, and Ljuba Tascheva

Bollinger+Grohmann Ingenieure, Westhafenplatz 1, 60327 Frankfurt am Main, Germany
{nhaschke,ahofbeck,ltascheva}@bollinger-grohmann.de

Abstract. This paper presents an approach for embedding data-based decisions into early design stages of processes in the AEC industry. In this approach, key performance indicators (KPIs) like life-cycle assessment characteristics (LCA) and recyclability categories are introduced to inform the users about the impact of their decisions. The developed concept establishes a platform-agnostic way using open-source tools and web technology to improve the applicability of LCA-analysis and the creation of multiple building variants. These tools were developed by the author in collaboration with Bollinger+Grohmann and the Bauhaus University Weimar to visualize the influence of design decisions and to provide a new workflow for LCA in the very beginning design stages.

Keywords: Data based decisions · Life-cycle-assessement · Web-framework

1 Introduction

Society in general and AEC in particular are facing major challenges. Climate emergency, resource scarcity and lagging digitalization are just a few to name. The invention of new tools or, more likely, the adaptation of existing concepts outside the industry are necessary to overcome these problems according to the principle: "The technology's figured out. The software's figured out. Processes are mostly figured out. We just have to readapt them to our industry." (Deutsch 2015, p. 32).

1.1 Climate Crisis

Compared to other industries, the construction sector has the strongest impact on the environment, as it accounts for 40% of global carbon dioxide emissions (World Business Council for Sustainable Development 2021, p. 5). While the building's usage has a major influence on that part, the production and construction process are responsible for over 50% of the building's lifecycle emissions (Gibbons and Orr 2020).

Considering the population growth and increasing wealth of emerging markets, it is necessary to implement new ways of thinking to address these issues. Since every building activity has a large environmental effect on its own, pragmatic approaches such as building less or building clever should always be prioritized.

Other parts of the solution to reach net zero 2050 are the invention of new and more environmentally sustainable materials and the development of new tools that exploit previously unused potential. The latter is the focus of this work.

C. Gengnagel et al. (Eds.): DMS 2022, *Towards Radical Regeneration*, pp. 260–268, 2023.
https://doi.org/10.1007/978-3-031-13249-0_22

1.2 Potential of LCA in Early Design Stages

According to the frequently cited MacLeamy curve, design decisions have the greatest influence in the early design stages of a building project. The further the planning process progresses, the greater are the effort and the potential costs associated with changes to the original design concept (Borrmann et al. 2015, p. 6). This applies also to LCA-analyses.

In conventional planning processes, the issue of LCA is incorporated into the design consideration at a point in time, when the building masses are already conceivable. These are then connected to their corresponding environmental indicators like Global Warming Potential values (GWP) to make assumptions about the building's ecological impact. One of the main reasons for LCAs elaborated in this way is to achieve sustainability certifications such as LEED or DGNB standards (Budig et al. 2021, p. 7). Thus, at this stage, the building process is usually already too advanced to derive efficient measures from these elaborations. But at the same time, these detailed input parameters are needed to make significant LCA assumptions.

1.3 Existing Tools

To accompany the planning process with LCA examinations, AEC firms have already a wide range of different options. The most common approach is the previously mentioned process of linking the derived building masses to the respective GWP using in house or provided excel sheets (Wood 2022). Beside the described disadvantage, this method is also prone to manual errors.

To overcome this pitfall, it is also usual to link BIM models to visual programming definitions in example Grasshopper (Budig et al. 2021, p. 8). The script can then generate a LCA from the given input. This approach has the disadvantage that the user must obtain some basic knowledge in the visual scripting environment.

A more sophisticated approach is taken by companies that developed tools like Arup Carbon, CAALA and OneClickLCA. These instruments offer the user a simple and intuitive interface that make the field of LCA accessible to a large number of architects and engineers.Arup Carbon uses Speckle to stream the BIM-model into a web-application, where users can access and easily create reports from the results (Brunn 2020). CAALA and OneClickLCA provide a stand-alone Rhino respectively Grasshopper plugin that connects the supplied 3D building model with environmental product declaration (EPD) databases (Apellániz et al. 2021, p. 2).

While all the mentioned tools have great utility in their intended use case, there are also some deficiencies in the early design stages. Despite their high information content, accurate BIM-models have the disadvantage that their creation is a laborious and time-consuming process.

This inflexibility makes it difficult to compare different design alternatives.

However, they are important in order to estimate the potential influence of a different structural material, a different structural system or changes in the column grid.

2 Methodology

Based on these considerations, the authors saw the urgency to develop a new concept to tackle the environmental challenges and to overcome the mentioned backlog.

To achieve this the developed tool would need to cover the following functionalities:

- A simple and intuitive user interface to enable everyone to make LCA-analyses
- Allowing to get real time feedback on desired performance indicators including sustainability criteria as well as KPIs such as, costs and total self-weight
- Get estimated LCA assumptions for the considered building project
- Allowing different input parameters like material, structural system, column grid, load selection, floor height
- A straightforward generation of multiple building variants to rate the influence of decisions

2.1 Web-framework

The building sector is a highly fragmented industry resulting in stagnant productivity and low sustainability. The digital transformation that has already been adopted in other industries is expected to solve these issues (Holzwarth et al. 2019, p. 1). This shift implies that off the shelf digital design tools are not any longer adequate.

Against this background and to overcome the problem of licensing and compatibility issues, the authors chose a modern way of a platform agnostic approach that is not embedded in any commercial software.

A web framework was developed to increase the accessibility of the created tool and to avoid interoperability issues of different software packages.

The ulterior motive was that anyone with a device and an internet connection can use the tool without having to install any software or plugin on his desktop computer.

The open-source 3D JavaScript library Three.js was used as a framework to handle the geometrical input and the rendering of the resulting building variant. Since Three.js is rather a graphics library than a modeling software, geometry operations that are usually provided by CAD-programs had to be implemented from scratch in Three.js. For the geometry input, the user has either the possibility to draw simple building outlines and cores direct in the browser or to upload them into the application. A JavaScript logic for the calculation algorithms on a Node.js backend was developed to create the desired performance indicators. The JavaScript framework Vue.js was used for the user interaction, while a database connection guarantees to get current data (Fig. 1).

2.2 Computational Design Concepts

The implemented logic of the web framework was created by using concepts of computational design. To extend the Three.js geometry classes, a BIM-like object-oriented approach was implemented to feed the parent class provided by Three.js with more information about the actual structural member.

Furthermore, using a parametric system allowed to generate different building variants from the input parameters. A parametric system is defined by clear rules and constraints that are arranged in a specific order (Bolpagni et al. 2022, p. 58). This enables the user to easily create different variants for the same considered project, without needing to have knowledge in parametric design. Whereas these parametric rules are determined by the developer, the user can explore the different options of the solutions space that the system offers.

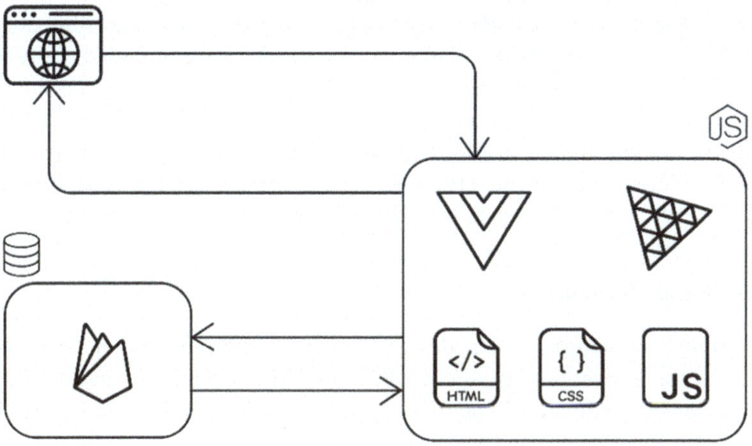

Fig. 1. Open-source web-framework of the developed tool

2.3 Predimension Algorithms

To create a quantity determination from the given input parameters, a predimensioning algorithm was developed that refers to the different selections of material and structural systems. In order to provide immediate feedback, FEM and other computationally intensive methods were avoided.

The predimension algorithm contains different approaches. For some members, the authors chose to implement estimated rules of thumb from academic literature, while others get dimensioned according to common structural proofs. Another way was to embed manufacturer tables into the program logic. These are then evaluated according to the entered load and geometry parameters. The described methods are complemented by data from projects that had been realized by Bollinger+Grohmann.

In total, the user can select between four different structural materials (concrete, steel, timber and a hybrid construction), whereas two choices of slab systems are available for every material (Fig. 2).

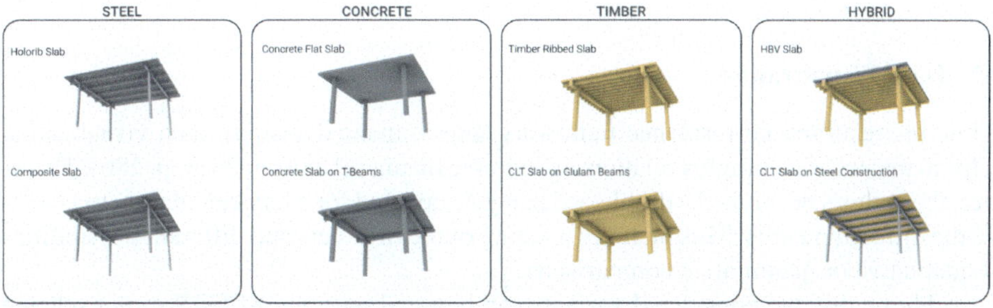

Fig. 2. Implemented structural systems for different materials

Since the tool targets at the early design stages of a project, the algorithm does not try to meet the exact dimension results. Rather, it aims at returning a reliable reference value to compare different variants within a considered project. This approach is adequate for the intended use case, if the inherent system is consistent (Toth et al. 2011, p. 531). To simplify the horizontal load transfer, the assumption was made, that the predimension algorithm considers actions from vertical loads, whereas horizontal loads are carried by reinforced concrete cores. This is applied to all considered building variants.

2.4 Performance Indicators

Once the structural members of a design option are determined, it is important to empower the user to evaluate these outcomes. For this reason, the authors introduced various performance indicators. The most important focus was to give feedback on the environmental impact of a building variant. Thus, the masses were connected to the embodied carbon factors (ECFs) of the respective material.

The user can choose between different ECF databases to consider different project locations. For this study, the ECFs proposed by the ICE and the ÖKOBAUDAT were used (Gibbons and Orr 2020).

Due to the location of this study in the field of structural engineering, the GWP refers exclusively to the building structure. Also, it considers the life cycle modules A1–A3 and C3–C4 for timber structures to prevent negative values for members made of timber. The GWP is not the only indicator to evaluate whether a building is sustainable or not. For example, steel may have a very high GWP, but has the advantage that it can be recycled almost completely. To take this fact into account, two different recyclability indicators were implemented.

The Material Recycling Content (MRC) shows to which parts a material can be recycled based on the actual processes. On the other hand, the Material End of Life (MEoL) indicator gives information about what is happening with the material at the end of the building's life cycle (Hillebrandt et al. 2018, p. 64).

In addition to that, information about the building's cost contrasts the sustainable indicators and the result panel shows other important values like the clear floor height, the structural ceiling height and the total surface area that might also have a big impact on the decisions to be made.

2.5 Data Visualization

While the resulting performance indicators help to make decisions, data visualization helps to understand complex relations in a simple manner (Deutsch 2015, p. 286). Therefore, the authors developed a dashboard that gets updated for every calculation to visualize the different results. This includes a listing of the raw facts and different possibilities of qualitative or quantitative comparisons.

Additionally, the considered variants can be rated with the SCORS rating scheme (Arnold et al. 2020) and the abstract value of embodied carbon is illustrated by the number of corresponding transatlantic flights. The final data visualization panel is a powerful tool to show the relevant building KPIs to clients and architects, so they can take a decision based on a holistic point of view. This is important, because there might not be a single optimal solution in the complex procedure of a construction process (Gholam 2020). The developed dashboard rather enables the user to compare a range of potential solutions (Fig. 3).

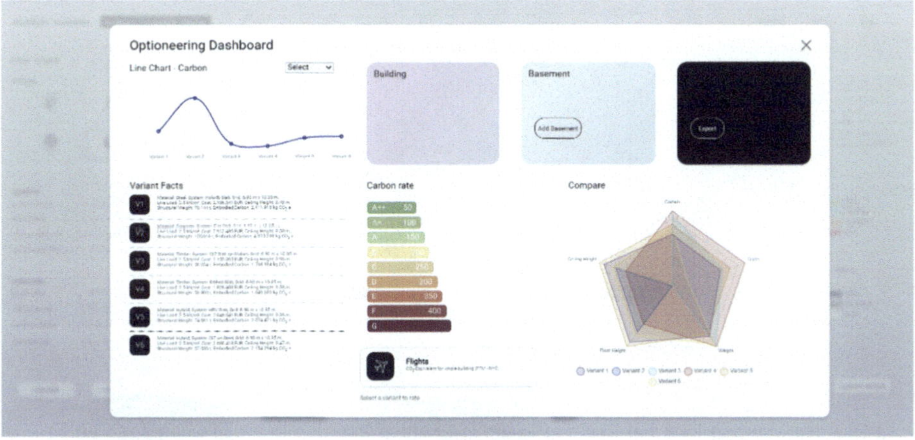

Fig. 3. Developed project dashboard to visualize design decisions

3 Results

In order to check whether the tool is suitable for the intended use case, it was tested in different scenarios.

Since the approach differs significantly from that of existing tools, which require more sophisticated digital models, another testing procedure was used. In the first phase, the tool was tested against other studies that elaborated the embodied carbon of different structures. Here, the developed tool produced similar results on a considered example of a generic six storey building (Roynon 2020) (Fig. 4).

The next step was to use the tool in the planning process of a real-life project by the engineering and design firm Bollinger+Grohmann.

The considered case-study is a six-storey research building that consists of two different building parts. While the structure of one building was predetermined by the activity of vibration-sensitive laboratories, the tool was used to examine different options for the office building. Since the column-grid should follow the predefined structure of the laboratory building, the investigated variants differed in material and structural system selections.

For testing the tool, a simplified floor plan consisting of the building outline and the building cores was uploaded into the web-application. Then, the input parameters were adjusted to fit the requirements of the laboratory building and all possible implemented options of materials and structural options were applied to the examined building. After evaluating the results of each design option, these were then limited to six final solutions.

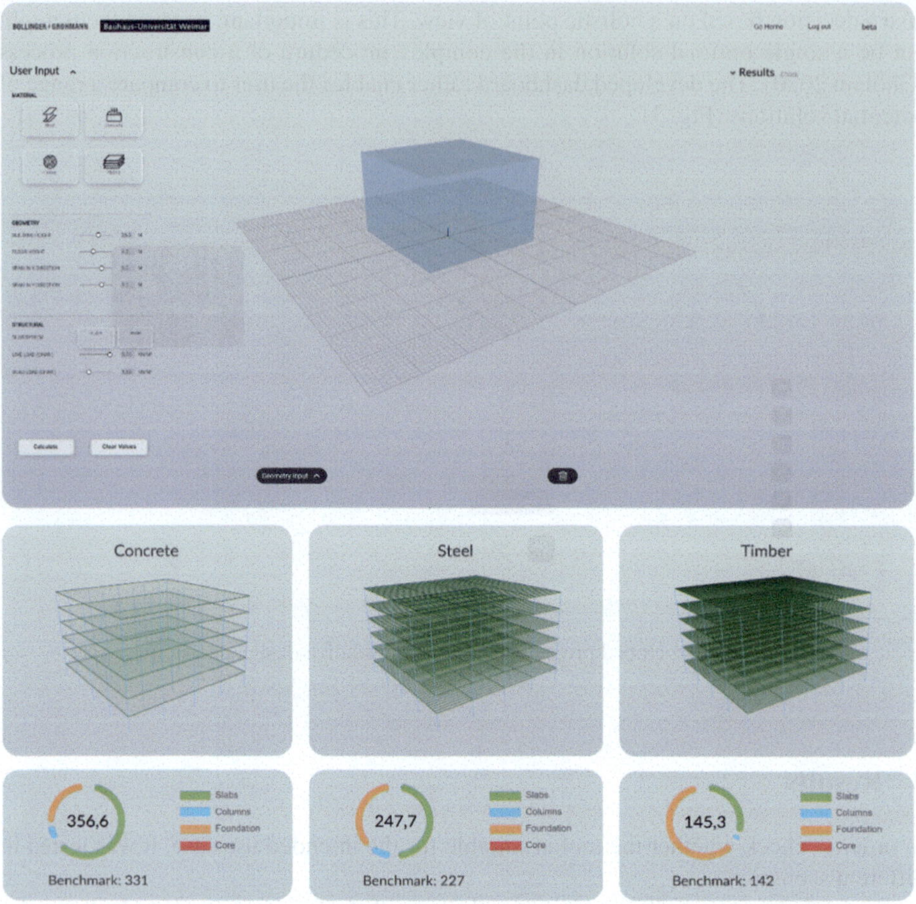

Fig. 4. Results of testing the tool against existing case studies (Roynon 2020)

As the structural embodied carbon and construction costs played an important role in the design process, the construction of a timber ribbed slab or a wood-concrete composite slab (HBV) that both have a low environmental impact while having low buildings costs are the proposed options from the developed design tool.

The final structure of the office building consists of HBV slabs, because other factors that influence design decisions like the susceptibility to oscillation, that are not regarded in the developed tool may have influences on the process as well (Figs. 5 and 6).

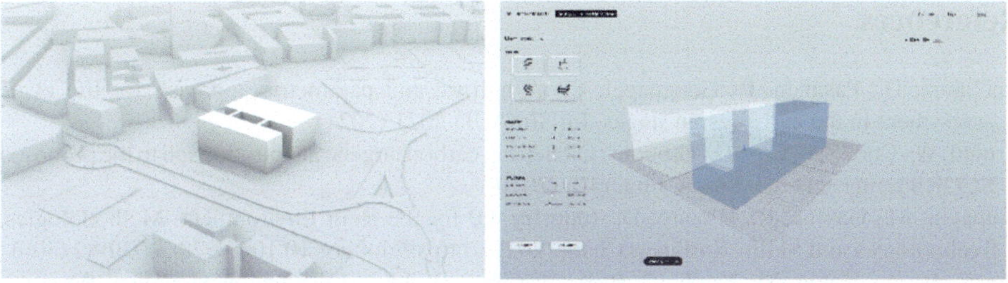

Fig. 5. Initial volumetric model and corresponding geometrical input in the web app

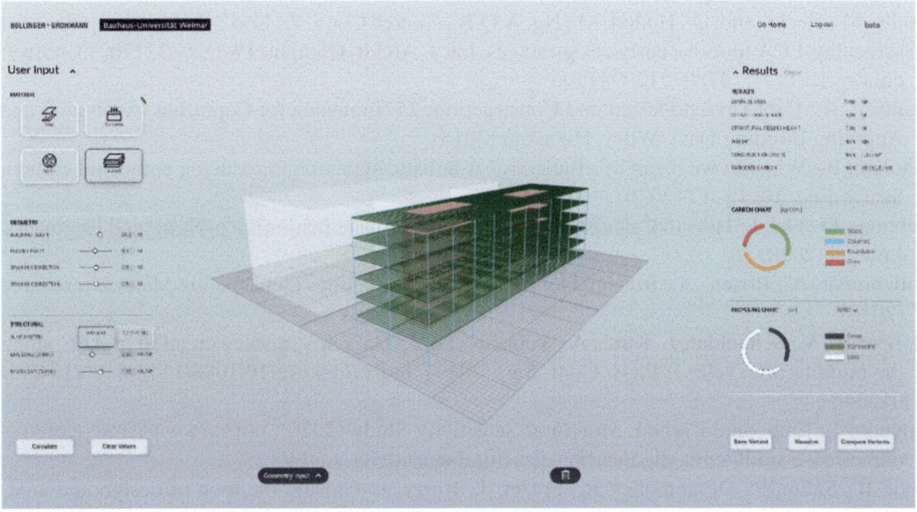

Fig. 6. Results of the finally chosen design variant

4 Conclusion

The described tests have shown that the developed concept offers a valuable tool to embed LCA in early design stages and to contrast sustainability criteria with other performance indicators. It can be stated that the tool produces a valuable design advice for the considered project and enables the user to get LCA estimates of different building alternatives within a short period of time. Here, the developed approach has great advantages over existing tools in terms of usability and time consumption. On the other hand, it does not provide equally detailed information of commercial LCA software, which is also not yet needed in this use case.

The chosen open-source web framework allows an easy way to add future functionalities. The tests have shown that the possibility to add basement levels and the export to commercial FE-software for future considerations might be desirable. In addition, it would be of great benefit to feed more existing data into the tool. In this way, AI algorithms might be the right way to process the quantities and to support the implemented algorithms.

References

Apellániz, D., Pasanen, P., Gengnagel, C.: A holistic and parametric approach for life cycle assessment in the early design stages. In: SimAUD 2021 (2021)

Arnold, W., Cook, M., Cox, D., Gibbons, O.: Setting carbon targets: an introduction to the proposed SCORS rating scheme. Struct. Eng. (10) (2020)

Bolpagni, M., Gavina, R., Ribeiro, D.: Industry 4.0 for the Built Environment: Methodologies, Technologies and Skills. Springer, Cham (2022). https://doi.org/10.1007/978-3-030-82430-3

Borrmann, A., König, M., Koch, C., Beetz, J. (eds.): Building Information Modeling. Springer, Wiesbaden (2015). https://doi.org/10.1007/978-3-658-05606-3

Brunn, T.: Building the right tools for a more sustainable future (2020). https://medium.com/arup-digital-news/building-the-right-tools-for-a-more-sustainable-future-268ae6935c51

Budig, M., Heckmann, O., Hudert, M., Ng, A.Q.B., Xuereb Conti, Z., Lork, C.J.H.: Computational screening-LCA tools for early design stages. Int. J. Archit. Comput. **19**(1), 6–22 (2021). https://doi.org/10.1177/1478077120947996

Deutsch, R.: Data-Driven Design and Construction: 25 Strategies for Capturing, Analyzing and Applying Building Data. Wiley, Hoboken (2015)

Gholam, B.: What do we mean by efficiency? A holistic approach to reducing embodied carbon. Struct. Eng. **10**, 14–17 (2020)

Gibbons, O., Orr, J.: How to Calculate Embodied Carbon. The Institution of Structural Engineers, London (2020)

Hillebrandt, A., Rosen, A., Riegler-Floors, P.: Atlas Recycling: Gebäude als Materialressource (2018)

Holzwarth, V., Schneider, J., Kunz, A., Vom Brocke, J.: Data driven value creation in AEC along the building lifecycle. J. Phys. Conf. Ser. (2019). https://doi.org/10.1088/1742-6596/1343/1/012046

Roynon, J.: Embodied Carbon: Structural Sensitivity Study (2020). https://www.istructe.org/resources/case-study/embodied-carbon-structural-sensitivity-study/

Toth, B., Salim, F., Drogemuller, R., Frazer, J., Burry, J.: Closing the loop of design and analysis: parametric modelling tools for early decision support. In: Herr, C.M., Gu, N., Roudavski, S., Schnabel, M.A. (eds.) Circuit bending, breaking and mending: Proceedings of the 16th International Conference on Computer-Aided Architectural Design Research in Asia (CAADRIA 2011), Circuit bending, breaking and mending, Newcastle. CAADRIA, Hong Kong, pp. 525–534 (2011)

World Business Council for Sustainable Development: Decarbonizing Construction: Guidance for investors and developers to reduce embodied carbon (2021). https://www.wbcsd.org/Programs/Cities-and-Mobility/Sustainable-Cities/Transforming-the-Built-Environment/Decarbonization/Resources/Decarbonizing-construction-Guidance-for-investors-and-developers-to-reduce-embodied-carbon

Robotic Wood Winding for Architectural Structures - Computational Design, Robotic Fabrication and Structural Modeling Methods

Georgia Margariti[1], Andreas Göbert[1], Julian Ochs[1], Philipp Eversmann[1], Felita Felita[3], Ueli Saluz[2], Philipp Geyer[2], and Julian Lienhard[1(✉)]

[1] Universität Kassel, Universitätsplatz 9, 34127 Kassel, Germany
lienhard@uni-kassel.de
[2] Leibniz University, Herrenhäuser Str. 8, 30419 Hannover, Germany
[3] Technische Universität Berlin, Straße des 17. Juni 152, 10623 Berlin, Germany

Abstract. Winding processes are known from the fiber composite industry for strength and weight optimized lightweight components. To achieve high resistance and low weight, mainly synthetic materials are used such as carbon or glass fibers, bonded with petrochemical matrices. For the construction industry, these additive processes present a very promising and resource-efficient building technology, yet they are still hardly used with sustainable materials such as natural fibers or timber.

The 3DWoodWind research prototype has developed a new generation of additive technologies to wood construction. The modular building system is built with a three-dimensional robotic winding process for material-efficient hollow lightweight components. An AI-controlled design logic enables the intelligent combination and design of modular components into multi-story structures, which may be used in the future to substitute solid wood panels and beams as well as concrete slabs and steel sections.

Our current research uses a continuous strip of thin timber veneer, which is a waste product from the plywood industry and therefore, presents a highly sustainable alternative to synthetic fibers usually used in winding, as well as solid timber products known in construction. The veneer's natural fibers are intact and continuous, and offer high tensile strength. In the presented project, three-dimensional winding processes were developed for material-efficient lightweight components made of wood. The demonstrator presents a modular column and ceiling system, which aims at large scale applications in multi-level structures. Having won an open national design competition for Germany's 'ZukunftBau' Pavilion, a first demonstrator is currently being built to be presented in May 2022, as part of the DigitalBau exhibition. The paper discusses all planning engineering and production processes in detail with particular emphasis on the machine-learning algorithm, which was trained during the design process to facilitate design iterations and future planning with this component-based building system.

Keywords: Additive manufacturing · Winding · FE-modeling · Machine learning

C. Gengnagel et al. (Eds.): DMS 2022, *Towards Radical Regeneration*, pp. 269–282, 2023.
https://doi.org/10.1007/978-3-031-13249-0_23

1 Introduction

Winding processes are known from the fiber composite industry for strength and weight optimized lightweight components. In order to achieve high resistance and low weight, synthetic materials are usually used, such as carbon or glass fibers bonded with petrochemical matrices. For the construction industry, these additive processes present a very promising and resource- efficient building technology, yet they are still hardly being used with sustainable materials such as natural fibers or timber.

Winding technologies of paper and veeners date back to the first half of the 20th century and have been gradually re-introduced in the last 20 years. Round wrapped columns were developed in Japan (Hata et al. 2001, Inaba et al. 2003). More recently, marketable round hollow profiles made of wound thin veneer layers were implemented by the company LignoTUBE (Beck 2014, Lignotube 2022). However, these are limited in terms of production technology to a circular, constant cross-section. Discontinuous processes were used to glue wooden sections in rings, for example, to produce poles or columns for the construction sector (Wehsener et al. 2014, Piao 2003). More recent studies show the winding of bamboo veneer with added resins for use in sustainable piping products (Chen et al. 2019). Recent research at University of Kassel explores textile tectonics in timber structures (Silbermann et al. 2019). In the NaHoPro project, the University of Kassel and the HNEE Eberswalde investigated the winding of veneer into circular profiles and subsequent forming processes into rectangular profiles (NaHoPro 2021). In the 3DWoodWind research project, these technologies have been further investigated for robotic free form winding (Goebert et al. 2022).

The design of such novel complex structures necessitates computer assistance during design. Recently, data-driven models utilizing surrogate modelling have been increasingly developed for energy performance prediction (Amasyali and El-Gohary 2018; Westermann and Evins 2019) and for structural design (Han and Liu 2020; Salehi and Burgueño 2018). These methods offer real-time feedback. During designing, not only pure analysis is of interest but also design space exploration (DSE), examining possible variations of a design configuration (Østergård et al. 2017). The ML-based surrogate methods allow for providing such comprehensive real-time assistance performing DSE and enabling causal reasoning embedded in digital design and modelling processes (Geyer et al. 2018; Chen and Geyer 2022).

With the realization of the 3DWoodWind research prototype, the above-mentioned technologies and design methods were implemented in a full-scale wood construction.

2 Computational Design and Manufacturing

2.1 Design Process

The 3DWoodWind Research Prototype consists of several winded components (Fig. 1): (a) A modular ceiling system made of equal, rectangular hollow profiles, which are supported by (b) mushroom-shaped columns. (c) Additional seating elements and tables were organized and placed freely according to the spatial division. The overall structure spreads out over 6 × 6 m with a clearance of 2.50 m. In total 41 ceiling modules with a

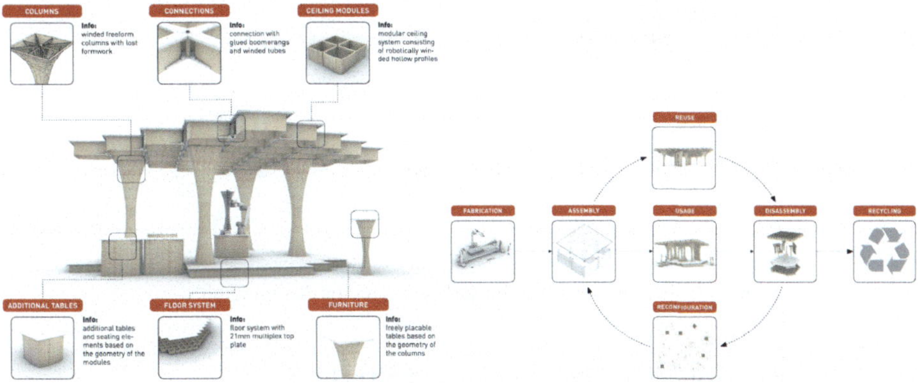

Fig. 1. Design process and modular component concept.

height of 310 mm and a width of 825 mm and four 2.80 m high freeform columns with varying cross-sections were used.

The components are designed in a modular concept, so that they can be reconfigured and reassembled in completely different scenarios. Apart from the actual application as a temporary exhibition pavilion, we developed a series of alternative scenarios with different requirements according to their configurations (Fig. 2).

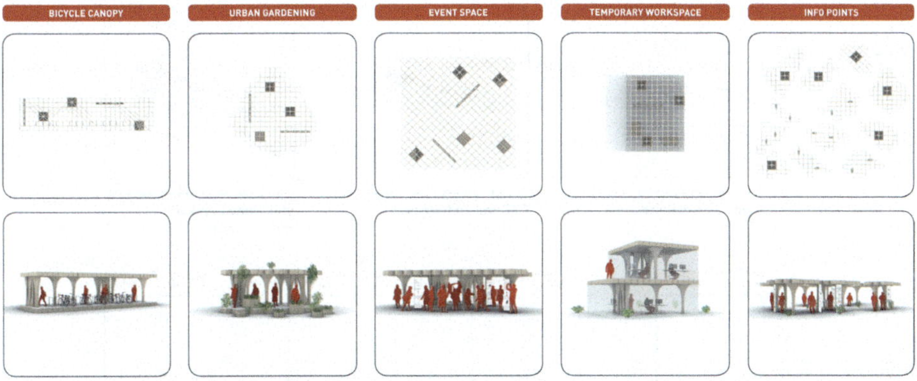

Fig. 2. Variation possibilities for the modular component concept.

2.2 Robotic Fabrication

For the realization of the winded ceiling modules and columns we used an ABB IRB4600-40/2.55 industrial robot arm with an ABB IRBP-L300-L4000 external workpiece positioner, on which the formwork was installed. A customized end-effector with a (a) filament guide, (b) stepper motor for extrusion, (c) extrusion roll and (d) cutting tool was designed for applying the veneer precisely on the formwork (Fig. 3).

In addition, we integrated a glue application system into the robotic winding process for a fully automated application of a defined amount of adhesive.

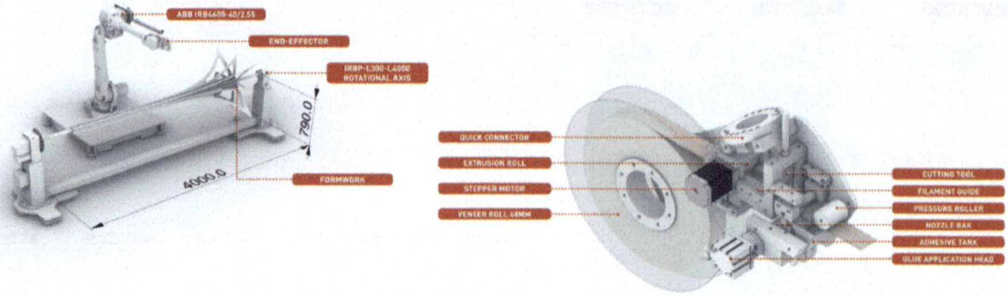

Fig. 3. Robotic setup (left) and end-effector design (right).

2.3 Component Geometry, Fibre Layout and Structure

Columns. Regarding the geometric possibilities of the fabrication technique, we investigated a variety of different component geometries. The principle of the geodesic line, which forms the basis for the robotic toolpath, immediately affects the design process. As the generated winding lines are densifying in areas with concave curvature and spacing out in convex areas, not only the material distribution is controlled by the geometry, but also the fibre directions are changing due to the various curvatures on the surface itself (Fig. 4).

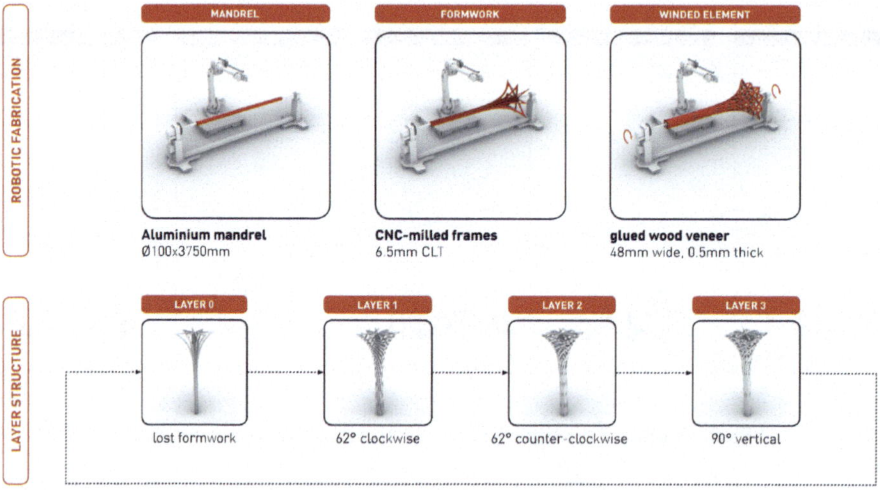

Fig. 4. Robotic fabrication and fibre layout of the columns.

Based on this principle, we investigated different cross-sections and evaluated their resulting fibre layouts according to the actual application possibilities and their design qualities. The best performing results were then selected and analyzed in detail according to their load-bearing behavior. The fibres were arranged in three directions: A first layer with 12 vertically placed fibres, followed by two layers of 8 clockwise and counter-clockwise oriented fibres. This pattern was shifted into the spaces and repeated (Fig. 5).

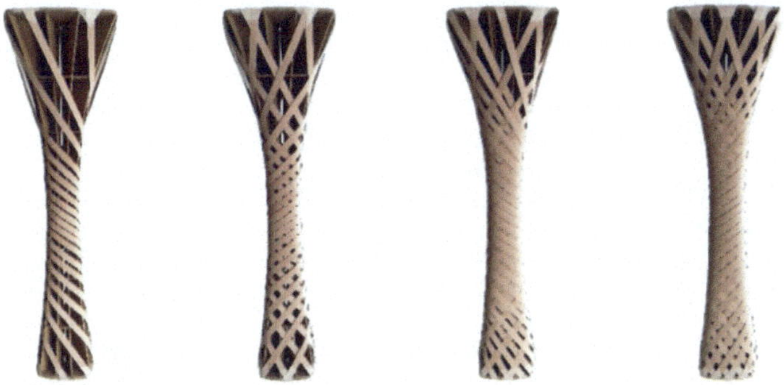

Fig. 5. Fibre layout and layer structure of the columns (from left to right): 8 strips clockwise (cw) and counter-clockwise (ccw), 8 strips 50% shifted cw and ccw.

Ceiling Modules. To ensure a sufficient contact pressure for the adhesive, we adjusted the rectangular shape of the modules by adding a slightly convex curvature to the component geometry. As a result of this, the veneer could be hold on tension during the winding process, so that the robot arm moved along a linear path parallel to the external rotational axis (Fig. 6). In doing so, the typical complex robot movements when winding rectangular shapes could be replaced with simple and fast motions, which led to a very high acceleration of the manufacturing process.

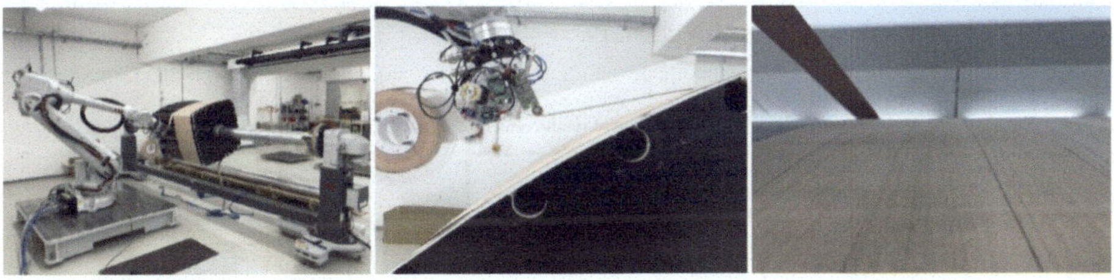

Fig. 6. Robotic winding process for the fabrication of the ceiling modules.

3 Structural Design and Simulation

3.1 Structural Scheme

The pavilion proposes a modular ceiling system supported by 3D wound columns. The composite ceiling consists of hollow 3D wound modules attached to a top plate and connected to each other at the bottom behaving like a hollow slab with structural height equal to the height of the modules (Fig. 7). Regarding the columns, apart from geometrical restrictions derived from the winding process of the outer surface, their shape is driven by the bending moments distribution of moment stiff columns at the top and bottom (Euler case 4). The columns consist of an inner plate structure which serves also as a lost formwork and to carry vertical loads. The exterior layer consists of the wound surface whose structural role is to stabilize the columns against buckling.

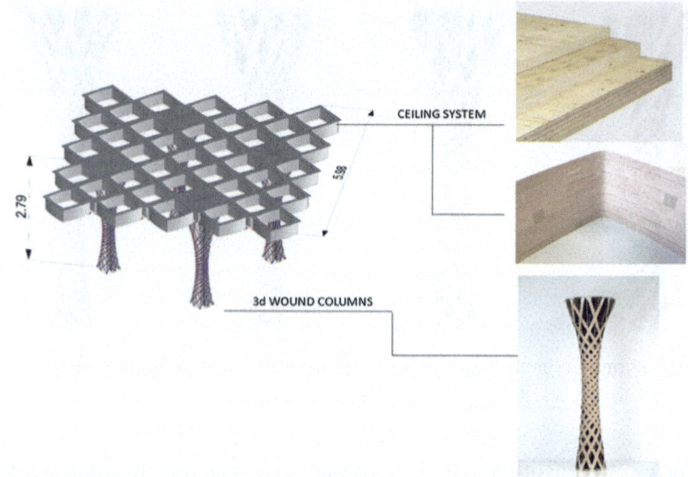

Fig. 7. Schematic plan view of the connection of the modules and resulting deformation.

3.2 Connections and Details

As in every modular system, the stiffness of the structure relies significantly on the type of connections between the structural elements. After a long investigation on the type of connections between the modules at the bottom (Fig. 8) it was proven sufficient that they will be connected only at the corners.

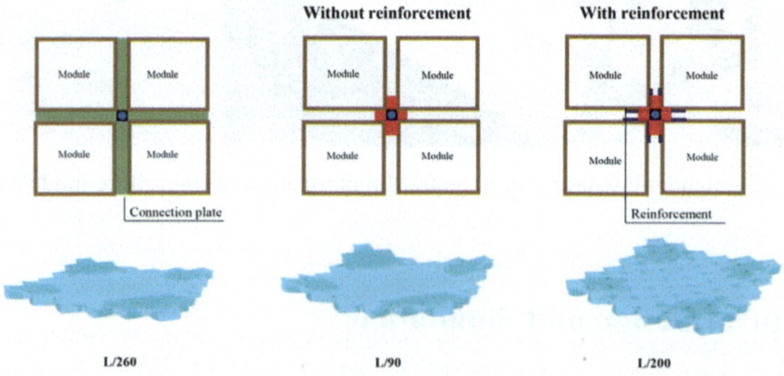

Fig. 8. Schematic plan view of the connection of the modules and resulting deformation.

A boomerang-shaped plate is glued at the corners of the modules and a star shaped plate connects the top and bottom of the ceiling through a steel rod (Fig. 9). The design of the detail aims to minimize material use, facilitate the assembly process and emphasize the structural role of the modules. At the corner area of high stresses, the boomerang element acts also as reinforcement of the modules. Finally, this detail offers the possibility to precamber the slab during assembly and minimize deflections under self weight.

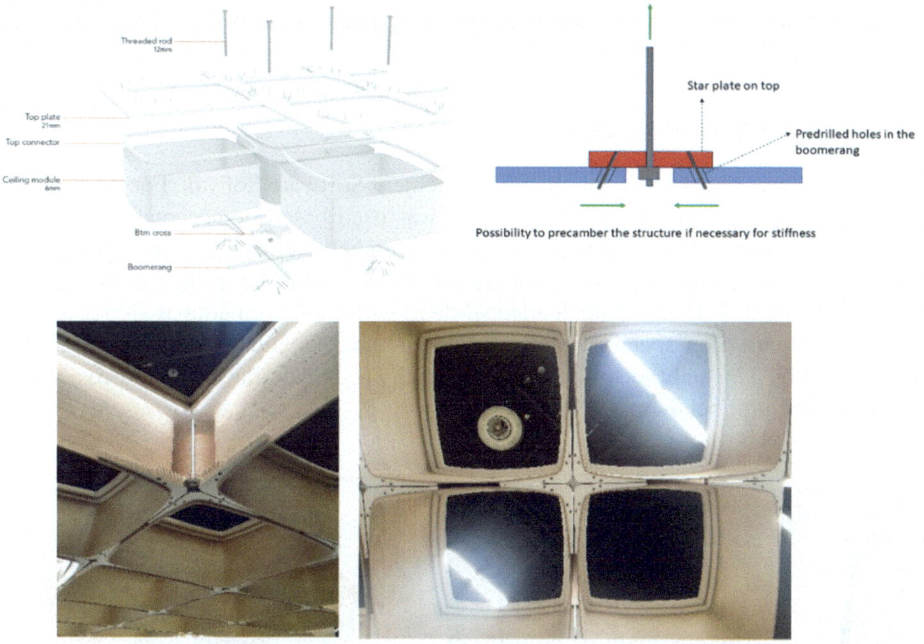

Fig. 9. Design and structural principle of the connection detail.

3.3 Material System

In order to simulate the structural behavior of the structure, material testing was conducted to derive the material properties of the laminate produced by the winding process of the veneer explained above. Fiber layouts with different layer structure and fiber orientation were tested as well as different glue types. The final layer structure includes 12 layers of 0.5 mm of veneer in angles of 0 degrees. For the orthotropy of this material,

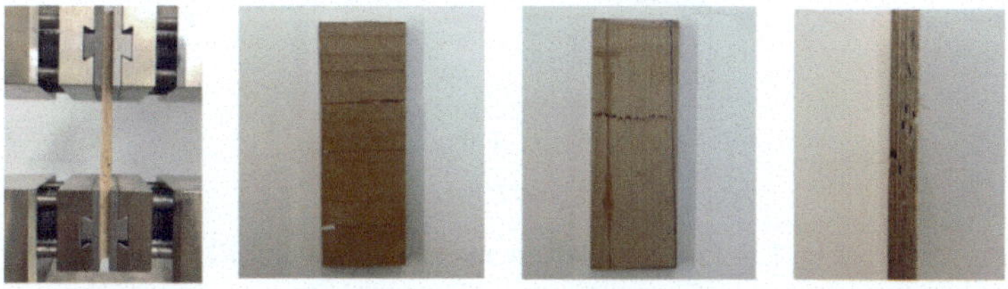

	$AN_{average}$	$JO_{average}$	AN_{sd}	JO_{sd}	AN	JO	
$E_{90,(1000)}$	925.13	1244.13	125.86	55.91	799.27	1188.22	N/mm^2
$F_{T,90}$	2314.88	4575.31	647.04	992.82	1667.85	3582.49	N
$f_{T,90}$	7.81	15.41	2.18	3.34	5.63	12.07	N/mm^2

Fig. 10. Test specimens with different glue. Table of the Young Modulus and Tensile strength of the module in the cross fiber direction.

tests for tension, compression and bending were carried out in two directions, parallel and perpendicular to the fibers (Fig. 10).

3.4 Global Design and Simulations

The pavilion was simulated using the Finite Element Software Sofistik. For the position of the columns, the machine learning model was used based on acceptable mass and deformation criteria. The modules are simulated as shell elements and the connections between them as spring elements. To define the stiffness of these springs a model in 1:1 scale was fabricated (Fig. 11) and loaded while measuring deformation to derive a load-displacement curve for specific points of the structure. An equivalent model was then simulated (Fig. 12), and the stiffness of the springs were calibrated in order to achieve the same structural behavior.

Fig. 11. Structural test – P_{max}: 175 kg each point.

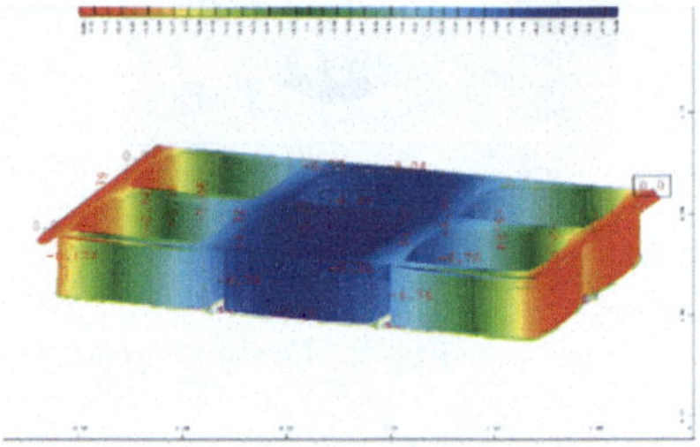

Fig. 12. Equivalent FEM model - Maximum deformation under 350 kg $\delta = 8.39$ mm.

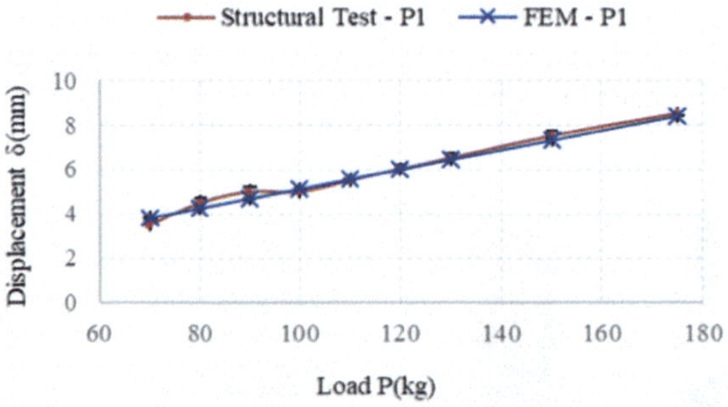

Fig. 13. Load displacement curves for point P1.

The diagram above (Fig. 13) shows the deformation of the structural test and the equivalent FEM model. The two curves are significantly close indicating the high precision of the simulation model to the real behavior of the system.

In addition, the global structural behavior is highly influenced by the exterior wound surface of the columns. The Fig. 14 below shows the first buckling modes with or without consideration of the wound surface.

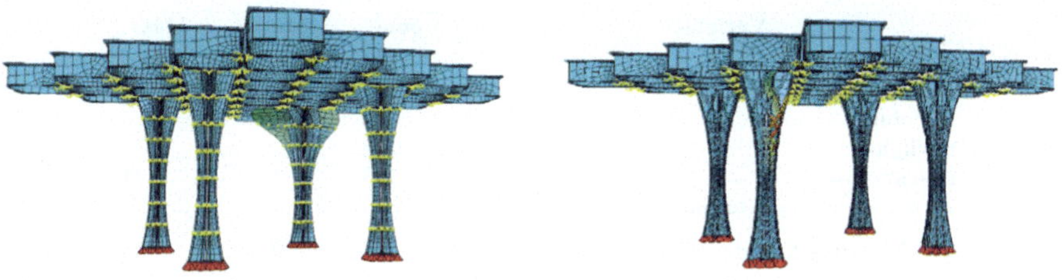

Fig. 14. (**a**)**.** Without wound surface: load factor: 1.08. (**b**)**.** With wound surface first: load factor: 2.28.

4 Machine-Learning-Based Dcsign Assistance

The complexity due to nonlinear material properties, energetic properties and systems in the presented wood wrap structure requires computationally intensive simulation models slowing down design processes. To overcome this bottleneck, an AI-based assistance system has been developed to support the design process based on simulated data sets in real time (Fig. 15).

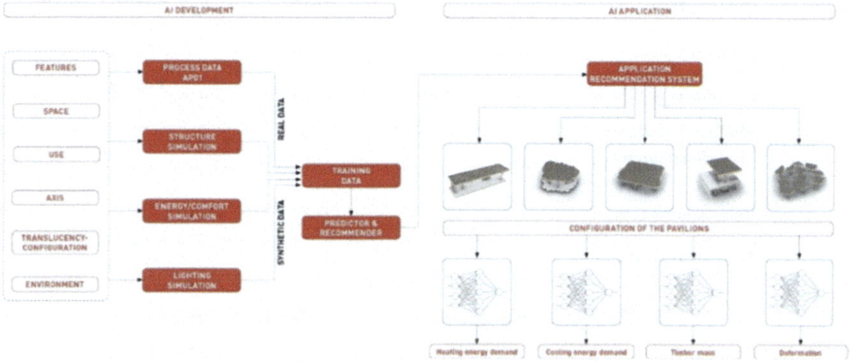

Fig. 15. Data pipeline for machine learning assistance.

4.1 Energy and Structure Simulation

Training data for energy and structure performance were with the variables and outputs shown in Figs. 16 and 17. For energy simulation, has been used to generate 243 data points. The parametric energy model was created using Energy Plus/Honeybee in Rhino and Grasshopper. The simulation generated 243 data points representing the variants of the design. The data set of 1659 data points for structure was generated using a high-fidelity Sofistik shell model and a fast-running parametric simulation using Karamba and Rhino/Grasshopper calibrated with the Sofistik model.

Variable	Count	Min	Max
WWR	243	0.5	0.9
Overhang	243	0.1	0.5
U-Value	243	0.6	1
Skylight	243	0.3	0.5
Orientation	243	0	180
Output			
Total Heating Energy (kWh/Annual)	243	6180.52	7787.40 3
Total Cooling Energy (kWh/Annual)	243	982.626 1	2452.89 7

Fig. 16. Variable and output range of energy data.

Variable	Count	Min	Max
Column A_x	1659	3.34108	7.59432 7
Column B_x	1659	3.34108	7.59432 7
Column C_x	1659	0	3.25269 1
Column D_x	1659	0	3.31633 1
Column A_y	1659	3.25269 1	7.59432 7
Column B_y	1659	0	3.69109 7
Column C_y	1659	3.25269 1	7.59432 7
Column D_y	1659	0	3.69109 7
Module Thickness	1659	5	9
Module Diameter	1659	500	1000
Output			
Deformation (cm)	1659	16.2752	2534.80 2
Timber Mass (kg)	1659	2.77869 6	9.04917 6

Fig. 17. Variable and output range of structural data.

4.2 Machine Learning Model

The machine learning model was trained using the Deep Neural Networks (DNN) method with SciKit-Learn and Keras/TensorFlow Python libraries. Several sets of hyperparameters were tested with up to 4 layers and 16 neurons, that resulted in poor fitting of energy prediction. For that reason, the energy model were built with 6 layers and 10 neurons for heating demand and with 5 layers and 10 neurons for cooling demand. Evidently, more layers and less neurons have more flexibility during the training and gave fairly good predictions. Several set of hyperparameters were also carried out for the structure model, with up to 6 layers and 25 neurons were tested. In the end, set of 3 layers and 35 neurons gave acceptable fitting for deformation and mass prediction. The data for each model is split with 80:20 ratio for training and test data (Figs. 18 and 19).

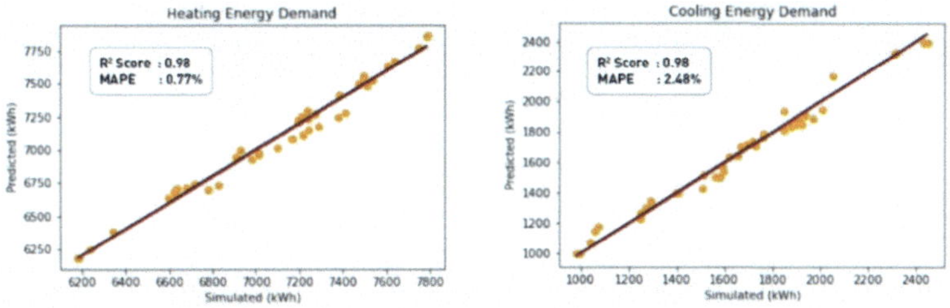

Fig. 18. Simulated ground truth and prediction of energy test data. Left: Heating energy demand; Right: Cooling energy demand.

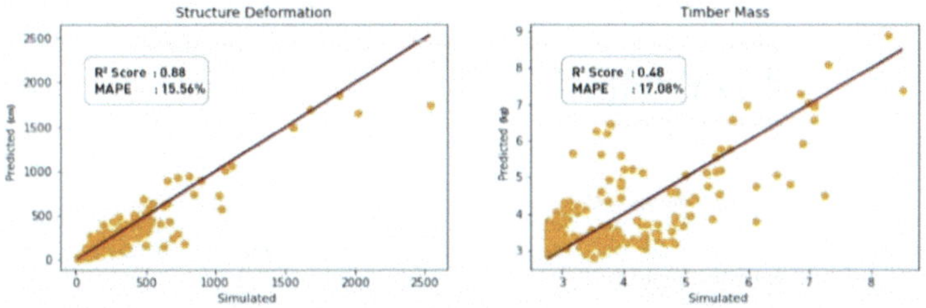

Fig. 19. Simulated ground truth and prediction of structural test data. Left: Deformation; Right: Timber Mass

Figures 20 and 21 show the prediction of the learned models against ground truth for prediction of the four performance indicators. Heating and cooling demand models achieved a good accuracy of a coefficient of determination R^2 of 0.98 both and mean absolute percentage error (MAPE) of 0.77% respectively 2.48%. For structure, the coefficient R^2 is lower equaling 0.88 and 0.48 and the MAPE is much higher equaling 15.56% and 17.08%. Whereas energy model shows satisfying performance for application as assistance model, the structural requires revision. As reason for unsatisfying low fitting of the structural model, we suspect much higher non-linearities combined with the fact the current model was trained with partially not fully-stressed configurations. A revised version of the model in future research will overcome these shortcomings.

4.3 Implementation

For design assistance, the pre-trained models were uploaded into Grasshopper and used as a surrogate model for an energy and structural simulation for interactive assistance in designing as demonstrated by a prototype GUI (Fig. 20). This environment allows interactive creation with real-time analysis of designs variations for this innovative timber structure.

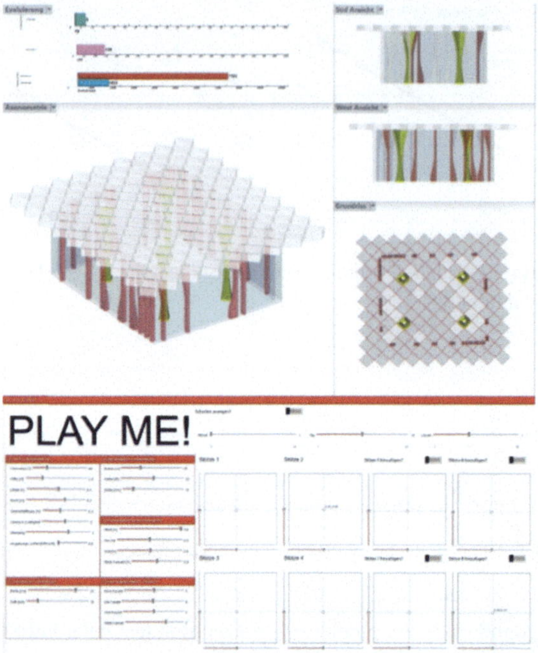

Fig. 20. Interactive AI assistance environment with real-time analysis.

Fig. 21. 3DWoodWind research prototype, presented at Digitalbau 2022 in Cologne.

5 Conclusion

The current work showed the complexities in working with material systems in the context of computer aided design-, engineering- and manufacturing (CAD, CAE, CAM) technologies. It was shown how a real-time assistance based on machine learning is setup that allows to support the design process for the novel construction method. The machine learning models were developed based on results from physical structural and energy simulation models, to predict structural performance, building physics, and material resource performance of parametrically in real time to all interactive design iterations. The problems in fitting the structural data to be enhanced with the next model generation showed that the methodology requires paying careful attention to the nature of the data and their underlying physical models in terms of non-linearity and boundary conditions.

Due to the short time span of this project, all disciplines were simultaneously developing their planning, simulation and production methods. Continuous back feeding of intermediate results into the workflow enabled fast design iterations to increase the overall performance of the system. Results of the AI based design tool, however, could not influence the final design of the first prototype and rather aims at the future development of the system. The good fitting especially of the energy predictions models highlights the potential of AI trained models and their design assistance based on simulated training data.

Finally, the research exemplifies how hollow lightweight components could save large amounts of material in timber construction, and serve as a substitute for solid wood, concrete or steel structures in the future. The design methodology exemplifies how time-consuming structural and climate simulations can be integrated in design processes using machine learning regression models.

Acknowledgements. The work presented in this paper was commissioned and founded by the German innovation program Zukunft Bau within the Federal Institute for Research on Building, Urban Affairs and Spatial Development (BBSR). The official title of this research project is "BBSR Research Prototype 2022".

References

Amasyali, K., El-Gohary, N.: A review of data-driven building energy consumption prediction studies. Renew. Sustain. Energy Rev. **81**, 1192–1205 (2018). https://doi.org/10.1016/J.RSER.2017.04.095

Beck, C., Taranczewski, R.: LignoTUBES–ein runder Furnierverbundwerkstoff für den ökologischen Leichtbau. ECEMP–European Centre for Emerging Materials and Processes Dresden (2014), 15 (1996)

Chen, X., Geyer, P.: Machine assistance in energy-efficient building design: a predictive framework toward dynamic interaction with human decision-making under uncertainty. Appl. Energy **307**, 118240 (2022). https://doi.org/10.1016/j.apenergy.2021.118240

Chen, F., et al.: Development of bamboo winding composite pipe (BWCP) and its compression properties. BioResources **14**(3), 5875–5882 (2019)

Geyer, P., Singh, M.M., Singaravel, S.: Component-based machine learning for energy performance prediction by MultiLOD models in the early phases of building design. In: Smith, I., Domer, B. (eds.) EG-ICE 2018. LNCS, vol. 10863, pp. 516–534. Springer, Cham (2018). https://doi.org/10.1007/978-3-319-91635-4_27

Göbert, A., Deetman, A., Rossi, A., Weyhe, O., Eversmann, P.: 3DWoodWind: robotic winding processes for material-efficient lightweight veneer components. Constr. Robot. (2022). https://doi.org/10.1007/s41693-022-00067-2

Han, X., Liu, J.: Rapid structural analysis based on surrogate models. In: Numerical Simulation-based Design, pp. 97–123. Springer, Singapore (2020). https://doi.org/10.1007/978-981-10-3090-1_6

Hata, T., Umemura, K., Yamauchi, H., et al.: 2001: Design and pilot production of a "spiral-winder" for the manufacture of cylindrical laminated veneer lumber. J. Wood Sci. **47**, 115–123 (2001). https://doi.org/10.1007/BF00780559

Inaba, D., Morita, M., Nakano, H., Takenaka, A., Kawai, S.: Continuous manufacture of cylindrical laminated veneer lumber. Wood Research Nr. 90, Kyoto 2003 (2003)

Lignotube: LignoTUBE – the new semi-finished product for lightweight construction. www.lignotube.com. Accessed 20 Jan 2022

NaHoPro: Rechteckige Konstruktions-Hohlprofile aus biobasierten Multimaterialsystemen als Substitution von Metallprofilen. https://www.fnr.de/ftp/pdf/berichte/22019018.pdf. Accessed 10 Dec 2021

Østergård, T., Jensen, R.L., Maagaard, S.E.: Early building design: informed decision-making by exploring multidimensional design space using sensitivity analysis. Energy Build. **142**, 8–22 (2017). https://doi.org/10.1016/j.enbuild.2017.02.059

Piao, C.: 2003: Wood Laminated Composite Poles. Louisiana State University, Baton Rouge, La (2003)

Salehi, H., Burgueño, R.: Emerging artificial intelligence methods in structural engineering. Eng. Struct. **171**, 170–189 (2018). https://doi.org/10.1016/j.engstruct.2018.05.084

Silbermann, S., Heise, J., Böhm, S., Eversmann, P., Klussmann, H.: Textile tectonics for wood construction. In: Hudert, M., Pfeiffer, S. (eds.) Rethinking Wood: Future Dimensions of Timber Assembly. Birkhäuser 2019 (2019)

Wehsener, J., Werner, TE., Hartig, J., Haller, P.: Advancements for the structural application of fiber-reinforced moulded wooden tubes. In: Aicher, S., Reinhardt, HW., Garrecht, H. (eds.) Materials and Joints in Timber Structures. RILEM Bookseries, vol. 9, pp. 99–108. Springer, Dordrecht (2014). https://doi.org/10.1007/978-94-007-7811-5_9

Westermann, P., Evins, R.: Surrogate modelling for sustainable building design – a review. Energy Build. **198**, 170–186 (2019). https://doi.org/10.1016/j.enbuild.2019.05.057

Extended Reality Collaboration: Virtual and Mixed Reality System for Collaborative Design and Holographic-Assisted On-site Fabrication

Daniela Mitterberger(✉), Evgenia-Makrina Angelaki, Foteini Salveridou, Romana Rust, Lauren Vasey, Fabio Gramazio, and Matthias Kohler

Gramazio Kohler Research, ETH Zürich, Zürich, Switzerland
mitterberger@arch.ethz.ch

Abstract. Most augmented and virtual applications in architecture, engineering, and construction focus on structured and predictable manual activities and routine cases of information exchange such as quality assurance or design review systems. However, collaborative design activities such as negotiation, task specification, and interaction are not yet sufficiently explored. This paper presents a mixed-reality immersive collaboration system that enables bi-directional communication and data exchange between on-site and off-site users, mutually accessing a digital twin. Extended Reality Collaboration (ERC) allows building site information to inform design decisions and new design iterations to be momentarily visualized and evaluated on-site. Additionally, the system allows the developed design model to be fabricated with holographic instructions. In this paper, we present the concept and workflow of the developed system, as well as its deployment and evaluation through an experimental case study. The outlook questions how such systems could be transferred to current design and building tasks and how such a system could reduce delays, avoid misunderstandings and eventually increase building quality by closing the gap between the digital model and the built architecture.

Keywords: Mixed reality · Virtual reality · Interactive design · Collaborative virtual environments · Remote collaboration · Immersive virtual environment

1 Introduction

In recent years, there have been remarkable advances in mixed-reality technologies for architecture, engineering, and construction (AEC). Most augmented and virtual applications in AEC focus on structured manual activities (Goepel and Crolla 2020; Jahn et al. 2019; Fazel and Izadi 2018), more routine cases of information exchange such as quality assurance (Dietze et al. 2021; Büttner et al. 2017), or design review (Liu et al. 2020; Zaker and Coloma 2018). Collaborative design activities or interdependent collaborative tasks, such as complex negotiation, task specification, and interaction, are, however, not yet sufficiently explored (Marques et al. 2021; Wang and Tsai 2011; Benford et al. 2001; McGrath and Prinz 2001).

C. Gengnagel et al. (Eds.): DMS 2022, *Towards Radical Regeneration*, pp. 283–295, 2023.
https://doi.org/10.1007/978-3-031-13249-0_24

Collaborative activities in the field of AEC can involve a plethora of different stakeholders with heterogeneous backgrounds and expertise. Furthermore, stakeholders can involve remote collaborators, ranging from on-site to off-site users. Especially for communication between remote users, knowledge transfer is critical for a successful collaboration, particularly for task decomposition, handover processes, and design revisions. Improved decision-making processes can enhance workflow efficiency and collaboration in the creative process as it supports the inclusion of expert knowledge.

Current computer-supported cooperative work systems (CSCW) focus on enhancing collaboration in AEC by providing users with diverse shared information. This information includes, for instance, access to shared digital context through common data structures and environments utilizing building information modeling (BIM) software such as *Autodesk Revit* or *ArcGIS*. Other systems provide access to shared administrative tasks such as project management platforms, e.g., *Microsoft* planning software or *Autodesk Navisworks*.

Current CSCWs are suitable for very distinct and asynchronous tasks that do not require extensive communication and collaboration between users. Their structure allows users to complete individual tasks and inform other users about their progress. Nevertheless, due to their task-specific structure, these platforms are relatively rigid and do not provide an environment that fosters an immersive communication and discussion platform between users. This lack of a communication environment can cause user frustration and inhibit creativity. Especially interwoven negotiated task activities require a more comprehensive range of communication between different stakeholders.

The research presented in this paper, "Extended Reality Collaboration" (*ERC*), aims to complement the functionalities of existing CSCW systems and groupware tools in AEC by providing workflows for not yet well-supported collaboration and communication tasks. This paper proposes a mixed-reality immersive collaboration system that enables bidirectional communication and data flow between on-site and off-site users, enabling them to operate together on a digital twin in a collaborative virtual environment (CVE).The workflow and functionalities of *ERC* have been applied and validated in an architectural scale prototype - a sticky note installation.

2 Background

Our work builds upon two general fields of research: collaborative virtual environments and augmented fabrication.

2.1 Collaborative Virtual Environments (CVE)

Churchill et al. (Churchill et al. 2001) define CVEs as distributed virtual systems that enable users to collaborate with a digital environment and with each other. Asymmetric CVEs (Grandi et al. 2019; Piumsomboon et al. 2017) support users with different input and visualization hardware, adapting to their various capabilities. *DollhouseVR* (Ibayashi et al. 2015) facilitates asymmetric collaboration between co-located users, one virtually inside the dollhouse using a head-mounted display (HMD) and the other using an interactive tabletop. Another asymmetric CVE of co-located users is *shareVR* (Gugenheimer et al. 2017), which uses floor projection and mobile displays with positional tracking to visualize a shared virtual world for non-HMD users. A system developed for geographically separated users is presented by Oda et al. (2015), which supports a remote expert to assist a local user. The results showed that a local user understood task instructions faster when the remote user wore a VR HMD and demonstrated the task in virtual space compared to written annotations. Commercial software such as *Wild* and *Iris VR* provide CVEs for multiuser object manipulation but do not link it with fabrication parameters and instructions and, therefore, miss out on streamlining the design and fabrication phase.

2.2 Augmented Fabrication

Augmented fabrication in AEC focuses primarily on guiding a craftsperson in a manual fabrication process (Nee et al. 2012). This guidance can be with audio instructions, projection mapping, or screen-based mixed-reality (MR). *Fologram* uses MR headsets to see virtual holographic 3D models in space and assist unskilled construction workers in complex fabrication tasks (Jahn et al. 2019). An example of a screen-based augmented-reality (AR) system is *Augmented Bricklaying* (Mitterberger et al. 2020). This system extends purely holographic AR with a context-aware AR system providing humans with machine precision by tracking objects in space. *IRoP* (Mitterberger et al. 2022) is a system that allows users to instruct robots via programming by demonstration and to preview generated designs on-site via projection-based AR. While the growing number of AR fabrication research shows the enormous potential of augmented fabrication, all the discussed systems are solely designed to be used in-situ. None of the above examples link a local user with a remote user.

3 Methods

Our *ERC* system aims to combine design and fabrication functionalities in a collaborative virtual environment and enhance communication between two geographically separated stakeholders. Consequently, the system not only converges on- and off-site activities but also integrates the processes of design development and physical fabrication into one virtual shared environment.

3.1 User Scenario

ERC involves at least two different stakeholders with different expertise that are in different locations; one user is on-site, and the other is off-site (see Fig. 1). The on-site user, "MR-User," is equipped with a MR headset, whereas the off-site user, "VR-User," utilizes a virtual-reality (VR) headset. The MR-User represents an expert construction worker, craftsperson, or construction site manager. The role of the MR-User is to provide site-specific data, insight knowledge, and instruct manual fabrication. The VR-User represents a stakeholder such as an architect or planner who navigates in a digital twin of the construction site. The role of the VR-User is to request and receive on-site information and feedback on the design to adjust the design accordingly. Furthermore, the VR-User provides different design options and supervises fabrication. Both users meet in the virtual space collaborating synchronously.

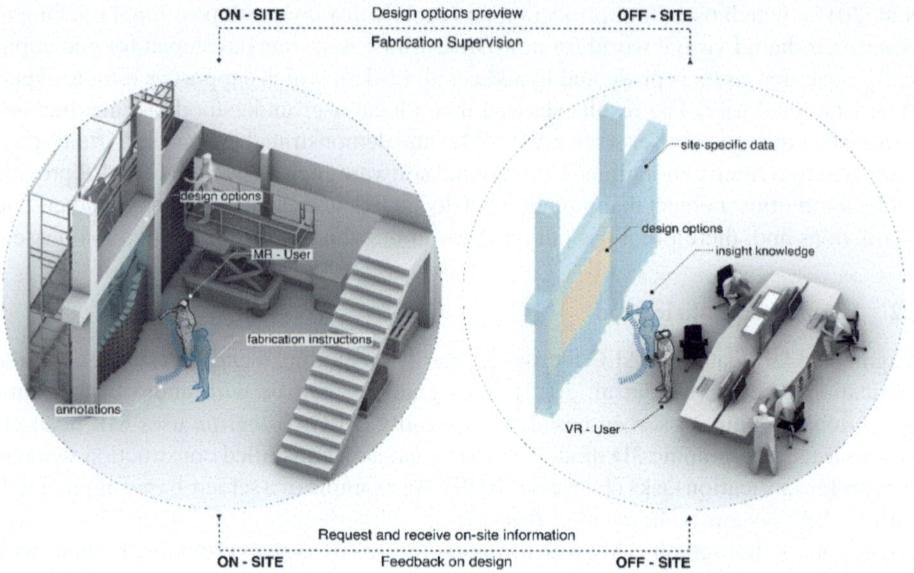

Fig. 1. User scenario showing the on- and off-site scenario with two distinct stakeholders

3.2 System Walkthrough

ERC is designed around two distinct phases, 1. Collaborative design phase, and 2. Augmented fabrication phase. In phase one, the VR-User (architect) and the MR-User (expert) evaluate the design options collaboratively. The MR-User creates the digital twin (see Fig. 2a), and then both users can meet in virtual space (see Fig. 2b). The MR-User sees the design options as holographs on-site, while the VR-User sees them in

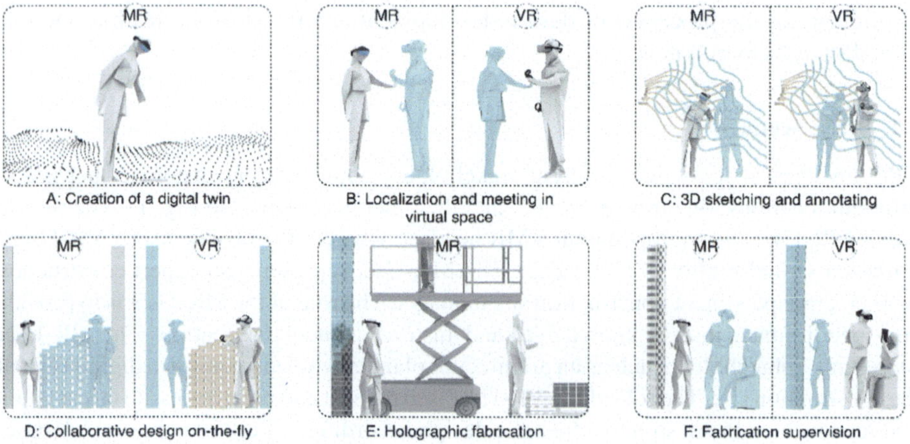

Fig. 2. System walkthrough.

the digital twin of the construction site. The collaborative design phase has two distinct features: 3D sketching and annotating (see Fig. 2c), and collaborative design on-the-fly (see Fig. 2d). Phase two allows the users to plan and fabricate the design and has two features: holographic fabrication (see Fig. 2e) and fabrication supervision (see Fig. 2f).

To illustrate a typical interaction, we consider the user scenario described in Sect. 3.1. The users follow a linear sequence of interactive design and fabrication sessions.

A - Creation of a Digital Twin

The creation of a shared digital twin model featuring both the construction site as well as as-built model can be done in two ways, resulting in meshes with different resolutions. The first option is asynchronous, creating a high-resolution point cloud using a Lidar scanner. The second option is synchronous using the spatial awareness system of the MR headset. This option can be accessed in *ERC* via the 'scanning' feature allowing the users to receive a current as-built mesh of the construction site with customized levels of detail. This feature consists of several interactive modes to further access and edit the generated spatial data. The MR-User can select and send meshes to the VR-User. Based on these meshes, the VR-User can adjust and update the design options.

B - Localization and Meeting in Virtual Space

Both users need to be localized in physical and virtual space to correctly send correlated spatial, geometric, and temporal data. Therefore, the local coordinate systems of the MR and VR spaces need to be aligned using relative transformation. The transformation requires the current position of each user relative to an origin frame. As an origin frame, the MR-User scans a referenced QR code in the physical space and then transmits the frame data to the VR-User. To share a mutual sense of presence, both users appear as avatars. The avatar position is updated in real-time and allows the users to communicate via hand movements and body motion trajectories.

C - 3D Sketching and Annotating

After localization and setting up a digital twin of the construction site, both stakeholders use a sketching and annotating feature to draw in 3D, highlight specific target areas or annotate existing designs (see Fig. 3). In this phase, both users can discuss potential design problems with the construction site's current as-built state.

D - Collaborative Design on-the-Fly

This feature allows users to preview and adjust a parametric design model on-the-fly in MR and VR and directly preview it as a hologram in-situ (see Fig. 4). The VR-User loads the parametric model using *Rhino.Inside*[1] and adjusts the parameters of the digital model according to the feedback of the MR-User. The VR-User has access to properties of the parametric model and can adjust these parameters in near real-time. Both users can sketch directly on the design options using the "3d sketch and annotate" feature.

[1] Rhino.Inside® is an open-source project which allows Rhino and Grasshopper to run inside other 64-bit Windows applications.

Fig. 3. 3D sketch and annotation feature. On the left is a first-person view of the MR-User watching the VR-User sketch. On the right side is a third-person view of the MR and VR-User drawing collaboratively within the VR space.

Fig. 4. Collaborative design on-the-fly feature. On the left is a first-person view of a hologram of the design on the installation site. On the right side is a third-person camera view of the MR and VR-User discussing the design in VR.

E - Holographic Fabrication

After deciding on a final design, the users switch from the interactive design phase to the fabrication mode (see Fig. 5). The fabrication mode can also include multiple other users as the system can be deployed on various augmented reality devices. The MR-User receives fabrication-specific information such as the holographic 3D model, estimated fabrication time, and the number of elements deployed. Furthermore, the MR-User can switch between fabrication sessions. These sessions are visualized in different colors representing the estimated daily working hours (see Fig. 6-2).

F - Fabrication Supervision

This mode allows the MR-User to store information about completed tasks, current fabrication sessions, and problematic areas. Furthermore, the VR-User can virtually join the fabrication session to supervise the process (see Fig. 6).

Fig. 5. A menu informing the MR-user about fabrication parameters and the holographic 3D model supporting fabrication.

Fig. 6. Fabrication supervision feature. On the left is a first-person view of the MR-User looking at the VR-User. On the right side is a third-person camera view of the MR and VR-User discussing the fabrication in VR. The different colored elements show the different fabrication sessions.

3.3 System Architecture

As displayed in Fig. 7, the system architecture consists of three main parts: (1) an on-site MR setup with a scanning system, (2) an online server, and (3) an off-site VR setup. The on-site MR setup consists of a laser scanning device (*Leica RTC 360*) providing high-resolution on-site scans, an MR-headset (*Microsoft Hololens2*), a laptop, and a WIFI router. The off-site hardware consists of a VR headset (*Oculus Quest 2*), a laptop, and a WIFI router.

The software setup is structured as follows. Two autonomous *Unity3D* applications were developed, one for MR and one for VR. The MR application uses the Mixed reality toolkit (*MRTK*) and *OpenXR* library to enable spatial awareness scanning and QR-code detection. The VR application is developed using the *OpenXR* library. Furthermore, *Rhinoceros3D*, *Grasshopper*, and Python are used to create algorithmic designs. *Rhino.Inside*, enables compatibility and bidirectional communication between external *Unity* processes and *Grasshopper*. The online communication is based on the Robot Operating System (*ROS*) (Quigley et al. 2009). The rosbridge package is used to access the publish-and-subscribe architecture of *ROS* and *ROS#* for the *Unity3D* applications.

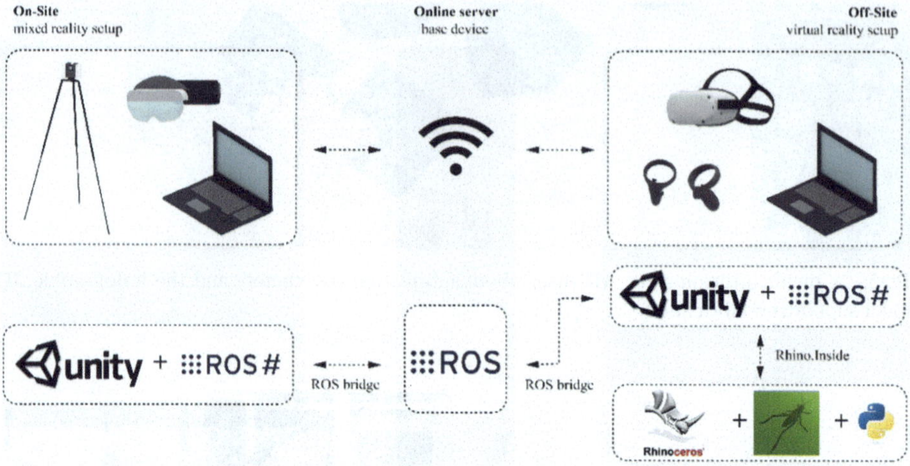

Fig. 7. System architecture

4 Case Study

To validate the feasibility of the proposed method and demonstrate the potential for a concrete fabrication system such as façade panels, we focused on one full-scale experimental implementation. For a user-friendly experience, the user interface design was based on each user's different roles and work packages (see Fig. 8). We used sticky notes as placeholders to showcase the various and complex types of information that can be exchanged between two geographically separated users. This information includes the position (P), rotation (a), size (f1), geometry (folding type) (f2), and color of each unit (see Fig. 9).

The total fabrication time was 26 h, whereas the interactive design was around 1 h. The final design was fabricated using two MR headsets, and a total of 4000 sticky notes were placed. The final design was split into distinct fabrication sessions of 60–90 min. We used an attractor-based approach for the computational design, which influenced the design depending on its location in space and its distance from physical boundaries (see Fig. 10). Specifically, the attractor's location (CP) changed the position (p), rotation (a), color, size and folding type (f) of the sticky notes. In our case study, the sticky note's location was projected onto the spatial mesh data (M) scanned by the MR User. This projection resulted in a precise position for each sticky note on the as-built data of the installation site. During the design phase, the VR-User moved the attractor as an interactive 3D prism in virtual space to control the design. The VR-User could adjust the design parameters collaboratively with the MR-User while the MR-User saw the different results as holographs in-situ. Furthermore, the MR-User could interact with the design via sketching to adjust the outline of the design. After agreeing on a final design, the MR-User fabricated the full-scale experimental implementation (see Fig. 11) while the VR-User supervised and informed the process.

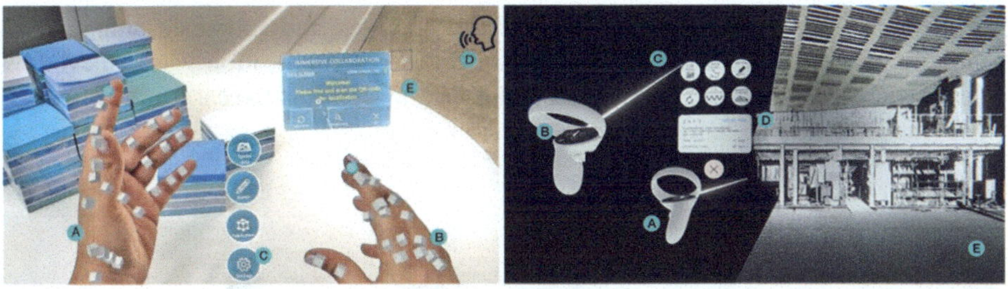

Fig. 8. Left: The MR-User interacts via hand tracking (A), gesture tracking (B), menu buttons (C), and voice commands (D). The MR-User sees an info window superimposed over their view (E). Right: The VR-User navigates the space and interacts via controllers (A) using controller buttons (B) and virtual menu buttons (C). The VR-User sees an info screen (D) and moves within a digital twin of the construction site (E)

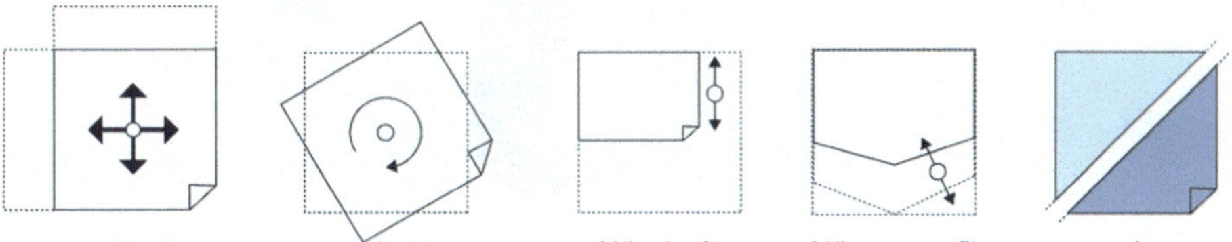

Fig. 9. Various and complex parameters that can be exchanged with the system

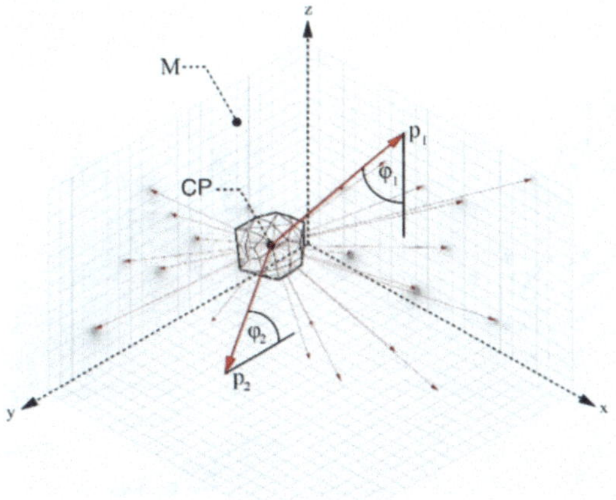

Design Paramters	
CP	centerpoint of attractor
p1, p2	intersection point
[φ1,φ2...]	angle
M	scanned mesh
P	sticky note position
a	sticky note rotation
f	sticky note folding size
color 0,1	sticky note color

Remapping values		
P	=	p
a	=	φ
f	=	remapped value of φ
φ < tolerance	=	color 0
φ > tolerance	=	color 1

Fig. 10. Computational attractor-based design logic and remapping values to determine sticky note position, rotation, folding size, and color

5 Results

Our *ERC* system allowed for an intuitive and real-time design interaction for users in different physical locations. The users had access to a full-scale impression of the architectural model augmented and contextualized by site-specific information. Both users collaboratively designed and fabricated a complex and full-scale architectural installation (see Fig. 12). Furthermore, personalized communication was achieved by creating avatars for all users. Implementing the *ERC* system and the case study provided us with insights into the hardware and software limitations.

Fig. 11. Photograph of the fabrication of the final physical installation.

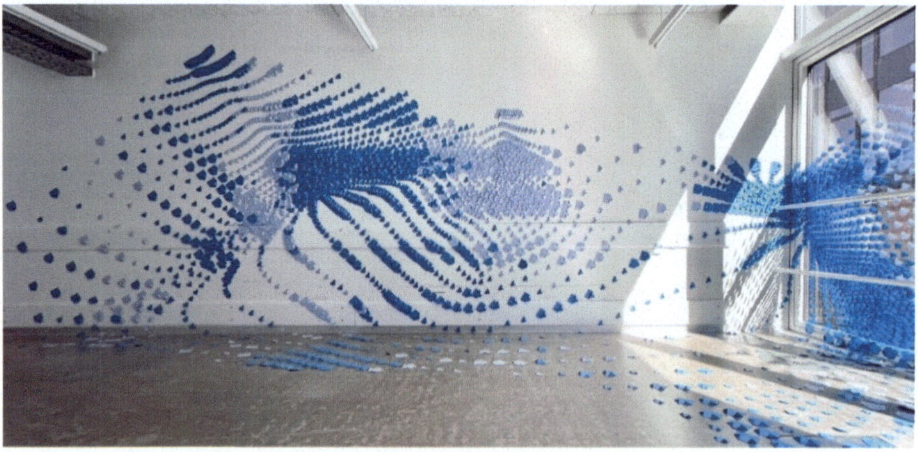

Fig. 12. Photograph of the final physical installation.

5.1 System Limitations

We experienced hardware limitations regarding the environmental scanning and localization as well as drift of the digital model (see Table 1). The main software limitations were delays and transmission speed, especially between the *Grasshopper* environment and the *Unity* interface with increased mesh count and internet connection speed. To avoid delays between the MR and VR-User, we used a mesh resolution of 20 triangles per cubic meter. Furthermore, the system still has a limited amount of drawing tools in the "3D sketch and annotation" feature. Extending the drawing tools would allow users to interact with a broader range of communication options. In noisy environments, it was difficult to get the other user's attention. Therefore, it would be essential to implement an "attention feature". Additionally, the current system lacks a "documentation feature" that would allow users to upload video, pictures, or voice memos to the digital model with associated location. Such a note collection could help on-site workers keep track of construction site notes and allow easier communication with off-site users. These notes could also be read asynchronously, allowing users to log into the system at different moments.

Table 1. Relation between QR Code placement and visibility and drift of the digital model. The QR code dimensions were 12.5 cm × 12.5 cm.

QR – code distance	Drift of the digital model
<0.45 m and in view	0.1–0.3 cm
~4 m and in view	1.5–2.3 cm
not in view	2–3 cm

6 Conclusion and Outlook

This research investigates the potential of collaborative design activities and how they could lead to better knowledge and information flows between on-site and off-site stakeholders during design and fabrication processes. The functionalities of the system were evaluated via a full-scale case study, aiming to define collaboration protocols and improve interaction and communication. Even though there are still limitations, this research shows the potentials of such a system to improve supervision and collaboration between on-site and off-site stakeholders, such as architects and construction supervisors, to support a paperless construction site. The key findings of this research are novel collaborative MR and VR interfaces, 3D workspace scenes with sufficient context-awareness, and a fabrication protocol that includes remote monitoring and planning. As an outlook, such a system could be applied towards detecting deviations between the as-built and the digital model in order to decrease project costs and building time. Such a system could be applied to real building scenarios, i.e., on-site construction meetings, custom interior designs, renovations, and complex building elements. *ERC* allows dispersed personnel to have more direct contact, thereby reducing problems of isolation and miscommunication. Furthermore, such a system could accelerate digital workflows and support a teleoperated construction site.

Acknowledgments. We want to thank Gonzalo Casas (ETH Zurich) for supporting the research on online communication. Furthermore, we would like to express our thanks to the Design++ initiative for giving us access to the Immersive Design Lab (IDL), equipment, and support throughout the project.

Author's Contribution. DM wrote the manuscript, did the conception of the work, and supervised the thesis. EA and FS developed the system as part of their MAS master thesis. EA contributed to research for the manuscript. RR wrote the manuscript, did the conception of the work, and supervised the thesis. LV was part of the supervision of the thesis. FG and MK contributed to the conception of the work. All authors reviewed the manuscript.

References

Benford, S., Greenhalgh, C., Rodden, T., Pycock, J.: Collaborative virtual environments. Commun. ACM **44**(7), 79–85 (2001). https://doi.org/10.1145/379300.379322

Büttner, S., et al.: The design space of augmented and virtual reality applications for assistive environments in manufacturing: a visual approach. In: Proceedings of the 10th International Conference on PErvasive Technologies Related to Assistive Environments, Island of Rhodes Greece, pp. 433–440 ACM (2017). https://doi.org/10.1145/3056540.3076193

Churchill, E.F., Snowdon, D.N., Munro, A.J. (eds.): Collaborative Virtual Environments: Digital Places and Spaces for Interaction. Computer Supported Cooperative Work. Springer, London, New York (2001). https://doi.org/10.1007/978-1-4471-0685-2

Dietze, A., Jung, Y., Grimm, P.: Supporting web-based collaboration for construction site monitoring. In: The 26th International Conference on 3D Web Technology, Pisa, Italy, pp. 1–8. ACM (2021). https://doi.org/10.1145/3485444.3495180

Fazel, A., Izadi, A.: An interactive augmented reality tool for constructing free-form modular surfaces. Autom. Constr. **85**, 135–45 (2018). https://doi.org/10.1016/j.autcon.2017.10.015

Goepel, G., Crolla, K.: Augmented reality-based collaboration - ARgan, a bamboo art installation case study. In: Holzer, D., Nakapan, W., Globa, A., Koh, I. (eds.) RE: Anthropocene, Design in the Age of Humans - Proceedings of the 25th CAADRIA Conference - Volume 2, Chulalongkorn University, Bangkok, Thailand, 5–6 August 2020, pp. 313–322. ACADIA (2020). http://papers.cumincad.org/cgi-bin/works/Show&_id=caadria2010_003/paper/caadria2020_426

Grandi, J.G., Debarba, H.G., Maciel, A.: Characterizing asymmetric collaborative interactions in virtual and augmented realities. In: 2019 IEEE Conference on Virtual Reality and 3D User Interfaces (VR), Osaka, Japan, pp. 127–35. IEEE (2019). https://doi.org/10.1109/VR.2019.8798080

Gugenheimer, J., Stemasov, E., Frommel, J., Rukzio, E.: ShareVR: enabling co-located experiences for virtual reality between HMD and Non-HMD users. In: Proceedings of the 2017 CHI Conference on Human Factors in Computing Systems, CHI 2017, pp. 4021–33. Association for Computing Machinery, New York (2017). https://doi.org/10.1145/3025453.3025683

Ibayashi, H., et al.: Dollhouse VR: a multi-view, multi-user collaborative design workspace with VR technology. In: SIGGRAPH Asia 2015 Emerging Technologies, SA 2015, pp. 1–2. Association for Computing Machinery, New York (2015). https://doi.org/10.1145/2818466.2818480

Jahn, G., Newnham, C., van den Berg, N., Iraheta, M., Wells, J.: Holographic construction. In: Gengnagel, C., Baverel, O., Burry, J., Ramsgaard Thomsen, M., Weinzierl, S. (eds.) DMSB 2019, pp. 314–324. Springer, Cham (2019). https://doi.org/10.1007/978-3-030-29829-6_25

Liu, Y., Castronovo, F., Messner, J., Leicht, R.: Evaluating the impact of virtual reality on design review meetings. J. Comput. Civ. Eng. **34**(1), 04019045 (2020). https://doi.org/10.1061/(ASCE)CP.1943-5487.0000856

Marques, B., Silva, S., Dias, P., Sousa-Santos, B.: An ontology for evaluation of remote collaboration using augmented reality (2021). https://doi.org/10.18420/ECSCW2021_P04

McGrath, A., Prinz, W.: All that is solid melts into software. In: Churchill, E.F., Snowdon, D.N., Munro, A.J. (eds.) Collaborative Virtual Environments. CSCW, pp. 99–114. Springer, London (2001). https://doi.org/10.1007/978-1-4471-0685-2_6

Mitterberger, D., et al.: Augmented bricklaying: human–machine interaction for in situ assembly of complex brickwork using object-aware augmented reality. Constr. Robot. **4**(3–4), 151–61 (2020). https://doi.org/10.1007/s41693-020-00035-8

Mitterberger, D., et al.: Interactive robotic plastering: augmented interactive design and fabrication for on-site robotic plastering, New Orleans. ACM (2022). https://programs.sigchi.org/chi/2022/program/content/68974

Nee, A.Y.C., Ong, S.K., Chryssolouris, G., Mourtzis, D.: Augmented reality applications in design and manufacturing. CIRP Ann. **61**(2), 657–679 (2012). https://doi.org/10.1016/j.cirp.2012.05.010

Oda, O., Elvezio, C., Sukan, M., Feiner, S., Tversky, B.: Virtual replicas for remote assistance in virtual and augmented reality. In: Proceedings of the 28th Annual ACM Symposium on User Interface Software & Technology, 405–15. UIST 2015. Association for Computing Machinery, New York (2015). https://doi.org/10.1145/2807442.2807497

Piumsomboon, T., Lee, Y., Lee, G., Billinghurst, M.: CoVAR: a collaborative virtual and augmented reality system for remote collaboration. In: SIGGRAPH Asia 2017 Emerging Technologies, Bangkok Thailand, pp. 1–2. ACM (2017). https://doi.org/10.1145/3132818.3132822

Quigley, M., et al.: ROS: an open-source robot operating system, 6 (2009)

Wang, X., Tsai, J.J.-H. (eds.): Collaborative Design in Virtual Environments. Intelligent Systems, Control and Automation: Science and Engineering, vol. 48. Springer, Dordrecht (2011). https://doi.org/10.1007/978-94-007-0605-7

Zaker, R., Coloma, E.: Virtual reality-integrated workflow in BIM-enabled projects collaboration and design review: a case study. Vis. Eng. **6**(1), 4 (2018). https://doi.org/10.1186/s40327-018-0065-6

Hosting Spaces

Encoding and Decoding Adaptive Digital Spaces Within a Reconfigurable Physical Pattern

Alexandra Moisi(✉), Nicolas Stephan, Robby Kraft, Mathias Bank Stigsen, Kristina Schinegger, and Stefan Rutzinger

Department of Design, i.sd | Structure and Design, University of Innsbruck, Technikerstrasse 21c, 6020 Innsbruck, Austria
alexandra.moisi@uibk.ac.at

Abstract. This paper presents a case study of an interactive and collaborative installation realized within a larger exhibition. The work mediates space and vision between human and machine and speculates on the role that spatial computing can play in architecture design. It explores the encoding and decoding of spatial experiences using a physical relief wall as a medium to encode and transfer data. By collaborating with a robot arm through an augmented layer, the users enable a feedback loop between the robotically assembled relief and a corresponding three-dimensional, digital layer.

The research intends to build upon already established relations between human perception and architectural composition and turn them into an operational logic that enables a more immediate experience and understanding of a hybrid space. It is an investigation into how our environments can become embedded with augmented information and how architecture can contribute to our digital future as a "hosting space".

The authors will discuss ideas and concepts related to theory, design, and technological implementations within the framework of the executed case study.

Keywords: Augmented Reality · Mixed Reality · Hybrid space · Bi-directional feedback · Phygital design · Hosting space

1 Introduction

Spatial computing, in combination with broad accessibility to Mixed Reality (MR) and Augmented Reality (AR) ready devices, allows digital interfaces to inhabit our physical environments. With the imminent arrival of web 3.0 and the Metaverse, there is a unique opportunity to rethink and reshape the framework for how design and architecture interact with new digital realities. This process creates new possibilities for design, enabling novel forms of social interaction between humans, machines, data, and our physical environment.

The following research aims at exploring novel ways of spatial interaction between physical space and digital content through architectural features, e.g., the figure/object (in a previous use case) or the relief (in the discussed work). As these architectural

C. Gengnagel et al. (Eds.): DMS 2022, *Towards Radical Regeneration*, pp. 296–304, 2023.
https://doi.org/10.1007/978-3-031-13249-0_25

motifs become "tangible" and operable, they constitute the actual interface of the digital experience and an intuitive way to reshuffle adaptive spaces.

It is an investigation into how architecture, as an integral part of our physical environment, can contribute to our digital future as a "hosting space." We argue that the digital and physical are now two dimensions of the same space - equally real and equally important. Consequently, architecture as the discipline of space creation has the potential to mediate between the physical and digital environment, to construct a negotiated middle ground - both hosting and being hosted.

1.1 Conceptual Background

As physical objects become embedded with digital information and deeper meaning, we look into ways of integrating them into one another, building bridges, and exploring ways of communication.

As humans, we can look at the physical space as the more easily accessible and understandable domain, whereas the digital space is more abstract and remote from our immediate understanding. The two domains could be seen as two different levels communicating to one another through bridges. The bridge that ties into the physical space, the more approachable and controllable domain to us, is an encoding bridge, whereas the opposite direction acts as a decoding one (Zafiris 2021).

In order to ensure the readability of the bridges, the design language of what we could call a bi-directional interface should be intuitive and homogenous.

To further explore this point, this paper will discuss a project which seeks to establish an interactive feedback loop between physical and digital aspects of architecture. The two sides are in a continuous dialogue, informing each other in a shared choreography. The prototype creates an immersive set-up, where various virtual and physical actions influence an adaptive hybrid space (Fig. 1).

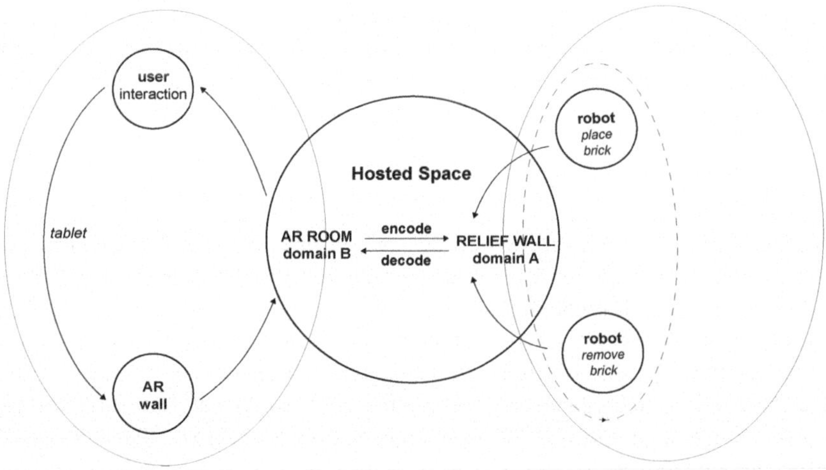

Fig. 1. The diagram shows the digital and the physical interactions in relation to the overall space – the hosted space.

1.2 Precedents

It is worth noting that so far, comparatively few attempts have been made by architects or designers to integrate AR into an architectural, urban, or cultural environment. Since its early days, the technology has been explored and developed by various researchers with an engineering background, such as Dr.-Ing. Wolfgang Höhl and M.Sc. Daniel Broschart (Höhl and Broschart 2015). They have provided solid foundational research in the field of AR and XR, primarily aimed at technical professionals, perhaps less approachable in a social and cultural context. Building on that, recent developments have also used the AR-enabled digital environment as an interface for both design and fabrication. Such setups lay the grounds for complex future human-machine collaboration and interaction (Giovanni et al. 2018).

In the gaming industry, one must take note of the massive social effects, both positive, and negative, that AR games such as PokemonGo and Minecraft Earth have had. It proved that AR has a massive potential to facilitate social events that can be shared with the people physically present. While "AR-enabled physical interaction" might initially sound like a contradiction, the technology has managed to partially reposition digital information from the arguably 2-dimensional space of the internet into our 3-dimensional surroundings. Additionally, by borrowing principles from games, meaning through "gamification", users can be motivated and involved enough to engage physically in various activities, whether they be actual games, design, or crafts (Robson et al. 2015).

However interesting and relevant these applications are, we believe that an important direction has been neglected - expanding our built environment with architectural digital content. Within the current discourse around the Metaverse, our own digital twins or avatars (Ensslin 2011) have taken the spotlight; architecture has been reduced to VR environments that are completely disconnected from the physical world we live in. Well known precedents for entangled physical and digital spaces are the "Fun Palace" by Cedric Price or the "Water Pavilion" by NOX. Although they differ in their formal approach, interaction between the user, information and architecture is a fundamental design concern. In the first, the building becomes provisional itself to be reprogrammed by its changing content (Mathews 2006), in the latter the pavilion's "motorgeometry" seamlessly integrates light and sound effects (Spybroek 1997; Görgül 2016).

1.3 Objectives

By bringing digital bits and pieces into our shared physical world, instead of cutting individuals entirely out of it and immersing them in the virtual, a shared social "phygital" space between humans is created.

The objective of this study is to explore this shared space, discuss how this bi-directional feedback between digital content and physical space can be informed through architectural motifs and elements, such as a relief, and how can architectural knowledge about perception and composition be turned into an operational logic that enables a more immediate understanding.

2 Methods

The installation was set up and tested in a four-month-long exhibition intended for a wider audience, including non-professionals. Our installation consists of the physical environment, a modular 3D-printed relief wall (composed of 72 3D-printed interchangeable parts), a UR robot arm moving on a linear axis, and an iPad running an AR environment. Without even picking up the iPad, users are invited to observe the physical part of the installation, and upon entering the AR environment, users not only uncover an entirely new aspect of the installation, but they are able to interact and influence both the digital and physical worlds (Fig. 2).

Fig. 2. The images show some of the many possible augmented spaces experienced through mobile devices

2.1 The Physical Space

One of the arguments that build on the idea of "phygital design" is that AR needs physical features and articulations. These features are still necessary from a technical point of view for the performance of SLAM (simultaneous localization and mapping) and stable spatial computing. The space was scanned with the Vuforia Area Target app on a Lidar-enabled iPad. In order to be able to anchor the digital content in the physical space, the exhibition area had to be well articulated and could not stay a classical "white cube" (O'Doherty 1986). Therefore, establishing a direct visual relationship between the elements of the installation, both physical and digital, between the "hosting" and "hosted" was key to our final proposal (Fig. 3).

2.2 The Relief Wall

The relief was chosen due to its ability to "compress" three-dimensional space in 2D; therefore, it can be seen as an early form of encoding of space, based on visual and spatial patterns. In contrast to numeric codes, the patterns and the specific aesthetics are

Fig. 3. The images are highlighting the design of the exhibition space, the white cube overwritten by content-related patterns. The patterns were used to map the space.

attuned to both human and machine perception. The wall acts as an interface and playful mediator of space.

The design of the relief wall focuses on modularity and high color contrasts to provide a range of differentiated recognizable perceptual patterns for human visitors (Arnheim 1969). Every individual brick is equipped with magnets and consists of five sub-components differentiated through shape and color (Fig. 4).

Fig. 4. The image shows a close-up of the intricate relief wall and the robot in process of removing one of the tiles in order to replace it

There are four different types of modules, varying in depth, complexity, distribution, and surface area of the colored sub-components. This variation aims to provide visual differentiation, which can be used to create diverse patterns and densities on the relief wall. The wall itself is equipped with yellow 3D printed connectors with embedded magnets. Each of these prints serves as a docking station for one relief module. When the robot places a module on the connector, the magnets attract each and "snap" the module into the correct position. If no component is placed, the shape and color of the connector signal its empty state to the onlooker.

There are 72 modules. Each slot can have five different states: empty (0), module A (1), module B (2), module C (3), module D (4). This enables high numbers of variability making the system suitable for grid-based computation of various kinds, such as two-dimensional cellular automata (Schiff 2008) or shape grammars (Stiny 2006). In the case of this exhibition piece, the states of the grid are congruent with the data of the computational model used to synchronize the physical space with its digital counterpart, aiming to provide a comprehensible visual reference of the interaction process.

2.3 Digital Interactions

In the center of the room, in front of the robot and the relief wall, sits a podium with an iPad, the window into the AR world. Initially, the AR environment is empty in all directions except for the robot and the relief wall, where the user will notice a precise overlay of one touchable tile on every relief wall module (Fig. 5).

Fig. 5. The images show snippets of how the users can engage with the digital overlay of the tiles. The dark blue highlighted tile has been selected and is queued to be replaced by the robot.

These digital overlays are interactable, and when the user taps on a relief tile on the screen, the tablet sends a request to the robot, and the request enters the command queue. Depending on if the robot is in a resting state or in a process, it will either move towards the tile, pick it off from the relief wall, and move it to another location or this new tile becomes the next step in the queue to be acted upon immediately after the completion of the current sequence.

As this real-world dance is spurred into action, a separate digital event begins to unfurl. Inside the AR environment, directly above the user, the digital 3D corresponding part of the relief wall tile that was selected takes on a much larger digital form and appears overhead. This digital relief tile is surrounded by four relief tiles, the same neighbors from the relief wall. All of these models are much larger, more detailed, and exhibit features inherited from their physical counterparts. Every 3D part is designed with fixed connecting points in position so that they can be reshuffled in multiple combinations, and still fit seamlessly. The center tile, directly above the user, stays put while the four neighbors begin a slow rotation, their outer edges easing downwards towards the user, eventually enclosing a cube space where the five tiles make up four walls and a ceiling. What was once an exhibition room is now a much smaller, vibrating, busy digital space filled with relief decorations probing inwards from five directions around the user (Fig. 6).

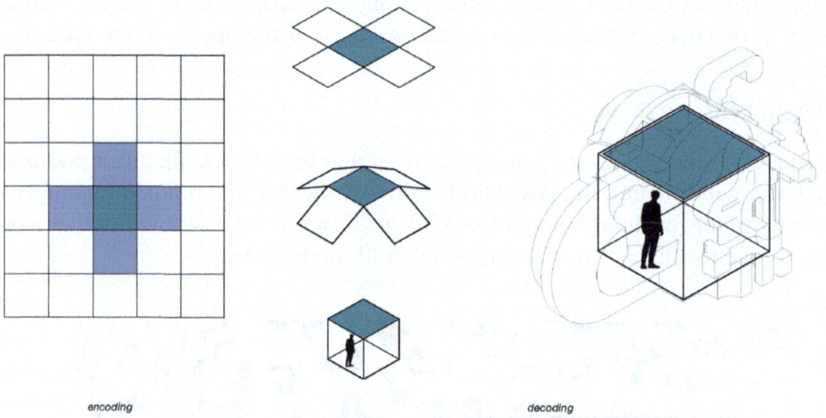

Fig. 6. The figure describes the digital interaction. On the left side, it represents the interaction with the relief wall, and on the right side, the decoding and final unfolding of the 3D augmented space.

2.4 Technical Implementation

The installation was driven by three computers. One server maintained a state representation of the board (relief wall, tile locations, and types) and handled HTTP requests from the other computers via Python and Rhino/Grasshopper. Another computer was running ROS on Linux and handled all lower-level commands with the robot. And finally, the iPad was running an AR app that was built and compiled in Unity. All communication occurred over a local area WIFI network, allowing the iPad the freedom of being wireless.

The technical challenge of communicating between three different computers, operating systems, and applications was simplified by trading JSON files back and forth, writing them to the file system, and watching the files for changes.

3 Visitors Feedback

According to feedback received, the coherence of the design among the different elements helped users better understand the installation and intuitively interact with it. A survey conducted within a small test group showed that visitors understood the installation as a whole, as well as the general connection between the physical and the digital. However, visitors found the coexistence of the two different systems more difficult to grasp, in which the robot assumes agency in the absence of human decision-making.

4 Conclusion

We argue that having a direct relationship between the context (hosting space) and content (hosted information) is essential in the design of phygital spaces.

Similarly, we hold a strong belief that the digital world needs a physical stage to perform on, not only for technical reasons but with a strong drive for creating new and exciting experiences and perceptions of space. Thus, we understand the relation of host and hosted as bi-directional. Therefore, design knowledge about architectural compositions (with all its fixed, reconfigurable, and informational or virtual elements) can be understood as a framework for a successful creation of phygital spaces. The deployment of known architectural elements diminishes thresholds for interaction. In the discussed use case, the relief allows the user to directly alter and change the AR experience in real-time by simply shifting or changing single tiles. By folding the tiles into an immersive space and scaling them up the relief becomes a 3D-space. The use case exemplifies how architectural elements could be reinterpreted with the help of AR, while being anchored in physical space and the result is a "multiscalar" experience. This will also allow to extend the design strategy of "hosting phygital spaces" to other contexts outside galleries.

We believe that the above-mentioned findings suggest there are vast opportunities to further explore this new, emerging field from various different angles: architectural, theoretical, and technical.

5 Outlook

The case study touches on multiple research questions related to the use of AR and MR in an architectural setting. On a theoretical level, it demonstrates the "natural communication model" of encoding and decoding two different dimensions, hosting and hosted spaces. On a technical level, it advanced knowledge and proved that a continuous, cyclic, and automated relationship can be established between humans and machines, digital and physical. From a designer's perspective, it is an early prototype, a speculation into how future "phygital" spaces could look like and function. The relief is also explored as a compressed space. Although almost 2D and usually integral to a wall or ceiling, we demonstrate how it can operate based on a "3D-ness" (Jacobi 1985).

This would also allow further application in architectural design processes, e.g. as a real-time modeling system where different scenarios of pre-programmed "types" could be arranged and experienced in real-time. In the future, we could see objects and architectural elements becoming embedded, not just with sterile layers of extra information, but also with stories, spaces and atmospheres. An intuitive connection between the two elements - physical and digital could promote a socially interactive, and participatory architecture that promotes creativity and gives users a sense of agency in space-making.

Acknowledgments. This work was produced at the Institute of Design, i.sd I Structure and Design, and was exhibited at the PhD-research exhibition "Potenziale 3" by the Faculty of Architecture, University of Innsbruck, and the AUT (Tyrolean Architecture Center). We thank the organizers and sponsors of the exhibition and Adam Geraia for technical and programming support in the development of the robotic communication system.

References

Arnheim, R.: Visual Thinking. University of California Press, Berkley (1969)

Ensslin, A.: Creating Second Lives. Avatar Needs and the Remediation of Architecture in Second Life. Routledge (2011)

Höhl, W., Broschart, D.: Augmented Reality im öffentlichen Raum. In: Proceedings Real Corp 2015 (2015)

Betti, G., Aziz, S., Ron, G.: Pop-up factory: mixed reality installation for the MakeCity festival 2018 in Berlin. In: Gengnagel, C., Baverel, O., Burry, J., Ramsgaard Thomsen, M., Weinzierl, S. (eds.) DMSB 2019, pp. 265–276. Springer, Cham (2020). https://doi.org/10.1007/978-3-030-29829-6_21

Görgül, E.: Space as a becoming: fresh water expo pavilion as a creative practice for an architecture to come. In: Design Research Society 50th Anniversary Conference, 2016 (2016)

Jacobi, F.: Zur Sprache des Reliefs. Dargestellt an ausgewählten Werken der Nationalgalerie. Staatliche Museen zu Berlin (1985)

Mathews, S.: The fun palace as virtual architecture. Cedric price and the practices of indeterminacy. J. Archit. Educ. ACSA **59**, 39–48 (2006)

O'Doherty, B.: Inside the White Cube. The Ideology of the Gallery Space. The Lapis Press (1986)

Papagiannis, H.: Augmented Human - How Technology Is Shaping the New Reality. O'Reilly Media, Sebastopol (2017)

Robson, K., Plangger K., Kietzmann, J.H., McCarthy, I., Pitt, L.: Is it all a Game? Understanding the principles of gamification. Bus. Horiz. **58**, 411–420 (2015)

Schiff, J.L.: Cellular Automata: A Discrete View of the World. Wiley, Hoboken (2008)

Spybroek, L.: Motor Geometry. TechnoMorphica (1997)

Stiny, G.: Shape: Talking About Seeing and Doing. The MIT Press, Cambridge (2006)

Vuforia: Vuforia Engine. AR SDK (2022)

Zafiris, E., Hovestadt, L., Bühlmann, V.: Natural Communication: The Obstacle-Embracing Art of Abstract Gnomonics. De Gruyter (2021)

Design, Control, Actuation and Modeling Approaches for Large-Scale Transformable Inflatables

Dimitris Papanikolaou(✉)

University of North Carolina at Charlotte, Charlotte, NC 28223, USA
dpapanik@uncc.edu

Abstract. Large-scale inflatables have a long history in the arts and architecture as quick, lightweight, and inexpensive means to create structures and environments of monumental scale. While several examples of inflatable installations are interactive, the design, control, actuation, and modeling techniques of transformable inflatables in large scales remains a relatively unexplored area of research. In the fields of human-computer interaction (HCI), design of interactive systems (DIS), and tangible, embodied, and embedded interaction (TEI), transformable inflatables are a more established area of research with applications primarily in object-based scales. Yet, important design, methodological, and technological differences make the direct application of existing research from small to large scales hard. In this paper, we introduce design, control, and actuation strategies for large-scale transformable inflatables for urban and architectural implementations and, through a case study, we discuss computational modeling, simulation, and experimental methods to study their behavior and performance.

Keywords: Transformable inflatables · Interactive systems · Shape-changing interfaces · Dynamic simulation

1 Introduction

1.1 Interactive Transformable Inflatables

Large-scale inflatables have a long history in the arts and architecture as quick, lightweight, and inexpensive means to create structures and environments of monumental scale. With the advent of digital technologies, artists, architects and designers increasingly explore interactive applications of large-scale inflatables that can actively change their shape based on environmental or anthropogenic stimuli. While several examples of inflatable installations are interactive, the design, control, actuation, and modeling techniques of interactive inflatables in large scales remains a relatively unexplored and actively ongoing area of design research. In the fields of human-computer interaction (HCI), design of interactive systems (DIS), and tangible, embodied, and embedded interaction (TEI), interactive inflatables are a more established area of research with multiple applications in shape-changing interfaces. Yet, important design, methodological, and

C. Gengnagel et al. (Eds.): DMS 2022, *Towards Radical Regeneration*, pp. 305–319, 2023.
https://doi.org/10.1007/978-3-031-13249-0_26

technological differences make the direct application of existing research from small to large scales hard. In this paper, we introduce design, control, and actuation strategies for large-scale transformable inflatables for urban and architectural implementations and, through a case study, we discuss computational modeling, simulation, and experimental methods to study their behavior and performance. We conclude with limitations, application areas, and directions for future work.

2 Background

2.1 Inflatables in Arts and Architecture

Pioneering examples of inflatables in the arts include the early works of Graham Stevens [1], Jeffrey Shaw [2], Tomas Saraceno [3, 4], Otto Piene [5], Doron Gazit [6] and, in architecture, the inflatable structures of Ant Farm [7] and of Jose Miguel de Prada Poole [8, 9]. Inflatable structures are, primarily, equilibrium forms and as such, most applications are design to be static after inflation. Several artists and designers have explored dynamic, non-equilibrium forms, such as the semi-inflated walkable membranes of Tomas Saraceno [10] and Jeffrey Shaw [2], yet, such examples are conceived and designed to be passively transformable, changing their shape based on how people or natural elements interact with them. Few artists and designers have explore actively transformable inflatables such as the giant performative flailing tubes (sky dancers) of Peter Minshall and Doron Gazit [11], debuted at the 1996 Olympics in Atlanta, which are activated by turbulent airflow from fans. Other artist and designers have explored interactive inflatable installations that actively transform their shape by mechanically circulating air between compartments through fans [12–14]. These applications provide limited control on the behavior of an inflatable and remain active only as long as fans provide airflow. Finally, another direction of actively transformable inflatables includes applications that actuate their shape with compressed air [15]. These applications constitute precisely controlled closed pneumatic systems that can retain their state even when airflow is not circulating, however, the energy, cost, and sophistication of their equipment and engineering limit their implementation to indoor installations. As such, large-scale shape-changing inflatable structures remain a relatively unexplored area of design research in arts and architecture.

2.2 Inflatable Shape-Changing Interfaces

Interactive pneumatic or inflatable systems gain increasing attention in the fields of HCI, DIS, and TEI mainly in areas of shape-changing interfaces design research. Shape and volume can change by moving air between compartments through air compressors and pumps and can they be preserved by blocking air through controllable valves. Examples in the literature of these fields include HydroRing [16], PneuSeries [17], Lifttiles [18], PneUI [19], aeroMorph [20], Printflatables [21], Pneuxels [6], JamSheets [22], AirPinch [23], tilePop [24], MultiPneu [25], Reflatables [26], Inflatable Mouse [27]. Most works focus on object scales, however, some works have expanded to room scales [18, 28, 29]. Most applications use active flow control while only a few use passive flow control [30].

2.3 Design, Control, Actuation and Modeling Challenges for Large-Scale Transformable Inflatables

Design, control, actuation and modeling for large-scale transformable inflatables have substantial challenges compared to their smaller-scale counterparts. 1) In outdoor public spaces, weather, wind load, public interaction, and safety requirements make controllability difficult and expensive compared to the predictable indoor conditions of object-based applications. 2) In public space scales, inflatable structures are almost exclusively actuated by airflow through fans or blowers that involve different physics and design principles than the high-precision digitally controlled air compression hardware (pumps, valves) of small-scale inflatable shape-changing interfaces. 3) Computational modeling of large-scale transformable inflatables is challenging due to the indeterminacy that semi-inflated structures have and due to the advanced mathematics involved in describing fluid behavior in dynamic contexts.

2.4 Contributions

This paper contributes novel methods to design, control, actuate and model large-scale transformable inflatables for architectural, urban, or territorial applications. 1) We present a design approach based on enclosing a transformable inflatable inside a constantly inflated external inflatable that provides structural stability from the outside and controllable conditions from the inside. 2) We present methods to passively control the internal inflatables by transferring air between compartments through a peristaltic sleeve damper that minimizes the amount of energy required to control them. 3) We present modeling, simulation, and experimental methods to study the design and behavior and we discuss their limitations. 4) We discuss potential application of large-scale networked interactive shape-changing inflatables.

2.5 Case Study

The design and prototypes presented in this paper are part of a study for a city-scale social telepresence art installation of wirelessly connected giant inflatable beacons placed at multiple neighborhoods of a city, that enable visitors in each neighborhood to physically experience the presence of visitors in other remote sites by exchanging air between the inflatables through a "closed system" metaphor (see Fig. 1). Each beacon features an emoji character and represents the genius loci of its location. Anyone can actuate any beacon by sending bursts of air to it using their mobile phones. Sending, however, a burst of air to a beacon requires removing the same burst of air from another beacon. Watching a beacon deflate signifies someone, somewhere, uses currently this air to inflate another beacon. Watching a beacon inflate signifies someone, nearby, uses currently this air to publicly manifest their presence in the observer's surrounding environment (see Fig. 2). While our contributions are generalizable, in what follows, we use this art project as a context for our research.

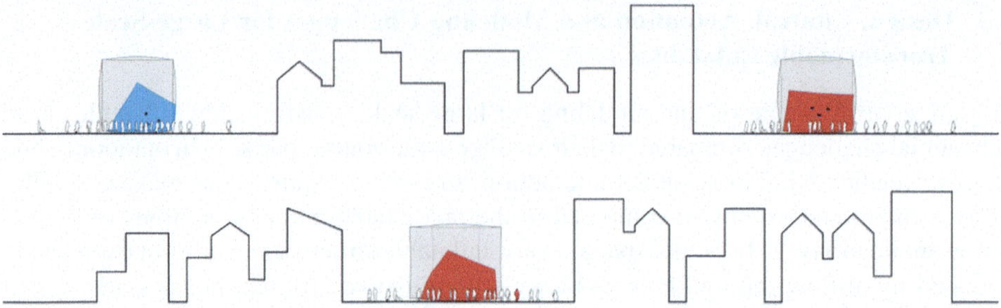

Fig. 1. Conceptual diagram of an example application for social telepresence. Users actuate giant inflatable beacons locally by pulling air from other beacons remotely. Air bursts between pairs of remote beacons cause beacons in some locations to marginally inflate whereas beacons in other locations to marginally deflate manifesting the presence of remote social groups.

Fig. 2. Photomontage of a semi-inflated inflatable beacon in a neighborhood.

3 Methods

3.1 Design Approach for Large-Scale Transformable Inflatables

We outline the following design principles for large-scale transformable inflatables. 1) A shape-changing inflatable is encapsulated inside a transparent constantly inflated external inflatable which provides structural support and protects the internal inflatable from environmental and anthropogenic conditions. 2) Shapes are changed by moving air between the two compartments. 3) Shapes are preserved by preventing air from moving between the two compartments. 4) Control of airflow is passive: the system does not know its state and the movements of air are not controlled or actuated digitally. 5) The only components that are digitally controlled are the relays that activate and deactivate the fans.

Our proposed physical system consists of the encapsulated inflatable interfaces, tubes, connectors, dampers, and anchoring methods. The encapsulated inflatable interfaces consist of an internal and an external inflatable bubble. The internal bubble can take practically any form whereas the external bubble acts as a bounding box for the internal one. Tubes move air between the inflatables and consist of two types. Soft tubes shrink if air pressure around them increases relatively to air pressure inside them, blocking thereby airflow. Hard tubes allow airflow through them independently of the relation between internal and external air pressures. Connectors are custom made interfaces between the inflatables, tubes, dampers, and fans. They include gaskets and flanges (the points where a tube penetrates or connects to the surface of an inflatable), and shoe-sockets for fans for placing them inside inflatables. Dampers (valves) control the flow of air through tubes or holes. Anchors provided fixed connections of the inflatable with the ground. For medium-size inflatables, sandbags can be places in their perimeter. For larger inflatables, sandbags must be combined with belts, cables, and anchors (screws or stakes) in the ground.

3.2 Passive Control and Actuation Methods

To control the inflatables, we wanted to be able to change their inflation state when a fan is activated and to preserve their inflation state when a fan is deactivated without, however, using additional energy other than that used for the fans and without having substantial reductions in airflow. To achieve this, we developed a design for a simple, low-cost, passive control damper that allows airflow above a threshold and blocks airflow below that threshold through peristalsis (contraction). The damper consists of a sufficiently long soft tube (sleeve), which is mounted at both ends to the outflows of two fans both of which are placed inside an inflatable (see Fig. 3).

The inflow of the deflating fan remains inside the inflatable whereas the inflow of the inflating fan is connected to a hard tube which exits the inflatable and links to its outside. When both fans are idle, the pressure inside the soft tube is lower than the pressure outside of it, causing the soft tube to contract and seal (the pressure inside an inflatable is always higher than the pressure outside of it due to the weight of the material of the membrane of the inflatable). When any of the two fans operates, it pushes air towards the interior of the soft tube which inflates it and allows the air to flow through it. In order to change the state of an internal inflatable, an opposite change must occur to the state of the external inflatable. The peristaltic damper with the two fans is placed inside the internal inflatable: one fan moves air from the external to the internal inflatable whereas the other fan moves air from the internal to the external inflatable. To inflate the internal inflatable, the first fan operates while the second fan is idle. To deflate the internal inflatable, the second fan operates while the first fan is idle. As air moves between the two inflatables, their volumes change reciprocally.

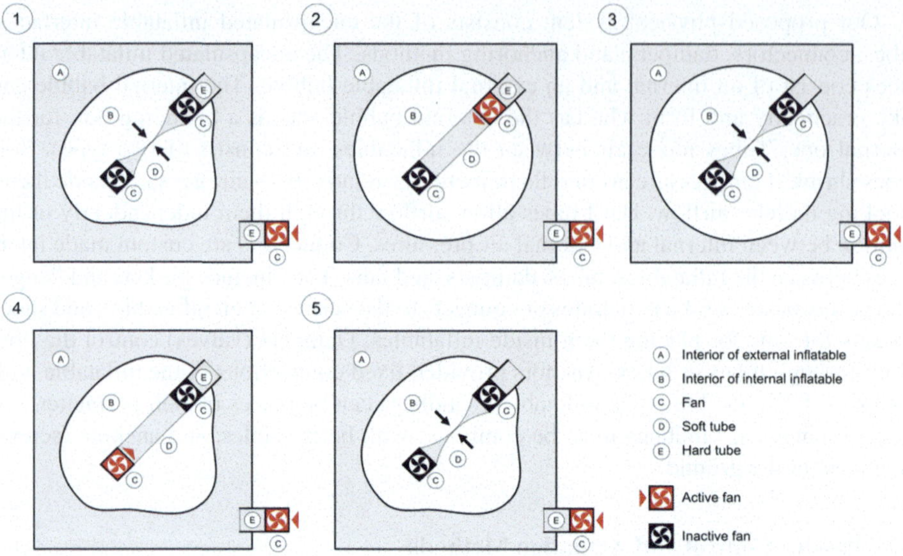

Fig. 3. Operation of control valves during incremental inflation and deflation. In all cases, the external fan keeps the interior of the external inflatable (A) under constant pressure: 1) The pressure of the interior of the external inflatable (A) compresses the interior of the internal inflatable (B) which in turn compresses the soft tube (D) interrupting the air flow and keeping the volume in B constant. 2) When inflow fan (E) in tube D operates, internal pressure in D increases enabling air to flow A to B (inflation). 3) When neither of the inflow or outflow fans of D operate, the internal pressure of A and B compress D sealing it. 4) When outflow fan of D operates, internal pressure in D increases enabling air to flow from B to A (deflation). 5) Same as 3.

3.3 Modeling and Simulation Methods

Design of transformable inflatables requires modeling various inflation states and the dynamic transition between them. Examples of questions during design process include: How to identify if the performance characteristics of a fan are sufficient for the desired outcome? How much time does it take for a transformation between two states to occur? Will material strength of a membrane be sufficient to withstand its surface tension during various degrees of inflation? How likely is for a deflating inflatable to be caught in the intake of a tube or a fan during deflation process? How to assess the performance of an inflation/deflation control strategy? To address such types of questions, a combination of modeling, simulation, and physical prototyping methods was used: physics-based simulation in Kangaroo for Grasshopper; system dynamics modeling of inflation and deflation in Vensim; and prototyping experiments.

Physics-Based Simulation in Kangaroo for Grasshopper

For geometric form exploration, we used Kangaroo Physics [31], a Grasshopper component for Rhinoceros 3D CAD modeling software [32]. Kangaroo is a particle springs system solver that simulates physics by converting a mesh into a combination of particles interconnected by springs, by applying physics goals (forces) and constraints, and by iteratively approximating their equilibrium by iteration and a threshold. On each timestep, the solver calculates the applied forces based on the deformation of springs, then it calculates the new velocities, and then it moves particles to the new positions. Once the difference in forces between two consecutive iterations is below a given threshold, the solver ends the iterative approach.

Figure 4 illustrates the Grasshopper definition that was developed with a description of each Kangaroo component and their interdependency. In order to simulate the contracting and expanding behavior of the peristaltic damper, we defined as separate meshes the inflatable cube and the tube within it and we applied different pressure loads to each of them. Figure 5 shows a simulation of the behavior of the peristaltic sleeve damper. Figure 6 shows a simulation of an inflatable in three different inflation states.

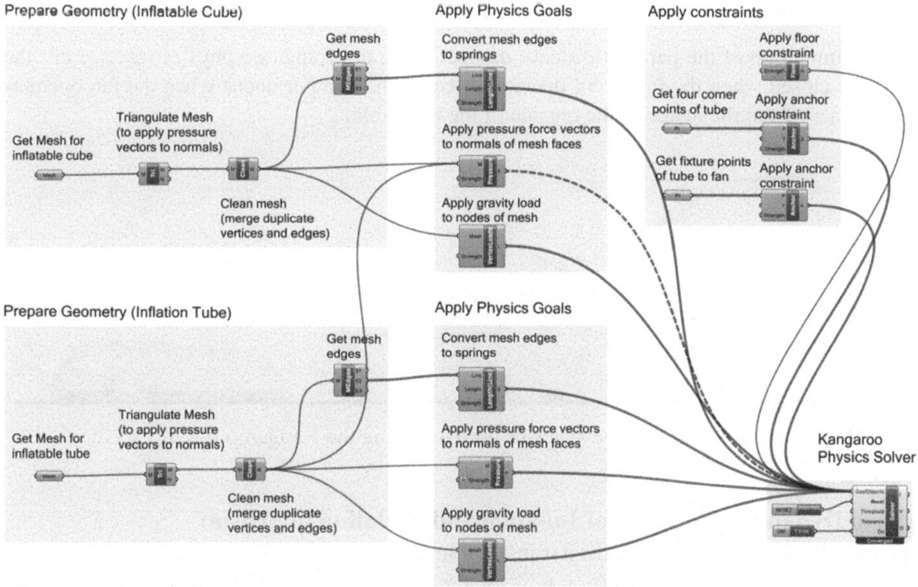

Fig. 4. Grasshopper definition of the simulation using Kangaroo components. The simulation model consists of three parts. First, input mesh geometries are used for both the inflatable cube and the inflation/deflation tube. These meshes are converted into nodes and edges. Then, the nodes and edges are converted into particles and springs and pressure and gravity forces are applied on them as goals. Then, anchoring and floor constraints are applied, also as goals. Finally, these goals are fed into the Kangaroo physics solver component which performs the simulation iteratively.

It is important to note that Kangaroo (like any other particle springs solver) simulates inflation physics by applying a pressure force vector to each triangulated mesh plane along its normal vector (a vector located at the center of a face perpendicular to its plane). This means that while Kangaroo can be used as a means to approximate the start and end state of a shape, develop intuition, and test a proof of concept, it cannot be used as a tool to accurately explore performance metrics such as timing of inflation.

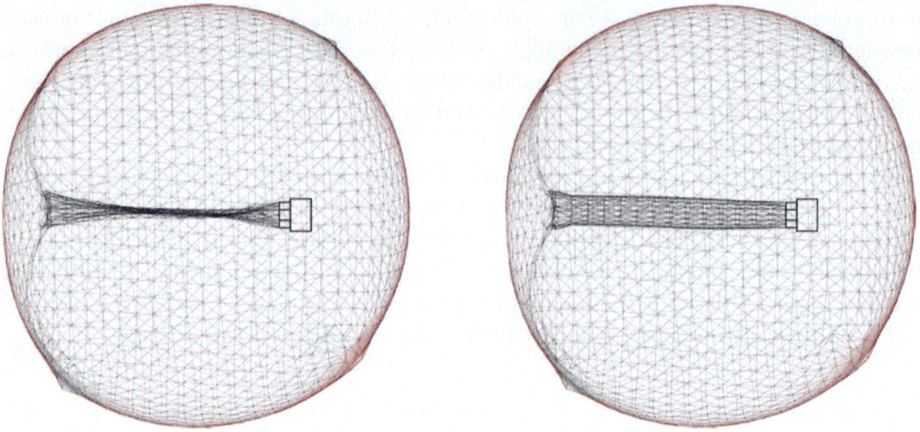

Fig. 5. Simulation of the peristaltic sleeve damper using the Kangaroo physics solver. Left: the damper is closed when the fan is not operating. Right: the damper opens when the fan operates and moves air from the inside to the outside of the inflatable.

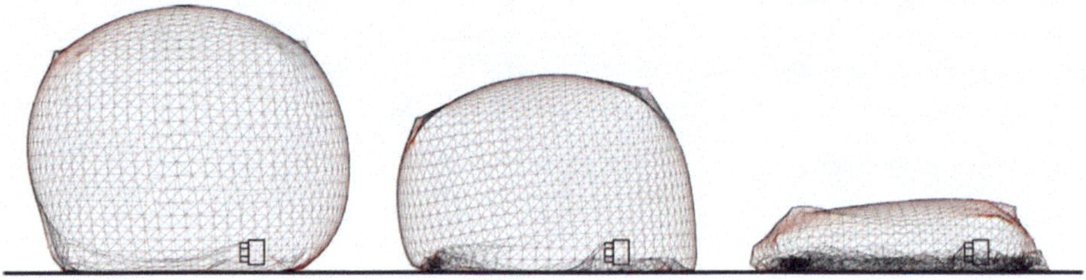

Fig. 6. Simulation of three states of inflation using the Kangaroo physics solver.

System Dynamics Modeling of Inflation and Deflation in Vensim

To model the time that it takes for a transformation of an inflatable between two different states based on a fan's airflow efficiency, the fan's static pressure, and the pressure applied to the air inside the inflatable because of the weight and thickness of its membrane material, we developed a compartment model that numerically simulates movement of a gaseous substance between the two inflatables through the tube. Since the compartment-based approach is shape generic, we used System Dynamics, a methodology of modeling and simulation of complex systems as combinations of stocks and flows [33]. To model the compartment system we used Vensim, a popular system dynamics modeling software [34] that allows a modeler to define mathematically relationships between components and simulate the dynamics of a system over time.

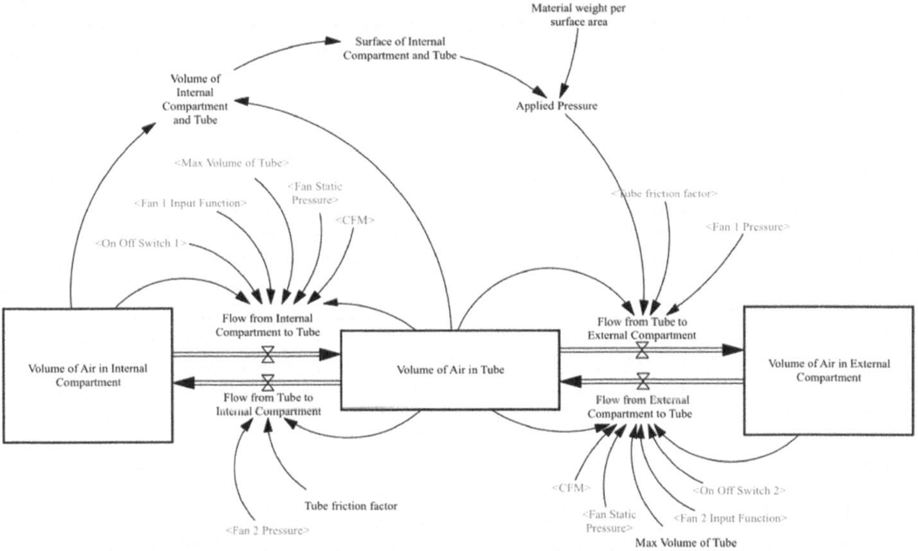

Fig. 7. System dynamics compartment model made in Vensim. The model models the changes in volumes of the inflatables as a result of exchange of air between them. The causal relationships of the model show that as the internal inflatable deflates, the pressure it applies to the tube decreases because the surface of material decreases. Thus, the rate with which the peristaltic damper contracts decreases.

We model the flow of air between the internal and the external inflatable through the tube as the exchange between three compartments (see Figs. 7 and 8). To simplify the modeling process, we assume that the pressure inside the tube is equal to the pressure inside the internal inflatable within which the tube is placed. This is because the pressure of the internal inflatable is applied to the outside tube and compresses it until the two pressures equalize. As a result, since flow between compartments is proportional to their relative pressure difference, we assume that the net flow between the internal compartment and the tube is zero (the flow from the compartment to the tube equals the flow from the tube to the compartment). In contrast, the pressure difference between the interior of the tube and the external inflatable (which is equal to the pressure applied inside the internal inflatable from its membrane's material weight) forces air to evacuate the tube and move inside the external inflatable. The flow from the tube to the external compartment is proportional to the pressure difference, the diameter of the tube, and inversely proportional to the length of the tube (friction). As air evacuates the tube, its volume decreases which further decreases the outflow rate until the tube collapses and prevents further air from escaping.

The model in Vensim in Fig. 7 captures the above relationships. The model allows a user to control the fans' input function (e.g., when a fan turns on or off), their throughput power (in cubic feet per minute or CFM), the maximum volumes of the internal and external inflatables and the tube, and the material weight, and to explore the dynamics of inflation and deflation in each of the three compartments.

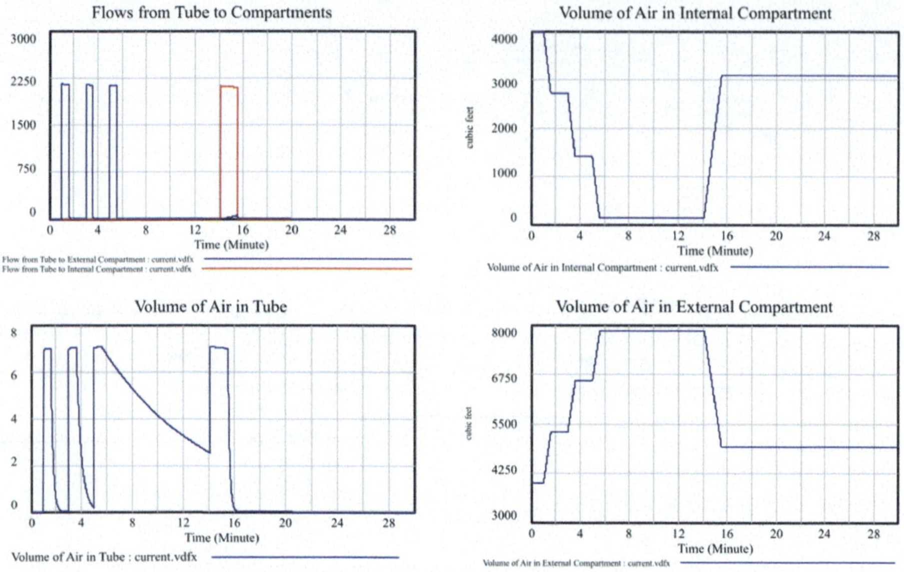

Fig. 8. Output graphs from numerical simulations in Vensim that show the dynamics of stocks and flows. The model simulates the outcome of three consecutive pulses of air from the internal to the external compartment followed by a longer pulse from the external to the internal compartment. Top left: flow rates from tube to external (blue) and from tube to internal (red) compartments measured in cubic-feet per minute. Top right, bottom left and bottom right: levels in the internal, tube, and external compartments measured in cubic-feet.

3.4 Prototyping and Experimental Testing

To test the design for the peristaltic damper and the behavior of the encapsulated inflatable during inflation and deflation, we developed a series of full-scale prototypes. Each inflatable interface consists of a $20' \times 20' \times 20'$ transparent external cube which contains a $17' \times 17' \times 17'$ opaque internal cube. The external cube remains constantly inflated through an industrial fan that produces 3,800 CFM of air flow. The internal cube is actuated by two 1-HP industrial fans with $18''$ diameter [35], each producing 5,800 CFM of air flow between the interior of the internal cube and the intermediary space between the internal and external cubes. The external cubes are made of a combination of 12-mil heavy-duty clear polyethylene tarp (poly-tarp) and 6-mil Tufflite IV (a high-performance polyethylene used in green houses) sheet, to ensure clear visibility of the internal cubes they contain. The internal cubes are made of 6 mil opaque polyethylene sheet. The poly-tarp and polyethylene sheets are either welded together with heat or they are bonded together with heavy-duty double-sided adhesive tapes. To keep the base flat, sandbags were used around the perimeter as anchor points.

Using the prototypes, we conducted a series of tests in which we evaluated the performance of the peristaltic sleeve damper, the behavior of the inflatables during inflation and deflation, the performance of material strength in combination to the fabrication and welding of the seams, and the actual timing of each transformation between different states.

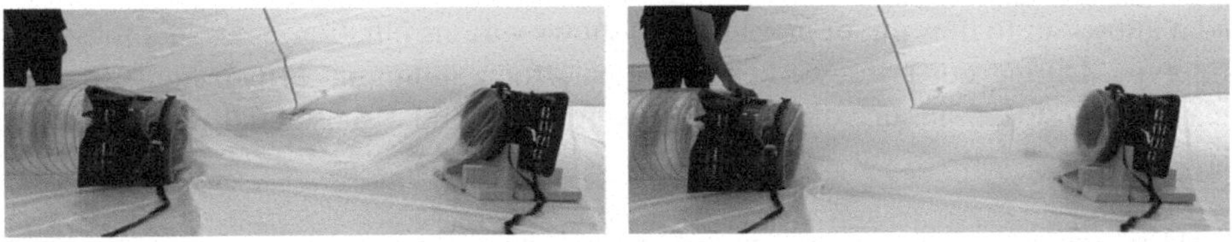

Fig. 9. Peristaltic sleeve damper. Left: contracted; the pressure inside the sleeve is lower than outside of it preventing air to flow through it. Right: expanded; the airflow of the activated fan increases the internal pressure in the sleeve and expands it allowing the air flow through it.

Fig. 10. Frames showing four inflation states of a full-scale prototype. Each state was able to retain its volume for more than two hours with about 5% loss even when almost inflated.

Figure 9 shows the behavior of the peristaltic sleeve damper inside the inflatable. The sleeve successfully contracts under the air pressure of the internal inflatable and seals the tube. When the fan is activated, the pressure inside the tube expands the sleeve and it allows air to flow out or in. Figure 10 shows various inflation states of a full-scale prototype. During our tests, we kept each state for a minimum of two hours and we measured the losses in air volume which were found to be substantially small (about 5%).

4 Discussion

We presented methods to design, control, actuate and model large-scale transformable inflatables for urban or territorial scales, primarily for the arts, with applications that may include public participation festivals, collaborative giant puppet festivals, collective telepresence applications, land art installations, and more. We furthermore demonstrated these methods with prototypes through a case study of a public art installation project. Other ongoing or future research steps include the development of the backend platform, development of methods for creating connectors, the design and development of additional passive control dampers, the development of alternative forms and geometries for the inflatable interfaces, and the development of the electromechanical infrastructure for actuating the fans, consisting of wireless-communication capable microcontrollers and relays.

A number of challenges are worth mentioning on the modeling side. While particle-spring simulation approaches are useful for approximating equilibrium forms during each state, they cannot model precisely the actual physics of the dynamics of air moving between compartments. Moreover, the initial mesh, created from the geometry of the cube, influences the final equilibrium form because it is used to generate the particles and springs and, thereby, it determines how loads, pressure, and tensile forces are distributed. Too dense meshes with low spring coefficients assimilate softer materials but increase computational complexity. On the other hand, too coarse meshes with higher spring coefficients assimilate thicker materials but they may fail to converge with an acceptable threshold. While system dynamics modeling is more accurate in modeling changes in volume because of transfer of air between compartments, a numerical simulation is not as accurate as a mathematical solution. On the other hand, modeling the dynamics of the system mathematically leads to unnecessary complexity. More accurate modeling approaches as presented in [36] are beyond the scope of this present study. Prototyping methods complemented modeling approaches with experimental measurements of inflation and deflation rates and volumes and they allow to verify results, calibrate models, and develop better intuition. To measure flow rates, we measured the time it took for changes in specific volumes. However, accurate results have difficulties because the actual behavior is influenced by prototyping imperfections. Despite these challenges, we believe interactive large-scale transformable inflatables is an exciting and promising emerging field of research in the arts, architecture, and human-computer interaction.

Acknowledgements. The design and research presented in this paper have been supported by the Knight Foundation through the "Celebrate Charlotte's Arts" grant and by the UNC Charlotte through a Faculty Research Grant. The following students have contributed to the research and prototyping (alphabetically): Elliot Ball, Alex Caviness, Manoj Deshpande, Hannah Duffey, Srinath Muralinathan, Matthew Robinson, Saquib Sarwar, Subham Sah, Nick Sturm, Aiden Williams, Ricky Young.

References

1. Graham, S.: Atmosfields, Centre Pompidou. https://www.centrepompidou.fr/en/ressources/oeuvre/cez5LMM. Accessed 21 Jan 2022
2. Airground: Jeffrey Shaw Compendium. https://www.jeffreyshawcompendium.com/portfolio/airground/. Accessed 20 Jan 2022
3. Poetic Cosmos of the Breath · STUDIO TOMÁS SARACENO, STUDIO TOMÁS SARACENO, 11 June 2018. https://studiotomassaraceno.org/on-the-poetic-cosmos-of-the-breath/. Accessed 15 Feb 2022
4. Museo Aero Solar: for an Aerocene era · STUDIO TOMÁS SARACENO, STUDIO TOMÁS SARACENO, May 21 2021. https://studiotomassaraceno.org/museo-aero-solar-for-an-aerocene-era/. Accessed 15 Feb 2022
5. Piene, O., Russett, R.: Sky, scale and technology in art. Leonardo **41**(5), 510–518 (2008). https://doi.org/10.1162/leon.2008.41.5.510
6. Gazit, D.: Doron Gazit Environmental artist, activist, Industrial Designer. Doron Gazit. https://www.dorongazit.com. Accessed 17 Jan 2022
7. Farm, A., Shapiro, M., Lord, C., Schreier, C.: Inflatocookbook: A Pneu-age Tech Book. Rip Off Press, Sausalito (1971)
8. de Prada Poole, J.M., et al.: José Miguel de Prada Poole: La arquitectura perecedera de las pompas de jabón, 1st edn. Recolectores Urbanos Editorial, Málaga (2019)
9. Prada Poole: The Perishable Architecture of Soap Bubbles, Centro Andaluz de Arte Contemporáneo (CAAC), Seville (2019). http://www.caac.es/english/exh/projects/frame_jmpp2019.htm. Accessed 17 Jan 2022
10. On Space Time Foam. STUDIO TOMÁS SARACENO, 17 Apr 2018. https://studiotomassaraceno.org/on-space-time-foam/. Accessed 17 Feb 2022
11. Gazit, D., Dranger, A.L.: Apparatus and method for providing inflated undulating figures, US6186857B1, 13 February 2001. https://patents.google.com/patent/US6186857B1/de. Accessed 13 Feb 2022
12. Fox, M., et al.: Bubbles, FoxLin Architects. https://foxlin.com/portfolio_item/bubbles/. Accessed 22 Jan 2022
13. HYPERBODY Research Group, Oosterhuis, K., Klevid, C.: Muscle NSA. http://www.hyperbody.nl/research/projects/muscle-nsa/index.html. Accessed 22 Jan 2022
14. Dickey, R.: Air hugs: a large-scale interactive installation. In: ACM SIGGRAPH 2019 Art Gallery, New York, July 2019, pp. 1–6 (2019). https://doi.org/10.1145/3306211.3320141
15. Lundén, E., Kauste, J.: Another Generosity. Lunden Architecture Co. https://www.lunden.co/research/another-generosity/. Accessed 13 Feb 2022
16. Han, T., Anderson, F., Irani, P., Grossman, T.: HydroRing: supporting mixed reality haptics using liquid flow. In: Proceedings of the 31st Annual ACM Symposium on User Interface Software and Technology, New York, NY, USA, October 2018, pp. 913–925 (2018). https://doi.org/10.1145/3242587.3242667
17. Chen, Y.-W., Lin, W.-J., Chen, Y., Cheng, L.-P.: PneuSeries: 3D shape forming with modularized serial-connected inflatables. In: The 34th Annual ACM Symposium on User Interface Software and Technology, New York, NY, USA, pp. 431–440. Association for Computing Machinery (2021). https://doi.org/10.1145/3472749.3474760. Accessed 17 Jan 2022
18. Suzuki, R., Nakayama, R., Liu, D., Kakehi, Y., Gross, M.D., Leithinger, D.: LiftTiles: constructive building blocks for prototyping room-scale shape-changing interfaces. In: Proceedings of the Fourteenth International Conference on Tangible, Embedded, and Embodied Interaction, New York, NY, USA, February 2020, pp. 143–151 (2020). https://doi.org/10.1145/3374920.3374941

19. Yao, L., Niiyama, R., Ou, J., Follmer, S., Della Silva, C., Ishii, H.: PneUI: pneumatically actuated soft composite materials for shape changing interfaces. In: Proceedings of the 26th annual ACM symposium on User interface software and technology, New York, NY, USA, October 2013, pp. 13–22 (2013). https://doi.org/10.1145/2501988.2502037
20. Ou, J., et al.: aeroMorph - heat-sealing inflatable shape-change materials for interaction design. In: Proceedings of the 29th Annual Symposium on User Interface Software and Technology, New York, NY, USA, October 2016, pp. 121–132 (2016). https://doi.org/10.1145/2984511.2984520
21. Sareen, H., et al.: Printflatables: printing human-scale, functional and dynamic inflatable objects. In: Proceedings of the 2017 CHI Conference on Human Factors in Computing Systems, New York, NY, USA, May 2017, pp. 3669–3680 (2017). https://doi.org/10.1145/3025453.3025898
22. Ou, J., Yao, L., Tauber, D., Steimle, J., Niiyama, R., Ishii, H.: jamSheets: thin interfaces with tunable stiffness enabled by layer jamming. In: Proceedings of the 8th International Conference on Tangible, Embedded and Embodied Interaction, New York, NY, USA, February 2014, pp. 65–72 (2014). https://doi.org/10.1145/2540930.2540971
23. Gohlke, K., Sattler, W., Hornecker, E.: AirPinch – an inflatable touch fader with pneumatic tactile feedback. In: Sixteenth International Conference on Tangible, Embedded, and Embodied Interaction, New York, NY, USA, February 2022, pp. 1–6 (2022). https://doi.org/10.1145/3490149.3505568
24. Teng, S.-Y., et al.: TilePoP: tile-type pop-up prop for virtual reality. In: Proceedings of the 32nd Annual ACM Symposium on User Interface Software and Technology, New Orleans, LA USA, October 2019, pp. 639–649 (2019). https://doi.org/10.1145/3332165.3347958
25. Zhao, X., et al.: MultiPneu: inflatable deformation of multilayered film materials. In: The Ninth International Symposium of Chinese CHI, New York, NY, USA, October 2021, pp. 158–161 (2021). https://doi.org/10.1145/3490355.3490517
26. Murayama, T., Yamaoka, J., Kakehi, Y.: Reflatables: a tube-based reconfigurable fabrication of inflatable 3D objects. In: Extended Abstracts of the 2020 CHI Conference on Human Factors in Computing Systems, New York, NY, USA, April 2020, pp. 1–8 (2020). https://doi.org/10.1145/3334480.3382904
27. Kim, S., Kim, H., Lee, B., Nam, T.-J., Lee, W.: Inflatable mouse: volume-adjustable mouse with air-pressure-sensitive input and haptic feedback. In: Proceedings of the SIGCHI Conference on Human Factors in Computing Systems, New York, NY, USA, April 2008, pp. 211–224 (2008). https://doi.org/10.1145/1357054.1357090
28. Deshpande, M., Sarwar, S., Mahdavi, A., Papanikolaou, D.: Pneuxels: a platform for physically manifesting object-based crowd interactions in large scales. In: Proceedings of the 2019 ACM International Joint Conference and 2019 International Symposium on Pervasive and Ubiquitous Computing and Wearable Computers, London, UK, pp. 9–12 (2019)
29. Swaminathan, S., Rivera, M., Kang, R., Luo, Z., Ozutemiz, K.B., Hudson, S.E.: Input, output and construction methods for custom fabrication of room-scale deployable pneumatic structures. Proc. ACM Interact. Mob. Wearable Ubiquitous Technol. **3**(2), 62:1–62:17 (2019). https://doi.org/10.1145/3328933
30. Paez-Granados, D., Yamamoto, T., Kadone, H., Suzuki, K.: Passive flow control for series inflatable actuators: application on a wearable soft-robot for posture assistance. IEEE Robot. Autom. Lett. **6**(3), 4891–4898 (2021). https://doi.org/10.1109/LRA.2021.3070297
31. Kangaroo3d. http://kangaroo3d.com/. Accessed 31 Mar 2022
32. R. M. & Associates: Grasshopper for Rhino 3D. www.rhino3d.com. https://www.rhino3d.com/6/new/grasshopper/. Accessed 31 Mar 2022
33. Forrester, J.W.: Principles of systems; text and workbook, chapters 1 through 10, 2nd. preliminary ed. Cambridge, Mass (1968)

34. Vensim. https://vensim.com/. Accessed 31 Mar 2022
35. XPOWER BR-450 Tube Man Blower» XPOWER Manufacture. https://xpower.com/shop/br-450-tube-man-blower/. Accessed 20 Jan 2022
36. Skouras, M., et al.: Designing inflatable structures. ACM Trans. Graph. **33**(4), 63:1–63:10 (2014). https://doi.org/10.1145/2601097.2601166

Timber Framing 2.0

An Interdisciplinary Study into Computational Timber Framing

Jens Pedersen[1,2(✉)], Lars Olesen[3], and Dagmar Reinhardt[4]

[1] Aarhus School of Architecture, Nørreport 20, 8000 Aarhus, Denmark
jenspedersen88@gmail.com
[2] Odico Construction Robotics, Oslogade 1, 5000 Odense, Denmark
[3] Brav Engineering, Kongensgade 66, 1, 5000 Odense, Denmark
[4] The University of Sydney, Sydney, NSW 2006, Australia

Abstract. Constructing buildings follow specific sequences and processes, which lends itself well to a combination of parametric workflows and fabrication technologies such as robotics and CNC machines. Architects and researchers have begun to adopt these workflows leading to the design and manufacturing of building systems for the AEC industry. This paper adds to this emerging field by discussing an interdisciplinary workflow and prototypical process developed for a construction system in which an industrial robot arm processes off-the-shelf standardised building components. The workflow combines a structural model with a fabrication model into one unified system, thus improving the manufacturing process from structurally valid building to the fabricated piece.

Keywords: Timber construction · Construction system · Robotic fabrication · Timber joints · Material tolerance

1 Preface

This paper presents an area of ongoing research aimed at extending the fabrication capabilities of an existing mobile robot unit (MRU). The paper presents a prototypical process that led to a modified timber framing construction method and gave indications as to how to design and modify the MRU. The project is part of an industrial Ph.D. study focusing on robotic timber fabrication and interface methods undertaken at Aarhus School of Architecture in collaboration with Odico Construction Robotics. Furthermore, the project was conceived in collaboration with Brav Engineering. The paper describes the prototypical stages that led to the first commercially robotically fabricated timber house in Denmark alongside the process of integrating the structural model with the fabrication model. The paper introduces the project context and state of the art for timber construction systems enabled through digital fabrication technologies (Sect. 2–2.1). Section 3 describes the overall system, the integration of structural and fabrication model (Sects. 3.1–3.2), and the fabrication setup (Sect. 3.3). Section 4 highlights three prototypes that led to the development of a construction system, which was used in the construction of a house in Odense, Denmark. Lastly, the paper discusses the findings and future work (Sect. 5) and concludes in Sect. 6.

C. Gengnagel et al. (Eds.): DMS 2022, *Towards Radical Regeneration*, pp. 320–331, 2023.
https://doi.org/10.1007/978-3-031-13249-0_27

2 Introduction

Digital fabrication has through the past years, seen a growing implementation within architectural fabrication, where off-site production facilities have fabricated infinitely varied elements for geometrically complex projects [1, 2]. These technologies also see regular use in the timber industry, where large CNC facilities, such as "Hundeggers" [3], fabricate architectural timber elements such as trusses or wall modules daily. However, since the 80's researchers and practitioners have attempted to migrate digital fabrication technologies to the construction site. Where the aim is not only to improve fabrication capabilities [4, 5], but also to improve on-site working conditions [6]. A contemporary example of such a system is the "*Factory on the Fly*" (FOTF) developed by Odico Construction Robotics [7], which functions as the research context for the PhD project. The FOTF is a modular architecture that can be modified to fit different needs, but what information is necessary to design the working environment for an FOTF system?

This project uses a "learning by doing approach" to better understand this problem. Here, the learning outcome from different prototypes is used to indicate what is needed for an initial design proposal for an on-site timber fabrication robot cell. This paper describes the process of going from prototypes, to extracting knowledge for further use. The starting idea for the first prototype, is found within complex geometrical projects, where simplicity of assembly is key. Simplicity in assembly could benefit conventional construction where adapting elements to one another is a high affordance task, or physically marking for elements positioning is time-consuming (Fig. 1).

Fig. 1. A photo from a site visit, where we can see, that drawings are made to indicate that a post is needed and where to position it. An area which can be subject to error.

The domain of timber joints is explored both as a resource and inspiration. The assembly of construction systems such as the half-timbered house closely resembles putting a puzzle together, whereby pieces can only be assembled in one way [8–10].

Therefore, this paper presents the process of developing a simplified variation of timber joints, aiming to simplify and decrease erection time of a timber-framed house.

This construction method is one of the most popular types of building methods for homes in the United States and was popularized in Denmark during the energy crisis around 1970 [11, 12]. The paper outlines how the structural and fabrication model can be combined and describes a robotic fabrication system consisting of an ABB IRB 6400-R industrial robotic arm, equipped with a rotary table, alongside workflow improvements that can be adopted for future research and projects.

2.1 State of the Art

Historically, timber has been used in documented construction systems since as early as the late 1600s. A half-timbered house is an example of a building system made of precise complex timber joints (mortise and tenon, scarf joints, or dovetails) that is numbered and can only be assembled in one way [8–10]. Designing bespoke pieces that can be assemble in one way is a concept widely adopted in the fabrication of complex geometries across research and academia. This concept is tested on single family homes in the construction system developed by EentilEen/OnetoOne [13]. Here, plywood sheets are cut on a CNC router, labeled, and assembled into plywood cassettes with finger joints (off-site). After assembly, the cassettes are sent to site for final assembly based on individual markings. Though little data is available about the system, especially machine time associated with fabricating a house using the method. It is assumed to be time-consuming due to the geometry of the joinery and the number of pieces per cassette. Thus, time spent per fabricated piece became point of departure for the system proposed in this paper.

As an example of research prototypes with timber components, recent research by Gramazio Kohler explores a timber construction system designed for dual robotic assembly. Here, timber is cut to size and positioned in modules by collaborating industrial robotic arms on a gantry system [14]. Assembled modules are sent to site for final assembly. Despite designing for robotic assembly, the researchers found that human intervention was necessary, to mitigate material impression such as warp or skew along elements [14]. This illustrates the need for incorporating assembly tolerances into the joinery system to mitigate material imperfections. Based on this precedent setup, the research defined the following project parameters:

1. The timber construction system and joinery process are based on standard, off-the-shelf timber, to simulate an industrial and conventional on-site construction framework.
2. The robotic fabrication should achieve minimal processing time pr piece to enable economic and affordable production.
3. Human intervention is integral for the assembly process.
4. The system should be able to respond to tolerances occurring in the fabrication and assembly process.

3 Computational Framework

The paper presents a computational framework that seamlessly allows transitioning from architectural design input to the structural- and fabrication model. The entire workflow is hosted through 3d modeling software McNeel Rhinoceros [15] and the parametric software extension Grasshopper 3D plugin [16], with further customised software extensions developed for this research. The flow from architectural design to fabrication within Grasshopper is outlined throughout this section and described in principle in Fig. 2.

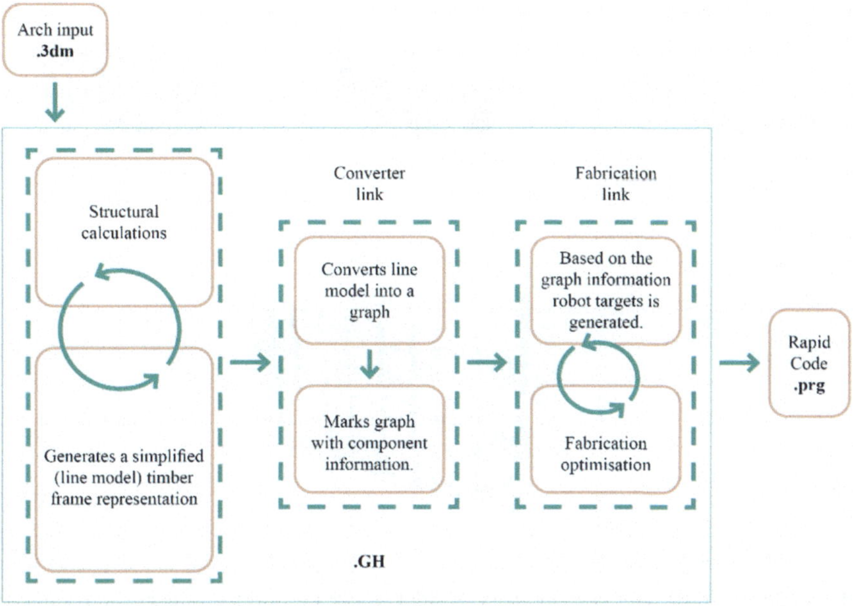

Fig. 2. From left to right describes the actions carried out through the computational framework. The highlighted file extensions specify which environment the actions occur in/result in.

3.1 Structural Model

Timber-framed structures follow specific rules to create repeating elements and seamlessly incorporate other materials such as insulation batts or chipboard material into the construction. In principle, the structure consists of a top and bottom beam, where posts are joined with a butt joint at 600 mm centers, allowing to insert insolation batts– or fixate plywood sheets to the posts (see Fig. 3). Therefore, the design of the structural model aims to solve a combination of a) resisting external loads and b) maintaining modularity for quick assembly on site.

The model is generated by tracing the external wall of an architectural drawing with a polyline in McNeel Rhino3D and importing it into Grasshopper. Here the polyline is broken into façade segments, and special cases at the corners are solved. Afterward, the remaining posts are positioned at the given modulus – this is visualised in Fig. 3 (a) – for visual clarity, we show timber elements instead of lines. Once this basic model is created, its sectional forces are extracted in Grasshopper to validate member sizes and connections. Windows and doors are implemented by positioning a box (Fig. 3(b)) along the polyline and tagging it as a door or window. This results in modifying existing posts and inserting additional posts on either side of the box (Fig. 3(c)). Once openings are positioned in the structure, the system rechecks the structural model, which adds extra beam elements where needed. This logic was reapplied when generating the floor beams (Fig. 3(c)).

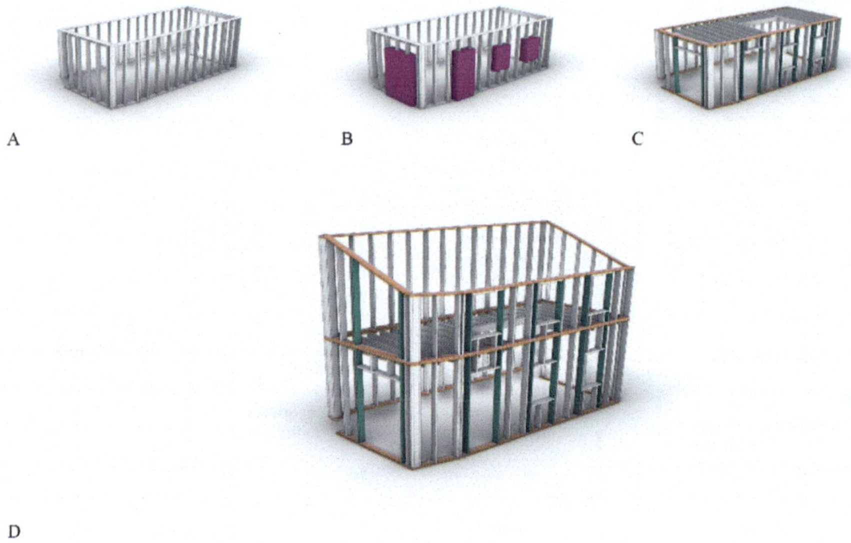

Fig. 3. A) Basic model with post and bars. B) Pink voids are inserted to generate openings. C) Model with post modified and new members inserted. D) a view of the entire structural model for the project described in Sect. 4.4.

3.2 Fabrication Model

Once a structurally valid result has been computed, the associated line model is passed to the first fabrication node. Here the line network is broken into a graph, using a brute force algorithm, where all lines intersect—performing this algorithm returns two layers of information; components identification and connectivity information (Fig. 3 and 4). Knowing what each component is allows to find identical elements and make fabrication batches that can be repeated, making for less robot programming. Additionally, the name and information of each component are marked on it, alongside any information of connecting elements – imagine the timber elements like puzzle pieces. The research team calculated minimal material use based on a Best-Fit bin packing algorithm [17]. Once the above steps have been computed, it is possible to use implicit information in the geometric model to generate robot targets using a bespoke Grasshopper plugin developed at Odico Construction Robotics.

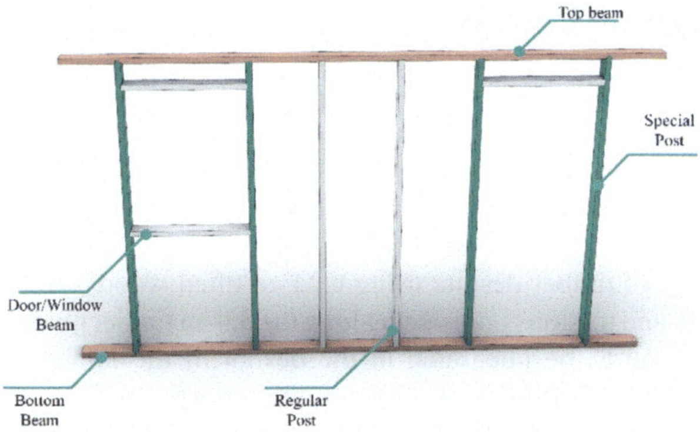

Fig. 4. Naming convention of the components of the marked graph, window/door openings have special posts due the added beams.

3.3 Fabrication Setup

The robotic setup used for this research consists of an ABB IRB 6400 R, equipped with spindle with a 25 mm routing bit, and linked to a rotary table where three timber elements can be fixed at a time. Timber was delivered in standard industrial lengths (5.4-m) and was precut to avoid spatial obstacles of the robot cell (walls, pedestal). Precut elements were fixated on the rotary table with clamps (Fig. 5, B) that helped to minimize impressions due to material imperfections (Fig. 5, A) – see Sect. 4.2 and 4.3. The robot would use the rotary table to trim pieces at both ends, which gave a fabrication sequence: mill details, trim one end, rotate the table, mill details, and finally trim the remaining end.

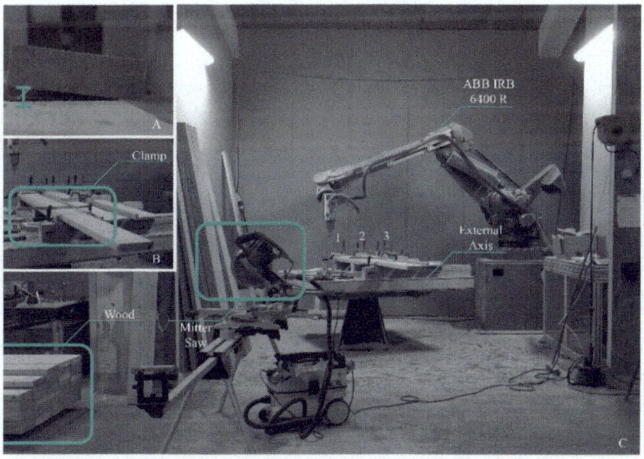

Fig. 5. Displays the robotic fabrication setup, where the position of various hand tools and the material has been optimized for smooth transitions between programs.

4 The Prototypical Process

The following section describes the five-week process of going from the first prototype to the finished construction system and full-scale commercial demonstrator (Sect. 4.4).

The research focused on processing off-the-shelf timber, which posed challenges in the form of material impressions and became the research problem space. Across several prototype studies, it was found that significant deviations in width or thickness could be found in each piece of timber despite using C24 certified wood[1]. Additionally, it was found that bending, and torsion could occur along the 5.4-m timber pieces. Consequently, these impressions had to be integrated in the design of the joinery strategy, and as mentioned into the design of the workstation of the robot (Sect. 3.3).

4.1 Prototype 01

The first prototype revolved around a simple mortise and tenon joint, that could be milled in one pass. Additionally, it was tested if it was possible to machine multiple parts simultaneously, as shown in Fig. 6, A. However, the tenon "migrated" from side to side due to thickness variations in the timber, making it tough to design the correct tenon. Therefore, one piece of timber was milled per slot for future prototypes. In this first prototype we used a tolerance of +0.5 mm, which proved insufficient to accommodate material inaccuracies.

Fig. 6. (A) shows the ambition of trying to machine multiple parts in one robot pass. (B) the first very simple prototype, and (C) a zoomed image of the mortise and tenon joint.

[1] C24, signifies the strength class of the wood. The higher the number, the stronger the wood. C24 is one of the highest ratings [18].

4.2 Prototype 02

The assembly of the second prototype was improved by increasing the tolerances in all joints from +0.5 to +2 mm, making it possible to accommodate material deviations. The focus was on testing a hybrid system between the mortise and tenon joints, and a simple notch that indicate the position of elements (Fig. 7, B & C). The idea of this hybrid was to use a sub-group of posts with the mortise and tenon strategy to form a stable frame into which the remaining posts could be inserted into the notches.

Some post elements of the prototype had torsion along them, making assembly difficult. However, it was possible to attach a clamp and twist the element(s) into place, but it became a concern for the overall project due to the affordances of on-site application. The prototype was deemed sufficient to continue development with the collaboration of two carpenters for refinement in prototype 03 (4.3).

Fig. 7. (A) second prototype, (B) & (C) show close ups of the new "notch" strategy tested during this prototype.

4.3 Prototype 03

For the third timber prototype, the mortise and tenon joint was abandoned since the tenon would restrict movement of the posts (Fig. 8, left). Free linear movement for the posts is necessary to maintain a straight internal wall due to width variations of the timber. A new element for this prototype became a strategy for handling floor separation, which led to development of a vertical beam that acted as a "comb". Floor beams rest in the comb, and a top comb is placed on top of the floor beams to adjust for torsion along the beams (see Fig. 8, right).

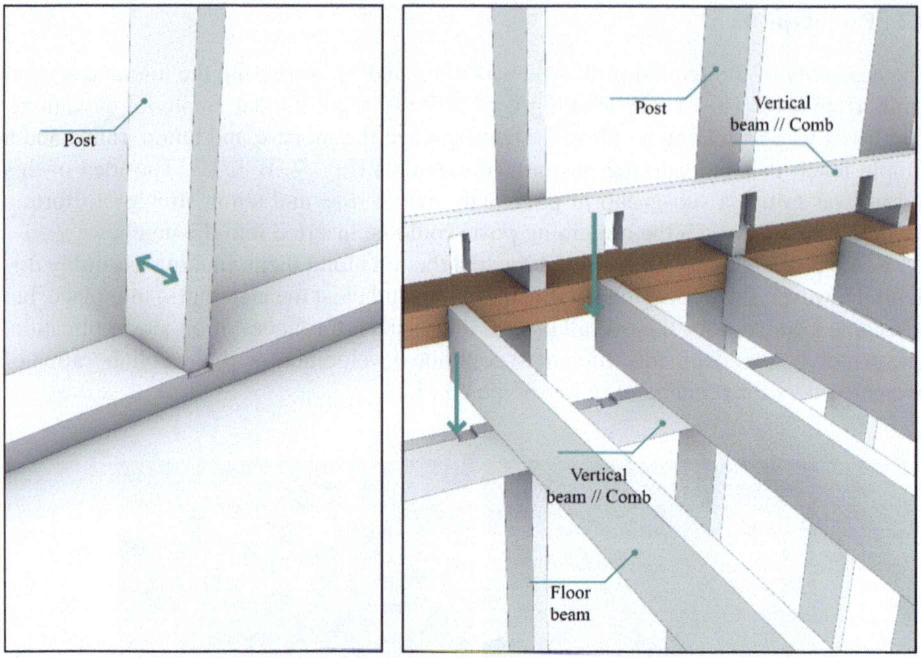

Fig. 8. Left) The post can move freely across the bottom/top beam, allowing to have a straight internal wall. But it results in a varied external wall, which is less noticeable. Right) a diagram highlighting the comb strategy.

Once the pieces for this prototype had been fabricated, they were grouped by type (Fig. 9, A), and two carpenters came and assembled the prototype. During the assembly process, the carpenters showed how to mitigate torsion in posts – a large wrench allowed them to twist elements into place (Fig. 9, B). This meant that the developed fabrication setup can overcome torsion along elements, if the elements are clamped properly, and is flat in the moment of fabrication.

4.4 From Prototype to Demonstrator: Olaf Ryes Gade

Based on the successful system developments across the prototypes and the collaborative improvement, the system was upscaled to a demonstrator: a timber frame structure for a small 50 m^2 two-story house. There were no changes to the system from prototype 03, which enabled a quick transition from prototyping to fabrication. The timber frame structure of this house was fabricated in four days, with the structure being erected in one day. This successfully demonstrated the system and workflow, with the additional finding that the floor separation strategy, made it simpler to fasten the moisture barrier and ensure a good seal (Fig. 10).

Fig. 9. (A) the pieces for the prototype, emulating how we could deliver them to site. (B) the wrench which allowed to untwist posts. (C) the entire prototype and (D) an image of the comb for floor separation.

Fig. 10. (A) image of the floor separation strategy, (B) complex corner and (C) the finished timber frame.

5 Discussion

The development across prototypes to finished timber frame enabled the research to undertake evaluations and iterative revisions. The main findings relate to material imperfections found in standardised industrial elements, integrating off-the-shelf elements in a fabrication pipeline, and establishing direct relationships between digital models and physical construction.

Notably, a review of parameters includes:

- *Processing time* - the processing time per piece was roughly 11.5 min, which includes precutting, fastening the timber with multiple clamps, executing the program, and taking the timber out. This leaves much to be desired, but simply removing the need for precutting and automating clamping would result in a dramatic speed increase.
- *Human Involvement* - developing the fabrication and construction system was centered around how two carpenters would work with it. Therefore, it would be of great interest to have more carpenters use the system, to see if alterations needs to be incorporated into the system.
- *Tolerance*s - some timber elements had deviations in thickness and width upwards of ±3 mm, which doesn't sound like much. Still, in a digital fabrication pipeline it is a lot. Therefore, all notches were cut wider by 4 mm, compared to the ordered material thickness of 45 mm. In the fabrication setup, it was tough to force timber flat/straight with clamps, an issue that need further development.
- *Off-the-shelf material* – the research used off-the-shelf timber and fabricated a house within an acceptable margin, which indicates that digital fabrication in construction may not need to be as accurate as assumed.

5.1 Future Work

The project findings highlight that the current hardware design needs to be redesigned, which makes it a point of interest as the research progress and aims to answer the second challenge as stated in the introduction (2.0 – *how is the work environment of the robot designed*). Therefore, future research aims to present a new hardware strategy that aims to solve the following:

- The station should minimize the need for handling timber.
- How to automate the fixation of timber elements
- Identify which robot can carry out the task
- The robot needs to be able to handle standard timber elements.
- Can the solution be mobile?

Concluding on these points will go toward describing the environment of a new Factory on the Fly solution aimed at timber framed fabrication. The solution should be able to decrease overall construction time, minimize errors, be material efficient and minimize heavy lifts for the construction workers. This will be presented in a future paper.

6 Conclusion

This paper has investigated the potential of robotic fabrication of simple timber joints and presented the successful development of a design-to-fabrication-to-construction work-flow from a first prototype to a finished demonstrator project in approximately five weeks. This was achieved by an agile interdisciplinary digital workflow that was integrated with an associated fabrication pipeline. Furthermore, it has showcased that it is possible to integrate off-the-shelf timber with digital fabrication systems and achieve acceptable tolerances despite material inaccuracies. This was only possible through a close relationship with the people carrying out work on-site while understanding their work processes and incorporating them into our way of thinking about fabrication.

References

1. Opus Dubai. https://odico.dk/en/references/opus-dubai-2. Accessed 11 June 2020
2. Stehling, H., Scheurer, F., Roulier, J., Gramazio, F., Kohler, M., Langenberg, S.: Bridging the gap from CAD to CAM: concepts, caveats and a new grasshopper plug-in. In: Fabricate 2014, DGO-Digital original, pp. 52–59. Negotiating Design & Making. UCL Press (2017). https://doi.org/10.2307/j.ctt1tp3c5w.10
3. Hans Hundegger AG: K2-Industry. https://www.hundegger.com/en-us/machines/joinery-machines/k2-industry. Accessed June 10 2022
4. Giftthaler, M., et al.: Mobile robotic fabrication at 1:1 scale: the In situ fabricator. Constr. Robot. **1**(1–4), 3–14 (2017). https://doi.org/10.1007/s41693-017-0003-5
5. Helm, V., Ercan, S.: In-situ robotic construction: extending the digital fabrication chain in architecture. In: ACADIA 12: Synthetic Digital Ecologies Proceedings of the 32nd Annual Conference of the Association for Computer Aided Design in Architecture (ACADIA), San Francisco 18–21 October 2012. CUMINCAD, 2012, pp. 169–176. (2012). ISBN 978-1-62407-267-3. http://papers.cumincad.org/cgi-bin/works/paper/acadia12_169
6. Yoshida, T.: A short history of construction robots research & development in a Japanese company. In: ISARC Proceedings, 5 October 2006, pp. 188–193 (2006)
7. Kahlen, J.: Factory on the fly. Odico (blog). https://odico.dk/en/factoryonthefly/.. Accessed 28 Aug 2020
8. Benzon, G.: Gammelt dansk bindingsværk. Det Benzon'ske Forlag, Bogense (1984)
9. Jensen, Chr.A.: Dansk Bindingsværk fra Renæssancetiden, dets Forhistorie, Teknik og Dekoration. Kunst i Danmark. Gad, Kbh. (1933)
10. Vejlby, U.: Bindingsværkshuset: renovering, fugtskader, isolering. 1. udgave, 2. oplag. Skovlænge, Søllested (1991)
11. Munch-Andersen, J.: Traeinformation: Traeskelethuse. Traeinformation, Kgs. Lyngby (2018)
12. Understand Building Construction: Timber Frame Construction | Wood Frame Construction | Timber Frame Homes | Timber Buildings. http://www.understandconstruction.com/wood-framed-construction.html. Accessed 21 Mar 2022
13. EEN TIL EEN – Arkitektur & Byggeri. http://eentileen.com/en/. Accessed 21 Mar 2022
14. Adel, A., Thoma, A.: Design of robotically fabricated timber frame structures. In: ACADIA 2018: Recalibration. On Imprecisionand Infidelity. Proceedings of the 38th Annual Conference of the Association for Computer Aided Design in Architecture (ACADIA), Mexico City, Mexico 18–20 October 2018, CUMINCAD, 2018, pp. 394–403 (2018). ISBN 978-0-692-17729-7. http://papers.cumincad.org/cgi-bin/works/paper/acadia18_394
15. Associates: Robert McNeel &. 'Grasshopper - New in Rhino 6'. www.rhino3d.com. https://www.rhino3d.com/6/new/grasshopper/. Accessed March 29 2022
16. Associates:Rhinoceros 3D'. www.rhino3d.com. https://www.rhino3d.com/. Accessed 29 Mar 2022
17. Johnson, D.S.: Near-Optimal bin packing algorithms. Thesis, Massachusetts Institute of Technology (1973). https://dspace.mit.edu/handle/1721.1/57819
18. Lavpris Træ: Hvad er styrkesortering? c18 - c24. http://lavpristrae.dk/Blog/information/hvad-er-styrkesortering-c18-c24Accessed 29 Mar 2022

Augmenting Design: Extending Experience of the Design Process with Glaucon, An Experiential Collaborative XR Toolset

David Gillespie(✉), Zehao Qin, and Martha Tsigkari

Foster + Partners, 22 Hester Road, London SW11 4AN, UK
dagillespie@fosterandpartners.com

Abstract. Architects are in the business of creating not only buildings but effectively experiences through the built environment. Historically, these experiences were only fully appreciated after the completion of the building or urban space. In the past couple of decades, innovation and technology have helped designers have a stronger understanding of how any built spaces would be occupied and experienced through the use of an array of tools and simulations that facilitated performance-driven design pipelines. Nevertheless, there is something very powerful around the idea of placing future users in the experience itself and allowing them to have a preview of how these spaces would look and feel in relation to themselves and their contexts.

To that end, the exponential development of augmented and virtual reality (combined referred to as extended reality (XR) environments) has provided the possibility to designers to do exactly that: create virtual environments, often overlayed on the physical space, that allowed architects, engineers, consultants and stakeholders to be able to experience in real-time how these spaces would look like and experiment in real-time with design changes and their effect they could have to the user's experience (physically and visually). To that end, this paper presents how technology has enabled a large architectural office to facilitate the experiential side of the design prior to the completion of a project, and how this has culminated in the development of a bespoke collaborative XR toolset called Glaucon. Glaucon's capabilities allow high fidelity virtual designs to be physically situated on the site and to experience it as if it were built. Implementing a collaborative toolset, Glaucon allows physically present and remote users to engage with design as an experience earlier in the process than has been traditionally possible through conventional means, increasing design participation and engagement.

Keywords: Immersive · Decision-making · User experience · Experiential storytelling · Participatory design · Efficiency · XR · Augmented reality · Experiential design

1 Introduction

Architectural design is concerned with designing human experience, most commonly manifest in the physical built environment. Through the design process, drawings, models

C. Gengnagel et al. (Eds.): DMS 2022, *Towards Radical Regeneration*, pp. 332–342, 2023.
https://doi.org/10.1007/978-3-031-13249-0_28

and 2D renderings are used to understand space. A full understanding of this experience often only begins during construction and becomes clear after a building is finished, thus limiting the potential for physical experience to make a meaningful contribution to the final product.

During the past few decades, the architectural, engineering and construction (AEC) industry has identified the potential that performance-driven design could have in tailoring the experience that the built environment could have for its users. Interactive interfaces and pipelines allowed users to test how well their designs performed, e.g. daylight, thermal comfort, wind or visual connectivity analyses, were developed and adopted in AEC pipelines to help designers make the right performance decisions early on. Still, the actual experiential side of what it means to be in that space was only possible through visualisations and movies. The growth of XR in terms of hardware (headsets and tablets) as well as software (Games Engines) has changed this, giving designers the opportunity to place themselves within an XR experience of their buildings to make experience-driven design decisions.

XR offers an opportunity to bring experience into the design process. Over the years, the authors' practice has developed XR tools enabling designers to experience their designs at early stages. Glaucon is an XR toolset developed by the authors that increases user ability to collaborate in the same space and remotely, connecting designer team workflows with experiential design tools and hardware. Glaucon has been used in several high-profile real-life projects within the practice with both internal and external users using a broad set of hardware types.

In this paper, the authors will show precedent technologies that were developed to facilitate the experiential side of design decision making and how these innovative approaches amplified its collaborative aspect. Leading to that, this paper demonstrates how these efforts have led to the development of Glaucon, what is the platform's architecture and how it has been used in the design process. The paper identifies several domains where Glaucon has been used to create a high degree of visual fidelity in a collaborative XR experiential design process and brought the experience of a built space into design processes, enhancing both design evaluation and iteration, in order to raise the quality of design outcomes.

2 State of the Art: The Development of XR and Its Incorporation in the AEC Industry

Virtual Reality is primarily seen as a platform for visualising digital or simulated environments (Machover and Tice 1994). VR (as a subset of immersive technologies and XR) can inherently benefit the AEC industry. XR technology centres on user experience and supports various types of communication and interactions in the design process, which has gradually become an area of increasing interest and development activities in the AEC industry. Research has shown that XR technology can enable better collaboration and communication between designers, stakeholders, managers and end-users (Bassanino et al. 2010; Fernando et al. 2013; Van den Berg et al. 2017) allowing users to evaluate and validate the proposed design (Bardram et al. 2002; Dunston et al. 2007;

Loyola et al. 2019), and it supports designers to collaboratively evaluate the functionality and usability of proposed environments (Whyte 2002; Hilfert and König 2016) and reduce cost and waste associated with physical mock-ups (Maldovan et al. 2006; Majumdar et al. 2006).

However, the applications of XR technologies in the AEC field often reveal the tension between the multi-disciplinary nature of the workflow and bespoke XR solutions for real-life projects (Van den Berg et al. 2017). Several studies have explored the possibilities of informing decision-making processes with XR HMD systems in the AEC industry from a technical point of view and concluded that realness could be extended for experiential congruence (Otto et al. 2003; Nikolić and Whyte 2021). Various existing XR systems are focusing on visual fidelity where pipelines for bridging CAD tools and visualisers are well developed but lacking the capability of enabling remote or local collaboration between designers and stakeholders and the possibilities of allowing users to partake in the design process across a board range of XR hardware systems and collaborate with others both physically and virtually (Enscape 2022; Resolve 2022; TheWild 2022). Also, many collaborative XR tools are being developed for remote collaboration in virtual environments (Spatial 2022; Horizon 2022). However, they have not been developed specifically for the AEC industry, where most interactions are developed solely for communication without considering interactions with the virtual context.

3 Experience and Decision Making in the Design Process in Practice

The authors are employees of an architectural practice that has been investigating the potential of XR and experiential feedback in the design process for over two decades. The authors' team has developed several immersive and interactive tools to enable designers to experience designs at 1:1 scale, thus facilitating a better understanding of geometries and spatial configurations. These interfaces and applications have been widely used in various real-life projects (from industrial design to city planning) and stages of the design process (conception to completion), offering an innovative approach to enhance the design process and workflows between design and support teams.

To that end, XR technologies have been used since their "commercial infancy" to facilitate design reviews of various projects. A stereo projector was used to immerse designers in their designs, allowing them to review boat hull geometry or even cockpit visibility while sailing the proposed design into the harbour. Similar, dome projectors were used for collaboratively evaluating architectural projects through the immersion of various stakeholders in the virtual space. The team had even backed VR HMD headsets in Kickstarter and used games engines to create VR walkthroughs (Fig. 1). These immersive experiences had proven essential for design decision-making, showcasing how the use of XR systems could – to an extent – replicate that of a physical mock-up. Nevertheless, these initial attempts with XR technologies allowed the user to experience the proposed environment in a predominantly stationary manner and were limited to specific interactions, originating from the nature of the hardware.

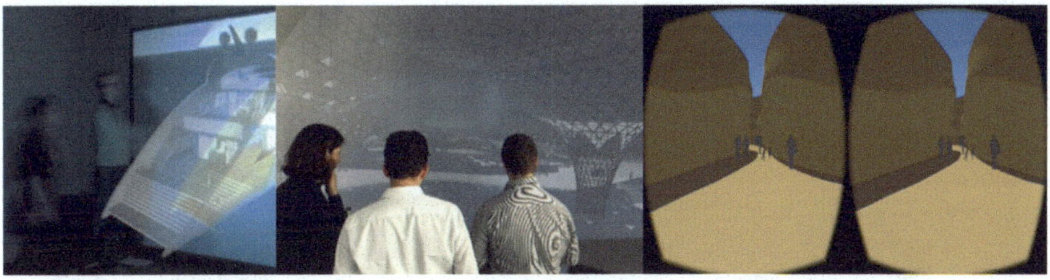

Fig. 1. Stereo projector, dome and VR use on prior projects

4 Glaucon – An Innovative XR Platform

Following the experimentation around various XR technologies and with the exponential evolution of both software and hardware, the authors have focused on developing an XR platform that could go beyond visualisations and allow for a fully immersive, collaborative virtual environment. Glaucon was developed to address a need to bring the experience of a built project directly to live site environments in order to understand what experiencing that space would be like more closely. This needed to be accessible irrespective of physical or virtual presence and across a broad range of XR hardware systems. The core application needed to be general-purpose to be easily reused or adapted for multiple project contexts. Additionally, bringing the collaborative nature to the heart of the virtual experience was key.

4.1 Methods

In order to develop this toolset, Glaucon was developed using a high-fidelity real-time graphics engine that provided online networking support together with a core API that could be readily built upon and extended using custom scripting and code. The toolset is built on top of a games engine, running on a high end gaming PC connected to a consumer grade VR headset with an additional camera to facilitate AR tracking (Gillespie et al. 2021). Versions have also been developed for iOS and PC streaming VR. Its common set of functionality as described below.

Graphical Fidelity
The application is built on top of a game engine (Epic Games Inc. 2022) that provides high-quality visual output across a broad spectrum of XR device types. This was the same video game engine employed by our practice's visualisation team to facilitate a smooth data pipeline when collaborating to create high-quality XR experiences.

Spatial Alignment
Tools were developed to enable virtual spaces to be aligned to their physical counterparts, establishing a correspondence between environments and allowing them to be augmented and extended virtually. These tools make use of computer vision and AR libraries to create image marker-based anchor points that could be positioned in the physical environment and mapped to corresponding virtual equivalents (Gillespie et al.

2021). When combined with spatial mapping technologies on most consumer-grade XR hardware, spaces could be aligned and tracked in real-time (Fig. 2).

Fig. 2. Spatial alignment establishes correspondence between reality and virtual environments

In some cases, device-specific implementations needed to be developed for specific hardware types. Some devices natively incorporated this functionality as standard (e.g. mobile AR), whilst others required additional hardware configuration and AR libraries or alternative input systems in order for the spatial alignment process to be carried out (Fig. 3).

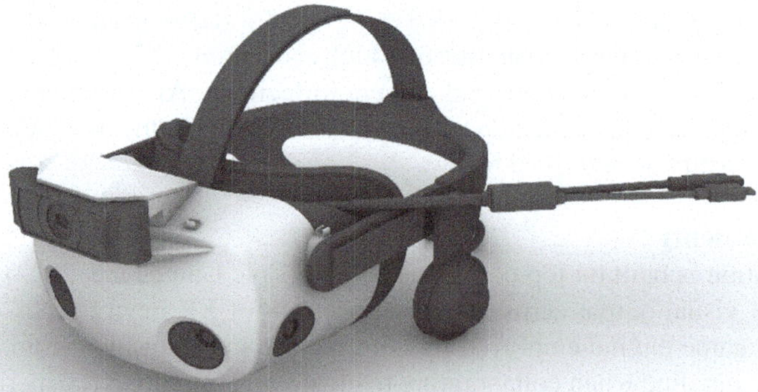

Fig. 3. Additional hardware added to VR headset to facilitate AR marker recognition

Collaboration

Glaucon supports collaboration between physically present and virtual users who can all see and speak with one other, and it includes a range of functions that enable users to share their experiences.

Collaborative functionality, already provided by the video game engine itself, was extended by implementing intuitive and straightforward interactivity together with a mode of user avatar representation appropriate for the review context. Basic human body language could be read in a manner corresponding with reality, inferred from user inputs (Gillespie et al. 2021). Project content forming the basis for collaboration could be loaded into virtual environments at any scale and linked design workflows.

Limitations in game engine's networking, such as data type replication and file size management, were overcome to allow collaborative virtual design environments of any scale to be created, saved and come back to as persistent experiences that could be revisited as part of a project design review process. A downloadable content pipeline (DLC) was also implemented, enabling third parties, such as visualisation teams using the same engine, to export high quality environments that Glaucon could load in.

Design Team Workflows
Glaucon was designed as a generic toolset to work alongside design team processes. The content loading system works with models exported from design applications, and a series of high-quality environments were implemented, providing several review contexts of varying scales for this.

Hardware
Glaucon has been designed to support a broad set of XR hardware types:

- 2D Desktop (PC)
- Tethered VR headsets (PC)
- Standalone VR and AR headsets (Streamed from PC)
- Web browser (Pixel Streamed from Server)
- Mobile AR Tablets (iPad)

Complexities developing for a broad set of target platforms were mitigated using a hardware agnostic SDK (Khronos Group 2022) and adopting XR streaming from PC servers for standalone headsets (NVIDIA Corporation 2022), which enabled consistency of XR experience across device types. Mobile AR was developed separately, allowing bespoke implementations of AR libraries and device hardware limitations to be managed.

4.2 Implementations

Glaucon has been used on several projects across different work stages in varying contexts. Three application domains have been identified, showcasing distinct scenarios where Glaucon has brought the experience of built space and experience to design processes.

Physical On-Site and Mock-Up Experiences
The ability to align physical and virtual experiences has enabled Glaucon to be used

to augment and extend the physical environment with the virtual at scale to experience projects as if they had been built in-situ.

Glaucon has been used to extend a physical mock-up virtually using a VR backpack PC and VR headset. It was able to extend a physical mock-up of half a lobby constructed in a warehouse (Gillespie et al. 2021). A full-scale experience of the environment was possible where physically, only half of the space could be constructed in plywood and plasterboard. The virtual allowed the extension of the physical to represent the built experience, adding materiality and views to city streets. It also allowed virtual design options to be cycled to aid decision-making (Fig. 4).

Fig. 4. Warehouse space used for XR mock-up

Glaucon has also been used to enhance the experience of a physical site to understand the contribution a design would make to the urban realm in-situ in a live urban context. A virtual model was developed to an enhanced level of detail to stand up to close scrutiny (Gillespie et al. 2021) which combined with the tactile nature of the site, the feel of the live urban context combined with the virtual to create an enhanced feeling of the proposed design within its live site context (Fig. 5).

Fig. 5. Onsite full-scale collaborative XR experience

Comparisons could be directly made between real and virtual by removing the headset, and the lead architect directly presented the project as a physically and virtually present user. A visitor experience was also developed that allowed the client to start their experience in the physical plaza and experience the visitor journey up the reception lift to a high-level viewing gallery, as if the project had been built (Fig. 6).

In this domain, all users were physically present, some but not all using XR hardware. Where users were represented as virtual avatars, this created a correspondence and sense of the present in physical and virtual spaces allowing the experience to be presented and discussed as if they had been physically built.

Fig. 6. Comparisons between reality and virtual experiences

Virtual Review Experience

Glaucon has been used in an entirely virtual context where users sought to experience or review an environment as if it were real. Multiple users who were not physically present could engage with the design and be part of the decision-making. These reviews have been primarily experienced with backpack VR in a similar manner to the onsite examples, and streaming VR has also been used successfully with a smaller number of users (Fig. 7).

By combining the experience of the space with the detailed design models and enabling multiple users to collaborate virtually, the impact of the detailed design, for example, in one case, a staircase in a residential project, and its contribution to the overall experience could be understood. Several environments have also been developed in conjunction with specialist teams to create the basis for a formalised review environment,

Fig. 7. Virtual review experience

allowing live project content to be loaded, cycled and review space states saved for immediate recall at a later point.

Mobile Experience
Glaucon has been used in a mobile context both as an onsite AR experience and entirely virtual use cases. It has been used to increase accessibility and application reach and to allow users who may not have access to or be comfortable using head-mounted XR hardware. Remote app deployment has been used to increase application reach to any user with a compatible device.

Fig. 8. Onsite mobile AR experience combining real and virtual experiences

Several projects have been developed using Glaucon's spatial alignment toolset to allow users to gain an understanding of a project as built, overlaying virtual with reality to create an augmented experience with the device acting as a "magic window". Virtual only experiences have also been developed where users have been able to experience a project using their mobile device as a window without needing to be situated on site. Collaboration between users has occurred physically through users showing others in the same space the view from the device (Fig. 8).

5 Conclusion

Glaucon has built upon prior precedent from within the practice and in the wider XR field allowing physical spaces to be enhanced through an XR experiential layer of immersive VR and tablet-based AR contexts. It is likely that given increased use as a design tool in practice, the impact of experience-driven decision making in this context will become more evident.

Glaucon has enabled design teams to understand and experience their designed spaces in a readily accessible and more immersive manner than previous tools have allowed, engaging with user experience and feel prior to the final built product. It has facilitated an enhanced level of collaboration across hybrid spatial environments, and through its approach to hardware development increased application reach both within and outside the practice, allowing virtual projects to be physically situated and experienced in-situ, has created a heightened sense of experience than has previously been possible.

Where alternatives to the experiences Glaucon enables are physical equivalents at full scale or achieved using costly mock-ups, tools like Glaucon have the potential to reduce waste from physical mock-ups whilst simultaneously allowing more options to be assessed at scale better informing any physical mock-ups constructed. Whilst the purpose of construction mock-ups is more than just visual, virtual mockups have a very clear economic benefit; in addition to that, Glaucon offers a higher degree of visual fidelity and situated physical experience than has been previously possible, which may therefore have a greater impact upon this process.

Glaucon's testing as the basis for a collaborative design review environment is an area of ongoing research. Where Glaucon offers an enhanced experience at scale and on site, its ability to directly load and configure varied review spaces and environments means that in similar ways that Glaucon offers benefits to the mock-up process, it could offer similar improvements to more traditional review environments and whilst this may not and arguably should not eliminate physical design review processes, it may enable them to increase in meaning and value when they do occur.

Given the rate of current XR hardware development, the opportunity to design hybrid spatial experiences and the spatial alignment technologies developed may increase Glaucon's value as a design tool. As virtual experiences themselves become key design outputs in a similar manner that BIM is and Digital Twins are becoming, future iterations of Glaucon or similar other tools may serve as test or deployment environments for augmenting and extending existing space using AR technologies to create hybrid spatial environments as part of a designed experience.

Acknowledgements. We would like to thank our colleagues, clients and collaborators for their contributions to the development and evaluation of this system, in particular Francis Aish for his contribution, and Gamma Basra and Ewan Couper from our Visualisation team.

References

Bardram, J.E., Bossen, C., Lykke-Olesen, A., Nielsen, R., Halskov, K.: Virtual video prototyping of pervasive healthcare systems. In: Proceedings of the 4th Conference on Designing Interactive Systems: Processes, Practices, Methods, and Techniques. London, England. ACM (2002)

Bassanino, M., Wu, K.-C., Yao, J., Khosrowshahi, F., Fernando, T., Skjærbæk, J.: The impact of immersive virtual reality on visualisation for a design review in construction. In: 14th International Conference Information Visualisation, pp. 585–589 (2010). https://doi.org/10.1109/IV.2010.85

Dunston, P.S., Arns, L.L., Mcglothlin, J.D., Lasker, G.C., Kushner, A.G.: An immersive virtual reality mock-up for design review of hospital patient rooms. In: 7th International Conference on Construction Applications of Virtual Reality, PA, USA. University Park (2007)

Enscape: Enscape - Real-Time Rendering and Virtual Reality (2022). https://enscape3d.com/. Accessed March 2022

Epic Games Inc.: Unreal Engine (2022). https://www.unrealengine.com/. Accessed 01 Apr 2022

Fernando, T., Wu, K.-C., Bassanino, M.: Designing a novel virtual collaborative environment to support collaboration in design review meetings. J. Inf. Technol. Constr. **18**,. 372–396 (2013)

Gillespie, D., Qin, Z., Aish, F.: Collaborative extended reality system for in-situ design review in uncontrolled environ-ments. ACADIA (2021)

Hilfert, T., König, M.: Low-cost virtual reality environment for engineering and construction. Vis. Eng. **4**(2) (2016)

Horizon: Horizon Worlds | Virtual reality worlds and communities (2022). https://www.oculus.com/horizon-worlds/. Accessed Mar 2022

Khronos Group: OpenXR Overview, March 31 2022. https://www.khronos.org/openxr/. Accessed 30 March 2022

Loyola, M., Rossi, B., Montiel, C., Daiber, M.: Use of virtual reality in participatory design. In: Architecture in the Age of the 4th Industrial Revolution. 37 Education and Research in Computer Aided Architectural Design in Europe and XXIII Iberoamerican Society of Digital Graphics, Porto, Portugal, pp. 449–454 (2019)

Hageman, A.: Virtual reality. Nursing **24**(3), 3–3 (2018). https://doi.org/10.1007/s41193-018-0032-6

Majumdar, T., Fischer, M.A., Schwegler, B.R.: Conceptual design review with a virtual reality mock-up model. In: Proceedings of the Joint International Conference on Computing and Decision Making in Civil and Building Engineering, Montréal (2006)

Maldovan, K.D., Messner, J.I., Faddoul, M.: Framework for reviewing mockups in an immersive environment. In: CONVR 2006: 6th International Conference on Construction Applications of Virtual Reality, Orlando, Florida, 6 (2006)

Nikolić, D., Whyte, J.: Visualizing a new sustainable world: toward the next generation of virtual reality in the built environment. Buildings **11**(11), 546 (2021)

NVIDIA Corporation: NVIDIA Cloud XR SDK (2022). https://developer.nvidia.com/nvidia-cloudxr-sdk. Accessed 29 March 2022

Otto, G., et al.: The VR-desktop: an accessible approach to VR environments in teaching and research. Int. J. Archit. Comput. **1**, 233–246 (2003)

Resolve: Resolve - Virtual Collaboration for Complex BIM (2022). https://www.resolvebim.com/. Accessed 10 June 2022.

Spatial: Spatial - Metaverse Spaces That Bring Us Together (2022). https://spatial.io/. Accessed March 2022

TheWild: The Wild - VR Collaboration for Architecture & Design Teams (2022). https://thewild.com/. Accessed 10 June 2022

Van den Berg, M., Hartmann, T., de Graaf, R.: Supporting design reviews with pre-meeting virtual reality environments. J. Inf. Technol. Constr. **22**, 305–321 (2017)

Whyte, J.: Virtual Reality and the Built Environment. Elsevier Science, Oxford (2002)

Design from Finite Material Libraries: Enabling Project-Confined Re-use in Architectural Design and Construction Through Computational Design Systems

Jonas Runberger[1](✉), Vladimir Ondejcik[2], and Hossam Elbrrashi[2]

[1] Chalmers Department of Architecture and Civil Engineering; Head of Dsearch, Digital Matter, White Arkitekter AB, Gothenburg, Sweden
jonas.runberger@white.se

[2] Architect and Developer at Dsearch, Digital Matter, White Arkitekter AB, Gothenburg, Sweden

Abstract. This paper presents two cases of material reuse in architectural practice, employing bespoke computational workflows. The first unpacks a complete cycle of from building demolition of an existing building in Järvsö, Sweden, through the design and fabrication of a temporary and mobile pavilion for use at a series of urban events, to its final destination as a wind shelter not far from the forests where the material was initially harvested. The second presents scaled approaches and methods to respond to the competition brief for the redevelopment of a partial urban block in Berlin, using the material stock from the existing building as a material library. Both cases have been developed in constrained project contexts where the resources of the material libraries have been limited and directly associated to the project – rather than being part of an open market of re-used resources – allowing data on availability, amount, and quality to be readily available. In this sense the approach can be defined as a project-confined re-use workflow, providing the opportunity to target the association between the design modelling environment and the material library.

Keywords: Material reuse · Computational design · Collaborative design · Circularity · Resource flows · Design system

1 Introduction

1.1 Background

The need for sustainable solutions within the AEC industry is ever more pressing (Ribeirinho 2020), and re-use of building elements from existing building stocks is emerging as an alternate approach to recycling (Josefsson and Thuvander 2020). This paper addresses the idea of project-confined material banks, where resources have been limited and directly associated to the task at hand, with data on availability, amount and quality at hand (Kozminska 2019). With simplified and pre-existing material banks there is no

C. Gengnagel et al. (Eds.): DMS 2022, *Towards Radical Regeneration*, pp. 343–359, 2023.
https://doi.org/10.1007/978-3-031-13249-0_29

need to document existing materials (Batalle Garcia 2021), and pressed development time following project conditions the design decision needs to be based on availability in the material bank, rather than on wide range of optional sources (Huang 2021). An important potential achievement is to provide designers with real-time data on material resources in order to provide a robust and durable design process (Batalle Garcia 2021, p. 9).

The paper uses two real-world cases where bespoke computational workflows have been developed as part of the design process, to identify key definitions, technical functionality, values provided and critical aspects relevant for architectural re-use design processes facilitated and enhanced through computational design. The first case ties into a cycle of reuse starting with the demolition of an existing lumber yard building in Järvsö, Sweden, through the design and fabrication of a pavilion. The second case presents approaches and methods in response to the competition brief for the redevelopment of a partial urban block in Berlin, using the material stock from the existing building as a material bank (Debacker 2016).

The two cases are steps in the continuous development of *PCARDS - Project Confined Automated Reuse Design Systems*, where automated workflows enabled by computational design facilitates creative design processes responding to functional and aesthetic needs to be associated to project-confined material banks. PCARDS regards situations where the material bank is complex, partially known, and/or dynamically changing over time, and where project specific conditions require bespoke semi-automated workflows. It also follows a reversed approach in which available components ultimately guide the design (Gorgolewski 2017, p. 118). The development is conducted by Dsearch, a computational design practice within White Arkitekter AB.

1.2 Key Definitions and Clarifications

A *Design Model* refers to one of potentially many digital models developed, and a *Script* refers to one of potentially many programmed functions associated to the Design Models[1]. A *Procedure* is a modularized part of a script serving a particular function. A *Computational Design System*, or a *Design System* for short, encapsulates the collected assets in terms of scripts, design models and workflows tailored to the task at hand. A distinction is made between the *User* of a Design System – typically an architect or engineer in a design team that integrates it into his/her practice – and the *Developer* that develops and maintains the design system and may or may not be an integral part of a design team. A *Use Case* indicates a particular well-defined task to be resolved by the design system, and a *Design Logic* refers to a specific design response to this task. Defined from a perspective external to the computational development, the Use Case can be seen as a function, objective or aspect of a building program needed to be fulfilled, and the Design Logic the conceptual, formal or technical response to this. In Case 1, a *Blank* refers to a raw oak beam as extracted from the source building, processed only in terms of cutting it to a useful length (removing deteriorated parts), while a *Beam* refers to an element of the proposed pavilion, cut to its final length.

[1] A majority of the Computational Design Systems developed by Dsearch use McNeel Rhinoceros 3D and Grasshopper, supported by Python scripts and an assortment of plugins.

A high degree of modularization and nesting of scripts makes them legible from a developer perspective – components within modules within segments within blocks (Magnusson et al. 2017). The term *Procedure* is here introduced to refer to a functional part of a script from an external perspective – a functional part that relates to a distinguishable task in the design process – and it would typically map perfectly to the block level of a script. The term Design System has previously been introduced as a more expansive assemblage including project bespoke methods and workflows as well as teams, policies, and contracts (Runberger and Magnusson 2015, Magnusson and Runberger 2016), but for the purposes of this paper the Design System represents the project specific scripts and workflows addressing the task at hand.

Fig. 1. The original Järvsö lumberyard barn, with some beams exposed.

2 Case 1 – The 200-Year Pavilion

2.1 Context, Use Case and Design Logic

There are several examples of computational design approaches for reuse of structural elements in steel and other materials (Brütting 2018, 2019). Structural timber elements are however often downcycled to non-structural elements and ultimately to incineration for energy production especially in areas where timber is abundant (Cristescu et al. 2020), following the theory of the cascading chain (Sirkin and ten Houten 1994). While it has been determined that reuse of timber for structural purposes depend on the development of new grading rules, it has also been suggested that new design principles that can allow the use of a variety of dimensions (in Case 1 restricted to lengths) are needed (Hradil 2014, pp. 33–36). As a response to this, the 200-year Pavilion project explored an opportunity to reuse oak beams from a 19th century lumberyard barn to be demolished in rural Järvsö (Fig. 1) for structural purposes, in the form of a pavilion developed in-house, and presented as part of an architectural festival hosted by White Arkitekter in 2019, setting a strict timeline for development.

Fig. 2. The beams of the material bank ready for inspection.

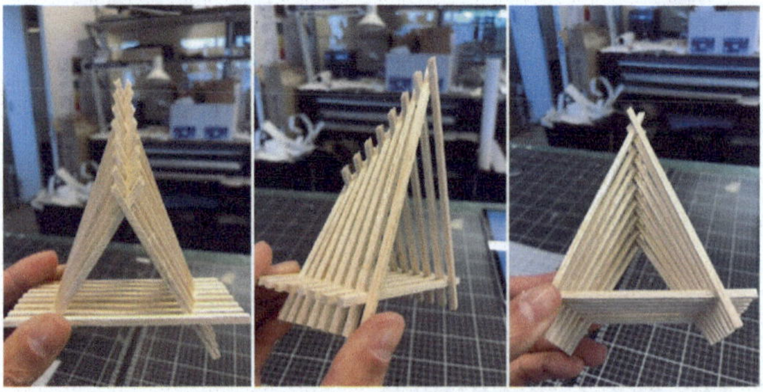

Fig. 3. Conceptual model showing the triangulated structural solution and the tapering effect.

While the potential length and number of beams was initially known, the specific condition of each beam was unknown until they were delivered on site a few days before the event (Fig. 2). The use case was defined as a pavilion exclusively constructed in timber only, with the reused material as main structure, to provide partial shelter, while taking ease of assembly, disassembly, and transportation into consideration. Beams of the specific dimension 190 × 190 mm were selected to provide a more rational structure, and a majority of the pin joints used were done in reused timber. Several concepts were considered through a sketching phase through drawings and physical models, leading to the decision of a singular design logic using a triangular shape that

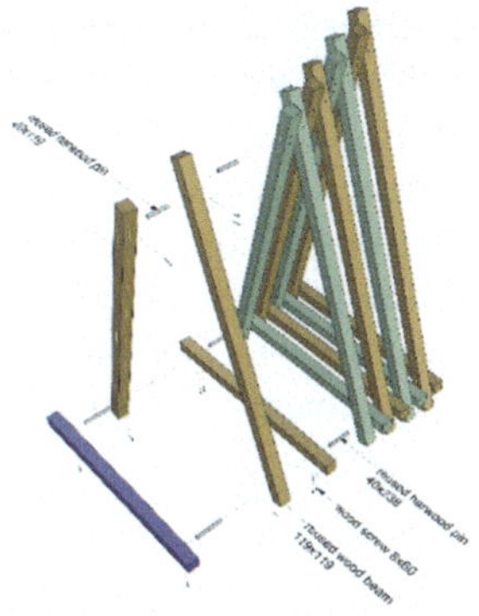

Fig. 4. Generated geometry with structural principle and joints.

Fig. 5. First installation at the architecture festival in Stockholm. Photo: Anders Bobert

Fig. 6. Final permanent installation of the 200-Year Pavilion in Järvsö. Photo: Elena Kanevsky

allowed a simple stable structure as the basis for the design system (Fig. 3). This logic was applied to three modules, each with a shifting cross section, and initially planned to be slightly different but ultimately instead repeated due to constrains in construction time. To facilitate the very short timeline between setting the final design parameters and completing construction where the final condition of the blanks could not be identifies until 2 days prior to construction completion deadline, the structural logic and all details were generated by the design system (Fig. 4). The final pavilion was presented at several temporary locations and is now permanently returned to Järvsö as a wind shelter for hikers (Fig. 5 and 6).

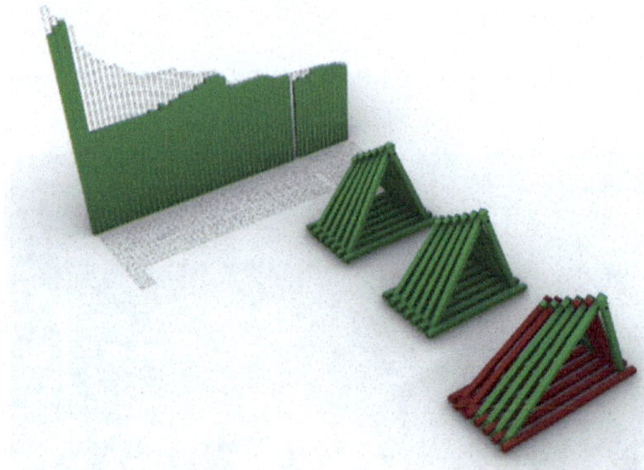

Fig. 7. Design model (right) and corresponding material bank (left), with colour coding indicating elements in design model missing in the material bank – red indicating missing elements.

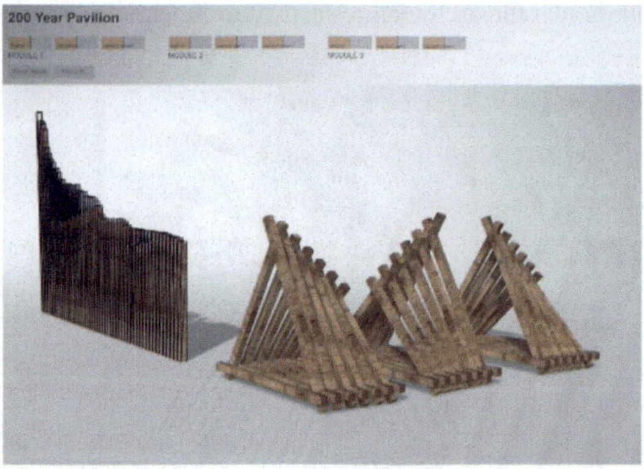

Fig. 8. User interaction enabled through online interface.

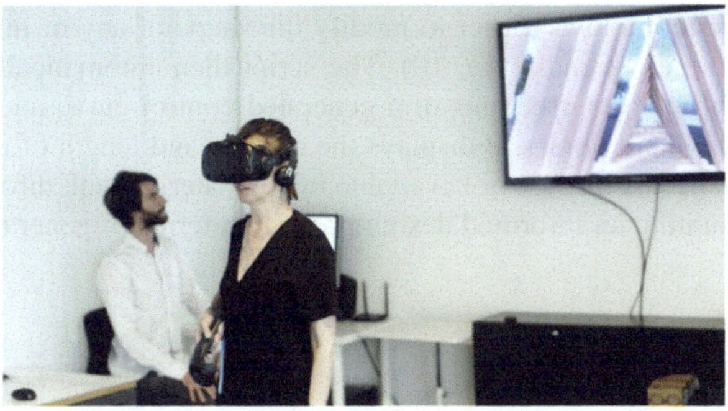

Fig. 9. Assessing design through VR.

2.2 The 200-Year Pavilion Design System

The scope of the design system was to enable a manual maximization of the size of the pavilion module, while a best beam to blank allocation optimization was automated in relation to the material bank. It provides real-time links between a dynamically adjustable material bank, and the parametrically adjustable design model (Fig. 7), allowing a user to engage directly with the design model and the script, or via an online interface (Fig. 8), and spatial qualities could be assessed through VR (Fig. 9)[2]. Given the very late update on the condition of the blanks, final finetuning could be conducted last minute before the production of fabrication documentation. While this documentation required a manual handling, the design system generated a data table with all blanks with mapped with beams, with ids and lengths for design reviews.

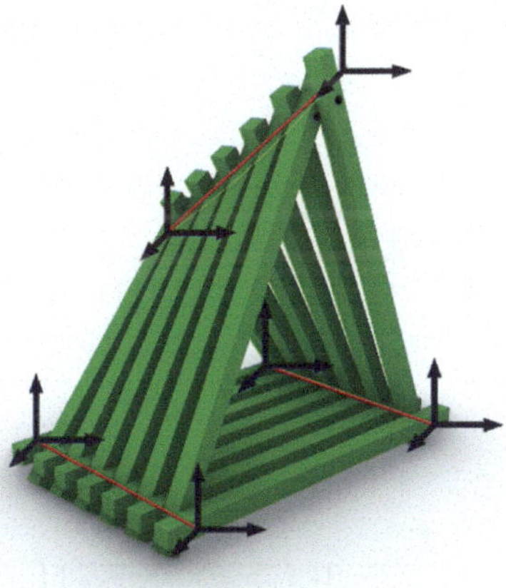

Fig. 10. Points for user manual manipulation in design model.

---

[2] Rhino Compute was used for the online interface, with the purpose of reducing the need for computational power or software license on behalf of the user. The VR setup was employing a semi-automated Rhino to Enscape workflow.

The design system allows a user to modify the shape of any of the pavilion modules by moving points at ends (Fig. 10). The script then automatically generates the data for beams through intersections of a generated control curve and maps this data on to the available blank library. It displays the amount and length of used and unused blanks, as well as beams that are not available in the material bank through colour coding (Fig. 7), facilitating an informed design process in terms of general form, resource

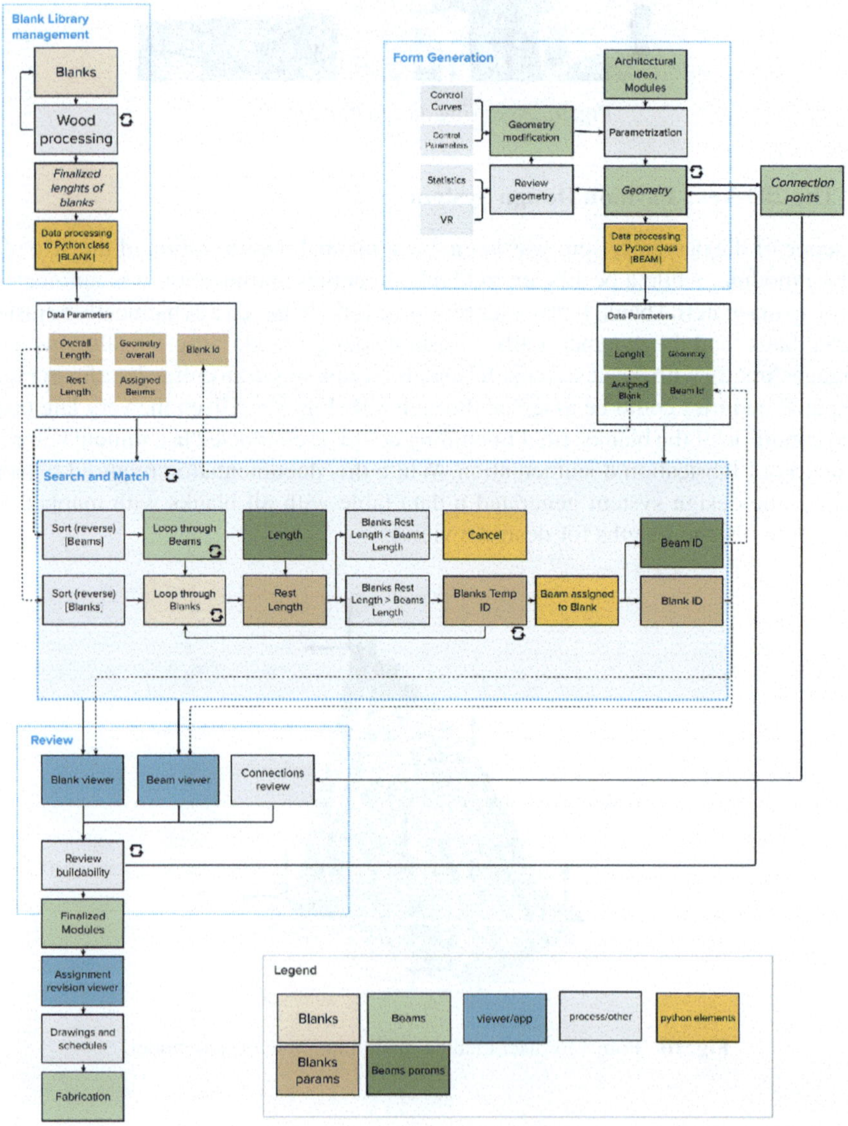

Fig. 11. Schematic overview over the design system script for Case 1

management and buildability. The connection logic is also displayed in terms of connection pegs, allowing informed revisions to the design configuration, both in terms of an intuitive understanding the limitations in beam length versus length of available blanks for resource management, and the restrictions in peg joint connections in relation to the beam overlap at the upper edge for buildability. Multiple versions of the design configurations can be saved through baking control curves to specific layers for later reinstatement.

Fig. 12. Design model and material bank with individual members colour coded after ids.

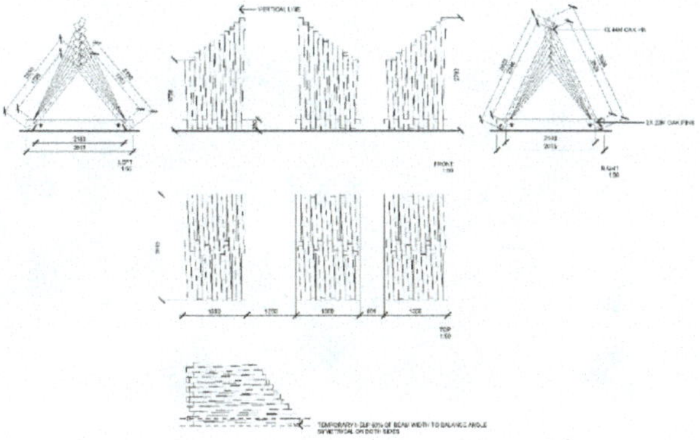

Fig. 13. Fabrication documentation.

The script is divided into four main functional procedures (Fig. 11):

- The *blank library management procedure* keeps track of blank lengths rounded to 1 cm. It also lists dents and damages along the length of the blanks, something which was not used downstream in the script but could in future development be used as an indicator for where to cut blanks into several beams.
- The *form generation procedure* automatically generates the selected design logic and all beams, based on three manually adjustable control curves located at the joints of the beams.
- The *search and match procedure* maps requested beams to the library of available blanks, pursuing the closest match in length, and mapping shorter beams to larger

blanks. This is facilitated through a simple but effective algorithm that sorts the numeric beam length data and available blank lengths. While a long blank may be divided into several beams, a beam longer than all blanks would result a colour indication of an unbuildable solution.

- The *review procedure* provides the visual feedback in terms of geometry for blanks in the material bank, as well as beams and joints (Fig. 7) in the design model, indicating availability through colour coding.

Within the script, data is handled through Python classes of two types – blank and beam, with interconnected data and simplified geometrical representations. Blank parameters include length, beam ids and length availability. Beam parameters include length, blank id and excess (indicating no availability in the material bank. The associated beams and blanks are also visualized to the user with ids and colour coding (Fig. 12). Once final design was set, fabrication drawings and schedules could be generated (Fig. 13).

3 Case 2 Karstadt am Hermannplatz Competition Entry

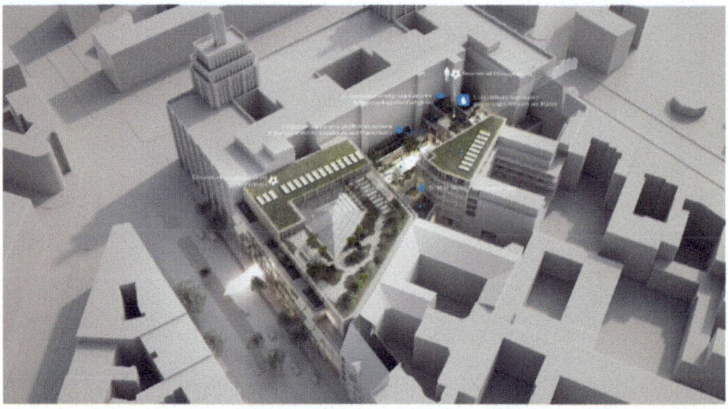

Fig. 14. Overview of Case 2 competition entry.

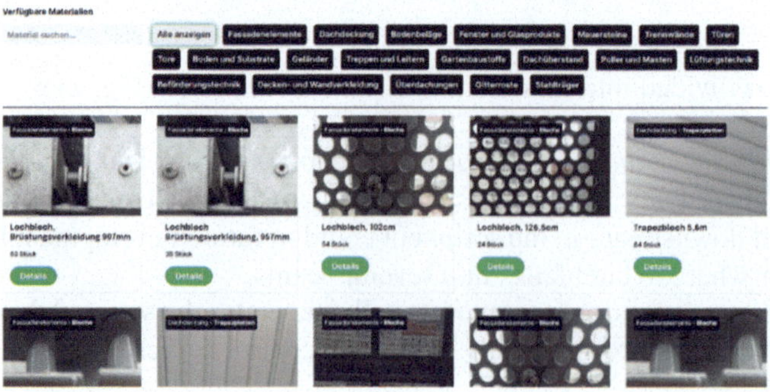

Fig. 15. Case 2 online material library, excerpt.

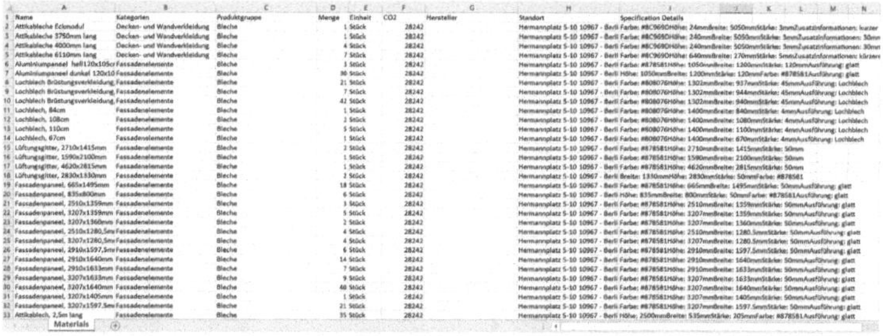

	Name	Kategorien	Produktgruppe	Menge	Einheit	CO2	Hersteller	Standort	Specification Details
2	Attikableche Eclonodul	Decken- und Wandverkleidung	Bleche	1	Stück	28242		Hermannplatz 5-10 10967 - Berli	Farbe: #8C969DHöhe: 24mmBreite: 5050mmStärke: 3mmZusatzinformationen: kurzer
3	Attikableche 3750mm lang	Decken- und Wandverkleidung	Bleche	1	Stück	28242		Hermannplatz 5-10 10967 - Berli	Farbe: #8C969DHöhe: 240mmBreite: 5050mmStärke: 3mmZusatzinformationen: 30mm
4	Attikableche 4000mm lang	Decken- und Wandverkleidung	Bleche	4	Stück	28242		Hermannplatz 5-10 10967 - Berli	Farbe: #8C969DHöhe: 240mmBreite: 5050mmStärke: 3mmZusatzinformationen: 30mm
5	Attikableche 6110mm lang	Decken- und Wandverkleidung	Bleche	7	Stück	28242		Hermannplatz 5-10 10967 - Berli	Farbe: #8C969DHöhe: 640mmBreite: 270mmStärke: 5mmZusatzinformationen: kürzere
6	Aluminiumpaneel hell120x105cr	Fassadenelemente	Bleche	3	Stück	28242		Hermannplatz 5-10 10967 - Berli	Farbe: #878581Höhe: 1050mmBreite: 1200mmStärke: 120mmAusführung: glatt
7	Aluminiumpaneel dunkel 120x10	Fassadenelemente	Bleche	30	Stück	28242		Hermannplatz 5-10 10967 - Berli	Höhe: 1050mmBreite: 1200mmStärke: 120mmFarbe: #878581Ausführung: glatt
8	Lochblech Brüstungsverkleidung,	Fassadenelemente	Bleche	21	Stück	28242		Hermannplatz 5-10 10967 - Berli	Farbe: #808076Höhe: 1302mmBreite: 937mmStärke: 45mmAusführung: Lochblech
9	Lochblech Brüstungsverkleidung,	Fassadenelemente	Bleche	7	Stück	28242		Hermannplatz 5-10 10967 - Berli	Farbe: #808076Höhe: 1302mmBreite: 944mmStärke: 45mmAusführung: Lochblech
10	Lochblech Brüstungsverkleidung,	Fassadenelemente	Bleche	42	Stück	28242		Hermannplatz 5-10 10967 - Berli	Farbe: #808076Höhe: 1302mmBreite: 940mmStärke: 45mmAusführung: Lochblech
11	Lochblech, 84cm	Fassadenelemente	Bleche	1	Stück	28242		Hermannplatz 5-10 10967 - Berli	Farbe: #808076Höhe: 1400mmBreite: 840mmStärke: 4mmAusführung: Lochblech
12	Lochblech, 108cm	Fassadenelemente	Bleche	2	Stück	28242		Hermannplatz 5-10 10967 - Berli	Farbe: #808076Höhe: 1400mmBreite: 1080mmStärke: 4mmAusführung: Lochblech
13	Lochblech, 110cm	Fassadenelemente	Bleche	5	Stück	28242		Hermannplatz 5-10 10967 - Berli	Farbe: #808076Höhe: 1400mmBreite: 1100mmStärke: 4mmAusführung: Lochblech
14	Lochblech, 67cm	Fassadenelemente	Bleche	1	Stück	28242		Hermannplatz 5-10 10967 - Berli	Farbe: #808076Höhe: 1400mmBreite: 670mmStärke: 4mmAusführung: Lochblech
15	Lüftungsgitter, 2710x1415mm	Fassadenelemente	Bleche	2	Stück	28242		Hermannplatz 5-10 10967 - Berli	Farbe: #878581Höhe: 2710mmBreite: 1415mmStärke: 50mm
16	Lüftungsgitter, 1590x2100mm	Fassadenelemente	Bleche	1	Stück	28242		Hermannplatz 5-10 10967 - Berli	Farbe: #878581Höhe: 1590mmBreite: 2100mmStärke: 50mm
17	Lüftungsgitter, 4620x2815mm	Fassadenelemente	Bleche	1	Stück	28242		Hermannplatz 5-10 10967 - Berli	Farbe: #878581Höhe: 4620mmBreite: 2815mmStärke: 50mm
18	Lüftungsgitter, 2830x1330mm	Fassadenelemente	Bleche	2	Stück	28242		Hermannplatz 5-10 10967 - Berli	Breite: 1330mmHöhe: 2830mmStärke: 50mmFarbe: #878581
19	Fassadenpaneel, 665x1495mm	Fassadenelemente	Bleche	18	Stück	28242		Hermannplatz 5-10 10967 - Berli	Farbe: #878581Höhe: 665mmBreite: 1495mmStärke: 50mmAusführung: glatt
20	Fassadenpaneel, 835x800mm	Fassadenelemente	Bleche	6	Stück	28242		Hermannplatz 5-10 10967 - Berli	Höhe: 835mmBreite: 800mmStärke: 50mmFarbe: #878581Ausführung: glatt
21	Fassadenpaneel, 2510x1359mm	Fassadenelemente	Bleche	3	Stück	28242		Hermannplatz 5-10 10967 - Berli	Farbe: #878581Höhe: 2510mmBreite: 1359mmStärke: 50mmAusführung: glatt
22	Fassadenpaneel, 3207x1359mm	Fassadenelemente	Bleche	5	Stück	28242		Hermannplatz 5-10 10967 - Berli	Farbe: #878581Höhe: 3207mmBreite: 1359mmStärke: 50mmAusführung: glatt
23	Fassadenpaneel, 3207x1360mm	Fassadenelemente	Bleche	2	Stück	28242		Hermannplatz 5-10 10967 - Berli	Farbe: #878581Höhe: 3207mmBreite: 1360mmStärke: 50mmAusführung: glatt
24	Fassadenpaneel, 2510x1280,5mi	Fassadenelemente	Bleche	4	Stück	28242		Hermannplatz 5-10 10967 - Berli	Farbe: #878581Höhe: 2510mmBreite: 1280.5mmStärke: 50mmAusführung: glatt
25	Fassadenpaneel, 3207x1280,5mi	Fassadenelemente	Bleche	4	Stück	28242		Hermannplatz 5-10 10967 - Berli	Farbe: #878581Höhe: 3207mmBreite: 1280.5mmStärke: 50mmAusführung: glatt
26	Fassadenpaneel, 2910x1597,5mi	Fassadenelemente	Bleche	6	Stück	28242		Hermannplatz 5-10 10967 - Berli	Farbe: #878581Höhe: 2910mmBreite: 1597.5mmStärke: 50mmAusführung: glatt
27	Fassadenpaneel, 2910x1640mm	Fassadenelemente	Bleche	14	Stück	28242		Hermannplatz 5-10 10967 - Berli	Farbe: #878581Höhe: 2910mmBreite: 1640mmStärke: 50mmAusführung: glatt
28	Fassadenpaneel, 2910x1633mm	Fassadenelemente	Bleche	7	Stück	28242		Hermannplatz 5-10 10967 - Berli	Farbe: #878581Höhe: 2910mmBreite: 1633mmStärke: 50mmAusführung: glatt
29	Fassadenpaneel, 3207x1633mm	Fassadenelemente	Bleche	9	Stück	28242		Hermannplatz 5-10 10967 - Berli	Farbe: #878581Höhe: 3207mmBreite: 1633mmStärke: 50mmAusführung: glatt
30	Fassadenpaneel, 3207x1640mm	Fassadenelemente	Bleche	40	Stück	28242		Hermannplatz 5-10 10967 - Berli	Farbe: #878581Höhe: 3207mmBreite: 1640mmStärke: 50mmAusführung: glatt
31	Fassadenpaneel, 3207x1405mm	Fassadenelemente	Bleche	1	Stück	28242		Hermannplatz 5-10 10967 - Berli	Farbe: #878581Höhe: 3207mmBreite: 1405mmStärke: 50mmAusführung: glatt
32	Fassadenpaneel, 3207x1597,5mi	Fassadenelemente	Bleche	21	Stück	28242		Hermannplatz 5-10 10967 - Berli	Farbe: #878581Höhe: 3207mmBreite: 1597.5mmStärke: 50mmAusführung: glatt
33	Attikablech, 2,5m lang	Fassadenelemente	Bleche	35	Stück	28242		Hermannplatz 5-10 10967 - Berli	Höhe: 2500mmBreite: 535mmStärke: 205mmFarbe: #878581Ausführung: glatt

Materials

Fig. 16. Case 2 Excel spreadsheet version of material library, excerpt.

3.1 Context, Use Cases and Design Logics

The Karstadt am Hermannplatz competition brief called for efficient and creative reuse, primarily using elements from the given material bank for the transformation of parts of a block in CITY (Fig. 14). The material bank included a large number of non-structural elements divided into 33 different categories, and a total of 228 types of (Fig. 15, 16), provided through a web-based platform based on a previous inventory, along with an associated Excel spreadsheet[3]. The design process included a range of use cases (interior ceilings, atrium glass partitions, façade cladding, solar shading, floor tiling etc.), each with one or several coupled design logics (in particular in terms of the ceilings). An initial challenge was to simplify the heterogeneous data structure of the material bank through computational means (Lokhandwala 2018, p. 2).

3.2 The Karstadt am Hermannplatz Design System

The scope was to provide users with efficient connectivity, utilization and interactivity with the material bank as part of the collaborative design process, in a way that combined qualitative design sensibilities with quantitative data feedback. Developed in parallel with the time-constrained competition timeline, it was presented as a conceptual workflow to the client but remained only partially functional and used by the design team. To facilitate collaboration, the design system combines several design models (structures, façades, interiors and landscape). The final competition entry included a full architectural design proposal, as well as a conceptual PCARDS workflow and examples of automated reuse design results. White Arkitekter was awarded second prize in the competition. If given the opportunity to continue with the resulting commission, Dsearch would have fully implemented the PCARDS workflows in the continued concept stage development.

[3] Provided by Concular GmbH.

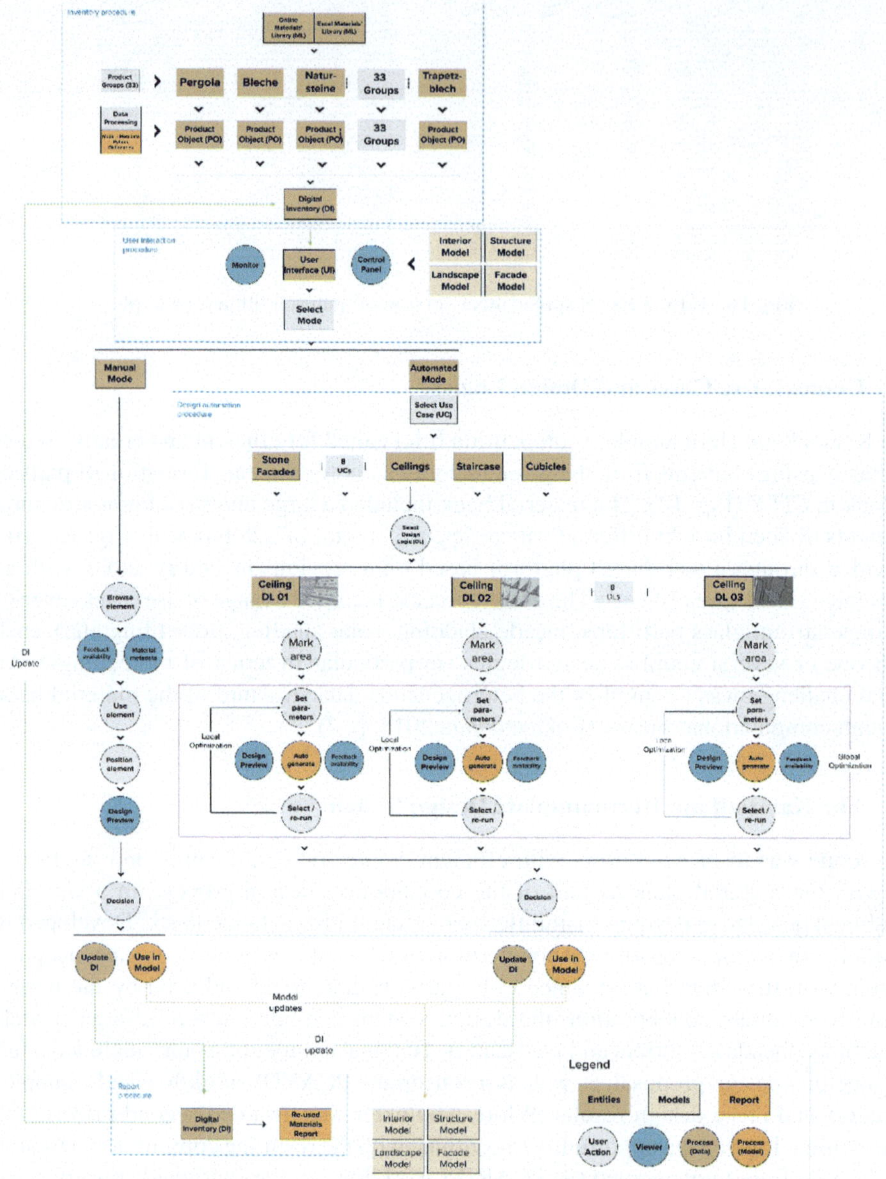

Fig. 17. Schematic overview over the design system script for Case 2

Fig. 18. Inventory Monitor (IM) and Control Panel (CP), showing amounts used of each material category.

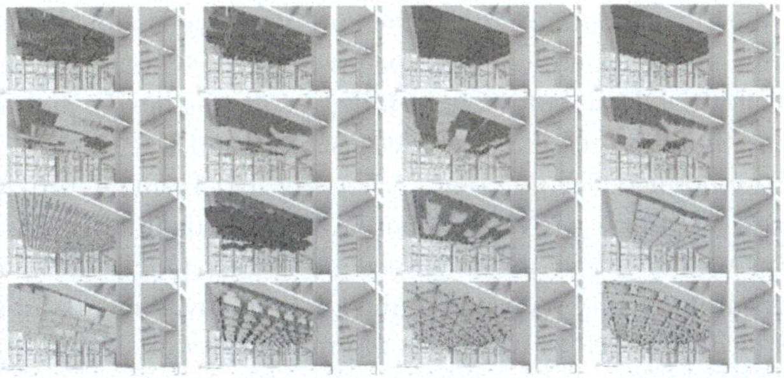

Fig. 19. Overview of 16 generated design logics for the same use case – interior ceilings.

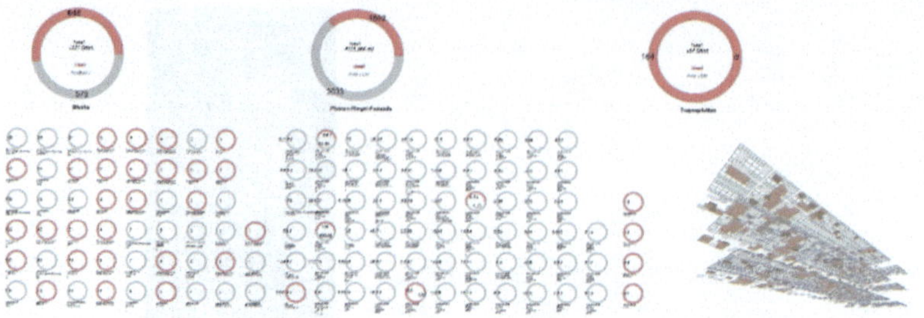

Fig. 20. Case 2 Use Case Report, showing amounts used of each material category.

The design system consists of four main functional procedures (Fig. 17):

- The *inventory procedure* reads the material bank from Excel, and a Digital Inventory (DI) is created, including geometries and their associated meta data as a digital twin library for the material bank. Meta data for the 33 distinct categories is interpreted and organized, and geometries corresponding to all types of elements are generated and stored together with meta data in 33 python dictionaries each representing one category, stored in the DI[4].
- The *user interaction procedure* provides access to the DI from all design models, allowing multiple users to independently and in parallel tap into the DI during the design phase through a custom developed User Interface (UI) with two main parts – the Inventory Monitor (IM) and the Control Panel (CP) (Fig. 18). The IM allows the users to check the availability of group(s) of objects from the DI before deployment in addition to totals charts as real-time heads-up graphical reports that work as customizable widgets to show the relevant data for the use case. The CP allows two modes of deploying elements from the DI. In Manual Mode, when specific materials/elements are selected, the corresponding geometries are displayed with the associated meta data, and can be manually copied in respective model, with an automatic update on used elements to the DI. In Automated Mode, it calls the design automation routine.
- The *design automation procedure* provides a selection of Use Cases (UCs), each with a selection of pre-defined Design Logics (DL) (Fig. 19). Each DL provides an iterative parameter-based workflow providing qualitative and quantitative feedback on local optimization as well as global use in all design models. Once a design is set, the use of elements is automatically recorded in the DI, again available to all users
- The *report procedure* provides two types of graphical reports can be extracted at any time. The PCARDS Score demonstrates the reusability performance of the design in reflection to the whole DI, including figures for total re-used elements, diversity of re-used elements, total re-used elements per category and other total figures featured in the inputted data such as carbon footprint (Fig. 18). The Use Case Reports are more specialized reports that detail the elements categories used in a specific use case (Fig. 20).

[4] Facilitated through the *Values toolkit*, a set of Python-based tools developed by Dsearch to support data management in projects by enabling the use of python dictionary data structures in the visual scripting environment of Grasshopper.

This parallel real-time design process enables a continuous negotiation of the use of elements between the design model users and allows extracting precise totals and detailed reports at any point, through real-time exchanges between the DI and all design models[5]. All elements deployed from the DI manually or automatically are made available in design models for presentation purposes (Fig. 21, 22 and 23) also associated with relevant meta data and could potentially be used in future BIM-model at detail design stage.

Fig. 21. Visualization of interior, featuring the interior ceiling use case with selected design logic for acoustic panels combining reused steel sheets with sound absorption materials.

Fig. 22. Visualization of courtyard featuring tiled paving use case with design logic for concrete tiles that are cut and reconfigured into a new pattern.

[5] Facilitated through Hops, a Grasshopper component referencing scripts across multiple models in parallel allowing heavy calculations without blocking user interaction.

Fig. 23. Visualization of external façade featuring façade cladding use case with design logic for natural stone cladding.

4 Conclusions

The two cases present two extremes of project confined reuse: a relatively simple material bank with partly unknown conditions being used for structural applications in timber, versus a well-known but diverse material bank being used for range of non-structural applications in a dynamically changing design model. In both cases the limitations and potentials of the material libraries guided the overarching design and development process through the iterative exploration of the potentials of the computational design system.

An important conclusion from the development of PCARDS in relation to the two cases is that there is potential in project-confined material banks, in that design and design system approaches can be closely tied to the available material and allow for rapid design processes. There is additional work to be done in terms of assessing the climatic impact of reusing material in this context – it could for instance be argued that using up more than needed material will prevent leftovers to transported (with negative environmental effect) or further degraded (limiting future use), if elements used in a project could be reused again in the future. Another important trajectory is to provide similar real-time links between project specific design systems to regional, national, or even international material banks, which will provide additional challenges in regard to data standards, grading of material, logistics and storage.

Acknowledgements. Elena Kanevsky (design lead case 1), Isabel Villar, Patsy Bellido, John Pettersson, Barbara Vogt and Fredrik Källström (design lead case 2), Max Zinnecker, Andreas Mitsiou, Niklas Eriksson, Martin Johnson, Frans Magnusson (Values toolkit), Concular GmbH, Karstadt Warenhaus GmbH.

References

Batalle Garcia, A., et al.: Material (data) intelligence: towards a circular building environment. CAADRIA (2021)

Brütting, J., et al.: The reuse of load-bearing components. IOP Conf. Ser.: Earth Environ. Sci. **225**, 012025 (2019)
Cristescu, C., et al.: Design for deconstruction and reuse of timber structures – state of the art review. InFutUReWood Report:1. RISE Report, May 2020 (2020)
Debacker, W. et al.: D1 Synthesis of the state-of-the-art. Report. BAMB Buildings as Material Banks (2016)
Gorgolewski, M.: Resource Salvation: The Architecture of Reuse. Wiley, Hoboken (2017)
Hradil, P., et al.: Re-use of Structural Elements: Environmentally Efficient Recovery of Building Components. VTT Technology, Espoo (2014)
Huang, Y., et al.: Algorithmic circular design with reused structural elements: method and tool. In: fib Symposium Proceedings, vol. 55 (2021)
Josefsson, T., Thuvander, L.: Form follows availability: the reuse revolution. IOP Conf. Ser. Earth Environ. Sci. (2020)
Kozminska, U.: Circular design: reused materials and the future reuse of building elements in architecture. Process, challenges and case studies. IOP Conf. Ser. Earth Environ. Sci. (2019)
Lokhandwala, Z.: Computational Design with Reused Materials. Thesis Project. Ecole des Ponts, Paris (2018)
Magnusson, F., Runberger, J.: Design system assemblages – the continuous curation of design computation processes in architectural practice. Prof. Pract. Built Environ. (2016). University of Reading
Magnusson, F., Runberger, J., Zboinska, M., Ondejcik, V.: Morphology & development: knowledge management in architectural design computation practice. eCAADe (2017)
Ribeirinho, M., et al.: The Next Normal in Construction: How Disruption is Reshaping the World's Largest Ecosystem, McKinsey & Company (2020)
Runberger, J., Magnusson, F.: Harnessing the informal processes around the computational design model. In: Thomsen, M., Tamke, M., Gengnagel, C., Faircloth, B., Scheurer, F. (eds.) Modelling Behaviour, pp. 329–339. Springer, Cham (2015). https://doi.org/10.1007/978-3-319-24208-8_28
SBE19 Brussels - BAMB-CIRCPATH: Buildings as material banks - a pathway for a circular future. 2019 IOP Conf. Ser. Earth Environ. Sci. **225**, 011001 (2019)
Sirkin, T., ten Houten, M.: The cascade chain: a theory and tool for achieving resource sustainability with applications for product design. Resour. Conserv. Recycl. **10**(3), 213–276 (1994)

Constructing Building Layouts and Mass Models with Hand Gestures in Multiple Mixed Reality Modes

Anton Savov(✉), Martina Kessler, Lea Reichardt, Viturin Züst, Daniel Hall, and Benjamin Dillenburger

ETH Zürich, Zürich, Switzerland
asavov@ethz.ch

Abstract. This paper presents a Mixed Reality framework for schematically defining building layouts and massing in multiple representations. Non-experts can use the framework to explore possible building configurations alone or in tandem with an architect. Our framework relies on a single-truth voxel matrix to track design changes and construct view-specific representations using the Marching Cubes and Marching Squares algorithms. We use only hand gestures for all design interactions instead of tangible objects or markers, to increase the mobility of users and make the application more accessible. The framework is tested in two prototypes for the HoloLens. The two prototypes have an objective to implement and test a variety of gestures for adding and removing volume, respectively area, from the designed building. The unified model representation across multiple MR views and interaction modes is the main contribution of this work and can be a valuable reference for the community developing applications of Mixed Reality in architecture. Additionally, we present a catalog of gesture-based interactions with the findings from our development process and the feedback from user studies.

Keywords: Architecture · Design · Hand gesture · Mixed Reality · Augmented Reality

1 Problem Statement

Today, aspiring homeowners can benefit from Mixed Reality (MR) to experience their future homes. At the same time, online configurators make it possible to define a home's layout and materials. While immersion in Mixed Reality helps non-experts judge the design and formulate their preferences more clearly, a configurator offers users a high level of control to repetitively change a design based on their preferences. Yet there is a lack of MR tools that allow for both control and immersion [1–3].

Creating immersive virtual experiences of a specific design requires specialized skills that not all architects can offer their customers. Creating an MR tool that lets users change the design on-the-fly is even more challenging because designing a building

C. Gengnagel et al. (Eds.): DMS 2022, *Towards Radical Regeneration*, pp. 360–373, 2023.
https://doi.org/10.1007/978-3-031-13249-0_30

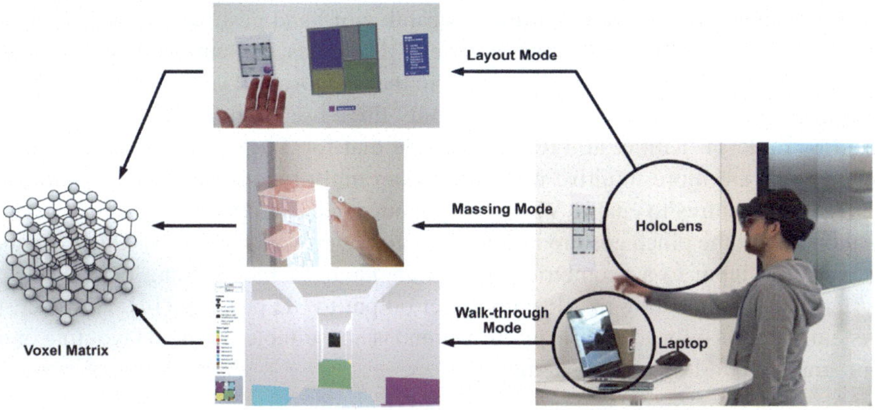

Fig. 1. A voxel matrix data model and the Marching Squares/Cubes algorithms enable various Mixed Reality interaction modes for creating schematic architectural designs.

encompasses two interdependent activities. One, shaping the building's volume to match site conditions and desired visual appeal and two, laying out a functional arrangement of the spaces needed by homeowners. There are MR precedents for laying out objects where users configure an arrangement of rooms or furniture [1, 4, 5]. There are also precedents where users model 3D shapes [6–8]. However, we have not identified precedents that support both laying out a building and 3D modeling, i.e., creating a building's massing.

We argue that this lack is due to the different requirements on the geometrical representations in both cases. Laying out rooms requires a representation consisting of 2D regions while 3D modeling requires a representation of solid primitives or boundary representation objects. In this paper, we present a geometrical data structure that has the potential to support both layout models and 3D models simultaneously (Fig. 1). We discuss two prototypical applications of this data structure in a Mixed Reality environment with four interaction modes: Layout, Walk-through, Massing at object scale, and Massing at 1:1 scale.

2 Background

Mixed Reality (MR) requires a media-specific mode of information representation that most likely differs from traditional, commonly used representations such as 2D drawings, 3D models, and physical models [9]. At the same time, design creation in MR requires new interactions [8, 10].

2.1 Gestures

We limit the scope of our work to using only gestural input, i.e. without tangible object instances. This allows our tests to be hardware and software light as the purpose is to identify the interactive and design potential of our proposed geometric data structure in various view modes.

Piumsomboon et al. (2013) present a catalog of hand gestures for 40 [11]. Besides pinch-like gestures to move, scale and orient 3D objects, the authors do not list specific design-related gestures. Of the 80+ works in the categories of Creation & Configuration [12] and Design Support & Review [3], only three use hand gestures for design tasks in MR (Table 1). Kim et al. (2019) conclude that for 3D manipulation gesture-based interactions are more intuitive and useful than multi-touch ones [13]. Fröhlich et al. (2018) use gestures to enable VR users to scatter trees and buildings [4]. However, the authors report the pinch gesture lacked precision. Our work overcomes this by mapping the gestural input to a geometric grid. Jailungka and Charoenseang (2018) present a system for 3D modeling in AR with direct output to 3D printing [14]. However, their system supports only geometric primitives and is not suitable for room layouting or more fine-grained building volume definition.

Table 1. Precedents that use hand gestures for design tasks in MR.

Gesture	Where	Used for
Pinch gesture	[4]	Scattering objects on a surface
Pinch gesture	[14]	Moving 3D object, or a vertex of a 3D object
Grab gesture	[14]	Rotating an object in 3D
Two-handed pinch gesture	[14]	Scaling an object
Pinch gesture	[13]	Orienting an object in 3D

2.2 Multiple Interaction Modes

MR can support both first-person and external, bird's eye views, respectively classified as an *egocentric* and *exocentric* frame of reference [15]. In their six-level framework of the gradient between physical and virtual realms Roo and Hachet (2017) treat the immersive mode is treated purely as a representational experience [16]. In our work, we aim to enable the user to edit the design in both exocentric and egocentric modes. The immersive experience of a walk-through in MR is poor due to the inconsistent overlaying of virtual objects over reality and non-photo-realistic rendering [3]. Therefore, we prototype a Walk-through Mode in a pure 3D environment.

Sandor and Klinker (2005) present a framework that is a mix of modeling-only and presentation-only modes [8]. However, their system, preceding today's integrated devices, is hardware heavy, with multiple custom developments and is only for 3D modeling, i.e. does not support layouting. Gai et al. (2017) present a system using a smartphone and custom-made physical elements to layout mazes and walk in them virtually [17]. However, it only offers the egocentric scale of interaction and layout configuration. Son et al. (2020) present a system that uses projection mapping and physical objects to retrieve and display floor plan layouts matching user-defined criteria [5]. However, it works only at the exocentric scale and the set up requires custom physical props.

2.3 Geometric Representations for MR in AEC

Previous works, such as Tilt Brush [18], Gravity Sketch [19], vSpline [20], and SandBOX [21], enable the creation of designs as abstract shapes, which an expert will need to reinterpret into an architecturally specific representation. We represent the user-created designs as isosurfaces over a voxel grid generated using the Marching Cubes, respectively the Marching Squares algorithms, which have been used for interactive architectural tools [22–25]. This delivers architecture-specific representations of the modeled design without an extra processing step, shortening the cognitive latency for the user [9].

3 Method

This paper presents two prototypes for the Microsoft HoloLens 2 that test the suitability of the Marching Cubes/Squares algorithm for capturing design changes and creating immersive representations. In the first prototype, the MR users can draw layouts of houses on any surface with their hands and can virtually walk in them to add windows and doors. The second prototype enables modeling the massing of a building. Two teams of four computer science students developed the prototypes in a master's level Mixed Reality course at ETH Zürich.

In user tests, participants were asked to perform a task in both MR prototypes. The task for the first prototype was to recreate and modify a given floor plan of a house (Fig. 2 left) using the Layout and the Walk-through Mode. In the second prototype, users were asked to model a building between two existing buildings represented as physical boxes (Fig. 2 right).

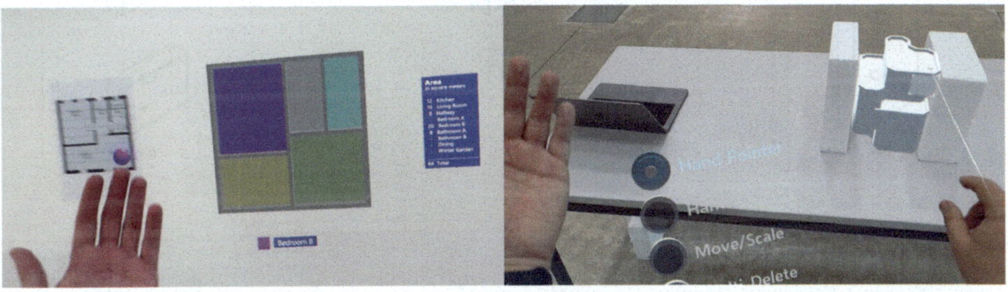

Fig. 2. User study tasks in the two MR prototypes.

4 Implementation

4.1 Model Representations

We store the current state of the design as a matrix of voxels that tracks which room a voxel is assigned to. In the 2D and 2.5D interaction modes we use the Marching Squares (MS) algorithm to automate the generation of the architectural layout representation. We use Marching Cubes (MC) in the Massing Mode.

Marching Squares. We have extended the MS algorithm to support different room types. To create a 3D mesh from the 2D data matrix, our algorithm places a wall between adjacent nodes with different room types (Fig. 3 and 4).

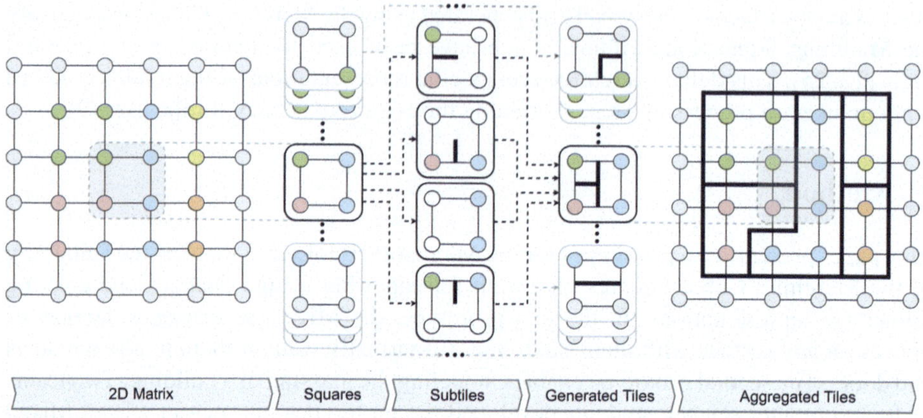

Fig. 3. From voxel to walls. The node values of each square are used to construct the tiles with the walls in the floor plan.

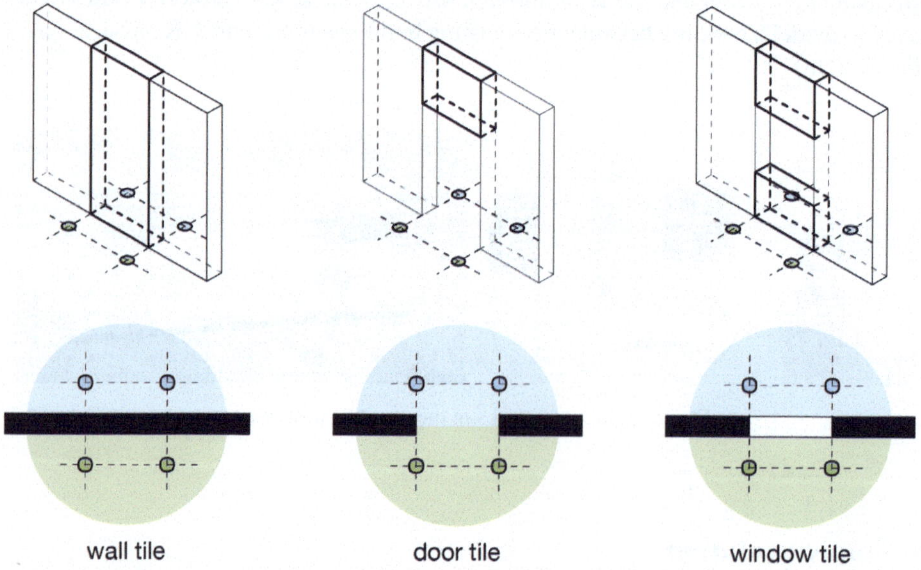

Fig. 4. The 2D and 2.5D tiles used to represent walls, doors and windows in the Layout Mode.

Marching Cubes. MC is a computer graphics algorithm that extracts isosurface meshes from a three-dimensional matrix data structure [26]. The algorithm replaces each matrix node with a mesh tile. Since we are dealing with architectural models, we use the approach described by Savov et al. (2020) to replace the original MC mesh tileset with a set of architectural mesh pieces [24].

Single Source of Truth, Multiple Representations. To synchronize the designs between devices we record user changes to the voxel matrix to a JSON file and push it to a web server (Fig. 5).

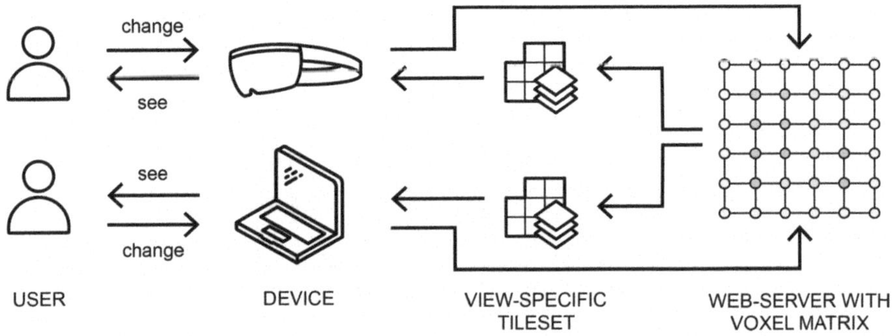

Fig. 5. Design changes are synced to a web-server. View-specific tilesets ensure the relevant information is displayed in the Massing, Layout and Walk-through modes.

4.2 Interaction Modes

The two MR prototypes explored the creation of designs in both *egocentric* and *exocentric frames* of reference [15]. Accounting for the two architectural representations, layout and massing, this gives four interaction modes (Fig. 6):

1. Layout Mode (exocentric) - where users can define the room layout.
2. Walk-through Mode (egocentric) - where users can add windows and doors.
3. Massing Mode (exocentric) - where users can define the shape of a building in a tabletop, birds-eye view.
4. Massing Mode (egocentric) - where users can preview and edit the shape of the building in near-real size.

Layout Mode. We developed the MR Layout Mode for the HoloLens 2 using Unity and the MRTK library. Wearing the HoloLens 2 headset, the user draws and edits floor plans with hand gestures on any flat surface.

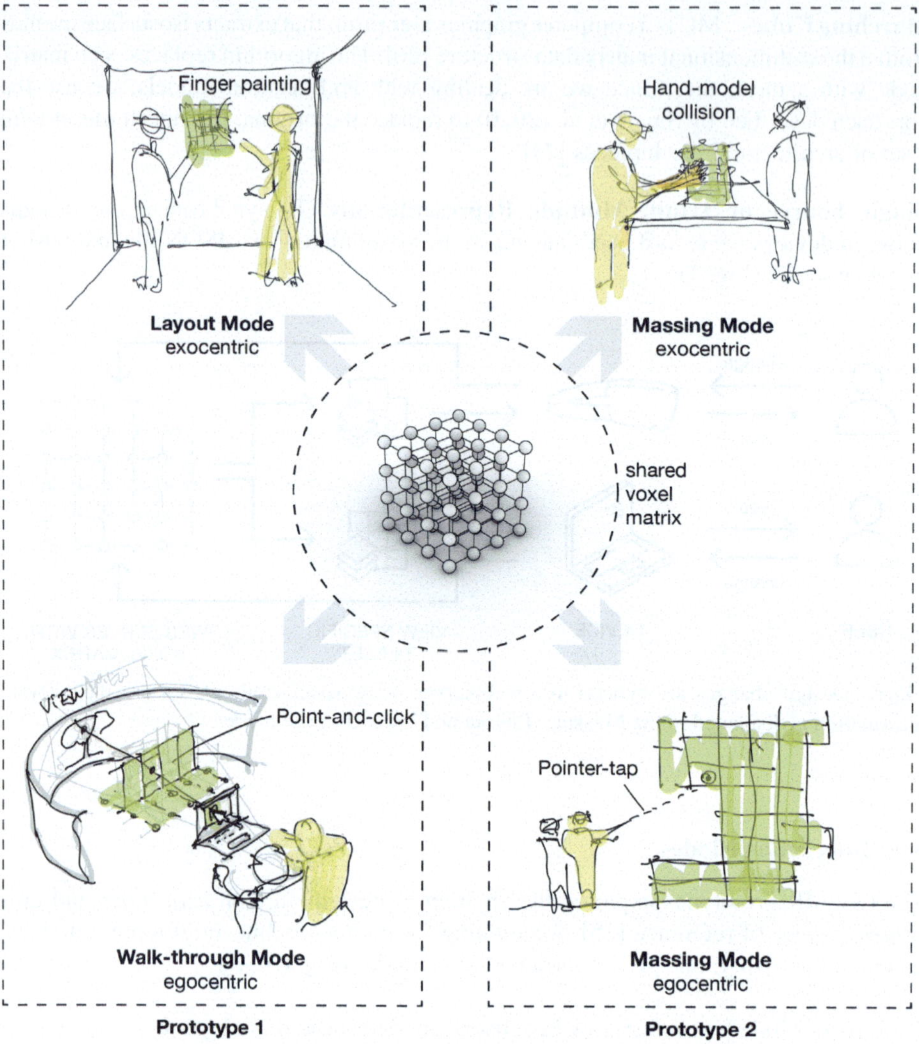

Fig. 6. The four interaction modes explored in this work.

To place the holographic floor plan grid on a surface, the user performs a *tap-to-place gesture*. The 2D grid automatically attaches to the available surfaces (Fig. 7 left). The user uses the *finger painting gesture* to create a layout on the grid, as shown in Fig. 7 (middle). To draw on the holographic grid, the user's index finger can maintain physical contact with the surface. To enable faster painting, the brush radius can be increased with a *two-handed pinch gesture* shown in Fig. 7 (right).

The *palm-up gesture* calls a hand-attached menu which lets the user reposition the workspace and choose the room they are painting with (Figs. 8 left, middle). The *area overview* illustrated in Fig. 8 (right) lists all room types of the current floor plan and their corresponding area in square meters.

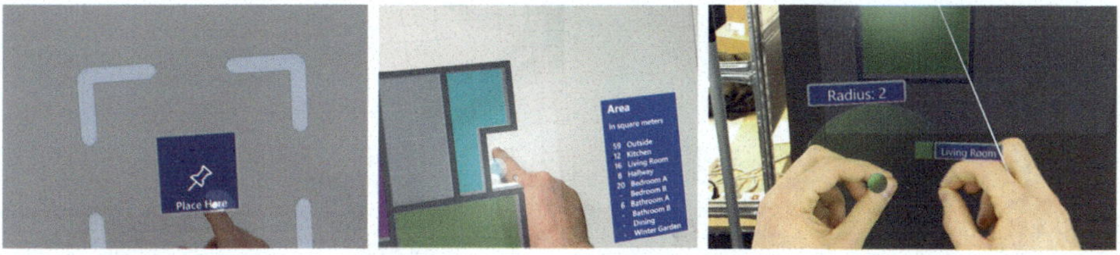

Fig. 7. Gestures in Layout Mode. (left) Tap-to-place. (middle) Finger painting. (right) Two-handed pinch.

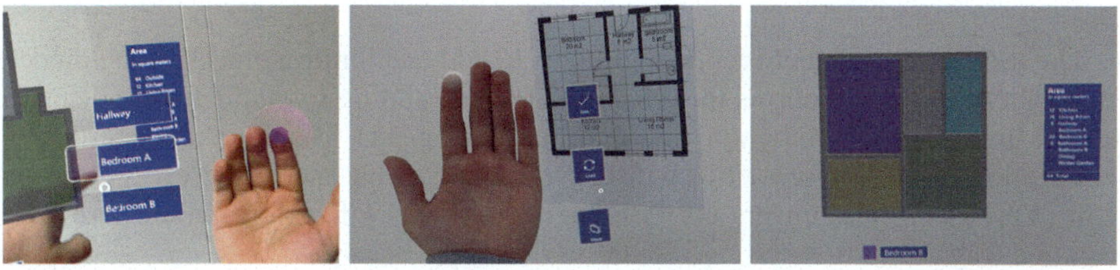

Fig. 8. UI elements. (left) Palm-up scroll wheel for changing the room type. (middle) Palm-up hand menu for synchronizing the design with the web server. (right) Area overview.

Walk-through Mode. The Walk-through Mode is a separate 3D application developed using Unity that runs on any computer and complements the Layout Mode (Fig. 9). It enables the user to experience the created floor plan and place doors and windows. A surrounding environment map adds realism as it shows the views. The wall modification type (door, window, wall) can be chosen and is applied to a wall when clicking on it. A mini-map helps the user to orient in the 3D environment.

Fig. 9. Screenshots from the Walk-through Mode. (left) the switch menu. (right) the walls with the newly created openings.

Massing Modes. The Massing Mode is in the second Mixed Reality prototype for the HoloLens 2 using Unity and the MRTK library.

The user can move and scale the model with the *single-hand* and *two-handed pinch gestures* respectively (Fig. 10). This gives users the freedom to shape the building as a scaled-down model (exocentric) or directly in its close to real-world size (egocentric).

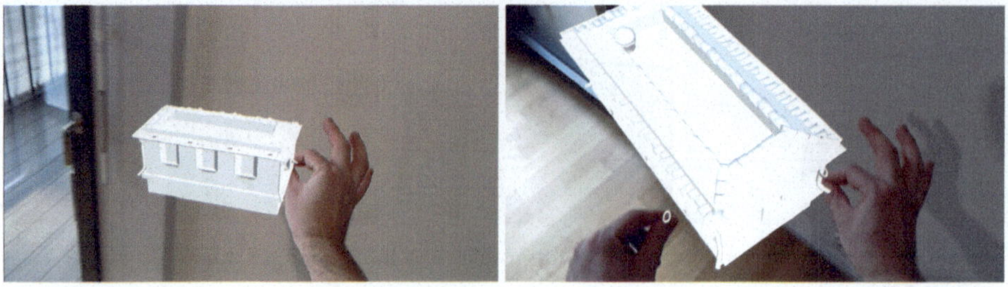

Fig. 10. Users move and scale the model with the one- and two-handed pinch gestures.

The user can switch between two gestures to add and remove blocks of the model. The *pointer-tap gesture* allows the user to shape the building's volume from a distance without 'touching' the hologram with their hands (Fig. 11). The *hand-model collision gesture* allows for direct interaction with the voxel grid. In this interaction, every collision between the user's index fingers and the holographic model will perform an add or remove operation (Fig. 12).

Both gestures described above can be used to add and remove blocks in *symmetric hand* or *asymmetric hand input*. In the *symmetric hand input*, both hands perform the same operation: adding or removing. Whereas in the *asymmetric hand input*, the right hand can be used to add blocks and the left hand can be used to remove blocks.

Fig. 11. Pointer-tap gesture interactions in the Massing Mode (exocentric and egocentric).

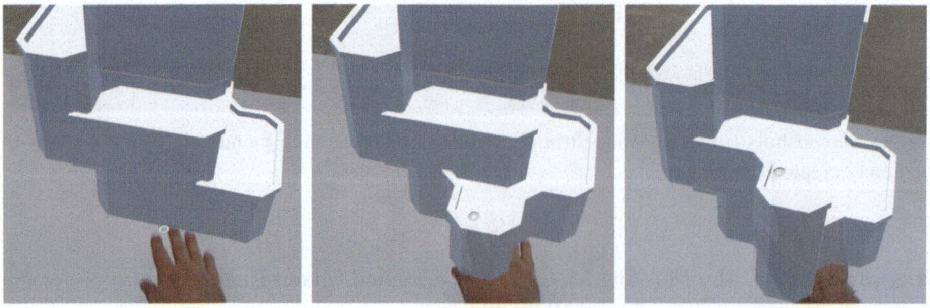

Fig. 12. Hand-object collision interaction in the Massing Mode.

5 Findings

The findings are based on interacting with the tools during development as well as by observing the users perform the two given tasks described in the chapter *Method*.

5.1 Design Representation

A benefit of using predefined tiles with the Marching Cubes and Marching Squares algorithms to construct the architectural geometry from the voxel values is that a scale is given through the details in the tiles. The same slice of the voxel space can be rendered using a different tileset as showcased in the Layout and Walk-through Modes. This enables design changes made in one mode to be immediately experienced in another.

5.2 View Modes

The Walk-through Mode allows multiple users to see the 3D model from the same point of view. At the same time, this setup allows that in future work, each of them can see the floor plan through their own HoloLens device and interact with it. However, the Walk-through Mode, making use of the first-person shooter controls, is challenging for non-gamers to use. A switch to a full immersion mode in Virtual Reality needs to be explored as a counter to that, but it would mean that the shared point of view will be lost.

5.3 Gestures and Interactions

Table 2 summarizes the gestures used in this work. Interactions based on hand collision with the virtual model were more intuitive across all tasks, in both the Layout and Massing Modes.

For the *tap-to-place gesture*, we tried two options for the workspace to follow the user to the new location. With *head tracking*, the user had to freeze their head to place the workspace, which was cumbersome, and often involuntarily shifted the workspace. With the *hand follow*, placing the workspace was more intuitive for the user.

The use of the *two-handed pinch gesture* for setting the brush size for layout painting proved well thought through and intuitive because it is physical [27] and reversible [11]. However, it happens that the user performs the gesture unintentionally, e.g. when holding a sheet of paper with both hands. A possible solution is to implement an eye-gaze condition, i.e., the gesture is only detected when the user looks at their hands.

The *pointer-tap gesture* makes editing life-size models from a single position possible. It works well for adding small corrections to a model as it offers precision. However, the repeated need to perform an air tap for every add/remove operation leads to reduced modeling speed and the gorilla arms syndrome [28].

The *hand-model-collision gesture* is optimal for editing from a short distance and quickly prototyping ideas. The intuitiveness of the gesture was very positively received by every user as they were able to create a large design change in minimal effort. Furthermore, there is less muscle fatigue because the users do not need to perform

Table 2. The hand gestures used in this work.

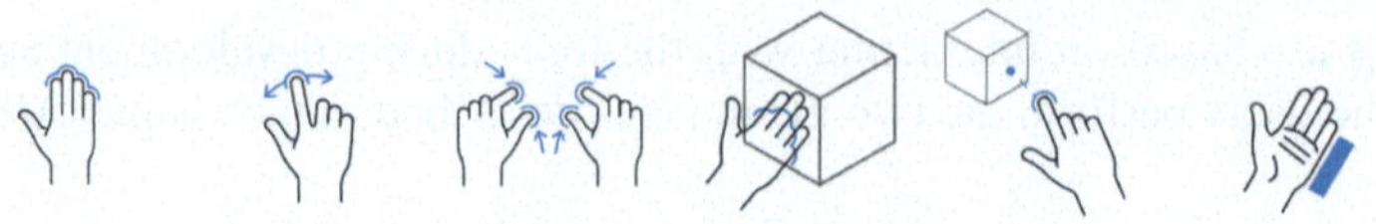

Tap-to-place	Finger painting	Two-handed pinch	Hand-model-collision	Pointer-tap	Palm-up
Used in: Layout Mode	Layout Mode	Layout Mode, Massing Mode	Massing Mode	Massing Mode	Layout Mode, Massing Mode
Used for: Workspace placement	To layout the rooms	Set brush size, Scale and Orient Workspace	Add/remove blocks from the design	Add/remove blocks from the design	Display tool menu
Pros: Intuitive in combination with hand follow	Intuitive, no fatigue	Intuitive	Natural flow, quick modeling	High precision modeling, no need to walk around model	Tools always available to user
Cons: Cumbersome when used with head-track follow	-	Accidental false detection when user's hands off camera view or holding an object.	Low precision	Arm fatigue, slow modeling speed	-

air taps all the time. The disadvantage of the gesture is the lack of precision as the users unintentionally add and remove blocks. To reduce this problem we only register collisions with the index fingers instead of all 10 fingers.

The *asymmetric hand input* allows the user to work on the model without having to use the hand menu often. However, the modeling process is not as intuitive as shaping a material such as clay, where one would usually use both hands at the same time. Therefore, we added the *symmetric hand input* option to use both hands for either adding or removing blocks. As a downside, the user needs to switch between adding and removing using the *palm-up hand menu*, which can distract from the original modeling task.

The multiple view modes require the users to change their physical location in the working room, moving between wall or table surfaces and the computer screen. Therefore in our prototypes, all tools are in the hand menus to ensure the user cannot get stuck with UI elements attached at an unreachable spot in the room [29].

5.4 Application Scenarios

The users find that feasibility studies and early-stage conceptual design are suitable use cases for the MR framework we propose. Potential scenarios for applying a multi-mode MR design tool based on the Marching Cubes data model are:

1. Future homeowners configure a single-family house or a cooperative apartment building on their own or together with an architect.
2. Real estate developers conduct a quick feasibility study for a plot on the market on their own or with an architect.

We do not see any indication of a direct value for architects working on their own on an early design concept in comparison to existing CAD and 3D tools. This is because the main benefit of Mixed Reality is its immersive quality that boosts a non-trained person's ability to imagine the design. Furthermore, the reliance on gestures as design input is intuitive for the non-trained user but cumbersome and inefficient for the professional designer. However, we see a great opportunity for architects to use the MR tool to communicate with their clients through quick iterative and immersive sketches.

6 Conclusion and Outlook

This paper presented a Mixed Reality framework for schematically defining building layouts and massing models in multiple representations. Non-experts can use the framework to explore possible building configurations alone or in tandem with an architect. The framework relies on a single-truth voxel matrix to track design changes and construct view-specific representations using the Marching Cubes and Marching Squares algorithms. Our goal was to avoid tangible objects and markers, and use only hand gestures for all design interactions, to increase the mobility of users and make applications more accessible.

In two prototypes we implemented and tested a variety of gestures for adding and removing volume, respectively area, from the designed building. The unified model representation across multiple MR views and interaction modes is the main contribution of this work and can be a valuable reference for the community developing applications of Mixed Reality in architecture. Additionally, we present a catalog of gesture-based interactions with the findings from our development process and the feedback from user studies.

Using a voxel matrix to store design configurations allows to plug in shape or layout autocompletion machine learning (ML) at a later stage. The ML model will fill the needed cells considering the desired room area and connectivity. Instead of painting rooms, users can be asked to define only the building's outlines and fitting designs can be retrieved from a database as in the work by Son et al. (2020) [5]. A useful addition to the framework will be a way to slice the building mass model and look into any floor's layout in the Layout Mode.

We modified the Marching Squares algorithm to support generating the tileset on the fly. This opens the possibility for increasing the design space of the whole system. The constraining factor of the orthogonality of voxel spaces can be overcome in future work by using irregular quad-mesh for the voxel grid.

The multi-mode Mixed Reality framework presented in this paper opens the door to integrating immersive experiences and building configurators for the benefit of including homeowners and other non-expert stakeholders in the early design stages.

Acknowledgments. The authors would like to thank the Mixed Reality course organizers Iro Armeni, Federica Bogo and Marc Pollefeys as well as the students who worked on the two projects: Giulia Argüello, Lukas Bösiger, David Enderlin, Adrian Hirt, Dominic Weibel.

References

1. Arrighi, P.-A., Mougenot, C.: Towards user empowerment in product design: a mixed reality tool for interactive virtual prototyping. J. Intell. Manuf. **30**(2), 743–754 (2016). https://doi.org/10.1007/s10845-016-1276-0
2. Caldas, L., Keshavarzi, M.: Design immersion and virtual presence. Technol. Archit. Des. **3**, 249–251 (2019). https://doi.org/10.1080/24751448.2019.1640544
3. Davila Delgado, J.M., Oyedele, L., Demian, P., Beach, T.: A research agenda for augmented and virtual reality in architecture, engineering and construction. Adv. Eng. Inform. **45**, 101122 (2020). https://doi.org/10.1016/j.aei.2020.101122
4. Fröhlich, T., Alexandrovsky, D., Stabbert, T., Döring, T., Malaka, R.: VRBox: a virtual reality augmented sandbox for immersive playfulness, creativity and exploration. In: Proceedings of the 2018 Annual Symposium on Computer-Human Interaction in Play. Association for Computing Machinery, New York, NY, USA, pp. 153–162 (2018)
5. Son, K., Chun, H., Park, S., Hyun, K.H.: C-space: an interactive prototyping platform for collaborative spatial design exploration. In: Proceedings of the 2020 CHI Conference on Human Factors in Computing Systems. Association for Computing Machinery, New York, NY, USA, pp. 1–13 (2020)
6. Arora, R., Habib Kazi, R., Grossman, T., Fitzmaurice, G., Singh, K.: SymbiosisSketch: combining 2D & 3D sketching for designing detailed 3D objects in situ. In: Proceedings of the 2018 CHI Conference on Human Factors in Computing Systems. Association for Computing Machinery, New York, NY, USA, pp. 1–15 ((2018))
7. Reipschläger, P., Dachselt, R.: DesignAR: Immersive 3D-modeling combining augmented reality with interactive displays. In: Proceedings of the 2019 ACM International Conference on Interactive Surfaces and Spaces. Association for Computing Machinery, New York, NY, USA, pp. 29–41 (2019)
8. Sandor, C., Klinker, G.: A rapid prototyping software infrastructure for user interfaces in ubiquitous augmented reality. Pers. Ubiquit. Comput. **9**, 169–185 (2005). https://doi.org/10.1007/s00779-004-0328-1
9. Du, J., Zou, Z., Shi, Y., Zhao, D.: Zero latency: Real-time synchronization of BIM data in virtual reality for collaborative decision-making. Autom. Constr. **85**, 51–64 (2018). https://doi.org/10.1016/j.autcon.2017.10.009
10. Trevisan, D., Vanderdonckt, J., Macq, B.: Analyzing interaction in augmented reality systems (2002). Undefined
11. Piumsomboon, T., Clark, A., Billinghurst, M., Cockburn, A.: User-defined gestures for augmented reality. In: Kotzé, P., Marsden, G., Lindgaard, G., Wesson, J., Winckler, M. (eds.) INTERACT 2013. LNCS, vol. 8118, pp. 282–299. Springer, Heidelberg (2013). https://doi.org/10.1007/978-3-642-40480-1_18
12. Kent, L., Snider, C., Gopsill, J., Hicks, B.: Mixed reality in design prototyping: a systematic review. Des. Stud. **77**, 101046 (2021). https://doi.org/10.1016/j.destud.2021.101046
13. Kim, M., Choi, S.H., Park, K.-B., Lee, J.Y.: User interactions for augmented reality smart glasses: a comparative evaluation of visual contexts and interaction gestures. Appl. Sci. **9**, 3171 (2019). https://doi.org/10.3390/app9153171
14. Jailungka, P., Charoenseang, S.: Intuitive 3D model prototyping with leap motion and microsoft hololens. In: Kurosu, M. (ed.) HCI 2018. LNCS, vol. 10903, pp. 269–284. Springer, Cham (2018). https://doi.org/10.1007/978-3-319-91250-9_21
15. Ware, C.: Chapter ten - interacting with visualizations. In: Ware, C. (ed.) Information Visualization, 3rd edn., pp. 345–374. Morgan Kaufmann, Boston (2013)

16. Roo, J.S., Hachet, M.: One reality: augmenting how the physical world is experienced by combining multiple mixed reality modalities. In: Proceedings of the 30th Annual ACM Symposium on User Interface Software and Technology. Association for Computing Machinery, New York, NY, USA, pp. 787–795 (2017)
17. Gai, W., et al.: Supporting easy physical-to-virtual creation of mobile VR maze games: a new genre. In: Proceedings of the 2017 CHI Conference on Human Factors in Computing Systems. Association for Computing Machinery, New York, NY, USA, pp. 5016–5028 (2017)
18. Google: Tilt Brush (2022)
19. Gravity Sketch Ltd.: Gravity Sketch (2022)
20. Arnowitz, E., Morse, C., Greenberg, D.P.: vSpline: physical design and the perception of scale in virtual reality. In: ACADIA 2017: Disciplines & Disruption, Proceedings of the 37th Annual Conference of the Association for Computer Aided Design in Architecture (ACADIA), Cambridge, MA 2–4 November 2017, pp. 110–117. CUMINCAD (2017). ISBN 978-0-692-96506-1
21. Psarras, S., et al.: SandBOX - an intuitive conceptual design system. In: Gengnagel, C., Baverel, O., Burry, J., Ramsgaard Thomsen, M., Weinzierl, S. (eds.) DMSB 2019, pp. 625–635. Springer, Cham (2020). https://doi.org/10.1007/978-3-030-29829-6_48
22. Hosmer, T., Tigas, P., Reeves, D., He, Z.: Spatial assembly with self-play reinforcement learning. In: Slocum, B., Ago, V., Doyle, S., Marcus, A., Yablonina, M., del Campo, M. (eds.) ACADIA 2020: Distributed Proximities/Volume I: Technical Papers, Proceedings of the 40th Annual Conference of the Association of Computer Aided Design in Architecture (ACADIA)]. Online and Global, 24–30 October 2020, pp. 382–393. CUMINCAD (2020). ISBN 978-0-578-95213-0
23. IndieCade Europe: Organic Towns from Square Tiles - a talk by Oskar Stålberg at IndieCade Europe 2019 (2020)
24. Savov, A., Winkler, R., Tessmann, O.: Encoding architectural designs as Iso-surface tilesets for participatory sculpting of massing models. In: Gengnagel, C., Baverel, O., Burry, J., Ramsgaard Thomsen, M., Weinzierl, S. (eds.) DMSB 2019, pp. 199–213. Springer, Cham (2020). https://doi.org/10.1007/978-3-030-29829-6_16
25. Stalberg, O.: Voxel house: breaking down the voxel house demo. Vertex **3**, 170–175 (2016)
26. Lorensen, W.E., Cline, H.E.: Marching cubes: a high resolution 3D surface construction algorithm. In: Proceedings of the 14th Annual Conference on Computer Graphics and Interactive Techniques, pp. 163–169. ACM, New York (1987)
27. Wobbrock, J.O., Morris, M.R., Wilson, A.D.: User-defined gestures for surface computing. In: Proceedings of the SIGCHI Conference on Human Factors in Computing Systems. Association for Computing Machinery, New York, NY, USA, pp. 1083–1092 (2009)
28. Hansberger, J.T., et al.: Dispelling the gorilla arm syndrome: the viability of prolonged gesture interactions. In: Lackey, S., Chen, J. (eds.) VAMR 2017. LNCS, vol. 10280, pp. 505–520. Springer, Cham (2017). https://doi.org/10.1007/978-3-319-57987-0_41
29. Azai, T., Otsuki, M., Shibata, F., Kimura, A.: Open palm menu: a virtual menu placed in front of the palm. In: Proceedings of the 9th Augmented Human International Conference. Association for Computing Machinery, New York, NY, USA, pp. 1–5 (2018)

Morphology of Kinetic Asymptotic Grids

Eike Schling[1(✉)] and Jonas Schikore[2]

[1] The University of Hong Kong, Pok Fu Lam, Hong Kong
schling@hku.hk
[2] Technische Universität München, Munich, Germany

Abstract. This paper investigates the kinetic behaviour of asymptotic lamella grids with variable surface topology. The research is situated in the field of semi-compliant grid mechanisms. Novel geometric and structural simulations allow to control and predict the curvature and bending of lamellas, that are positioned either flat (geodesic) or upright (asymptotic) within a curved grid. We build upon existing research of asymptotic gridshells and present new findings on their morphology. We present a digital and physical method to design kinetic asymptotic grids. The physical experiments inform the design, actuation strategy and kinetic boundaries, and become a benchmark for digital results. The kinetic behaviour of each sample is analysed through five stages. The digital models are used to calculate the total curvature at every stage, map the energy stored in the elastic grids and predict equilibrium states. This comparative modelling method is applied to seven asymptotic grids to investigate transformations and the impact of singularities, supports and constraints on the kinetic behaviour. Open grids without singularities are most flexible and require additional, external and internal constraints. The cylindrical typology acts as a constraint and creates symmetric kinetic transformations. Networks with one, two and four singularities cause increasing rigidity and limit the kinetic transformability. Finally, two prototypical architectural applications are introduced, an adaptive shading facade and a kinetic umbrella structure, that show the possible scale and actuation of kinetic designs.

Keywords: Asymptotic networks · Semi-compliant mechanism · Kinetic behaviour · Comparative modelling

1 Introduction

Transformable structures are 4-dimensional and offer to design through time, beyond the static, and adapt to environmental conditions, structural influences or user's needs. We can distinguish between conventional rigid-body mechanisms utilizing hinges or telescopes, and compliant mechanisms (Howell 2002), which utilize the elastic properties of the material to perform a smooth change in curvature and store strain energy. This research is focused on a hybrid (semi-compliant) typology of kinetic grid mechanisms (Schikore et al. 2020), coupling elastic slats with scissor joints in a doubly-curved quadrilateral grid. The paper combines insights from architectural geometry and structural engineering (Fig. 1).

C. Gengnagel et al. (Eds.): DMS 2022, *Towards Radical Regeneration*, pp. 374–393, 2023.
https://doi.org/10.1007/978-3-031-13249-0_31

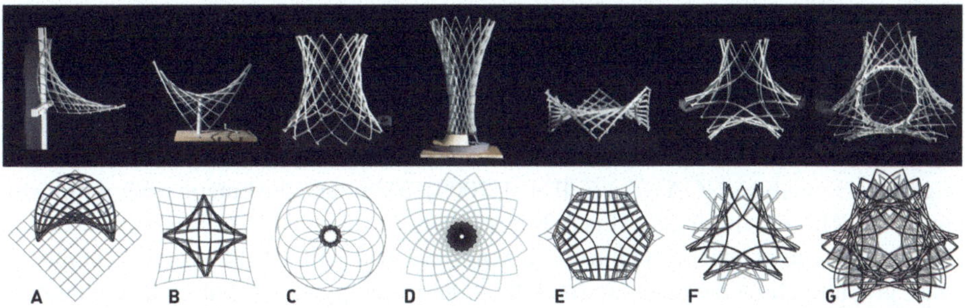

Fig. 1. Seven experimental asymptotic grids with variable topology. A) Regular grid (used for adaptive shading façade), B) Hypar grid, C) Regular rotational grid, D) Rotational grid with support constraints (Kinetic Umbrella), E) Schwarz D grid with one singularity, F) Tricylinder with two singularities, G) Quadcylinder with four singularities.

There is beautiful accordance of the mechanical behaviour of elastic slats in a grid, with the geometric properties of curves on a surface. Some of which have been described as early as 1897 by the mathematician Sebastian Finsterwalder (Finsterwalder 1897). This approach has gained momentum through the field of Architectural Geometry (Pottmann et al. 2007), which – among many other insights - brought forward methods to design doubly curved grids from developable strips (Tang et al. 2016). The approach is based on the theory of the *curvature of curves on a surface.* As long as the curves are attached continuously to the surface, and are measured in respect to the surface normal, their three local normal curvature k_n, geodesic curvature k_g and geodesic torsion τ_g can be calculated and correspond to the local deformation of an elastic profile (bending around **y**, bending around **z**, and torsion around **x**) within a gridshell (Schling and Barthel 2020). This knowledge has been applied to architectural construction by designing lamella gridshells along the asymptotic curves of anticlastic surface (Schling et al. 2017). Such asymptotes exhibit zero normal curvature ($k_n = 0$) and can thus be assembled from exclusively straight slats. The slats are positioned upright in the grid and adhere to the design geometry solely by twisting and bending around their weak axis. During the construction of the first steel prototype (Fig. 2), it was discovered that the asymptotic lamella grid was sufficiently constrained to deform predictably, following a semi-compliant mechanism (Schling et al. 2018). The complete lamella grid was assembled flat and transformed into the designated design shape purely by push of the hand.

Fig. 2. Kinetic behaviour of the first asymptotic steel grid. The lamellas are assembled flat and transform into the design shape simply by a push of the hand. They are constrained by the strong axis of lamellas and the scissor joints.

The phenomenon of controlled deployable gridshells has enjoyed some attention in the computational design community (Soriano et al. 2019; Isvoranu et al. 2019; Haskell et al. 2021). Both G-shells and X-shells focused on geodesic lamellas, tangential to the design grid. (Schikore et al. 2020) developed a typology, linking the kinetic behaviour to the three curvatures (k_n, k_g and τ_g) and corresponding profile axis (**x**, **y** and **z**) and used iso-geometric analysis (IGA) (Cottrell et al. 2009) to simulate the kinetic grids within the smooth NURBS environment. This publication also evaluated the necessary constraints to control a regular and a cylindrical asymptotic grid and introduced the curvature-square graph to allow designers to track the kinetic energy stored within throughout its transformation and predict its natural equilibrium state of minimum energy.

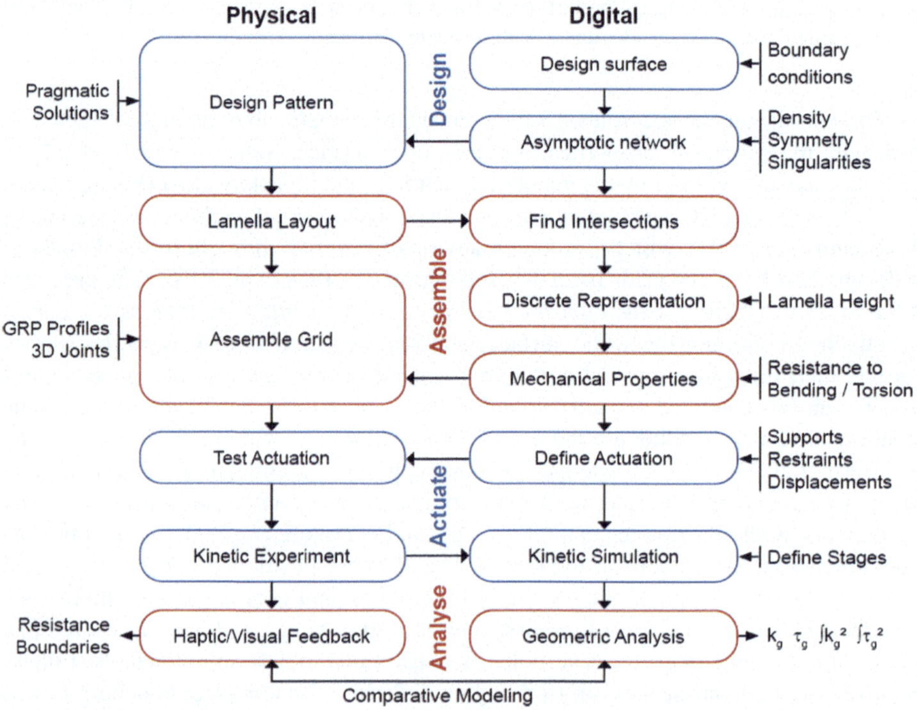

Fig. 3. Comparative Modelling. A hybrid approach is used, in which physical experiments and digital simulations are conducted in parallel to inform each other.

Contributions. The goal of this research is to extend the design language of kinetic asymptotic grids, by systematically investigating the impact of grid morphology on kinetic behaviour. The research is using a hybrid method of digital and physical experiments to design and evaluate 7 networks. Its feasibility for architecture is demonstrated through two design implementations. The work was first presented as an interactive digital and physical exhibition (https://eikeschling.com/2021/09/02/pmq-exhibition-kinetic-grid-structures/) in Hong Kong in September 2021 (Schling and Schikore 2021).

The paper presents a method to build accurate physical models of elastic lamellas grids from GRP and 3D printed joints (Sect. 2.1), digitally design comparative asymptotic networks and simulate their kinetic behaviour with the ideal mechanical properties (Sect. 2.2). Both physical and digital models are actuated through five stages to determine the natural boundaries and behaviour of semi-compliant mechanisms. Mapping the total curvature gives insights into the structures' residual stresses and likely equilibrium states. In Sect. 3, this method is applied to 7 open and closed asymptotic networks of incremental complexity, increasing the number of singularities and cylinders: A regular grid (used later for the adaptive shading façade) (A), a hypar grid (B), a regular rotational grid (C), the rotational grid (a scaled model of the Kinetic Umbrella) (D), a Schwarz D grid with one singularity (E), a tricylinder grid with two singularities (F) and a quad-cylinder with three singularities (G). Finally, in Sect. 4, we introduce two prototypical architectural applications, an adaptive shading facade and the Kinetic Umbrella that show the possible scale and actuation for kinetic designs.

2 Comparative Modelling

This study is based on a hybrid physical and digital exploration (Fig. 3). This comparative modelling strategy allows input and feedback from both worlds at any time of the investigation.

2.1 Physical Modelling

Design and Assembly. The physical experimental models are designed and constructed from glass fibre reinforced plastic lamellas (1 × 10 mm) and joined laterally on two levels using 3D-printed sleeves. M3 steel bolts (20 mm) create the scissor joints, which allow rotation around the normal axis. The joints are located with a 3 mm offset to the theoretical intersection points due to the lateral connection. The positions are marked by hand and the 3D joints are threaded and glued onto the glass fibre lamella. The physical models are not replicas of a digital design but were used actively to find suitable networks, adjust their density and learn about the assembly process. Samples A and C were designed without any digital help based on pragmatic regular flat and cylindrical patterns.

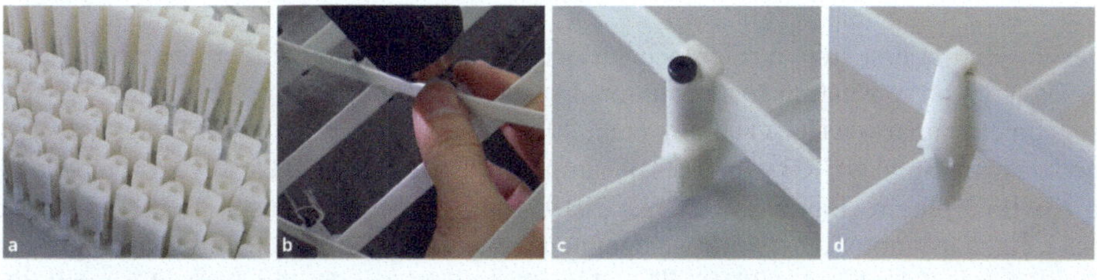

Fig. 4. The GRP lamellas were connected by hand (b) with 3d printed sleeves and M3 bolts (c). The sleeves are batch-printed (a) with a Form 3 Stereolithography (SLA) printer. A second, rigid sleeve (d) is used as internal restraint.

Actuation. The elastic grid is used to test various supports, constraints and displacements that control the kinetic behaviour. These physical tests are the basis for the digital simulations and development of actuation systems. Once a controlled kinetic behaviour is identified, the morphology is explored to its natural boundaries, which are marked by one of three occurrences (see also Fig. 7, Fig. 8, Fig. 9):

- **Planarity.** Sample A, B, C, and E are bounded by a planar state, which marks a point of symmetry in the kinetic behaviour. From here, the inverse transformation can be expected.
- **Self-collision.** Sample C, D, F and G display a limit state, in which lamellas collide and hinder further movement.
- **Resistance.** All kinetic transformations (except for C) were naturally limited by the resistance of the material (torsion or bending). These boundaries were later adopted in the digital simulation.

The maximal movement is actuated by hand and systematically documented in plan and elevation.

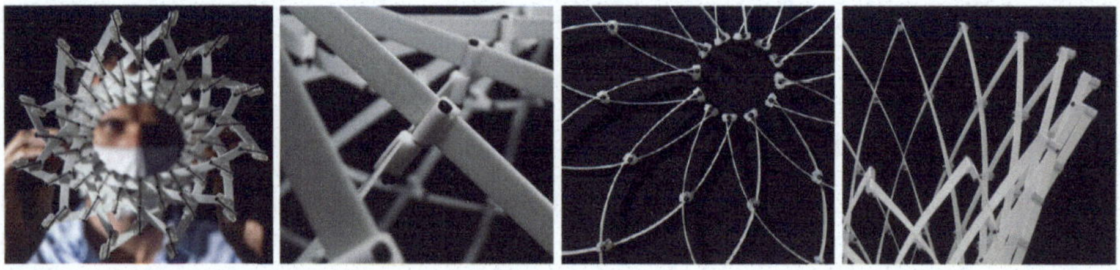

Fig. 5. The GRP models are tested extensively to determine suitable supports, constraints and displacements that control the kinetic behaviour.

2.2 Digital Design and Simulation

These physical experiments are simulated digitally based on the surface and network geometry, the stiffness parameters, support, restraints and actuation forces.

Design. In the first step, a reference surface is generated using either geometric operations (e.g. curve rotation for a rotational surface) or numerical optimization. We use isogeometric analysis (IGA) (Bauer et al. 2016; Oberbichler et al. 2019; Bauer 2020) to model the minimal NURBS surfaces of samples E, F and G. The IGA solver is embedded in Kiwi!3D plugin (Bauer and Längst 2019) for Grasshopper. Based on the reference surface, the asymptotic network is generated using the Bowerbird Pathfinder plugin for Grasshopper (Oberbichler 2019). The algorithm calculates the direction of asymptotic curves at any point on the surface and iteratively finds the path of zero normal curvature to draw two families of curves (Schling et al. 2017). Defining a homogenous network with appropriate density is straightforward for regular open surfaces (A, B) or rotational surfaces (C, D). In the case of complex surfaces with singularities (E, F, G) we use principal curvature lines, which bisect the asymptotic network, to define regular intersection points (Schling and Wan 2022).

This *initial design model* of surface and network can be used to create architectural visualizations, model the lamella geometry, define offsets, and find the intersection of curves. The length of curves and distance between intersections is the only digital data needed to mark straight strips of material and build a physical model.

Assembly. The initial design model is supplemented with ideal mechanical properties. For this purpose, we create an abstract discrete model of polygonal strips, which represent the intersections and lamellas. Each intersection is represented by a line i_i normal to the design surface which sits centered on the intersection point pt_i. Consecutive intersection-lines ($i_0, i_1, i_2, \ldots$) are connected at top and bottom with lines t_i and b_i, creating polylines t and b above and below the asymptotic curves a. Together t, b and i form strips of quads, resembling the elastic lamellas (Fig. 6).

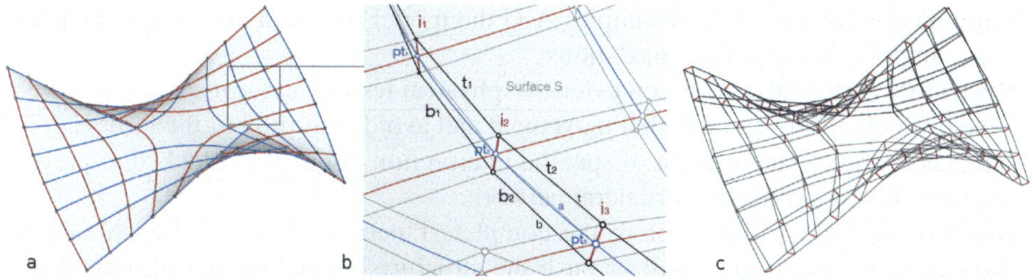

Fig. 6. Digital modelling. The initial, smooth design model (a) is rationalized (b) into a discrete polyline model (c) that allows embedding the mechanical properties.

We use the particle-spring solver Kangaroo 2 (Piker 2015) to embed simple optimization goals which resemble the ideal stiffness of profiles and rotation of joints. There are 6 *mechanical properties* (4 hard goals and 2 soft goals) that need to be embedded in the model.

Hard goals

- Resistance against **axial forces** (compression and tension): All lines i_i, t_i and b_i are restricted to their initial length.
- Resistance against **shear of the strong axis** of the lamella: The corners of the rectangle created by i_i, t_i, i_{i+1}, and b_i are restricted to 90 degrees.
- Resistance against **bending of the strong axis** of the lamella: The first two goals (length and angle) ensure that the two polylines t and b remain parallel and of the same length, and the quad-strip cannot curve up or down.
- **Scissor joints** with rotation limited to the normal axis: The points in which lines touch are treated as hinged connections by Kangaroo2. The intersection lines act as coupling of intersecting quad-strips and naturally limit the rotation to the normal intersection axis.

Soft goals

- Resistance against **torsion** of the lamella: The angle between consecutive intersection lines (i_i, i_{i+1}) is drawn towards 0. This motivates the intersection lines to become parallel and resist torsion of the polygon strip.
- Resistance against **bending of the weak axis** of the lamella: The angle between consecutive lines (t_i, t_{i+1}) and (b_i, b_{i+1}) is drawn towards 0. This motivates the quad-strips to become straight, i.e. minimize their geodesic curvature.

Constraints and Actuation. Once all mechanical properties are embedded, the Kangaroo solver is started, and the asymptotic grid will start transforming in real-time to assume a new equilibrium shape that optimizes the soft properties of minimal torsion and minimal geodesic curvature.

Additional external and internal constraints may be applied to control the kinetic movement. We divide these constraints into three groups:

- **Supports (external):** For all samples A-G the model was held in a symmetry plane to avoid global rotations or translations.
- **Restraints (internal):** Based on extensive physical testing additional restraints were added, which ensure a controlled movement and avoid distortion of the grid. E.g. for samples A and B, the rotation of specific intersection points was blocked to prevent in-plane shearing of the quadrilateral network.
- **Displacements:** To fully explore the complete kinetic movement of each sample, displacements are used to pull or push the structure beyond its equilibrium shape. These loads mimic either the external hand gestures or internal actuation systems used during physical experimentation.

This method is flexible and allows to carry out and adjust *kinetic simulation* on the fly to closely match the physical behaviour of networks.

Analysis. The kinetic behaviour is recorded in five stages. We can derive valuable geometric information from this sequence: Measuring the geodesic curvature k_g and geodesic torsion τ_g give insight on where the highest strain occurs and how they are distributed over the grid. The internal strain energy of the grid structure $\Pi_{i(t)}$ can be approximated using the total sum of curvature-squared (CS) and the beams stiffness parameters:

$$\Pi_{i(t)} = \frac{GI_T}{2} \int_c \tau_{g(t)}^2 ds + \frac{EI_y}{2} \int_c \kappa_{g(t)}^2 ds \tag{1}$$

The progression of the total CS sum throughout the transformation offers to predict the equilibrium state[1] and design the kinetic behaviour by adjusting the profile stiffness parameters. To measure these curvature values, the discrete model is transferred back

[1] This is valid, if the deformability of the grid is highly constrained, and thereby stiffness parameters do not affect the structure's transformation path.

to a smooth NURBS geometry, by interpolating the intersection points pt_i to recreate curves and surfaces with sufficient accuracy. We use the Plugin Bowerbird to analyse each "curve on surface", measure k_g and τ_g, create their total CS sums and map the results over five stages. The *CS-graphs* (Fig. 7, Fig. 8, Fig. 9) show the interpolated curve of the five CS results for geodesic torsion (in red) and geodesic curvature (in blue). Marked with a green line is the natural equilibrium state observed in the physical models. This state can theoretically be moved to any state within the green area, by adjusting the proportion of torsional to bending stiffness in the profile.

3 Morphology

We designed 7 open and closed asymptotic networks, to systematically investigate their kinetic behaviour (Fig. 7, Fig. 8 and Fig. 9). The data reveals fundamental relationships of the topology of networks with the kinetic boundaries, necessary constraints, equilibrium states and actuation methods.

Kinetic Boundaries. All open networks (A, B, and E) exhibit a boundary in the form of planarity. The planar state acts as symmetry plane from which an inverse kinetic behaviour is possible. The CS-graph can be mirrored at this state. Rotational grids (C and D) may also display a state of planarity where torsion is zero (C), and geodesic curvature becomes maximal. This creates a snap through effect (indicated with a green circle in the CS-graph). Any cylindrical network (C, D, F, G) will further exhibit a self-collision when the cylinders are closed and the grid approaches the cylinder axis. Other self-collisions were preceded by the resistance of the material.

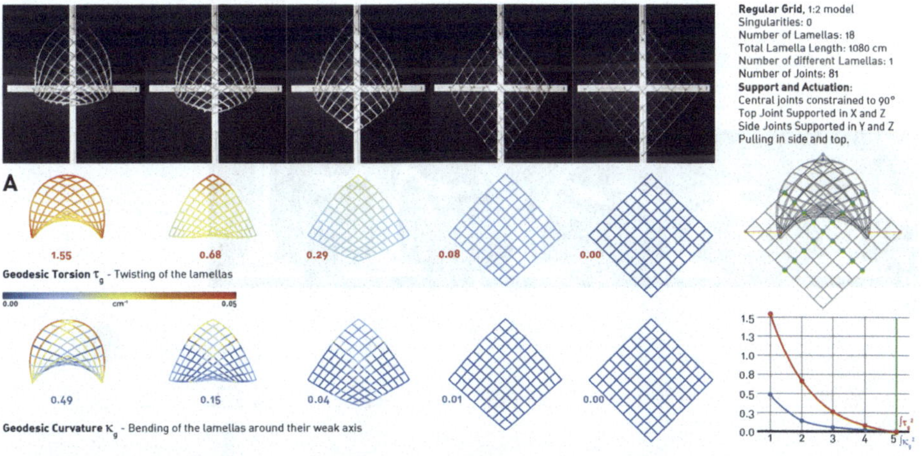

Fig. 7. Kinetic analysis for regular grid A. The grid is constrained by rigid joints (green dots). The outer corners are pulled inwards (orange lines) to actuate the double-curved shape. The planar state is the natural equilibrium (green line in CS-graph) where both geodesic torsion and curvature are zero. The graph can be mirrored here.

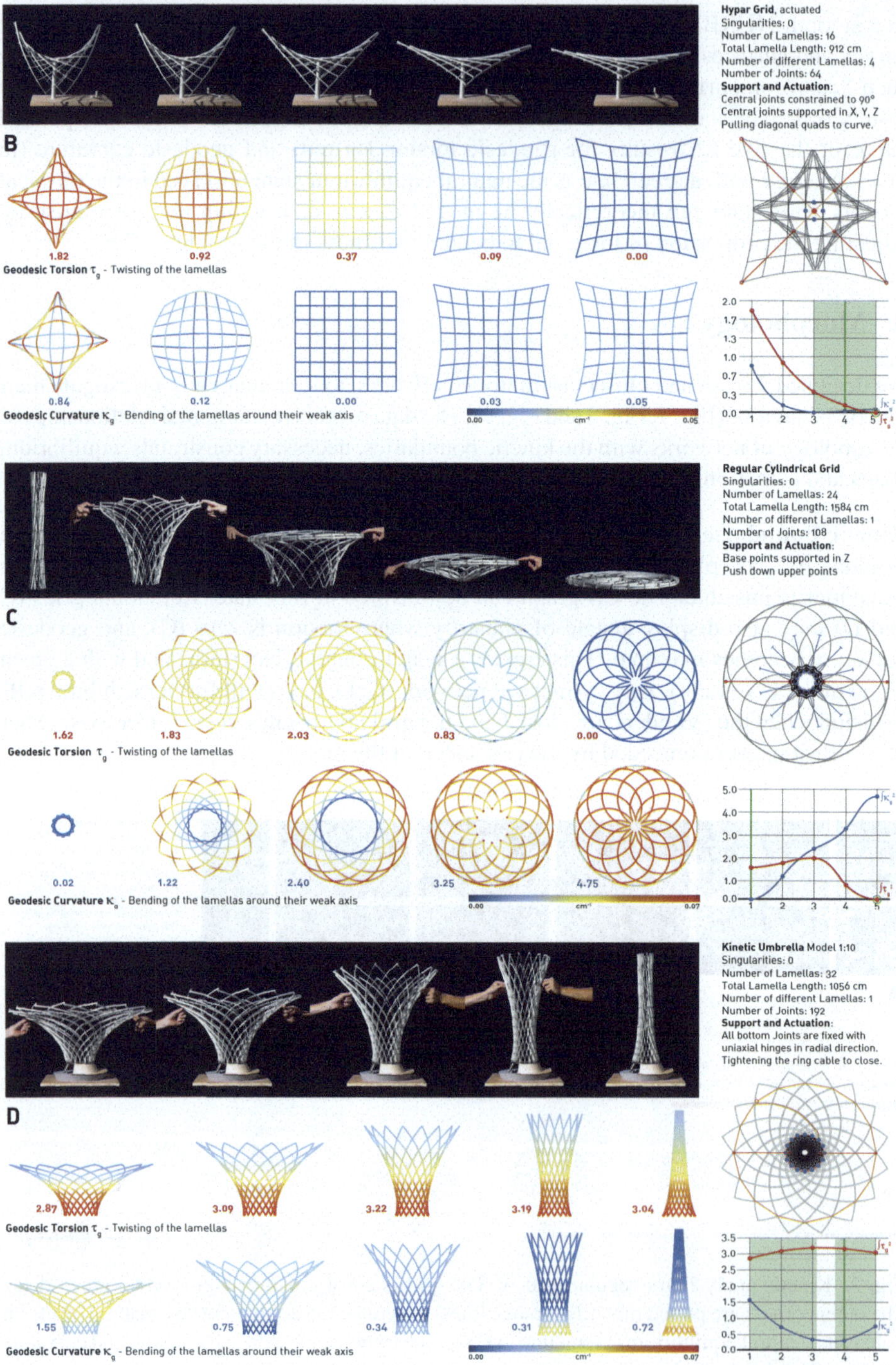

Fig. 8. Kinetic analysis showing the physical and digital models, constraints and CS-graph.

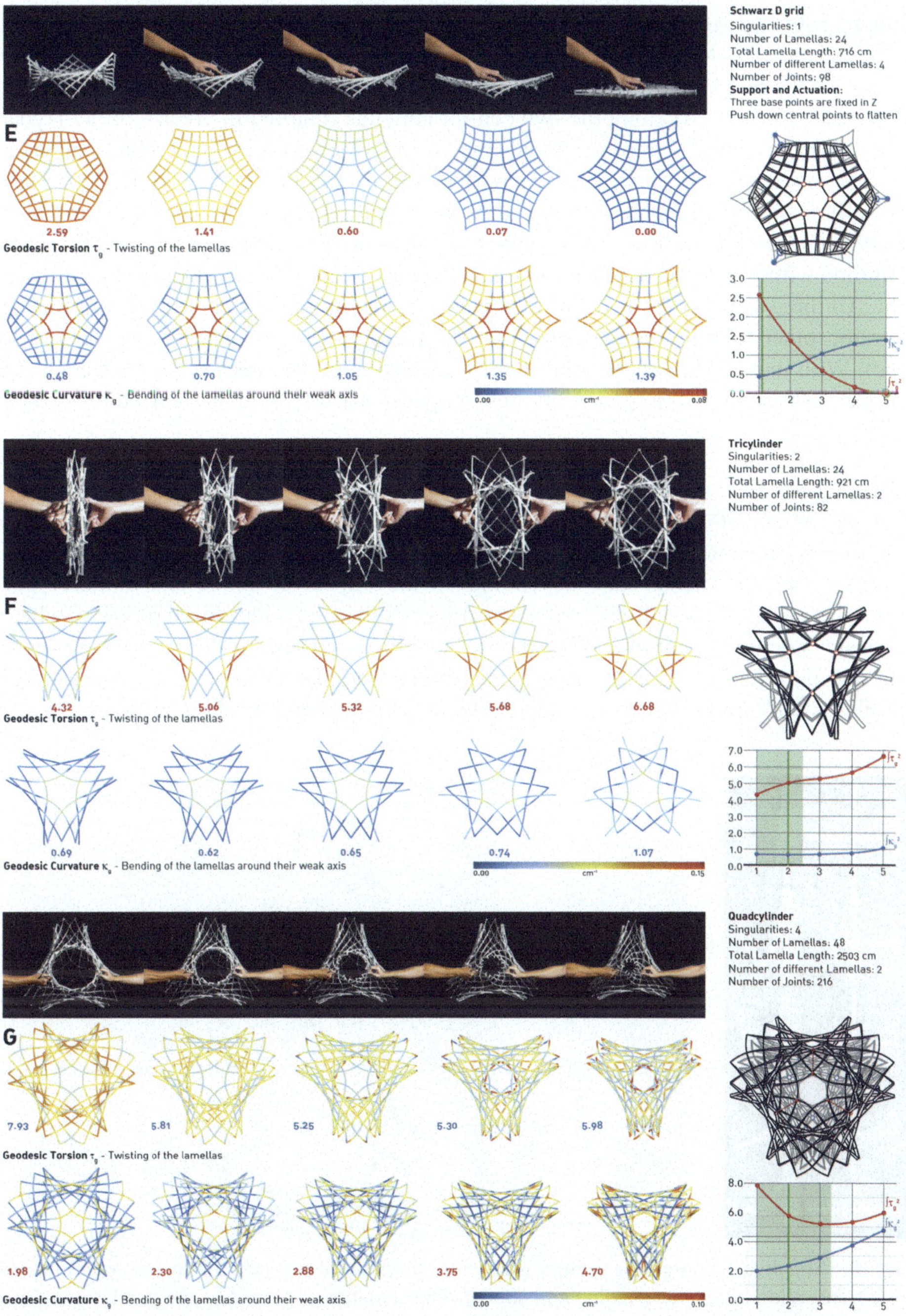

Fig. 9. Kinetic analysis showing the physical and digital models, constraints and CS-graph.

Constraints and Actuation. Open grids without singularities are initially unstable and require external and internal constraints to avoid a kinematic scissor/shear movement. In the case of the regular grid, the controlled kinetic transformation was achieved by adding rigid joints (see Fig. 4) along the two central lamellas (marked in green). In the Hypar model (B), the central square was fixed (Fig. 5, B) to prevent shearing. Cylindrical grids are less susceptible to shearing and tend to keep a rotational symmetry (C, D). They naturally shorten and increase their radius when pulled apart (C). The transformation can be controlled by adding pinned supports at one end (D). Singularities (E, F, G) act as restraints within a grid (like rigid joints) and create a controlled symmetric transformation. All three samples (E, F, G) were actuated by hand (marked in red) without further restraint, other than the natural supports (the table they were standing on, marked in blue). With increasing singularities, their transformation becomes more restricted. All samples presented in this paper, are design symmetrically, which contributes to their controlled movement.

Equilibrium. The natural equilibrium state (in which the energy is minimal) is dependent on the specific stiffness parameters of lamellas. For the regular open grid (A) planarity is the state of this equilibrium. For irregular networks (B), this equilibrium is offset from the planar state depending on the impact of geodesic curvature. The minimum-energy state of rotational networks tends to be near the closed position, where geodesic curvature is minimal. Singularities inevitably embed geodesic curvature in the grid and introduce reciprocity of bending and torsion. This increases the spectrum of possible minimum-energy states, depending on torsional and bending stiffness.

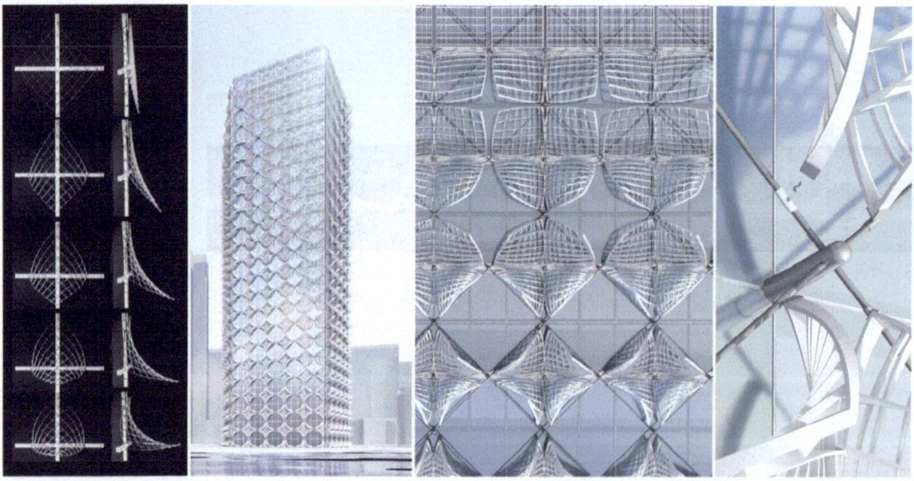

Fig. 10. Adaptive shading. The regular grid, system A, was developed for an integrated façade system that allows dynamic shading adjustment. The actuation is guided by diagonal rails.

4 Architectural Applications

The experiments are directly applicable to architectural practice. System A was further developed into an **adaptive shading system** (Fig. 10) to be integrated into existing building facades. The GRP lamellas are actuated by shifting the two diagonal corners inward transforming the regular grid into a doubly-curved canopy. The mechanism can be locked in any position to adjust to the sun angle.

Fig. 11. The Kinetic Umbrella was completed in August 2021. The 6 m tall structure is actuated via a winch at the base and transforms from a cylinder into a blossom-like canopy of 8 m diameter.

The **Kinetic Umbrella** (Schikore et al. 2020) was completed in August 2021 at the at the “Kreativquartier” in Munich. The rotational asymptotic grid is based on system D. It was covered with an adaptive system of ribbons to create shading. The structure performs a reversible semi-compliant transformation from a slender cylinder of 6 m height (closed) to a funnel shape of 8 m diameter (open). Figure 11 shows the structure and shadow effect in five stages of transformation. In contrast to the scaled model D, this large-scale application naturally tends to open due to gravity, and can thus be actuated reversely by pulling the a ring cable and closing the cylindrical grid. The kinetic structure (Fig. 12) consists of two layers of 16 8 × 80 mm GRP slats, connected with aluminium joints and steel bolts (a). A textile cover of 10 mm ribbons provides shadow and adapts to the grid structure’s transformation (a, b, c). Three circular fixed cables (yellow) are locking the transformation in the open state (b, d). The transformation is actuated by an additional, circular cable (red), leading down to a winch (f). The lamellas are connected via uniaxial hinges (e) to an octagon steel base that sits on concrete foundation bodies (d, f). Timber panels, covering the concrete and steel base, accommodate circular seating (f).

Fig. 12. Components and details of the Kinetic Umbrella. a: The lamellas are joined with aluminium profiles on two levels. b/d: Three ring cables (yellow) secure the opened structure. c/d: A fourth ring cable (red) actuates the closing movement. e: Uniaxial hinges constraint the base but allow a controlled movement. f: The red cable is pulled by a winch that is attached to the concrete base. Timber panels provide seating.

5 Conclusion

The hybrid approach of physical and digital experiments offers a broad investigation of complex kinetic systems while ensuring a realistic evaluation and application of results. The digital method proposed is fast and flexible, and accurately simulates the kinetic morphology of complex grids. The analysis is purely geometrical, as no actual material properties or profile dimensions are given. Nonetheless, measuring the geodesic curvature and geodesic torsion allows the prediction of residual bending and torsional stresses. Mapping the total curvature-squared gives insights into the likely minimal-energy states of the grid. This method is applied to seven asymptotic networks, and systematically investigates the impact of topology, singularities and support conditions on the kinetic behaviour. The data reveals strategies to constrain the transformation of asymptotic grids. Open grids without singularities are most flexible and require additional constraints. Cylindrical grids offer a more directed movement and can be controlled through pinned supports. Singularities act as internal constraints and create symmetric kinetic transformations. However, with an increasing number of singularities and cylinders, the movement becomes more restricted.

Two prototypical kinetic designs, an adaptive shading facade and a kinetic umbrella structure show possible architectural applications, including actuation system and detailed constructive solutions. The use of slender lamellas offers elastic transformation around the weak profile-axis while maintaining high structural stiffness of the strong axis. For large-scale application, the effect of gravity influences the minimum-energy state, favouring geometries with less elevated mass.

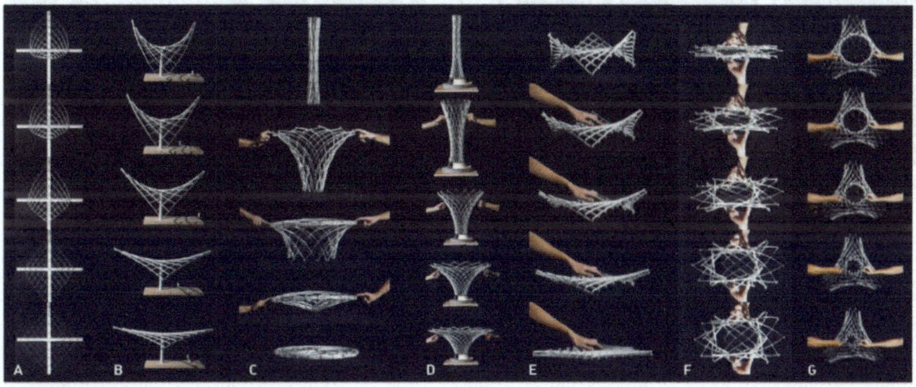

All actuated models in five phases.

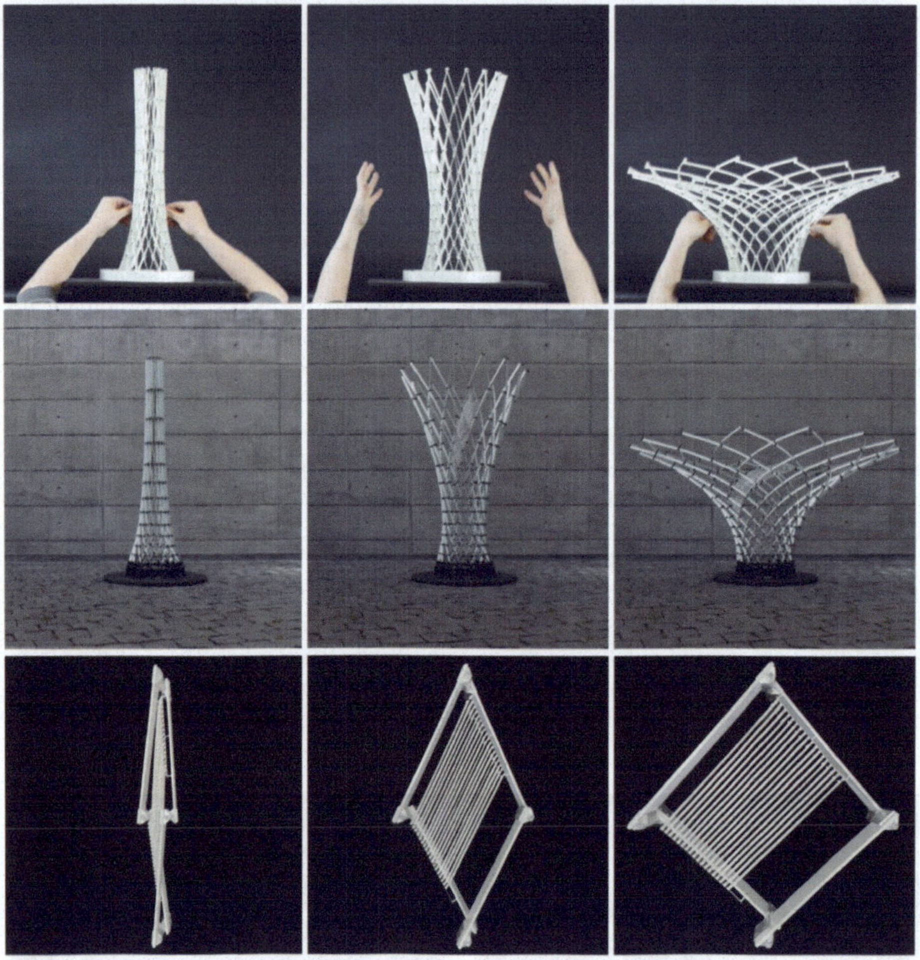

Physical models of the Kinetic Umbrella. Top, 1:10, Middle 1:3, Bottom 1:1

Close-up view of umbrella model

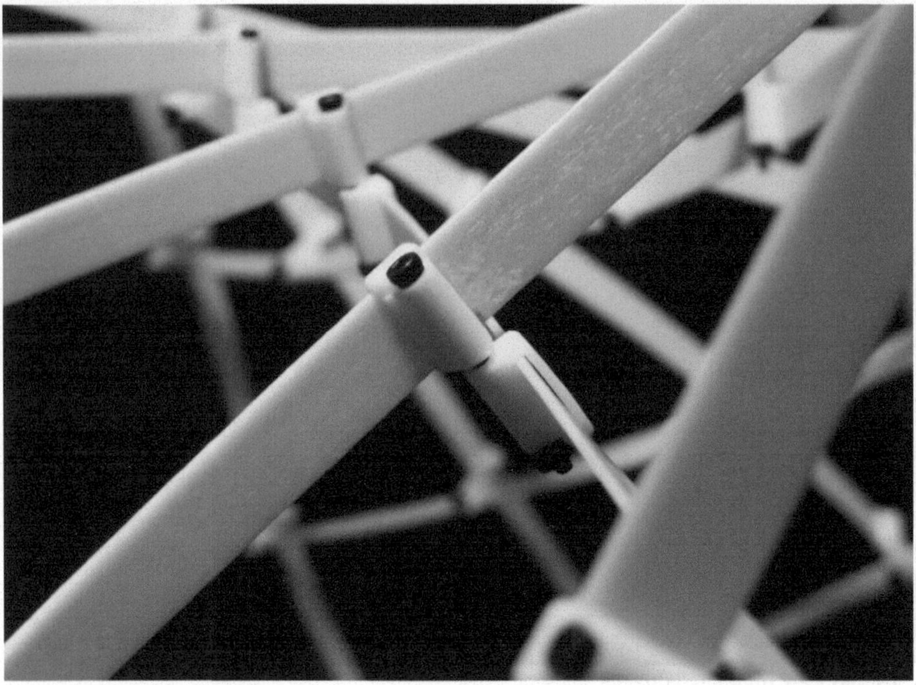

Close up of grid and joints

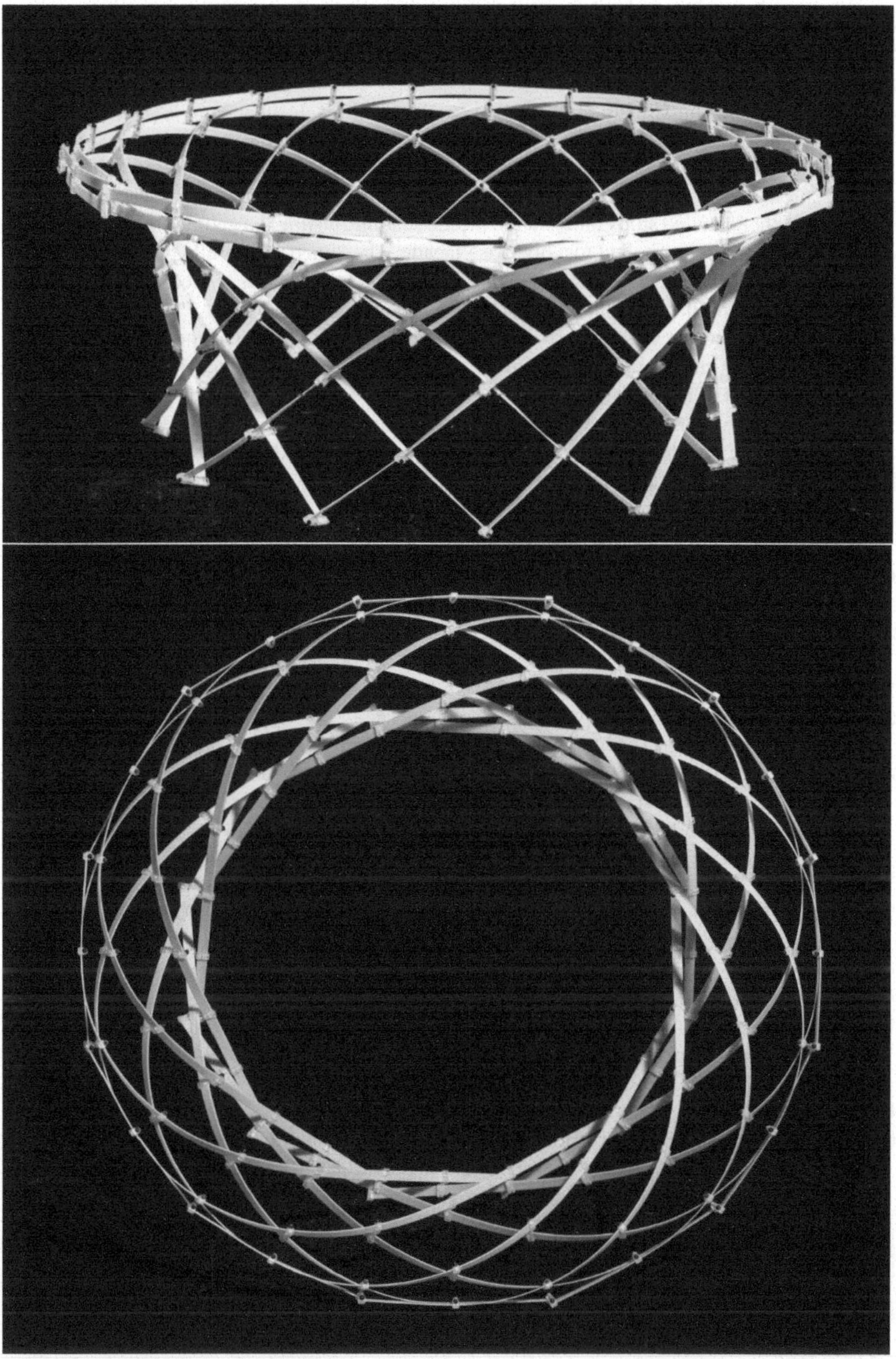

Alternative stable state of regular rotational grid.

Acknowledgements. The morphological research was funded by the HKU DoA Seed Fund, Repetitive structures. We thank the research team at HKU for their tireless work throughout the summer of 2021 to build, test and improve seven complex kinetic lamella models (Jan Yip Choy, Ka Mak Fan, Hao Feng Chuah, Hing Fung Li). We thank Zongshuai Wan and Jacky Chu for their support in construction and design and setup of the interactive exhibition. We are grateful to Jason Ji Xiang for his professional photography of the models, and to Alyssa Williams and Yongki Sunarta for filming the kinetic behaviour and workflow throughout the research. We thank Subin Park for her detailed study on dynamic façade shading, testing the geometric boundaries and actuation necessary to be applied to a high-rise façade. The Kinetic Umbrella project was funded the Dr. Marschall Stiftung and the STO Foundation. The Umbrella was designed by Jonas Schikore in collaboration with Eike Schling, Pierluigi D'Accunto, Muye Ma, Clemens Lindner and Tao Sun. Material and construction was supported by our industry partners Metal Manufacturer BRANDL from Eitensheim, Aluminium manufacturer FACTUREE from Berlin, and Ropes and Ribbons manufacturer GEPOTEX from Emskirchen. We thank everyone involved in the prefabrication assembly erection and maintenance of the Kinetic Umbrella (Frauke Wilken, Tao Sun, Clemens Lindner, Frederic Chovghi, Fabian Matella, Maria Rau, Sarah Sendzek, Sebastian Hoyer, Merlin Bieling, Sanziana Maximeasa, Sebastian Dietrich).

References

Bauer, A.M., Breitenberger, M., Philipp, B., Wüchner, R., Bletzinger, K.-U.: Nonlinear isogeometric spatial Bernoulli beam. Comput. Methods Appl. Mech. Eng. **303**, 101–127 (2016). https://doi.org/10.1016/j.cma.2015.12.027

Bauer, A.M., Längst, P.: Kiwi3d! Meshfree, Isogeometric FE Analysis integrated in CAD. Chair of Structural Analysis. TU Munich (2019). https://www.kiwi3d.com/. Accessed 27 May 2020

Bauer, A.M.: CAD-integrated isogeometric analysis and design of lightweight structures. München: Lehrstuhl für Statik, Technische Universität München (Schriftenreihe des Lehrstuhls für Statik TU München, Band 43) (2020)

Cottrell, J.A., Hughes, T.J., Bazilevs, Y.: Isogeometric Analysis. Toward Integration of CAD and FEA. Wiley, Chichester, Hoboken (2009)

Finsterwalder, S.: Mechanische Beziehungen bei der Flächen-Deformation. GDZPPN00211626X. In: Deutsche Mathematiker-Vereinigung (ed.) Jahresbericht der Deutschen Mathematiker-Vereinigung, vol. 6, pp. 43–90. Teubner (6), Göttingen (1897). https://www.digizeitschriften.de/dms/img/?PID=GDZPPN00211626X. Accessed 13 Feb 2019

Haskell, C., Montagne, N., Douthe, C., Baverel, O., Fivet, C.: Generation of elastic geodesic gridshells with anisotropic cross sections. Int. J. Space Struct. **36**(4), 294–306 (2021). https://doi.org/10.1177/09560599211064099

Howell, L.L.: Compliant Mechanisms. Wiley, New York (2002)

Isvoranu, F., Panetta, J., Chen, T., Bouleau, E., Pauly, M.: X-shell pavilion. a deployable elastic rod structure. In: Carlos Lázaro, K.-U., Bletzinger, E.O. (eds.) FORM and FORCE 2019, pp. 606–613. International Centre for Numerical Methods in Engineering (CIMNE), Barcelona (2019)

Oberbichler, T.: Bowerbird. Blug-In for Grasshopper, Munich (2019). https://github.com/oberbichler/Bowerbird. Accessed 19 Nov 2020

Oberbichler, T., Bauer, A.M., Goldbach, A.-K., Wüchner, R., Bletzinger, K.-U.: CAD-integrierte Analyse im Entwurfsprozess. Bautechnik **96**(5), 400–408 (2019). https://doi.org/10.1002/bate.201800105

Piker, D.: Kangaroo. Live Physics Enging (2015). https://www.grasshopper3d.com/group/kangaroo. Accessed 31 July 2018

Pottmann, H., Asperl, A., Hofer, M., Kilian, A: Architectural Geometry. Bentley Institute Press, Exton (2007)

Schikore, J., Schling, E., Oberbichler, T., Bauer, A.M.: Kinetics and design of semi-compliant grid mechanisms. In Baverel, O., Pottmann, H., Mueller, C., Tachi, T. (eds.) Advances in Architectural Geometry 2020, Paris, pp. 108–129 (2020). https://thinkshell.fr/wp-content/uploads/2019/10/AAG2020_06_Schikore.pdf

Schling, E., Barthel, R.: Repetitive structures. In: Gengnagel, C., Baverel, O., Burry, J., Ramsgaard Thomsen, M., Weinzierl, S. (eds.) DMSB 2019, pp. 360–375. Springer, Cham (2020). https://doi.org/10.1007/978-3-030-29829-6_29

Schling, E., Hitrec, D., Barthel, R.: Designing grid structures using asymptotic curve networks. In: de Rycke, K., et al. (eds.) Design Modelling Symposium Paris 2017. Humanizing Digital Reality, pp. 125–140. Springer, Singapore (2017). https://doi.org/10.1007/978-981-10-6611-5_12

Schling, E., Kilian, M., Wang, H., Schikore, J., Pottmann, H.: Design and construction of curved support structures with repetitive parameters. In: Hesselgren, L., Olsson, K.-G., Kilian, A., Malek, S., Sorkine-Hornung, O., Williams, C. (eds.) Advances in Architectural Geometry 2018, 1st edn., pp. 140–165. Klein Publishing, Wien (2018)

Schling, E., Schikore, J.: Kinetic grid structures. exhibition at PMQ, 14 September–5 October 2021. With assistance of Jan Yip Choy, Ka Mak Fan, Hao Feng Chuah, Hing Fung Li, Zongshuai Wan, Jacky Chu, Muye Ma, HKU, HKU Architecture Gallery at PMQ, Hong Kong (2021). https://eikeschling.com/2021/09/02/pmq-exhibition-kinetic-grid-structures/

Schling, E., Wan, Z.: A geometry-based design approach and structural behaviour for an asymptotic curtain wall system. J. Build. Eng. **52**, 104432 (2022). https://doi.org/10.1016/j.jobe.2022.104432

Soriano, E., Ramon, S., Boixader, D.: G-shells. Flat collapsible geodesic mechanisms for gridshells. In: Carlos Lázaro, K.-U., Bletzinger, E.O. (eds.) FORM and FORCE 2019, pp. 1894–1901. International Centre for Numerical Methods in Engineering (CIMNE), Barcelona (2019)

Tang, C., Kilian, M., Bo, P., Wallner, J., Pottmann, H.: Analysis and design of curved support structures. In: Adriaenssens, S., Gramazio, F., Kohler, M., Menges, A., Pauly, M. (eds.) Advances in Architectural Geometry 2016. ETH. 1. Auflage, pp. 8–23. vdf Hochschulverlag, Zürich (2016). http://www.dmg.tuwien.ac.at/geom/ig/publications/2016/curvedsupport2016/curvedsupport.pdf

FibreCast Demonstrator

New Ways of Connecting Timber Elements with Fixed and Detachable Interlocking Connections Made of Reinforced Polymer Concrete

Michel Schmeck[1](✉), Leon Immenga[2], Christoph Gengnagel[1], and Volker Schmid[2]

[1] Konstruktives Entwerfen und Tragwerksplanung (KET), UdK Berlin, Hardenbergstraße 33, 10623 Berlin, Germany
schmeck@udk-berlin.de

[2] Entwerfen und Konstruieren – Verbundstrukturen, TU Berlin, Berlin, Germany

Abstract. Efforts toward the circular economy necessitate new approaches for the construction industry. Repair and repurposing of structures through easily disassembled connections facilitate maintenance, preservation and concept changes. Decreasing availability of resources requires the efficient, material-conserving use of the materials employed.

The decisive factor for the material consumption of timber construction elements are mostly the loads in the nodal area. This is the deciding factor for the required cross-section. As a result, wooden structures are oversized in wide areas in order to withstand the loads that occur in the connection. Inspired by approaches to building in existing structures with polymer concrete, a cast connection is created that can be cast directly against the timber to be joined. The resulting filigree and partially detachable connections are reinforced with different reinforcement concepts, thus increasing the load-bearing capacity to above the level of the timber to be connected.

In a pavilion, the performance of the novel connections will be demonstrated for the first time on an architectural scale, showcasing the application potential of the connection method.

Keywords: Timber-connections · Timber-joinery · Polymer-concrete · Pavilion · Deployable structure · Topology optimization · Additive manufacturing · Rapid prototyping

C. Gengnagel et al. (Eds.): DMS 2022, *Towards Radical Regeneration*, pp. 394–410, 2023.
https://doi.org/10.1007/978-3-031-13249-0_32

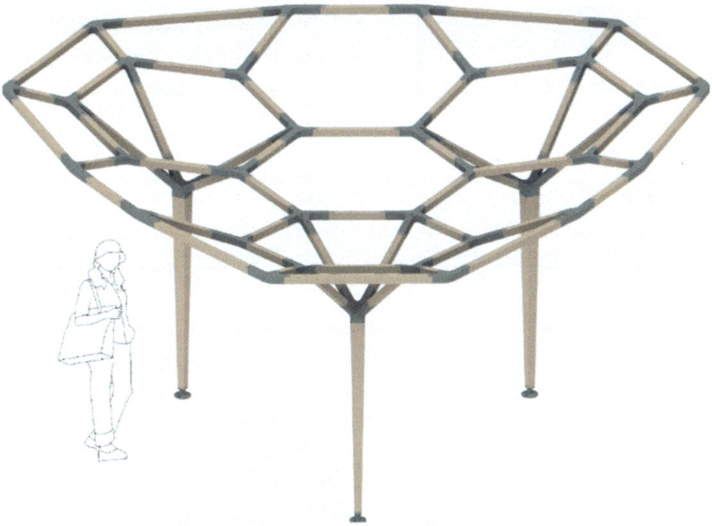

1 Introduction

Light lattice girders or slim lattice shells made of hollow or solid steel profiles are used for the majority of filigree facade and roof constructions. The use of these structural elements made of metallic materials has various disadvantages: a relatively high use of energy in production, a high heat transfer coefficient and thus the risk of thermal bridges as well as the need for sometimes expensive corrosion protection measures. To avoid these disadvantages, components made of fiber-based materials are used, such as full cross-sections and profiles made of timber and timber-based materials. However, their extensive use in construction has so far been limited by the lack of a suitable connection technology.

Usually, in the building industry linear load-bearing elements made of timber are connected with the help of metal, like pin-shaped fasteners (screws, nails, dowels). The force is thus transmitted in discrete areas of the support elements that locally lead to high stress concentrations. As a result, the dimensions of the structures are often defined by the connection technology and thus lead to oversizing of the beam's cross-sections outside the connection areas. Efficient, glued connections, on the other hand, cannot be produced on site [7, 8] Another disadvantage in terms of sustainability is their irreversibility [9].

In order to promote an economically and ecologically sensible use of rod-shaped structural elements made of natural and artificial fiber composite materials, a new type of connection technology was developed. The use of (fiber) reinforced and mineral-filled polymer cast parts represents a completely new approach. The basic ideas here are the new possibilities for effective use of material- and form-fit connections as well as a high degree of freedom in terms of their shape. For example, with a toothed casting geometry, a uniform transmission of force between the parts to be joined is possible. At the same time, the detachability of the connection is guaranteed [2, 5] (Fig. 1).

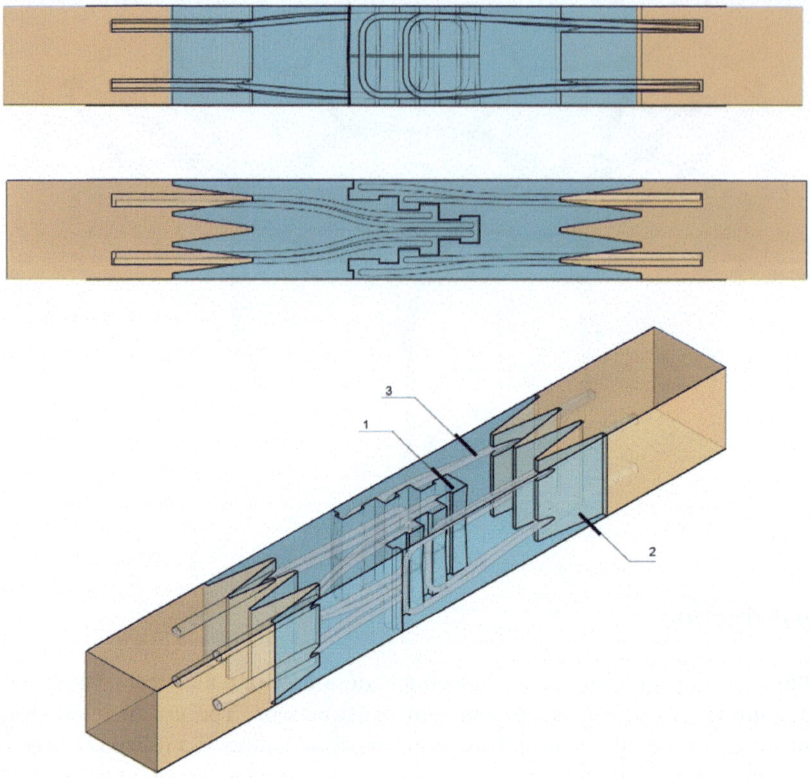

Fig. 1. Design principle: timber beams connected by a cast-node: 1. Form-fit connection 2. Finger joint connection provides the main force transfer between the wood and the polymer concrete 3. Internal reinforcement, anchored in the timber-beam

2 Design

Within the project a two form-fit and a robust, permanent connection type are developed and investigated. The form-fit connections will be presented in detail. Mechanical characteristics, which form the basis of the node design, were determined in a series of tensile, compression and bending tests, whose results were published earlier [3, 4, 6]. The design goal is to increase the load-bearing capacity of the connection to the level of undisturbed timber C24 ($f_{m,mean} = 37{,}1$ MPa). The dimensions of all profiles are square 50/50 mm. Design, force-flow and reinforcement are the key factors in achieving this goal.

2.1 Design of the Joints

Form-fit connections are widely used in steel-composite constructions to transfer forces between different structural components, such as bridge deck and tension cables [1]. Each joint consists of two interlocking parts. Male and the female part slide into each other and can easily be disassembled. The forces are transferred to the timber rods via a reinforced finger-joint [3, 5].

2.2 The 3-Tooth-Connection with Fibre Reinforcement

The tooth geometry was developed based on the ratios of the bearable shear and pressure stresses of the polymer concrete. According to the test results the compression-to-shear ratio is 5:1. The tooth geometry has to follow the same ratio, resulting in a 2 mm wide and 10 mm long tooth. The force is therefore applied over the tooth width of 2 mm by compressive stresses and transmitted over the tooth length of 10 mm by shear stresses in the cross-section. In order to equalize the stress distribution along the row of teeth, the central tooth is 50% higher and therefore more rigid, so that force inserted by each tooth becomes the same along each row of teeth. In the final version the geometry is realized as a hybrid of half male and half female (Fig. 2).

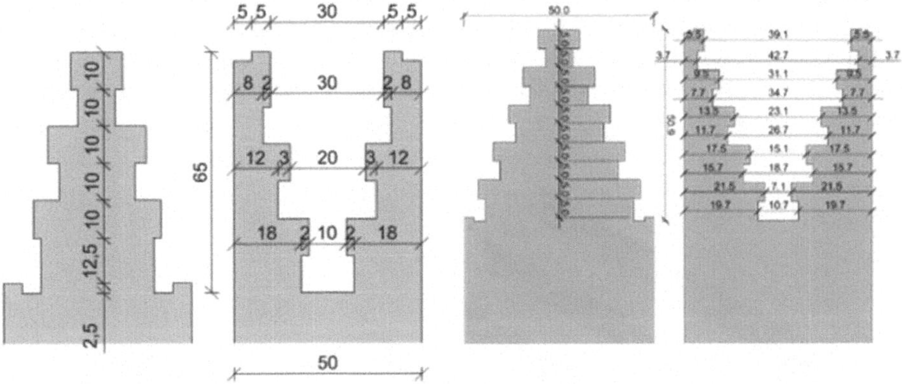

Fig. 2. Geometry of the 3-tooth-design (left); Geometry of the 5-tooth design

In order to increase the load bearing capacity of the connection different strategies of (fibre-) reinforcement were investigated and evaluated for performance and usability. The geometry of the reinforcement is based on FEA-force-flow analysis and on the failure mechanisms of the mechanical tests. While force-flow analysis is an effective tool to understand the internal reaction of the connection, the post-crack behavior can be observed in detail in time-laps recordings of the physical tests. However the combination of these two methods and a number of iterations led to a comprehensive observation of the occurring mechanisms before and after concrete cracks. The resulting novel micro carbon-fibre-reinforcement proved to be very effective to stabilize the geometry even after severe cracks occurred.

2.3 The 5-Tooth-Connection with 3D-Printed Metal Reinforcement

The 5 Tooth-geometry is designed for metal reinforcement. The shear to compression ratio is no longer the driving design. Steel as an isotropic material is capable of tension, compression and bending. The load is still inserted via the 2 mm teeth, but instead of shear, the mechanics of bracket-reinforcement can be applied. This was used to shrink the tooth size by 50% while increasing the number of teeth from 3 to 5 teeth.

3D-Printing Metal-Reinforcement

The reinforcement was printed in a cooperation with Fraunhofer IPK Berlin in a SLM process. Using the SLM (selective laser melting) process allows the implementation of geometries that would not be possible with subtractive or formative manufacturing processes. Due to the high temperatures of the melting process thermal deformation occurs. In order to limit this, support structures are used over the whole area in order to quickly dissipate the heat from the melting process into the base plate. The printed parts are 'welded' to the base plate. This supporting structure has to be laboriously removed in post-processing.

Optimizing the Metal Geometry

In the beginning a choice had to be made between a real 3d-analysis and a multi-layer 3d analysis. The advantage of the 3d systems provided by ANSYS or Autodesk's Fusion 360 allow for quick results but are limited to just one material, thus ignoring the load bearing capacity of the polymer-concrete. The decision was made in favour of a customized 2d-approach (Fig. 3).

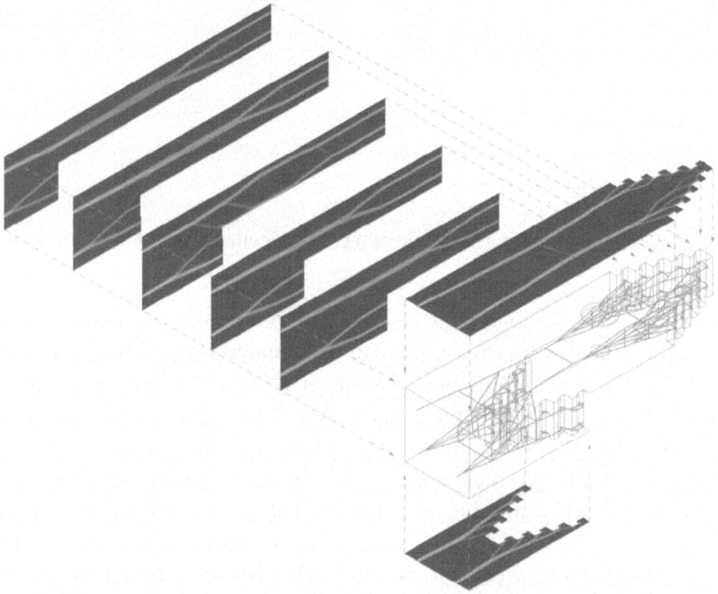

Fig. 3. In a hybrid process the results of multi-layer 2d topology optimisation, analytic truss models and production-constraints are balanced to create a satisfying solution for the 3d printed steel reinforcement.

To optimise the topology of the reinforcement a multi-layer 2d-analysis was conducted in order to better understand and control the results in comparison to a 3d approach. The setup was built upon Rhino and Grasshopper, enabling a customized workflow. The results of the force-flow analysis are interpreted, combined and if necessary adopted to meet the structural requirements, as well as the requirements brought in by the SLM process. Ideally the metal reinforcement would only transfer tension forces, but local effects (micro brackets) and the necessary anchoring length of the reinforcing elements lead to compression and bending forces as well. The resulting 3d framework can be transferred directly to an FEA environment. In the following analysis, concrete struts can be added to the model as coupling elements during the calculation and a spatial truss can be generated and dimensioned without mayor bending distortions (Fig. 4).

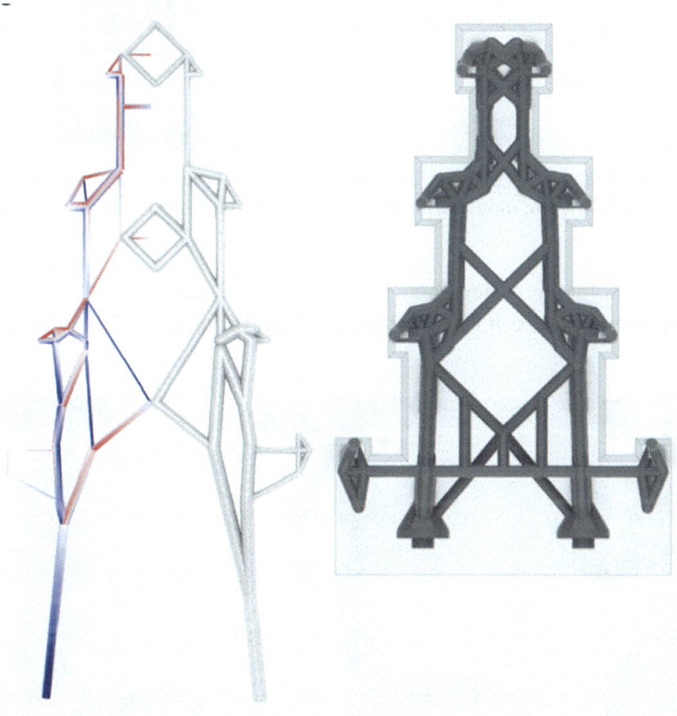

Fig. 4. 3D-FE analysis and generation of cross-sections in an early design stage with lots of bending distortions (left), final iteration optimised for structural and process requirements (right)

2.4 Fixed Nodes

Combining multiple timber beams in one node raised the necessity of a fixed node, capable of the of the forces occurring in a grid structure. The node needs to be compatible with both metal- and fibre-reinforced tooth-connectors. The omission of geometric details increases its robustness and thus extends its potential for use in a range of practical construction applications. Dimensioning of the reinforcement can be conducted with ordinary hand calculation methods The simplicity of the node is also reflected by the design of the reinforcement, which consists of only one roving of impregnated carbon fibres, spiraling upwards and thus creating 4 layers of reinforcement (Fig. 5).

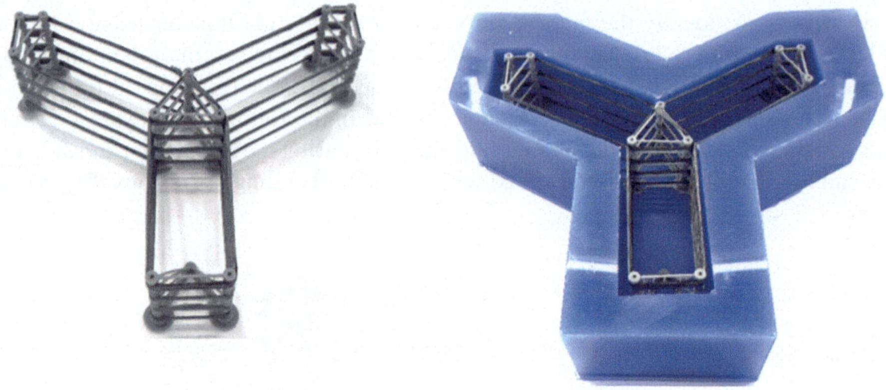

Fig. 5. Fixed node design for 3 beams (left) and its 4-layer fibre reinforcement consisting of just 1 carbon roving (left) Silicone mould with fibre reinforcement prior to the casting process (right).

The Y-node geometry is designed for the physical testing of the nodes structural capacity and can be adopted for more than 3 rods to be connected (Fig. 6).

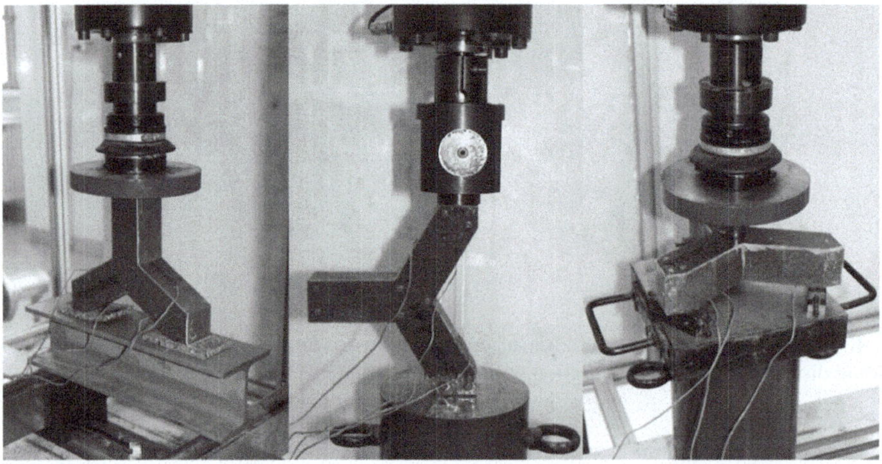

Fig. 6. Y-node in physical testing (left to right): positive bending moment, negative bending moment, in-plane bending moment

3 Production

Applying means of rapid prototyping allows for processes enabling high precision, a minimum degree of robustness while still integrating a certain amount of manual labor to compensate for exceptions in early design stages. The developed processes for the concrete casting and the production steel and fibre reinforcement are described in the following.

3.1 Production of Moulds

Silicones for mould production are available with different mechanical properties. While soft silicones are easy to unform, versions with a higher hardness are more stable and allow repeatedly and precise casting results. The master-model, based on the CAD geometry is produced with a stereolithography SLA 3D printer. The cured resin is compatible with silicone; partially described crosslinking disturbances on the surface of the UV-resin based master forms were not observed. An addition-curing silicone rubber (platinum silicone) with a Shore hardness of 45 ShA is used; a release agent is not required to produce the mould. The moulds will then be reused for small batches of casting polymer-concrete up to 15 times. As a release agent, Vaseline dissolved in petroleum-ether has proven to be very effective and easy to process (Fig. 7).

Fig. 7. CAD geometry (left), 3d printed master piece (centre), extraction of the masterpiece from the silicone mould (right)

3.2 Production of Steel Reinforcement

The metal reinforcement is made from tool steel in the SLM process. The metal reinforcement is manufactured from tool steel type 1.2709 with a (theoretical) tensile strength of up to 2200 MPa (depending on elements diameter and surface defects) using the SLM process. The actual tensile strength could be tested to 1150 MPA for 1.5 mm square profile and 1450 MPa for a round 3 mm cross-section. The tensile specimens were printed together with the first reinforcement specimen so that their results could only be taken into account in later designs.

Before the metal reinforcement is produced, the support structures are defined in preprocessing. Ideally, functional surfaces are omitted in order to reduce the effort required for postprocessing. The minimum angle of branching elements must not exceed 45° in the build-up direction.

Before production, conflicts are identified, for example, the minimum angle of branching elements must not exceed 45° in the build-up direction. When defining the support structures, functional surfaces are ideally omitted to reduce the effort required for post-machining. In post-processing, the components are separated from the build platform, washed in an ultrasonic bath, support structures (manually) removed, sand-blasted. Thermal annealing at 600 °C over 24 h reduces internal stresses and can double the tensile strength of the components. The final correction of any component protrusions then takes place before the reinforcement cages are installed in the formwork. Machining by hand can take up to 30–60 min per cage. The forces of the reinforcement in the interlocking area are transferred to four M6 threaded rods with a screw connection and these are then anchored in the timber profile via slotted holes (Figs. 8 and 9).

Fig. 8. Additive manufacturing: printing chamber, gas is ventilating through the circular openings from the right, to the left (left), printed parts on the building plate after cleaning (centre), metal saw, removing the parts from the base-plate

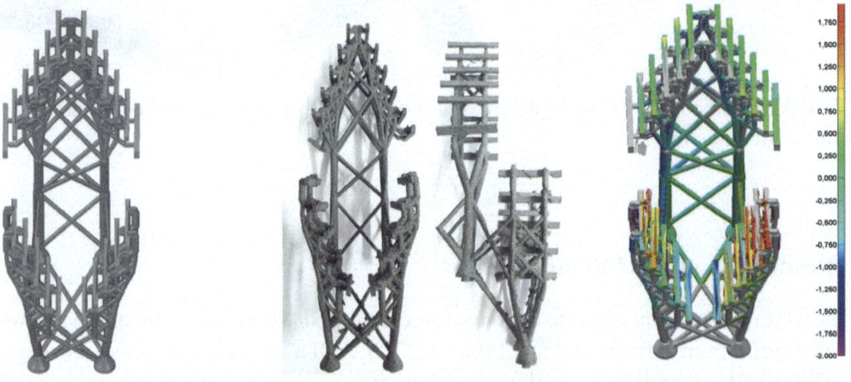

Fig. 9. Precision: CAD (left), top and side view of the printed piece (centre), 3d scan false colors display (right) showing deviation from the CAD-geometry up to −2/+1.75 mm

3.3 Production of Fibre Reinforcement

For fibre reinforcement, carbon roving of 800-3300tex (tensile strength 4000 MPa, Young's modulus 240 GPa) are soaked in a slow-curing epoxy resin matrix. A pot life of 300 min allows storage of soaked rolls in the freezer and use over a period of about 2 weeks. The fibres are wetted on one side with resin and then deflected several times over a set of 6 20 mm rolls in order to ensure complete impregnation of the fibres. The impregnated filament is wound onto a roll. The different reinforcement cages can then be wound without keeping minimum radii. The designed geometry is lasered onto a wooden plate and provided with deflection pins, which later serve as spacers in the formwork. The reinforcement is always designed so that a cage can be wound in one operation with a continuous fibre bundle (Fig. 10).

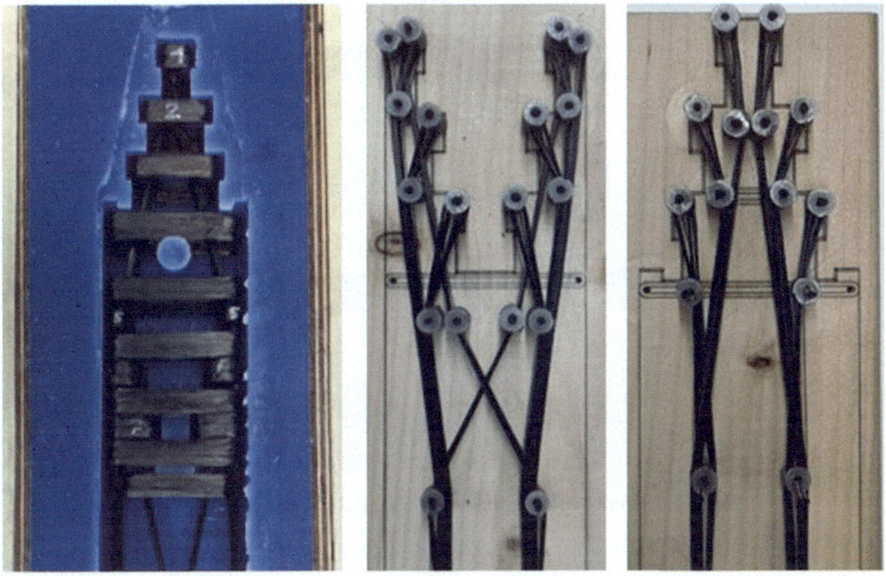

Fig. 10. Carbon roving based reinforcement for a non-detachable-shaped connection (left) and the male/female version of the 3-tooth connection (right). The rovings are impregnated with slow curing epoxy resin to enable longer processing time. The reinforcement cages are manufactured in a winding process on custom-made 3D printed winding pins, which also serve as spacers for the subsequent casting process.

4 Physical Testing

To evaluate the load-bearing and deformation behaviour of the connection node, 4-point bending tests were performed on the reinforced interlocking geometry. The polymer concrete test specimens were enhanced with 3D-printed reinforcement made of high-strength tool steel.

4.1 Test Setup

The tested beams have a cross-section of 50 mm × 50 mm and a total length of 950 mm. The span in the 4-point bending test was selected to be 900 mm in accordance with the requirements of DIN EN 408. The distance between the load application points of the testing machine used (Toni Technik Type 1225) is a maximum of 280 mm, so that the distance between the support and load application points is 310 mm on each side. The test setup for the 4-point bending tests is shown in Fig. 11.

Fig. 11. 4-pint-bending test setup (Toni Technik Type 1225)

4.2 Test Specimen 5-Tooth-Connection with Steel Reinforcement

The connection area is located in the middle between the load application points and thus free of shear force and under constant moment load. The connections are tested standing and lying (90° rotation around the longitudinal axis of the beam). The concrete age on the test day is 7 days (±1 day).

4.3 Test Procedure

In accordance with the specifications of DIN EN 12390-13, three preloading cycles are performed before the actual test until failure of the specimens. The loading is displacement-controlled with a loading speed of 0.60 mm/min in the servo-hydraulic testing machine. In addition to the testing machine's internal transducers (force and displacement), strain gauges with a gauge length of 20 mm are applied to measure the strains on the top and bottom of the polymer concrete beams.

The test program includes a total of 11 bending tests on the reinforced 5-tooth geometry. Two versions of the 5-tooth reinforcement were investigated, first the variant 5Z1.1 later, the adapted variant 5Z1.2

4.4 Test Results and Observations

As in classical reinforced concrete construction, distinct cracking can be observed before the actual failure of the test specimens, which is accompanied by a significant drop in the component stiffness. In all tests, the first crack occurs after passing through the three preloading cycles during the final loading ramp. The initial cracking moments are of a similar magnitude for all tests (0.439–0.572 kNm), with the largest initial cracking moments observed in the tests with standing teeth orientation. A clear correlation between initial cracking moment and bending axis or reinforcement variant cannot be established.

Fig. 12. 5-tooth connection after physical failure: full separation after failure of reinforcement, tested in vertical orientation (left); tested in horizontal orientation, numrous cracks, failure by teeth slipping on th elower, tensile side.

With the exception of one test specimen, all test specimens fail due to a bending tensile failure in the middle of the connection node. At this point, the concrete cross-section is already reduced by the bolt hole. The cause of the bending tensile failure is the cracking of the 3D-printed steel reinforcement, which branches in this area (stress jump), resulting in a complete separation (Fig. 12 left). Due to the brittle material behaviour of the high-strength tool steel, the failure occurs abruptly, without advance notice through large plastic deformations. A different failure mechanism can be observed with the specimen in horizontal orientation. Here, the failure occurs by shearing or slipping of the toothing on the lower row of teeth subjected to tensile stress (Fig. 12 right).

The breaking moments of the test specimens with vertical and horizontal tooth orientation are similar. This corresponds to the project objective of withstanding the full moment load in both bending axes.

The adjustment of the reinforcement cage after the tests of variant 5Z1.1 resulted in the desired effect: In the vertical and horizontal test specimen orientation, a bending load capacity of reinforcement variant 5Z1.2 can be measured that is clearly above the values measured for variant 5Z1.1. The increase in the flexural strength is about 24 percent for the horizontal and 41 percent for the vertical orientation.

The initial moments of fracture, the moments of failure and the performance compared to the referenced C24 timber are summarized in the following table (Table 1):

Table 1. Initial moments of fracture, the moments of failure and the performance compared to the referenced C24 timber

Geometry tested	Reinforcement version	Reinforcement material	Orientation	Number of specimen	mean cracking moment M_{cr} [kNm]	mean failure moment M_u [kNm]	performnce over reference C24 timber [%]
5-Tooth	5Z1.1	3D printed steel	y-axis (vertical)	3	0,490	0,825	106,8
5-Tooth	5Z1.1	3D printed steel	z-axis (horizontal)	3	0,478	0,931	120,4
5-Tooth	5Z1.2	3D printed steel	y-axis (vertical)	3	0,559	1,161	150,3
5-Tooth	5Z1.2	3D printed steel	z-axis (horizontal)	2	0,512	1,154	149,3
3-Tooth	3Z2.4	CFRP	y-axis (vertical)	3	0,487	0,634	82,1
3-Tooth	3Z2.4	CFRP	z-axis (horizontal)	3	0,637	1,158	149,8

The aim of the FibreCast project is to develop a filigree and at the same time detachable node made of polymer concrete that has the same load-bearing capacity as the (undisturbed) wood cross-section connected to it. For a C24 solid timber cross-section of 50/50 mm with an average bending tensile strength of $f_{m,mean} = 37.1$ N/mm^2 ($f_{m,k} = 24$ N/mm^2), the maximum bending moment that can be absorbed can be calculated as 0.773 kNm. This reference value is clearly exceeded by all test specimens in the 4-point bending tests on the reinforced gear geometry.

The fibre-based reinforcement used in the 3-tooth geometry is characterized not only by its corrosion resistance but also by its lower dead weight and significantly higher tensile strength compared to steel reinforcements. In addition, the manufacturing method developed allows a relatively free configuration similar to that of printed metal reinforcement. The manufacturing time also undercuts the effort required for post-processing the metal reinforcement.

The experiments on the fiber-reinforced 3-tooth reinforcement have not yet been completed at the time of manuscript submission. Due to the small spread of the results so far, at least a indicative evaluation can be presented in advance. The test setup and execution remain unchanged (Fig. 13).

Fig. 13. 3-tooth geometry with fibre reinforcement at failure. High density of cracks and a widening of the female body, which leads to a sliding of the toothing on the (lower) tension side.

5 Demonstrator

While in ordinary timber structures the joints are the structural weak points, in the concept shown, it is the other way around. The connections have a very high load-bearing capacity, similar to the connected timber profiles. This allows particularly slim profiles without an accumulation of material in the node area, thus allowing for a design freedom only known from welded aluminum and steel connections. This allows the use of a hexagonal lattice in which all beams are subjected to bending in two axis (Figs. 14 and 15).

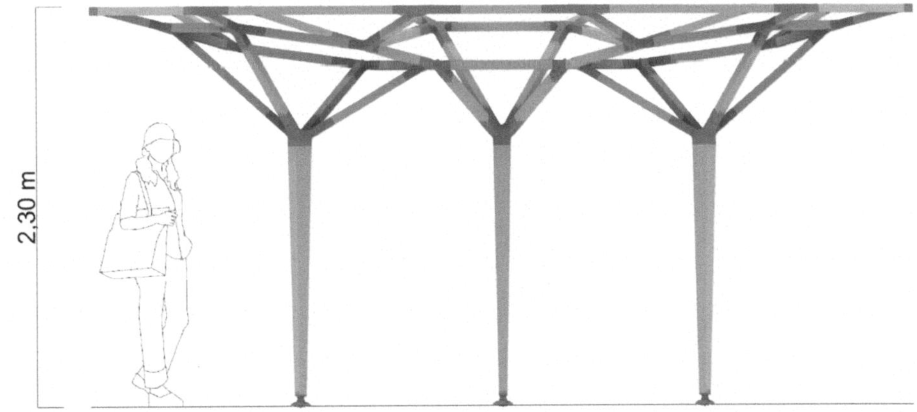

Fig. 14. Prototype demonstrating the architectural potential of the developed connections. The occurring bending forces can be transmitted by the high performance nodes.

The structure consists of 3 identical parts reducing complexity and allowing to define processes that can be repeated up to three times. For segmentation, structural groups of different sizes can be produced by varying the use of detachable and fixed nodes. While larger segments go hand in hand with a reduced number of elements, large segments require complex formwork scaffolds to accurately produce the nodes at the correct angle (Fig. 16).

The form fit connections have a gap of 0.3 mm between male and female parts allowing for rapid assembly without tools. The 6th degree of freedom is finally fixed with a single bolt per node.

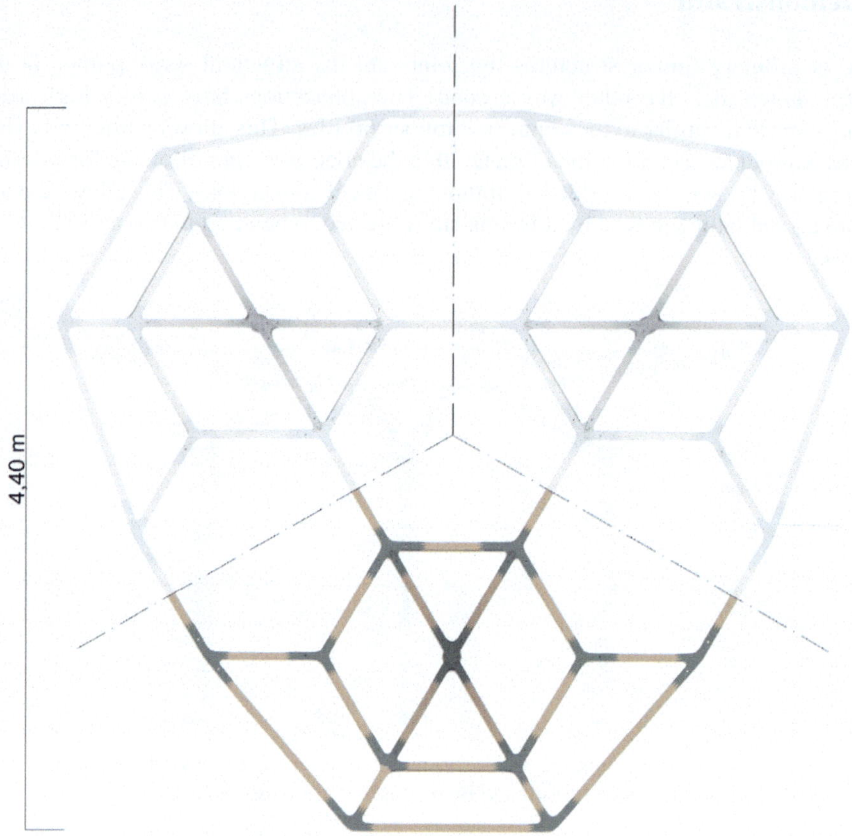

Fig. 15. Top view: triple symmetry of the pavilion

Fig. 16. Form-fit connections of the pavilion combined with a robust, branching node into one complex polymer concrete unit.

6 Outlook

The development of a high-performance, (fibre-) reinforced cast connection made of polymer concrete for timber construction can be successfully applied. The high performance of the connection up to above the load-bearing level of the connected timber components enables much slimmer construction elements and new design freedom. Further development to reduce the complexity of the detachable connection and their production as well as their size is desirable.

Classical concepts of fire protection, such as over dimensioning or cladding of the profiles, are not applicable in the intended application as slender façade elements. The structure has no fire resistance capability.

However, in combination with higher strength timber, the developed connection system has the potential to replace metal structures. This is for temporary structures such as tent structures and facade systems. In this way, the developed connection system can contribute to resource-conserving construction.

Acknowledgements. We are thankful for the Deutsche Forschungsgemeinschaft (DFG, German Research Foundation) for funding our research in the framework of the project "Fibre Cast - lightweight, high performance castings made from FFRP (filled fibre reinfoced polymer) for lightweight structures made from FRP and timber" (GE 1643/2–1 AOBJ: 624629) and "FibreCast 2.0 – Material singularities and multidimensional joints in FFRP" (GE 1643/2–3 AOBJ: 660407). We thank the Sto-Stiftung for financially supporting the realisation of the research demonstrator. We thank the company Bennert for their material supply.

References

1. Schlaich, J., Schmid, V., Schlaich, M.: Stahlverbundbrücken – neue Erfahrungen. Die Entwicklung von Verbindungen mit Zahnleisten; Bauingenieur Band 77, März 2002. Springer-VDI-Verlag, Düsseldorf (2002)
2. Arendt, St., La Magna, R., Quinn, G., Schmeck, C., Gengnagel, C., Schmid, V.: Properties and applications of polymer concrete for timber constructions. In: HCU Hamburg (Hrsg.) Proceedings of the IASS 2017: Interfaces, Hamburg, S.1–9 (2017)
3. Schmid, V., Arendt, S., Gengnagel, C., Schmeck, M.: Analysis and design of interlocking connections made of polymer-concrete for lightweight constructions. In: IASS (Hrsg.) Proceedings of the IASS Symposium 2018 Schweden (2018)
4. Schmid, V., Arendt, S.: Neue Ergebnisse zu epoxidharzbasiertem Polymerfeinbeton mit und ohne GFK- Bewehrung, Bautechnik 96, pp. 472–484. Ernst & Sohn (2019)
5. Schmeck, M., Arendt, S., La-Magna, R., Schmid, V., Gengnagel, C.: Neues Verbindungskonzept aus bewehrtem Polymerfeinbeton für Faserverbund- und Holzwerkstoffe. Ernst & Sohn, Bauchtechnik 96, Heft 6 (2019)
6. Immenga, L., Schmeck, M., Schmid, V., Gengnagel, C.: Teilflächenbelastung von epoxidharzbasiertem Polymerfeinbeton. Beton und Stahlbetonbau (2022). https://doi.org/10.1002/best.202200013
7. da Silva, L., Adams, R.: Techniques to reduce the peel stresses in adhesive joints with composites. Int. J. Adhes. Adhesives **27**, 227–235 (2007)
8. Glos, Horstmann: Strength of glued lap timber joints. In: CIB – W18A Meeting 22, pp. 7–8 (1989)

9. Steiger, R.: Faserverstärkte Kunststoffe in Holztragwerken: Entwicklungen zur Erweiterung der Anwendungsmöglichkeiten. IV. Tragfähigkeit und Verformung einer Zugstosses mit eingeklebten CRP-Lamellen – Einfluss von Materialsteifigkeit, Lastdauer und Temperatur., EMPA Abt. Holz, (No. Forschungsbericht 115/44), Dübendor (2001)
10. Kirlikovali, E.: Polymer concrete composites – a review. Polym. Eng. Sci. **21**(8), 507–509 (1981)
11. Ohama, Y.: Recent progress in concrete-polymer composites. Adv. Cem. Based Mater. **5**(2), 31–40 (1997)
12. Reis, J.M.L.: Fracture and flexural characterization of natural fiber-reinforced polymer concrete. Constr. Build. Mater. **20**(9), 673–678 (2006)
13. Rautenstrauch, K., Kästner, M., Jahreis, M., Hädicke, W.: Entwicklung eines Hochleistungsverbundträgersystems für den Ingenieurholzbau. Bautechnik **90**(1), 18–25 (2013)
14. Schober, K.-U.: Untersuchungen zum Tragverhalten hybrider Verbundkonstruktionen aus Polymerbeton, faserverstärkten Kunststoffen und Holz. Doctoral thesis, Bauhaus-Universität Weimar (2008)
15. Diederichs, U., Haroske, G., Krüger, W., Mertzsch, O.: Tragwerke aus Polymerbeton. Bautechnik **79**(5), 306–315 (2002)
16. Serrano, E.: Adhesive joints in timber engineering - modelling and testing of fracture properties. Doctoral thesis, Lund University (2000)
17. Linne, S.: Lösbare kraftschlüssige Verbindungen für modulare Bauwerke aus Faserverbundkunststoffen. Doctoral thesis, Bauhaus-Universität Weimar (2010)

Augmented Reuse

A Mobile App to Acquire and Provide Information About Reusable Building Components for the Early Design Phase

Bastian Wibranek[1] (✉) and Oliver Tessmann[2]

[1] University of Texas at San Antonio, 501 W. Cesar E. Chavez Blvd., San Antonio, TX 78209, USA
bastian.wibranek@utsa.edu

[2] Digital Design Unit (DDU), Technical University of Darmstadt, El-Lissitzky-Straße 1, 64287 Darmstadt, Germany
tessmann@dg.tu-darmstadt.de

Abstract. Construction materials are one of the main contributors to the global waste production. Compared to other industries, the reusability of building materials and components is hard to implement due to each project's individual properties and the difficulty of sharing information across the various stakeholders. In order to foster the reuse of building components, the gap between the existing building stock and the design phase of new buildings has to be minimised by bringing suppliers' data about the existing stock closer to the designers. This research illuminates how to provide relevant information from material passports and integrate them into the design environment. We compared nine passports and extracted relevant variables for the early design phase. Additionally, an augmented reality measurement app enables quick capturing and data exchange of materials and components from existing buildings. Finally, a compression-only design scheme is proposed to simplify the load capacities of the reused concrete components from an existing building. By providing information about existing materials and components in the strategically important role of the designer, reuse could be enhanced for a more sustainable built environment based on circular construction.

Keywords: Design with digital and physical realities · Circular construction · Augmented reality · Building material platform

1 Introduction

Buildings produce large amounts of waste during construction, their life cycle, and after demolition. Especially the usage of construction materials is estimated to contribute drastically to the greenhouse gas emission of buildings between 2020–2060 if the construction industry does not start implementing more material-efficient strategies (Zhong et al. 2021). To reduce material consumption in construction, we need to prevent obsolete building components from turning into waste. With their strategically important position within the production chain, architects can help establish systems in which the existing building stock may be turned into valuable material for future buildings in the sense of a circular economy for construction (Fig. 1).

C. Gengnagel et al. (Eds.): DMS 2022, *Towards Radical Regeneration*, pp. 411–423, 2023.
https://doi.org/10.1007/978-3-031-13249-0_33

Fig. 1. Future designs could guide the demolition of a building

However, implementing a circular construction system depends on the generation and availability of relevant information. Especially component reuse is hard to implement as design processes usually start with a white sheet of paper and abstract geometry models, grids, and ordering systems that are not defined by available materials in the first place. Thus, the circular economy approach is hard to implement in the early design phases (Honic et al. 2019), making it even harder to implement the reuse of parts in the later phases. Today, designers have limited access to information about the existing building material stock, let alone a geometric representation (Rašković 2020).

One approach to tackle this information gap is the idea of a passport for building components that is currently developed in academia and practice (e.g., Heinrich and Lang 2019; Groh and Dubnik 2019). Material passports can enable circular economy business models in construction by sharing information about used materials and components (Jensen and Sommer 2016). Therefore, it is crucial to distribute this information between the various stakeholders, including suppliers, deconstruction companies, builders, subsequent users, and designers (Debacker et al. 2017). In addition, especially in cases without original documentation, technologies to capture the as-built status of a building, such as 3D scanning, might provide relevant geometrical information.

It was shown that the design process can be conceptually and technically adapted for reuse by switching inputs and outputs of a design, rendering reusability as an operational constraint for structural designs (Fivet and Brütting 2020). Researchers started to implement tools for predicting the impact of early-design decisions on the circularity performance of buildings in the design environment (Heisel and Nelson 2020). Additionally, the intelligent dismantling of buildings can be beneficial for resource recovery from construction and demolition waste (Ghaffar et al. 2020). Reclaimed materials can significantly change the design and construction process but require match-making between the existing material stock and a new design (Gorgolewski 2008).

In this paper, we investigate the barriers regarding the reuse of existing building materials from the designer's perspective through digital technologies. We identify information that needs to be accessible to close the gap between the existing construction material stock and the early building design phase. To generate geometrical data about existing building components, we propose the usage of an augmented reality measurement app that streams the 3D data directly through an online platform into a conventional CAD design environment. We developed a matchmaking process between an existing building's linear concrete components, such as beams and columns, and a compression-only design scheme. Due to this simplified load case, we reduced the structural effort for the reused elements. Although we limited the scope of this study to geometrical information, it still renders a possible path towards a circular construction system based on digital technologies capable of informing designers to reuse building components.

2 Methods and Data

The research presents a prototypical investigation into the entire production sequence involved in reusing a building for future construction. Therefore, the research methodology is structured into three main areas: (i) data acquisition and structuring, (ii) design methods and algorithms, and (iii) inventory matchmaking between available and necessary resources.

2.1 Data Structuring and Acquisition

For this study, nine reference material passports were analysed from different resources from both commercial and research areas. Nine existing passports were compared, and 32 variables were collected and compiled into the overall Table 1. Similar variable names such as price or value were combined to narrow this list during this compilation. From these existing variables, the most relevant for the design phase were extracted and implemented in an Android mobile application and web platform (Fig. 2).

Table 1. All identified variables are listed, making a total of 32 from all the reference passports.

9
Reference Passports
Webgis
Circular Cloud
Simplified Metabolic
BAMB
BIM-Based for Recycling
Harvest MAPS
Wood Frame Brazil
Madaster
BIM Archicad

Abbreviation Name	Expected Useful Life	Manufacture Energy Demand	String Description
Acoustic Ratings	Function	Material	Thermal Mass
Area	File	Name	Thermal Transmittance
Building Name	Fire Rating	Owner	Thickness
Country	GTIN Number	Coordinates / Level	Type of union or connection
Cost	Image	Projected Area	Volume
Defects	Length	Quantity of same element	Weight at time 0
Density	Manufacturer	Reuse Potential	Weight currently

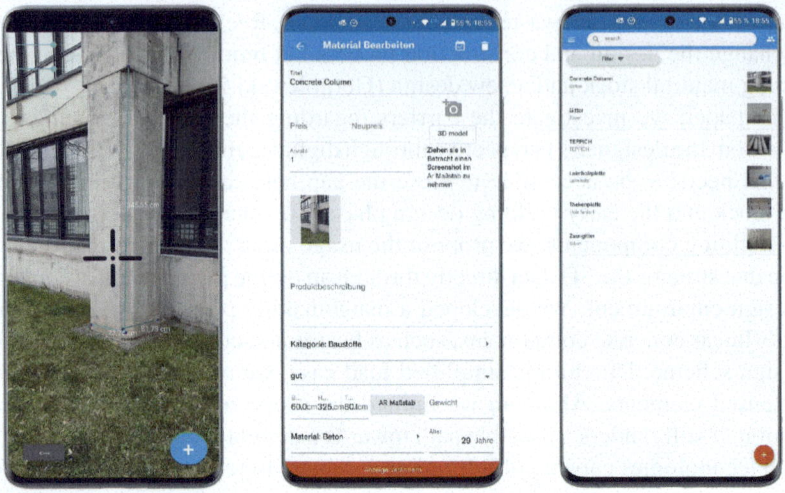

Fig. 2. The Android application and its three primary modalities, the augmented reality ruler for taking measurements (left), the material dashboard to fill in the information (middle), and the browser for available materials with filter options (Image: Bastian Wibranek)

The acquisition of building materials should happen in stages; first, a rough estimation, providing an overview of availability, and second, a realistic and holistic capturing of the entire stock. For the first stage, we propose a mobile app with photo capturing and an AR measurement tool to quickly identify what is available at the location. The application for mobile devices feeds captured data into an online platform to share available building components. It utilises the AR core for Android, specifically the AR Ruler, for linear measurements of building component dimensions (Fig. 3). The database of each building component includes image uploads, location capturing, material, condition, weight, age, availability dates, and measurements. The augmented reality measuring tool captures the dimensions and generates 3D data of components and materials. All gathered information is stored on a web server implemented in Firebase and can directly stream into the design environment Rhino 3D via a Grasshopper script (Fig. 3).

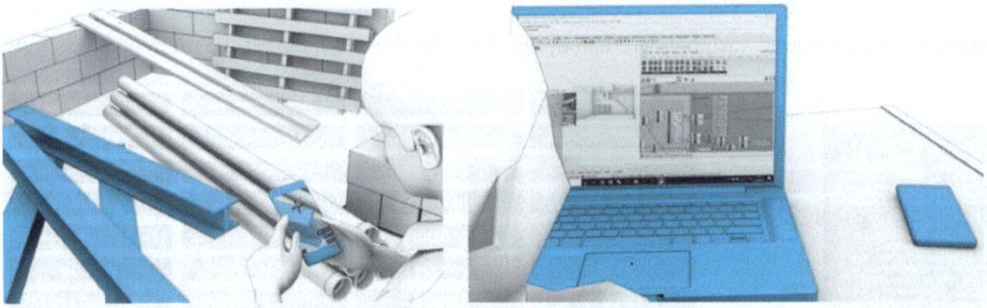

Fig. 3. The mobile app for measuring and capturing building components and direct stream of information into Rhino 3D (left) (Image: Julia Kuhn)

Furthermore, an entire building can be captured using 3D scanning procedures such as Lidar 3D scanning or photogrammetry and remodelled into an as-built BIM model (Nahangi et al. 2014). Therefore, we conducted lidar scans inside the building, generating a 3D point cloud model. These clouds were arranged and managed through the Grasshopper plugin Volvox (Evers and Zwierzycki 2016). Additionally, we used drone photogrammetry to capture the exterior of the building, resulting in a 3D mesh. Finally, the point cloud model and 3D mesh were combined in one model, and the concrete components such as slaps, columns, and facade elements were remodelled as solid boundary representations (B-rep) in Rhino (Fig. 4). The transformation from point cloud to B-rep went through a manual modelling process, as fully automated procedures do currently not exist to the best of our knowledge.

Fig. 4. The 3D mesh model captured through photogrammetry overlayed with a surface representation of individual components (right) and an initial stock representation of the concrete elements grouped by type (left). (Image: Nastassia Sysoyeva and Lucia Martinovic)

2.2 Design Algorithm Development

We focused on the available concrete columns to narrow the design space for our design explorations. Based on this choice, we developed a vector active bridge design scheme in which only compression forces are at play. This choice limits the structural behaviour to one load-bearing case, addressing difficulties in assessing rebars. The resulting three-dimensional thrust network has no moment resistance at the joints and was created using graphic statics. In addition, we implemented the PolyFrame Grasshopper Plug-In into our design algorithm to create Polyhedral Reciprocal Diagrams for compression-only structural forms (Nejur and Akbarzadeh 2021) to inform the arrangement of the existing columns.

The algorithm takes vertical loads as applied forces at the top external nodes of the structure and four support loads as boundary conditions as input to calculate the compressive members along the axis (Fig. 5). The area of the polyhedral faces correlates with the forces along the axis. We tested several subdivisions of the vector active network to identify a load case that meets the cross-sections of the concrete columns (Fig. 6). The final subdivision is based on the highest internal forces, meeting the available concrete members in the existing building (Akbarzadeh et al. 2017).

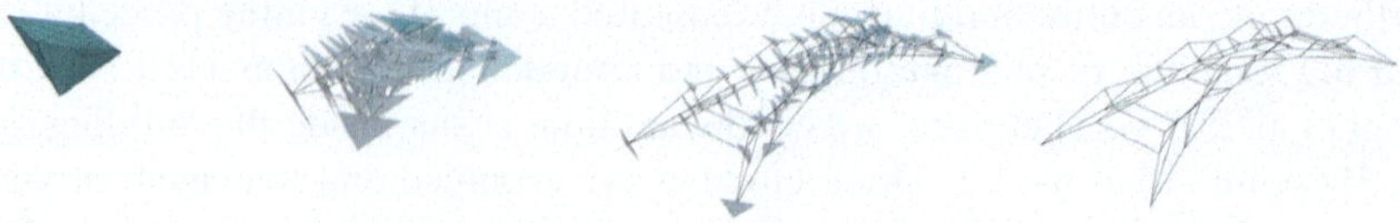

Fig. 5. The reciprocal relationship between the polyhedral force diagram and the form diagram for a bridge structure (Image: Jianpeng Chen, 2020).

Fig. 6. Three different subdivisions of the bridge structure result in various area sizes of the force polyhedral. (Image: Jianpeng Chen, 2020)

2.3 Inventory Matchmaking

Based on the bridge design and the existing building, the two must be aligned by conducting an inventory matchmaking. Therefore, the existing building components are linked with the stack, providing locational information and fixed ids for each component (Fig. 7). Based on the extracted stack, a match between the stack and future design must be drawn. In this process, the existing building is analysed for the typology of existing columns to comply with the bridge scheme. As a result of this process, the conventional inventory, as seen in Fig. 4, is reassessed to adjust the member lengths of the concrete

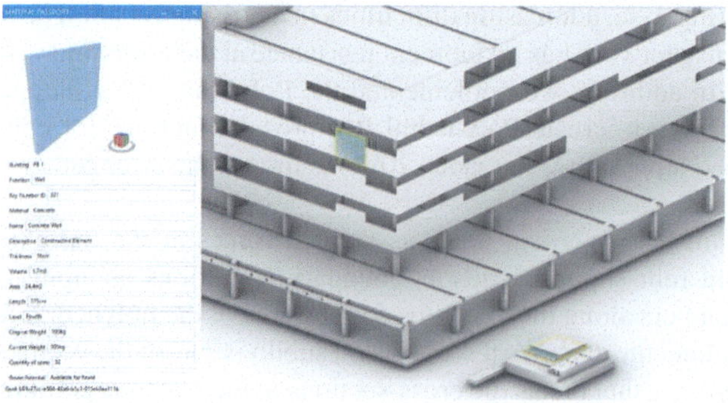

Fig. 7. The passport interface in Rhino 3D and the linkage of in-building position and disassembled stack (Image: Martin Avendaño, 2020)

columns across building stories, resulting in more extended components than the storey heigh columns (Fig. 8).

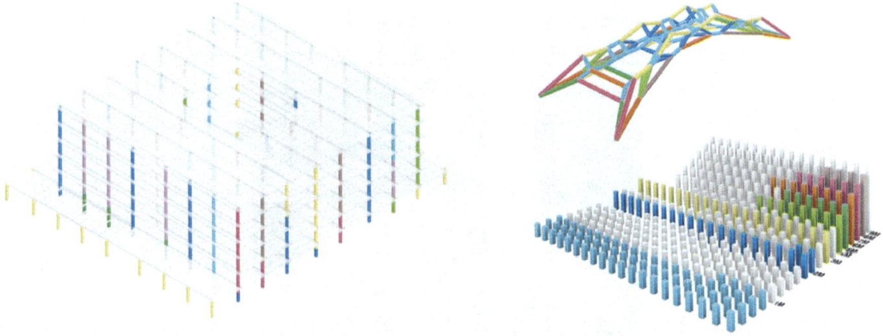

Fig. 8. The reassessed mapping of column length in the building (left) and the resulting types in the stock for the bridge design (right) (Image: Jianpeng Chen, 2020)

2.4 Connections

The structural system is designed to avoid shear and bending forces and only transfer compression forces through dry joining. The geometrical solution for joining several components in a node is generated using the Voronoi algorithm, as presented in Fig. 9. First, the algorithm creates cells where each point inside the cell is closest to the initial seed point. Next, these points are defined for each axis per node, defining cutting faces between the individual components. Finally, a joining piece is created parallel to the cut faces and extruded along the components' outer edges, providing stability against slippage.

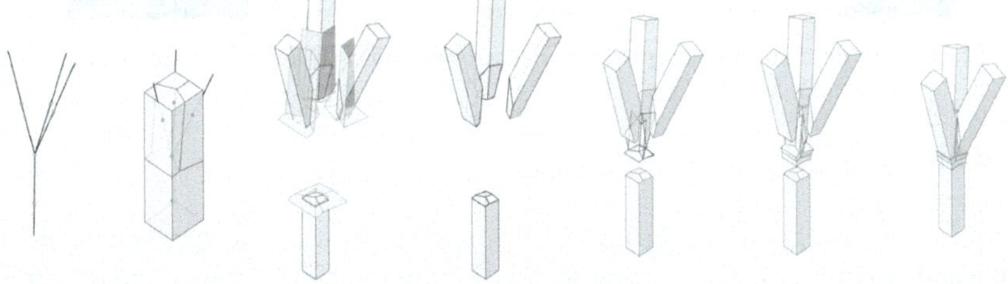

Fig. 9. The geometrical modelling of the joint configuration from left to right: the axial arrangement, Voronoi cells (Image: Jianpeng Chen, 2020)

The concrete components must be extracted from the building by introducing simple cuts orthogonal to their axial orientation (Fig. 10). In a second step, the components must be cut according to their arrangement and nodal neighbours. We prototyped the cutting process on a small-scale robotic setup, using a table saw and wood parts (Fig. 11).

Fig. 10. Test sawing of a concrete column for reuse (Image: Anne Wagner)

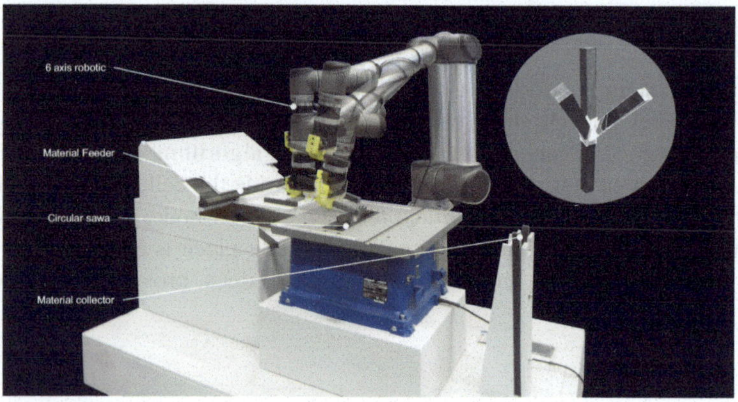

Fig. 11. The robotic cutting setup with the robot in motion (Image: Jianpeng Chen, 2020)

3 Results

The application was easy to use through a self-explanatory user interface and did not need special training (Fig. 12). The geometrical data combined with images and descriptions of materiality, location, and age presents relevant data for the early design phase. The captured building component was immediately streamed into the Rhino user interface, simplifying the acquisition to design workflow; see video link in the list of references (Große-Lohmann 2020).

Fig. 12. Capturing a concrete column with an AR-ready Android mobile phone (Image: Max Eschenbach, 2021)

In total, 138 out of the 400 available stock members were used for the final design of the bridge structure. The members included nine of the available ten proposed types with a cross-section of 40 cm by 40 cm, ranging in length from 1 m to 8 m. A geometrical solution was created for all nodal connections based on the Voronoi algorithm.

Fig. 13. The 138 components and the 52 connectors for the model (Image: Jianpeng Chen, 2020)

Finally, a model in the scale 1:50 was built to test the dry fitting joining system. The model consisted of 3D-printed components (Fig. 13). The only joints that were screwed were the ones touching the ground; the rest of the model was assembled without adhesives or additional fixation and is 48 cm wide and 180 cm long (Fig. 14).

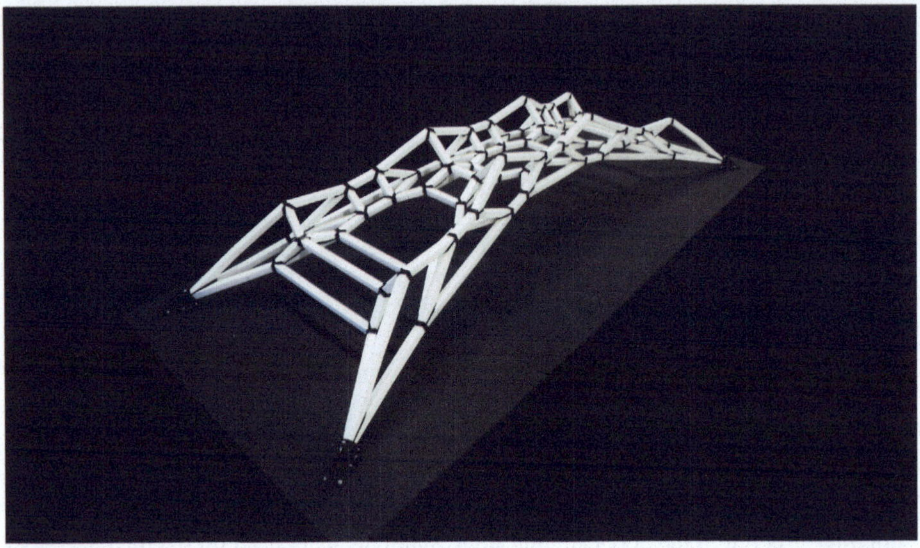

Fig. 14. The assembled model without adhesives was fixed to a ground plate. (Image: Jianpeng Chen, 2020)

Additionally, based on the remaining column components, two more design schemes were developed; a pavilion and a tower, highlighting the ease provided by the design algorithm (Fig. 15). The automated design algorithm enables quick iteration based on the existing stock. Moreover, the dry joined assembly technique turns future deconstruction into a reversed assembly.

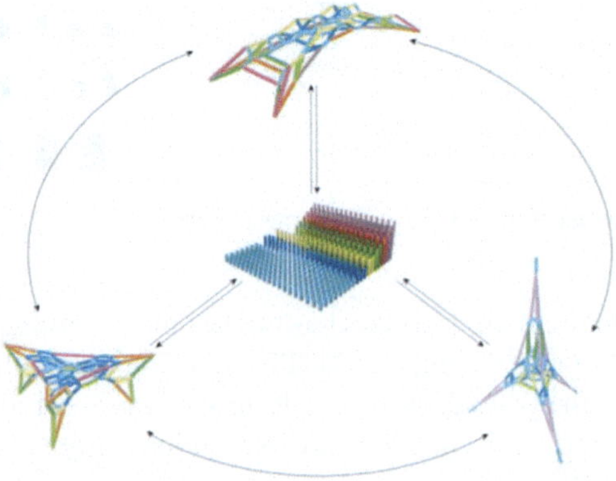

Fig. 15. Three structures mapped onto a constraint building inventory (Image: Jianpeng Chen, 2020)

4 Discussion

The work presented in this paper introduces a novel methodology integrating data acquisition and structuring, inventory matchmaking, and design for deconstruction. The proposed procedure highlights the impact of design decisions on the deconstruction approach by readjusting the extraction of building materials based on a design scheme. The workflow demonstrated how an existing building could serve as the input donor for an approximate target design (Fig. 16).

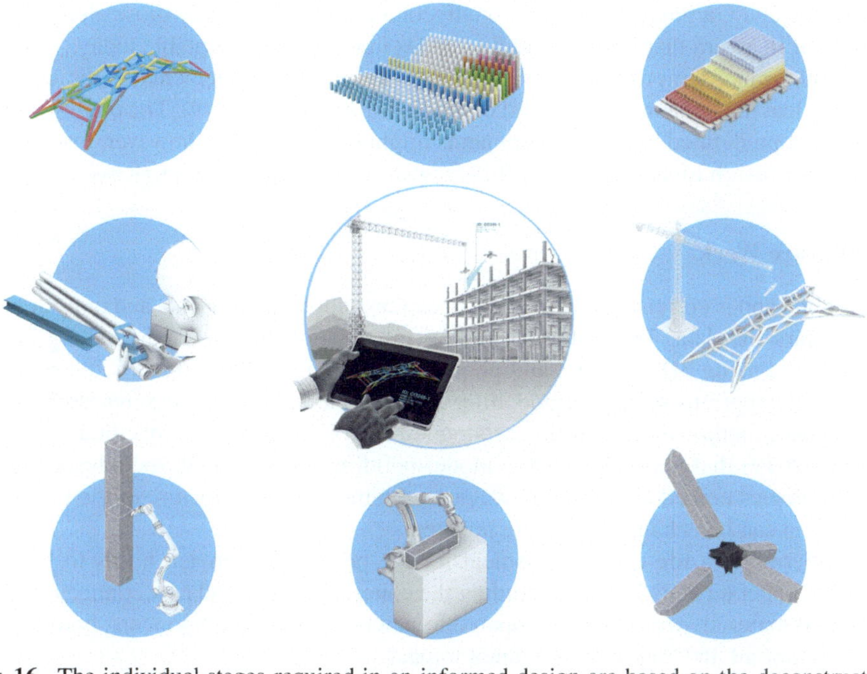

Fig. 16. The individual stages required in an informed design are based on the deconstruction of an existing building. Clockwise from top left: the mobile data acquisition, streamed into the design environment, matchmaking between design and stock, deconstruction, logistics, component preparation, joining, assembly, and the augmented deconstruction (centre) (Image: Bastian Wibranek, 2022)

The existing building components force designers to develop design protocols to address the available stock and its non-ideal structural capacities. Our design for the new structure is based on Graphic Statics, generating a compression-only structure, thus reducing the necessity of steel reinforcement. Additionally, the dry-joining of the concrete components with their joints is intended to make future deconstruction easier through simplified disassembly.

4.1 Limitations

The presented research scope consciously excludes a series of highly relevant as the building component inspection is reduced to visual and geometric aspects. However, a thorough investigation of the structural capacities of existing elements is critical to their application in future building construction. Especially during the deconstruction, the components may become damaged or destroyed. Therefore, protocols for the inspection and quality assessment of reused components are inevitable. For example, the concrete elements would need to be visually inspected for cracks, chipping, and signs of corroding reinforcement using non-destructive evaluation methods, such as Ferroscanning (Balayssac and Garnier 2017). Additionally, one might choose only those elements for reuse built in an era that follows building codes that are still acceptable today.

At this stage, the proposed design focused on the linear elements of the existing building, leaving most material stored in the slabs unutilised (Fig. 9). The combinatorial problem of predefined building components might be a limitation; however, researchers on discrete design tools reformulate these constraints as opportunities (Rossi 2021).

4.2 Future Work

The mobile app is currently only feasible to capture single components and only streams rudimentary 3D data. The as-built Building Information Model of an existing building captured using 3D scanning and photogrammetry is still preliminary and requires manual interventions such as remodelling complex parts. Nevertheless, researchers are investigating methods such as fitting catalogue-based CAD parts in point clouds (Klawitter et al. 2019) that may improve the model-building process. With the advancements in technologies embedded in mobile devices, the measurement feature of the proposed application can be enhanced in future updates.

Furthermore, a life cycle assessment for the proposed approach needs to be conducted to understand its impact, including the concrete cutting and new connectors. The reclaimed materials will still be compared to newly produced components, both from a price standpoint and their environmental impact.

5 Conclusion

The described methods reduce the information gap between the physical construction materials and the early design phase. With designers at the forefront of such strategic decisions about what materials to use and how to arrange these, it is crucial to ease the integration of existing components to tackle material scarcity in construction. In this context, the proposed mobile app provides a tool integrated with a web platform and the design environment Rhino 3D. Furthermore, the research suggests sourcing materials from obsolete buildings based on a new design. As a result, a future deconstruction may be informed by the designer's intention to reuse the locally available building inventory. The presented digital technologies can contribute to more sustainable consumption in construction, focusing on reducing waste through prevention, reduction, recycling, and reuse by informing designers.

Acknowledgements. The authors thank Jianpeng Cheng for the work during his Master Thesis BEndless ReAssembly. We also want to thank the team of Material for All for their contribution to the app development. Parts of the research in this paper were developed within the research project Fertigteil 2.0 (Precast Concrete Components 2.0), funded by the Federal Ministry of Education and Research Germany (BMBF) through the funding measure Resource-efficient circular economy - Building and mineral cycles (ReMin).

References

Balayssac, J.P., Garnier, V.: Non-destructive testing and evaluation of civil engineering structures. In: Non-Destructive Testing and Evaluation of Civil Engineering Structures (2017). https://doi.org/10.1016/c2016-0-01227-5

Debacker, W., Manshoven, S., Peters, M., Ribeiro, A., De Weerdt, Y.: Circular economy and design for change within the built environment: preparing the transition. In: International HISER Conference on Advances in Recycling and Management of Construction and Demolition Waste (2017)

Evers, H.L., Zwierzycki, M.: VOLVOX. Food4Rhino (2016). https://www.food4rhino.com/app/volvox. Accessed 21 May 2022

Fivet, C., Brütting, J.: Nothing is lost, nothing is created, is reused structural design for a circular economy. Struct. Eng. **98**, 74–81 (2020)

Ghaffar, S.H., Burman, M., Braimah, N.: Pathways to circular construction: an integrated management of construction and demolition waste for resource recovery. J. Clean. Prod. (2020). https://doi.org/10.1016/j.jclepro.2019.118710

Gorgolewski, M.: Designing with reused building components: some challenges. Build. Res. Inf. (2008). https://doi.org/10.1080/09613210701559499

Groh, J., Dubnik, P.: BIM as a tool to implement circular economy into construction projects' life-cycle. Aalborg University (2019)

Heinrich, M., Lang, W.: Materials passports - best practice. In: Innovative Solutions for a Transition to a Circular Economy in the Built Environment (2019). https://doi.org/10.5281/ZENODO.2556515

Heisel, F., Nelson, C.: RhinoCircular: development and testing of a circularity indicator tool for application in early design phases and architectural education. In: AIA/ACSA Intersections Research Conference: CARBON (2020)

Honic, M., Kovacic, I., Rechberger, H.: Concept for a BIM-based Material Passport for buildings. IOP Conf. Ser. Earth Environ. Sci. (2019). https://doi.org/10.1088/1755-1315/225/1/012073

Klawitter, D., Bringmann, O., Tonn, C.: Generic fitting in point clouds. In: Cocchiarella, L. (ed.) ICGG 2018. AISC, vol. 809, pp. 832–841. Springer, Cham (2019). https://doi.org/10.1007/978-3-319-95588-9_70

Große-Lohmann, F.: MFA-Material für Alle App Präsentation @ LUMINALE Frankfurt 2020 on Vimeo (2020). https://vimeo.com/401345201. Accessed 9 June 2022

Nahangi, M., Yeung, J., Brilakis, I., Haas, C.: State of research in automatic as-built modelling state of research in automatic as-built modelling, January 2015 (2014). https://doi.org/10.13140/2.1.2163.4885

Rašković, M., Ragossnig, A.M., Kondracki, K., Ragossnig-Angst, M.: Clean construction and demolition waste material cycles through optimised predemolition waste audit documentation: a review on building material assessment tool. Waste Manag. Res. **38**, 923–941 (2020)

Rossi, A.: Wasp v0.4.013: Discrete Design for Grasshopper (2021). https://www.food4rhino.com/en/app/wasp. Accessed 21 May 2022

Zhong, X., et al.: Global greenhouse gas emissions from residential and commercial building materials and mitigation strategies to 2060. Nat. Commun. (2021). https://doi.org/10.1038/s41467-021-26212-z

Design for Biosphere and Technosphere

Self-interlocking 3D Printed Joints for Modular Assembly of Space Frame Structures

Pascal Bach[1](✉), Ilaria Giacomini[1], and Marirena Kladeftira[1,2]

[1] Institute of Technology in Architecture, Eidgenössische Technische Hochschule Zürich, 8093 Zurich, Switzerland
pascal-bach@gmx.ch, kladeftira@arch.ethz.ch
[2] Digital Building Technologies, Zurich, Switzerland

Abstract. This paper presents a novel system of 3D printed self-interlocking space frame structures that are designed to facilitate automatic assembly using robots or drones. The research focuses on fundamental geometrical investigations of connection mechanisms enabled by additive manufacturing (AM) and their computational framework. It seeks to find out in which way AM can advance the design of space frame structures in order to enable automation in the AEC industry.

The developed system consists of bespoke 3d printed connections and carbon fiber tubular members harvesting the geometric freedom that AM allows in order to encode multiple details. The novelty of the method lies in the customization of the joints to enable a hybrid scheme of standard and automated assembly. The system operates in a two-step process: humans assemble light rigid modules in a prefabrication facility and later these modules are assembled on-site in a quick fashion using mobile robots.

The paper describes multiple investigations of connection mechanisms and joint designs that were tested through physical prototypes. The investigations focus on different self-interlocking mechanisms that address local demands in the structural system. Finally, the prototypes presented are assembled simulating the robotic unit due to the short span of the project.

Keywords: Space frames · Joints · Robotic assembly · Additive manufacturing · Self-interlocking

1 Introduction

Space frame structures proliferated in the construction industry during the last half of the 20th century thanks to their ability to cover large spans with little material. With the advent of new building materials and new construction technologies, these structures offer a combination of lightness, cost and time efficient construction.

Standardized lightweight structures consist of repetitive joints and off-the-shelf linear elements that generate rigid three-dimensional modules, obtaining repetitive and similar geometries. As computer-aided design provides opportunities to architects and

C. Gengnagel et al. (Eds.): DMS 2022, *Towards Radical Regeneration*, pp. 427–441, 2023.
https://doi.org/10.1007/978-3-031-13249-0_34

engineers to create bespoke or freeform structures, existing space frame systems are being challenged.

Nowadays technological innovations like additive manufacturing (AM) enable the materialization of complex geometries encoding multiple information. In addition to AM, architecture today is influenced by other technologically advanced systems, such as robotic construction and automated assembly. Specifically, AM paired with automated assembly can exploit the advantages of a fast, prefabricated and on-site construction approach at a maximum level.

In parallel, the AEC sector is facing a shift towards sustainable construction methods, as our industry is responsible for almost 40% of the global CO2 emissions [1]. As a response to the UN sustainable development goals and specifically action 11 for achieving sustainable cities and communities [2] this paper proposes a construction scheme that addresses economy of material, reduction of cost, production and on-site time, as well as circularity achieved through modular and disassemblable structures.

The research presented in this paper describes a novel connection system that is based on AM and facilitates a two-step assembly of space frame modules. The system relies heavily on prefabrication and is conceptualized for future transfer to a fully automated assembly using mobile robotic units.

2 State of the Art

2.1 AM Bespoke Connections for Lightweight Structures

A number of projects have investigated the opportunities that AM offers for lightweight structures by customizing the connection elements. The Smart Nodes [3] is a system where bespoke 3D printed metal joints were designed using topology optimization, following a similar principle to the ARUP joint [4]. In the Digital Metal pavilion [5] 200 custom metal joints are cast in 3D printed molds but offer limited resolution. The Ultra- light network [6] used Fused Deposition Modelling (FDM) to fabricate bespoke joints that integrate LED lights controlled by sensors. The Digital Bamboo [7], features bespoke joints for bamboo produced with MultiJet Fusion that integrate cable fittings and connection to the envelope.

2.2 Robotic Assembly of Spatial Structures

Several projects in academia address the topic of robotic automated assembly, on-site or in a prefabricated environment. The Dfab House [8], where bespoke timber frame modules are assembled by two robotic arms fixed to a gantry system in a human-machine collaborative process. In parallel, the Cooperative spatial assembly [9], demonstrated the in-situ assembly and connection of metal space frame structures by welding together linear discrete elements. Another research uses Wire Arc Additive Manufacturing (WAAM) technology to generate in-place detailing in a similar setup [10].

In this context, this paper explores the opportunities at the intersection of design, fabrication and freedom for bespoke joints that account for sequential self-interlocking assembly. The research focuses on ultra-light prefabricated modules composed of 3D printed nylon connections and carbon fiber rods with the goal to achieve automated on-site robotic assembly in the future.

2.3 Objectives and Contributions

This paper investigates bespoke self-interlocking dry connection details suitable for automated robotic assembly. The features of these connections allow for the design of dynamic and versatile modular shapes that can be easily assembled and disassembled, but can also be reconfigured and extended. The following questions are addressed:

- What types of different connections are relevant for robotic assembly and which computational method is suitable to design and print such connections?
- How can 3D printing facilitate the automated assembly of lightweight modules?
- What are the qualities of aggregating 3D printed spatial frame structures?

3 Methods

Nowadays it is possible to develop freeform space frames by designing both the overall form and the connections of the linear elements in a single digital workflow (see Fig. 1). In parallel, AM allows an efficient fabrication of the resulting complex joints. The methods that follow describe the computational and design workflow developed to facilitate this link.

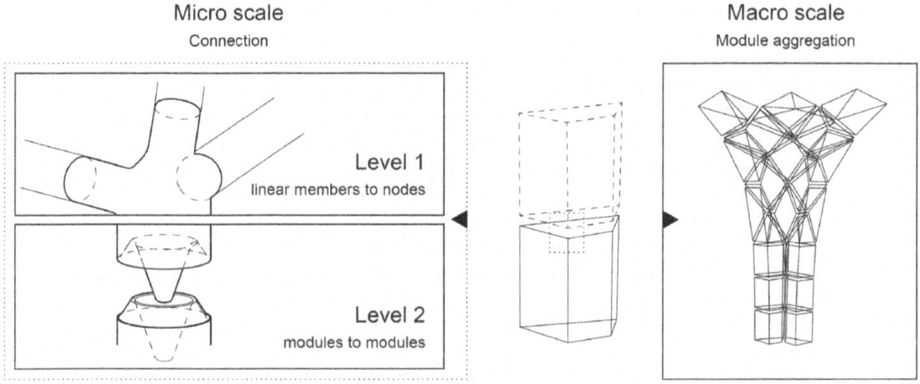

Fig. 1. Diagram of the computational workflow.

The design process starts with the subdivision of complex spatial structures into bespoke modules specifically designed for an automated assembly. The fabrication process consists of printing the connection detail and then aggregating the modules in a larger structure:

- Phase I - 3D printing the connections and cutting the linear elements to the right length.
- Phase II - humans pre-assemble, in a prefabricated environment, discrete bespoke modules.
- Phase III - robots assemble the modules which are aggregated into a larger space frame structure.

3.1 Materials and Technology

This research demonstrates how by resolving details can help in building large aggregations thanks to a bottom-up process: starting from the development of the micro scale of the connection detail it is possible to create bespoke modules that constitute a space frame structure. The systems investigated here are still experimental in nature and so a choice was made to use ultralight materials, such as carbon fiber tubes and nylon, to minimize the weight of the modules.

The powder-based HP Multi Jet Fusion (MJF) 3D printer has been selected as the most promising technology to print complex lightweight connections yet integrating high resolution surface details that would be impossible to achieve with conventional fabrication methods [7]. Thanks to its multi-agent printing process it is possible to achieve a detailed surface finish and unidirectional mechanical strength.

3.2 Macro Scale - Aggregation of the Structure

A computational approach is explored to define the design of the overall structure: the input is a NURBS surface and subsequently the depth of the space frame is defined. The object is discretized into polyhedral modules using a custom subdivision algorithm. The topology of the structure and its weight are parameters of the design and are explored through different configurations of polyhedral patterns (see Fig. 2).

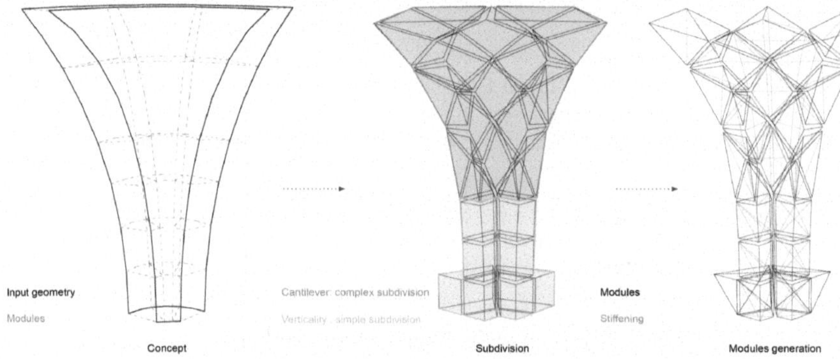

Fig. 2. Computational method to design the spatial structure.

The design of the structure is informed by possible scenarios of assembly (see Fig. 3):

1. members are connected punctually,
2. one linear member is doubled or
3. planar connection between two faces.

The punctual connection was rejected due to the introduction of high momentary forces in the connection. As a result, the structure should only include edge-to-edge and face-to-face aggregations.

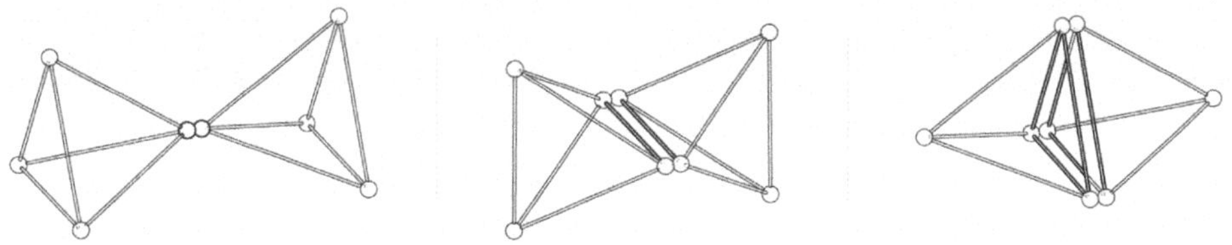

Fig. 3. From left to right: node-to-node, edge-to-edge, face-to-face.

3.3 Micro Scale: Level 1 - Connecting Linear Members to Joints

This connection emerges from the development of two parts. The first secures the linear element to the joint (level 1) and the second connects two or more modules (level 2) with an interlocking mechanism.

The linear members are manually assembled in a prefabricated environment. The following three main types have been investigated (see Fig. 4):

- threaded connections (A, B, C, H, I)
- attaching the linear element with additional fasteners (D, F, G, J)
- snap fit connections (E, K, L) (see Fig. 5a, b)

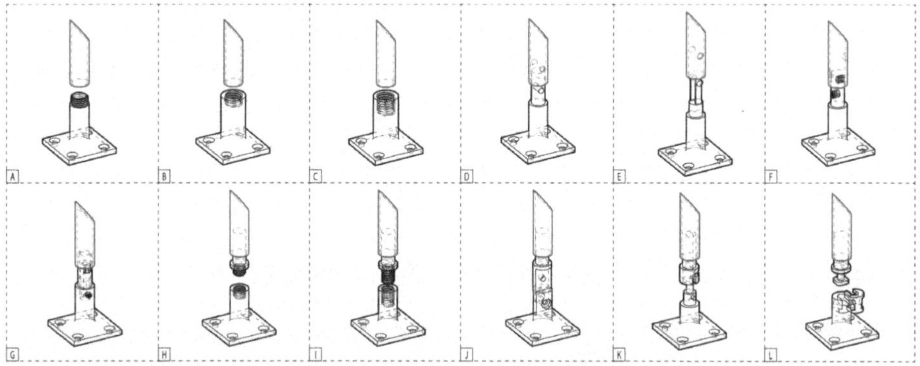

Fig. 4. Design study of linear member to joint connections - level 1.

Connections type D, F, G, J need extra fasteners, increasing the time and the precision needed to assemble a stable connection.

The strongest and simplest connection to assemble is the thread. Its biggest advantage is the ease of assembly and disassembly by rotation of members, as well as tolerance compensation in the axial direction. In order to fasten the linear member between two connected joints a right thread in one end and a left thread in the other end are needed (see Fig. 5).

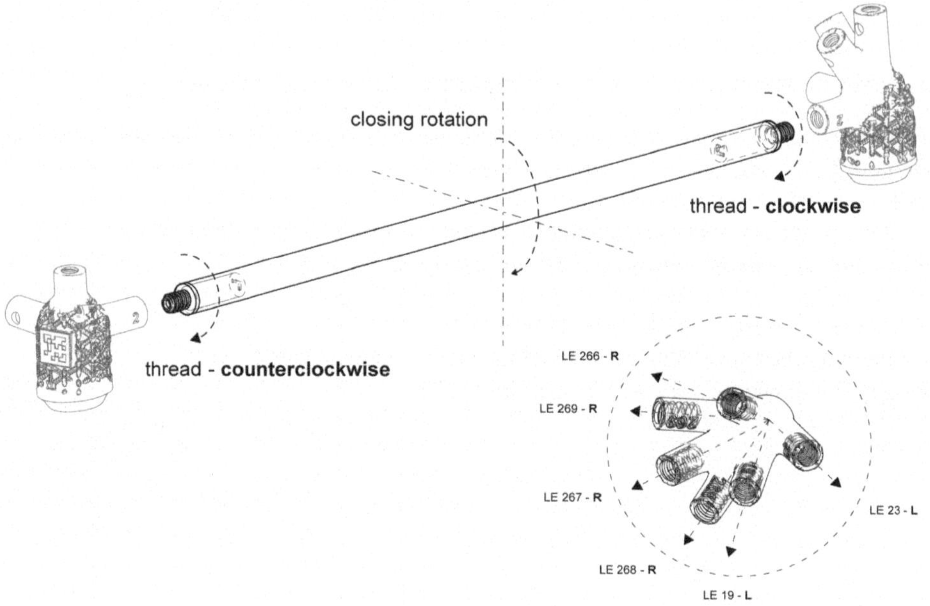

Fig. 5. Overview of the connection system.

3.4 Micro Scale: Level 2 - Connecting Modules to Modules

For the module connections, different systems were evaluated like rotary, clamping, slip or self-interlocking joints. In addition, add-on features such as magnets, screws or brackets have been investigated to enhance the performance of each connection mechanism.

Three types of self-interlocking mechanism were analyzed for the second level (see Fig. 6):

- Unidirectional connections: these are of secondary nature to the principal locking system and the locking mechanism is placed at the direction of assembly (A, B).
- Multi-part locking connections: need to be locked with a pin, another module or a rotational movement (C, D, E, F).
- Sliding connections: the locking mechanism is placed perpendicular to the force flow and could be locked additionally with a fastener (G, H, I).

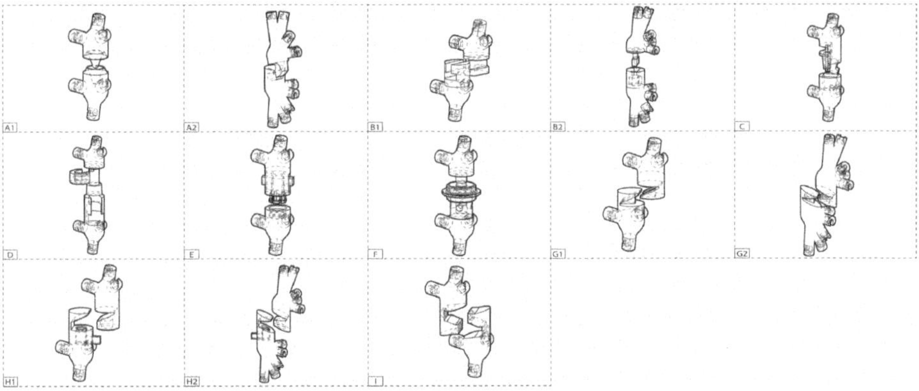

Fig. 6. Design study of module-to-module connections - level 2.

The types C and E were identified as the most promising because of the guaranteed stiffness, the ease in assembly and disassembly as well as the possibility to implement their mechanisms for multi-module connections.

Type C (see Fig. 7) is an advanced snap-fit system locked with a pin directly printed inside the joint. Although the connection is stable and very stiff once the pin is locked, this mechanism needs a lot of force in the vertical direction to be assembled and therefore was deemed unfitting for robotic assembly.

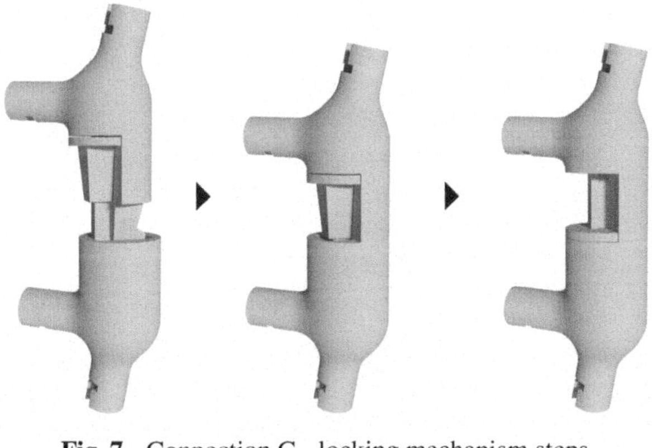

Fig. 7. Connection C - locking mechanism steps.

In connection type E (see Fig. 8) the upper part locks in the lower one and is then secured with a pin held in place thanks to a small magnet. As soon as the upper part locks into the lower one, another stronger magnet initiates the pin to slide, locking the entire connection. The integration of magnets does not have any structural or load-bearing purpose; they are used as an alignment feature.

After studying the connection between two discrete modules, the same principles were transferred to more complex cases where three or more modules meet at one point.

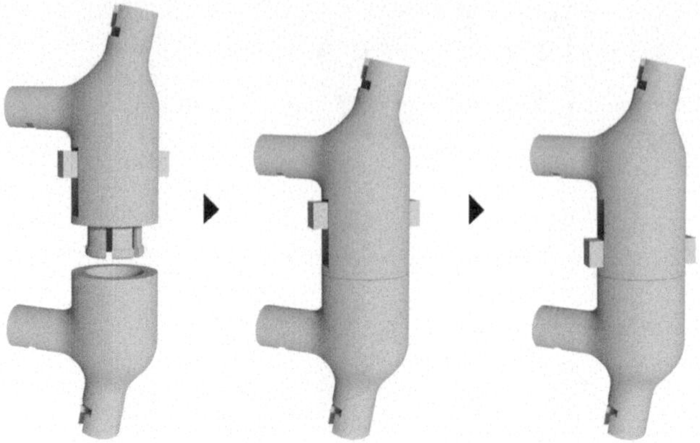

Fig. 8. Connection E - locking mechanism steps.

The principles are established with four modules and are based on the following rules (see Fig. 9):

- Splitting the basic locking mechanism in 4 ways (A).
- Doubling the mechanism by adding one locking mechanism in between. This allows the modules to be locked sequentially (B).
- The module assembled locks the previous one thanks to locking slots (C).
- The connection is split perpendicular to the assembly vector and guarantees a vertical sequential assembly (D).

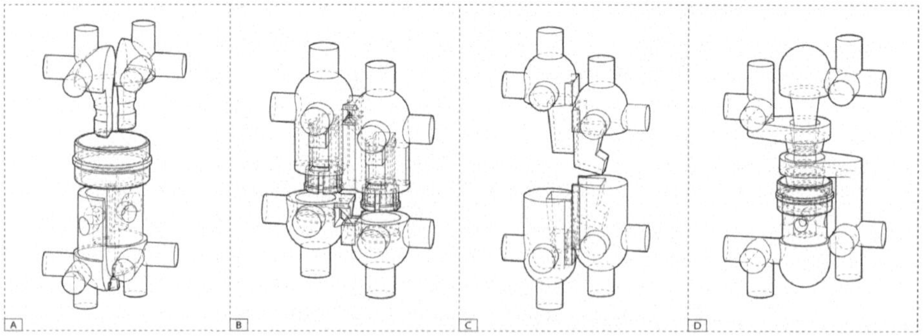

Fig. 9. A. splitting mechanism, B. doubling mechanism, C. self-interlocking mechanism, D. layering mechanism.

These mechanisms were tested in simulations and physical prototypes. The interlocking mechanism (C) is evaluated as the most promising option since it guarantees an easy but stiff connection for an automated (dis)assembly. Moreover, the mechanism works without any additional fasteners, decreasing the number of steps necessary. In fact, every module is secured by the one placed after it, which creates a chain of locking sequences. The interlocking system requires a strict order of (dis)assembly.

The assembly sequence is based on the height and number of modules of the structure. Starting from the bottom corner and moving in continuous linear motions (see Fig. 10), it is possible to assemble the structure module by module. It is fundamental to define the division of the connection depending on the assembly orientation of each module.

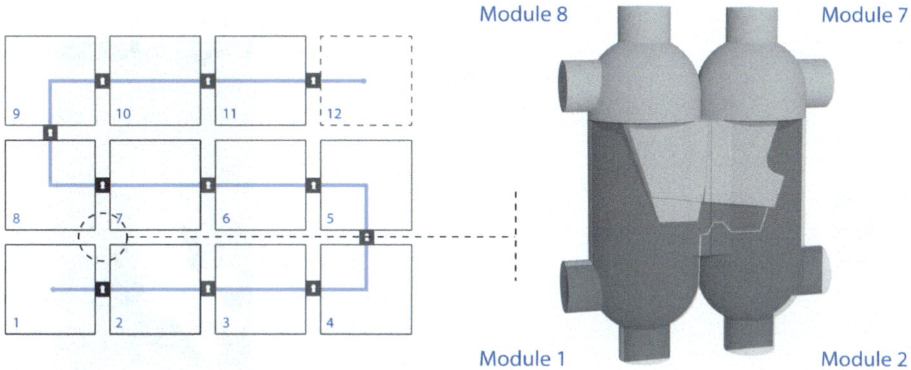

Fig. 10. Assembly logic

3.5 Robotic Assembly-Specific Features

To take full advantage of AM and make the connections as efficient as possible, different infill structures were investigated to reduce the weight and material consumption of each connection. A spatial lattice structure reduced the weight more than 50% and was implemented into the generation algorithm (see Fig. 11). A further investigation will be to analyze the ratio of structural performance to weight.

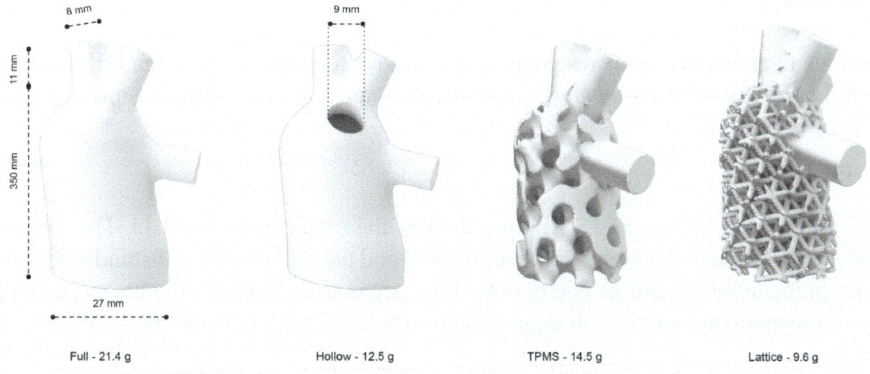

Fig. 11. Weight study with different infill structures.

Due to the use of bespoke members and connections, the assembly process becomes exponentially complex as the size of the structure increases. To simplify the process the number of each joint and the rotation direction of the thread is engraved. On the joint's surface an AprilTag - a visual fiducial system - is implemented. It allows to encode multiple information, such as the index, rotation and position of the module in space and among other modules. AprilTags are designed for small data payload, between 4 and 12 bits, for easier camera detection and from longer ranges (see Fig. 12).

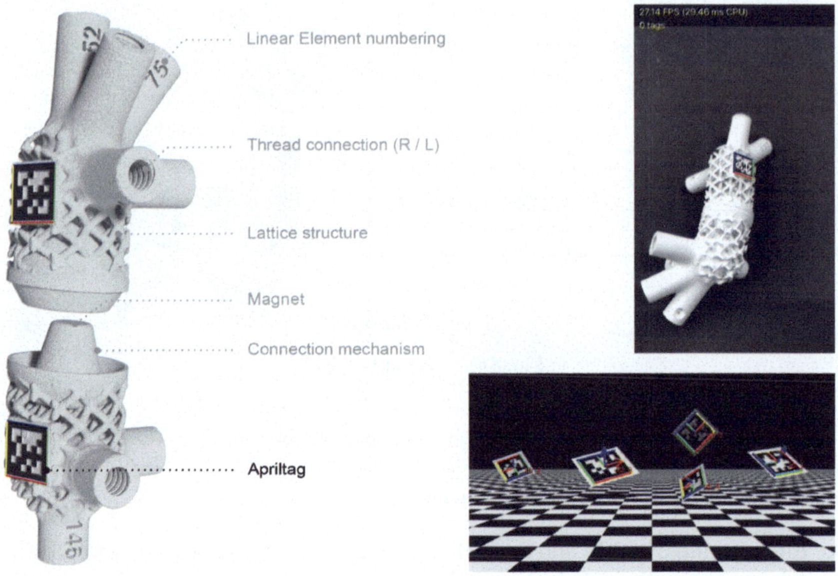

Fig. 12. Additional features on the connection.

4 Prototypes

Different mechanisms were tested in 1:1 scale prototypes to investigate the ease of assembly and disassembly as well as the strength and stability of each connection.

4.1 Prototype 1

In the first prototype (see Fig. 13) a pin secures the connection (level 1). This pin was found unstable due to the high tolerances needed and the inability to withstand forces that are not in the axial direction. The two modules are instead connected with an advanced snap-fit mechanism locked with a pin which is directly printed inside the joint (level 2).

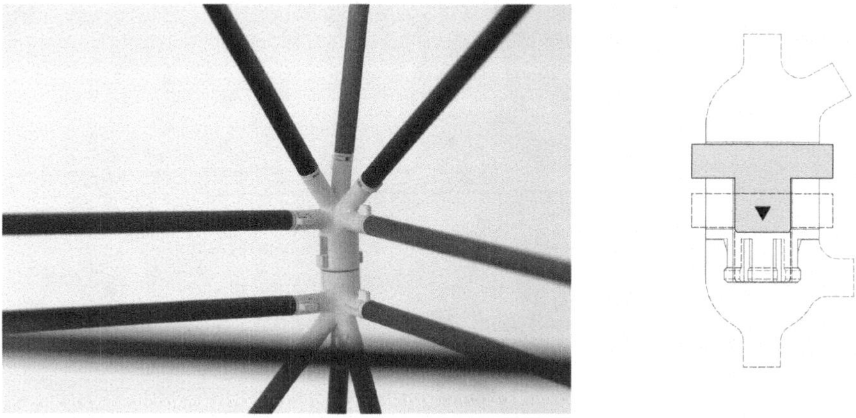

Fig. 13. Connection mechanism of prototype 1.

4.2 Prototype 2

In the second prototype (see Fig. 14) magnets were implemented to facilitate the assembly of the modules with guided movement. The connection tested in the second prototype is a self-interlocking mechanism of type C explained in Sect. 3.4. It became clear that the pin locking the mechanism could become part of the connecting module.

Fig. 14. Self-interlocking connection of prototype 2.

4.3 Prototype 3

The third prototype (see Fig. 15) identified the stability of the modules in the cantilevering part of the structure. To evaluate the aggregation of a much more complex system, the assembly orientation was analyzed and consequently the connection was segmented for multiple modules. To achieve a flawless assembly of the four modules, the connection is divided according to the average vector defining the assembly direction of the modules.

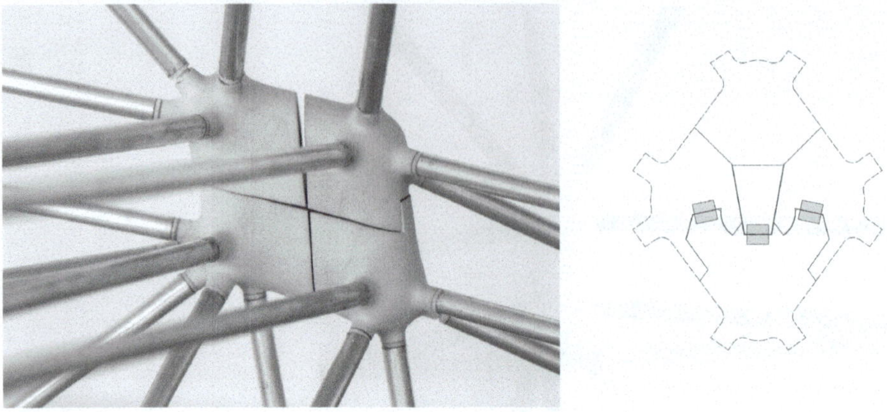

Fig. 15. Multi-module connection of prototype 3.

4.4 Final Demonstrator

After a careful evaluation of the prototypes, the best combination of features was identified: quick (dis)assembly, number of parts required, stiffness and stability of the connection as well as the difference forces that each mechanism responds better to. All these traits led to the build-up of a final large-scale prototype of a space frame structure (see Fig. 16).

Fig. 16. Final demonstrator

The number of modules to connect pinpointed the two types of connections needed to guarantee a locked stable structure.

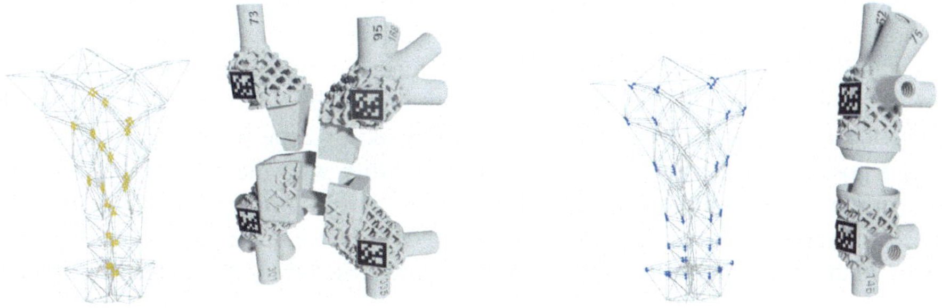

Fig. 17. Multi module connection and compression connection.

The best combination of investigated features was used to generate the joints for the final demonstrator (see Fig. 17). AM allowed the encoding of all details in one connection: the locking mechanism and the inlets for the magnets, the threaded detail, the AprilTag for robotic assembly and information for human pre-assembly. The result is a space frame structure of 3.5 m high and consists of 180 m of glass fiber rods and 287 single connections with a total weight of only 18 kg.

5 Discussion

AM allows the fast fabrication of bespoke joints while encoding multiple information. In combination with automated assembly with advanced mobile robotic units, the developed system could respond to current topics that our profession is facing, such as economy of material, cost and time on the construction site.

Space frames are high-performance construction systems that allow for lightweight buildings where materials are used optimally for axial loading. Changing the paradigm of current construction processes we must also address other aspects beyond material efficiency. The authors see promise in the proposed research as it allows for a high degree of prefabrication and digital control. The two-step assembly process targets precise prefabrication in a controlled facility, where humans or machines can perform quality control, while the modular system guarantees efficient assembly on-site. Although the second step of robotic assembly was manually simulated due to the short timeframe of this project, the system was developed in consultation with robotics engineers to respect the paylimits, path planning limitations and tolerances that are relevant for robotic assemblies. Therefore, the authors believe that the method is transferable to a robotic unit with minor modifications, although further research is needed to validate the findings.

Furthermore, the proposed scheme creates the basis for material circularity as it is entirely disassemblable and elements can be reused or recycled. The modularity of the system in discrete rigid modules is key in achieving both transportability and automated assembly as, if taken into account in the design phase, the structure can be self-supported to further eliminate the need for any scaffolding and not limit robotic access.

The authors see the implemented connection as one suitable possibility but not the sole one, thus the research aims to display a variety of evaluated joints with different features. The computational framework that has been developed guarantees a quick evaluation of suitable connections. Taking into consideration the form and the forces that the structure needs to bear, it is possible to generate infinite configurations.

A mobile robotic unit that was planned for further application of the research was being developed in parallel by the Robotic Systems Lab (RSL) at ETH Zurich. The choice of carbon fiber rods was driven by the need to experiment with the lightest material possible as the aim was to test the feasibility with the robotic arms without exceeding the limited payload of a UR5. However, the system developed is not limited to the use of carbon fiber, further research is needed to identify the intersection of material choices so that the materiality of the joint can serve structurally the material system.

6 Conclusion

This paper showcases the potential of self-interlocking lightweight structures assembled by unique modules. The system is designed to facilitate, in future research, automated assembly with a mobile robotic unit. The material and the technology that has been presented allows to decrease human labor on site, which is one of the most expensive and unsafe factors in the construction industry.

The ability to easily assemble and disassemble complex structures creates the opportunity to reuse most components of complex systems and extend their usage into multiple life cycles for a sustainable future. This is made possible thanks to AM, which allows the fabrication of connections that can be adapted to bespoke geometries and to the material chosen.

Acknowledgements. This work was conducted during the individual thesis of the MAS DFAB 2020–2021. The research was situated within the larger research stream "Additively manufactured bespoke joints" and supported by the NCCR Digital Fabrication, funded by the Swiss National Science Foundation (NCCR Digital Fabrication Agreement #51NF40- 141853).

Pascal Bach and Ilaria Giacomini were involved in the conceptualization, investigation, visualization, and writing of the original draft. Marirena Kladeftira is the main researcher of the stream "Additively manufactured bespoke joints" and was responsible for the conceptualization and supervision of the thesis as well as writing and reviewing the manuscript.

References

1. United Nations Environment Program, and Global Alliance for Buildings and Construction: 2020 Global Status Report (2020). https://wedocs.unep.org/handle/20.500.11822/34572
2. Guterres, A.: The Sustainable Development Goals Report 2020, United Nations Publication Issued by the Department of Economic and Social Affairs, pp. 1–64 (2020)
3. Crolla, K., Williams, N.: Smart nodes, a system for variable structural frames with 3D metal-printed nodes. In: Proceedings of the 34th Annual Conference of the Association for Computer Aided Design in Architecture, ACADIA 2014 (2014)

4. Galjaard, S., Hofman, S., Perry, N., Ren, S.: Optimizing structural building elements in metal by using additive manufacturing. In: International Association for Shell and Spatial Structures, IASS 2015 (2015)
5. Aghaei Meibodi, M., Giesecke, R., Dillenburger, B.: 3D printing sand molds for casting bespoke metal connections. In: Intelligent and Informed - Proceedings of the 24th International Conference on Computer-Aided Architectural Design Research in Asia, CAADRIA 2019, pp. 133–142 (2019)
6. Raspall, F., Amtsberg, F., Banon, C.: 3D printed space frames. In: Proceedings of the IASS Symposium 2018, August 2018
7. Kladeftira, M., et al.: Digital Bamboo. A digital ultra-light bio-structure. In: Hybrids & Haecceities - Proceedings of the 41st Annual Conference of the Association for Computer Aided Design in Architecture, ACADIA 2022 (2022)
8. Adel, A., Thoma, A., Helmreich, M., Gramazio, F., Kohler, M.: Design of robotically fabricated timber frame structures. In: Proceedings of the 38th Annual Conference of the Association for Computer Aided Design in Architecture, ACADIA 2018 (2028)
9. Parascho, S., Kohlhammer, T., Coros, S., Gramazio, F., Kohler, M.: Computational design of robotically assembled spatial structures. In: AAG - Advances in Architectural Geometry, pp. 112–139 (2018)
10. Ariza, I., et al.: In Place Detailing. Combining 3D printing and robotic assembly. In: On Imprecision and Infidelity - Proceedings of the 38th Annual Conference of the Association for Computer Aided Design in Architecture, ACADIA 2018 (2018)

Matter as Met: Towards a Computational Workflow for Architectural Design with Reused Concrete Components

Max Benjamin Eschenbach[1(✉)], Anne-Kristin Wagner[2], Lukas Ledderose[3], Tobias Böhret[4], Denis Wohlfeld[4], Marc Gille-Sepehri[5], Christoph Kuhn[2], Harald Kloft[3], and Oliver Tessmann[1]

[1] Digital Design Unit (DDU), Technical University of Darmstadt, El-Lissitzky-Str. 1, 64287 Darmstadt, Germany
eschenbach@dg.tu-darmstadt.de

[2] Entwerfen und Nachhaltiges Bauen (ENB), Technical University of Darmstadt, El-Lissitzky-Str. 1, 64287 Darmstadt, Germany

[3] Institute of Structural Design (ITE), TU Braunschweig, Pockelsstrasse 4, 38106 Braunschweig, Germany

[4] FARO Europe GmbH, Lingwiesenstraße 11/2, 70825 Korntal-Münchingen, Germany

[5] THING TECHNOLOGIES GmbH, Am Kronberger Hang 8, 65824 Schwalbach am Taunus, Germany

Abstract. Over the past decades computational design, digital fabrication and optimisation have become widely adapted in architectural research, contemporary practice as well as in the construction industry. Nevertheless, current design and fabrication process-chains are still stuck in a linear notion of material use: building components are digitally designed, engineered, and ultimately materialised by consumption of raw materials. These can be defined as *digital-real* process chains. But these parametric design logics based on mass customisation inhibit the reuse of building components. In contrast to these predominant and established *digital-real* process chains, we propose a *real-digital* process chain: departing from our real, already materialised built environment. We digitise and catalogue physical concrete components within a component repository for future reuse. Subsequently, these components are reconditioned, enhanced if necessary and transitioned into a modular building system. The modularised components are then recombined to form a new building design and, eventually, a new building by combinatorial optimisation using mixed-integer linear programming (MILP). An accompanying life cycle assessment (LCA) complements the process and quantifies the environmental potential of reused building components. The paper presents research towards a feasible workflow for the reuse of structural concrete components. Furthermore, we suggest a digital repository, storing geometric as well as complementary data on the origin, history, and performances of the components to be reused. Here, we identify core data to be integrated in such a component repository.

Keywords: Component reuse · Circular design · Life cycle assessment · 3D scanning · Concrete dry-joints

C. Gengnagel et al. (Eds.): DMS 2022, *Towards Radical Regeneration*, pp. 442–455, 2023.
https://doi.org/10.1007/978-3-031-13249-0_35

1 Introduction

1.1 Background

Digital-real process chains enable us to design, plan, optimise and build complex structures, even from a large number of individualised components. While the advantage of such parametrically designed, engineered, and optimised structures is clearly that fewer raw materials are consumed in fabrication, the impact of these savings may be diminished or even eliminated by the use of energy-intensive manufacturing processes, complex assembly procedures and what can be called an *over-individualisation* of building components that inhibits component reuse. More recently these parametric design and construction logics based on digital mass customisation of components are being challenged by the evolution of combinatorial as well as discrete design and modelling paradigms [1]. These methods rely on discrete parts, the ability to recombine existing building components [2] and on designing and building with *matter as met*[1] [3]: creating structures from heterogeneous natural materials [4] or reusing material formerly considered waste, like scrap or rubble [5]. These *real-digital* process chains act on the basis of the real, already materialised components of our built environment and proceed to recombine them in a digital manner. Such approaches have been studied and applied in several case-studies before: projects like *Mine The Scrap* [6], *Unmaking Architecture* [7] and *Reform Standard* [5] demonstrate the potentials of the approach. The 2021 publication of *The Cannibal's Cookbook* [8] may even serve as a manifesto for this practice, suggesting *cyclopean cannibalism* as a "construction technique consisting of the re-appropriation of pre-existing building stock for the assembly of 'new' buildings" [8].

1.2 Workflow

We propose a *real-digital* process chain for the reuse of concrete components (Fig. 1). We start by digitising concrete components from obsolete buildings using 3D scanning. The components are then harvested, their performances inspected, they are reconditioned using cutting robots and thus transformed into new precast concrete components (PCC). A combinatorial modelling pipeline recombines these discrete components into dry-jointed structures that serve as both, buildings for today and component repositories for future buildings, making PCCs building components decoupled from the lifespan of a single building, circulating through time and space.

[1] The expression *matter as met* is used as a translation of the french *matériau de rencontre*, adapted from a lecture by Yves Bréchet. In the official translation by Liz Libbrecht, the expression is translated to *materials discovered by chance.*

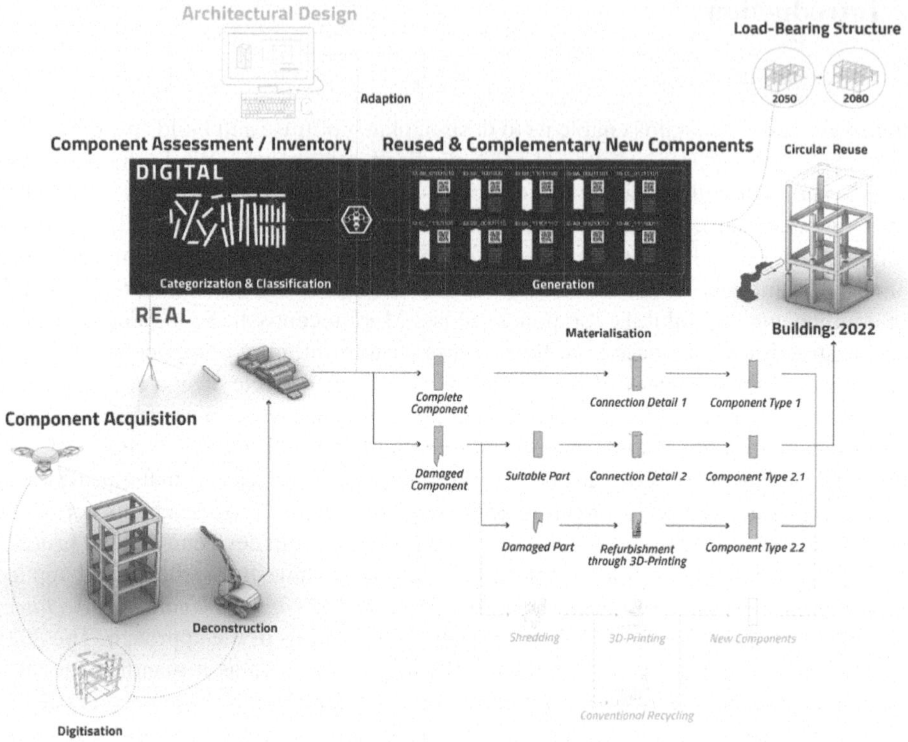

Fig. 1. Diagram depicting all key subprocesses of the proposed *real-digital* process chain.

1.3 Component and Data Repository

Our case study on the reuse of concrete components enables us not only to develop a seamless process and evaluate it, but also to further research the establishment of the component repositories. Developing a process to enable a loop of reuse within construction does not only entail the building made with reused concrete components to double as a storage for material resources. It also entails the creation and storage of an array of data relative to the reuse of existing building components once the life of the building they are part of is over. Our research therefore builds on our process to assess the kind of information that is necessary and how it can be gathered as we move through the process.

2 Component Digitisation, Abstraction and Feature Extraction

The process starts with the digitisation of existing concrete components in obsolete buildings. Here, we used a FARO® Focus Laser Scanner in conjunction with a FARO Freestyle 2 handheld scanner. The stationary laser scanner captures the key features of components. These scans are then enhanced using the handheld scanner, reducing shadows, and capturing further details. All scans were registered using FARO SCENE software, resulting in one single point cloud. For the identification of relevant components within the point cloud and their subsequent abstraction to surface geometry, two approaches are explored:

1. The point cloud is processed with FARO As-Built™ software by one-click selection and trimming of relevant regions (Fig. 2) using the approach presented by Tonn et al. [9]. After the abstraction, a DXF model is exported for further processing.

Fig. 2. Processing of the point cloud resulting from merging all captured 3D scanning data; Left side: One-click selection and growing of regions. Right side: Abstraction and trimming of regions against each other.

2. The merged point cloud is exported to the E57 file format and imported into Rhinoceros® [10]. It is then processed in Grasshopper® using the Cockroach plug-in [11–14]. Specifically, it is downsampled and processed using statistical outlier removal (SOR). Normals are computed, the point cloud is clustered and processed using a random sample consensus (RANSAC) algorithm for identifying planar regions. These are then trimmed against each other, resulting in an abstract surface geometry (Fig. 3).

Through experimentation with both methods, FARO As-Built™ software has proven to be more stable and intuitive. While it is possible to achieve a sufficiently abstracted surface model using Grasshopper and Cockroach, it requires substantial trial and error as well as manual post-processing in Rhinoceros. In contrast, the user interaction and one-click selection of the FARO As-Built™ software yields better abstraction results in a shorter amount of time due to also leveraging the user's knowledge, perception, and human intuition. The assessment of these two approaches has thus enabled us to identify how to perform the first step of our process, gathering geometrical data as the basis of our repository and for use in our design process.

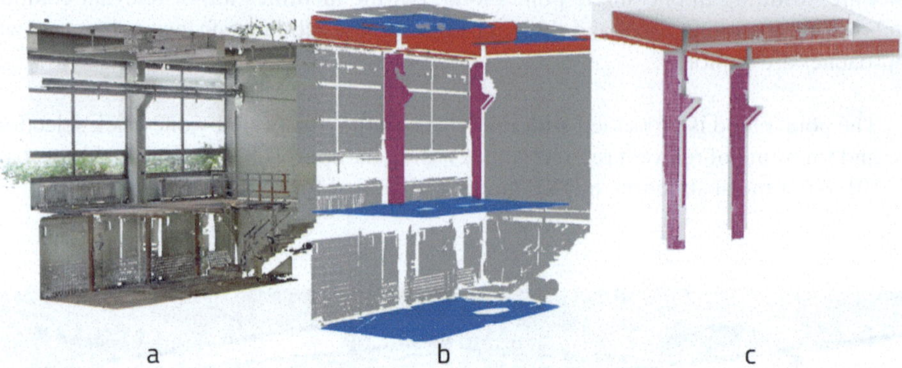

Fig. 3. Point cloud resulting from merging all captured 3D scanning data (a); Clustering and colour-coding of key components: columns (magenta), beams (red) and floor (blue) (b); Trimming of identified regions and polygonal surface reconstruction of components (c).

3 Harvesting of Components and Quality Inspection

The digitised components are harvested by sawing at key points to disassemble the obsolete structure into individual components for reuse (Fig. 4). Components are kept as large as possible to avoid limitations during recombination. Harvested components are categorised according to the time of their initial production: in Germany, all elements are expected to adhere to building codes (after 1943 DIN 1045 [15], and since 2004 DIN EN 1992 [16]) that clearly define the amount of reinforcement and concrete quality, allowing certain assumptions to be made regarding the performances of the components to be reused. However, the elements must be subjected to tests to assess their potential for reuse. Mettke's research work [17] provides useful guidance on how the inspection of concrete components can be carried out: in a multi-stage decision-making process, the concrete components are subjected to test criteria both before and after disassembly, transport, and possible storage. These concern the concrete quality, reinforcement condition, structural-physical properties, geometry, and aesthetics. The analyses range from an initial assessment of the condition based on visual inspection and information from documents to sampling and detailed technical examinations of the building material (e.g., compressive strength tests).

Components not fit for reuse can be inspected using destructive methods and thus complement the information on material performance. Since tests can be both time and money consuming, we propose the repository to host two sections relative to the diagnosis. The first one, mandatory, relates to the history of the component, indicating where it was situated in the obsolete building and how it was used. The second, optional, will comprise the results of the material diagnosis assessment. By relying on these two sections, requirements or suggestions can be made depending on the reuse scenario. E.g., in the case where only the first section is filled, the new PCCs are preferably used in a similar fashion to the original use-case: columns will be reused as columns, beams will be reused as beams.

Fig. 4. Selective disassembly using diamond-fitted concrete saw as a proof of concept of harvesting concrete components for reconditioning and reuse.

4 Recombination of Components Using Combinatorial Optimisation

The repository now hosting data relative to the geometry, origin, history and the performances of the components, allows us to perform the next step: recombination of components in a new building design. Several suitable tools and modelling methods have emerged for discrete design, i.e., WASP [18] or Monoceros [19] for discrete aggregation. Furthermore, methods focused on building from a stock of heterogeneous components by combinatorial optimisation have also been integrated and applied successfully in previous projects [2, 7, 20]. Recently Brütting et al. [21] have detailed an approach for the reuse of reclaimed steel elements by formalisation as a cutting stock problem (CSP) and optimisation using mixed integer linear programming (MILP). Combinatorial optimisation using MILP is especially flexible since it allows for the integration of additional constraints based on data from the repository, e.g., structural and LCA data.

As a result of experimentation with all mentioned approaches, MILP turned out to be the most flexible and thus was integrated into our pipeline. The Hops plug-in [22] was used in conjunction with the ghhops-server package [23], enabling a seamless interface between Grasshopper and the MILP solver Gurobi [24]. The program is written in Python 3.8 and based on the formalisation presented by Brütting et al. [21]. It is adapted by introducing additional constraints, ensuring matching cross-sections of columns and beams during the CSP optimisation. Demand and stock are defined in Grasshopper, the data is then passed to Gurobi via Hops. Once the optimal assignment is found, the result is passed back to Grasshopper via the same Hops component (Fig. 5).

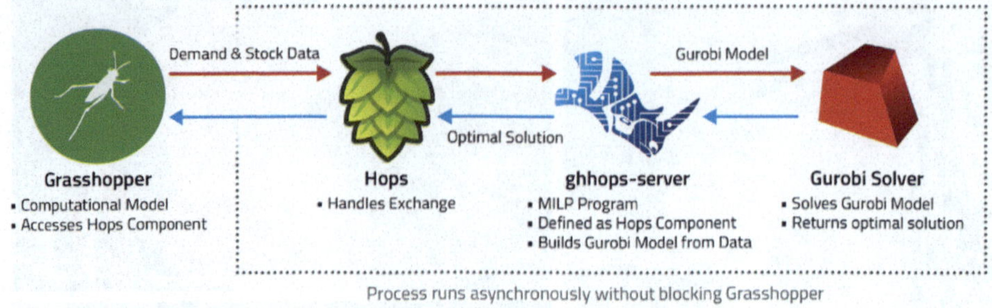

Fig. 5. Diagram depicting the computational modelling pipeline, interfacing the Gurobi optimisation solver with Grasshopper through Hops and ghhops-server.

To tackle the challenge of recombining components, combinatorial optimisation is not only effective but necessary to optimise the potentials of structures made from reused components, especially when multiple constraints need to be fulfilled. While existing computational design tools for aggregation of discrete components [18, 19] work very well with purely geometric constraints, additional development is needed to integrate numerical conditions and constraints into the aggregation process. Optimisations based on the Hungarian method have shown great potential when working with heterogeneous components and natural materials. However, the Hungarian method requires a cost matrix with a single cost value per possible assignment and thus either requires elaborate weighting of the cost coefficients or will suffer from significant limitations when working with complex constraints.

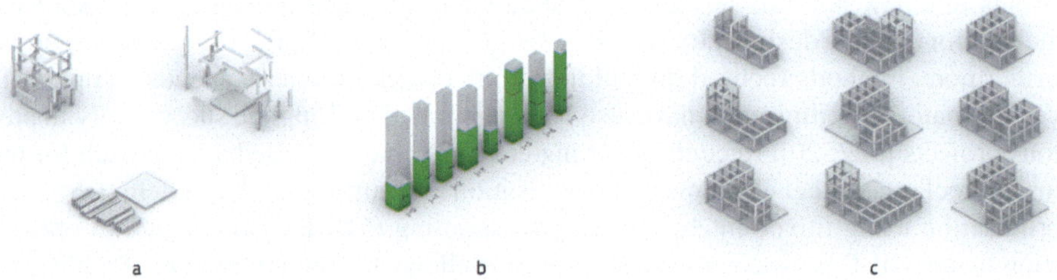

Fig. 6. Schematic visualisation of digital disassembly based on abstracted surface geometry acquired from point clouds (a), modularisation and combinatorial optimisation of the resulting CSP formalised as MILP program (b) and digital recombination of modularised components (c).

We were able to overcome the limitations of the other approaches using MILP. Our developed modelling pipeline works seamlessly and yields optimal results consistently and efficiently. This allows recombination of the PCCs into new skeleton structures, utilising the gathered data from our component repository (Fig. 6).

5 Reconditioning and Dry-Jointing of Reclaimed Concrete Components

Non-destructive harvesting of concrete structures enables the production of resource-efficient concrete components. The quality of the joints is particularly important for the on-site assembly of subtractive manufactured components and will play a crucial role in the successful practical implementation of circular reuse. The objective of our component investigations is to study the mechanisms contributing to the load-bearing capacity of the joint region of subtractive manufactured concrete components, which will be combined adaptively (using post-tensioning), considering the different concrete qualities and specific load situations. Three main types of connections are used in the construction industry today: dry joints, adhesive joints, and mortar joints. Adhesive joints and mortar joints are generally suitable as joints for PCCs but have the disadvantage that they must harden. However, dry joints can be integrated directly into the subtractive process by milling or sawing with diamond tools. They also simplify the future dismantling of segmental structures, a key issue to keep costs of reuse processes low enough for them to be competitive.

The Digital Building Fabrication Laboratory (DBFL) at the ITE provides the authors with a fabrication centre including a freely movable robot as well as a CNC sawing and milling unit, which enables the subtractive machining of reclaimed concrete components at full scale and the production of high-precision joints. Figure 7 illustrates the essential CNC-based process steps for the production of a dry-jointed, segmented beam from a reclaimed concrete component. Concrete properties and the location and characteristics of the reinforcement encountered are of critical importance here.

During this research project, the extensive findings on the manufacturability of dry joints [25–27] are to be transferred to the application of reused concrete components (Fig. 7a–c). In tests on concrete components from existing buildings, segmented beam joints have been produced with four different prototype dry joints, including joints with steel sleeves, turnbuckles for prestressing and transferring tensile forces (Fig. 7d–f), as well as hinge joints and the integration of channels for post-tensioning. The focus was on optimising the sawing and milling parameters and tool wear for aged, 50-year-old concrete, as well as the achievable dimensional accuracy compared to the digital twin. These tests demonstrated the rational and cost-effective robot-aided producibility and possible reconditioning of reclaimed concrete components with a dimensional accuracy of a few tenths of a millimetre. The data obtained on energy consumption and tool wear were integrated into the LCA.

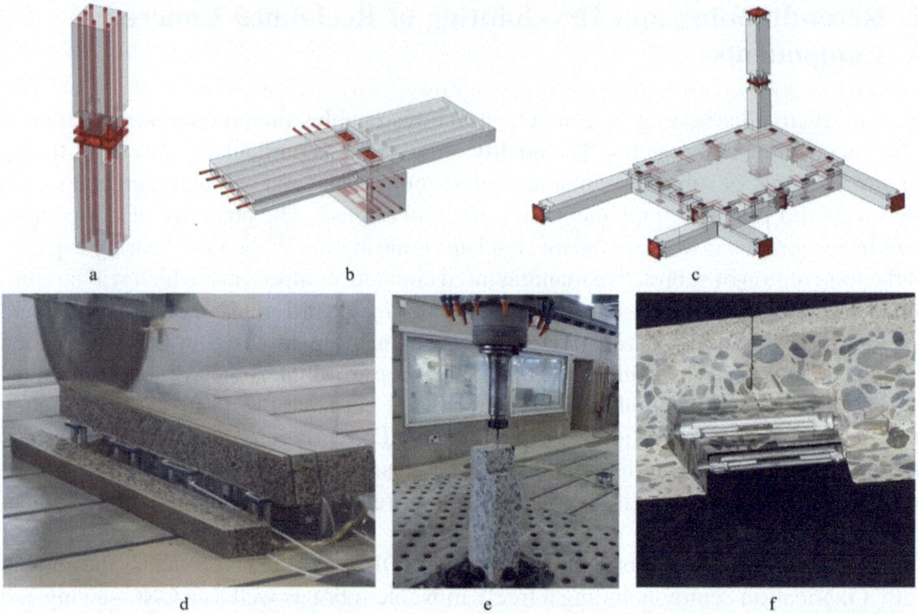

Fig. 7. Design of dry joints using integrated steel components with glued-in rebars: column to column (a), floor to beam (b), section of the assembled structure (c); From reclaimed concrete components to novel segmented concrete structures: CNC sawing of a concrete component into the required blanks (d), CNC milling of the joint geometry (e), finished, dry-jointed segmented beam with turnbuckles in the tension zone (f).

6 LCA of Reusing Building Components

The aim is to quantify the environmental impact and resource use of the reused concrete components and to compare it with the impact of conventional concrete production with and without a proportion of recycled components (conventional concrete and recycled concrete). For this calculation a LCA based on DIN EN 15804 [28] and DIN EN ISO 14040 [29] is used. The system boundary includes the life cycle phases of demolition/deconstruction, recycling, and reconditioning for reuse as well as the production of the raw materials and the construction of the new components (Fig. 8). These phases are divided in several processes and for each process the energy consumption and the use of resources (e.g., production of the concrete raw materials) are inputs for the LCA calculation. For each process the environmental emissions are calculated as output. All processes contain the amount of energy and material needed for a specific unit (functional unit) of the final product. The functional unit serves as a unit of comparison, in our case 1 m^3 of concrete as a component.

The input data for the energy consumption and the mass of the reused concrete component (scenario A) is first estimated based on literature sources [30] and conservative empirical values. No real processes have been determined at this stage, preliminary data and estimations are used. After recording the real processes, the data will be adjusted, i.e., the energy consumption of the sawing process is to be examined in detail. The

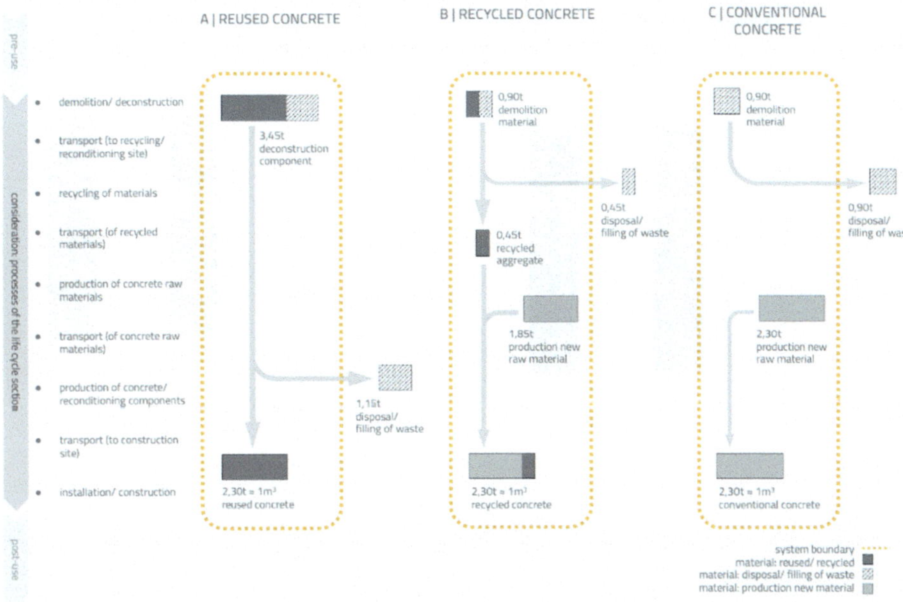

Fig. 8. System boundaries and mass balance of the life cycle processes considered, own representation based on preliminary data as well as Heyn and Mettke [30].

input data for the energy consumption and production of the concrete raw materials for the recycled concrete (scenario B) and conventional concrete (scenario C) is based on the research of Heyn and Mettke [30]. Environmental impact and resource use are determined based on the described inputs and the ecological database Ökobaudat [31].

The LCA results show the global warming potential and non-renewable primary energy demand of the three scenarios: reused concrete (scenario A), recycled concrete (scenario B) and conventional concrete (scenario C) (Fig. 9). The reused component shows comparatively the lowest values: under the model conditions, reuse can save about 70% of the greenhouse gas emissions (carbon dioxide equivalent, CO_2e) and about 50% of the non-renewable primary energy (megajoule, MJ). The largest ecological savings are attributed to the process "production of concrete raw materials". The production of cement requires a high energy input and causes more than 90% of the greenhouse gas emissions of recycled and conventional concrete production. By reusing the concrete components, no new cement is needed, and the environmental impact is significantly reduced. The emissions and resource use due to the additional efforts of other processes generated by reuse (e.g., deconstruction, reconditioning, and installation) are low compared to the production of the raw materials. It becomes clear that the reuse of concrete components offers a great opportunity to reduce the ecological footprint. Furthermore, our LCA lays out the base for an environmental assessment of the components to be performed and integrated as complementary data in the repository.

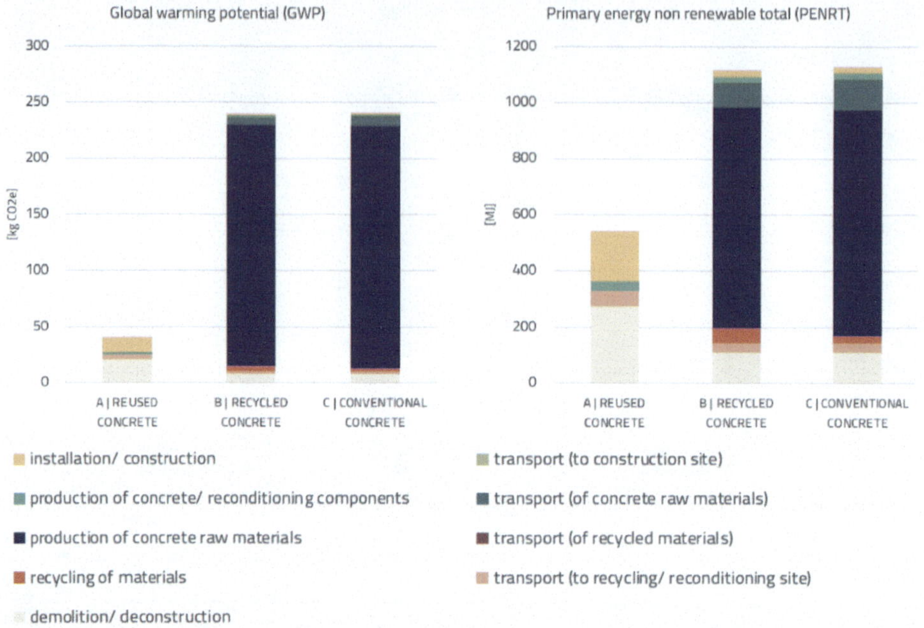

Fig. 9. Results: global warming potential (GWP) and primary energy non-renewable total (PENRT) for the processes and scenarios considered.

7 Conclusion and Outlook

Our workflow enables reuse, reconditioning, and recombination of existing concrete components from obsolete buildings: A digital repository of building components is created based on geometry from 3D scanning and point cloud processing. The repository is further informed by data on building history and from concrete quality inspection. Harvested concrete components can be reconditioned using state of the art robotics. Dry-jointed connections provide a feasible way of assembly. MILP can optimise the recombination of existing components and can be integrated seamlessly into a CAD modelling environment. Our LCA shows that an approach based on the reconditioning, recombination and circular reuse of existing concrete components is not only technically feasible but also effective.

However, further research and development in all key areas is required to arrive at a seamless process: point cloud processing can be improved to create simplified BIM models (Scan-to-BIM) for our repository. Regarding the LCA, the estimated data for the reuse of components needs to be measured and adapted to real processes, especially the sawing of concrete is to be examined for optimisations (e.g., environmental impacts, waste reduction). Furthermore, only one life cycle of the component was calculated here, several life cycles have to be considered and calculated to complete the evaluation. Various scenarios and sensitivity analyses (e.g., component storage, different transport distances, etc.) can provide additional information about the opportunities and challenges of working with reused components. Integration of LCA into a digital model can be achieved using existing plug-ins, e.g., as presented by Apellániz et al. [32]. Component specific LCA data can also be integrated into the combinatorial optimisation.

The results presented in this paper constitute intermediate findings towards a feasible workflow for the reuse of structural concrete components and identifying core data that has to be integrated in a component repository. A full-scale demonstrator made from PCCs will be built in 2023. This research, i.e., the presented workflow and the component repository, can inform the conditions of industrialisation but is just one building block embedded within several aspects that require fundamental change. To make the process chain economically viable, a series of socio-economic and building code related aspects need to change as well.

Acknowledgments. The research in this paper was developed within the research project *Fertigteil 2.0* (Precast Concrete Components 2.0), funded by the Federal Ministry of Education and Research Germany (BMBF) through the funding measure *Resource-efficient circular economy - Building and mineral cycles (ReMin)* on TRL 3.

References

1. Tessmann, O., Rossi, A.: Geometry as interface: parametric and combinatorial topological interlocking assemblies. J. Appl. Mech. **86** (2019). https://doi.org/10.1115/1.4044606
2. Huang, Y., Alkhayat, L., De Wolf, C., Mueller, C.: Algorithmic circular design with reused structural elements: method and tool. In: Proceedings of the International Fib Symposium on the Conceptual Design of Structures, pp. 457–468 (2021). https://doi.org/10.35789/fib.PROC.0055.2021.CDSymp.P056
3. Bréchet, Y.: The science of materials: from materials discovered by chance to customized materials: inaugural lecture delivered on Thursday 17 January 2013. Collège de France (2015). https://doi.org/10.4000/books.cdf.3641
4. Wibranek, B., Tessmann, O.: Digital rubble - compression-only structures with irregular rock and 3D printed connectors. In: Proceedings of the IASS Annual Symposium 2019 – Structural Membranes 2019, Barcelona (2019)
5. Huang, C.-H.: Reinforcement learning for architectural design-build. In: Proceedings of the 25th International Conference of the Association for Computer-Aided Architectural Design Research in Asia, Hong Kong, pp. 171–180 (2021)
6. Certain Measures: MINE THE SCRAP (2015). https://certainmeasures.com/MINE-THE-SCRAP. Accessed 10 June 2022
7. Marshall, D.: Unmaking architecture: holding patterns for misfit matter. Master thesis, MIT (2019). https://files.cargocollective.com/c374561/ThesisBook.pdf. Accessed 10 June 2022
8. Clifford, B.: The Cannibal's Cookbook: Mining Myths of Cyclopean Constructions. Goff Books, Novato (2021)
9. Tonn, C., Schmidt, H., Bringmann, O., Klawitter, D.: Abstracting and trimming reality - one-click-region-growing and surface trimming in point clouds. In: Computing for a Better Tomorrow - Proceedings of the 36th ECAADe Conference, Lodz, Poland, vol. 1, pp. 257–262. CUMINCAD (2018)
10. Robert McNeel & Associates: Rhinoceros® 7 & Grasshopper®. Software (2021). https://www.rhino3d.com/. Accessed 10 June 2022
11. Vestartas, P., Settimi, A.: Cockroach: a plug-in for point cloud post-processing and meshing in Rhino environment, EPFL ENAC ICC IBOIS. Software (2020). https://github.com/9and3/Cockroach. Accessed 10 June 2022

12. The CGAL Project: CGAL User and Reference Manual, 5.4. edition. CGAL Editorial Board (2022). https://doc.cgal.org/5.4/Manual/packages.html. Accessed 10 June 2022
13. Zampogiannis, K., Fermuller, C., Aloimonos, Y.: cilantro: a lean, versatile, and efficient library for point cloud data processing. In: Proceedings of the 26th ACM International Conference on Multimedia, MM 2018, pp. 1364–1367. ACM, New York (2018). https://doi.org/10.1145/3240508.3243655
14. Zhou, Q.-Y., Park, J., Koltun, V.: Open3D: a modern library for 3D data processing (2018). https://doi.org/10.48550/arXiv.1801.09847
15. DIN 1045, Tragwerke aus Beton und Stahlbeton: 2., überarb. Aufl. Fraunhofer IRB-Verl.; Beuth. Praxis Bauwesen, Stuttgart (2005). ISBN: 3410160000
16. DIN EN 1992-1-1:2011-01, Eurocode 2: Bemessung und Konstruktion von Stahlbeton- und Spannbetontragwerken - Teil 1-1: Allgemeine Bemessungsregeln und Regeln für den Hochbau; Deutsche Fassung EN 1992-1-1:2004_+ AC:2010. Beuth Verlag GmbH, Berlin. https://doi.org/10.1109/5.771073
17. Mettke, A., et al.: Schlussbericht zum Forschungsvorhaben Rückbau industrieller Bausubstanz – Großformatige Betonelemente im ökologischen Kreislauf. Brandenburgische Technische Universität (2008). https://www-docs.b-tu.de/ag-baurecycling/public/Forschungsberichte/0__Einfuehrung.pdf. Accessed 10 June 2022
18. Rossi, A.: Wasp - Discrete Design for Grasshopper. Software (2022). https://github.com/ar0551/Wasp/. Accessed 10 June 2022
19. Pernecký, J., Tóth, J.: Monoceros: A Wave Function Collapse plug-in for Grasshopper, Subdigital. Software (2021). https://github.com/subdgtl/Monoceros. Accessed 10 June 2022
20. Schein, M., Eschenbach, M.B., Fromm, A.: Tailored structures - parametrics for sustainable constructions. In: Inspiring the Next Generation - Proceedings of the International Conference on Spatial Structures 2020/21 (IASS 2020/21 Surrey 7), Guildford, UK, pp. 243–254 (2021). https://doi.org/10.15126/900337
21. Brütting, J., Vandervaeren, C., Senatore, G., De Temmerman, N., Fivet, C.: Environmental impact minimization of reticular structures made of reused and new elements through Life Cycle Assessment and Mixed-Integer Linear Programming. Energy Build. **215**, 109827 (2020). https://doi.org/10.1016/j.enbuild.2020.109827
22. Robert McNeel & Associates: Hops. Software (2022). https://github.com/mcneel/compute.rhino3d/tree/master/src/hops. Accessed 10 June 2022
23. Robert McNeel & Associates: ghhops-server. Software (2022). https://github.com/mcneel/compute.rhino3d/tree/master/src/ghhops-server-py. Accessed 10 June 2022
24. Gurobi Optimization LLC: Gurobi Optimizer Reference Manual (2022). https://www.gurobi.com. Accessed 10 June 2022
25. Mainka, J., Lehmberg, S., Budelmann, H., Kloft, H.: Non-standard Fügeprinzipien für leichte Bauteile aus UHPFRC. In: Beton- und Stahlbetonbau, Nr. 11, vol. 108, pp. 763–773. © Ernst & Sohn Verlag für Architektur und technische Wissenschaften GmbH & Co. KG, Berlin (2013). https://doi.org/10.1002/best.201300055
26. Lehmberg, S., Ledderose, L., Wirth, F., Budelmann, H., Kloft, H.: Von der Bauteilfügung zu leichten Tragwerken: Trocken gefügte Flächenelemente aus UHPFRC. In: Beton- und Stahlbetonbau, Nr. 12, vol. 111, pp. 806–815. © Ernst & Sohn Verlag für Architektur und technische Wissenschaften GmbH & Co. KG, Berlin (2016). https://doi.org/10.1002/best.201600053
27. Baghdadi, A., Heristchian, M., Kloft, H.: Connections placement optimization approach toward new prefabricated building systems. Eng. Struct. (2021). https://doi.org/10.1016/j.engstruct.2020.111648
28. DIN EN 15804: Nachhaltigkeit von Bauwerken – Umweltproduktdeklarationen – Grundregeln für die Produktkategorie Bauprodukte; Deutsche Fassung EN 15804:2012+A2:2019 + AC:2021 (2022)

29. DIN EN ISO 14040: Umweltmanagement – Ökobilanz – Grundsätze und Rahmenbedingungen (ISO 14040:2006 + Amd 1:2020); Deutsche Fassung EN ISO 14040:2006 + A1:2020 (2021)
30. Heyn, S., Mettke, A.: Ökologische Prozessbetrachtungen-RC-Beton (Stofffluss, Energieaufwand, Emissionen). zum Forschungsprojekt: Einsatz von Recycling-Material aus mineralischen Baustoffen Zuschlag in der Betonherstellung. Brandenburgische Technische Universität (2010). http://www.rc-beton.de/vortraege_pdfs/Stofffluss-Energieaufwand-RC-Beton101102.pdf. Accessed 10 June 2022
31. Bundesministerium für Wohnen, Stadtentwicklung und Bauwesen: ÖKOBAUDAT-Release 2021-II Internet Database (2021). https://www.oekobaudat.de. Accessed 10 June 2022
32. Apellániz, D., Pasanen, P., Gengnagel, C.: A holistic and parametric approach for life cycle assessment in the early design stages. In: Proceedings of the Symposium on Simulation for Architecture and Urban Design (SimAUD) (2021). https://www.oneclicklca.com/wp-content/uploads/2021/05/SimAUD-2021-Camera_Ready-LCA_early_design_stages.pdf. Accessed 10 June 2022

Lightweight Reinforced Concrete Slab

Georg Hansemann(✉), Christoph Holzinger, Robert Schmid, Joshua Paul Tapley, Stefan Peters, and Andreas Trummer

Institute of Structural Design ITE, Graz University of Technology, Graz, Austria
georg.hansemann@tugraz.at

Abstract. Although increased efforts have been made in recent years to reduce the environmental impact in construction, greenhouse gas emissions in this sector are at record levels worldwide. In addition to reducing energy consumption and using environmentally friendly materials, the optimization and further development of conventional construction methods could make a significant contribution to achieving the given climate goals. In this regard, concrete, the most commonly used building material worldwide, plays a major role. Additive fabrication offers promising perspectives to change conventional reinforced concrete construction. 3D concrete printing does not require complex formwork construction and it makes targeted and economical fabrication of small quantities of concrete possible. This raises the question of whether the use of 3D printing technologies can reduce the CO_2 emissions of the construction industry and whether the digital planning and production used in the manufacturing process represents an economical alternative on the construction site compared to conventional construction methods [1]. Previous work shows that material savings of 30% to even 70% are possible by using this technology [2]. However, these values do not usually correspond to the CO_2eq saved. The calculations often do not take into account the high cement value of the printed concrete. In order to be able to exploit the potential of the technology and enable successful, large-scale use, it is essential to use a print material that meets the ecological and economic requirements [3].

This paper investigates the use of this new technology to produce a lightweight concrete slab using printed voids and additional in-situ concrete. It provides information on the design, the entire planning, the construction on site and the implementation in the context of a real construction project - a 100 m^2 slab with 130 3D-printed voids. The research project provides information on the functionality, economic efficiency, CO_2 savings, practicality and applicability of the new fabrication technology from the digital design to the production of the prefabricated parts in the production facility as well as the reinforcement and concreting of the structure on site. The project in Lunz am See proves that additive fabrication can be applied in construction practice in a timely manner and that the technology is suitable as a supplement to and further development of conventional construction methods. The wide-span slab construction is representative for a sustainable attitude towards the use of reinforced concrete, which, relies on economical, digital fabrication methods for saving resources.

Keywords: 3D concrete printing · Lightweight construction · CO_2 reduced construction · 3D-printed slab · Sustainable design

C. Gengnagel et al. (Eds.): DMS 2022, *Towards Radical Regeneration*, pp. 456–466, 2023.
https://doi.org/10.1007/978-3-031-13249-0_36

1 Introduction

Climate change and the achievement of the goals set in the Paris Climate Agreement of 2015 are one of the central challenges facing our society. We are already living with a multitude of resulting consequences and are still emitting proportionally far too much CO_2 for successful mitigation of the temperature rise. A closer look at the distribution of global energy-related CO_2 emissions shows that the building sector is one of the largest emitters with 38% [4].

While some savings have been achieved in recent years within the operation of buildings, there is a need for action in the area of grey emissions, which includes the production of building materials and the construction itself. In concrete construction, the load-bearing structure is responsible for a large proportion of the grey emissions due to the extensive use of concrete and reinforcement; in terms of load-bearing structure, this relates mainly to slab and roof structures. Thus, in the context of sustainability, the focus shifts to the reinforced concrete flat slab [5].

In this paper, the research group provides an answer to the question of how a lightweight ceiling construction can be realized with the aid of digital and parametric planning, calculation and production methods. In particular, the aim is to develop a practical construction method that saves as much CO_2 as possible while using modern approaches for planning and production [6, 7].

Fig. 1. Drone footage of Schloss Seehof in Lunz am See. Annex at the eastern wing during construction.

The described project is intended as a contribution to reducing concrete consumption in slab structures by using thin-walled mineral-based formwork produced by additive fabrication. These 3D-printed voids are positioned on the slab formwork and thus the concrete not involved in load transfer is reduced in the tension zone. This construction method of a mass-reduced slab, developed at the Institute of Structural Design (ITE) of the Graz University of Technology, was applied to a construction project at Schloss Seehof in Lunz am See, Austria (Fig. 1). The individual construction phases are shown below and form the basis for the development of a sustainable and competitive construction method.

2 Methods

2.1 Design

Hans Kupelwieser, Austrian sculptor, graphic artist, photographer and media artist, commissioned an extension and annex at the lower part of the eastern wing of Schloss Seehof in Lunz am See, Austria. The annex included a 100 m^2 room in masonry construction, which will serve as a studio. The design was developed by ITE in cooperation with Kupelwieser and it represents a combination of design and structural optimization in order to reach the mentioned goals [8] (Fig. 2).

The entire process, from initial design to production planning and the production itself, is based on a single, parametric model that has been carried from the beginning to the end of the project. Not only are the print paths and machine codes for the 3D-printed voids generated from this slab model, but it also serves as the basis for structural design and analysis. The meshing of the structural FEM model is as automated as the calculation itself. The results are in turn used for reinforcement planning in this 3D model. In consequence, a time-consuming manual reinforcement design for the inclined slab with slightly curved and length-variable beams could be parameterized and thus accelerated many times over. At the beginning of the iterative workflow the parametric design model, created by using the algorithmic visual scripting editor grasshopper3D, was very basic. The geometry was reduced to the essentials and related to the external dimensions of the structure and the axis lines of the ribs to be designed (Fig. 3). The shape determination concerning the number, course and curvature of the ribs could be determined in a cycle of influencing factors such as statics, size of the subdivisions, slab height and expected concrete savings. The information obtained about the structure in the first step, such as the required dimensions of the ribs and component heights, was adopted and defined the outer contours of the individual void bodies.

Fig. 2. Top view of the mass-reduced slab design

In a further step, the actual geometry of the 3D printed formwork could now be developed. The geometry was constantly checked for its load-bearing capacity in the pouring process. A single load and the concreting pressure were simulated as load cases in the analyses, and the stresses occurring in the printed formwork skin influenced the design of the void bodies, which led to an optimization of the structure and lightweight properties [9]. The selection of the right material thickness and suitable radii in the corner areas enabled a reduction of the stresses.

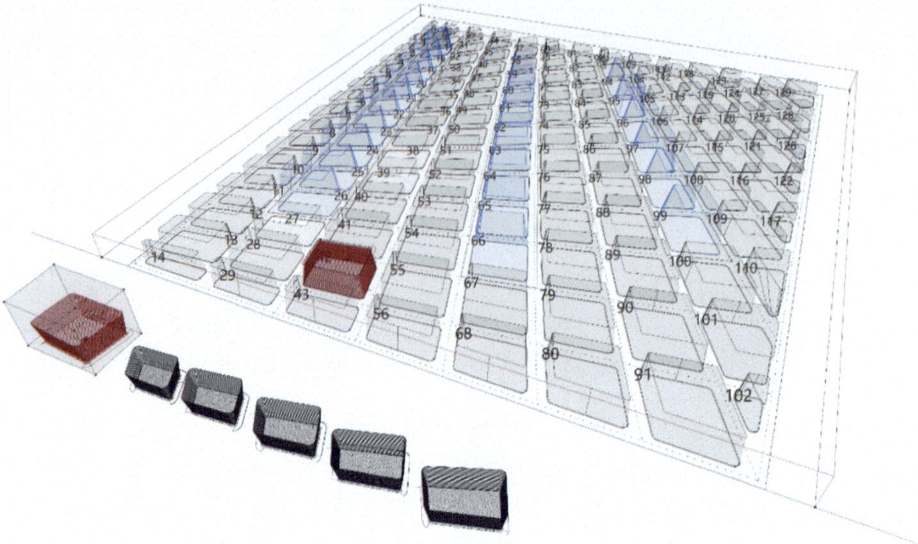

Fig. 3. Parametric 3D model in Rhino grasshopper

2.2 Additive Fabrication of the Voids

For the project, 130 different voids were produced using the 3D concrete printing system from Baumit in the Robot Design Lab at Graz University of Technology (Fig. 4). The individual prefabricated elements are used as lost formwork, they are unreinforced and do not perform any structural functions. Furthermore, the print-material acts as a part of the concrete cover. The elements weigh between 4 and 90 kg and reach dimensions between 20 cm × 15 cm × 25 cm (l/w/h) and 125 cm × 105 cm × 45 cm (l/w/h) while skylight elements reach a height of up to 70 cm. The travel speed or Tool Centre Point speed (TCP speed) of the robot during printing was 300 mm/s, resulting in an effective printing time of 18 h to produce all the void elements. 5.6 t of the Baumit printing material (Printcrete 230 N) specially adapted to the printing system was used in total. The cross section of the Printed Concrete Line (PCL) was set to 8 mm × 18 mm [10].

The box-shaped voids were printed upside down and are assembled from a flat base/ lid, and tapered walls (Fig. 4). In order to avoid defects in the lid and to prevent predetermined breaking points, the two layers of the flat base were built up in cross-ply layers. The lid is extended over the side walls to obtain a clean appearance on the inside. The resulting cantilever enables better handling on site and simplifies the lifting and positioning of the elements.

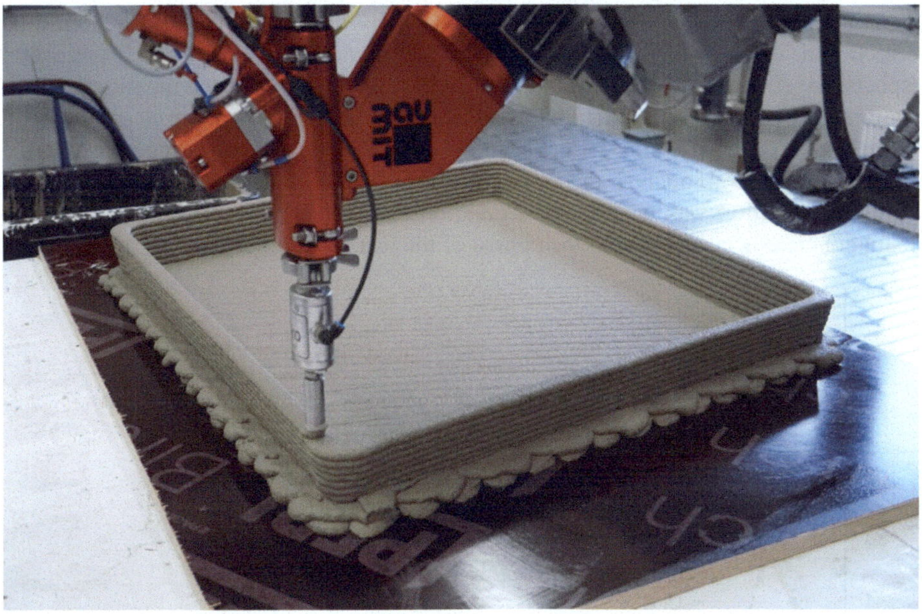

Fig. 4. Production of a void in the robot design laboratory of the Graz University of Technology using the Baumit concrete printing system (Bauminator)

As a proof of concept, a large-scale test was carried out in a former research project at the Laboratory for Structural Engineering at Graz University of Technology to demonstrate the consistency and robustness of the new construction method (Fig. 7). A lightweight slab segment with a length of 8 m was produced and loaded with different load scenarios (Fig. 5).

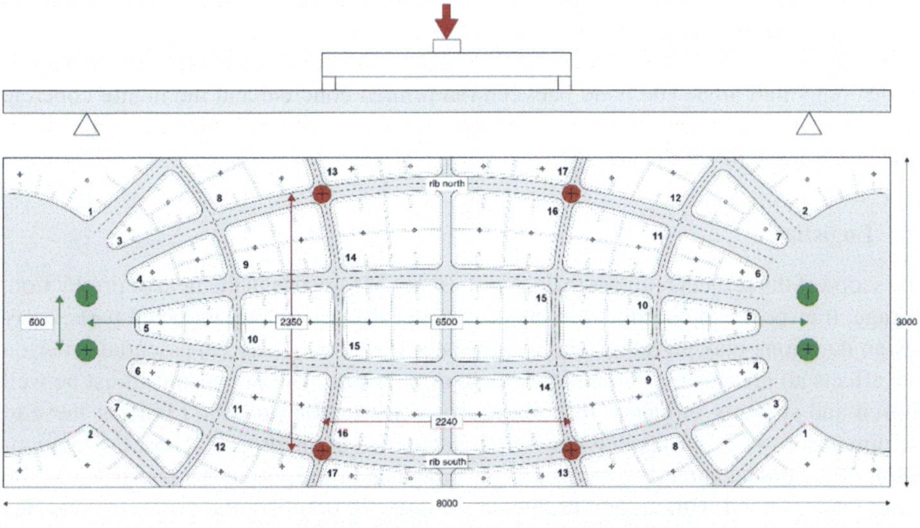

Fig. 5. Schematic of the experimental setup

Thus, on the one hand, the load-bearing capacity of the overall structure could be verified and, on the other hand, the serviceability could be demonstrated in several respects. At serviceability load, a maximum deformation of 9.6 mm was measured, which is within the permissible limits (Fig. 6). The maximum crack width at serviceability load was measured to be 0.3 mm. After reaching the design load for bearing capacity, the test was continued to analyse the behaviour of the void elements in the structure. An interesting question was whether delamination between the printed concrete and the in-situ concrete occurred during the course of the test.

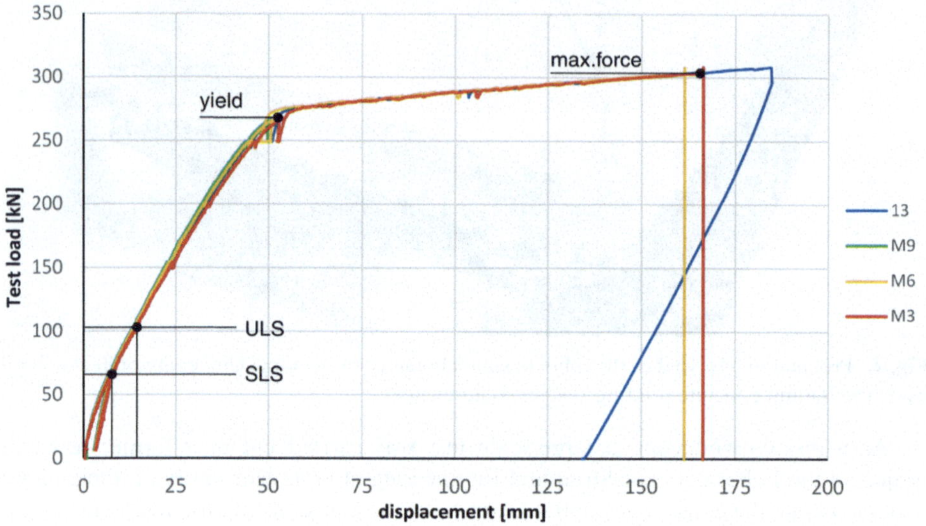

Fig. 6. Force deformation curve

The test specimen was loaded up to 4.5 times the serviceability load or 3 times the design load, resulting in a deformation of 200 mm. At this point, there was no failure of the structure but the stroke of the test equipment was over.

Despite this enormous deformation, there was no delamination or other abnormalities in the compound area. The bond between the printed concrete and the in-situ concrete can be regarded as quasi-monolithic, which was also particularly evident in the well-distributed crack pattern [11].

2.3 Logistics

The scope of the project requires to deal with issues of logistics in the field of production, storage, transport and coordination on site. In a project using prefabricated parts, early design decisions play an important role for later logistics issues. An essential criterion that affects all areas is the choice of element dimensions. The dimensions must be well chosen and strongly influence the subsequent construction process. Limiting these to the dimensions of a Euro pallet was a great advantage in the project. Thus, the manipulation of the bodies directly after printing, storage and transport to the construction site could be organized without any problems. In order to optimize the processes and due to the quantity of prefabricated parts produced, digital aids from production planning

Fig. 7. Large scale test of a lightweight slab segment with 3D printed voids

were also used such as nesting tools and organising and labelling the voids regarding the laying sequence and fitting process on the construction site. The shape of the 130 individual parts is not suitable for stacking. For space-saving storage of the elements, a suitable stacking or shelving system is recommended, which would also optimize transport capacity utilization (Fig. 8).

Fig. 8. Voids placed on the slab formwork and reinforcement

2.4 Sustainable Construction Method

The selected slope direction for drainage favours the ratio of beam heights to the respective spans, so the reinforcement tonnage can be reduced by 30% compared to an equivalent monolithic flat slab. In this comparison, the new construction method also achieves a concrete saving of 40%.

When carrying out a life cycle assessment, however, it must be considered that the printing concrete used for the voids has a much higher CO_2eq than the normal concrete used for the loadbearing structure due to its high cement ratio. The material saving therefore does not correspond to the CO_2 saving, which in this case is approx. 30%.

For a generative design, regarding a sustainable and resource-saving construction, the material-specific volumes of the parametric 3D model were deposited with the corresponding CO_2eq values and the concrete saving potentials of the 3D printed void bodies were graphically displayed [12].

3 Discussion

According to the authors of this paper, a major advantage of digital calculation, design and fabrication methods is the possibility to design load-bearing concrete components in a much more differentiated way. A good and urgently needed motivation for a paradigm shift in this context is the timely exploitation of all potentials for a more economical use of materials for reinforced concrete structures, of course without violating the necessary and reasonable basic rules of economic efficiency. Only if these are adhered new opportunities for value creation arise from them.

This project shows that digital planning methods with all their possible complexity can be integrated into a real construction process, which can even profit from it. New approaches are necessary, architects, planners, contractors and clients must be convinced to move away from their conservative approaches and give new technologies a chance in the broad application, not only in the academic environment [13].

The presented project of a 40% lighter slab (Fig. 9) is representative for a multitude of further possibilities. Following a similar method, reinforced concrete wall elements and foundation components are currently being researched. It is the responsibility of all building professionals, from clients, approving institutions and planners to the contractors, to demand and exploit the possibilities mentioned in this article in the future. Other research tasks are already defined and in progress, such as printing methods with integration of reinforcement and modification of the mineral printing material for further gradation and differentiation of load-bearing cross-sections. Meanwhile, it will be interesting to see how long it will take before the inevitable consequences of climate change fundamentally change the specifications for the construction of our load-bearing structures.

Fig. 9. Ceiling after removing the formwork

References

1. Anton, A., Reiter, L., Wangler, T., et al.: A 3D concrete printing prefabrication platform for bespoke columns. Autom. Constr. **122**, 103467 (2021). https://doi.org/10.1016/j.autcon.2020.103467
2. Block, P., Rippmann, M., van Mele, T.: Compressive assemblies: bottom-up performance for a new form of construction. Archit. Des. **87**, 104–109 (2017). https://doi.org/10.1002/ad.2202
3. Dey, D., Srinivas, D., Panda, B., et al.: Use of industrial waste materials for 3D printing of sustainable concrete: a review. J. Clean. Prod. **340**, 130749 (2022). https://doi.org/10.1016/j.jclepro.2022.130749
4. Global Alliance for Buildings and Construction: Global Status Report for Buildings and Construction: towards a zero-emissions, efficient and resilient buildings and construction sector. UN Environment Programme, Nairobi (2020)
5. Weidner, S., Mrzigod, A., Bechmann, R., et al.: Graue Emissionen im Bauwesen – Bestandsaufnahme und Optimierungsstrategien. Beton- und Stahlbetonbau **116**, 969–977 (2021). https://doi.org/10.1002/best.202100065
6. Schmid, R., Hansemann, G., Autischer, M., et al.: Adaptive foam concrete in digital fabrication. In: Buswell, R., Blanco, A., Cavalaro, S., et al. (eds.) DC 2022, vol. 37, pp. 22–28. Springer, Cham (2022). https://doi.org/10.1007/978-3-031-06116-5_4
7. Anton, A., Jipa, A., Reiter, L., et al.: Fast complexity: additive manufacturing for prefabricated concrete slabs. In: Bos, F., Lucas, S., Wolfs, R., Salet, T. (eds.) DC 2020, vol. 28, pp. 1067–1077. Springer, Cham (2020). https://doi.org/10.1007/978-3-030-49916-7_102
8. Hansemann, G., Schmid, R., Holzinger, C., et al.: Additive fabrication of concrete elements by robots: lightweight concrete ceiling. In: Fabricate, pp. 124–129 (2020)

9. Nguyen-Van, V., Panda, B., Zhang, G., et al.: Digital design computing and modelling for 3-D concrete printing. Autom. Constr. **123**, 103529 (2021). https://doi.org/10.1016/j.autcon.2020.103529
10. Hansemann, G., Schmid, R., Holzinger, C., et al.: Lightweight reinforces concrete slab. In: CPT, pp. 69–73 (2021)
11. Peters, S., Trummer, A., Hansemann, G., et al.: Gedruckte Schalungen für den Stahlbeton-Leichtbau. Detail **2020**, 14–16 (2020)
12. Agustí-Juan, I., Habert, G.: Environmental design guidelines for digital fabrication. J. Clean. Prod. **142**, 2780–2791 (2017). https://doi.org/10.1016/j.jclepro.2016.10.190
13. Adaloudis, M., Bonnin Roca, J.: Sustainability tradeoffs in the adoption of 3D Concrete Printing in the construction industry. J. Clean. Prod. **307**, 127201 (2021)

Growth-Based Methodology for the Topology Optimisation of Trusses

Christoph Klemmt(✉)

University of Cincinnati, Cincinnati, USA
christoph@orproject.com

Abstract. This paper presents a novel methodology for a growth-based topology optimization of trusses. While most methods of topology optimization are based on voxel grids that result in free-form volumes, the topology optimization of trusses exists as subtractive methods that start with a large number of initial beams. This method instead commences with a minimal amount of beams. The model is iteratively refined by node repositioning and node division according to structural forces to arrive at a complex truss. Case studies of a cantilever and a table show the results and reduction in mass achieved by the algorithm.

Keywords: Topology optimization · Truss · Growth · Michell cantilever

1 Introduction

In order to reduce the energy used in the production of load-bearing system, the development of lightweight structures for architectural applications is paramount, achieved by tools of Topology Optimization (TO). This paper presents a novel methodology for the TO of bending-stiff trusses that is based on principles of growth by node division. Inspired by the way that organisms grow by processes of cell division, including the growth of their structural systems such as leaf veins or trabecular bone, an initial simple truss is continuously refined by the insertion of new nodes and beams.

The most widely used methods of TO are based on a 3D grid of voxels and result in free-form solid volumes, which are often 3D-printed or CNC-milled (Aage et al. 2017). While the CNC-milling causes large amounts of removed waste material, 3D-printing is often praised for its minimal use of material for the production of a part. However, 3D-printing has been identified to cause extensive waste via failed test prints, and most of the commonly used materials are unsustainable thermo-polymers. Even bioplastics such as polylactic acid (PLA) do not biodegrade when placed in a landfill, while large-scale metal or cement printing require large amounts of energy in the production. It is therefore important to develop methods of TO for the generation of structures that can easily be constructed from components such as timber beams, or standard linear extrusions that have a possibility to be reused.

The proposed methodology is therefore based on linear beam elements, which are more easily applicable to architectural structures than the free-form geometry of a voxel-based TO, which requires further rationalisation for constructability. Most TO methods

C. Gengnagel et al. (Eds.): DMS 2022, *Towards Radical Regeneration*, pp. 467–475, 2023.
https://doi.org/10.1007/978-3-031-13249-0_37

of trusses are subtractive in nature, such as the Ground Structure method, whereby an initially large set of elements is iteratively reduced (Dorn et al. 1964; He and Gilbert 2015). Our method instead commences with an initial simple network of beams that is continually adjusted and refined according to its structural performance (Fig. 1).

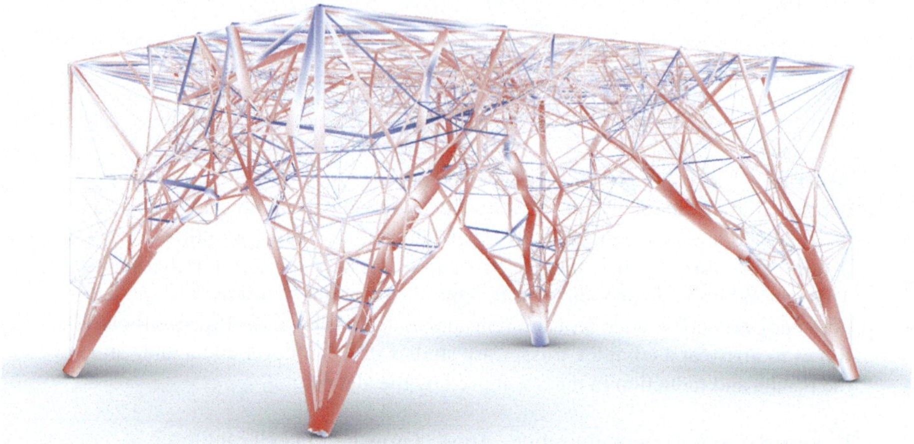

Fig. 1. 3D roof structure generated by the proposed methodology for TO of trusses.

2 Related Work

2.1 Growth Simulations

Computational simulations of growth processes have a long history, with early examples such as Cellular Automata or Diffusion Limited Aggregation (Gardner 1970; Witten and Sander 1981). As computational possibilities increased, more recent attempts focus on modelling growth at the level of individual cells (Lomas 2014; Andrasek 2016). This is pursued in the fields of design and architecture with the aim to create digital art, either in the form of videos that illustrate the development, or by generating specific geometries that may become sculptural art or architectural geometry.

In the sciences, computational growth modelling has been a method in developmental biology in order to gain an increased understanding of the processes that shape organisms and to test hypotheses relating to their formation (Walpole et al. 2013; Palm and Merks 2015). In oncology, simulations of tumour development are used as a predictive tool, as well as a method to predict the effects of therapeutic treatments (Shirinifard et al. 2009, Bearer et al. 2009, Neufeld et al. 2013).

2.2 Topology Optimization

The most common method for TO is based on a 3-dimensional grid of voxels that encompasses the design domain, with loads, supports and boundary conditions applied at individual voxels (Rozvany 2001; Xie and Steven 1993; Bendsøe and Sigmund 2013). Iteratively, a Finite Element Analysis (FEA) is calculated for the grid of voxels, with solid voxels added to the structure and underutilized voxels being removed. This results in a solid volume of voxels that is suitable for a load transfer, commonly with the solid smoothed into a free-form geometry. The method is used extensively, especially in areas such as mechanical engineering, where the resulting parts are 3D-printed, CNC-milled, or cast with CNC-milled moulds.

While this works well for small-scale parts that are made in one piece, larger-scale objects and structures are commonly assembled from discrete parts, including building structures, bridges, or the frames of transport vehicles. For such structures, the voxel-based approach leads to the problem of a discretization of the free-form surface, which requires postprocessing and may yield non-optimized assemblies (Subedi et al. 2020). Consequently, TO has so far only found limited applications at the large scale. An alternative are methodologies of TO of trusses, which are based on networks of linear beams. Most of those are subtractive in nature, such as the Ground Structure method, whereby an initially very large set of beams is iteratively reduced according to the FEA (Dorn et al. 1964; He and Gilbert 2015). Its biggest drawback is the computational cost of the large amount of initial beams, and different methods for a post-processing of the resulting geometries have been proposed. Nevertheless, the method is so far used far less widely that the voxel-based TO.

Also additive methods of a TO of trusses have been proposed, whereby iteratively more beams are inserted into an initial simple truss. However, in the literature so far those methodologies have remain 2-dimensional, with examples in 3D consisting of no more than 7 beams (Martínez et al. 2007, Kwok et al. 2016). The algorithm presented here therefore constitutes the first functioning implementation of an additive, growth-based TO of trusses.

3 Methodology

3.1 Repositioning

The algorithm commences with an initial truss, global loading and boundaries conditions that the structure is to support, and possibly a design domain that the structure has to remain within. Iteratively an FEA is calculated, and based on this, decisions are taken at the local level of each node that guide its behaviour of repositioning, as well as the possibility of a node division. Two major behaviours of repositioning for a node have been developed:

Firstly, a straightening of adjacent elements that both act under either compression or tension, ForceAxialAngle. This is achieved by identifying two elements that meet at a node at an angle $> 0.5\ \pi$, and by moving the node along the angle bisector of those two beams. An increased strength of the forces acting in the elements leads to a larger movement of the connecting node. The behaviour is applied at each node for the beams that act strongest in compression as well as in tension.

Secondly, an arrangement of two beams with one acting in compression and one in tension, whereby those either attempt to achieve an orthogonal, or alternatively an acute angle, ForceOffAxialAngle. The behaviour moves the node in the opposite direction of the angle bisector of the two elements that act strongest in compression and tension. An increased strength of the forces acting in the elements leads to a larger movement of the connecting node.

Those two behaviours are combined with a spring force that attempts to equalize the lengths of two equally loaded beams that are aligned with each other, ForceSpring. This only acts on beams that are both in compression or both in tension, by moving the node away from the closer and towards the further of the two beam endpoints.

Based on those behaviours, the new positions of the nodes are calculated. In order to avoid an undesirable intersection of elements, the nodes are further prevented from crossing their surroundings.

3.2 Division

Every set number of iterations, the strongest loaded node of the network is identified for division. Certain conditions will prevent a node from dividing and select the next node instead, for example if any of the surrounding nodes will have too many connections after the division, or if the division results in irregularities within the network. A new node is inserted adjacent to the dividing node, and beams are established to connect it to the surrounding network.

The execution of the algorithm is terminated either after a set number of divisions, or if a maximum permissible displacement is achieved. In a further step, underutilized elements are removed, and the sizes of the elements are adjusted utilizing a commercially available cross-section optimizer (Preisinger 2013, Preisinger 2016).

In order to provide results that are comparable, the mass of the structure is kept constant, so that its maximum and mean node displacements can be evaluated. A decrease in the displacements then equals a desired decrease of the mass of the structure while remaining within a maximum permissible displacement. For this work, the mass rather than the maximum displacement was kept constant as a model can easily be scaled to it.

4 Results

The methodology was applied to a variety of case studies in 2D and 3D. A specific aim was the verification of the resulting geometries on a quantitative level in regards to the achieved displacements of the structure, and on a qualitative level in comparison to known solutions for architectural structures. Investigations therefore focussed on the case study of the Michell/Prager cantilever, and on the case study of a simple bridge or table structure.

4.1 Michell/Prager Cantilever

The Michel/Prager cantilever is a horizontally symmetric structure with two supports on the left and one downwards acting load on the right. The optimal mass distribution for this case study was calculated by Michell, and discretized into linear elements by Prager (Michell 1904; Prager 1977). The growth simulation starts with a simple truss that connects the loads and supports around one centrally placed node (Fig. 2). Different setting were investigated, with successful results for the strengths ForceSpring 0.015, ForceAxialAngle 5.0 and ForceOffAxisAngle 2.0. The simulation was terminated after 200 divisions.

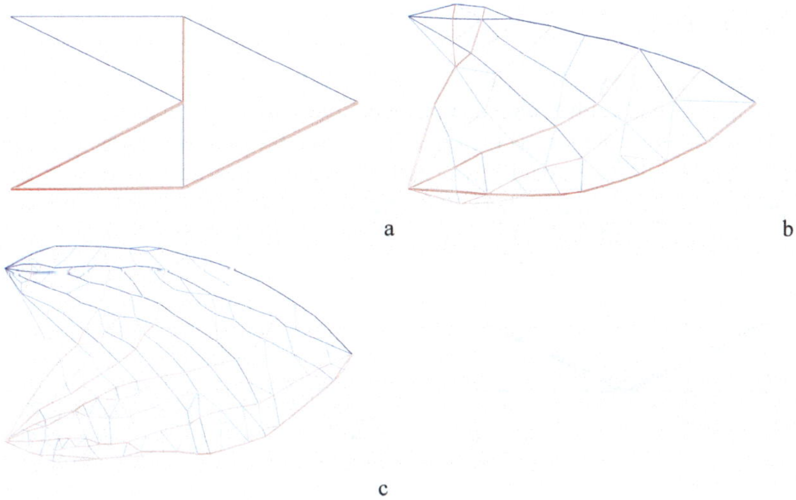

Fig. 2. Development of the model during its growth at divisions 0, 40 and 200. Blue in tension, red in compression.

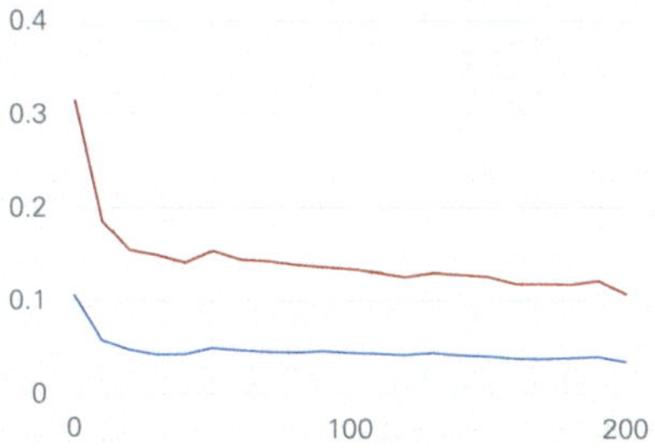

Fig. 3. Development of the maximum displacement (red) and mean node displacement (blue) during the simulation.

As shown in Fig. 3, on a quantitative level the displacements of the structure significantly decrease during the course of the simulation. This shows an improvement of the structural performance at a constant mass, which can be translated into a reduced mass for a constant permissible displacement. The resulting structure is therefore shown to require less mass, less cost and less energy consumption in its material production.

On a qualitative level, the resulting geometry has close similarities to the Michell/Prager cantilever. Those include curved, outwards-bulging beams that connect the supports to the load and that intersect at near orthogonal angles. The structure is near symmetric horizontally, with the element density increased at the loads. Major differences result from the triangulation of the structure, whereas the Prager cantilever is quadrangulated, as well as the iterative development of the growth structure, which necessarily results in deviations when a single node divides without its neighbours or mirrored nodes dividing simultaneously.

4.2 Table Structure

The table or bridge structure has two supports at the bottom corners and supports a downwards acting load applied along a horizontal line above. For materials that act better in compression than in tension, as the material chosen for this case study, the formation of an arch is expected. As the cantilever example, the initial structure connects the loads and supports around a central node (Fig. 4). Settings used were ForceAxialSpring 0.15, ForceAxialAngle 5.0 and ForceOffAxisAngle 2.0, terminated after 200 divisions.

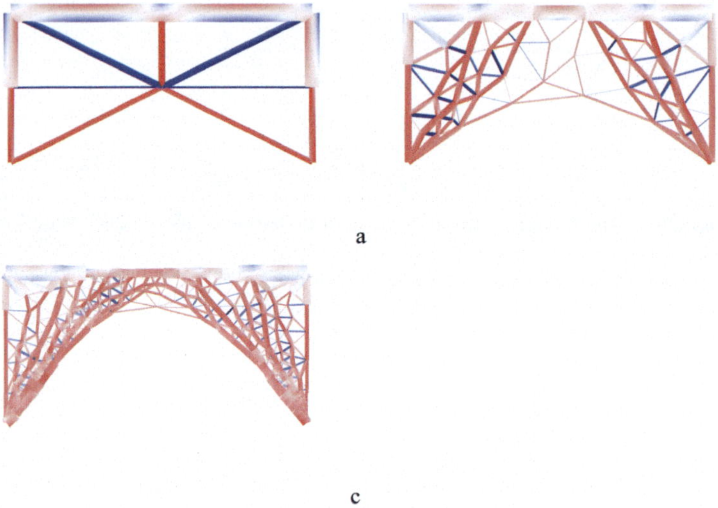

Fig. 4. Development of the model during its growth at divisions 0, 40 and 200. Blue in tension, red in compression.

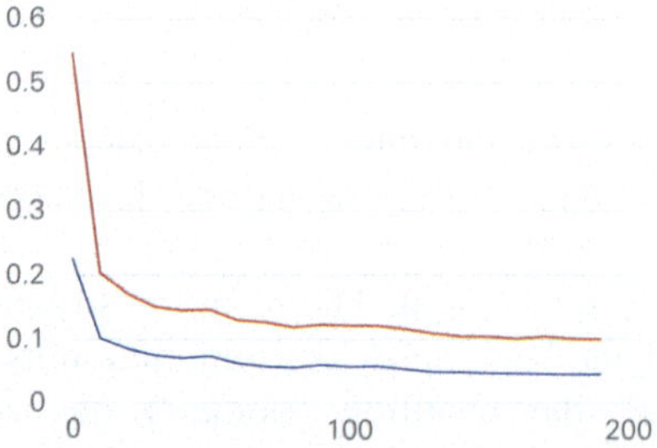

Fig. 5. Development of the maximum displacement (red) and mean node displacement (blue) during the simulation.

Figure 5 shows the improvement in performance during the course of the simulation, similar to the Michell/Prager cantilever example. The continuous decrease of the maximum and mean displacements at a constant mass result in a reduction of mass, cost and energy consumption when staying within a defined maximum displacement.

At first linear elements start to form from the supports towards the top load (Fig. 4b). The expected formation of compressive arches then clearly emerges in Fig. 4c. Similar to the Michell/Prager truss example, not a perfect symmetry, but a near symmetry can be observed between the left and right sides of the structure.

A 3D version of the table structure was calculated with the values of ForceAxialSpring 0.005, ForceAxialAngle 0.025, and no ForceOffAxialAngle (Fig. 1). As in the 2D version, arches are forming that start at the supports and branch out to support the load above.

5 Evaluation

Quantitatively, both examples clearly show a significant decrease of the maximum and mean displacements of the structures during the course of their development. The topology of the resulting structures is therefore significantly optimized in comparison to the initial structures. However, while this proves the algorithm's ability for optimization, the logic has been developed heuristically, and there is currently no proven optimum for the strength values of the repositioning behaviours. Future work should therefore seek a mathematical foundation for the setting of those parameters, or devise a method for their optimization.

Qualitatively, the resulting geometries closely match the expected forms based on know results for the case studies. However, while the overall appearance matches the expected results, there are clear deviations on the detail levels, such as a near symmetry instead of a perfect symmetry for perfectly symmetric start conditions. This can be explained by the iterative approach of the algorithm, whereby always only one node divides, resulting in an imbalance between the two mirrored halves of the structure that leads to continuing deviations. However, as the algorithm was inspired by the cell-based growth of organisms, the similarity to the near symmetry found in natural structures such as leaf veins is appreciated.

In terms of fabrication, the proposed methodology has the advantage over voxel-based approaches that the resulting geometries are trusses and can be easily constructed as such from discrete parts. The beams can be made from readily available standard extrusions, utilizing re-usable elements or timber components. More complex is the manufacturing of the large number of unique nodes, where currently several beams meet at non-repetitive angles. A simplification has been implemented within the algorithm by limiting the number of beams that can meet at a node, and by setting a minimum distance between nodes. Attempts at limiting the beams to meet a node at discrete angles have not yet yielded functioning results. Further work should therefore be carried out to limit node complexity.

It is also important to note that currently, the models are completely triangulated, or in 3D tetrahedralized. This is necessary in the current implementation so that each node is aware of its surrounding nodes to prevent the occurrence of beam intersections, even though some of the connecting beams may not be utilized structurally at all. Therefore prior to the removal of underutilized elements, the resulting geometries are always triangulated. Alternatives to this full triangulation could be explored if other methods are used to prevent an intersection of beams.

6 Conclusions

The proposed methodology of TO for trusses based on growth by node division has been shown to successfully generate optimised lightweight structures. It forms a novel tool for structural optimisation, specifically aimed at trusses that are constructed from discrete parts, leading to a reduction of material cost, energy consumption and resources. Unlike most voxel-based approaches to TO that rely on 3D-printing or CNC-milling for manufacturing, the methodology is specifically useful for the fabrication of structures from standard extrusions or timber beams, which at the end of the structure's lifecycle allow for a re-use or decomposition respectively. Future work should address improvements to the algorithm that reduce possible errors and unsuitable results, as well as buildability by limiting the angles at which beams can meet at a node.

Acknowledgements. This project was carried out as part of a doctorate under the supervision of Klaus Bollinger at the University of Applied Arts Vienna.

References

Aage, N., Andreassen, E., Lazarov, B.S., Sigmund, O.: Giga-voxel computational morphogenesis for structural design. Nature **550**(7674), 84–86 (2017)

Andrasek, A.: Xenocells: in the mood for the unseen. Archit. Des. **86**, 90–95 (2016)

Lomas, A.: Cellular forms: an artistic exploration of morphogenesis. In: SIGGRAPH Studio (2014)

Bearer, E.L., et al.: Multiparameter computational modeling of tumor invasion. Cancer Res. **69**(10), 4493–4501 (2009)

Bendsøe, M.P., Sigmund, O.: Topology Optimization: Theory, Methods, and Applications. Springer, Heidelberg (2013)

Dorn, W., Gomory, R., Greenberg, H.: Automatic design of optimal structures. J. De Mécanique **3**(1), 25–52 (1964)

Gardner, M.: Mathematical Games: the fantastic combinations of John Conway's new solitaire game "Life." Sci. Am. **223**, 120–123 (1970)

He, L., Gilbert, M.: Rationalization of trusses generated via layout optimization. Struct. Multidiscip. Optim. **52**(4), 677–694 (2015). https://doi.org/10.1007/s00158-015-1260-x

Kwok, T.H., Li, Y., Chen, Y.: A structural topology design method based on principal stress line. CAD Comput. Aided Des. **80**, 19–31 (2016)

Martínez, P., Martí, P., Querin, O.M.: Growth method for size, topology, and geometry optimization of truss structures. Struct. Multidiscip. Optim. **33**(1), 13–26 (2007)

Michell, A.G.M.: The limits of economy of material in frame-structures. Lond. Edinb. Dublin Philoso. Mag. J. Sci. **8**(47), 589–597 (1904)

Neufeld, E., Szczerba, D., Chavannes, N., Kuster, N.: A novel medical image data-based multiphysics simulation platform for computational life sciences. Interface Focus **3**(2), 4 (2013)

Palm, M.M., Merks, R.M.H.: Large-scale parameter studies of cell-based models of tissue morphogenesis using CompuCell3D or VirtualLeaf. In: Nelson, C.M. (ed.) Tissue Morphogenesis. MMB, vol. 1189, pp. 301–322. Springer, New York (2015). https://doi.org/10.1007/978-1-4939-1164-6_20

Prager, W.: Optimal layout of cantilever trusses. J. Optim. Theory Appl. **23**(1), 111–117 (1977)

Preisinger, C.: Linking structure and parametric geometry. Archit. Des. **83**(2), 110–113 (2013)

Preisinger, C.: Parametric Structural Modeling - Karamba - User Manual for Version 1.2.2 (2016)

Rozvany, G.I.N.: Aims, scope, methods, history and unified terminology of computer-aided topology optimization in structural mechanics. Struct. Multidiscip. Optim. **21**(2), 90–108 (2001). https://doi.org/10.1007/s001580050174

Shirinifard, A., Gens, J.S., Zaitlen, B.L., Popławski, N.J., Swat, M., Glazier, J.A.: 3D multi-cell simulation of tumor growth and angiogenesis. PLoS ONE **4**(10), 10 (2009)

Subedi, S.C., Verma, C.S., Suresh, K.: A Review of methods for the geometric post-processing oftopology optimized models. J. Comput. Inf. Sci. Eng. **20**(6), 12 (2020)

Walpole, J., Papin, J.A., Peirce, S.M.: Multiscale computational models of complex biological systems. Annu. Rev. Biomed. Eng. **15**, 137–154 (2013)

Witten, T.A., Jr., Sander, L.M.: Diffusion-limited aggregation, a kinetic critical phenomenon. Phys. Rev. Lett. **47**, 1400 (1981)

Xie, Y.M., Steven, G.P.: A simple evolutionary procedure for structural optimization. Comput. Struct. **49**(5), 885–896 (1993)

RotoColumn

A Continuous Digital Fabrication Framework for Casting Large-Scale Linear Concrete Hollow Elements

Samim Mehdizadeh[1](✉), Adrian Zimmermann[2], and Oliver Tessmann[1]

[1] Digital Design Unit (DDU), Technical University of Darmstadt, Darmstadt, Germany
Mehdizadeh@dg.tu-darmstadt.de

[2] Institut für Werkstoffe in Bauwesen (WiB), Technical University of Darmstadt, Darmstadt, Germany

Abstract. The research project RotoColumn aims to produce hollow, vertical concrete elements in various shapes as self-carrying permanent formwork for load-bearing concrete elements. The work is motivated by the urgent need to improve how we build with concrete. The project conceives the digital design simulation, robotic fabrication technologies, and advances in concrete rheology as the drivers for improvements in state of the art in construction industry. This paper unfolds the sequencing of individual methods as a part of the larger research goal, described through the digital simulations and different series of large-scale physical prototypes, in collaboration with the Industry partner. The project RotoColumn is a part of the research trajectory Rotoform, a dynamic robotic casting method for manufacturing hollow concrete building elements.

Keywords: Design simulation and robotic fabrication · Flexible formwork · Sequential dynamic casting · Rotoform · Digital concrete

1 Introduction

The building industry has the highest material consumption among all the sectors worldwide. Especially concrete construction is responsible for a considerable part of solid waste as formwork material and carbon emission for producing cement. The amount of cement produced in 2014 is three times more than in 1995 (Kromoser and Kollegger 2017); on the other hand, the technology of the prefabricated concrete elements does not show a significant change over the last 50 years. It is crucial to rethink building methods with concrete to use the minimum amount of the material with the most durable structures. Kromposer addresses three different strategies (Kromoser et al. 2019). A) structural design and form optimized elements, B) compact concrete material, which uses fly ash and other industrial waste, C) building light-weight with concrete and rethinking the reinforcement material. The awareness of the destructive consequences of the ubiquitous use of concrete is increasing (Watts 2019), and a lot of recent research addresses this problem. Therefore, we aim at using advances in digital technologies to innovate

C. Gengnagel et al. (Eds.): DMS 2022, *Towards Radical Regeneration*, pp. 476–486, 2023.
https://doi.org/10.1007/978-3-031-13249-0_38

concrete construction. A team of researchers and practitioners from the construction industry (Spannverbund GmbH./Prof. Stefan Böhling) developed a 1:1 scale prototype to bridge the gap between the innovative research field and the actual state of the building industry by:

Fig. 1. A six-axis robotic arm is a rotoforming a twisted hyperelastic membrane formwork and wooden scaffolding.

- Reducing concrete consumption through lightweight, hollow concrete elements.
- Innovating efficient material allocation through rotational robotic trajectories in conjunction with concrete rheology.
- Saving material through membranes as formwork, held in place by reusable timber scaffolds, assembled and re-assembled by robots.
- Construction efficiency through automation, prefabrication, and modularity.
- The prototyping exploration is based on the coalescence of advanced technologies such as digital simulation, robotics, and material science (Fig. 1).

2 Background

The emergence of digital technologies and utilization of (CAD/CAM) design for fabrication methods increased the possibilities for materializing complex and individual geometries and reduced material consumption (Wangler et al. 2016). Combining digital fabrication with the precise control of concrete flow characteristics is the aim of the

RotoColumn project. The challenges of digital concrete are seen in the following topics: textile form finding (Ahlquist and Menges 2013), fabric formwork, digitally-driven material placement, and concrete rheology. In this context Andrew Kudless investigates the relationship between fabric as concrete formwork and digital form-finding tools as the design driver (Kudless 2011). The enormous potential of innovative fabric formwork for reducing the material waste of formwork is explored at various research institutes (Veenendaal et al. 2011). The interconnection of robotic positioning and concrete rheology has been shown in projects such as Smart Dynamic Casting at ETH Zürich (Lloret et al. 2015). The research trajectory Rotoform contributes to the digital fabrication with concrete research area (Buswell et al. 2020) as a formative process.

2.1 On Rotoform

RotoColumn is a project within a larger research trajectory called RotoForm, conducted by the Digital Design Unit (DDU) at TU Darmstadt. RotoForm is based on Rotomoulding, a fabrication method used in the plastic industry to make large-scale hollow bodies such as kayaks and water tanks: Low amounts of liquid material or granules are cast into a heated mold that slowly rotates (Wang et al. 2011). The material then disperses along the mold surface and adheres to the formwork creating an inner cavity. RotoForm transfers this technique to the concrete and building industry realm (Tessmann and Mehdizadeh 2019a). We pour a minimum amount of concrete to only cover the surface of the formwork through rotation, creating hollow elements with inner cavities (Tessmann and Mehdizadeh 2019b). This method reduces material consumption and the hydrostatic pressure of liquid concrete exposed to the formwork, enabling the utilization of lightweight hyperelastic membrane as a formwork without deviation and deformation. In the RotoColumn project, we aim at fabricating individual hollow concrete elements by utilizing computational design and simulation methods and robotic fabrication. The target of the research project RotoColumn in this stage is to materialize the self-carrying permanent concrete formwork. The inner cavity of the object allows for the integration of the reinforcement rebars and entirely casting on the construction site.

3 Methods and Data

The RotoColumn project is based on a fabrication-aware design process in which material properties and fabrications constraints are digitally simulated in the early design process. The formerly linear design process turns into a feedback loop and allows to explore solution spaces constrained by materials and processes (Fig. 2).

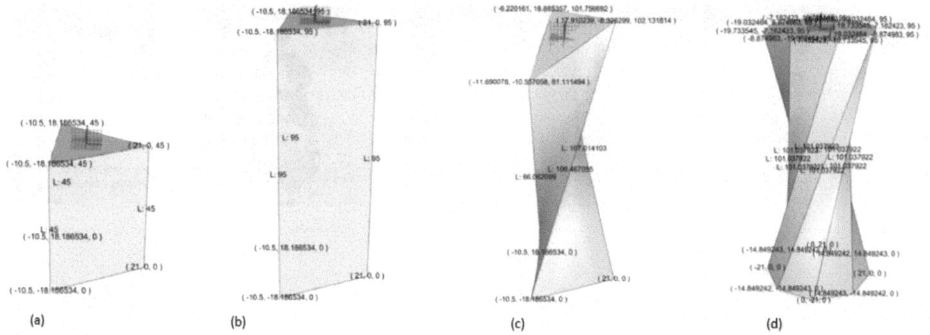

Fig. 2. The parametric model and the digital mass-spring simulation define the spectrum of possible twisted geometries for the formwork system. (a) length; (b) twist angle; (c) tilt angle between the caps; (d) number of faces

3.1 On the Formwork System for the RotoColumn

This project identifies two groups of experiments and prototypes based on geometrical constraints of the hyperelastic membrane and the scaffold system. Both experiments target a continuous digital fabrication workflow comprised of the following steps:

1. Digital simulation
2. Formwork system and formation
3. Scaffold system and assembly procedure
4. Robotic trajectory and concrete flow characteristic

3.2 Hyperelastic Formwork System

A parametric design model defines a spectrum of possible twisted geometries (Fig. 2). A digital simulation with a particle-spring model in Kangaroo II (Piker 2013) generates a tailoring pattern for the planar hyperelastic membrane before stretching. After the tailoring and attaching phase, a six-axis robotic arm stretches the membrane formwork and precisely positions it according to the parametric design model. Subsequently, the reusable timber scaffold bars that fix the membrane position are assembled manually (Fig. 3.a). A 3D scan shows the deviations of the physical prototype from the digital simulation.

The second group of the formwork systems focuses on the hyperelastic membrane and adaptive scaffolding system. The adaptive scaffold is a cylinder-shaped cage comprised of plywood caps, 3d printed connectors, radial aluminum rails, and geometric articulation elements. This group of prototypes allows for achieving geometric complexity without tailoring the membrane. Instead, it reaches the complexity with a set of articulative elements attached to the surface of the membrane. The cutting pattern is generated with the same simulation system. Subsequently, we attach the membrane to

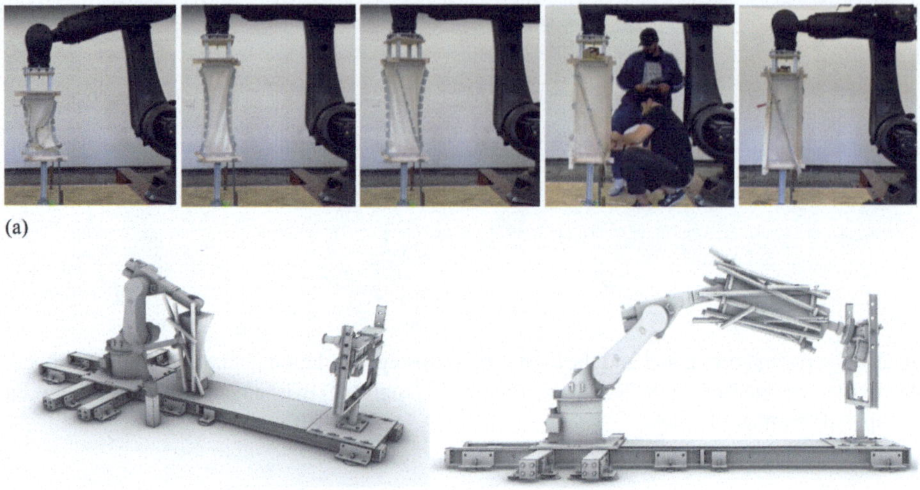

(a)

(b)

Fig. 3. Robotic membrane stretching and positioning; assembly of formwork timber material as scaffolds for the hyperelastic membrane formwork system. (a) human robot collaborative process; (b) L: complete automated process from formwork stretching with a second robot; R: robotic rotoforming with an industry robot and anchor point.

the scaffold and the articulation elements to the radial aluminum rails (Fig. 4). After positioning the anchor points with marker-based holographic projection and the Microsoft-HoloLens, head-mounted augmented reality glasses. We use a Holographic projection add-on for rhino and grasshopper CAD software Fologram (Vermisso, et al. 2019). Using the markers enables the positioning sequence guided by the digital model. The surface articulation is limited to the elasticity, shore parameter, and thickness of the hyper-elastic membrane.

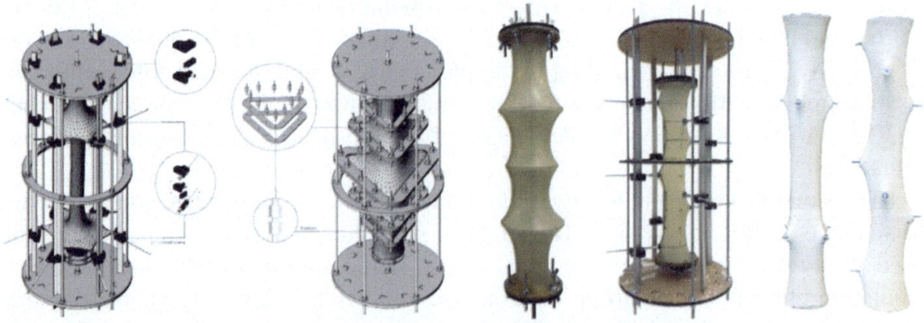

Fig. 4. The adaptive scaffolding and the clamp system with the prestressed latex on an adaptive scaffolding with the articulation anchors and compression rings.

3.3 The Challenges of Rotoforming Concrete

The prototyping process consists of correlating the two main challenges: defining the rotation trajectory and controlling the material flow characteristic of concrete. These two challenges are solved with a series of experiments and prototypes. The first series of experiments define the relation of two rotational axes in conjunction with material flow. In contrast to the early research on hollow nodal elements in the RotoForm Project (Tessmann and Mehdizadeh 2019a), the geometric proportion of linear elements requires asynchronous relation and sequence between the rotational axis. Therefore, we simulated the material flow in the formwork with a simplified rigid body model with Kangaroo in Grasshopper. The two-axis are rotating, and the flow of a rigid body wall in the formwork draws a trajectory. The trajectory with the longest path is the best solution for the geometry. We conduct the same with a robotic arm moving a transparent box with a ball inside on a small scale. Then we arranged a set of experiments with the small-scale transparent framework and replacement material inside to test the physical distribution inside the framework. These tests allowed us to define rough variables and a numeric model for the rational axis.

3.4 The Machinery and Robotic Trajectory

To scale the material system for automated industrial use cases, we defined two sets of machines; a two-axis rotational rotoforming called a "rock n roll" machine (Wang et al. 2011) and a six-axis robotic arm. The "rock n roll" machine is one of the earliest devices used in the plastic industry for rotational molding. The design involves one axis for rocking action and the other perpendicular axis with 360° rotation. The "rock n roll" machine was initially used to produce low-cost products because the device's volume is not limited to the plastic heating cage. We adopted this machine design to produce concrete elements with a size of up to 120 cm in length. Two independent axes powered by stepper motors enable us to control the relationship between the rotation ratio between the axes independently (Fig. 5). However, this machine design has limited rocking angles.

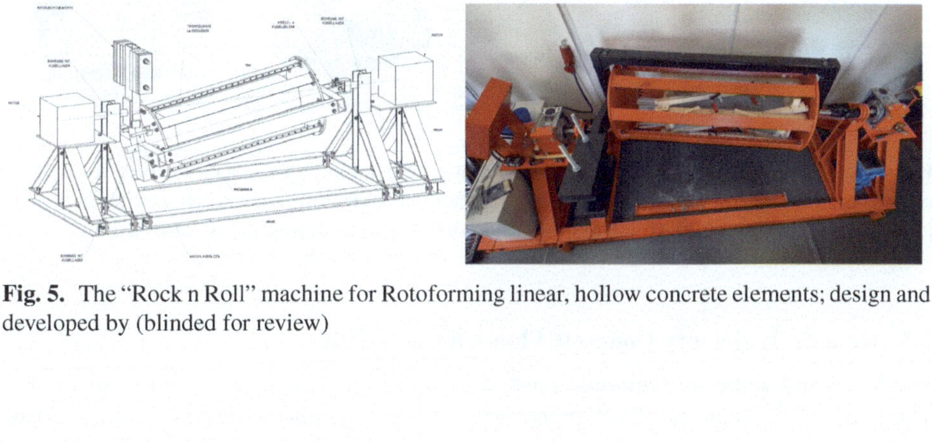

Fig. 5. The "Rock n Roll" machine for Rotoforming linear, hollow concrete elements; design and developed by (blinded for review)

We set up a KUKA robotic arm with a pivotal joint in front of it to overcome the limited rocking angles. The distance between the Pivotal Joint and Robot Flange is variable due to the robot's position. The pivotal joint enables the materialization of long cantilevers of 1 to 3 m and with up to double weight the maximum payload capacity of the robotic arm (in this case, 300 kg). Moreover, the robotic arm can perform different motion patterns and sequences, which allows for defining any set of motions with differentiated trajectories and velocities the robot setup enables us to rotoform elements with significantly more complex geometry and distribute material within uneven formwork to make hollow elements with variable wall thickness. In the next stage, we applied the numeric model results of a rotoforming test with small robots *Universal Cobot UR10* to the robotic setup with a *KUKA-KR300-R2500* industrial robot (Fig. 6). In order to focus on the development of the robotic movement and trajectories we conducted several tests with the concrete replacement material Acrystal (a water-based acrylic resin polymer) that cures faster and has a higher ductility, compared to concrete, temporarily excluding challenges concerning the concrete rheology. The robotic arm slowly rotates and positions the mounted formwork to distribute the material. Gravity is the only force that lets the concrete flow inside the formwork. The early test results showed that the robotic setup with Acrystal is applicable for rotoforming the 120 cm long linear hollow elements. The robotic setup enables the continuous and automated digital fabrication process from robotic formwork positioning, timber scaffold assembly to the robot positioning of timber assembly. In the next phase of research, we aim to closely link the robotic trajectory to the material flow characteristics of concrete.

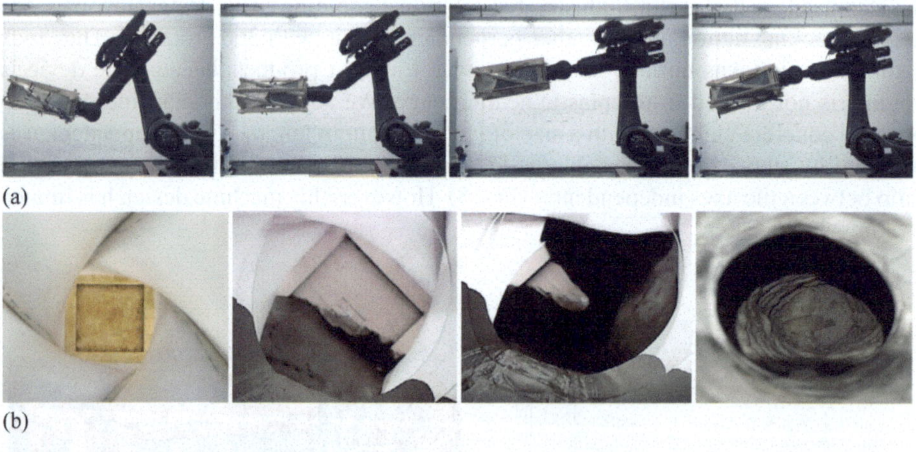

Fig. 6. (a) The third stage of the continuous robotic fabrication process. Rotoforming a 1 m long twisted column in hyperelastic membrane and timber scaffold with a KUKA robot (without pivotal joint) (b) inside view of the concrete flow inside the formwork during the robotic casting.

3.5 Robotic Trajectory Concrete Flow Characteristics

We developed a special hydraulic paste for rotoforming large-scale hollow elements. The focus of the material development was on rheology and setting of the fresh paste as well as reaching a homogeneous and aesthetically attractive surface; we developed a single rotational axis machine to test each mix design before scaling it up for the final linear elements. The upscaling was the most challenging part because of the larger formwork diameter. After pouring material inside the formwork, the robots start to rotate the formwork slowly for 30 min until the material is viscous enough to stop the rotoforming process.

The hydraulic paste flows uniformly under its own weight at a suitable rotation speed, thereby adheres to the hydrophobic formwork membrane and stops flowing after 30 ± 5 min. The first paste layer has a thickness of 5 ± 2 mm, and after each sequential iteration, more layers can be added to reach the final desired thickness. The paste reaches a final compressive strength of 70,4 ± 2,3 [N/mm^2] as measured on standard mortar prism with a homogeneous and smooth surface. This strength is within the range of a high-performance concrete (HPC). The laboratory tests went through an iterative approximation to adjust the final mix design. The mixture contains Portland cement (CEM I 42,5 R), fly ash, silica fume, and various admixtures. Fly ash and silica fume enable to control the fresh paste rheology, increasing the final strength, reducing lime efflorescence, and optimizing the CO2 footprint. Admixtures were required to further adjust flowability, adhesion to the hydrophobic formwork membrane, and initial setting. The CO2 emitted during the production of the Portland cement is partly re-bound by carbonation.

In this context, it is advantageous that the hollow elements are in direct contact with air and that their walls are relatively thin, usually less than 15 mm. (Fig. 7) shows the carbonation depth of a two-years-old hollow body stored in air as measured with phenolphthalein. The original grey colored area indicates the carbonated outer zone with a thickness of about 2 mm and the violet colored area the not yet carbonated core. On this basis, it can be predicted that a common hollow body with a wall thickness of 10 mm, whose walls are in contact with air on both sides, will be completely carbonated after approx. 12.5 years, which should be significantly shorter than the service life of a column (Stark and Wicht 2013).

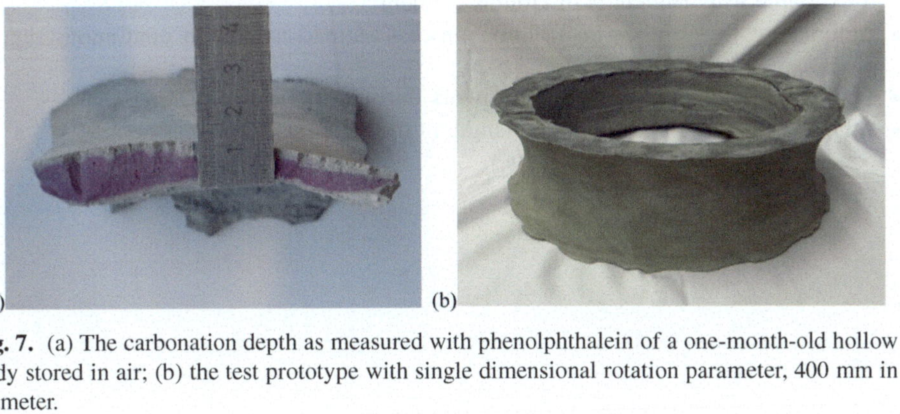

(a) (b)

Fig. 7. (a) The carbonation depth as measured with phenolphthalein of a one-month-old hollow body stored in air; (b) the test prototype with single dimensional rotation parameter, 400 mm in diameter.

4 Findings

Within the research project RotoColumn we developed a digital process for producing hollow, precast concrete columns that consist of digital simulation, a continuous robotic fabrication process, a special hydraulic HPC-like paste, and a sustainable membrane formwork system. Applying the fabrication constraints to the design simulation enables a fully automated and digital continuous fabrication process in the following steps: fabric formwork positioning timber scaffold assembly and rotoforming.

The research is exemplified by two series of prototypes, consisting of a twisted column made from three stacked concrete hollow elements (Fig. 8) and three stacked round sections of highly articulated columns. The digital design simulation framework enables interactive design workflow and offers a vast range of design possibilities. The prototypes demonstrate that the specially developed HPC-like paste for this research works for rotoforming linear elements of up to 120 cm in length and 50 cm in diameter. The elements have high visual and tactile surface quality. The 3D scan of the prototypes shows that the geometry of the cast object has a minimal deviation of 3 mm from the initial membrane formwork. The assembly strategy allows for building complex geometries with offcut timber material as for scaffolding and natural hyperelastic latex membrane as formwork. The structural tests show that the paste reaches 70,4 $\pm$ 2,3 [N/mm^2] compressive strength as measured on standard mortar prisms. Throughout this research project, we solved the challenge of migrating rotomoulding from plastic materials into the production of concrete elements by scaling the procedure and developing a concrete like material suitable for robotically rotated casting. We identified the following tasks for the future of research to apply this research for building more significant linear hollow elements.

- Integrating the reinforcement inside the concrete Hollow Formworks and casting them with Concrete
- Integrating the Reinforcement fort the Hollow Elements to avid the Buckling
- Integrating the digital simulation of material flow and defining the robot Trajectory for variable wall Thickness of Hollow element
- Automating the Timber Scaffolding process step to achieve a continuous digital process in our future work (Fig. 3.b).
- Minimizing the vibration in the formwork during robotic rotoforming.
- Adapting the paste mixture to the length of the elements.

Fig. 8. The final prototype of RotoColumn with twisted geometry and made from three stacked concrete hollow elements.

Acknowledgements. The RotoColumn research project was developed by DDU in close collaboration with the spannverbund GmbH., represented by Prof. Stefan Böhling, Jan Müller and Thomas Engel. The material development was conducted by Adrian Zimmermann of the Institute of Construction and Building Materials (WiB) at the TU Darmstadt, headed by Prof. Dr.ir. Eddie Koenders. The Institute of Structural Mechanics and Design (ISMD) represented by Prof. Ulrich Knaack and Amir Chaddah consulted in the field of membrane formwork. The Prototypes has been built with the technical Support of TU Darmstadt Students Danial Ahmad and Yasin Taha Yürur, Elisabeth Hagenthau and Janine Junen. The Materials has been sponsored by Sika

GmbH. And Dyckerhoff GmbH. The research project was funded by the "ZIM" program. (Zentrales Innovationsprogramm Mittelstand), which means "Central Innovation Programme for small and medium-sized enterprises (SMEs)". A funding program of the Federal Ministry for Economic Affairs and Climate Action.

References

Ahlquist, S., Menges, A.: Frameworks for computational design of textile micro-architectures and material behavior in forming complex force-active structures (2013)

Buswell, R.A., et al.: A process classification framework for defining and describing Digital Fabrication with Concrete. Cem. Concr. Res. **134**, 106068 (2020)

Hawkins, W.J., et al.: Flexible formwork technologies–a state of the art review. Struct. Concr. **17**(6), 911–935 (2016)

Kromoser, B., Kollegger, J.: Aktives Verformen von ausgehärteten Betonelementen zur Herstellung von räumlich gekrümmten Betonflächen. Beton-Und Stahlbetonbau **112**(2), 106–115 (2017)

Kromoser, B., Preinstorfer, P., Kollegger, J.: Building lightweight structures with carbon-fiber-reinforced polymer-reinforced ultra-high-performance concrete: research approach, construction materials, and conceptual design of three building components. Struct. Concr. **20**(2), 730–744 (2019)

Kudless, A.: Bodies in formation: the material evolution of flexible formworks (2011)

Lloret, E., et al.: Complex concrete structures: merging existing casting techniques with digital fabrication. Comput. Aided Des. **60**, 40–49 (2015)

Piker, D.: Kangaroo: form finding with computational physics. Archit. Des. **83**(2), 136–137 (2013)

Stark, J., Wicht, B.: Dauerfähgkeit von Beton, 2nd edn., p. 121. Springer, Heidelberg (2013). https://doi.org/10.1007/978-3-642-35278-2

Tessmann, O., Mehdizadeh, S.: Rotoform: realization of hollow construction elements through roto-forming with hyper-elastic membrane formwork. In: Proceedings of the Symposium on Simulation for Architecture and Urban Design, April 2019, pp. 1–7 (2019a)

Tessmann, O., Mehdizadeh, S.: Hollow-Crete. In. In: Gengnagel, C., Baverel, O., Burry, J., Ramsgaard Thomsen, M., Weinzierl, S. (eds.) DMSB 2019, pp. 474–486. Springer, Cham (2019b). https://doi.org/10.1007/978-3-030-29829-6_37

Veenendaal, D., West, M., Block, P.: History and overview of fabric formwork: using fabrics for concrete casting. Struct. Concr. **12**(3), 164–177 (2011)

Vermisso, E., Thitisawat, M., Salazar, R., Lamont, M.: Immersive environments for persistent modelling and generative design strategies in an informal settlement. In: Gengnagel, C., Baverel, O., Burry, J., Ramsgaard Thomsen, M., Weinzierl, S. (eds.) DMSB 2019, pp. 766–778. Springer, Cham (2019). https://doi.org/10.1007/978-3-030-29829-6_59

Wang, Y., Zhang, K., Dai, Y.C., Liu, J., Zhang, Y.Y.: State-of-the-art of rotational moulding technique and its application. Appl. Mech. Mater. **80**, 980–984 (2011)

Wangler, T., et al.: Digital concrete: opportunities and challenges. RILEM Tech. Lett. **1**, 67–75 (2016)

Watts, J.: Concrete: the most destructive material on Earth. Guardian **25**, 1–9 (2019)

Statistically Modelling the Curing of Cellulose-Based 3d Printed Components: Methods for Material Dataset Composition, Augmentation and Encoding

Gabriella Rossi[1(✉)], Ruxandra-Stefania Chiujdea[1], Laura Hochegger[1], Ayoub Lharchi[1], John Harding[2], Paul Nicholas[1], Martin Tamke[1], and Mette Ramsgaard Thomsen[1]

[1] Centre for IT and Architecture (CITA), Royal Danish Academy, Copenhagen, Denmark
garo@kglakademi.dk
[2] University of Reading, West Berkshire, UK

Abstract. The Machine-Learning models thrive on data. The more data available, or creatable, the more defined is the problem representation, and the more accurate is the obtained prediction. This presents a challenge for physical, material datasets, specifically those related to fabrication systems, in which data is tied to physical artefacts which necessitate fabrication, digitisation and formatting to be used as input for predictive models.

In this paper we present a design-based methodology to producing a material dataset for statistically modelling the curing of cellulose-based 3d-printed components, as well as associated methods for geometric data encoding, tolerance-informed data augmentation and statistical modelling. The focus of the paper is on the digital workflows and considerations for dataset composition - the material case of 3d-printing cellulose is secondary. We use a built 3d-printed demonstrator wall as a material dataset, through which we generate datapoints that stem from a real design-scenario and inform the fabrication model. Using a feature-engineering approach, select geometrical features are encoded numerically. We perform statistical analysis on the data, and test different shallow models and neural networks. We report on the successful training of a Polynomial Kernel Ridge Regressor to predict the vertical shrinkage of the pieces from wet print to dry element.

Keywords: Machine learning · Dataset · Sensing · Data augmentation · Biomaterials · 3d printing

1 Introduction

Sustainability challenges our standard architectural material practice. It not only challenges our choice of materials, but also how we understand their properties and behaviours during their lifespans, and how we engineer fabrication processes for them [1]. Biogenic materials, while providing a smaller embodied carbon footprint, come

C. Gengnagel et al. (Eds.): DMS 2022, *Towards Radical Regeneration*, pp. 487–500, 2023.
https://doi.org/10.1007/978-3-031-13249-0_39

with increased anisotropy and heterogeneity, which makes them difficult to control, when compared to standard industrialized materials [2]. Biogenic materials are composites - part fibres, part polymer chain - and often have a hygroscopic relationship to water. This is of particular importance when developing 3d-printable slurries. The water that is included to control the rheology of the slurry is later removed from the system through drying processes resulting in a material system in which properties and geometries transform. Controlling the discrepancy between wet and dry state is the key challenge when 3d printing with biopolymer slurries for architectural applications [3, 4].

This drying transformation can be difficult to formalize using analytical models, in which all occurring mechanistic processes of surface evaporation and material compaction need to be described. To represent the material heterogeneity, temporal behaviours and scale dependent dynamics at play it would necessitate a complex model with many parameters and high degrees of freedom to explain the complex interactions between the printed piece and its specific curing environment [5, 6]. In this research, our objective is predictive rather than explanatory [7], i.e. to predict the final geometrical output, rather than explain the underlying causalities of the drying phenomena, which relies on a large number of parameters (geometric, environmental, material mix etc.). Existing state of the art looks at the behaviour [8]. However, these findings can only be implemented in highly simplistic geometric primitives not useful to an architectural design scenario, or an integrated design to fabrication process. In our work we use a different modelling paradigm, based on numerical statistical models to infer correlation. Here, we examine how machine learning can act as a support for adaptive fabrication towards material control [9]. We aim to infer, with sufficient precision the effects of the drying process on the geometry of the printed parts, by using time-stamped data collected during the drying process. The learnt prediction is instrumentalized to establish the relationship between digitally designed and physically fabricated object, in terms of dimensional tolerance.

2 State of the Art

2.1 Dataset Composition

With a machine learning approach, the complexity of the modelling exercise shifts from being solely about the model, to equally importantly being about the dataset it uses to train. Certain rules need to be followed when designing datasets that successfully encapsulate the features and space of the problem they intend to model. Important dataset qualities are that the data should be Independent and Identically Distributed (IID), feature signals should be clear, feature representation should be balanced, and the more datapoints, the better the algorithm should perform [10].

When making material data sets, in which datasets are drawn from physical artefacts achieving a sufficient quantity of data can become a significant problem. Making experiments and digitizing the outcomes is time and resource consuming. This problem of material datasets is particularly relevant for disciplines such as medicine and biology [11] material science [12], as well as for digital fabrication [13]. For digital fabrication-based datasets, this problem is further complicated by the need to correctly represent the design space breadth and relevance. The datapoints should both be able to embody pertinent fabrication parameters linked to the specifics of a given material system and represent valid designs while the sampling range of the datapoints should allow to interpolate the prediction to future unknown samples. This presents a generalization vs specificity dilemma – where does one draw the line of what the model can or cannot predict? For instance, if the model is trained cylindrical samples, then predicting waves-form samples becomes beyond its capacity [14].

2.2 Dataset Encoding

A second important consideration is that the amount of data needed for training is directly correlated to the complexity and category of model [15]. The choice of model category is directly dependent on the type and dimensionality of data to be used. Determining the level of abstraction of the problem representation is crucial for a successful encoding of the dataset. This level of abstraction is not obvious. Physical samples can be represented in different ways with more or less direct correlation between the physical object and its datapoint descriptors. For 3d prints, this can be the print bead, the toolpath curve, or a point cloud scan of the printed artefact. This choice is subsequently reflected on the model where one that handles numerical input is not only less complex than a model that handles point clouds but is also less data hungry.

Additionally, some data types lend themselves to augmentation better than others. For example, while meshes and networks are immutable, image-based data, can be easily augmented using scaling, shearing, cropping, mirroring and rotating [16, 17]. This works well for surface or depth-based representation but is not suited for 3-dimensional prints. Numerical methods of synthetic dataset augmentation as bootstrapping [18], The Synthetic Minority Over-sampling TEchnique (SMOTE) [19], and Kohonen Map [20] interpolations, risk producing data points that are not desirable to be designed or possible to be fabricated.

3 Methods

In this paper a design-based methodology is presented in order to produce a material dataset for statistically modelling the curing of cellulose-based 3d-printed components, as well as associated methods for geometric data encoding, tolerance-informed data augmentation and statistical modelling.

We use a built 3d-printed demonstrator wall as a dataset, to generate datapoints that stem from a real design-scenario. Rather than adopting a deep-learning approach using raw point-cloud data, a feature-engineering based approach is adopted where select geometrical features are encoded numerically. This allows us to perform statistical analysis on the data, and to be able to test different types of shallow models and neural networks, which present adequate complexity for the amount of data we have available. The data is collected using a tracking system, it is subsequently compiled and augmented using tolerance-informed gaussian noise. The dataset is then used to train a Polynomial Kernel Ridge Regressor [21] to predict the vertical shrinkage of the pieces from wet print to dry element. We report on the different stages below.

3.1 Sample Design, Making and Curing

In our previous research paper [4], we have established geometry as the critical driver for surface evaporation. To study the relationship between geometry and drying transformation, our dataset -demonstrator wall design- creates a balanced representative sampling of geometric features such as spread and height of the pieces and thickness of the cross section. While previous research explores circular geometries that are stacked in a column assembly [22, 23], we are interested in a wall assembly made of interlocking polygonal components, as they offer the opportunity to study assembly tolerances as well as design functionalization. The design is developed in Rhino/GH and integrates printing toolpath with reinforcing crossings and male/female joints. The usage of a parametric algorithms and data tree structure allows to keep track of all digital geometric features during the dataset compiling (Sect. 3.2). The components are robotically 3d printed in our biomaterial recipe [24]. We track 18 front wall blocks from the front wall up to 160 h into their drying using an 8-camera Optitrack Prime x13 motion capture system. Every edge is tracked using 2 reflective markers (Fig. 1b–c). We use Motives [25] loopback streaming for marker position streaming using a customized NatNet c++ client [26]. Values are uploaded every 5 min to our InfluxDB [27] cloud database for storage of time-series data.

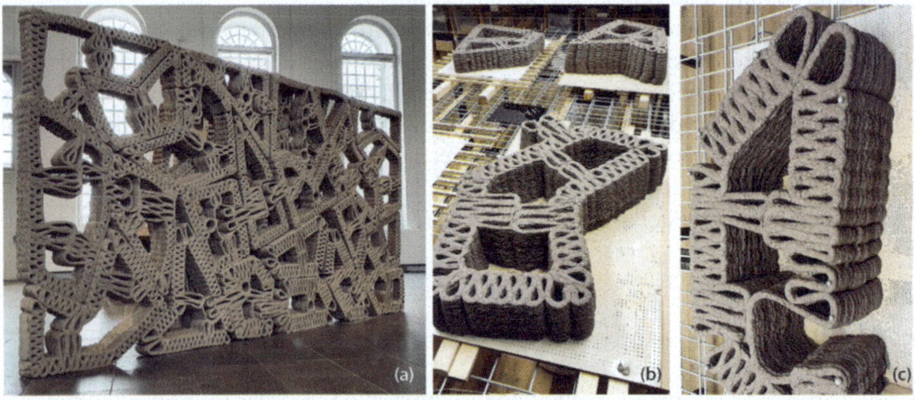

Fig. 1. (a) Demonstrator wall 3 m long × 1 m wide and 1.8 m high with the frontal 18 components serving as dataset for shrinkage prediction. (b) Curing room setup drying 4 blocks at a time monitored through an 8-camera optitrack system. (c) Reflective markers are placed on each edge of the component and tracked by the cameras with their position regularly uploaded to the cloud database.

3.2 Dataset Compilation

At the end of the drying period for each batch of blocks, we use a custom workflow using influxDB API [28] to run queries and compile a Pandas [29] DataFrame containing the xyz positioning of the reflective markers over the elapsed curing week. The .csv file is then processed in Grasshopper, where the Motion Capture information is matched with the existing data tree of digital components. The reflective markers are repositioned in space, and a timestamped polygonal outline of the component is geometrically rebuilt at T_S "start of sensing" – when the component is wet, and at T_e "end of sensing" – when the component is dry. This allows us to visualize the shrinkage and warpage that has occurred over drying, and extract metrics such as lengths, angles, perimeters and area at those timestamps. Additionally, we match the physical *as printed* features, with the existing *digitally designed* features, that we retrieve from the parametric model data tree. These features include the print component, the thickness of the component outline, and whether the edges are touching across the component (termed *crossing*) or whether the edge presents a joint. They are semantically characterising the edges, and have been observed to influence on the shrinkage behaviour. In order to maximise our dataset size, rather than treating every component as a datapoint, we consider every edge of each component as a separate datapoint. Therefore, we are able to extract 87 data points from 18 components. The dataset can be found here [30] and presents 20-dimensional feature vector for the 87 entries (Fig. 2).

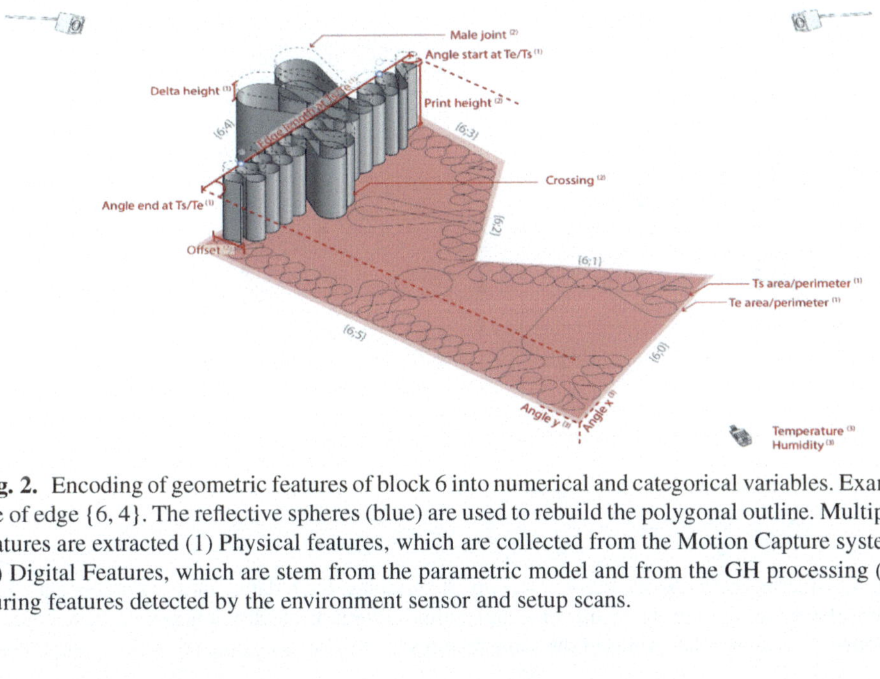

Fig. 2. Encoding of geometric features of block 6 into numerical and categorical variables. Example of edge {6, 4}. The reflective spheres (blue) are used to rebuild the polygonal outline. Multiple features are extracted (1) Physical features, which are collected from the Motion Capture system (2) Digital Features, which are stem from the parametric model and from the GH processing (3) Curing features detected by the environment sensor and setup scans.

3.3 Exploratory Data Analysis (EDA)

Predicting the curing behaviour of the print requires identifying variables that would allow to geometrically rebuild the transformation. Our previous findings [4] reveal that the magnitude of transformation is most marked in the vertical direction, rather than in-plane. This hypothesis is further confirmed by empirical observation of the components as they cure. Therefore, as a starting point, we assume that the height difference between the initial wet and final dry position of the reflective markers best captures the shrinkage of the material overall. As per best practice [31], we isolate this vertical shrinkage termed Delta Height a as response variable and perform graphical and statistical exploratory analysis on the data with Seaborn [32], in order to identify contributing correlations between it and the input variables. Delta height ranges from 21 mm to 45 mm (Fig. 3). As expected, we observe that edges from the same block shrink differently depending on their categorical features, hence their importance.

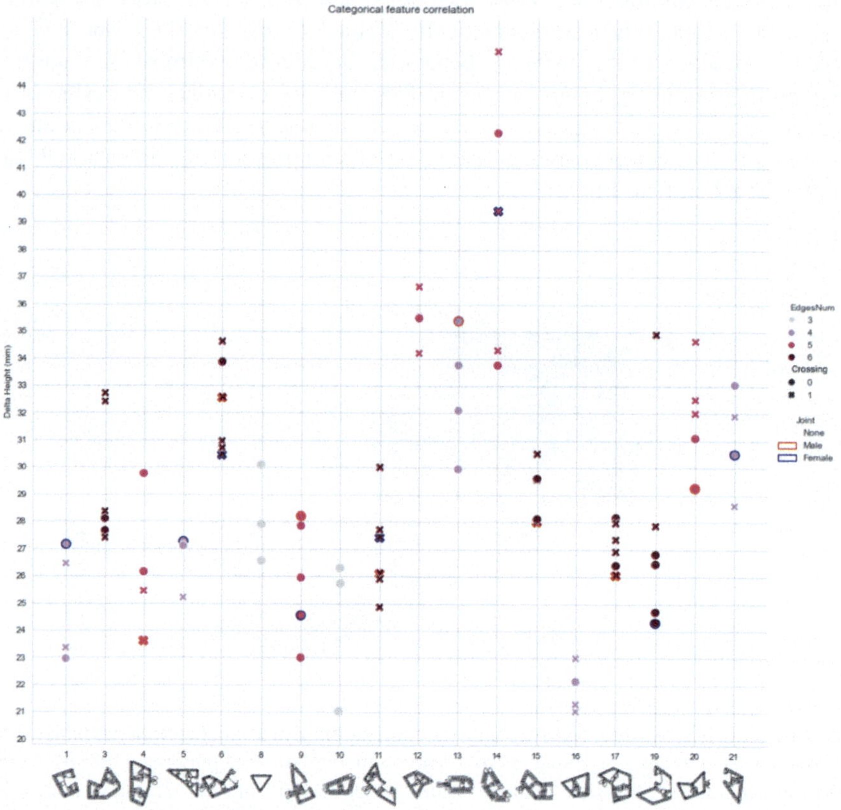

Fig. 3. Isolating *Delta height* across categorical variables. Each datapoint is a component edge, where changes of marker shape, fill color, and outline color indicate different geometric features of the edge. We can see that points of the same features tend to behave similarly within components.

Pair plots of the numerical features (Fig. 4) showcase correlations with the print height, offset and angle start, but no/weak correlation with print area and print perimeter and edge length. Given the reduced nature of the sample size, further feature distillation to avoid overfitting is necessary. To this effect a Principal Component Analysis (PCA) model was generated using Scikit-Learn [21] to reduce the dimensionality of the data and observe the main variance directions. Analysing the weighting of the variables to the Principal Components (PCs), points to variables that can be collapsed due to their high correlation. It was found that the first 5 principal components explained over 85% of the data variance, meaning for later experiments (see Sect. 4). Additionally, the initial features could be reduced from 20 features, to the 8 with most diverging eigenvectors, namely: 'Offset', 'PrintHeight', 'WetLength', 'WetAngStart', 'WetArea', 'FemMal?', 'Cross?' as input variables and 'DeltaHeight' as response variable.

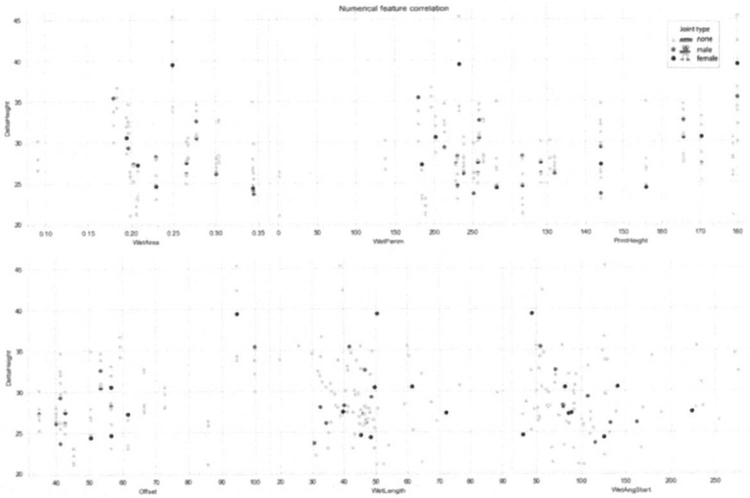

Fig. 4. Delta height plotted in y-axis, against multiple numerical variables in x-axis, with joint type indicated as hue

3.4 Data Augmentation

Due to the small sample size, dataset augmentation was adopted help to increase model training data. Established methodologies for numerical dataset augmentation have proven effective to improve model training [33], however given the physical nature of our samples, specific attention must be placed on the type of augmentation. Kohonen Maps for instance, being a relative mapping algorithm, risks producing unbalanced samples, SMOTE algorithm for regression, assumes that outliers are visible to the dataset, and oversamples at average values, which alters the real distribution of the phenomena and therefore undesirable in this case.

In contrast, it was proposed to add gaussian noise to our dataset as a method of augmentation. While literature suggests that noise could add robustness to the training process [34], it is often applied infinitesimally after data normalization and standardisation - for instance with pixel images. In this study, we add gaussian noise to the variables before standard scaling. This allows for the noise to be informed by the tolerances of the real physical sample. We term it tolerance-informed gaussian noise augmentation. Here for each variable, we define a range and standard deviation that stems from real-life inaccuracy of measurement, or discrepancy between digital and physical, and that is acceptable for an architectural application. Using SciPy [35] Truncated Normal Distribution we generate 4 noisy copies of the data, which are appended to the dataset, therefore going from 87 entries to 524 in total. We observe that this operation does not distort the statistical distribution of the data (Fig. 5).

```
NoiseDomain = {"Offset" : [0, 1, -2, +2], "PrintHeight":
[0,1, -2, 2],"WetLength" : [0,0.5, -1, 1],"WetAngStart" :
[0, 0.125, -0.25, 0.25],"WetArea": [0, 0.0005, -0.001,
0.001],"DeltaHeight" : [0,0.5,-1,1]}
```

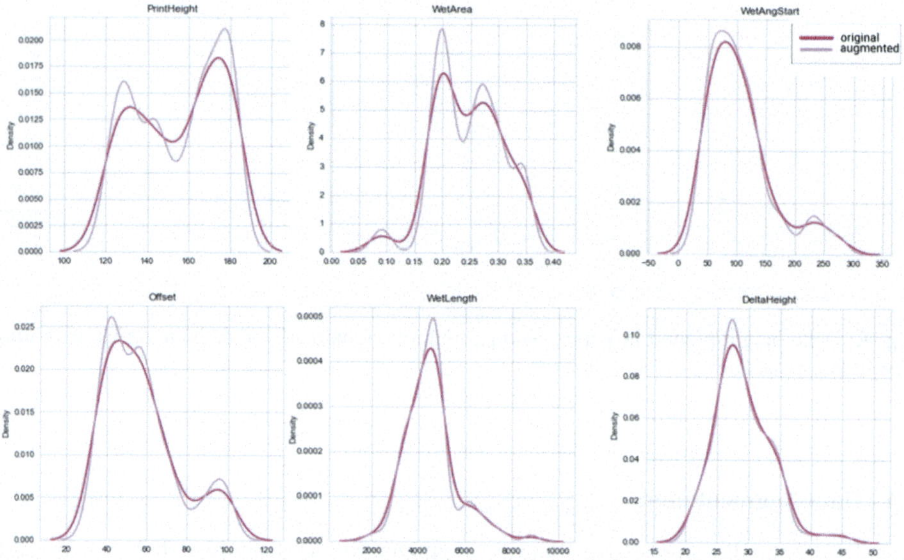

Fig. 5. Normalized KDE plot showcasing the distribution of the variables before and after tolerance-informed gaussian augmentation. The mean and deviation of the variable appear maintained, yet some local changed peaks are created in the curves

4 Predictive Model Results

We split our dataset into train and test. In order to ensure equal representation of numerical and categorical features across the two sets (Fig. 6), we split our PCA-reduced dataset into 8 clusters using Scikit-Learn's [21] K-Means algorithm, and we randomly sample 10 entries out of each cluster.

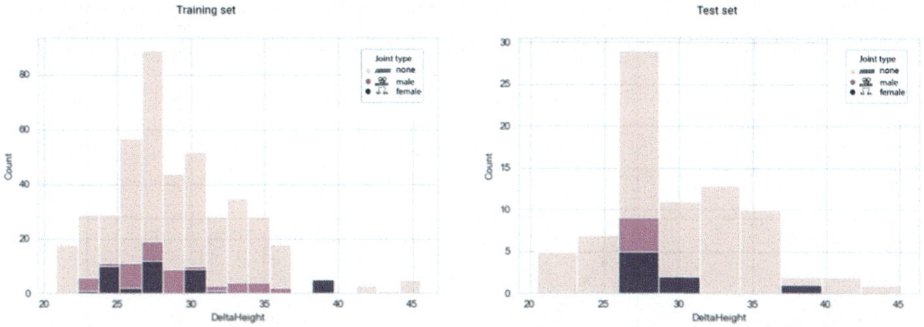

Fig. 6. Response variable distribution, coloured by categorical feature "Joint" among Train and Test sets. This showcases a balance in the dataset split granted by the k-means clustering approach

Models of increased degrees of freedom are tested for compatibility with our problem: Scikit Learn's [35] Linear Regressor and Polynomial Kernel Ridge Regressor, XGboost tree [36] and a simple Artificial Neural Network (ANN) [37]. Mean Squarred error is our loss, with the response variable Y kept in original scale. We test three version of our dataset:

Raw Data: original dataset:15 features, 71 Train/16 Test
PCA reduced data: First 5 PC factors, 71 Train/16 Test
Gaussian Augmented data: 5x entries, 8 features, 427 Train/80 Test

The performance matrix (Fig. 7) shows that while an acceptable training accuracy can be achieved using Scikit Learn's Polynomial Kernel or a XGBoost model, test accuracy are very poor across the board. This means that the model is overfitting and unable to generalize. These symptoms persist with the PCA-reduced data, pointing to the fact that the poor predictive performance could be due to the number of samples. Indeed, when testing with the gaussian augmented data, we see an overall improvement of the performance across all models, yet we achieve a great performance using the polynomial kernel ridge regression, which hints to the fact that the model might present the correct degree of freedom to model the shrinkage, unlike the ANN which might be too complex. We achieve a training loss of 0.27 mm and test loss of 0.38 mm. This gladly fits into fabrication tolerance requirements. Predictions are visualized against real measured shrinkage (Fig. 8).

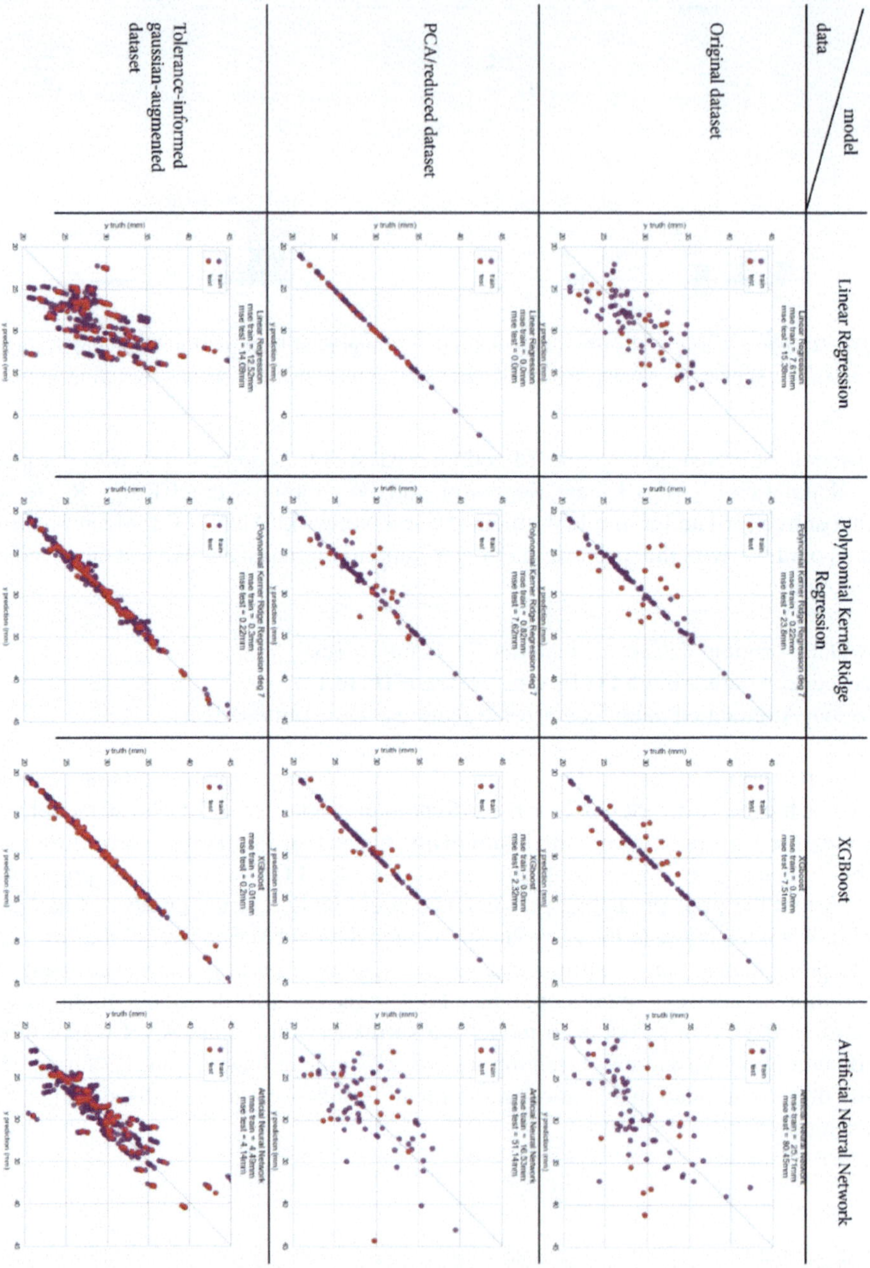

Fig. 7. Predictive model performance matrix. Plotting the dataset's yTruth against the model's yPredict provides a quick way to visualize its performance: The closer the scatter of points is to the diagonal line, the more accurate the predictions are. However, perfect diagonal alignment is often symptom of model overfitting, as observed with the Linear model combined with the PCA data, or with the train data and the XGboost models

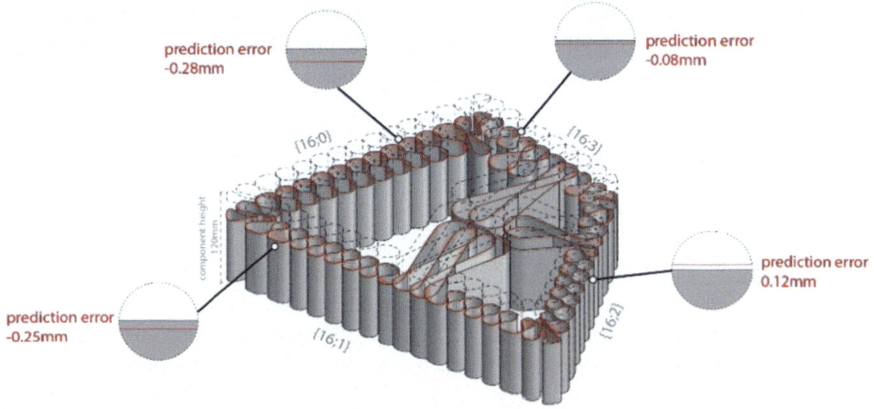

Fig. 8. Visualizing prediction error on component 16. The feature vector of each edge is sent to the model, and the delta height is returned (dashed: wet print height, black: measured dry height, red: predicted dry height).

5 Conclusions

In this paper we presented a methodology for physical dataset composition, augmentation and encoding for predicting the vertical shrinkage behaviour of 3d-printed cellulose slurry. Adopting a design-functionalized approach to composing datasets, rather than a random generator approach, filters out invalid design possibilities and hones the training of the model towards datapoints that are relevant for the fabrication system.

By choosing a processed feature extraction approach coupled with simple ML models, in contrast to raw rich data approach and complex models, we target our model towards the application scenario directly, rather than needing post processing for fabrication-relevant consequences extraction. This approach however requires a thorough knowledge of the modelled process, a good intuition about the parameters that are influencing the phenomenon, and a good design of the sensing workflow and a solid digitization pipeline from design to production.

Our findings showcase, that even with simple statistical models, the dataset size is a persistent bottleneck for obtaining accurate predictions out of physically made datasets. Our tolerance-informed gaussian augmentation methodology has proven successful in increasing prediction accuracy. The enabler of such is the understanding of the physical process and tolerances associated with making and sensing, and their incorporation in the creation of synthetic data points. This opens the door to more ML-controlled fabrication processes.

Our immediate next step is to move from single response-variable prediction to multiple response variable prediction, namely the dry state of the edge (length and angles with neighbours). We will maintain the established method of tolerance-informed gaussian augmentation and test it with both shallow models and autoencoder neural networks. Finally, feedback and integration of the prediction will be deployed back into the design environment and be leveraged to compensate the print for accurate manufacturing.

Acknowledgements. This project is funded by Independant Research Fund Denmark (DFF) PROJECT NUMBER 9131-00034B "Predicing Response", in collaboration with Anders Egede Daugaard and Arianna Rech (Denmark Technical University) and John Harding (University of Reading).

References

1. Thomsen, M.R., Tamke, M.: Towards a transformational eco-metabolistic bio-based design framework in architecture. IOP Bioinspir. Biomim. (2022)
2. Pradhan, R.A., Rahman, S.S., Qureshi, A., Ullah, A.: Biopolymers: opportunities and challenges for 3d printing. In: Sabu, T., Gopi, S., Amalraj, A. (eds.) Biopolymers and their Industrial Applications: From Plant, Animal, and Marine Sources, to Functional Products, pp. 281–303. Elsevier (2021). Accessed 31 Mar 2022
3. Dritsas, S., Vijay, Y., Teo, R., Halim, S., Sanandiya, N., Fernandez, J.G.: Additive manufacturing with natural composites. In: Intelligent & Informed, Proceedings of the 24th International Conference of the Association for Computer-Aided Architectural Design Research in Asia (CAADRIA 2019), Hong Kong, pp. 263–272 (2019)
4. Rossi, G., et al.: A material monitoring framework: tracking the curing of 3d printed cellulose-based biopolymers. In: Realignments: Towards Critical Computation (2022)
5. Faircloth, B., Ramsgaard Thomsen, M.: Rocking the cradle of the mechanistic model. In: CITA Complex Modelling. pp. 174–179. Riverside Architectural Press, Toronto (2020)
6. Defraeye, T.: Advanced computational modelling for drying processes – a review. Appl. Energy **131**, 323–344 (2014)
7. Baker, R.E., Peña, J.-M., Jayamohan, J., Jérusalem, A.: Mechanistic models versus machine learning, a fight worth fighting for the biological community? Biol. Lett. **14**, 20170660 (2018)
8. Katekawa, M.E., Silva, M.A.: A review of drying models including shrinkage effects. Drying Technol. **24**, 5–20 (2006)
9. Tamke, M., Nicholas, P., Zwierzycki, M.: Machine learning for architectural design: practices and infrastructure. Int. J. Archit. Comput. **16**, 123–143 (2018)
10. Koch, S., et al.: ABC: a big CAD model dataset for geometric deep learning. In: 2019 IEEE/CVF Conference on Computer Vision and Pattern Recognition (CVPR), Long Beach, CA, USA, pp. 9593–9603. IEEE (2019)
11. Calimeri, F., Marzullo, A., Stamile, C., Terracina, G.: Biomedical data augmentation using generative adversarial neural networks. In: Lintas, A., Rovetta, S., Verschure, P.F.M.J., Villa, A.E.P. (eds.) ICANN 2017. LNCS, vol. 10614, pp. 626–634. Springer, Cham (2017). https://doi.org/10.1007/978-3-319-68612-7_71
12. Vanpoucke, D.E.P., van Knippenberg, O.S.J., Hermans, K., Bernaerts, K.V., Mehrkanoon, S.: Small data materials design with machine learning: when the average model knows best. J. Appl. Phys. **128**, 054901 (2020)
13. Ramsgaard Thomsen, M., Nicholas, P., Tamke, M., Gatz, S., Sinke, Y., Rossi, G.: Towards machine learning for architectural fabrication in the age of Industry 4.0. In. J. Archit. Comput., 147807712094800 (2020)
14. Vijay, Y.: Advanced manufacturing with natural materials. University of Technology and Design, Singapore (2018)
15. Verleysen, M., François, D.: The curse of dimensionality in data mining and time series prediction. In: Cabestany, J., Prieto, A., Sandoval, F. (eds.) IWANN 2005: LNCS, vol. 3512, pp. 758–770. Springer, Heidelberg (2005). https://doi.org/10.1007/11494669_93

16. Rossi, G., Nicholas, P.: Re/learning the wheel: methods to utilize neural networks as design tools for doubly curved metal surfaces. In: ACADIA 2018: Recalibration. On Imprecision and Infidelity: Proceedings of the 38th Annual Conference of the Association for Computer Aided Design in Architecture, Mexico City, pp. 146–155 (2019)
17. Zwierzycki, M., Nicholas, P., Ramsgaard Thomsen, M.: Localised and learnt applications of machine learning for robotic incremental sheet forming. In: De Rycke, K., et al. (eds.) Humanizing Digital Reality, pp. 373–382. Springer, Singapore (2018). https://doi.org/10.1007/978-981-10-6611-5_32. Accessed 31 Mar 2022
18. Nevitt, J., Hancock, G.R.: Performance of bootstrapping approaches to model test statistics and parameter standard error estimation in structural equation modeling. Struct. Equ. Model. **8**(3), 353–377 (2001)
19. Torgo, L., Ribeiro, R.P., Pfahringer, B., Branco, P.: SMOTE for regression. In: Correia, L., Reis, L.P., Cascalho, J. (eds.) EPIA 2013. LNCS (LNAI), vol. 8154, pp. 378–389. Springer, Heidelberg (2013). https://doi.org/10.1007/978-3-642-40669-0_33
20. Van Hulle, M.M.: Self-organizing maps. In: Rozenberg, G., Bäck, T., Kok, J.N. (eds.) Handbook of Natural Computing, pp. 585–622. Springer, Heidelberg (2012). https://doi.org/10.1007/978-3-540-92910-9_19. Accessed 31 Mar 2022
21. Pedregosa, F., et al.: Scikit-learn: machine learning in Python. J. Mach. Learn. Res. **12**, 2825–2830 (2011)
22. Dristas, S., Vijay, Y., Halim, S., Teo, R., Sanandiya, N., Fernandez, J.G.: Cellulosic biocomposites for sustainable manufacturing. In: Burry, J., Sabin, J., Sheil, B., Skavara, M. (eds.) Fabricate 2020: Making Resilient Architecture, pp. 74–81. UCL Press, London (2020). Accessed 17 Nov 2020
23. Goidea, A., Floudas, D., Andréen, D.: Pulp Faction: 3d printed material assemblies through microbial biotransformation. In: Burry, J., Sabin, J., Sheil, B., Skavara, M. (eds.) Fabricate 2020: Making Resilient Architecture, pp. 42–49. UCL Press, London (2020)
24. Rech, A., et al.: Predicting Response: aste-based biopolymer slurry recipe for 3d-printing. Zenodo (2021). https://zenodo.org/record/5557218. Accessed 8 June 2022
25. Natural Point: Motive. OptiTrack. http://optitrack.com/software/index.html. Accessed 31 Mar 2022
26. NatNet SDK - Stream motion tracking data across networks. OptiTrack. http://optitrack.com/software/natnet-sdk/index.html. Accessed 31 Mar 2022
27. InfluxDB: Open Source Time Series Database. InfluxData. https://www.influxdata.com/. Accessed 31 Mar 2022
28. influxdb-client-python. InfluxData (2022). https://github.com/influxdata/influxdb-client-python. Accessed 31 Mar 2022
29. Reback, J., et al.: pandas-dev/pandas: Pandas 1.0.3. Zenodo (2020). https://zenodo.org/record/3715232. Accessed 31 Mar 2022
30. Rossi, G., et al.: Predicting Response: 3d printed biopolymer block dataset. Zenodo (2022). https://zenodo.org/record/6631767. Accessed 10 June 2022
31. Tukey, J.W.: Exploratory Data Analysis. Addison-Wesley Pub. Co., Reading (1977)
32. Waskom, M.L.: Seaborn: statistical data visualization. J. Open Source Softw. Open J. **6**, 3021 (2021)
33. DeVries, T., Taylor, G.W.: Dataset augmentation in feature space. arXiv:1702.05538 [cs, stat] (2017)
34. Goodfellow, I., Bengio, Y., Courville, A.: Regularization for deep learning. In: Deep Learning. MIT Press (2016)
35. Virtanen, P., et al.: SciPy 1.0: fundamental algorithms for scientific computing in Python. Nat. Methods **17**, 261–272 (2020)

36. Chen, T., Guestrin, C.: XGBoost: a scalable tree boosting system. In: Proceedings of the 22nd ACM SIGKDD International Conference on Knowledge Discovery and Data Mining, pp. 785–94. ACM, New York (2016). http://doi.acm.org/10.1145/2939672.2939785
37. TensorFlow. Zenodo (2022). https://zenodo.org/record/5949169. Accessed 31 Mar 2022

Design-to-Fabrication Workflow for Bending-Active Gridshells as Stay-in-Place Falsework and Reinforcement for Ribbed Concrete Shell Structures

Lotte Scheder-Bieschin (Aldinger)[1(✉)], Kerstin Spiekermann[1], Mariana Popescu[2], Serban Bodea[1], Tom Van Mele[1], and Philippe Block[1]

[1] Institute of Technology in Architecture, Block Research Group, ETH Zurich, Stefano-Franscini-Platz 1, 8093 Zurich, Switzerland
aldinger@arch.ethz.ch

[2] Faculty of Civil Engineering and Geosciences, Delft University of Technology, Stevinweg 1, 2628 Delft, The Netherlands

Abstract. Facing the challenges of our environmental crisis, the AEC sector must significantly lower its carbon footprint and use of first-use resources. A specific target is the reduction of the amount of concrete used. Funicular structures that base their strength on their structurally-informed geometry allow for material efficiency. However, a bottleneck for their construction lies in their costly and wasteful formworks and complex reinforcement placement.

This research presents an alternative flexible formwork system consisting of a bending-active gridshell falsework and fabric shuttering for ribbed funicular concrete shells. The falsework becomes structurally integrated as reinforcement and is designed as two connected layers offering shape control and sufficient stiffness to support the wet-concrete load.

The paper focuses on the development of a design-to-fabrication workflow and a graph-based data structure for gridshell falsework and reinforcement in the computational framework COMPAS. The implementation utilises, customises and creates packages for the form finding of the ribbed shell with TNA and the gridshell with FEA.

The research is based on a demonstrator realised in the context of the Technoscape exhibition at the Maxxi Museum in Rome, Italy. The computational workflow was used to design this system and translate it for materialisation. The demonstrator serves as proof-of-concept for the novel material-efficient construction system. Its key to efficiency lies in the structurally-informed geometry for both the formwork and the resulting ribbed concrete shell.

Keywords: Bending-active gridshell · Flexible formworks · Integrated formwork · Reinforcement · Ribbed concrete shell · COMPAS framework · FEA · Double-layered gridshell

C. Gengnagel et al. (Eds.): DMS 2022, *Towards Radical Regeneration*, pp. 501–515, 2023.
https://doi.org/10.1007/978-3-031-13249-0_40

1 Introduction

The critical impact of the construction sector on our environment demands a fundamental rethinking of design and construction practices towards more sustainable solutions. This is of particular relevance for reinforced concrete; as the most widely used construction material, a major contributor to global CO2 emissions and resource depletion (Lehne and Preston 2018). Conventional construction practices often rely on its excessive use to provide structural strength. In contrast, strength can be achieved through structurally-informed geometry like thin doubly-curved concrete shells with local stiffening articulations (Block et al. 2020). However, their complex shapes require bespoke formwork systems typically produced with CNC-milled foam or timber, which can be costly, wasteful, and limited to high-tech construction contexts (García de Soto et al. 2017). Additionally, the manufacturing and installation of custom reinforcement is cost- and labour-intensive (Waimer et al. 2019).

This paper presents an alternative flexible formwork solution employing geometric stiffness similar to tensile flexible formwork systems (Popescu et al. 2021). It utilises active bending, a deliberate-deformation method to elastically bend slender straight elements into curved geometries without formwork. It allows the construction of lightweight, material-efficient, self-contained and self-supporting structures that can be flat-packed for transport and easily deployed (Lienhard 2014). The presented construction system is a formwork system consisting of a bending-active gridshell falsework and fabric shuttering for ribbed funicular concrete shells. The strained gridshell is designed as a double-layered structure to ensure sufficient stiffness and shape control. This falsework becomes structurally integrated inside the ribbed concrete shell as reinforcement. Therefore, complex shell reinforcement does not have to be bent in and placed in an extra tedious construction step.

The focus of this paper is a design-to-fabrication workflow and data structure for gridshell falsework in the computational COMPAS framework (Van Mele et al. 2017–2022). This is required for the complex topology and geometry of the double-layered structure and for an integrative design approach. It enables the reconciliation of the form-finding geometry for the bending-active falsework and funicular shell, the simultaneous consideration of fabrication design for both, and the separate structural design with common focus on stiffness and shape control (Fig. 1).

The research is based on a demonstrator being built in the context of the Technoscape exhibition at the Maxxi Museum in Rome, Italy. Section 3 describes the system design; Sect. 4 presents the modelling workflow for such systems based on the case study, and, Sect. 5 touches on the materialisation and suitability assessment.

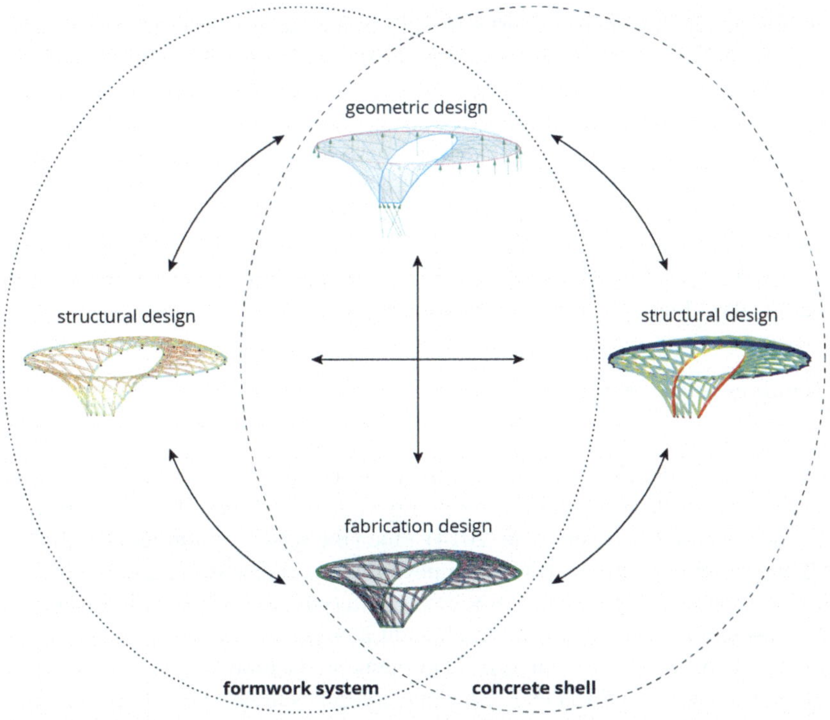

Fig. 1. Integrative design approach.

2 Background

2.1 Funnel-Shaped, Ribbed, Funicular Concrete Shells

Strength through geometry can be achieved with funicular compression-only structures through form finding with Thrust Network Analysis (TNA) (Block 2009). Van Mele et al. (2012) extended TNA to account for tensile members by lifting the necessary condition for compression-only TNA of convex, non-overlapping edges in the force diagram and allowing edges to flip orientation to close the force diagram. This enabled Rippmann and Block (2013) to explore a variety of efficient and expressive funnel shapes designed as funicular ribbed shells with tension rings.

2.2 Flexible Formwork Systems and Structurally-Integrated Falsework

Tensile flexible formworks are self-supporting, lightweight, waste-saving and compactly-packageable alternatives to conventional formworks for complex geometries (Hawkins et al. 2016). Mendez Echenagucia et al. (2019) and Popescu et al. (2021) utilised textile shuttering on cable-net falsework for expressive shells. However, these tensioned systems are limited to anticlastic curvature, and the tensile reaction forces necessitate large boundary structures.

Bending-active flexible formworks are self-contained, not needing extensive boundary support, and allow both syn- and anticlastic curvatures. Cuvilliers et al. (2017) demonstrated the use of a strained gridshell as falsework with a passive textile shuttering layer. Popescu et al. (2018) introduced stiffening articulation in both bending-active formwork and resulting shell. Both examples give the outlook to structurally integrate the gridshell falsework in hybrid composite action with the resulting concrete shell, whereas Hack et al. (2020) developed a construction system of passively bent rebar that serves as formwork and reinforcement for non-standard geometries. Already the pioneer Nervi (1956) had combined formwork and reinforcement in his Ferro-cement invention to overcome their bottleneck for complex geometries.

2.3 Bending-Active Structures

Bending-active structures suffer the dilemma of requiring low stiffness for the elastic forming process and sufficient stiffness to withstand external loads. Thus, gridshells such as the Multihalle Mannheim are often designed as double-layered systems with shear connectors (Adriaenssens et al. 2014) – then the splines' minimum bending radius is small due to small spline sections with nevertheless, larger structural height.

The form finding of strained gridshells is commonly performed with dynamic relaxation for fast simulation or with finite element analysis (FEA) for accurate numerical modelling and analysis. Bellmann (2017) implemented a built-in function into the commercial FEA-software SOFISTIK (2020) that computes internal bending stresses based on curved input splines' initially straight, unstressed states.

2.4 Computational Frameworks

The modelling of gridshells with complex topology and geometry is challenging with visual programming software that typically stores the data in tree structures without their connectivity. Instead, the Python-based, open-source framework COMPAS for research in computational architecture, engineering and construction (Van Mele et al. 2017–2022) provides a versatile graph-based data structure that can be customised and interface with various COMPAS packages, as in Mendez Echenagucia et al. (2019).

3 System Design

The presented construction system undergoes three main construction stages: the assembly of the bending-active gridshell falsework, the mounting of the fabric formwork and the in-situ concreting of the ribs of the ribbed funicular concrete shell structure (Fig. 2). The falsework stays in place after casting and is structurally integrated into the concrete as reinforcement. The textile of the faces in between the ribs that stabilises the shuttering during casting can either be removed or remains as an architectural feature. Alternatively, it could also be used as shuttering for casting a thin continuous shell layer. The demonstrator will not include the in-situ concreting step for logistical and sustainability reasons.

The geometry of the case study exhibits a funicular funnel shape with a droplet-shaped opening inspired by soap film explorations by Otto (1988). The ribbed concrete shell structure is a design descendent and homage of the funicular rib vaults by Rippmann and Block (2013), using a pattern reminiscent of Pier Luigi Nervi's Palazzetto dello Sport, which is in close vicinity. The rib pattern is key to the structure's aesthetics as it highlights the force flow and is informed by the funicular structural logic. The rib's cross-sections are equilateral triangles with sectional dimensions of 18 cm (Fig. 2 – right). The cross-section is consistently translated along the rib curves, following the normal orientation of the surface into torsional triangular prisms. The global structure's circular outer diameter and maximum height measure 9 m and 3.5 m, respectively. It is supported by conventional scaffolding props along its perimeter.

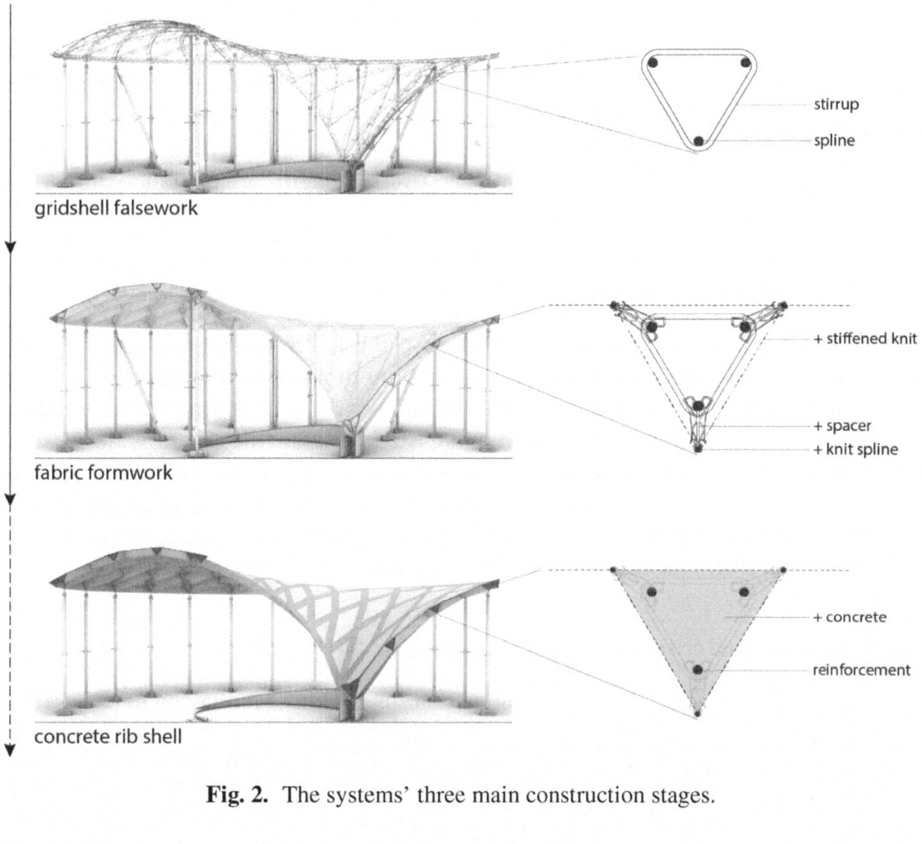

Fig. 2. The systems' three main construction stages.

3.1 Falsework Plus Reinforcement: Bending-Active Double-Layered Gridshell

The gridshell falsework (Fig. 2 – top) is the support structure for the wet-concrete load and defines the global shape. It is a double-layered system to ensure sufficient stiffness and shape control and subsequently serves as reinforcement. The falsework is materialised using conventional steel rebar, and constructed using typical techniques of the concrete industry. The mechanical properties of reinforcement steel are just within the range of suitable materials for bending-active structures that must offer a high ratio of flexural strength to stiffness (Lienhard 2014). However, compared to typical materials for strained gridshells like timber or fibre-reinforced polymers, steel offers workability with a high plasticity that permits the irreversible bending of stirrups.

A trio of continuous 10 mm rebar splines form a rib and cross another rib's splines with hinged conditions. To allow for the required minimum curvature of the global geometry the spline's highly slender sizing of 10 mm is at its maximum. Consequently, the structural height must be provided by the double-layered system. The three splines are connected with 6 mm rebar stirrups to space the splines and with a pair of inclined 6 mm stirrups as shear connectors to stiffen and lock the gridshell into its doubly-curved shape (Fig. 12). The crossing connections in between the splines are made using wire twist-ties. These crossing locations and the splines' overall lengths are extremely important as they dictate the global geometry.

3.2 Formwork Shuttering: Knitted Textile

The falsework supports and defines the shape of the fabric formwork shuttering into the prismatic profiles (Fig. 2 – middle). To ensure minimum concrete coverage, custom spacers are clipped onto the splines and hold thin rods at their extremities to shape the fabric. The fabric formwork is made of a CNC-knitted textile that provides custom features and pockets for the rods as in Popescu et al. (2021). Alternatively, the shuttering could be made of woven textiles with sewn pockets, which would demand more manual operations and no high technology. For both, the non-prestressed fabric must be stiffened with a resin or cement paste prior to concrete casting.

3.3 Concrete Shell Structure: Ribbed Funicular Shell

Each trio of splines with fabric wrapping is cast into a reinforced-concrete (RC) rib of the funicular ribbed shell structure (Fig. 2 – bottom). The geometry is designed as a compression-only vault with a continuous tension ring along the perimeter balancing the thrusts under self-weight (Fig. 7). However, for the non-funicular live loading cases, the concrete ribs and nodes must withstand bending moments, especially because the topology lacks bracing triangulation. The rebar splines and stirrups provide the required bending-stiff, non-discrete RC ribs. Thus, the dual role of the gridshell as falsework and reinforcement is critically important.

4 Computational Design Modelling Methods

The integrative design of all construction stages and design constraints for both the formwork system and the concrete shell demands an elaborate computational modelling workflow from geometric to structural to fabrication design (Fig. 3). The implementation therefore employs existing COMPAS packages, customises them, and develops new packages.

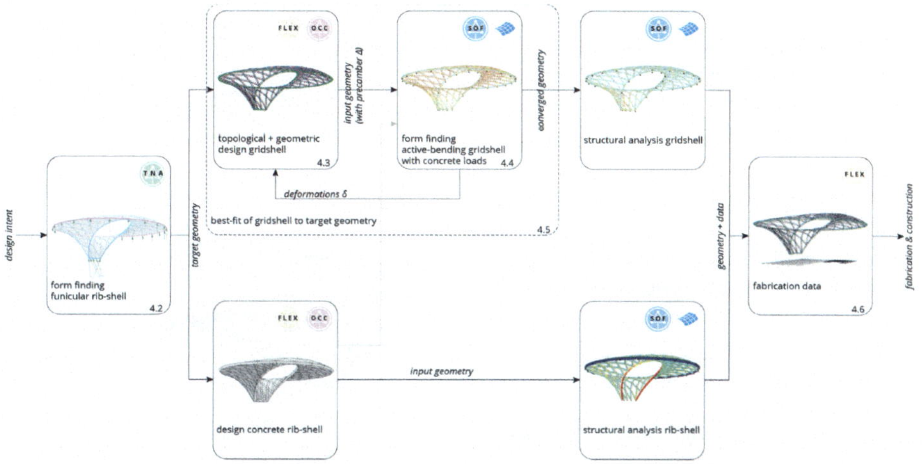

Fig. 3. Computational design-to-fabrication workflow.

4.1 Assembly Graph Data Structure

A custom package named compas_flex was developed for versatile data handling of such flexible formworks and reinforcement of complex geometry and topology with numerous different parts and connections. It inherits and customises the assembly graph data structure from the COMPAS library and stores a unique network, mesh or curve in each graph node and their respective connectivity in its edges. All parts and edges of the assembly have a unique identifier key and can be called by their attributes such as position, size, or type (Fig. 4). Separate networks guarantee spline continuity, and the graph edges store the hinge conditions in their attributes (Fig. 5). Furthermore, virtual connections between spline networks and the ribbed shell mesh store how different construction stages relate.

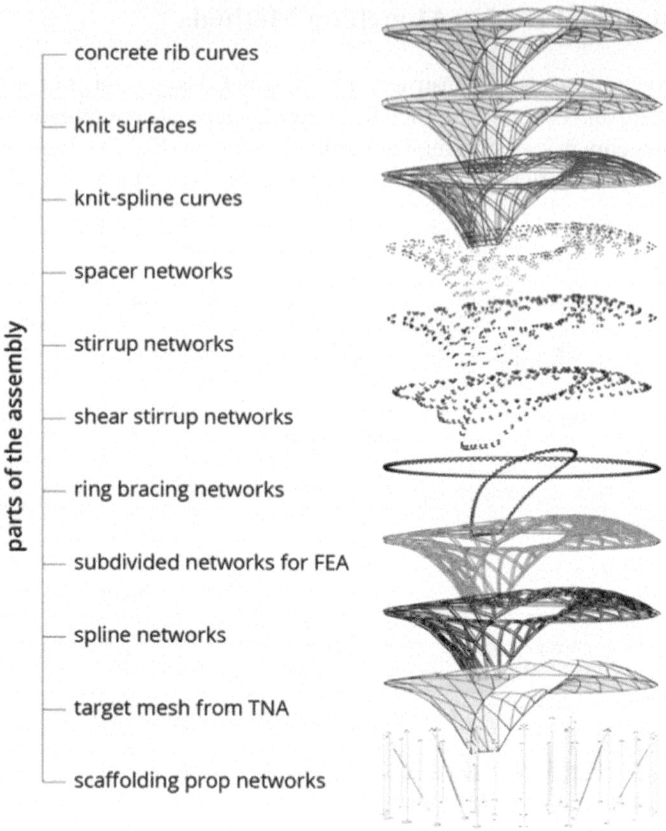

Fig. 4. Parts of the assembly graph by type.

4.2 Form Finding of Ribbed Concrete Shell

The workflow initiates with the form finding of the funicular shell using the compas_tna package. For the case study, the input pattern is generated from 48 radial splines that cross along five equally spaced hoops (Fig. 5 – left). In the form diagram, the tensile outer ring's edge attributes are set to tension (Fig. 5 – middle). Tension causes flipping edges to close the force diagram (Fig. 5 – right). Five vertices of the inner ring and all vertices of the outer ring are designated as supports after the horizontal equilibrium is found to ensure exclusively vertical reaction forces under the design loads. The resulting thrust network finds vertical equilibrium in the desired funnel with a droplet-shaped opening (Fig. 6 – left).

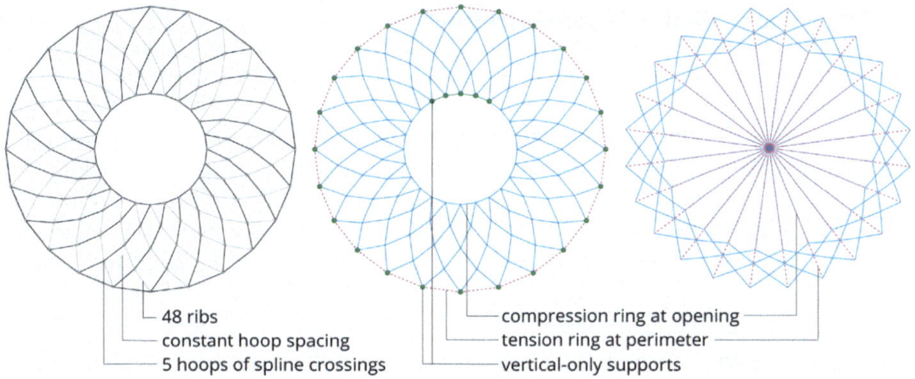

Fig. 5. Input pattern, form diagram, and force diagram.

Even though the equilibrium shape is independent of the shell's material density, the loading distribution is decisive for the funicular geometry, differing from a continuous shell to a ribbed shell by up to 8 cm (Fig. 6 – left). This is due to the varying ratio of ribbed shell openings to tributary vertex area, as the size of the faces and the skewness vary. Compas_tna is implemented for continuous shells, and thus its function to update loads in each iteration for finding vertical equilibrium is retrofit for ribbed shells. All mesh face-edge polygons are offset. Then, for each vertex, the tributary area of the offset edges is computed using the cross vector of the face centroid and adjacent half-edges, and thereafter subtracted from the sum of vertex area of the continuous shell (Fig. 6 – right).

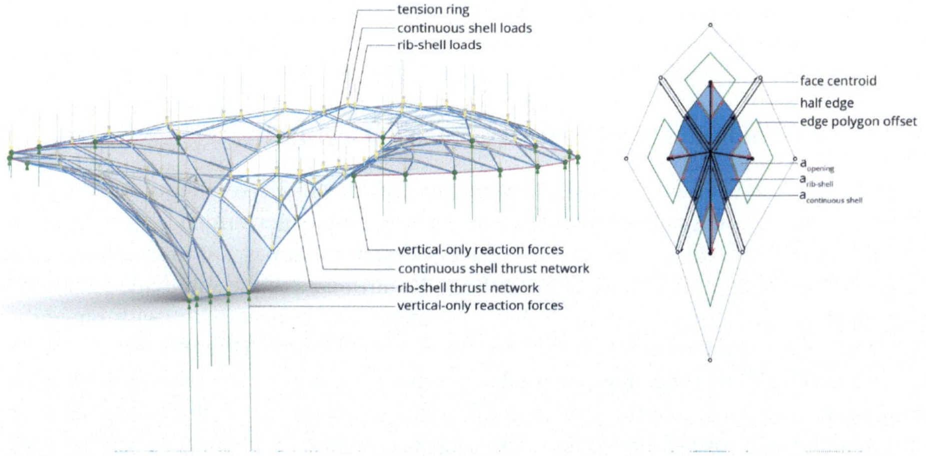

Fig. 6. Thrust networks of continuous versus ribbed shell (left) due to difference of tributary area distribution (right).

4.3 Modelling of Double-Layered Gridshell

The resulting funicular shell is the target geometry for the topological and geometric modelling of the double-layered, bending-active gridshell (Fig. 7). First, the assembly data structure stores the separate ribs as networks (generated from the mesh edges) with main attributes {direction, is_ring, at_base} and connectivity to each other and the mesh. Second, with the compas_occ package, a Python-based COMPAS binding for OpenCascade3D (2022), the NURBS curves defining the ribs are modelled to generate tangential frames that are normal to the target geometry to generate the spline trios. Third, the resulting networks are added as nodes to the assembly graph with their complex connectivity, especially on the upper layer where four splines cross per node. The spline geometry is corrected such that the splines intersect at their closest midpoint, and the networks are updated accordingly. Finally, the splines are subdivided into dense networks along reconstructed NURBS curves for the FEA input, preserving the initial coarse network nodes for connectivity. Stirrups (in blue) and inclined shear stirrup pairs (in pink) are modelled along frames and added to the assembly graph with their specifics in attributes and connectivity to the splines.

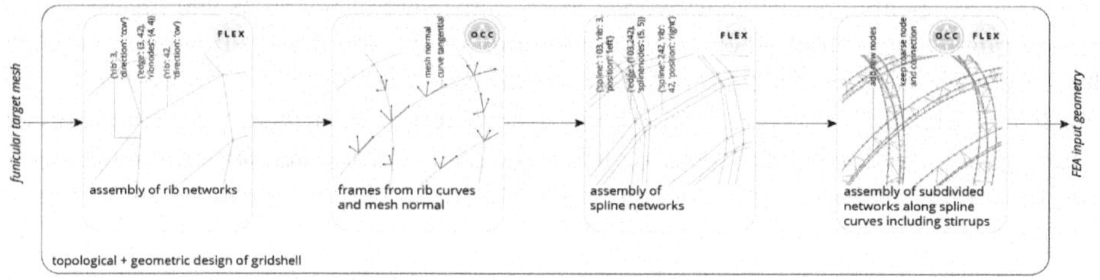

Fig. 7. Modelling of gridshell.

4.4 Form Finding and Structural Simulation of Gridshell

The spline's lengths and crossing positions are the shape-decisive input for the form finding of the bending-active structure. The form finding is performed with SOFISTIK through a direct interface from a custom package named compas_sof to the commercial FEA software (Fig. 8). Compas_sof translates the attributes of the subdivided networks into input parameters of sectional values, geometry and connectivity (with hinged or stiff couplings) into the.dat-file format for the SOFISTIK input (Fig. 8 – left).

In the SOFISTIK procedure, internal active-bending stresses are introduced with the built-in function such that the system equilibrates with non-linear, third-order analysis into its deformed form-found shape. This step is computed without the shear stirrups; they are added in a second step with wet-concrete load to match the physical assembly process (Fig. 8 – middle).

The FEA results are read out directly from the SOFISTIK.cdb-database by the compas_sof package and stored in the attributes of the parts and connections. The geometry in COMPAS is then updated with the deflections, and the shape difference (Fig. 8 – right) is evaluated and minimised prior to computing the fabrication data (Sect. 4.5). The same procedure allows setting specific loading distributions for concreting sequences and wind loads for the structural analysis (Aldinger et al. in preparation).

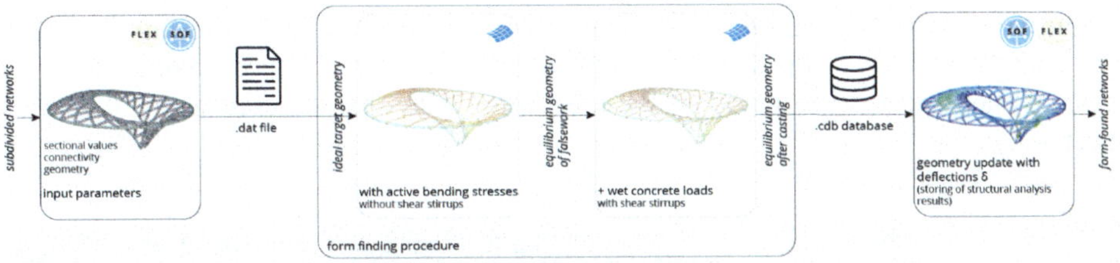

Fig. 8. Form finding of gridshell.

4.5 Translation to Fabrication Data

The fabrication data is generated from the geometry of compas_flex. From the assembly nodes, it exports the splines' unstressed lengths with their key and naming convention for cutting; from the assembly edges, the crossing locations and connecting element's key for marking (Fig. 9). Further, the surfaces of the textile shuttering are generated with compas_occ and numbered based on the connectivity of the half-edge mesh data structure. Subsequently, these serve for the knit pattern generation (Popescu et al. in preparation).

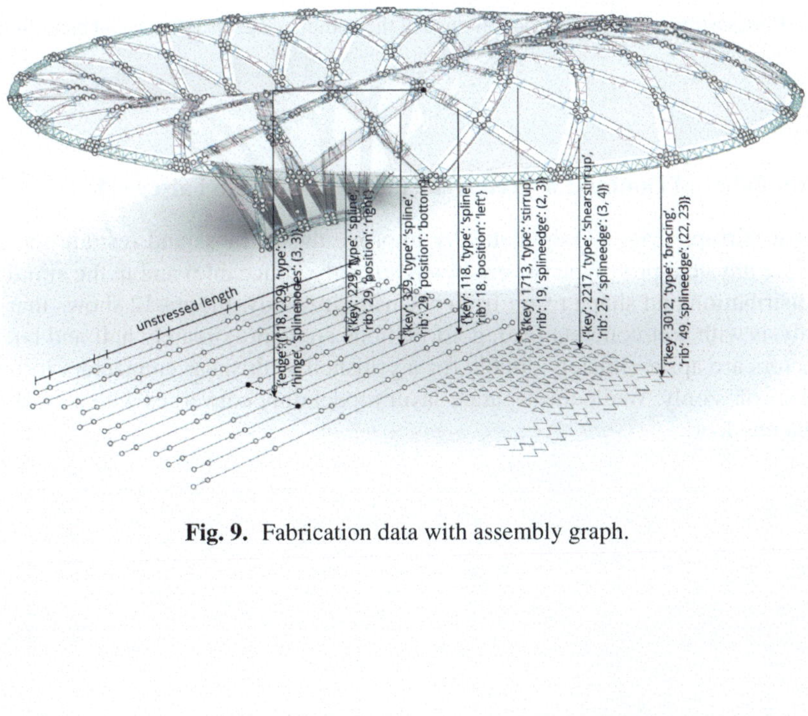

Fig. 9. Fabrication data with assembly graph.

5 Results

5.1 Materialisation and Construction

A pavilion demonstrator validates the materialisation of the falsework and reinforcement system (Fig. 10). The handling is labour-intensive but feasible for unskilled labour. The complexity lies in the correct assembly order and precise positioning of crossings, labelled and marked based on the fabrication data of unstressed segment lengths (Sect. 4.5).

Fig. 10. Demonstrator in the construction stage of the completed gridshell made of bending-active splines.

5.2 Suitability of Double-Layered Gridshells as Concrete Falsework

The shear stirrups (Fig. 11) dramatically improve the stiffness and restrain the shape in both the physical prototype (where it was strongly noticeable) and in the simulation. Their distribution and shape result from a sensitivity study. Figure 12 shows that in an FE-analysis with wet-concrete load, deformations are approximately half and buckling load factors are approximately double for a gridshell with shear connectors instead of normal stirrups only. While the separate-layer model only converges with a third of the wet-concrete load.

Fig. 11. Stirrup and shear stirrups of the gridshell.

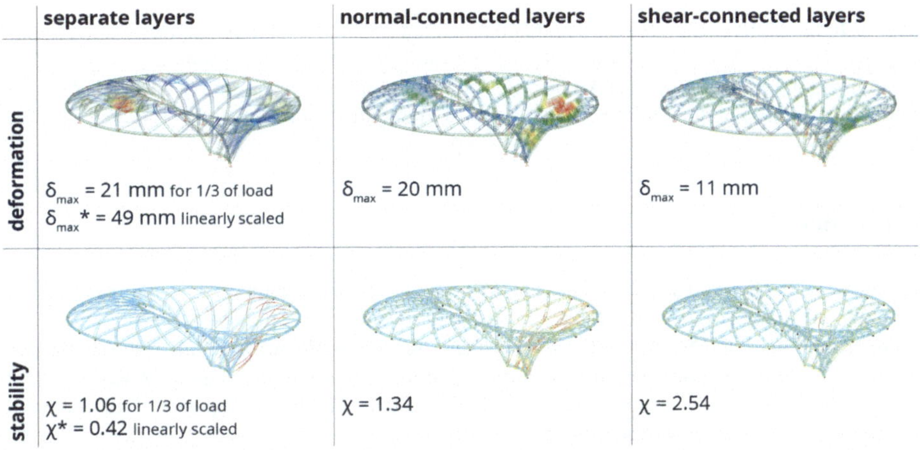

Fig. 12. Comparison of deformation and buckling load factors of double-layered gridshell under wet-concrete load.

6 Discussion and Outlook

Activating the structural height of the double-layered gridshell with shear connectors makes the proposed system a valid falsework solution. However, the sensitivity lies in the global shape accuracy toward spline lengths and crossing positions.

Further research will investigate the structural analysis with sensitivity studies and concrete loading sequence, the physical assembly logistics, and the demonstrator's construction, quantifying tolerances and stiffening (Aldinger et al. in preparation). The demonstrator validates the system design (Bodea et al. in preparation) with the knit (Popescu et al. in preparation) and the design-to-fabrication workflow as well as demonstrates the potential of the construction system for the efficient shaping of syn- and anticlastic-curved ribbed shells.

The computational development with COMPAS allows the use and customisation of existing packages as well as the creation of new packages resulting in a reproducible, comprehensible and generalisable workflow for such construction systems. The assembly data structure offers high control of the geometric, structural and fabrication design with complex topology and geometry.

The computational workflow allowed the form finding and design of a structural and construction system so that the key to efficiency for both the flexible formwork and the funicular concrete shell lies in their structurally-informed geometry. Thus, unlike the conventional CNC-milled moulds, the formwork system is less wasteful and not limited to high-tech construction contexts. Even though it is labour-intensive, no additional reinforcement installation step is required. As an outlook, such a construction system could find application in bridges, vaulted roofs or vaulted rib-stiffened floors with an additional continuous shell layer. All this could make the system applicable to a wide range of construction contexts and mitigate the environmental crisis by reducing the material consumption of both formwork and concrete structure.

Acknowledgements. The authors would like to express their gratitude to their industry supporters Debrunner Acifer Bewehrungen, Gisler Bewehrungen and Doka Schalungstechnik. The research was supported by an ETH Architecture & Technology fellowship and by the NCCR Digital Fabrication, funded by the Swiss National Science Foundation (Agreement # 51NF40-141853).

References

Adriaenssens, S., Block, P., Veenendaal, D., Williams, C.: Shell Structures for Architecture: Form Finding and Optimization. Routledge Taylor & Francis Group, Abingdon (2014)

Aldinger, L., Popescu, M., Bodea, S., Spiekermann, K., Van Mele, T., Block, P. (in preparation)

Bellmann, J.: Active bending starting on curved architectural shape. In: Bletzinger, K.-U., Onate, E., Kröplin, B. (eds.) 8th International Conference on Textile Composites and Inflatable Structures - Structural Membranes 2017 (2017)

Block, P.: Thrust network analysis: exploring three-dimensional equilibrium. Ph.D. dissertation, Massachusetts Institute of Technology, Cambridge, USA (2009)

Block, P., Van Mele, T., Rippmann, M., Ranaudo, F., Calvo Barentin, C., Paulson, N.: Redefining structural art: Strategies, necessities and opportunities. Struct. Eng. **98**(1), 66–72 (2020)

Bodea, S., Popescu, M., Aldinger, L., Spiekermann, K., Van Mele, T., Block, P. (in preparation)

Cuvilliers, P., Douthe, C., Du Peloux, L., Le Roy, R.: Hybrid structural skin: prototype of a GFRP elastic gridshell braced by a fiber-reinforced concrete envelope. IASS J. **58**(1), 65–78 (2017)

García de Soto, B., et al.: Productivity of digital fabrication in construction: cost and time analysis of a robotically built wall. Autom. Constr. **2018**, 297–311 (2017)

Hack, N., et al.: Structural stay-in-place formwork for robotic in situ fabrication of non-standard concrete structures: A real scale architectural demonstrator. Autom. Constr. **115**, 103197 (2020). https://doi.org/10.1016/j.autcon.2020.103197

Hawkins, W.J., et al.: Flexible formwork technologies - a state of the art review. Struct. Concr. **17**(6), 911–935 (2016). https://doi.org/10.1002/suco.201600117

Lehne, J., Preston, F.: Making concrete change; innovation in low-carbon cement and concrete. Technical report, Energy Environment and Resources Department, London (2018)

Lienhard, J.: Bending-Active Structures: form-finding strategies using elastic deformation in static and kinematic systems and the structural potentials therein. Ph.D. dissertation, Universität Stuttgart (2014)

Mendez Echenagucia, T., Pigram, D., Liew, A., Van Mele, T., Block, P.: A cable-net and fabric formwork system for the construction of concrete shells: design, fabrication and construction of a full scale prototype. Structures **18**, 72–82 (2019)

Nervi, P.L.: Structures. F.W. Dodge Corp., New York (1956)

OpenCascade3D: Open Cascade 3D Technology (2022). https://dev.opencascade.org/doc/overview/html/

Otto, F.: IL 18. Seifenblasen, Stuttgart (1988)

Popescu, M., Reiter, L., Liew, A., Van Mele, T., Flatt, R., Block, P.: Building in concrete with an ultra-lightweight knitted stay-in-place formwork: prototype of a concrete shell bridge. Structures **14**, 322–332 (2018)

Popescu, M., et al.: Structural design, digital fabrication and construction of the cable-net and knitted formwork of the KnitCandela concrete shell. Structures **31**, 1287–1299 (2021)

Popescu, M., Aldinger, L., Bodea, S., Spiekermann, K., Van Mele, T., Block, P. (in prepartion)

Rutten, D., McNeel, R.: Grasshopper3D. Seattle, Robert McNeel & Associates (2007)

Sofistik: Sofistik: structural engineering software (2020). https://www.sofistik.com

Rippmann, M., Block, P.: Funicular funnel shells. In: Gengnagel, C., Kilian, A., Nembrini, J., Scheurer, F. (eds.) Proceedings of the Design Modeling Symposium Berlin 2013, Berlin, pp. 75–89 (2013)

Van Mele, T., Lachauer, L., Rippmann, M., Block, P.: Geometry-based understanding of structures. J. Int. Assoc. Shell Spat. Struct. **53**(4), 285–295 (2012)

Van Mele, T., et al.: COMPAS: a framework for computational research in architecture and structures (2017–2020). http://compas.dev

Waimer, F., Noack, T., Schmid, A., Bechmann, R.: From informed parametric design to fabrication: the complex reinforced concrete shell of the railway station Stuttgart 21. In: Design Modelling Symposium Berlin, pp. 390–400 (2019)

Redefining Material Efficiency
Computational Design, Optimization and Robotic Fabrication Methods for Planar Timber Slabs

Kristina Schramm[1(✉)], Carl Eppinger[1], Andrea Rossi[1], Max Braun[2], Matthias Brieden[2], Werner Seim[2], and Philipp Eversmann[1]

[1] Institute for Architecture, Chair of Experimental and Digital Design and Construction, University of Kassel, Universitätsplatz 9, 34109 Kassel, Germany
schramm@asl.uni-kassel.de

[2] Institute for Structural Engineering, Chair of Timber Structures and Building Rehabilitation, University of Kassel, Kurt-Wolters-Straße 3, 34125 Kassel, Germany

Abstract. Available building materials are scarcer than ever before. The shortage of materials influences also timber construction, which has been experiencing a revival in the last decades, due to the material's excellent reputation as a sustainable resource. The rising motivation to build more sustainably demands large material quantities that the market can hardly supply. Hence, strategies to increase the efficiency of material usage are needed. Conventionally, material efficiency is equated exclusively with the reduction of the total amount of material used. However, a more holistic approach that considers not only the total quantities but also the dimensions and material grading could offer novel strategies to improve material usage and reduce waste. A fundamental shift in the design and construction of timber building elements is required with a particular focus on strategies enabling the reuse and recycling of small-scale timber components and on joining methods that enable the elimination of adhesives and thus allow for disassembly. This research, therefore, proposes a novel system of hollow timber slabs comprised of multiple layers consisting of an internal layer of small-scale beams in optimized locations connected to two outer plates through wood dowels. To design and fabricate these pure timber slabs the application of computational design and optimization methods to identify ideal material layouts and the use of automated robotic assembly processes to simplify production are required.

Keywords: Material efficiency · Timber construction · Material layout · Robotic additive manufacturing

1 Introduction

The construction industry is facing a shortage of materials. It is also noticeable in timber construction, which has experienced an upswing in recent decades as the material enjoys an excellent reputation as a sustainable resource. The quest for more sustainable solutions in this industry requires large quantities of material that the market can hardly supply and which endanger the renewability of forests (Caulfield 2020). Therefore, strategies to

C. Gengnagel et al. (Eds.): DMS 2022, *Towards Radical Regeneration*, pp. 516–527, 2023.
https://doi.org/10.1007/978-3-031-13249-0_41

increase the efficiency of material usage are needed that are not exclusively equated with the reduction of the total amount of material used but, more holistically, also consider the dimensions and material grading. This could offer novel strategies to improve material usage and reduce waste.

In exploring such possibilities, it is important to keep in mind considerations regarding both the possibility of reuse of timber elements, as well as their end-of-life scenarios. Most timber elements are still simply being burned at the end of their life cycle, as the extensive use of adhesives makes the elements hardly separable for recycling. Addressing this, a fundamental shift in the design and construction of timber building elements is required with a particular focus on strategies enabling the reuse and recycling of timber components and on joining methods that enable avoiding adhesives and thus allow for disassembly.

As surface-based slab elements for buildings need to respond to a range of criteria that require an ideal combination of structural capacity, material usage, acoustics, and building technology integration, they appear to be the ideal candidates for this study. Hence, this research proposes a novel system of hollow timber slabs, consisting of a layered construction of an internal layer of short beams, connected to two outer plates through wood dowel connections. The construction of these seemingly massive, yet hollow elements not only utilizes smaller amounts of material compared to massive CLT building elements, but their cavities also offer the potential for insulation infill and building technology integration (Orozco et al. 2021).

Designing and fabricating these elements requires the application of computational design and optimization methods to identify ideal layouts for given requirements, as well as the use of automated robotic fabrication processes to simplify production (Fig. 1).

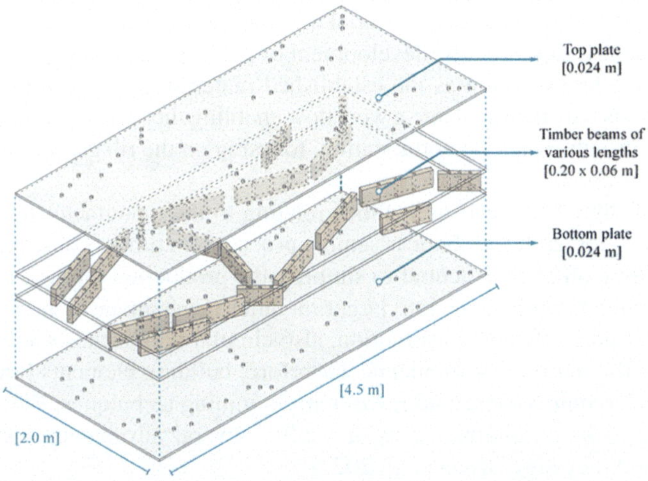

Fig. 1. Exploded isometric perspective of the proposed system: planar timber slab consisting of two outer plates and an internal layer of beam components.

1.1 Related Work

With a few exceptions, the majority of computational methods for material distribution rely either (A) on the automated generation of an ideal layout or (B) on the optimization of a predefined, generally uniform layout.

Recently, the generation of an ideal material layout (A) through topology optimization (TO) of the full volume of a building element has been investigated, as it offers great potential to achieve significant material reductions, but it also ensures an efficient distribution of the material. This approach appears to be particularly suitable for building elements from concrete as no further discretization is necessary to manufacture such elements (Jipa and Bernhard 2016).

Experiments on using TO achieve an ideal material layout for timber structures have been conducted as well. However, as the resulting geometry requires further discretization, a subsequential finite element analysis to discretize the geometry for the location and orientation of discrete components has been used to generate a layout for such elements (Naboni and Kunic 2019).

Other projects have been focusing on the optimization of a uniform material layout (B), such as in the case of a concrete slab whose structural system consisted of curved ribs that were based on a uniform 2-dimensional grid. As the slab had to meet multiple geometric and structural criteria, the exact position of the curved ribs was subsequentially generated using a multi-objective evolutionary optimization algorithm (Meibodi and Leschok 2018).

Dawod et al. used a 2d topology optimization algorithm with a predefined grid to maximize the stiffness while minimizing material usage, through optimization of the material distribution of a high-resolution structural beam element made from continuous wood filament (Dawod et al. 2020).

As described, most research on material distribution has mainly focused on either of the two approaches. However, the development of a thin-vaulted, unreinforced concrete floor represents one exception as the established material layout generation and optimization methods are integrated into a workflow enabling the form-finding of a structural rib layout and its subsequential optimization, focusing on the rib geometries' shape and size (Liew et al. 2017).

In terms of digital fabrication methods relating to the manufacturing of such optimized structures, research on robotic assembly processes in timber construction has been conducted as they offer the potential to simplify the production through the automated assembly of smaller components into larger modules (Eversmann et al. 2017).

Besides resource-efficient construction, also circular approaches need to be considered, enabling the reuse of components. Therefore, building elements have to become separable which requires the rethinking of current joining techniques. Wood-wood connections seem to be an alternative as they offer several advantages such as ease of disassembly and recycling (Ruan et al. 2022).

LIGNO ceiling elements represent a good example of hollow building elements made from massive timber components. They offer the possibility for large spans but also convince with their inherent structural and serviceability qualities. However, their reliance on adhesives to join the individual components together makes them unsuitable for recycling or reuse (LIGNO TREND 2022).

2 Methods

This paper presents a study on the combination of automated layout generation methods for material distribution with multiple optimization strategies for its usage and structural performance. In conjunction, to enable the fabrication of the resulting highly irregular structures, a timber-based robotic additive manufacturing and assembly process was developed (Fig. 2).

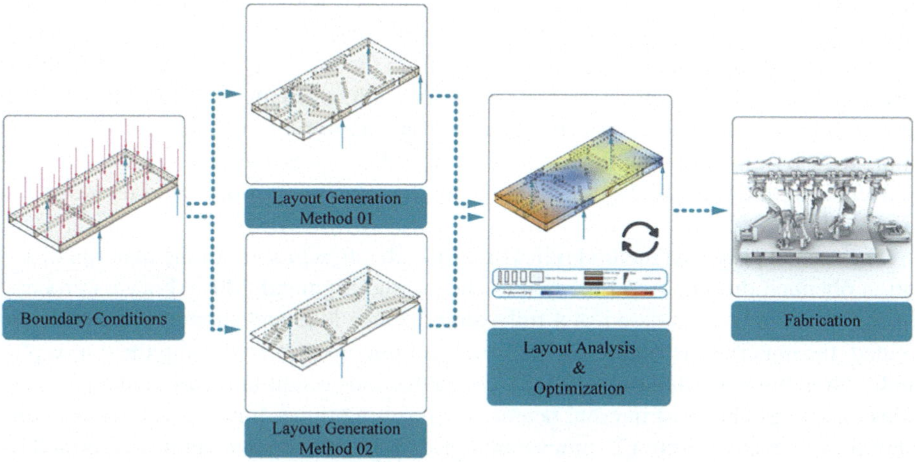

Fig. 2. Overall diagram of the developed continuous workflow linking together material layout generation, layout analysis & optimization and robotic assembly.

2.1 Layout Generation

We developed two computational layout generation methods for a structurally efficient material distribution of beams within the internal layer of the proposed construction system. Both methods enable automatic layout generation, while constantly providing feedback on the material usage and compliance with structural performance criteria, through the integration of displacement analysis.

The first automated layout generation method is based on tracing principal stress trajectories, as they appear as a suitable option to locate and orient material in areas where high stresses occur. Hence, the slab element is considered as a single shell element and subsequently analyzed through a finite element analysis (FEA) using the Karamba3d plugin for Grasshopper (Preisinger et al. 2022). The principal stress trajectories in the first direction for the upper and lower layer of the shell are generated in response to a selected support scenario and the corresponding load conditions. For each support scenario, an initial set of beams is predefined as a boundary condition. These beams are placed to prevent very high shear forces or bending moments at the supports. As the internal layer of the ceiling element currently consists of a single layer of beams, the obtained first principal stress vectors and the associated values of both, the top and the bottom layer of the analyzed shell, are assigned to a 2-dimensional point grid. This value grid is then used to create the layout for the placement of the beams. In the first step, the points with the highest tension and compression values from the upper layer are selected, as well as the associated vectors in direction of the first principal stress direction. To enable a more comprehensive, yet irregular distribution, each value is multiplied by a random number in a range of 0 to 1 and subsequently compared to a predefined threshold number. Once the value exceeds this threshold, its associated point is selected as a possible position for a beam. At the current stage of the method development, only single-length beams are considered. The beams are then placed and aligned with the vector directions, and iteratively checked for collisions with the previously placed ones. The same procedure is performed for the obtained values and vectors from the lower layer of the shell analysis. Lastly, a final collision check is run to test possible overlaps between the first and the second set of beams. Interfering beams from the non-priority set can thereby be eliminated. By changing the maximum considered compression and tension values or by decreasing the probability threshold value, the total amount of material placed can be controlled.

The second proposed method relies on a discretization process of the material distribution obtained through a TO method (Bendsoe and Sigmund 2004). For this process, the slab is initially considered as a full volume. Once load and support conditions are defined, the material distribution within such volume is optimized using the tOpos plugin for Grasshopper (Bialkowski 2017). Given that the current fabrication methodology relies on a single layer of internal beams, the resulting three-dimensional material distribution is converted into a 2-dimensional grid. Such conversion can be performed by either averaging all values for grid points with the same XY location, or by selecting the highest values among them. The resulting 2d values map is used to seed the placement of the beams. This is achieved by first calculating a material gradient vector field, which for each point defines the direction in which the material density values are increasing. Elements are then iteratively placed on the grid points with the highest values, and oriented according to the material gradient direction. If beams with variable lengths are desired, the algorithm grows the length of the beam in pre-defined steps along the material gradient direction, stopping when the material density values start to decrease. Once one element is placed, all points within it are disabled for further placement. An additional offset of the beam can be defined to disable further points around it, enabling the definition of minimum distances between elements, which is relevant for fabrication. The total amount of material placed can be controlled by either defining the total number of elements to be placed or the total length of all placed beams.

2.2 Layout Analysis and Optimization

All resulting layouts are assessed through an isogeometric finite-elements analysis (IGA), implemented through the Kiwi!3D plug-in for Grasshopper (Längst et al. 2017). Plates and beams are defined as shell elements with a given thickness and material. Couplings between beams and plates are set as hinged connections, to account for the use of dowels. We considered a dead load of 1.5 kN/m^2, a live load of 3.5 kN/m^2, and the elements' self-weight given its materials. Currently, we only considered uniform loads. Studies on the impact of point loads on the material layout and structural performance will be conducted in future research. The maximum displacement in the vertical direction is assumed as the parameter for evaluation.

The resulting information is used for optimization studies. The layout of beams is assumed as fixed, but optimization is used to select the most suitable sizing and material for each beam. For sizing, each beam is allowed to vary its width in 2 cm steps in a range from 4 cm to 12 cm. For materials, timber with different gradings is used. The optimization algorithm can either optimize one of the two parameters or both in parallel. Two different approaches have been investigated. The first one is a single-objective optimization with the goal of minimizing displacement. This relies on Galapagos, the built-in genetic algorithm for Grasshopper (Rutten 2013). While the goal of this paper is not to exclusively focus on maximum structural performance, this study allowed us to assess the impact of sizing and material choice on the structural performance of the slab element.

A second approach focused on the exploration of interactions between multiple objectives. This was implemented using Octopus, a multi-objective optimization plug-in for Grasshopper (Vierlinger 2018). In this case, displacement was not included as optimization goal but was rather used as a hard constraint, discarding any solution which would not satisfy the allowed limit. Looking at parameters that are related to sizing and material of the elements, different optimization goals were defined, such as:

- Reduction of the total material volume.
- Reduction of the total costs of the structure, by using lower grade timber where possible.
- Reduction of the global warming potential (GWP) of the used materials.

2.3 Fabrication

To fabricate the generated layouts, we defined a robotic fabrication system, which could adapt to a large variety of different geometries, joining methods and sizes. This required rethinking the common requirements of a robotic work cell and end-effectors. The cell was arranged to allow easy reconfiguration fitting different component dimensions. This was achieved by creating modular platforms which were used as pick-up and work areas. As the domain of beam lengths is variable, the gripper had to be adaptable to a range of sizes. The tools for joining were designed to allow for variable joint geometry. In the fabrication process, the robot places beams and plates without fixation. Wood nails for fixation were used to allow precise pre-drilling of the holes and insertion of the wood dowels. While the nail gun and the drill are conventional manual tools attached to and actuated by the robot, the tool for dowel insertion had to be developed. This reproduced the manual process of hammering dowels using a pneumatic piston, using high force to insert the dowel through several impacts of the piston (Fig. 3).

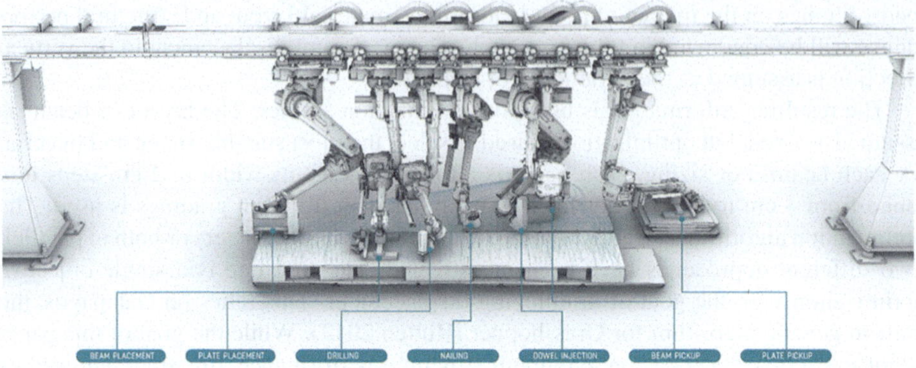

Fig. 3. Diagrammatic perspective view of the robotic fabrication setup showing the individual steps of the assembly process.

3 Results

We tested layout generation and optimization methods for a variety of support conditions and assessed their performance in terms of maximum displacement and material savings in comparison to conventional solutions. We further explored optimization approaches and their impact on variables such as cost, material volume and GWP. To validate the feasibility of the proposed wood-wood connection and related robotic fabrication process, we fabricated full-scale prototypes in our robotic laboratory.

3.1 Layout Generation

See Fig. 4.

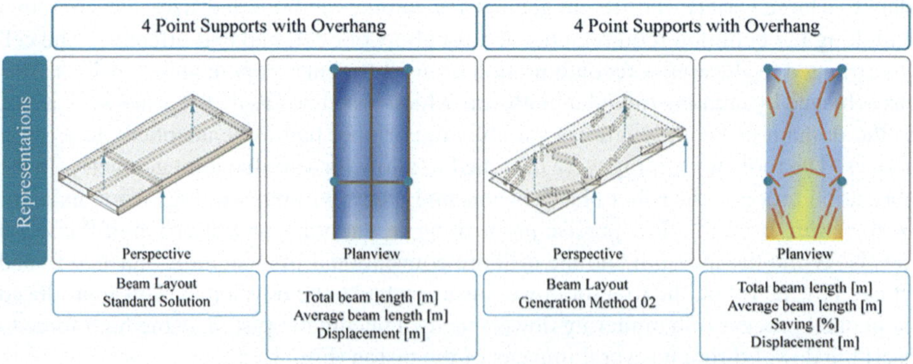

Fig. 4. Explanatory diagram of representations

Comparing generated layouts with conventional solutions allows for evaluating the advantages and drawbacks of the different approaches. It must be noted that, in most cases, our approaches cannot match the displacement results of conventional solutions. This is because the emphasis was not on reaching the minimum possible displacement, but rather on reducing the overall amount of material used and on reducing the average length of the beams used while maintaining the displacement below the maximum recommended limit of l/250 (in the case of a 4.5 m slab, the maximum allowed displacement is 0.018 m).

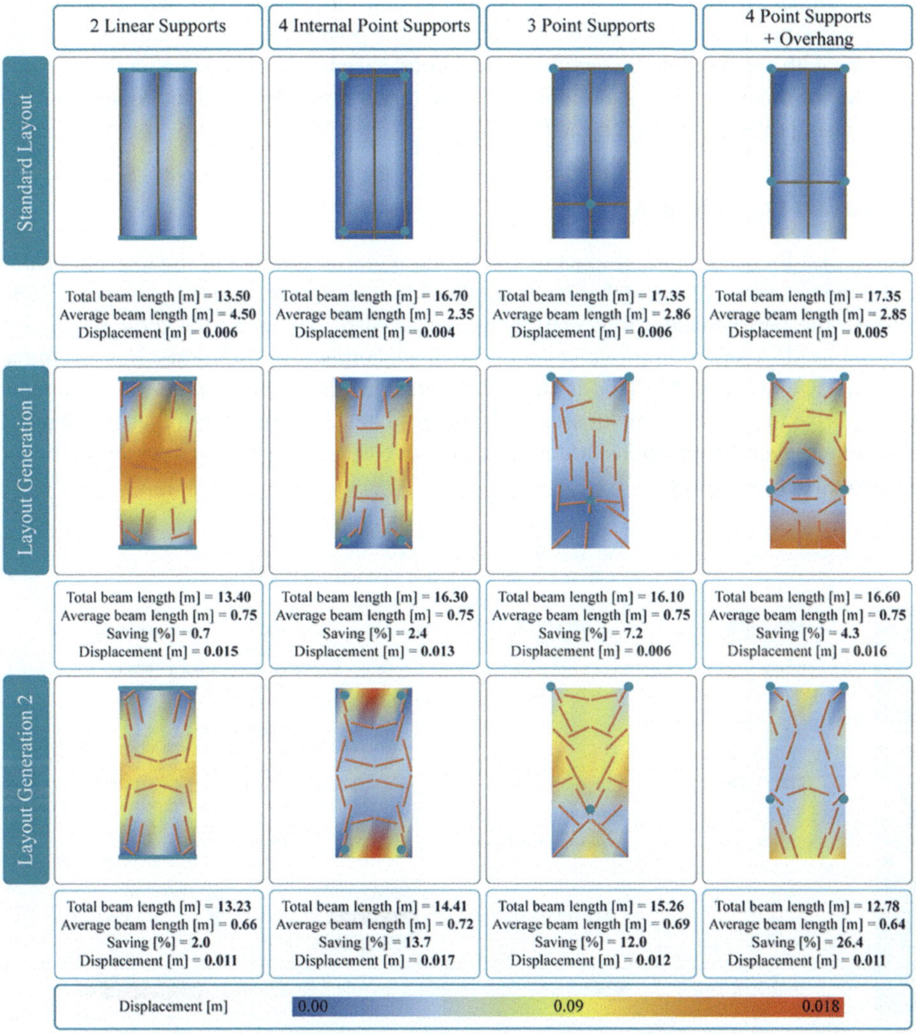

Fig. 5. Comparison of the results from the developed material layout generation methods with conventional beam layouts for several support scenarios, showcasing the potential to reduce the overall amount of material of the beams and the possibility to decrease their average lengths while still complying with the structural requirements.

A summary of the results is provided in Fig. 5. While in the case of two edge supports our methods provide very small material savings, in other cases with more complex configurations they provide significant material reductions, ranging between 2% and 26%. Furthermore, they provide a considerable reduction in the need for larger beams. In particular, the first layout generation method, based on stress-lines tracing, matched the desired performance using exclusively elements with a length of 0.75 m. The TO-based approach used elements with a variable dimensional range, but it can achieve higher overall material savings. In the first case, improved performance and material savings could be achieved by including the possibility for the algorithm to use elements with varying lengths, while in the second case it would be beneficial to include an option of placing additional elements in the larger areas left without beams, as well as increasing overlaps between the beams.

3.2 Layout Optimization

First studies focused on the impact of sizing and material grading of the internal beams on the performance of the slab. Using single-objective optimization, it was possible to improve the performance only marginally, while also increasing the total volume of material used. Average improvements of the displacement for different cases are in the range of 5% (see Fig. 6). This is mainly because, at the current stage, the generated layouts drive the majority of stresses in the outer plates, lowering the impact of changes in the inner layer. More studies are required to develop layouts where stresses are better distributed across layers.

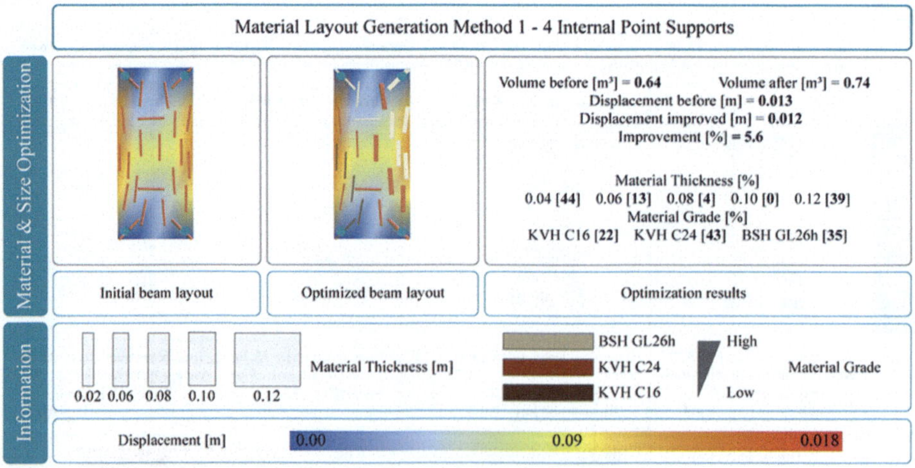

Fig. 6. Overview of the results from the performed material & size optimization on one layout for the 4 internal point supports scenario. As the displacement of the slab is only improved by 0.001 m, further studies on suitable combinations of material and sizing to improve other parameters than structural ones will be conducted in future research.

Given the only marginal improvements in structural performance possible, further studies focused on utilizing the optimization process to find suitable combinations of materials and sizing which would improve other parameters beyond structural ones. Using a multi-objective approach, it was possible to reduce the total volume of material used and its cost, while also improving the overall GWP value (see Fig. 7).

While current results are based on simplified assumptions, they nevertheless demonstrate the potential of linking efficient layout generation methods with multi-objective optimization approaches, allowing to explore a variety of solutions where specific material properties and dimensions can be leveraged to achieve a variety of requirements.

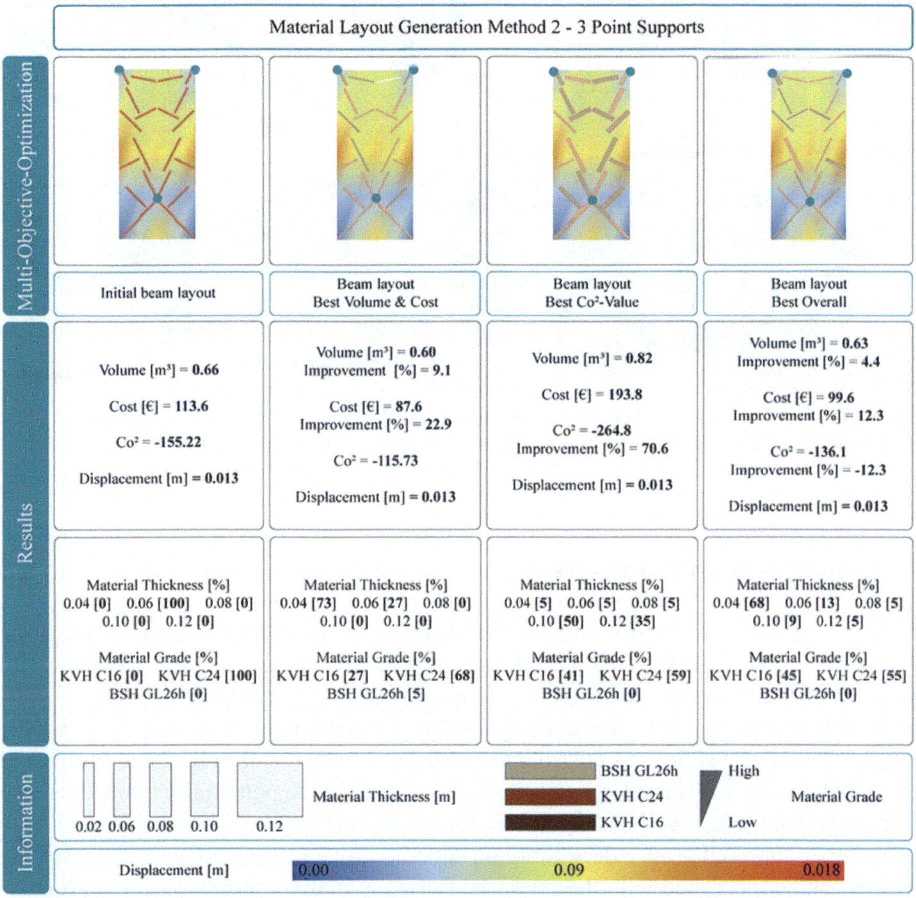

Fig. 7. Table: the results from the multi-objective-optimization on one layout for the 3 point support scenario indicate the possibility to reduce the total amount of material used and related to this also its costs as well as improving the total GWP value of the slab.

3.3 Fabrication

Using the described robotic fabrication methods, we were able to manufacture several large-scale (1 × 4,5 m) prototypes, demonstrating that not only the automated assembly of complex geometries is feasible, but also fully automated, adhesive-free and reversible joining methods can be integrated. This required overcoming several challenges regarding nailing and dowel insertion, such as replacing the conventional drilling bit with a wood auger and developing a drying process to ensure the ability to insert dowels despite their dimensional variability (Fig. 8).

Fig. 8. From left to right – top: pick and place procedure of the beams. Laying the top plate without any fixation – bottom: nailing the plates for fixation; drilling; injecting dowels

4 Conclusion and Outlook

The results from the developed computational design and optimization methods illustrate the flexibility of the established workflow to respond to multiple support scenarios with the possibility of reducing material in comparison to standard hollow timber floor slabs while still complying with the defined structural performance criteria. By using the proposed layered system, it is possible to break away from the need for large beams and the reliance on non-reversible gluing processes, as well as to potentially allow the reuse of material from construction demolition, the use of offcuts from production or even to differentiate material grades depending on specific project requirements. The resulting layouts can be fabricated with the proposed fully-automated robotic assembly process.

Further research on more detailed FEA modeling and structural testing of the behavior of the wood dowel joints to determine the load-bearing capacity as well as the serviceability of the slabs is necessary. In particular, given the material reductions achieved, evaluation of the impact of vibrations in the resulting lightweight slabs should be considered. Moreover, we are planning to integrate fabrication-related parameters into the design and optimization workflows to include limitations and constraints based on the robotic setup and investigate the design of layouts based on given material inventories.

Overall, the proposed methods enable a more holistic understanding of material efficiency, augmented by the combination of computational design processes and robotic fabrication. The further steps of this research will look into using such an approach enabling the flexible reuse of building elements in a circular process, further reducing the impact of timber construction.

Acknowledgements. Redefining Material Efficiency is part of the research project Robotic Timber, funded by the Deutsche Forschungsgemeinschaft (DFG). We would like to thank Georgia Margariti from the Chair of Structural Design (TWE), University of Kassel, for her support.

References

Bendsoe, M.P., Sigmund, O.: Topology Optimization: Theory, Methods, and Applications. Springer, Heidelberg (2004). https://doi.org/10.1007/978-3-662-05086-6

Bialkowski, S.: TOpos - GPGPU accelerated structural optimisation utility for architects. In: Proceedings of the 35th ECAADe Conference, Rome, Italy, vol. 1, pp. 679–688 (2017)

Caulfield, J.: A New Report Predicts Significant Demand Growth for Mass Timber Components. Building Design + Construction, 5 July 2020. https://www.bdcnetwork.com/new-report-predicts-significant-demand-growth-mass-timber-components

Dawod, M., et al.: Continuous timber fibre placement. In: Gengnagel, C., Baverel, O., Burry, J., Ramsgaard Thomsen, M., Weinzierl, S. (eds.) DMSB 2019, pp. 460–473. Springer, Cham (2020). https://doi.org/10.1007/978-3-030-29829-6_36

Eversmann, P., Gramazio, F., Kohler, M.: Robotic prefabrication of timber structures: towards automated large-scale spatial assembly. Constr. Robot. **1**(1–4), 49–60 (2017). https://doi.org/10.1007/s41693-017-0006-2

Jipa, A., Bernhard, M.: 3D-printed stay-in-place formwork for topologically optimized concrete slabs. In TxA Emerging Design + Technology (2016)

Längst, P., Bauer, A.M., Michalski, A., Lienhard, J.: The potentials of isogeometric analysis methods in integrated design processes. In: Proceedings of IASS Annual Symposia, vol. 2017, pp. 1–10. International Association for Shell and Spatial Structures (IASS) (2017)

Liew, A., López López, D., Van Mele, T., Block, P.: Design, fabrication and testing of a prototype, thin-vaulted, unreinforced concrete floor. Eng. Struct. **137**, 323–335 (2017). https://doi.org/10.1016/j.engstruct.2017.01.075

LIGNO TREND: Deckenkonstruktionen Made of LIGNO Bedeutet Mehrwert (2022). https://www.lignotrend.de/produkte/konstruktion/deckenbauteile

Meibodi, M.A., Leschok, M.: Smart slab. Computational design and digital fabrication of a lightweight concrete slab. In: Proceedings of the 38th Annual Conference of the Association for Computer Aided Design in Architecture, Mexico City, Mexico, pp. 434–443 (2018)

Naboni, R., Kunic, A.: A computational framework for the design and robotic manufacturing of complex wood structures. In: Blucher Design Proceedings, pp. 189–196. Editora Blucher, Porto (2019). https://doi.org/10.5151/proceedings-ecaadesigradi2019_488

Orozco, L., et al.: Design methods for variable density, multi-directional composite timber slab systems for multi-storey construction, p. 10 (2021)

Preisinger, C., Vierlinger, R., Tam, M.: Karamba3D (version 2.2.0) (2022). https://www.karamba3d.com/

Ruan, G., Filz, G.H., Fink, G.: Shear capacity of timber-to-timber connections using wooden nails. Wood Mat. Sci. Eng. **17**(1), 20–29 (2022). https://doi.org/10.1080/17480272.2021.1964595

Rutten, D.: Galapagos: on the logic and limitations of generic solvers. Archit. Des. **83**(2), 132–135 (2013)

Vierlinger, R.: Octopus (version 0.4) (2018). https://www.food4rhino.com/en/app/octopus

Strategies for Encoding Multi-dimensional Grading of Architectural Knitted Membranes

Yuliya Sinke(✉), Mette Ramsgaard Thomsen, Martin Tamke, and Martynas Seskas

CITA - Center for IT and Architecture, Royal Danish Academy, Philip de Landes Alle 10, 1435 Copenhagen K, Denmark
ybar@kglakademi.dk

Abstract. This paper introduces the knit as a special case of functionally graded materials. The ability to specify knitted material both structurally and materially allows for multi-dimensional grading of the knit. This ability to functionally grade knit is defining for knitted material's relevance to architecture and construction, as it allows for combining performance multiple criteria in one material by smoothly transitioning between the properties in a single manufacturing act. However, the scaling of knit to architectural scale brings computational challenges for automating material specifications. This paper investigates the methods for solving these computational challenges when operating with multi-dimensional material gradings that increase the scale and resolution of knitted textiles and therefore increase the complexity of their encoding.

The method for multi-dimensional material grading of CNC-knitted membranes is developed through the project of Zoirotia, a large-scale textile installation (Fig. 1). Zoirotia is made of 88 unique membranes, where each is uniquely graded at the levels of both material structure and yarn composition. This results in surfaces of varied expansion properties and color transitions across the entire structure, achieved by changing yarn morphology and material. The project is understood as a testing ground for solving the challenge of large volume material graded specification through simplified numerical data and the process of binarisation. A dithering technique is used to translate rich design-driven gradients for material specification into binary gradients for CNC fabrication. This paper presents the gradient numerical map as a solution for handling the material specification spanning across the multi-element structures.

Keywords: Multi-dimensional grading · CNC-knitting · Computation · Digital fabrication · Functionally graded materials · FGMS

1 Introduction

1.1 Background

In architecture, lightweight composite materials contribute to resource and energy saving by combining different properties towards improved performance and reduced weight of the structure. Here, several homogeneous matters are fused together to achieve a

C. Gengnagel et al. (Eds.): DMS 2022, *Towards Radical Regeneration*, pp. 528–542, 2023.
https://doi.org/10.1007/978-3-031-13249-0_42

multi-functional composite material. However, these types of fused composites create an undesired accumulation of internal material stress at the point of material sudden shift (Fig. 2) (Shinohara 2013).

Fig. 1. Multi-dimensional grading of architectural knitted membranes in Zoirotia, 2021.

Functionally graded materials (FGMs) can be understood as a novel class of composite materials characterized by the gradual variation in composition and structure throughout a surface or volume (Ichiro and Yoshinari 1997). Here, the smooth transition between the properties reduces the thermal and residual stress concentrations found in traditional composites. Furthermore, FGMs optimize material deployment, reducing unnecessary material deposition, while concentrating it in the places of structural utilization.

The usefulness of FGM composites was recognized by Bever and Duvez already in 1972. However, this has had a limited impact on the industry due to a lack of suitable production methods. Early fabrication methods were based on the manual sequential casting of matrix-based materials (Pajonk et al. 2022), which was slow and imprecise. The emergence of additive manufacturing, supported by computational interfaces, has led to new fabrication processes of digitally controlled material deposition enabling the production of FGMs of higher grading complexity (Richards and Amos 2014; Grigoriadis 2015).

In this paper we introduce knit as a special class of FGMs. Knit, being constructed by an accumulation of stitches to form a textile, is an additive manufacturing technique. The advancements in CNC knitting technologies has enabled the fabrication of textiles of higher material complexity with a fine level of material differentiation (Ramsgaard Thomsen et al. 2016). Here, the shift of stitch structure and material composition allows

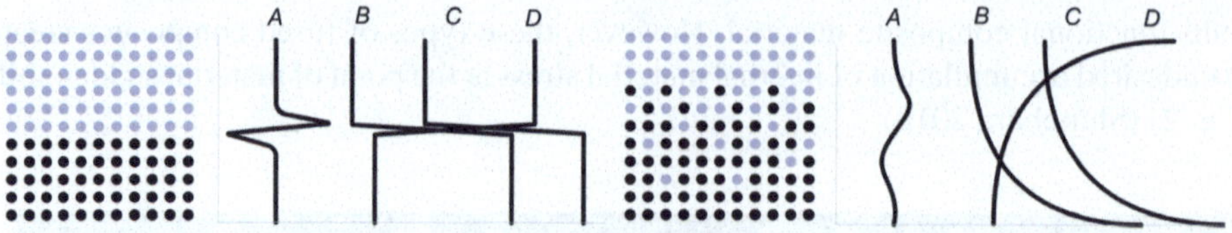

Fig. 2. Comparison between composite (left) and functionally graded materials (right) composition, metal+ceramics: A - thermal stress, B - thermal expansion, C - thermal conductivity, D - heat resistivity. Original image by Shinohara (2013), reworked by Sinke 2022.

the definition of knit as a technique allowing for a multi-dimensional material grading, which extends the definition of FGMs.

This paper presents the conceptualization of a new class of multi-dimensional FGMs for architectural application. By using the installation Zoirotia as a material case, we examine knit as a functionally-graded additive manufacturing technique, that allows for the making of differentiated architectural membranes. By controlling the design and specification of the material composition and structure at the micro-scale of the stitch we can steer the membrane performance at the macro-scale, smoothly incorporating varied surface properties without the need for cutting and sewing as done with woven textile materials for architectural membranes.

1.2 Problem Definition and Goals

Where knit allows for micro-scale control of the textile this also compounds the problem of specification. Textile traditions are frequently based on patterns that repeat across the surface allowing for relatively compact specification data. When appropriating knit for architectural scale and engaging in performance-driven functional grading, the specification data is radically extended. To enable the design and production of complex and highly differentiated knitted structures we, therefore, have to develop means of translating design intent into compact encodings, described with a small number of parameters that can be steered towards high-performance multi-layered designs.

Multi-dimensional grading is often handled independently for each parameter of change, as they are informed differently. This naturally introduces parallel layers of multi-grading representations and presents the challenge of overlapping these into a single binary mono-dimensional file for fabrication. These tasks are not trivial, as reflect the ongoing challenges of computer science regarding modelling complex systems and managing big data (Stanley and Miikkulaninen 2002; Doursat et al. 2012).

The work presented in this paper, prototypes and tests efficient methods for multi-dimensional material grading of multiple criteria into a single fabrication file. The presented method describes a workflow for interfacing and combining the specification of two independent grading criteria and its further adaptation for fabrication. The method furthermore develops ways of defining local grading strategies from global design intent, while keeping the high resolution for fabrication specifications. We introduce the application of these approaches through the material case of Zoirotia, a large tensile membrane structure commissioned by the ZKM museum (Zentrum fuer Kunst und Medien), Karlsruhe, 2021–22.

2 State of Art

2.1 Functionally Graded Structures

Functionally graded materials (FGMs) belong to a relatively new class of materials characterized by gradient variation of micro-structure and mechanical properties. They consist of two or more constituent phases with a continuously variable composition creating properties that change spatially within the structure, in order to design components optimized for specific applications (Maalawi 2018). They solve a number of engineering problems, eliminating the stress singularities occurring at the sudden shift of materials, like in traditional composites. The emergence of digitally informed additive manufacturing has enabled more detailed material specifications of FGMs, allowing for smoother transitions between material graded properties.

FGMs are studied and applied within fields where it is crucial to provide great material performance. In the medical industry, additively manufactured FGMs are applied for producing biological tissues and organ transplants, successfully replicating the original heterogeneous nature of these (Watari et al. 2004). In aerospace, the thickness of the metal parts for airplanes can be functionally graded to achieve the best mechanical performance and improve aerodynamics (Saleh et al. 2020). In mechanical engineering geometrical optimization is used to materially grade mechanical joinery in order to reduce material deposition while keeping the performance intact (Maalawi and Badr 2009).

2.2 FGMs in Architecture and Construction

The principles of lightness, high performance, and shape optimization, have high relevance to architectural construction. FGMs for architectural application have been investigated by combining several materials (Grigoriadis 2015; Herrmann and Sobek 2017; Duque Estrada et al. 2020) and grading the density of the structural composition (Nicholas and Tamke 2012; Soldevila et al. 2015; Tish et al. 2018; Lesna and Nicholas 2020). These demonstrate smooth material property transitions based on required structural performance or design indent as well as the feasibility of adopting FGM strategies in architecture and construction.

Following the classification for FGMs (Loh et al. 2018) the examples mentioned above can be classified as either *mono-material* FGMs or *multi-material* FGMs. In mono-material FGM a single matter is used to form a graded material. The grading is achieved through gradual structural modification, smoothly transitioning between the properties by changing the density of the material. In multi-material FGM several different materials are combined in a single additive process to create objects with varied properties. Both strategies for mono- and multi-material grading are mono-dimensional as they operate with *one* parameter of change, either change of material density or the change of material types. This is primarily because of the fabrication limitation of additive manufacturing of 3d printing as this technology doesn´t permit a multi-criteria change during the fabrication (Oxman 2010).

2.3 Knit as Multi-dimensional FGM

Knit is inherently both multi-material and multi-structural, which enables the fabrication of multi-dimensional knitted FGMs. In knit, fine material structures can be achieved, as the yarn is manipulated between the needle beds. Several yarns can be intertwined simultaneously, combining different material properties in one surface. As such, knit allows for *multi-dimensional* material gradation, both at the level of the structure (stitch type) and matter composition (yarn type), thereby combining in itself concepts of mono- and multi-material FGMs.

The recent interest in knit as an architectural material instrumentalises this ability to design the textiles specifically for performance and initiated the investigations of knit application as an alternative for structural membranes and formwork. The principles of *structurally graded surface* have been examined through the mono-material grading, where a single yarn, entwined into various stitch types, is used (Ahlquist 2015; Deleuran et al. 2015; Ramsgaard Thomsen et al. 2016, 2019a, b; Tan and Lee 2018). These surfaces result in homogeneous matter composition but vary structurally across their surface, which results in differentiated density and expansion properties across the membrane. The principles of *materially graded surface* have been examined in multi-material grading strategies, where multiple yarns of varied matter origin (e.g. elastic cotton+stiff polyester etc.) are combined, using the same stitch type (Ramsgaard Thomsen et al. 2016; Ramsgaard Thomsen 2019a, b; Anishchenko 2021). These textiles often result in a structurally homogeneous surface with a heterogeneous material composition, which can be reflected in color or texture variation.

A few research outputs explore multi-dimensional aspects of knit specification (Hörteborn and Zboinska 2021; Liu et al. 2020; Tan 2021) by incorporating multi-colored structurally differentiated surfaces. However, the potential for finer material gradings and transitions between the properties is not fully explored as distinct property transitions are observed in these precedents.

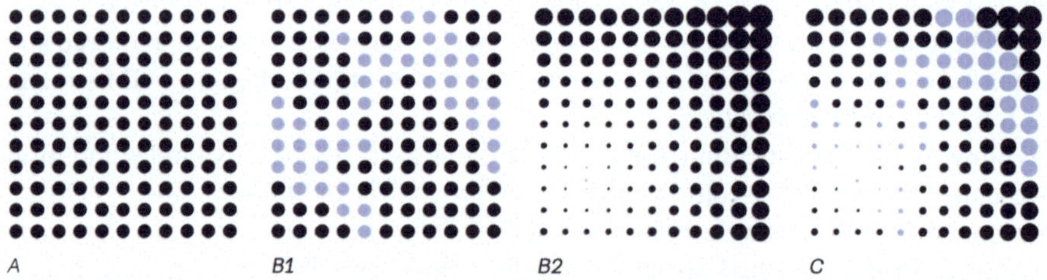

Fig. 3. A - non-graded material, B - mono-dimensional grading (B1 - same stitch varied color. B2 - same colour, varied stitch). C - multi-dimensional grading combining (varied stitch and color). Important to note that within this classification the dimension is implied in a meaning of a parameter, rather than as a direction of grading as used by Hascoet et al. (2011).

3 Multi-dimensional Knitted Membranes of Zoirotia

3.1 Core Parameters of Zoirotia Material Grading

This paper presents Zoirotia, a large-scale tensile installation, that uses multi-dimensional grading, extending the modes of material differentiation towards smooth transitions between the properties. The installation is made of 88 individual unique membranes, where each is graded in a bespoke manner for two purposes (Fig. 4):

- Structural grading for three-dimensionality and structural performance. It influences the expansion of the material and is achieved by combining two stitch types of varied locking point lengths: open loose double jacquard and dense double jacquard stitches.
- Material grading of different yarn colors for visual expression (via double jacquard technique) creates transitions in hue, across a large body of the installation and contributes to the chromatic perception of the design.

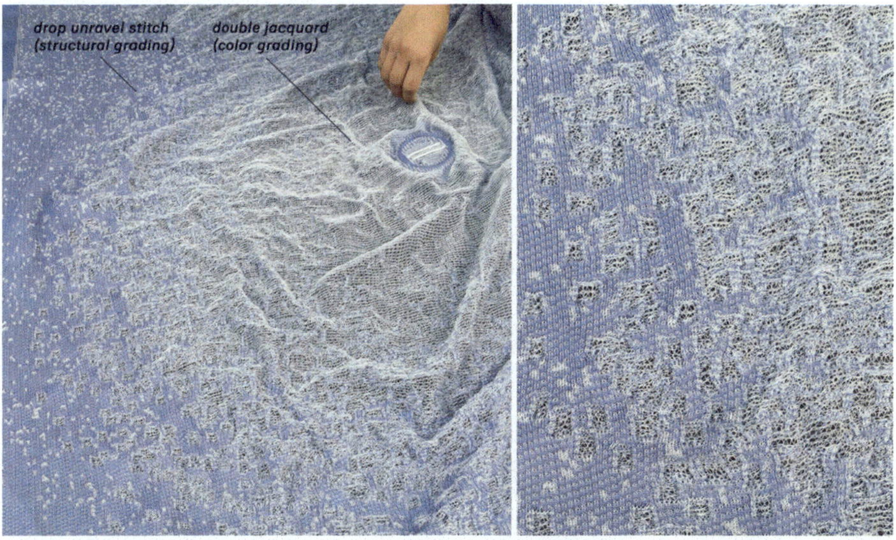

Fig. 4. Prototyping membrane shows multi-dimensional grading through the double jacquard technique (color and structure)

Here, the morphology of stitches and yarn color are the two core parameters of surface differentiation, simultaneously existing within each membrane. However, these parameters of material grading are informed and operate differently, which requires different modes for their representation, reflected in two individual parallel models: one for structural grading and another for color grading. Both models serve the purpose of material grading design and generate material specification data reflected in a form of color gradients, which is then modified for fabrication purposes (Fig. 5).

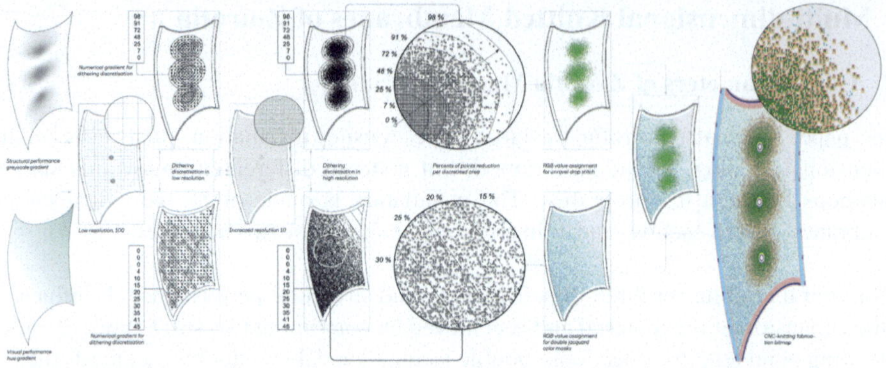

Fig. 5. Membrane surface programming through dither gradient patterns

3.2 Informing the Grading Models

The fundamental differences between these models lay in the modes how they are informed. The structural grading model is informed by geometrical analysis, where the form-found surface is evaluated for identification of areas of larger displacement (or stretch). The structural textile units are form-found using a spring-based simulation, where the mesh is simulated as homogeneous mesh in order to represent the behavior of the non-differentiated surface. This allows to identify the areas of larger expansion and where the specification with a larger stitch type would be required in order to achieve desired textile expansion into a three-dimensional shape when under tension. The calculation of mesh displacement is done between the final equilibrium mesh and the stage of form-finding, before prestressing the surface with the cabinets. The displacement distance values are remapped onto a planarized monochrome grayscale gradient that

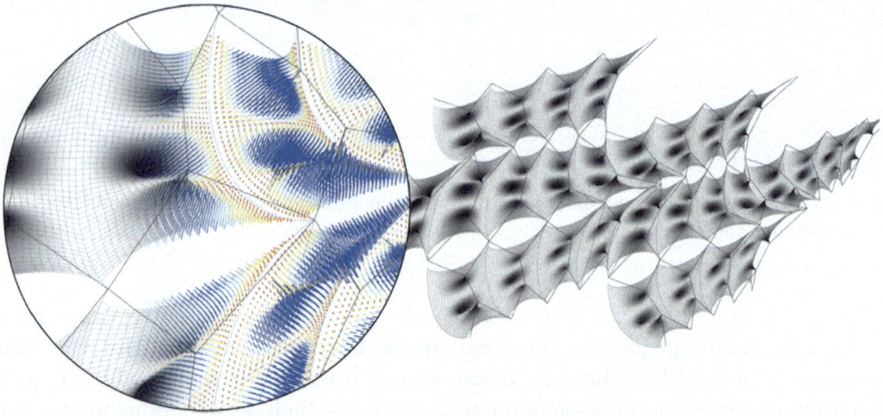

Fig. 6. Form-finding of the facade of the structure showing the displacement degree of the meshes represented through the monochrome gradient of the meshes

informs the stitch type concentration density over another (Fig. 6). Later, these gradients are modified in order to be read by the knitting machine for fabrication, which is described further in more detail.

The material grading model is informed by the design intent of combining multiple colors within the body of the installation. The ambition to achieve a vivid color expression is strongly connected to the knitting technique of double jacquard, which is used to introduce the color to the patch. Given the structural grading complexity already present in the membrane the color transition per patch was limited to two colors. This implies the nature of color distribution across the structure through the linear gradient, which allows to keep the dual-color shift per membrane (Fig. 7).

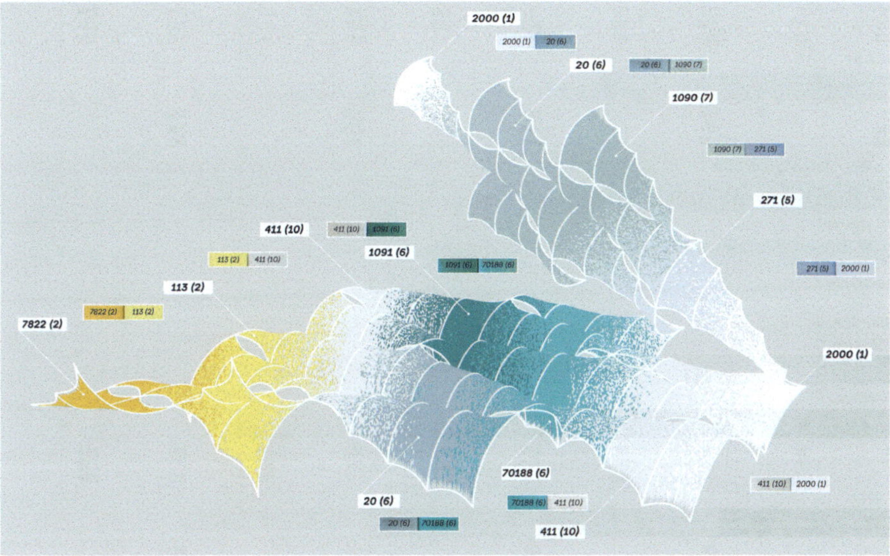

Fig. 7. Axonometric diagram with remapped colors and yarn color code labels

3.3 Challenges of Knit Multi-dimensional Grading

The differences between design and fabrication environments possess a challenge of data translation. The gradient in the digital design environment, representing the material gradation, is constructed through an almost indefinite range of colors when transitioning from color A to B, while the CNC knitting is a binary technique, operating with a definite number of stitch types and yarn carriers (hence the colors). This implies a necessity of translation of a continual material gradation, represented through these rich color gradients into binary material gradients achieved through a *dithering* technique. In computer graphics, dithering is an image processing operation used to create the illusion of color depth in images with a limited color palette (Fig. 8). The colors that are not available in the palette are approximated by a diffusion of colored pixels within the available color range (Furht 2008). In Zoirotia only two stitch types and only two color yarns per patch are used to achieve the local structural differentiation accompanied by the global hue material transition across the entire structure, total expressing gradients between 12 colors.

The dithering of colors through the density gradation, rather than through the additional intermediate colors, allows for maintaining the effect of a smooth transition through the optical effect of blending while using a low number of instrumental colors.

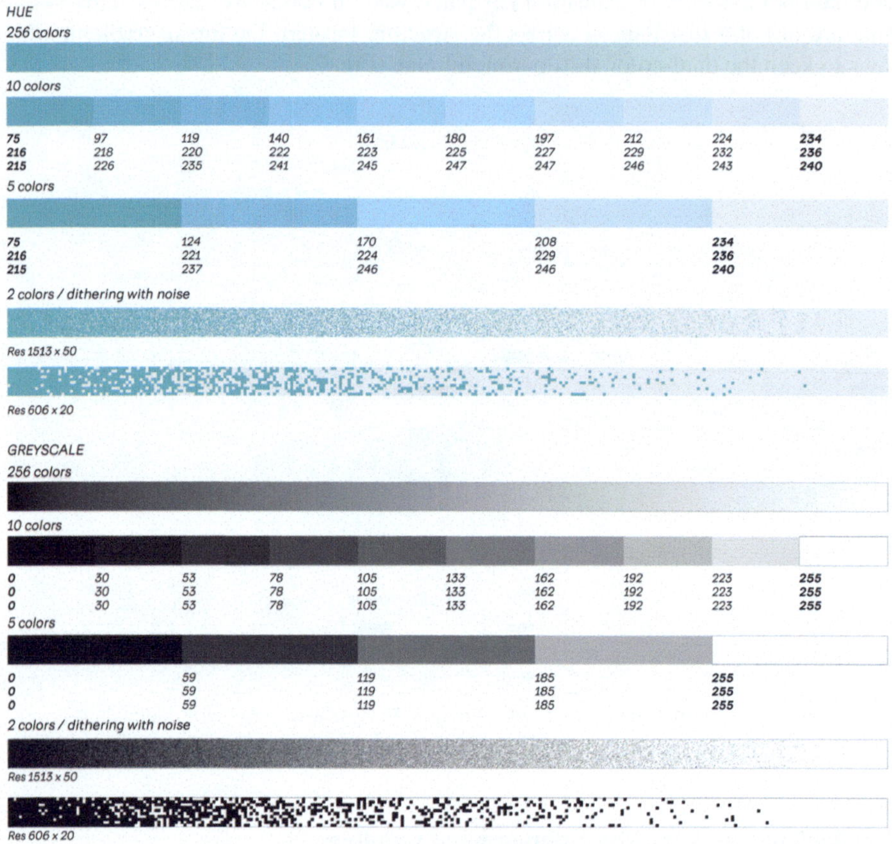

Fig. 8. Gradient representation through the varied amount of colors/resolution, for color and structural change.

In Zoirotia, the material gradings are coded within the grid-based system of points, representing the knit stitches. The dithering gradual transition is achieved through the reduction of points density, which is informed by the numerical value, representing the number of removed points per the subdivision area, introduced at the range of the gradient. For the structural grading, these are the circular areas around the pulled cones, while for the material grading these are diagonal parallel subdivisions, informed by the vector of color transition. The random function for the selection of removed points is applied, which results in the noise-like appearance of the gradient and is believed to provide a smooth blending between subdivided areas of gradient densities (Fig. 5).

The varied scales of multi-dimensional material grading pose the challenge of their resolution, compatibility and combination of these independent gradings. Zoirotia is made of 88 individual membrane units and is structurally graded within each single membrane, which allows for a high resolution of specification as computationally affordable. In contrast, the color grading of Zoirotia spans across the entire structure and is designed globally, as the overall gradient is constructed from multiple gradients of individual membrane patches. Here, the use of high resolution for material grading of color is computationally expensive when used across such a large model.

In order to ensure computational feasibility and accuracy in transferring the property density data from the global design down to the patch scale when shifting between the scales, a method of numerical gradient map is used (Fig. 11). For that, an auxiliary gradient map model is constructed, where the numerical value is used to describe surface parameters of dither density. This way we elevate in scale and do not grade directly on the stitch level, but rather through subdivision areas of the entire structure (Fig. 10). Here, the center of each subdivision area generates a vertical line, that differs in its height based on the proximity to the areas of the most color intensity, assigned manually by attractor points (Fig. 9). The height values of these lines (domain 0–100) correspond to the intensity (%) of the dither per subdivision area. The percent value describes the amount of present/removed points per subdivision area. This results in a noise dithering pattern, that smoothly transitions between subdivision areas. As a result, the independent numerical gradient values are extracted from this map into an individual list of gradient numbers, that are used to grade membranes locally at high resolution, combining structural and hue differentiation within one membrane.

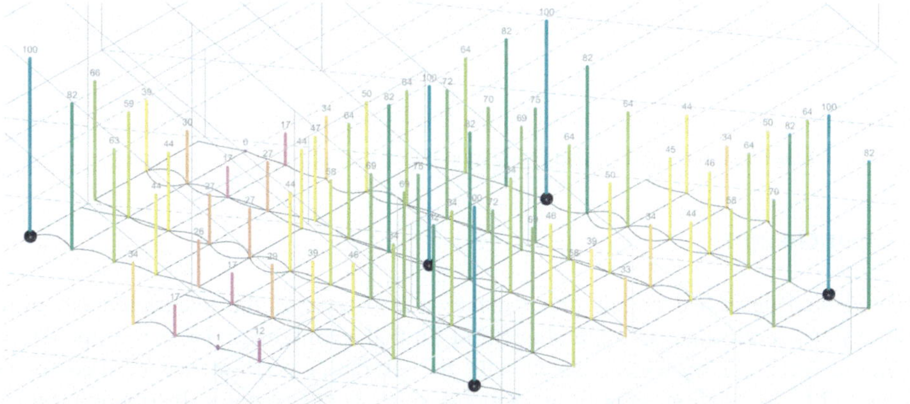

Fig. 9. Auxiliary model for generating color height-map gradient landscape

Fig. 10. Auxiliary model for generating height-map gradient landscape

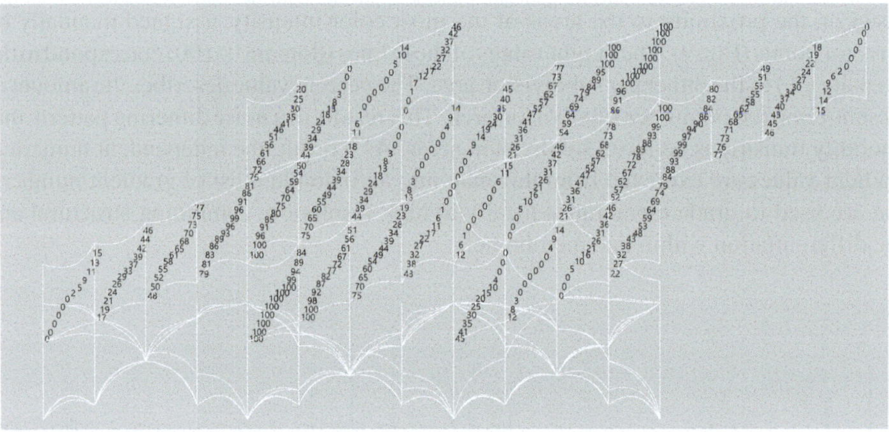

Fig. 11. Numerical gradient map model of the facade part of the structure.

4 Conclusion

The method described in this paper makes a step towards the integration of multi-dimensional performances of architectural knitted membranes. Here the membrane material local grading is correlated with the structural performance of membranes informed by simulation, as well as global design intends of color distribution, which introduces the knit as a multi-dimensional functionally graded material. Through the use of advanced CNC-knitting manufacturing and computational interfaces, knitted material is graded finely at the level of stitch, transitioning membrane structural and material properties in a gradient manner (Fig. 12).

The computational challenges associated with this process are discussed and methods for solving these are presented. Through the process of binarisation by using the dithering technique only two structural stitch types and only two yarn colors per patch are combined to achieve a very continuous merge of properties on structural and material levels at the scale of the single patch as well as across the entire structure. Here, a random-based density reduction dither pattern is generated, through the reduction of pixel density derived through the function of the curve. However, alternative dither patterns could be potentially tested as they might contribute with the novel surface expressions as well as better performing properties of the material. The translation of rich gradients is done by binarisation over pixel reduction through curve function and concentric zones subdivisions. However, in further work, a direct link between the monochrome-rich gradient to the dither reduction percentages should be explored as believed to be more integral.

Additionally, a numerical gradient method is used to link scalar differences when converting from the lower resolution in design environments - to high-resolution for fabrication.

Fig. 12. CNC-knitting bitmaps demonstrating the multi-dimensional gradients for structural and color transitions

The overall color gradient appearance was successful, although some imperfections in the technique and the method were observed. The nature of the double jacquard technique creates inverted colors at the backside of the membrane (Fig. 13). This reversed gradient becomes more visible when the high contrast yarn colors are used. In addition, the translucency of the fabric drastically affects the perception of the color gradient in multi-directional lighting conditions. Here the actual perception of hue transition in the physical structure differs from the smooth diagram from the digital environment of not transparent meshes used in the design phase (Fig. 7). An intarsia approach (in which yarns are fully replaced across the textile) for color gradient might have been a more successful knitting technique, as it has the same color appearance from both sides of the textile.

On the overall note, the degree of grading increases with the number of overlayed properties of material differentiation. While a higher degree of grading adds additional richness to the fabric it simultaneously increases material intricacy and hence the complexity of functional grading material models. Therefore, the investigation into these models and their improvement is considered a worthwhile direction for further research investigations.

Fig. 13. Photo of the finished structure (view from the facade part)

References

Ahlquist, S.: Integrating differentiated knit logics and pre-stress in textile hybrid structures. In: Thomsen, M., Tamke, M., Gengnagel, C., Faircloth, B., Scheurer, F. (eds.) Modelling Behaviour. Springer, Cham (2015). https://doi.org/10.1007/978-3-319-24208-8_9

Anishchenko, M.: Bespoke knitted textiles for large-scale architectural projects. In: Paoletti, I., Nastri, M. (eds.) Material Balance: A Design Equation, pp. 75–82. Springer, Cham (2021). https://doi.org/10.1007/978-3-030-54081-4_7

Bever, M., Duvez, P.: Gradients in composite materials. Mater. Sci. Eng. **10**, 1–8 (1972)

Deleuran, A.H., et al.: The tower: modelling, analysis and construction of bending active tensile membrane hybrid structures. In: Proceedings of the International Association for Shell and Spatial Structures (IASS) (2015)

Doursat, R., Sayama, H., Michel, O.: Morphogenetic engineering: reconciling self-organization and architecture. In: Doursat, R., Sayama, H., Michel, O. (eds.) Morphogenetic Engineering: Toward Programmable Complex Systems. Springer, Heidelberg (2012). https://doi.org/10.1007/978-3-642-33902-8_1

Duque Estrada, R., et al.: Spatial winding: cooperative heterogeneous multi-robot system for fibrous structures. Constr. Robot. **4**, 205–215 (2020). https://doi.org/10.1007/s41693-020-00036-7

Furht, B.: Dithering. In: Furht, B. (ed.) Encyclopedia of Multimedia, pp. 198–199. Springer, Boston (2008). https://doi.org/10.1007/978-0-387-78414-4_18

Grigoriadis, K.: Material fusion: a research into the simulated blending of materials using particle systems. IJAC **13**(3–4), 313–333 (2015)

Hascoet, J.Y., Muller, P., Mognol, P.: Manufacturing of complex parts with continuous functionally graded materials (FGM). University of Texas at Austin (2011)

Herrmann, M., Sobek, W.: Functionally graded concrete: Numerical design methods and experimental tests of mass-optimized structural components. Struct. Concr. **18**(1), 54–66 (2017)

Hörteborn, E., Zboinska, M.A.: Exploring expressive and functional capacities of knitted textiles exposed to wind influence. Front. Archit. Res. **10**(3), 669–691 (2021)

Lesna, J.M., Nicholas, P.: De gradus - programming heterogeneous performance of functionally graded bio-polymers for degradable agricultural shading structures. In: CAADRIA 2020: RE: Anthropocene, Bangkok, Thailand, pp. 383–392 (2020)

Liu, Y., Chai, H., Yuan, P.F.: Knitted composites tower. In: RE: Anthropocene, Proceedings of the 25th International Conference of the Association for Computer-Aided Architectural Design Research in Asia, CAADRIA, Hong Kong, p. 10. Association for Computer-Aided Architectural Design Research in Asia (2020)

Loh, G., Pei, E., Harrison, D., Monzón, M.: An overview of functionally graded additive manufacturing. Addit. Manuf. **23**, 34–44 (2018)

Maalawi, K., Badr, M.: Design optimization of mechanical elements and structures: a review with application. J. Appl. Sci. Res. **5**, 221–231 (2009)

Maalawi, K.: Modeling and applications of FGMs in aerospace structures. J. Aeronaut. Aerosp. Eng. **7** (2018)

Nicholas, P., Tamke, M.: Composite territories: engaging a bespoke material practice in digitally designed materials (2012)

Oxman, N.: Structuring materiality: design fabrication of heterogeneous materials. Archit. Des. **80**(4), 78–85 (2010)

Pajonk, A., et al.: Multi-material additive manufacturing in architecture and construction: a review. J. Build. Eng. **45**, 103603 (2022)

Ramsgaard Thomsen, M., et al.: Knit as bespoke material practice for architecture. In: ACADIA 2016: Posthuman Frontiers (2016)

Ramsgaard Thomsen, M., et al.: Systems for transformative textile structures in CNC knitted fabrics – Isoropia. In: Proceedings of the TensiNet Symposium 2019 Softening the Habitats, pp. 95–110 (2019a). https://doi.org/10.30448/ts2019.3245.08

Ramsgaard Thomsen, M., Karmon, A.: Listener: a probe into information based material specification. In: Where Art, Technology and Design Meet (Studies in Material Thinking), vol. 07, pp. 1–10 (2012)

Ramsgaard Thomsen, M.: Predicting and steering performance in architectural materials. In: Sousa, J.P., Xavier, J.P., Castro Henriques, G. (eds.) Architecture in the Age of the 4th Industrial Revolution - Proceedings of the 37th eCAADe and 23rd SIGraDi Conference, University of Porto, Porto, Portugal, 11–13 September 2019, vol. 2, pp. 485–494. CUMINCAD (2019b). http://papers.cumincad.org/cgi-bin/works/paper/ecaadesigradi2019_150. Accessed 12 June 2022

Richards, D., Amos, M.: Designing with gradients: bio-inspired computation for digital fabrication. In: ACADIA 2014: Design Agency, Los Angeles, USA, pp. 101–110 (2014)
Saleh, B., et al.: 30 years of functionally graded materials: an overview of manufacturing methods, applications and future challenges. Compos. Part B Eng. **201**, 108376 (2020)
Shinohara, Y.: Functionally graded materials. In: Somiya, S. (ed.) Handbook of Advanced Ceramics, Chap. 11.2.4, 2nd edn., pp. 1179–1187. Academic Press, Oxford (2013)
Soldevila, L.M., Royo, J.D. and Oxman, N.: Form follows flow: a material driven computational workflow for digital fabrication of large scale hierarchically structured object, p. 9 (2015)
Stanley, K.O., Miikkulainen, R.: Evolving neural networks through augmenting topologies. Evol. Comput. **10**, 99–127 (2002)
Tan, Y.Y., Lee T.L.: The flexible textile mesh - manufacture of curved perforated cladding panels. In: Fukuda, T., Huang, W., Janssen, P., Crolla, K., Alhadidi, S. (eds.) Learning, Adapting and Prototyping - Proceedings of the 23rd CAADRIA Conference, Tsinghua University, Beijing, China, 17–19 May 2018, vol. 2, pp. 349–358 (2018)
Tan, Y.Y.: Graded Knit Skins - design and fabrication of curved modular facade cladding panels. In: PROJECTIONS - Proceedings of the 26th CAADRIA Conference, CAADRIA, Hong Kong, pp. 733–742 (2021)
Tish, D., Schork, T., McGee, W.: Topologically optimized and functionally graded cable nets. New approaches through robotic additive manufacturing. In: ACADIA 2018: Re/Calibration: On Imprecision and Infidelity, Mexico City, Mexico, pp. 260–265 (2018)
Watari, F., et al.: Biocompatibility of materials and development to functionally graded implant for bio-medical application. Compos. Sci. Technol. **64**(6), 893–908 (2004)

Deep Sight - A Toolkit for Design-Focused Analysis of Volumetric Datasets

Tom Svilans(✉), Sebastian Gatz, Guro Tyse, Mette Ramsgaard Thomsen, Phil Ayres, and Martin Tamke

CITA Centre for IT and Architecture, Royal Danish Academy, Architecture, Design Conservation, Philip de Langes Allé 10, 1435 Copenhagen, Denmark
tsvi@kglakademi.dk

Abstract. In response to global challenges of resource scarcity, increasing attention is being paid to bio-based materials - a domain that covers familiar materials such as timber and emerging materials such as bio-plastics and mycelium composites. The ability to observe, analyse, simulate, and design with their interior heterogeneity and behaviour over time is a necessity for a bio-based and cyclical material practice and opens a deep reservoir of creative and technical innovation potentials within architecture and aligned design practices.

This paper describes a research inquiry which seeks to integrate volumetric material data acquired through non-intrusive methods into materially-led digital design workflows. The inquiry is developed as a set of computational tools and approaches to architectural modelling, and demonstrated through three main material tracks: structural glue-laminated timber assemblies, mycelium composites, and bio-luminescent bacteria substrates. Each addresses the acquisition, analysis, and simulation of deep volumetric material data at different scales and in different deployment contexts. In doing so, we demonstrate a novel shift in the digital modelling of bio-based architectural materials and set out its implications for new design practices that deeply embed the individuality and temporality of materials.

We contribute a perspective on the possibilities afforded by a volumetric modelling approach to bio-architecture and a computational framework for operating with volume data of heterogeneous materials.

1 Introduction

In response to global challenges of resource scarcity, increasing attention is being paid to bio-based materials - a domain that covers familiar materials such as timber and emerging materials such as bio-plastics, mycelium composites and novel classes of living materials. The ongoing transition to a bio-based paradigm in construction fundamentally challenges existing building practice - especially on a material level. A common characteristic of bio-materials is their heterogeneous and variegated structure, and innate temporality. Shaped by growth cycles and formed by their environment, bio-based materials are characterised by differentiated internal structures that perform differently under external forces.

C. Gengnagel et al. (Eds.): DMS 2022, *Towards Radical Regeneration*, pp. 543–555, 2023.
https://doi.org/10.1007/978-3-031-13249-0_43

Modelling this material heterogeneity challenges traditional surface-based architectural modelling techniques and instead requires a different interfacing with a varying volumetric material continuum. Whereas surface-based techniques operate on a binary understanding of inside and outside, a changing material continuum is described in gradients and transitions - some sharp and abrupt, others gradual.

This paper examines how volumetric representations can allow for a rethinking of how material assemblies can be observed, analysed, simulated, designed and fabricated. It presents a set of studies into emerging bio-based material practices that ask how volumetric modelling can support the representation of material heterogeneity, its analysis, and temporal behaviour - and how these representations support the design of novel assemblies, functional properties and morphologies.

These studies have helped to evolve a computational toolkit - *Deep Sight* - which integrates and appropriates established tools from medicine and visual effects for representing, analysing, and manipulating heterogeneous volume data into architectural design modelling and fabrication workflows. The studies demonstrate the current capabilities of *Deep Sight* and point to avenues of future development and extension.

2 State of the Art

The medical field has pioneered developments in volumetric data acquisition and processing, with focus on two primary objectives - accelerating automated image handling and data management, and developing feature-specific workflows for quantitative information extraction. Current methods of non-intrusive acquisition include Computed Tomography (CT), Confocal Microscopy and Magnetic Resonance Imaging (MRI). These methods were developed as a way to reconstruct three-dimensional representations of soft tissue structures (Beckmann 2006), characterised by their variable densities and entangled make-up. Computationally, this has led to efforts to develop shared tools for analyzing information gathered via these methods (McCormick et al. 2014). A somewhat similar effort can be seen in the visual effects (VFX) industry, with the standardization of formats for volumetric data and their usage in digital content creation (Museth 2013).

The study of variegated tissue structures in medicine have informed efforts in the field of architectural design and computation. The limits of surface representations to describe material variability in design modelling have led to new toolkits for generating multi-channel voxel-based representations based on data structures and techniques in medical imaging (Michalatos and Payne 2013; Michalatos and Payne 2016). The maturing of 3d printing technology has also created a focus on material variation and the variegation of material distribution, requiring the specification of properties throughout the entire volume of building elements, not just their boundary representations (Bernhard et al. 2018). Coupled with structural simulation tools, this has allowed the specification and steering of material deposition strategies towards particular design objectives (Hanna and Mahdavi 2004; Oxman 2010, 2012; Palz 2012). In manufacturing, a similar eschewing of traditional surface-based model representations leads to possibilities in composing complex material structures with 3d printing that avoid the traditional separation and sequencing of distinct production steps (Regli et al. 2016). These efforts are not relegated to 3d printing: similar approaches to functional and volumetric aggregation of material are done at larger scales with robotic timber assemblies (Naboni and Kunic 2020).

Together, these efforts can be characterised by their use of heterogeneous volumetric representations and digital simulation to *specify* material properties and tailor their internal material composition for some additive process. Other implementations of volume data structures in design modelling environments are focused on the extraction of surface-based geometry as its final output.

The same medical imaging techniques are also deployed in other industries, such as in forestry and sawmilling. Specifically, the CT-scanning of wood logs and boards has become a promising area of research for mitigating the unpredictable nature of the forest resource and therefore gaining a more precise understanding of its behaviours and capabilities (Freyburger et al. 2009; Chiorescu 2003; Huber et al. 2021; Yu et al. 2007; Shuxia et al. 2007). This has typically been focused on increasing the value yield of sawn boards from harvested logs by optimizing the sawing pattern based on their internal composition and detected features such as knots. In this case, the use of volumetric representation is to *acquire* and *analyse* a detailed material map of a bio-based resource in order to inform further production steps.

3 Methods

Our work investigates how to integrate volumetric material data acquired through non-intrusive methods or constructed through simulations into materially-led digital design workflows. The inquiry is instigated as a set of computational tools and developed further into prototypical modelling workflows in three material cases: structural glue-laminated timber assemblies, mycelium networks, and bio-luminescent bacteria substrates. Each addresses the acquisition, analysis, and simulation of deep volumetric material data at varying scales and in different deployment contexts.

The timber case study explores the mapping of prescribed performance demands in designed glulam assemblies across different found material qualities in CT-scanned sawn logs. The mycelium case study employs the skeletonization of sparse volume datasets to extract topological features of the hyphal network in a fungal colony to support functionalisation as a computing circuit. Finally, the bio-luminescence case study, questions how volumetric representations can be understood as a matrix for characterising the dynamic interactions between living systems and their environment over time. The three case studies demonstrate a novel shift in the digital modelling of bio-based architectural materials and set out its implications for new design practices that deeply embed the variance and temporality of materials.

Together, the case studies demonstrate the current capabilities of the *Deep Sight* toolkit as well as possible avenues for extension.

4 Case Studies and Results

4.1 Material Quality Mapping in Structural Glulam Assemblies

The *RawLam* project uses industrial CT-scanning of felled logs to derive a mapping of timber quality that drives the composition of curved glue-laminated beams (Tamke et al. 2021a; b). The scans - provided by partner Microtec - are enriched with metadata and important log features such as knots which create a detailed 3d map of the raw material resource. A series of physical prototypes and a full-scale structure are constructed to test, refine, and deploy a design modelling workflow based on these acquired material mappings. A tight informational link is therefore created between the raw forest resource and the design performance demands of individual glulam elements - integrating a wider breadth of the digital timber value chain (Svilans et al. 2022).

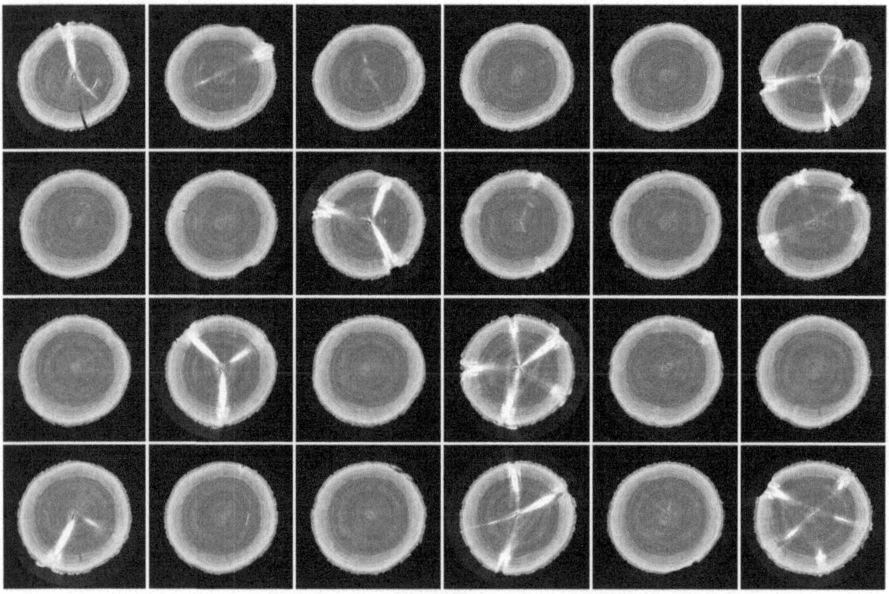

Fig. 1. Transverse slices through the CT-scan data of a log.

The key step in this project is to interface the CT-scan datasets acquired from the scanners in the sawmill with design modelling environments, allowing them to inform the further material specification of the designed prototypes (Fig. 1). The *Deep Sight* toolkit is therefore initiated as a means to translate the scan formats into a more widely-used and appropriate file format - OpenVDB (Museth 2013) - for volumetric data storage and querying.

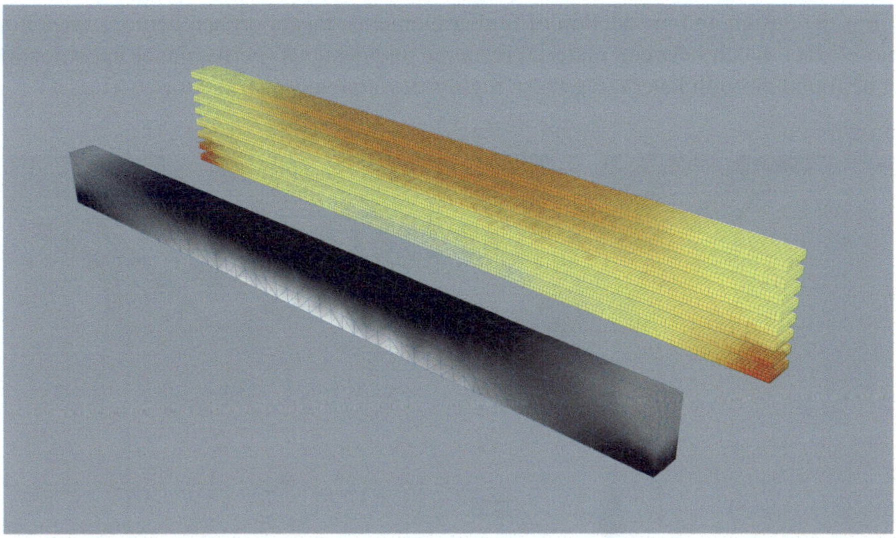

Fig. 2. Simulated stresses on the beams (bottom left) are translated into performance demand mappings on each lamella in the glulam beam (top right).

The physical prototypes are modelled using a separate toolkit for the design and material specification of glulam beams (Svilans 2021). A workflow is devised to augment this material specification with queries from the CT-scan data. The glulam model is layered with different types of performance demands, corresponding with the need for higher- or lower-quality material: proximity to the joints requires stronger material; areas of curvature should have more continuous grain areas; and a simulated stress mapping from a finite-element analysis adds a further layer of high- and low-performance demands (Fig. 2). A heuristic attempts different distributions of the lamellas from the glulams in the logs: they are digitally "placed" in particular spots of the CT-scan data in attempt to match their performance demands with the underlying material quality distribution of the logs. A comparison between their performance demands and the underlying material map of the log yields an overall fitness value for each lamella which the heuristic attempts to improve through subsequent allocation iterations.

The CT-scan data is further used to generate iso-surface boundaries to get the boundary representation of the log for use in more traditional modelling workflows. Slices of this boundary representation as well as data about knot positions and sizes are used to align production data with the physical material to ensure traceability throughout the fabrication process. A physical demonstrator is constructed to test the feasibility of this tree-to-product workflow (Fig. 3).

The use of acquired CT-scan data of timber elements therefore allows a volumetric tailoring of material to be expanded beyond only a designed specification, to include a consideration of the detailed properties of the specific material being used. This also demonstrates how this volumetric approach does not entirely replace or exclude traditional modes of surface modelling: conversely, it allows a broad range of surfaces to be extracted and defined from the volume gradient.

For the design and production of timber elements, this approach demonstrates how a more exact match between material resource and designed performance specification can be found through leveraging these highly-detailed volumetric datasets (Fig. 4).

Fig. 3. The RawLam 3 demonstrator.

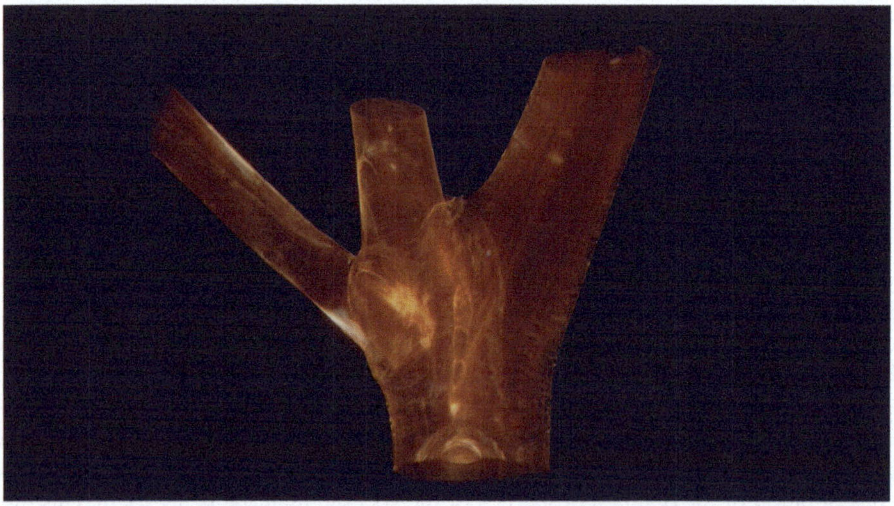

Fig. 4. A 3d visualization of a CT-scanned tree branch.

4.2 Hyphal Network Feature Extraction in Fungal Colonies

The *Fungar* project investigates the architectural use of mycelium composites as a combined construction and computing material (Adamatzky et al. 2019). Drawing upon recent findings in the literature that demonstrate bio-electric spiking behaviour in response to environmental stimulation in both the fruiting bodies and mycelium of fungi (Adamatzky 2018; Dehshibi et al. 2021), we hypothesise implementing a data processing circuit within the fungal colony.

The plausibility of this hypothesis is grounded in findings from the field of Unconventional Computing - a branch of computer science that seeks to apply general computing principles to novel substrates (Adamatzky et al. 2007). For example, within the Unconventional Computing literature, it has been shown that topological features of protoplasmic networks in slime mould (*Physarum polycephalum*) can be mapped to logical functions and processing demonstrated (Adamatzky and Schubert 2014). These principles can be extended to other kinds of fluid-based networks.

The initial acquisition of the fungal network representation, and its analysis to determine a 3d graph, has been previously reported in the literature (Adamatzky et al. 2021; Beasley et al. 2021). Here, we focus on the contribution of a network analysis method to the *Deep Sight* toolbox. This is implemented using the Python library *networkX*.

The graph is constructed as an undirected weighted graph, where weights represent edge lengths and therefore act as a proxy for electrical resistance. However, it should be noted that these are edge lengths of the rationalised graph rather than actual lengths of the fungal hyphae, which, in reality, are rarely piece-wise linear between branching points. This marks an area of future improvement for our graph reconstruction methods.

A data processing circuit was designed and its functionality validated using an online circuit simulation tool (Fig. 5). The circuit takes a sequence of binary stings encoding [P, R, O, J, E, C, T] and converts them to output strings encoding [null, F, U, N, G, A, R]. The circuit employs logic OR gates and Half-Adders (shown as 'circuit' in the figure), which combine logic AND and logic XOR to sum the inputs and provide a carry.

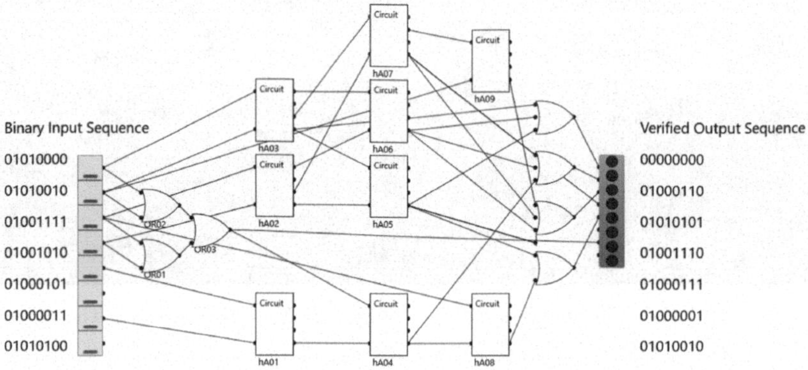

Fig. 5. A prototypical circuit design comprising OR gates and Half-Adders to perform data processing. A sequence of input strings is defined on the left and the circuit produces the outputs on the right.

The 3D graph is analysed to map all start/end nodes and the degree (number of connecting edges) of all branching nodes. Figures 6 and 7 show the identified degree 3 nodes (mapping to logical OR) and degree 5 nodes (mapping to Half-Adders), respectively.

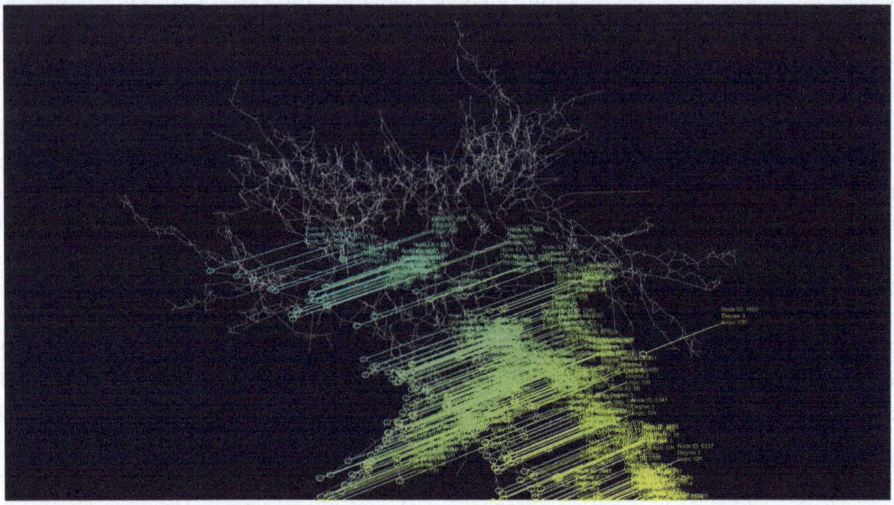

Fig. 6. Partial visualisation of all degree 3 branches identified in the fungal network. The topology of these nodes maps to the Boolean OR.

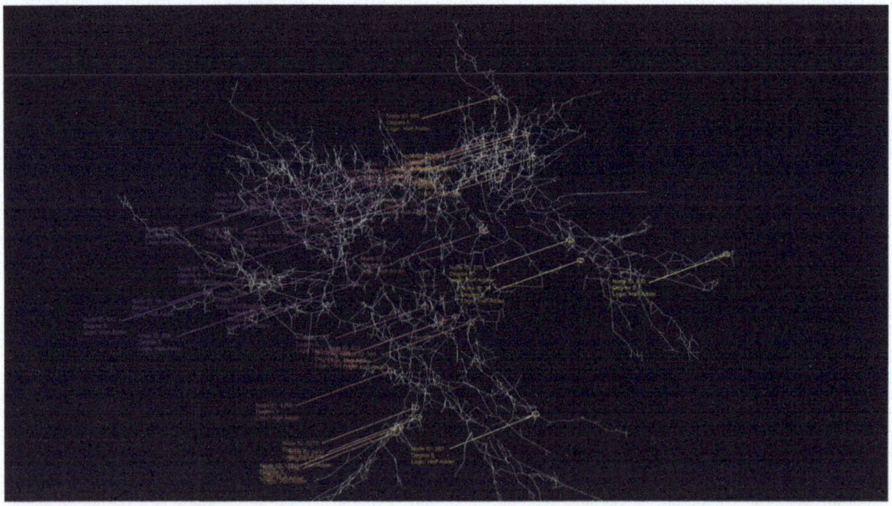

Fig. 7. Visualisation of all degree 5 branches identified in the fungal network. The topology of these nodes maps to a Half-Adder circuit.

This data set can then be interrogated to find routes through the network that map topological features to the required functionality of the circuit design, using the shortest paths possible. Figure 8 visualises the circuit found within the fungal network.

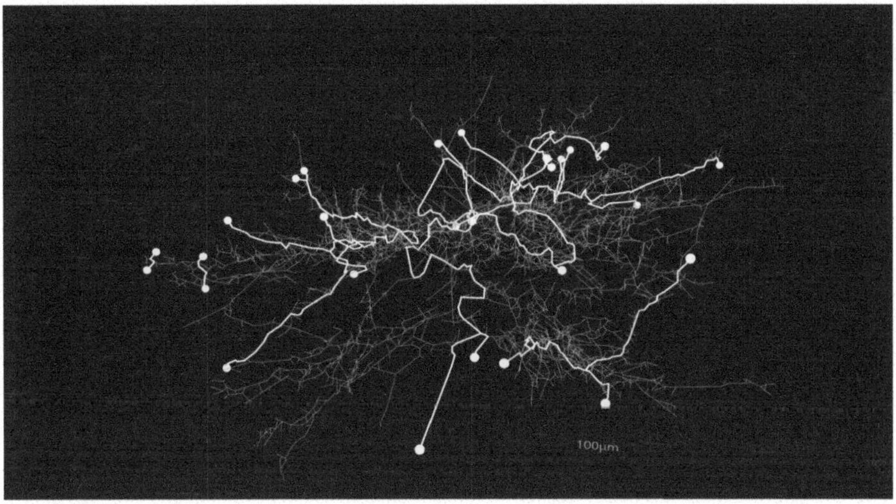

Fig. 8. Mapping of the data processing circuit design into the fungal network.

Whilst this investigation is highly speculative, even practically unfeasible, it is theoretically plausible. However, the contribution here is not found in the use-case, but the addition of analytical methods that extend the *Deep Sight* toolbox. The ability to interrogate network features of volumetrically segmented data provides a further dimension of insight into heterogeneous study targets.

4.3 Time-Based Simulation for Living Organisms

The last case examines future perspectives by which to develop the *Deep Sight* toolkit. Here, we employ volumetric modelling for time-based simulation of living organisms and employ an agent-based system to pass information across the voxel structure. This probe into future development stages uses our project *Imprimer la Lumière* into bioluminescent bacteria as a material case.

Imprimer la Lumière asks how the dynamics of living materials; their propagation, metabolising, and expiration can be functionalised as an architectural light source. It examines how the design of topologically-complex 3D-printed bacterial growth medium can steer the propagation and performance of bioluminescent bacterial colonies (Ramsgaard Thomsen et al. 2021; Tyse et al. 2022) (Fig. 9). As part of the investigation, we examine how the bacterial colonies can be simulated to enable the prediction of light performance. Our strategy is based on a volumetric model in which the behaviour of individual bacteria and their interactions with their containing environment are simulated using a time-based agent system. This study offers a venue for extending the *Deep Sight* toolkit from static models of heterogeneous media to evolving representations of time-based phenomena.

The simulation offers ways by which to test the correlation between environmental factors and performance. Through experiments we have observed that the form of the medium, the surface to volume ratio, and the integration of moisture-capturing pools affect the luminescence, propagation, and lifespan of the bacteria as they shape the influence of key environmental factors including availability of nutrients, salinity, alkalinity, and oxygen. Across their life spans, they deplete their environment and change their performance. The environmental change is furthermore dominated by changes in humidity. As time passes, and as bacterial colonies propagate, the medium dries making the habitat less liveable for the bacteria.

Fig. 9. 3D printing micro architectures (from left): slicing preview, printed agar, glowing bacteria.

Our simulation allows us to model these interactions. Each voxel contains a set of gradient variables that are recalculated in each time step in response to neighbouring voxel values. A part of these variables describe the environmental condition - humidity, nutrient level, and oxygen - which are calculated in respect to the physical properties of gravity, diffusion, and evaporation. The other part describes a bacteria count occupying the voxel within the given time state. The voxels interact with each other through their 26 immediate neighbours. For each time step of the simulation the physical conditions and bacteria count of each voxel are updated by using their previous conditions and their neighbours' conditions as input (Fig. 10).

Fig. 10. Stages of the simulation: printed geometry, adding liquid where the broth was poured, voxelizing, simulating, post production blurring in order to align simulation with photography of bacteria.

The resulting simulation is evaluated against the time-stamped recording of the live control experiment. Comparing the simulation to our recordings we can see that the emergent behaviour of the volumetric simulation correlate to the material experiments. Where the time sequencing of the simulation is slightly compressed we can see that the changes in luminosity follow similar patterns to that of the simulation (Fig. 11).

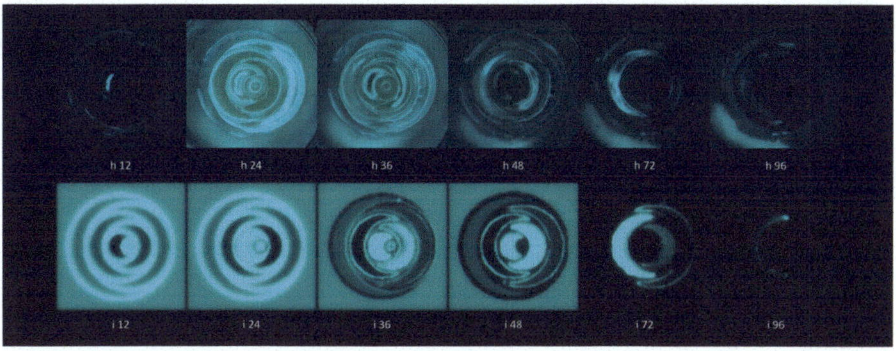

Fig. 11. Comparison of time-lapse photographs (top row) and simulation (bottom row).

The study allows us to estimate how topology impacts luminosity and life span and gives guidelines to how future architectural design environment would be able to functionalise or steer a living material substrate. As a guide to the future perspective of *Deep Sight*, the study presents a framework to more deeply embed time-steps and time-changes in a heterogeneous material substrate. The ability to integrate these temporal and interactive dimensions are central to a bio-based material paradigm in which material state change is always formed by its response to environment in time.

5 Conclusion

We contribute a perspective on the possibilities afforded by a volumetric modelling approach to bio-architecture and a computational toolkit for operating with volume data of heterogeneous materials. The case studies reveal the implications of making high-quality interior data of bio-materials accessible within design-oriented digital workflows. Access to the composition of raw timber logs permits the precise mapping of timber quality across designed architectural elements, in response to particular design and performance goals. Mapping the topology of fungal colonies allows properties of their hyphal networks to be mapped, evaluated and selected to functionalise them towards data processing objectives. Simulating the growth of living bacteria substrates makes their predictive use in architectural construction both feasible and steerable.

Together, these case studies demonstrate a path towards a design modelling approach that allows us to describe material as a changing continuum that is transformed by environmental interactions. Here, these graded volumetric representations embrace notions of material continuum, heterogeneity, and - ultimately - a material depth rather than delineation.

Acknowledgements. The authors are grateful for the collaboration and support from the involved partners. The timber case study - *RawLam* - is supported by collaborations with Microtec, Norra Timber, Luleå University of Technology (LTU), Aarhus School of Architecture, Vævestuen Savværk, and Umeå Bildmuseet. The mycelium case study was supported by University of Utrecht and the Unconventional Computing Lab (UWE Bristol). The work contributes to the project *Fungal Architectures* which is funded under the European Union's Horizon 2020 research and innovation program FET OPEN "Challenging current thinking" under grant agreement No 858132. *Imprimer la lumière* is a cross-disciplinary collaboration between CITA (Centre for IT and Architecture, KADK) and Soft Matters group, Ensadlab (ENSAD). The project benefits from the support of the Institut Français du Danemark (2018), the Danish Arts Foundation (2019), Eur-Artec (2020), Agence Nationale pour la Recherche (2021–25) and the European Research Council (ERC) under the European Union's Horizon 2020 research and innovation programme (Grant agreement No. 101019693).

References

Adamatzky, A.: Towards fungal computer. Interface Focus **8**(6), 20180029 (2018)

Adamatzky, A., et al.: Fungal electronics. Biosystems **212**, 104588 (2021)

Adamatzky, A., Ayres, P., Belotti, G., Wösten, H.: Fungal architecture position paper. Int. J. Unconv. Comput. **14**, 397–441 (2019)

Adamatzky, A., Bull, L., De Lacy Costello, B.: Unconventional Computing 2007. Luniver Press, Beckington (2007)

Adamatzky, A., Schubert, T.: Slime mold microfluidic logical gates. Mater. Today **17**(2), 86–91 (2014)

Beasley, A.E., Ayres, P., Tegelaar, M., Tsompanas, M.-A., Adamatzky, A.: On electrical gates on fungal colony. Biosystems **209**, 104507 (2021)

Beckmann, E.C.: CT scanning the early days. Br. J. Radiol. **79**(937), 5–8 (2006). https://doi.org/10.1259/bjr/29444122

Bernhard, M., Hansmeyer, M., Dillenburger, B.: Volumetric modelling for 3D printed architecture. In: AAG - Advances in Architectural Geometry, pp. 392–415 (2018). https://research.chalmers.se/en/publication/504188

Chiorescu, S.: The forestry - wood chain: simulation technique, measurement accuracy, traceability concept. Doctoral thesis, Luleå University of Technology (2003). http://epubl.luth.se/1402-1544/2003/03

Dehshibi, M.M., et al.: Stimulating fungi pleurotus ostreatus with hydrocortisone. ACS Biomater. Sci. Eng. **7**(8), 3718–3726 (2021)

Freyburger, C., Longuetaud, F., Mothe, F., Constant, T., Leban, J.-M.: Measuring wood density by means of X-ray computer tomography. Ann. For. Sci. **66**(804) (2009). https://doi.org/10.1051/forest/2009071

Hanna, S., Mahdavi, S.H.: Modularity and flexibility at the small scale: evolving continuous material variation with stereolithography. In: Fabrication: Examining the Digital Practice of Architecture: Proceedings of the 23rd Annual Conference of the Association for Computer Aided Design in Architecture and the 2004 Conference of the AIA Technology in Architectural Practice Knowledge Community, pp. 76–87 (2004)

Huber, J., Ekevad, M., Broman, O.: Using computed tomography data for finite element models of wood boards, April 2021. https://doi.org/10.23967/wccm-eccomas.2020.355

McCormick, M., Liu, X., Ibanez, L., Jomier, J., Marion, C.: ITK: enabling reproducible research and open science. Front. Neuroinform. **8** (2014). https://doi.org/10.3389/fninf.2014.00013

Michalatos, P., Payne, A.: Monolith: the biomedical paradigm and the inner complexity of hierarchical material design. In: Complexity & Simplicity - Proceedings of the 34th eCAADe Conference, vol. 1, pp. 445–454 (2016)

Michalatos, P., Payne, A.O.: Working with multi-scale material distributions. In: ACADIA 2013: Adaptive Architecture - Proceedings of the 33rd Annual Conference of the Association for Computer Aided Design in Architecture, pp. 43–50 (2013)

Museth, K.: VDB: high-resolution sparse volumes with dynamic topology. ACM Trans. Graph. **32**(3) (2013). https://doi.org/10.1145/2487228.2487235

Naboni, R., Kunic, A.: A computational framework for the design and robotic manufacturing of complex wood structures. In: 3 (Carpo 2017), pp. 189–196 (2020). https://doi.org/10.5151/proceedings-ecaadesigradi2019_488

Oxman, N.: Structuring design fabrication of heterogeneous. Archit. Des. **80**(4), 78–85 (2010)

Oxman, N.: Material computation. In: Sheil, B. (ed.) Manufacturing the Bespoke: Making and Prototyping Architecture, pp. 256–265. Wiley, New York (2012)

Palz, N.: Emerging Architectural Potentials of Tunable Materiality through Additive Fabrication Technologies I Norbert Palz (2012). Academia.edu

Ramsgaard Thomsen, M., Tamke, M., Mosse, A., Tyse, G.: Designed substrates for living architecture performance - Imprimer la Lumière. In: Tadeu, A., de Brito, J. (eds.) Proceedings of CEES 2021 - Construction, Energy Environment & Sustainability, Itecons, University of Coimbra, Coimbra, Portugal (2021)

Regli, W., Rossignac, J., Shapiro, V., Srinivasan, V.: The new frontiers in computational modeling of material structures. CAD Comput. Aided Des. **77**, 73–85 (2016). https://doi.org/10.1016/j.cad.2016.03.002

Shuxia, H., Lei, Y., Dawei, Q.: Application of X-ray computed tomography to automatic wood testing. In: Proceedings of the IEEE International Conference on Automation and Logistics, ICAL 2007, pp. 1325–1330 (2007). https://doi.org/10.1109/ICAL.2007.4338775

Svilans, T.: GluLamb: a toolkit for early-stage modelling of free-form glue-laminated timber structures. In: Proceedings of 2021 European Conference of Computing in Construction (2021 EC^3) (2021)

Svilans, T., Tamke, M., Ramsgaard Thomsen, M.: Integrative strategies across the timber value chain. In: Proceedings of the 5th International Conference on Structures and Architecture (ICSA) (2022)

Tamke, M., Svilans, T., Gatz, S., Ramsgaard Thomsen, M.: Tree to Product - Prototypical workflow connecting Data from tree with fabrication of engineered wood structure - RawLam. In: Proceedings of WCTE 2021a (2021a)

Tamke, M., Svilans, T., Gatz, S., Ramsgaard Thomsen, M.: Timber elements with graded performances through digital forest to timber workflows. In: Behnejad, S.A., Parke, G.A.R., Samavati, O.A. (eds.) Proceedings of the IASS Annual Symposium 2020/21 and the 7th International Conference on Spatial Structures (2021b)

Tyse, G., Tamke, M., Ramsgaard Thomsen, M., Mosse, A.F.: Bioluminescent Micro-Architectures: Planning Design in Time, an Eco-Metabolistic Approach to Biodesign. Archit. Struct. Constr., 1–9 (2022). https://doi.org/10.1007/s44150-022-00038-9

Yu, L., Han, S., Qi, D., Gu, H.: Automatic and fast testing of wood density based on computed tomography. In: 2007 IEEE International Conference on Control and Automation, ICCA 2007, pp. 2560–2565 (2007). https://doi.org/10.1109/ICCA.2007.4376824

Spatial Lacing: A Novel Composite Material System for Fibrous Networks

Xiliu Yang[1,2](✉), August Lehrecke[1], Cody Tucker[1], Rebeca Duque Estrada[1,2], Mathias Maierhofer[1], and Achim Menges[1,2]

[1] Institute for Computational Design and Construction (ICD), University of Stuttgart, Stuttgart, Germany
xiliu.yang@icd.uni-stuttgart.de

[2] Cluster of Excellence Integrative Computational Design and Construction for Architecture (IntCDC), University of Stuttgart, Stuttgart, Germany

Abstract. This paper presents a fibre composite material system, *Spatial Lacing*, inspired by the traditional craft of Bobbin Lace. The system utilises parallelised, coordinated fibre-fibre interactions to create nodes in a spatial network, combining the design space of lattice- and surface-based structures. The nodes retain topology during transformation between flat-packed and tensioned states, thus enhancing logistical flexibility for curing and deployment. The system is developed from the micro (fibre nodes and local structural behaviours), meso (fibre topologies and component types) and macro (global design and computational workflow) levels. A notation system defined based on elementary lacing actions informs a graph-based modelling method to represent fibre geometries and integrate fabrication information. The design and construction of a 2.4 m physical artefact demonstrate the *Spatial Lacing* system, which showcases unique fibre tectonics unachievable by existing production methods. Through the transfer of a craft process into the realm of computational design and spatial fibre composites, this work aims to expand the design and fabrication space of fibre systems in architecture.

Keywords: Fibre composites · Spatial Lacing · Computational design · Craft

1 Introduction

This research presents a spatial fibre composite system that extrapolates from the traditional 2D craft of Bobbin Lace and extends its logic to 3D for components at architectural scale. The paper is structured as follows. First, contexts and state of the art methods that inform the *Spatial Lacing* concept are introduced. Second, material system developments are presented, progressing from notation system to combinatorial node type analysis. Fibre topologies and corresponding component types and potential applications are then discussed, followed by an overview of modelling approaches that facilitate global design and fabrication. Finally, the demonstrator (Fig. 1) and its design-to-fabrication workflow is presented, and the benefits, shortcomings and future directions of research are discussed.

C. Gengnagel et al. (Eds.): DMS 2022, *Towards Radical Regeneration*, pp. 556–568, 2023.
https://doi.org/10.1007/978-3-031-13249-0_44

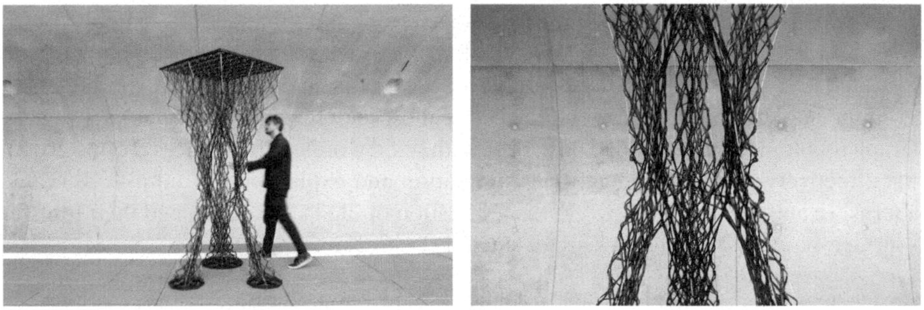

Fig. 1. Spatial lacing demonstrator.

1.1 Fibre Systems

Accounting for 38% of global CO2 emissions, the construction industry faces pressing needs to reinvestigate typical construction methods and materials (United Nations Environment Programme 2020). A promising alternative are fibre systems, which whether knit, woven, or wound, attain structure through customised fibre-level material arrangement. They are prevalent in nature where raw material is expensive and material efficiency is a fundamental design driver. Almost all load bearing structures in nature can be seen as fibre composites, where a high tensile strength material is embedded in a matrix that surrounds and supports it (Menges and Knippers 2015). Fibre systems are lightweight, efficient, and highly tunable, making them relevant for architecture where most projects are unique and require structurally performative solutions (Prado et al. 2014).

1.2 Textiles

Textiles are a subset of fibre systems made by interlocking yarns, where structural capacities are enabled by sequences of fibre interactions (Abrahart and Whewell 2021). In woven textiles, warp and weft yarns form interlaced patterns accurately controlled by CNC looms, but the product is constrained to machine sizes and fibre directionality is limited to fixed configurations. Knitted textiles are more elastic, as a single continuous yarn forms courses in series (Bilisik et al. 2016). They are highly programmable and can form 3D shapes but are limited to a pixel-based programming logic. Additionally, 2D textiles require templating and assembling to achieve 3D forms, which could result in a mismatch between force flow and fibre directions. In comparison, 3D textiles such as 3D braiding, require little post-processing, as fibres are pulled through a shape-defining profile to create densely packed objects (Ma et al. 2017). This process, however, is limited by the horn gear matrix and packing profile. One textile technique that has the potential to overcome these limitations (e.g., machine size, grid-based logics, discrete assembly) is *Bobbin Lace*.

1.3 Bobbin Lace

Bobbin Lace is a traditional craft dating to the 1500's, which involves braiding and twisting lengths of thread over a template that defines the overall pattern (Fuhrmann 1985).

Complex patterns emerge from simple rules governing fibre interactions, namely Twist, Cross, and Pin. Classic notations consist of written lacing instructions combined with illustrated fibre paths. Mathematical Bobbin Lace (Irvine and Ruskey 2014) brought the technique into the computation realm, where patterns are formally notated as a 2-regular digraph topologically embedded on a torus with braid words at the vertices (Irvine 2016). This effectively allows computational simulation and exploration of infinite workable patterns. In authors' previous work, the adaptation of 2D lace patterns and their transfer to surface-based 3D geometries are discussed (Lehrecke et al. 2021).

1.4 Filament Winding

Filament Winding (FW) differs from textile techniques in that it places fibres in desired locations without additional forming, and lays fibres sequentially instead of in parallel. While traditional FW winds fibres around a mandrel, Coreless Filament Winding replaces the mandrel with a winding frame, enabling the creation of large components without costly formwork (Reichert et al. 2014). While desired geometry is achieved through layering fibres, material interactions are limited to surface contact. Spatial Winding extends the notion of winding points with a type of fibre interaction, by wrapping fibres around wound ones, enabling the creation of space frames (Duque Estrada et al. 2020). This evolution of winding techniques shows a process whereby component geometry decouples from fixed frames, and fibre interactions increase geometric possibilities. *Spatial Lacing* builds upon this by introducing the parallel fabrication logic of textiles to FW to create topology-preserving, multi-fibre interactions at nodes, and further decrease reliance on frames.

1.5 Digital Craft

Manual craft procedures provide opportunities to discover how time-tested techniques of fibre arrangement, which historically enabled unique, customizable textile objects, could be evolved towards a new architectural material system that embodies a "textile logic" in structure, design, and representation (Thomsen et al. 2012). Though in the post-industrial era many manual crafts are relegated to minority practice or replaced by automation, the intersection of craft and computational technologies has been conducive to reincorporating forgotten techniques, sustaining cultural knowledge, and discovering new applications (Noel 2020). This research engages in parallel investigations through craft-based and computational methods.

2 Method

2.1 Notational System

While Bobbin Lace has been applied to surface-based, hand-wound architectural components (Eberfeld 2020) and performance-adapted fibre composites (Lehrecke et al. 2021), this research extrapolates the craft into a truly 3D system capable of fabricating spatial structures. 2D patterns are represented based on fibre count, lacing actions, and graph denoting fibre paths (Fig. 2b and c). By including spatial actions (such as changing the lacing plane and splitting and merging bobbins pairs) and extending the graph to 3D (Fig. 2d), the adapted notation system facilitates the spatial expansion of design possibilities to both surface and lattice geometries.

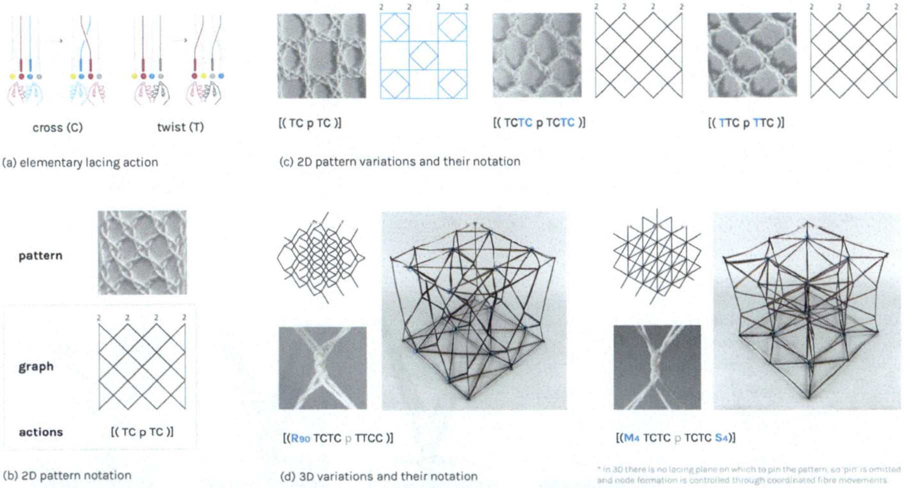

Fig. 2. (a) Elementary bobbin lace actions (b) elements of the 2D notation system (c) 2D pattern variation and lace artefacts (d) extended 3D actions and lattice artefacts.

2.2 Node Studies

Different combinations of Cross (C) and Twist (T) result in various fibre interactions (Fig. 3a). Given combinatorial possibilities of the two actions, the resulting nodes can be evaluated based on four fabrication and structural criteria.

- **Action complexity** counts the number of T and C actions required to fabricate the node. A number between 4 and 6 creates reasonably complex interaction.
- **Node adjustability** evaluates the ease of positional adjustment during fabrication as determined from physical models. Nodes that slide easily are eliminated.
- **Fibre-fibre contact** denotes number of fibre interaction points in the node, which directly relates to degree of bonding after resin application. Nodes with less than 2 contacts are eliminated.
- **Bending radius** is observed based on local geometry from physical models. An action repeated twice creates an apex along fibre paths, which is unfavourable.

These criteria are used to refine the selection to three nodes for further study, namely CTCTC, CCTTCC, CTC (Fig. 3b).

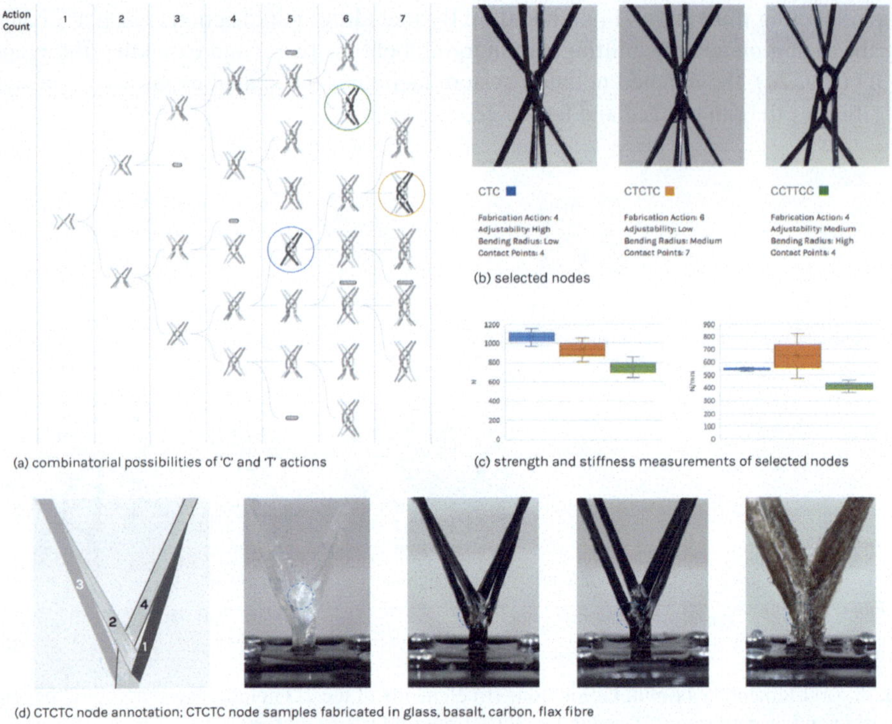

Fig. 3. Node studies (a) combinatorial map (excluding combinations where an action repeats more than twice) (b) selected nodes (c) strength and stiffness measurements (d) notation and material samples.

Node Types. To establish relative performance characteristics of the three node types, each node was fabricated twice using TEX3200 carbon fibre infused with EPIKOTE Epoxy Resin and subject to compression test with a Zwick/Roell Z100 testing machine. In the CTCTC and CCTTCC samples, failure occurs by first a buckling deformation in one of the fibre segments, and subsequently delamination and fracture at the entry point between the fibre segment and node bundle. One of the CTC nodes failed in the same manner, while the other buckled midway through the node. The strength and stiffness of three nodes are shown in Fig. 3c. CTC showed the highest strength, medium stiffness, and was selected for application in surface-based designs, where its tendency to slide is not a critical issue. The stiffest CTCTC node is used in lattice sections due to its ability to hold position. The weakest CCTTCC node is eliminated.

Materials. Various materials were surveyed for lacing, considering fabrication factors such as friction between bundles, resin absorption, tendency to fray, and fibre cross section. Compression tests were conducted using the same setup to observe the impact of material choice on node performance and failure modes. Four materials are tested through a CTCTC node infused with epoxy resin. Carbon, basalt, glass, and flax fibre roving were used to create two samples each in amount comparable to each other in dry fibre weight. Five out of eight specimens fractured at fibre 3 (Fig. 3d), as previously described. Dry carbon fibre was chosen for prototyping due to its superior performance as well as moderate friction and fraying tendency, but natural materials like basalt are also viable for exploration.

2.3 Fibre Topology

Spatial Lacing produces fibre structures by creating a network of nodes in space; the organisation and valence of these nodes determines the topology of the component. Through craft experimentations, three primary classes of topology are identified, and their relationship with component type, cross section, and application in architecture are described below.

Lattice. When nodes are connected in a space-filling manner, the resultant organisation draws from periodic lattices. The component cross section is a single fibre bundle (Fig. 4a), where the density and connectivity of fibres can be tailored to force directions while maintaining thread count. Compared to existing methods for composite space frame structures (Duque Estrada et al. 2020; Francom and Jensen 1999) this method generates a richer set of geometries and node conditions with minimal increase in setup complexity, as no fixed winding points are needed beyond the starting positions. This strategy can be used to create box trusses or more heterogeneous lattice distributions, contributing to potentially more efficient material utilisation for long-span structures.

Surface. The lattice component can be scaled up by expanding a single fibre bundle into a circular fibre array, forming hollow tubes and branching structures. This strategy can be generalised to other surface-based designs, where nodes are connected and flow along any surface that can be appropriately tiled in 2D. Rather than assembling 2D pieces to achieve an enclosed shape (Lehrecke et al. 2021), freeform geometries can be laced directly in space (Fig. 4b). The variety of surface patterns and configurations enable composite elements ranging from modular spatial nodes which form larger networks, to continuous monolithic shell structures.

Hybrid. When lattice and surface topologies are combined, a hybrid condition emerges which lends the system to more complex geometries and larger scale: surface fibres define the boundary and lattice fibres infill and strengthen the geometry. Here the component cross section becomes a tube with spatial connections within (Fig. 4c). In comparison to architectural components produced by FW, geometries are no longer limited to hollow makeups. The core becomes a new design space, where internal fibres can be tuned to geometric and structural constraints. This could lead to components capable of larger spans and resistant to a broader range of load cases. The demonstrator presented is based on this topology.

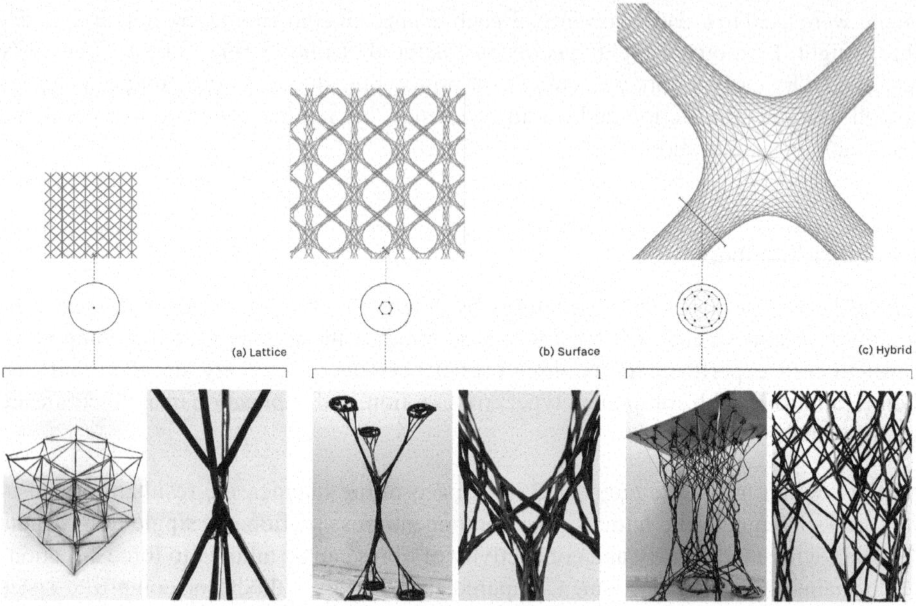

Fig. 4. Lattice, surface, and hybrid fibre topologies.

2.4 Computational Modelling

The main goal of computational modelling is to translate design intent to fibre geometries and fabrication instructions, in a manner that is cohesive with system logic. Lace structures in 3D are abstracted as Directed Acyclic Graphs where nodes on the graph are objects containing parameters such as the length and vector of connected fibre segments, as well as computed fabrication and structural constraints. This node object together with surrounding fibres form a spatial unit (Fig. 5) which in the lattice case can directly tile a bounding volume through regular or semiregular tessellation. Such a voxel-based approach is easy to generate and fabricate but is limited in geometric freedom. To overcome these limitations, hexahedral meshing presents a promising means to generate volumetric cubic tiles which conform to any given boundary conditions (Corman and Crane 2019).

For surface-based structures, the modelling method follows braiding logics where braided preforms are defined as meshes containing information on fibre interaction at each face (Zwierzycki et al. 2017). With this logic, surface-based topologies can be modelled using low resolution meshes derived from hollow tubes defined around edges of a macro graph. The hybrid topology extends this approach, where each edge in the macro graph represents a spatial lattice instead of a hollow tube (Fig. 5). This strategy uses tileable blocks, similar to an approach outlined by Wu et al. for modelling 3D woven structures (Wu et al. 2020). The main difficulty is dealing with the intersections between two or more blocks in a network while maintaining correct fibre distribution. There are infinite possibilities when distributing fibres between blocks and automating their solutions lies outside the scope of this work.

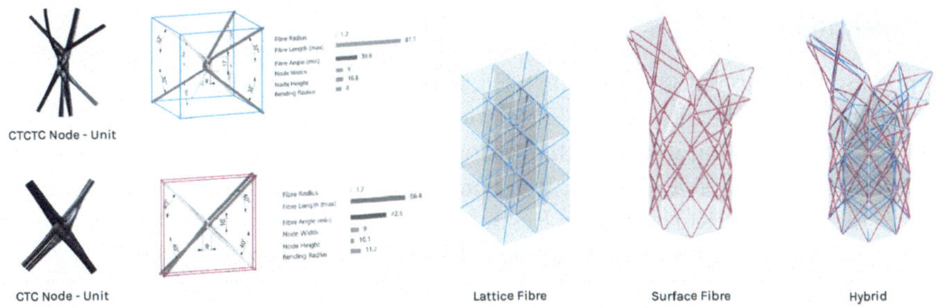

Fig. 5. Physical nodes (left), digital nodes as spatial units embedding information (middle), and fibres annotated on a branching geometry (right).

3 Demonstrator

The material system is tested through a physical demonstrator at architectural scale. The aim is to validate the design and fabrication process and showcase how the system can blend various topologies and maintain form while transitioning between flat and tensioned states. The final 2.4 m tall artefact demonstrates the *Spatial Lacing* concept and showcases unique topologies unachievable by existing production methods.

3.1 Computational Design

The computational workflow used to design the demonstrator is shown in Fig. 6. Given a 1.5 m $\times$ 1.5 m $\times$ 3 m boundary condition and an evenly distributed vertical load, Topology Optimisation is used to generate a preliminary geometry, from which a macro graph model is extracted. Fibre densities are assigned to graph edges based on size of the branch and span. Each edge is then populated with spatial units, where an octahedral lattice repetition with CTCTC lattice nodes and CTC surface nodes are applied. Transitions between units are precalculated to ensure symmetrical fibre distribution. Subsequent local adaptations of the computed graph network to specific strains are not considered, though it presents an important area for future development. The network is relaxed in a particle-spring simulation to estimate the fabricated form. The final design consists of 384 fibres transitioning through multiple hybrid configurations.

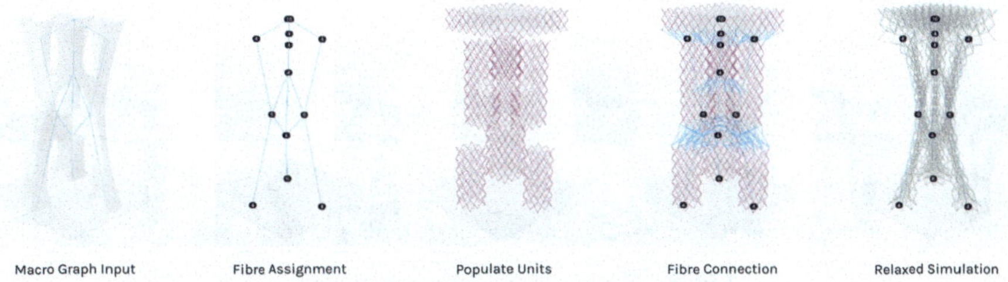

Fig. 6. Demonstrator computational workflow.

3.2 Fabrication

Each node in the graph model embeds fabrication instructions which can be carried out by a craftsperson or integrated with a robotic fabrication system (discussed in a future paper). The 1:1 demonstrator took three people 90 h to lace, and the entire fabrication process was completed over 7 days. It is constructed with 3500 g of TEX3200 carbon fibre and 4500 g of EPIKOTE Epoxy Resin from Lange+Ritter. Given the sheer number of fibres, bobbins are simplified to bundles affixed with clothes pins (Fig. 7).

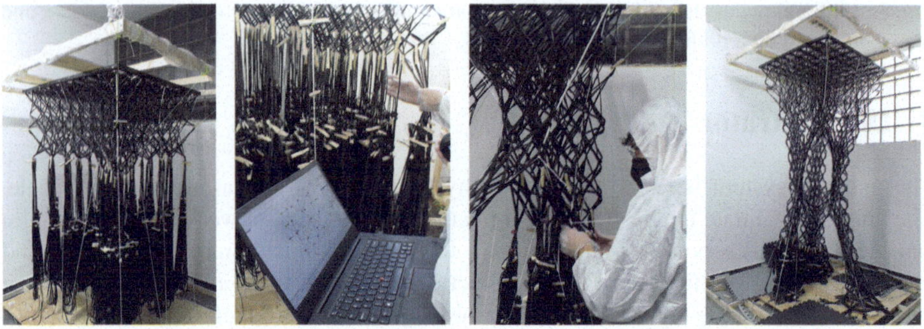

Fig. 7. Fabrication process

Before lacing starts, all fibres are fixed onto a starting plane, which consists of two plates milled to shape. Carbon fibres are cut to length and fed through pre-drilled holes on the first plate, which is screwed against a second plate to clamp the fibres. The top section is laced as a sparse lattice with 384 separate fibres before branching into four quadrants, where four fibres merge into one bundle, reducing total bundles to 96. As lacing continues, the starting plane is raised via a Z axis. When the structure is complete, fibres are cut, fed through the bottom plate, and tensioned to reduce slack. After the bottom plate is screwed, the structure is taken off the Z axis, collapsed into a flat state and inserted into a vacuum bag. Vacuum Assisted Resin Transfer Moulding (VARTM) is applied to infuse the component, and the saturated structure is reattached to the Z axis and tensioned back to its original height (Fig. 8). After 48 h of curing at room temperature, the structure is complete.

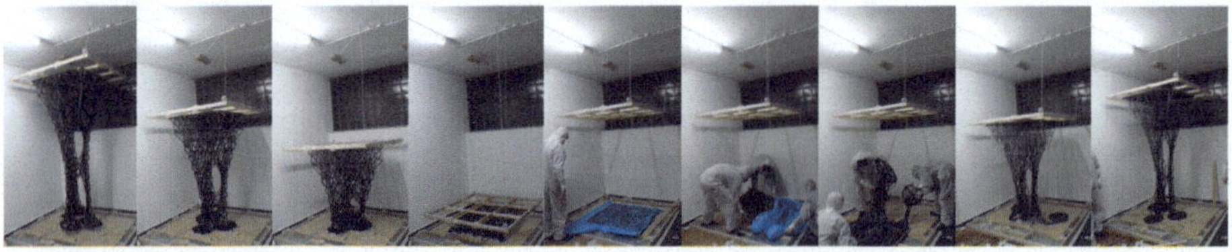

Fig. 8. Flat-packing and re-tensioning.

4 Conclusion

This research proposes a novel lacing system that expands the design and fabrication space of fibre structures in architecture. *Spatial Lacing* makes use of interlocking fibre connections derived from the traditional craft of Bobbin Lace, which allow fibres to retain topology when collapsed, thus introducing a key flexibility advantage. The component can be prefabricated with dry fibres, infused with resin, and transported to site in a frozen, flat-packed state, then deployed and cured in-situ. This has the potential to reduce transportation emissions. To improve the system's carbon footprint and recyclability, natural fibre and resin should be considered.

By combining lattice-based and surface-based topologies, hollow components can be strengthened through internal lattice connections unachievable by existing FW methods. This enables potentially higher performance, lightweight systems that can sustain high loads with less material, thus reducing material footprint. However, further studies on global and local structural performance are needed. Modelling and computational design methods should also be improved upon, with particular attention to the precision and dynamic nature of the material system, as well as strategies for graded, local adaptations.

The labour-intensive fabrication process is a major shortcoming which, combined with the need for parallelised fibre manipulation, create an opportunity for multi-agent robotic fabrication. The system can scale with adaptations in cross section, meaning systems at different scales can be abstractly represented with the same graph. This presents an interesting challenge: while modelling at the fibre level allows better control, modelling at the macro graph level allows more intuitive operations at runtime. These modelling challenges are synergistic with fabrication approaches, where the former is conducive to robotic fabrication while the latter could be carried out by craftspeople assisted via AR technologies.

The focus on industrialised, automated systems of fibre manufacturing risks overlooking potential benefits inherent to some lesser-known crafts of fibre arrangement. Under the theme of radical regeneration, examination of historical techniques can yield novel and materially efficient designs that blend technological advancements with the wisdoms of century-old crafts. Fibre structures, built upon a rich history of craft traditions and inherent tunability for efficient material distribution, is a promising avenue to explore environmentally and socially sustainable means of architectural production. By leveraging craft as a lens and method of inquiry, these lightweight systems can hopefully find new ways to transcend existing limitations in architecture.

Acknowledgements. Xiliu Yang, August Lehrecke, and Cody Tucker contributed equally to this research within their ITECH Master's thesis at the University of Stuttgart. The research was supported by DFG under Excellence Strategy - EXC 2120/1-390831618 through the IntCDC Master's Thesis Grant. The authors thank Marta Gil Pérez and Prof. Dr.-Ing. Jan Knippers for their guidance, as well as Autodesk, Lange+Ritter and Deutsche Basalt Faser GmbH for their generous support.

References

Abrahart, E.N., Whewell, C.S.: textile | Description, Industry, Types, & Facts. Encyclopedia Britannica (2021). https://www.britannica.com/topic/textile

Bilisik, K., Karaduman, N.S., Bilisik, N.E.: 3D fabrics for technical textile applications. In: Jeon, H.-Y. (ed.) Non-Woven Fabrics. InTech (2016). https://doi.org/10.5772/61224

Corman, E., Crane, K.: Symmetric moving frames. ACM Trans. Graph. **38**, 87:1–87:16 (2019). https://doi.org/10.1145/3306346.3323029

Duque Estrada, R., Kannenberg, F., Wagner, H.J., Yablonina, M., Menges, A.: Spatial winding: cooperative heterogeneous multi-robot system for fibrous structures. Constr. Robot. **4**, 205–215 (2020). https://doi.org/10.1007/s41693-020-00036-7

Eberfeld, N.J.: Computing embodied effort in the constructible design space of bobbin lace. Master thesis, Massachusetts Institute of Technology (2020)

Francom, L., Jensen, D.: Three-dimensional iso-truss structure. US5921048A (1999)

Fuhrmann, B.: Bobbin Lace: An Illustrated Guide to Traditional and Contemporary Techniques. Dover Books on Needlepoint, Embroidery, Dover, New York (1985)

Irvine, V.: Lace tessellations: a mathematical model for bobbin lace and an exhaustive combinatorial search for patterns. Ph.D. dissertation, University of Victoria (2016)

Irvine, V., Ruskey, F.: Developing a mathematical model for bobbin lace. J. Math. Arts **8**, 95–110 (2014). https://doi.org/10.1080/17513472.2014.982938

Lehrecke, A., Tucker, C., Yang, X., Baszynski, P., Dahy, H.: Tailored lace: moldless fabrication of 3D bio-composite structures through an integrative design and fabrication process. Appl. Sci. **11**, 10989 (2021). https://doi.org/10.3390/app112210989

Ma, P., Jiang, G., Gao, Z.: 11 the three dimensional textile structures for composites. In: Advanced Composite Materials: Properties and Applications, pp. 497–526. De Gruyter Open (2017). https://doi.org/10.1515/9783110574432-011

Menges, A., Knippers, J.: Fibrous tectonics: fibrous tectonics. Archit. Des. **85**, 40–47 (2015). https://doi.org/10.1002/ad.1952

Noel, V.A.A.: Situated computations: bridging craft and computation in the Trinidad and Tobago carnival. Dearq, 62–75 (2020). https://doi.org/10.18389/dearq27.2020.05

Prado, M., Dörstelmann, M., Schwinn, T., Menges, A., Knippers, J.: Core-less filament winding. In: McGee, W., de Ponce Leon, M. (eds.) Robotic Fabrication in Architecture, Art and Design 2014, pp. 275–289. Springer, Cham (2014). https://doi.org/10.1007/978-3-319-04663-1_19

Reichert, S., Schwinn, T., La Magna, R., Waimer, F., Knippers, J., Menges, A.: Fibrous structures: an integrative approach to design computation, simulation and fabrication for lightweight, glass and carbon fibre composite structures in architecture based on biomimetic design principles. Comput. Aided Des. **52**, 27–39 (2014). https://doi.org/10.1016/j.cad.2014.02.005

Thomsen, M.R., Bech, K., Krisjana, S.: Textile logics in digital architecture. In: Achten, H., Pavlicek, J., Hulin, J., Matejovska, D. (eds.) Physical Digitality. Presented at the Physical Digitality: 30th eCAADe 2012, Prague, pp. 621–628 (2012)

United Nations Environment Programme: 2020 Global Status Report for Buildings and Construction - Executive Summary (2020). https://wedocs.unep.org/xmlui/handle/20.500.11822/34572

Wu, R., et al.: Weavecraft: an interactive design and simulation tool for 3D weaving. ACM Trans. Graph. **39**, 1–16 (2020). https://doi.org/10.1145/3414685.3417865

Zwierzycki, M., Vestartas, P., Heinrich, M.K., Ayres, P.: High resolution representation and simulation of braiding patterns (2017)

Design for Humans and Non-Humans

Investigating a Design and Construction Approach for Fungal Architectures

Phil Ayres(✉), Adrien Rigobello, Ji You-Wen, Claudia Colmo, Jack Young, and Karl-Johan Sørensen

Centre for Information Technology and Architecture (CITA), Royal Danish Academy, Copenhagen, Denmark
phil.ayres@kglakademi.dk
https://kglakademi.dk/cita

Abstract. The design research presented in this paper grounds itself in a tradition of seeking new architectural form from the affordances and proclivities of new materials. We report on the developmental stages of a construction concept that involves the growing of mycelium-based composites within stay-in-place scaffolds produced using Kagome weaving techniques. We demonstrate how speculative design is used to generate hypotheses - testable design statements - for directing empirical investigation, and how results drive the progression of the design inquiry and its associated digital design tools. Our core contribution is to expose new design pathways that operate reciprocally between material, tectonic and spatial exploration. We argue that such reciprocity is a prerequisite for supporting the invention of new architectural forms, vocabularies and systems.

Keywords: Biohybrid architecture · Mycelium composites · Kagome weaving · Biofabrication · Living architecture · Living materials

1 Introduction

Mycelium-based composites (MBC) are a relatively new class of biodegradable materials generally derived from the fungal colonisation of lignocellulosic substrates. Such substrates can be readily found as agricultural and land-management waste streams, making MBC an exemplary embodiment of circular economy principles [1]. They also exhibit many properties suited for application within building construction [2]. This makes them an attractive and necessary target for research, with disruptive potential against the backdrop of challenges facing the construction industry, particularly in relation to resource scarcity for the production of conventional materials.

Much of the research in this rapidly expanding arena is motivated by the worthy ambition of substantiating mycelium composites as viable replacements for materials with less positive environmental credentials due to, for example, reliance on non-renewable resources and/or having high embodied energy. Within an industry context, the aim of the replacement paradigm is to support incorporation into existing construction systems and practices. However, it can be argued that operating within a paradigm of 'replacement'

C. Gengnagel et al. (Eds.): DMS 2022, *Towards Radical Regeneration*, pp. 571–583, 2023.
https://doi.org/10.1007/978-3-031-13249-0_45

risks constraint within an established repertoire of expectations and design thinking. It also risks missing other interesting properties exhibited by these materials - especially in their living state - such as regeneration, adaptation, decision making, reproduction and resource balancing.

The research reported on in this paper, grounds itself in a tradition of seeking new architectural form from the affordances and proclivities of new materials. We develop a construction concept that involves the growing of mycelium based composites within stay-in-place scaffolds produced using Kagome weaving techniques. The research operates across various scales of thinking and engagement, constructing bilateral relations of influence between material composition, architectural tectonics and spatial configuration. We demonstrate how speculative design generates hypotheses - testable design statements - for directing empirical investigation, and how results not only refine the design inquiry, but become instrumentalised within digital design tools. This reciprocal and iterative approach creates a rich design space for architectural investigation leading to novel outcomes and producing research results within material, tectonic and spatial spheres of design activity.

2 State of the Art

The most common approach to MBC product production is through moulding and assembly as discrete units [3, 4]. In general, once colonisation has reached satisfactory levels, units are denatured by heat-treatment thereby preserving functional properties. As an alternative, the denaturing process can be avoided, leaving units hydrated and biologically active so that the parts can fuse once assembled - assuming cultivation conditions are kept favourable [5]. Particularly within the architectural research community, the scope of production approaches has been enriched in recent years, to include monolithic production [6], 3d printing [7, 8] and hybrid production techniques involving fusing of discrete blocks and shaping of the living composite into geometric design targets using robotic wire-cutting [9]. In Fig. 1 we present an overview matrix of projects within this field, as represented in the literature. In the research reported here, we develop architectural proposition that combines monolithic and unit based approaches, with extensions to the state-of-the-art in terms of a combined stay-in-place mould and reinforcement strategy for monolithic production. Standard protocols are followed for discrete unit production.

Prior to engagement with production methods, MBC material-level specification must be determined. This opens a vast and under-explored MBC design space between parameters of substrate composition and the wealth of widely distributed saprotrophic (dead wood decomposing) fungi that can be used for binding. Substrate structure can take the form of wood shavings, dusts, straws, shives, husks and even include proportions of non-organic constituents - all of which have been investigated in the literature and demonstrated to impact functional properties. Desired properties can therefore be targeted by curating aggregates for their geometry, size, nature and distribution. As an extension of this mode of aggregate curation, in which the aggregate is generally homogeneous, we have previously demonstrated that the introduction of structuring natural fibres and organic textiles - that is, designing and curating a heterogeneous aggregate structure with bias towards orientated fibres - can have significant effect on compressive and flexural mechanical behaviour [10, 11]. A further dimension of MBC property tuning is by supplementing the chemical profile of the substrate to modify the mycelium expression [12].

The research described in this paper offers new ways of engaging with this material aspect of the state-of-the-art by seeking ways of instrumentalising material-level MBC specification as an integrated part of the architectural design workflow, thereby contributing to broader efforts at enhancing design engagement across scales [13].

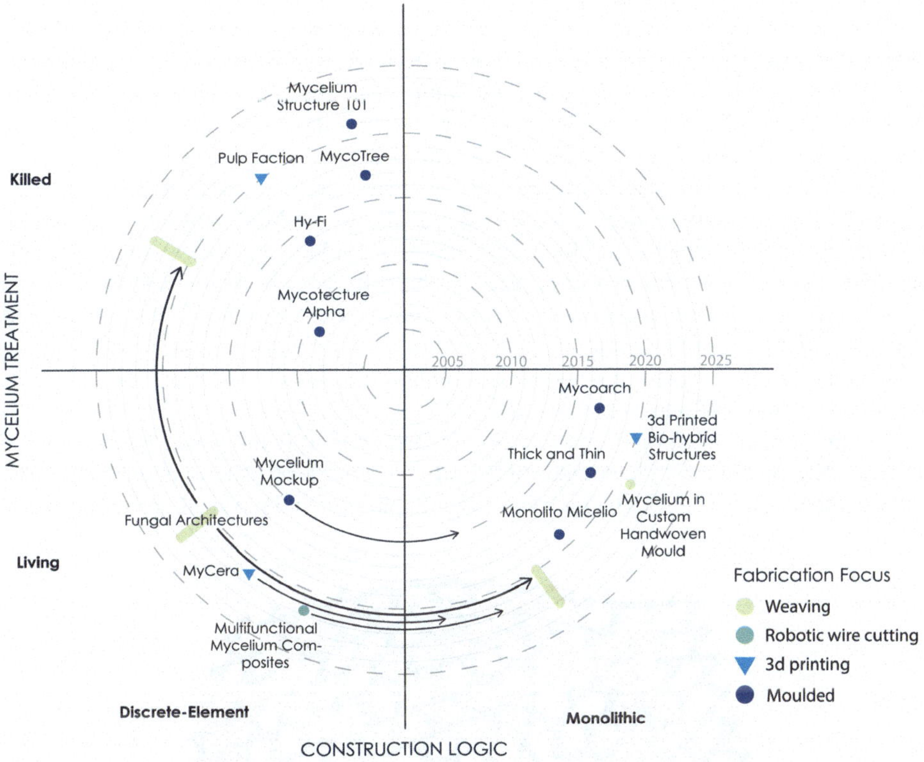

Fig. 1. A mapping of the state-of-the-art in architecturally focused projects employing MBC, reported in the literature.

3 Methods

The overarching aim of investigating how novel materials can inform the synthesis of new architectural form, and vice versa, necessitates a plurality of research approaches and methods that oscillate between the foundational, use-inspired and applied [14].

As such, we employ speculative design proposition as a method of hypothesis building by identifying specific testable statements/conditions from the design proposition.

We then develop experimental setups for investigation and integrate results into subsequent iterations of design proposition. This can occur at the scale of material design through to full architectural proposition. Here, we report upon the use of this iterative research approach to investigate the development of a construction method that seeks to combine MBC with triaxial Kagome weaves. This novel combination is motivated by the fact that MBC production is generally predicated on the use of moulds to constrain the growth phase to a desired shape, and that Kagome weaves can approximate any shape describable by a manifold mesh whilst being producible with straight strips of material. Therefore, we hypothesise that the weave can act as a combined stay-in-place mould, reinforcement and nutritional supplement.

With regard to the Kagome weave, we have previously demonstrated the digital instrumentalisation of topological principles underlying Kagome patterns and their relation to the generation of local surface curvature [15, 16]. We extend this work with the development of a digital design workflow allowing the design and investigation of topologically principled weave patterns for arbitrary manifold geometries.

The base geometry is randomly generated to create many regions of stiffening double curvature. A secondary weave is introduced that acts as the MBC stay-in-place mould, reinforcement and supplement. This weave adopts a similar but higher resolution geometry and intersects the primary weave (Fig. 2). The difference between these two weaves creates a rich a varied set of conditions interior and exterior conditions, clearly exhibited in the scaffold prototype shown in Fig. 3. Within the digital design space, both geometries can be locally adapted in response to simulation feedback of the parameters under investigation until objectives are satisfied.

We present, compare and evaluate design iterations to demonstrate the process of design development through methods of hypothesis generating using speculative design, and targeted inquiries through empirical experimentation and physical prototyping.

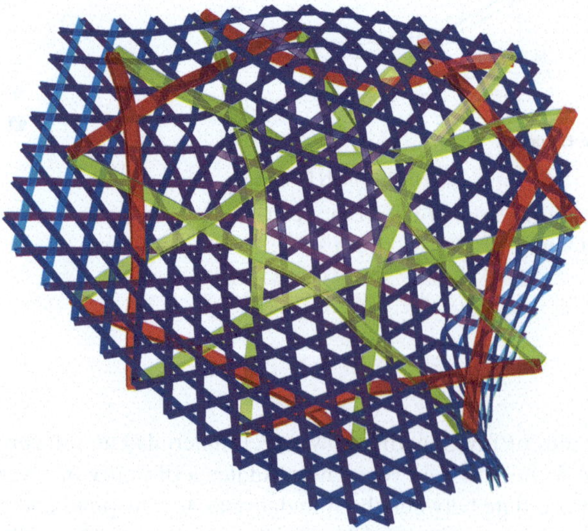

Fig. 2. Digital model of a weave fragment combining a structural Kagome grid-shell layer and a higher density Kagome layer to receive mycelium composite.

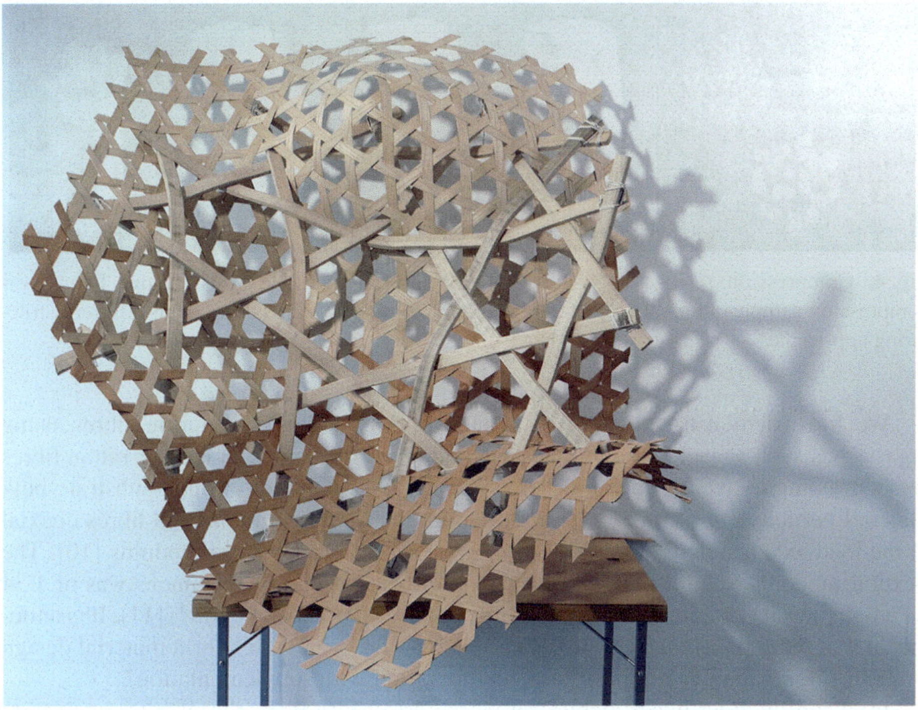

Fig. 3. Physical scaffold prototype of the digital model shown in Fig. 2

4 Case Studies and Results

4.1 MBC Composition Studies

Based on a material engineering review of the enzymatic activity of ligninolytic fungi [12], we have identified three principal material design strategies for MBC systems: *densification* (by dense packing, cold or hot-pressing), *composition* (by introducing structuring elements, or modifying particles and/or fibre properties), and *supplementation* (targeting mycelium properties based on chemical tuning of the substrate). Of these three strategies, we have conducted an experimental series targeting composition, which remains the least investigated strategy in the MBC literature. Drawing upon insights from the field of synthetic composite design, in which fibre composition is a primary design vector for functional property tuning, we have reported on a substrate composition approach of orientated fibres in combination with different particle sizes and geometries for the bulk MBC volume.

Fig. 4. Compression series. (left) Bulk volume substrate particle sizes. (right) Four post-compression samples illustrating different mechanical properties resulting from different structuring approaches.

We have demonstrated the significance of introducing structuring fibres using approaches such as jacketing with hessian and reinforcement with reed or rattan fibres for both compressive and flexural behaviour (Fig. 4). With an optimal substrate bulk material particle size of 0.75–3.0 mm, the introduction of common reed fibres coaxial to the load axis resulted in a 2.77-fold increase of the compressive modulus [10]. The flexural modulus of a longitudinally rattan-reinforced group of specimens was of 1.34 GPa for a density of 249.48 kg/m^3 (control: 192.71 MPa, 232.24 kg/m^3) [11]. It is anticipated that these preliminary results can be improved upon with hybrid material design strategies, for example, composition in combination with supplementation.

In an on-ongoing experimental series we are determining the effect of substrate composition on thermal performance. Three series of MBC panels representing different substrate densities and particle geometries are being evaluated according to ISO-9869 [17]. Early results indicate that the medium density substrate is the best performing. Thermal conductivity values, derived from the U-value experimental results, are in the range 0.0352−0.0516 W/mK, which approaches the thermal conductivity of standard commercial mineral wool products ($\approx$0.035 W/mK). The experimental plan, protocols of preparation and production, results and evaluation will be the subject of a forthcoming paper.

The results of these investigations into the functional properties of MBC should be understood as validating courses of action for tuning performance, rather than results indicating optimal performance - that is to say, they can likely be improved upon but are satisfactory for sanctioning a period of informed design speculation through which we can cycle back into MBC material design with more refined performance demands.

4.2 Steps Towards Weave and MBC Integration

Early probing experiments sought to establish plausibility for the construction concept of utilising Kagome weaves as a stay-in-place mould, reinforcement and feed-stock supplementation. A Y-branch component with branches $\approx$65 mm diameter and comprising a rattan weave with a series of valence 7 singularities to achieve the morphology, was successfully cultivated. Here, our measures of success include consistency of fungal growth, incorporation of the Kagome scaffold by the mycelium and minimisation of contamination. The visual record provides evidence that these criteria were met (Fig. 5(left)).

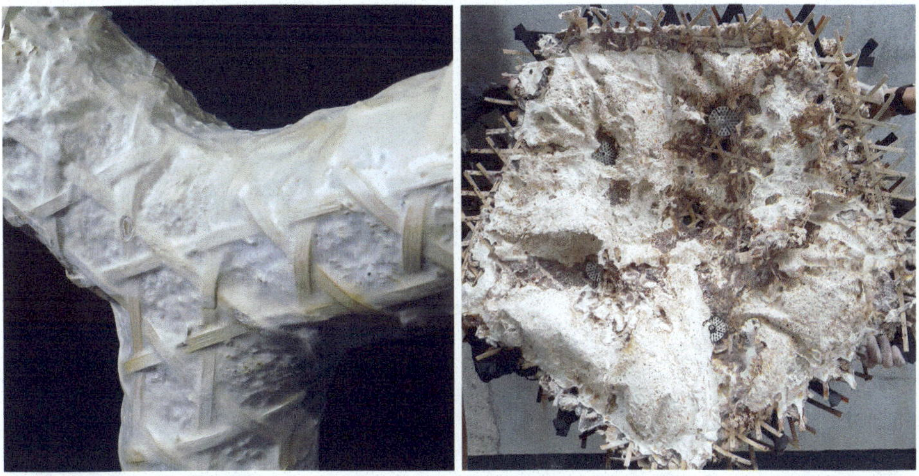

Fig. 5. Preliminary prototypes of Kagome weave and MBC integration. (left) A Y-branch component with good integration of the rattan weave within the mycelium skin. (right) A synclastic fragment of an enclosure surface composed as a sandwich of weaves - rattan on external faces and a structural carbon-fibre weave, internally bound. The MBC in this case is much less conforming, but spatially suggestive.

Building off this result, a subsequent probing experiment tackled a larger scale ($\approx$1 m diameter) assembly for a notional enclosure condition. This presented new functional demands and design challenges. For example, the self-weight of the MBC would be greater and new loading cases had to be considered. Where rattan weaving material had been sufficient for the Y-branch component, at this scale it would not provide adequate support. To resolve this, the design concept was modified to include a stiffening carbon fibre (CF) weave. This was developed as a sandwich construction with the CF weave enclosed in MBC and rattan weaves on external faces to contain it. Two prototypes were fabricated exhibiting synclastic and anticlastic curvature respectively. The results demonstrated the challenges of scaling up with poor consistency of colonisation and significant degrees of contamination. However, the lack on conformity, exhibited through both pocketing and excessive material build-up, suggested creative spatial and tectonic opportunities that could be investigated further (Fig. 5(right)).

Articulating these newly found qualities commenced with an initial design hypothesis of two weaves interfering to create a heterogeneous set of spatial conditions whereby regions of the primary weave can be expressed internally, externally or contained. An arbitrary perlin-noise surface was generated to act as the target geometry of the primary structural weave. A secondary ‘enclosure’ weave geometry was generated with a higher frequency value. To achieve a level of geometric coherence between the weaves, a wave summation principle was employed to bring the secondary weave into alignment with the structural weave.

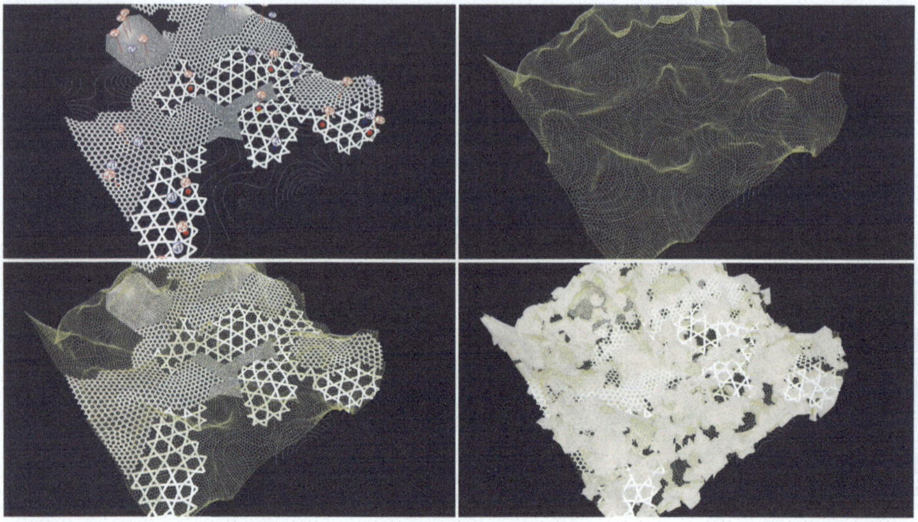

Fig. 6. The buildup of primary and secondary weaves, articulating the spatial qualities of the prototype shown in Fig. 5(right). The frame, bottom-right, shows a 'maturing' stage design hypothesis with MBC colonising the weaves.

Fig. 7. The (left) A valence 5 (note the pentagon, lower center) beech weave using timber members of 40 × 20 mm cross-section. (right) Detail of the interlacing density achieved with 60 × 30 mm solid beech members. Note the plastic deformation of the members, achievable whilst they have a high moisture content.

Furthermore, using the loop subdivision methods on the mesh faces of the weave representation, we hypothesised that discrete changes in weave density for the structural weave could be achieved, embedding the possibility of altering weave density towards functional objectives in later design iterations (Fig. 6). From this design hypothesis, an experimental plan was devised to empirically determine weave parameters (most importantly, minimum reciprocal triangle size) given solid beech and ash weavers of compressed timber with cross-section 40 × 20 mm and 60 × 30 mm respectively. The timber was provided in maximum lengths of 3 m due to processing limits. The timber was firstly heat-steamed and then compressed, rendering the material significantly more pliable whilst it retains a high moisture content.

Two weaves were produced using the 40 × 20 mm timber - a valence 5 (yielding a snyclastic surface, see Fig. 7(left)) and a valence 7 (yielding an anti-clastic surface). A single valence 5 weave was produced using the 60 × 30 mm timber (see detail, Fig. 7(right)). All weaves were assembled without the use of, or subsequent need, for mechanical or chemical fixings. This exploits the inherent jamming properties of the reciprocal triangles in the Kagome pattern, and the plastic deformation of the timber into sinusodial geometry to realise material interlacing. An impromptu load-test was conducted without boundary restraints on the valence 5 weave composed of 60 × 30 mm weavers. A load of ≈250 kg was applied around the weave's singularity and comfortably accommodated.

It was found that the relationship between material cross-sectional dimension and reciprocal triangle dimension would not allow discrete changes in weave density - disproving our hypothesis. However, it was found that a continuous change of weave density could be supported. Based on measurements of the minimum achievable reciprocal triangle dimension across the material cross-sections utilised, Eq. (1) was determined from the geometric relations presented in Fig. 8 and incorporated into the design workflow to refine our weave design approach. The minimum edge length of the reciprocal triangle for a given material cross-section is determined by:

$$\sqrt{2r^2 - (2r - T)^2} + \left(\tan A * \frac{w1}{2} + \frac{w2}{2\sin A}\right) + \left(\tan B * \frac{w1}{2} + \frac{w3}{2\sin B}\right) \tag{1}$$

This equation is incorporated into the digital tools to support the modelling of continuous changes in weaver density which can be associated to different model features and design objectives; two cases are shown with density changes driven either through distance to an attractor point (Fig. 9, left), or mean curvature analysis (Fig. 9, right).

With these refinements instrumentalised, a new design hypothesis was developed (Fig. 10). In this case, a context was chosen to force the consideration of spatial conditions relative to context. Entrances, extended thresholds, internal courtyards, ground sculpting and spatial organisation were now explicitly considered. In addition, the parallel material understanding developed for MBC functional performance tuning, permits geometric conditions to be assessed and specified according to intended performance requirements as suggested in Fig. 11.

Further iterations of MBC design in conjunction with spatial design developments will focus on additional refinements informed through specific structural, spatial, environmental and functional objectives.

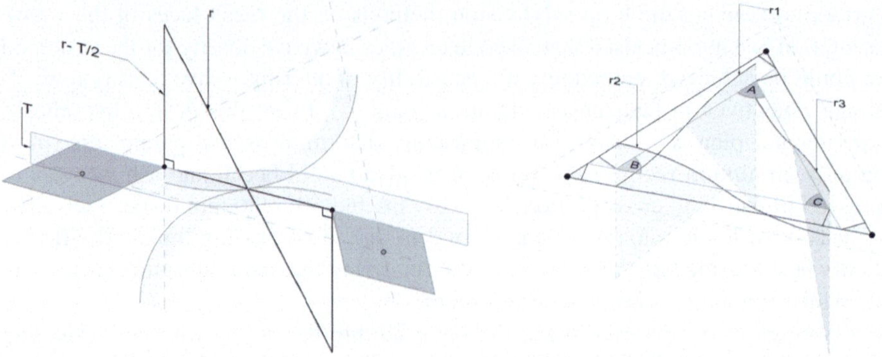

Fig. 8. Description of geometric relationships employed to construct an equation relating material cross-section dimensions to the minimum dimensions of a woven reciprocal triangle.

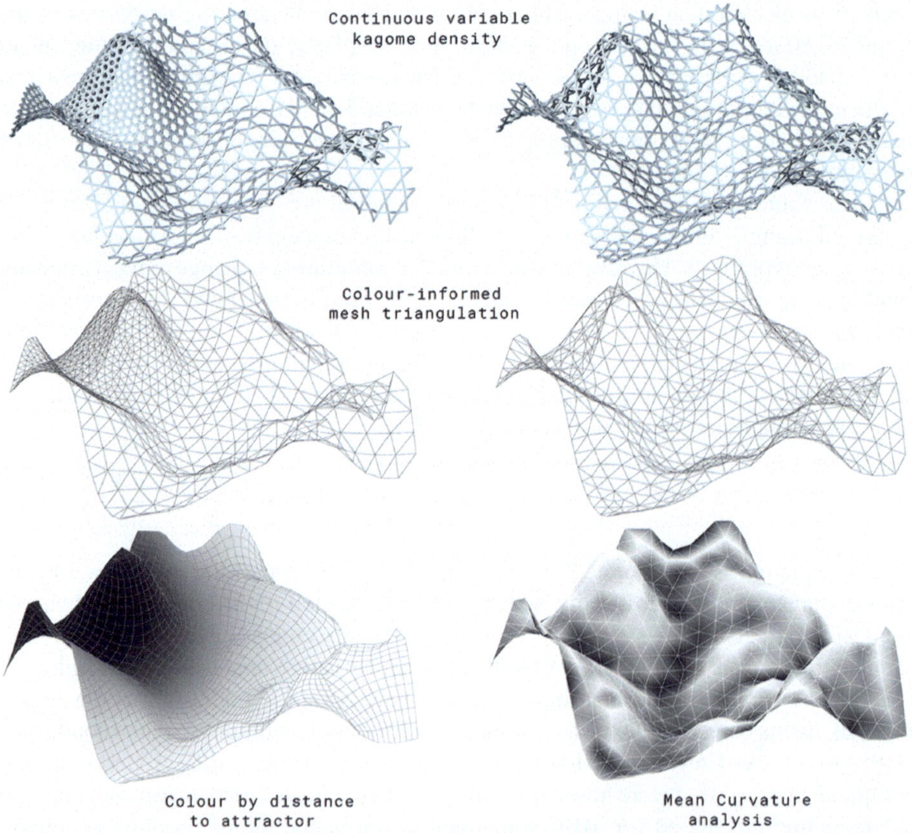

Fig. 9. Refined weave density approach for continuous change within a range parameterised by the material cross-section dimensions used. On the left, the weave density is informed by distance to an attractor point; on the right, the density is informed by mean curvature analysis of the design surface.

Fig. 10. A refined design hypothesis informed through empirical investigations, and the instrumentalisation of their findings within a digital design environment.

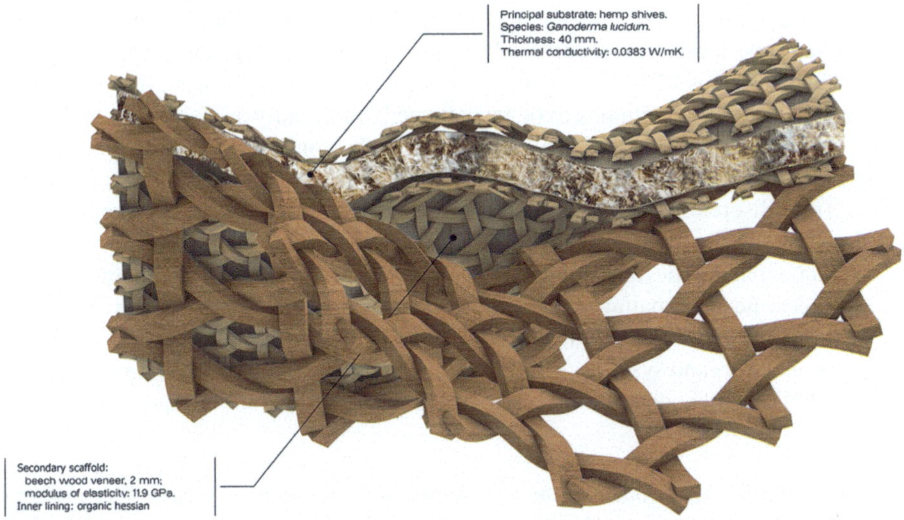

Fig. 11. Detail of the refined design hypothesis with material composition and functional performance specifications derived from design objectives. These specifications will drive further iterations of MBC material investigation.

5 Discussion

The field of MBC design research is currently flourishing, as evidenced by a burgeoning corpus of literature. The thriving community of researchers responsible for this are collectively enriching MBC practice and expanding the 'Ashby map' of functional performance through design, engineering and cultivation strategies, productively enhancing the possible use-cases and application domains of these materials. However, where new materials offer a starting point for investigating new architectural expressions, this extended sphere of exploration remains under-explored. Where the linear design pattern of developing structural form from the proclivities of functional material performance has been productively investigated [4], the work presented here expands the design repertoire by exercising a parallel and reciprocal spatial and material inquiry. Having demonstrated that MBC functional properties can be designed and refined through, among other approaches, substrate structuring strategies, design intent can be used as a driver for designing material performance, offering enhanced and enriched modes of design engagement.

6 Conclusion

We have demonstrated the refinement of design intent through a combination of design-led hypothesis building and empirical testing that bilaterally couples material and spatial inquiry within the rapidly expanding field of MBC related to architecture. Within this work, an initial intention of developing structurally performing MBC bound by a stay-in-place Kagome weave has progressed to the contribution of a more nuanced system that integrates various MBC composition strategies with a load-bearing timber kagome gridshell. This system continues to be investigated empirically at increasing scale.

We have shown how this inquiry also results in the contribution of enhanced digital tooling, as empirical investigations provide both qualitative (weave density strategies moving from discrete to continuous) and quantitative (scaling of weave reciprocal triangles relative to material cross-section) insights that can be instrumentalised. This supports an expanded design inquiry in which speculation and development can drive material, tectonic and spatial investigation in productive, reciprocal ways. We argue that such reciprocity is a prerequisite for supporting the invention of new architectural forms, vocabularies and systems. This has been demonstrated here in the novel architectural vocabulary and construction system based on MBC in combination with Kagome weaving.

Acknowledgments. This work contributes to the Fungal Architectures project. This project is funded by the European Union's Horizon 2020 research and innovation programme FET OPEN "Challenging current thinking" under grant agreement No. 858132.

References

1. Appels, F.V.W., et al.: Fabrication factors influencing mechanical, moisture-and water-related properties of mycelium-based composites. Mater. Des. **161**, 64–71 (2019)
2. Jones, M., Mautner, A., Luenco, S., Bismarck, A., John, S.: Engineered mycelium composite construction materials from fungal biorefineries: a critical review. Mater. Des. **187**, 108397 (2020)

3. Nagy, D., Locke, J., Benjamin, D.: Computational brick stacking for constructing free-form structures. In: Thomsen, M., Tamke, M., Gengnagel, C., Faircloth, B., Scheurer, F. (eds.) Modelling Behaviour, pp. 203–212. Springer, Cham (2015). https://doi.org/10.1007/978-3-319-24208-8_17
4. Heisel, F., et al.: Design of a load-bearing mycelium structure through informed structural engineering. In: World Congress on Sustainable Technologies (WCST-2017), pp. 45–49 (2017)
5. Dahmen, J.: Mushroom furniture. University of British Columbia (2016). https://doi.org/10.14288/1.0343116
6. Dessi-Olive, J., et al.: Monolithic mycelium: growing vault structures. In: 8th International Conference on Non-Conventional Materials and Technologies Construction Materials and Technologies for Sustainability, Nairobi, Kenya (2019)
7. Colmo, C., Ayres, P.: 3D printed bio-hybrid structures-investigating the architectural potentials of mycoremediation. In: Anthropologic, eCAADe Conference, pp. 573–582. CUMINCAD (2020)
8. Goidea, A., Floudas, D., Andréen, D.: Pulp Faction: 3D printed material assemblies through microbial biotransformation. In: Fabricate 2020, pp. 42–49. UCL Press (2020)
9. Elsacker, E., Søndergaard, A., Van Wylick, A., Peeters, E., De Laet, L.: Growing living and multifunctional mycelium composites for large-scale formwork applications using robotic abrasive wire-cutting. Constr. Build. Mater. **283**, 122732 (2021)
10. Rigobello, A., Ayres, P.: Compressive behaviour of anisotropic mycelium-based composites. Sci. Rep. **12**(1), 1–13 (2022)
11. Rigobello, A., Colmo, C., Ayres, P.: Effect of composition strategies on mycelium-based composites flexural behaviour. Biomimetics **7**(2), 53 (2022)
12. Rigobello, A., Ayres, P.: Design strategies for mycelium-based composites. In: Satyanarayana, S.K., Deshmukh, T. (eds.) Fungi and Fungal Products in Human Welfare and Biotechnology. Springer (2022, in press)
13. Faircloth, B., et al.: Multiscale modeling frameworks for architecture: designing the unseen and invisible with phase change materials. Int. J. Archit. Comput. **16**(2), 104–122 (2018)
14. Stokes, D.E.: Pasteur's Quadrant: Basic Science and Technological Innovation. Brookings Institution Press, Washington, D.C. (2011)
15. Ayres, P., Martin, A.G., Zwierzycki, M.: Beyond the basket case: a principled approach to the modelling of Kagome weave patterns for the fabrication of interlaced lattice structures using straight strips. In: Advances in Architectural Geometry 2018, pp. 72–93. Chalmers University of Technology (2018)
16. Ayres, P., You-Wen, J., Young, J., Martin, A.G.: Meshing with Kagome singularities: topology adjustment for representing weaves with double curvature. In: Advances in Architectural Geometry, pp. 188–207. Ponts Chaussées (2021)
17. Rasooli, A., Itard, L.: In-situ characterization of walls' thermal resistance: an extension to the ISO 9869 standard method. Energy Build. **179**, 374–383 (2018)

Demonstrating Material Impact

A Computational Design Framework Promoting Environmental Justice

Elizabeth Escott, Sabrina Naumovski, Brandon M. Cuffy, Ryan Welch, Michael B. Schwebel, and Billie Faircloth(✉)

KieranTimberlake, 841 N American Street, Philadelphia, PA, USA
bfaircloth@kierantimberlake.com

Abstract. As designers normalise the practice of BIM-LCA, connecting models to environmental impact data, their interaction with the underlying datasets presents an opportunity to evaluate building materials through an environmental justice (EJ) lens. This proof-of-concept method demonstrates the potential importance in reviewing environmental impact data for the local impacts, such as Photochemical Ozone Creation Potential (POCP). When the method is applied to a building project case study, an identical product manufactured at two or more sites is shown to have vastly different impacts on the local communities, illustrating the need for a closer look at manufacturer-provided environmental impact data.

1 Introduction

Embodied carbon (EC) modelling of building materials using life cycle assessment (LCA) is an established method for design teams to engage in global climate action by decarbonising the materials supply chain. Using building information modelling (BIM), modellers have assessed whole buildings including sub- and super-structure, enclosure and interior finish systems (Bruce-Hyrkäs et al. 2018, Al-Ghamdi and Bilec 2017). Model types include early-phase part-to-whole assessments, materials formulation and component optioning, assembly and detail optimisation, and end-of-project benchmarking (Feng et al. 2021, Al-Ghamdi and Bilec 2017). As designers normalise the practice of BIM-LCA, connecting models to environmental impact data, their interaction with the underlying datasets presents an opportunity to evaluate building materials through an environmental justice (EJ) lens.

The US Environmental Protection Agency (EPA) defines EJ as "the equitable treatment and meaningful involvement of all people regardless of race, colour, national origin, or income, with concern to the development, implementation, and enforcement of environmental policies." In the US, Black, Indigenous, and People of Colour (BIPOC) and poor populations accrue a higher share of pollution risk due to land-use policy placing manufacturing plants and material disposal locations in proximity to BIPOC communities (Ash and Boyce 2018). Spatialising the impacts improves understanding of site-specific toxicity and the effects on local communities (Castellani et al. 2016; Azapagic et al. 2013).

C. Gengnagel et al. (Eds.): DMS 2022, *Towards Radical Regeneration*, pp. 584–600, 2023.
https://doi.org/10.1007/978-3-031-13249-0_46

In current BIM-LCA practice, designers associate building elements with lifecycle inventory (LCI) data to understand the environmental impacts of building materials, with a particular focus on global-scale impacts such as EC. Designers access LCI data through Environmental Product Declarations (EPDs), manufacturer's reports created according to product category rules (PCRs). For a given product, a manufacturer discloses the expected emissions across the product lifecycle in stages, including the product manufacturing stage (A3), at geographically locatable facilities (ISO 2006). Whereas global-scale impacts, such as EC and ozone depletion potential (ODP), are understood as cumulative emissions, local-scale impacts, such as eutrophication, acidification, and photochemical ozone creation potential (POCP), are more challenging. Contextualising the consequences of local-scale impacts requires a deeper understanding of the ecological and demographic differences in the receiving environments (Nitschelm et al. 2016; Seppälä et al. 2006; Huijbregts et al. 2000). Given the distributed nature of the building materials supply chain, the complexity of building materials, and the multi-scalar linkages between ecosystems, the impacts of our building activities are both global and local.

This paper demonstrates a proof-of-concept (POC) methodology for computational designers to extend underlying LCI datasets, hypothesising that when we mosaic these with other forms of demographic and health data, we can establish a complementary pathway for integrating EJ concerns into design decision-making. We base our approach on the developing field of GIS-LCA (Li et al. 2021; Megange et al. 2020; Bulle et al. 2019) and our firm's development of the first BIM-LCA tool (Bates et al. 2013). Our research expands upon these modelling methods, incorporating conventions for geospatial modelling of ecosystems and sociological data into a computational framework. Our approach results in a practice that couples materials selection with environmental impact data, community health data, and building product manufacturing sites. While our POC demonstrates the importance of making EJ considerations visible, this framework is currently limited by the lack of transparency in EPD LCI data and the uneven availability of EPDs within manufacturing based on geography and material.

2 Model Schema

Our modelling framework promotes the following EJ calculation methodology. We begin with established methods: generating a Bill of Materials (BOM) for a building using BIM-LCA; linking materials found within the BOM to a range of product-specific EPDs; pulling manufacturing location and pollution data for stage A3 from selected EPDs. We represent communities adjacent to manufacturing sites by incorporating a *Demographic Index,* calculating a spatialised sensitivity factor identifying historic inequity factors, including race, colour, and income. We add spatially explicit probability of adverse health effects of exposure based on public health data, termed the *Vulnerability Index.* We layer a location-based diffusion of POCP pollutants (from EPDs), termed the *Pollution Index.* Finally, we combine the three indexes to create an EJ score for the selected building products, bringing attention to EJ considerations within the building materials supply chain (Fig. 1).

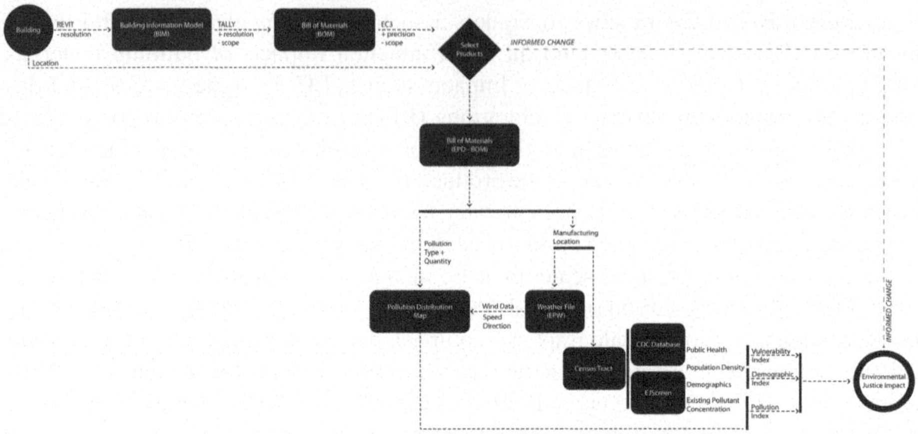

Fig. 1. The overall approach to approximating EJ considerations incorporating a building's BOM, the manufacturing location of selected building products, and community population and health data.

2.1 Representing the Building Materials (BIM-LCA to BOM)

To create a BOM for a selected building, we begin with an Autodesk Revit® BIM. We use Tally®, a BIM-LCA application, to quantify the materials associated with modelled elements and add the necessary quantities of unmodeled materials, such as adhesives, coatings, finishes, and studs (Bates et al. 2013). As BIM-LCA practice does not yet account for materials associated with MEP systems, casework, and site work, there are restrictions to creating a complete BOM, placing initial limitations on material manufacturing sites assessed for EJ considerations during the design process.

2.2 Representing Environmental Impacts (BOM to EPD)

As Tally links to generic LCI data for construction materials, its underlying data presently lacks the information necessary to investigate EJ issues requiring product-specific environmental impacts and locations of manufacture. To link the building's BOM to manufacturer-specific LCI data, we export a BOM from Tally directly into the Embodied Carbon in Construction Calculator (EC3). Although EC3 is one of the largest EPD databases available, assessing a BOM for EJ issues using EC3 may be limited by the following: EPD availability for a material category; under-representation of manufacturing regions; or non-disclosure of a manufacturing site on an EPD.

We proceed with a detailed examination of the EPDs hosted in the EC3 database through the public API. Our selected set of EC3 entries contains products from nine MasterFormat Divisions: Concrete, Masonry, Metals, Wood/Plastic/Composites, Thermal and Moisture Protection, Openings, and Finishes. As impacts cannot be compared between characterisation schemes, we filter our EPDs to those following US EPA TRACI 2.1, which organises impact quantities per kg of material according to ODP, EC, acidification, eutrophication, smog formation potential, and PCOP. With the final filtered list of

BOM-related EPDs, we are ready to examine community health and EJ considerations within the building materials supply chain.

2.3 Representing the Community and Community Health

1. Demographic Index (Fig. 2)

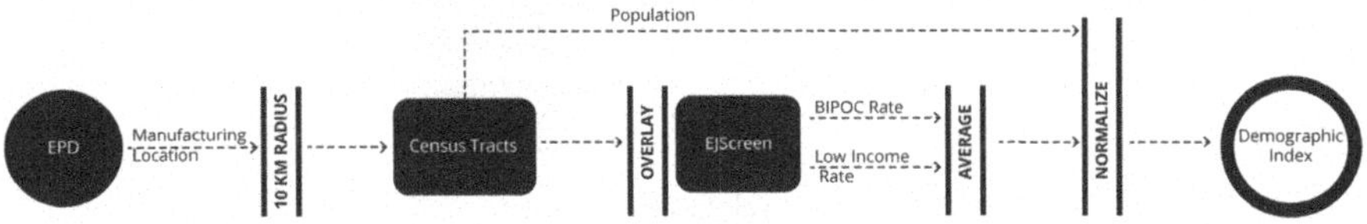

Fig. 2. Modelling the demographic index includes manufacturing location and population characteristics included in census data.

The Demographic Index, published in the US EPA's EJScreen database, combines two socioeconomic indicators: low income and people of colour (Environmental Protection Agency 2022, Federal Register 1994). The demographic index is referenced in this study to identify the potential for any disproportionate burden on areas with more highly populated levels of people of colour and low-income residents (Environmental Justice Indexes 2022).

2. Vulnerability Index (Public Health) (Fig. 3)

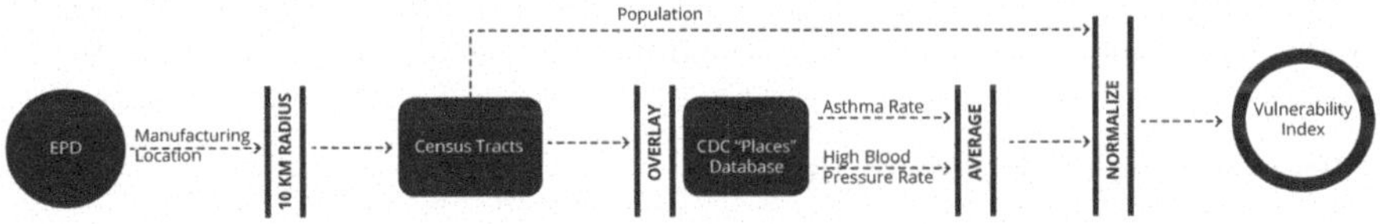

Fig. 3. The calculation method for the Vulnerability Index includes the manufacturing location, public health data, and population density.

The US Center for Disease Control and Prevention (CDC) "Places" database (CDC 2021) tracks several EJ public health indicators. After mapping these health impacts to pollutants of concern, we determined POCP to be a primary pollutant of concern for EJ health metrics (Fig. 4).

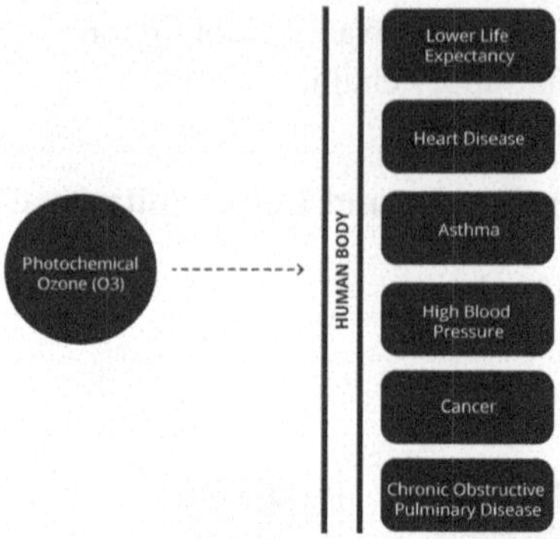

Fig. 4. Public health metrics related to photochemical ozone creation

Within the related health impacts, we conducted Pearson Correlation testing to remove unintentional bias creation from using overly correlated variables, tracking against life expectancy:

$$\rho_{X,Y} = \frac{\text{cov}(X, Y)}{\sigma_X \sigma_Y}$$

Equation 1. Pearson Correlation Coefficient Equation

This analysis determined high blood pressure would serve as a proxy for both heart disease and COPD, and asthma was sufficiently independent to maintain as a separate indicator. As with the method for the Demographic Index, we used an elementwise mean per census tract to generate the vulnerability index impact vector (VIP). We multiply the VIP by the population of its respective census tract to create a population-based metric matching the format of the Demographic Index and reflecting differences in expected exposure dependent on population density around the manufacturing location (Fig. 5).

2.4 Representing the Environment

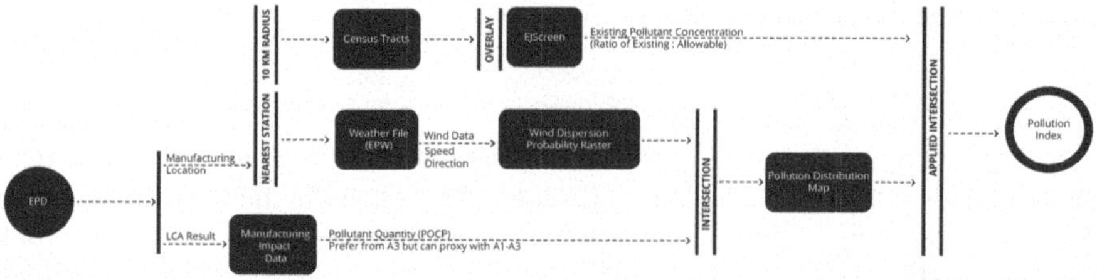

Fig. 5. Modelling method for the pollution index begins with EPD location and pollutant data, combined with existing pollution concentration and factors in wind dispersion of the factory pollution.

1. Pollution Quantity and Capacity

We use measured O_3 concentrations at each manufacturing plant to determine the likelihood that the environment can absorb the POCP emissions associated with the manufacturing of the product without exceeding established safe pollutant concentration levels. The ratio of the metered level of pollutant to the known allowable concentration gives this resilience:

$$C = \frac{POC}{A}$$

$C :=$ *pollution capacity*
$POC :=$ *recorded ground level ozone in parts per billion*
$A :=$ *allowable ground level ozone for "good air quality" rating, currently established as 54 ppb* (EPA 2015, p. 2)

Equation 2. Pollution Capacity

2. Wind Multiplier Value

To model the relative spatial distribution of EPD-reported POCP, we applied a wind-driven diffusion model with a 10 km cut-off radius, beyond which emissions were assumed to be sufficiently dispersed (Browning and Lee 2017). To efficiently approximate plume modelling, we used typical meteorological year (TMY) weather data from the closest airport (Ladybug-Tools 2022) to define hourly wind conditions for each site. We assumed steady-state conditions to establish a straightforward calculation of hourly concentrations. Because solar radiation and warm temperatures are factors in POCP formulation, we limited time sampling to daylight hours between spring and fall equinoxes. We calculated the resulting concentrations of time-integrated spatial distribution of POCP using the 2D diffusion equation:

$$C(\overline{x}, h) = \frac{1}{4\pi D} \sum_{i=1}^{n} e^{\frac{-\left|\overline{x}-\overline{u}_h \Delta t_{i,n}\right|^2}{4D\Delta t_{i,n}}}$$

$\overline{x} :=$ *location of site* (x, y) *relative to factory at* $(0, 0)$
$h :=$ *hour of simulation*
$D :=$ *2D diffusion coefficient* $(150\,\mathrm{m}^2/\mathrm{s})$
$n :=$ *number of discrete samples per hour*
$\overline{u}_h :=$ *wind velocity at hour* $h(m/sec)$
$\Delta t_{i,n} :=$ *seconds elapsed at sample* i $(3600\ i/n)$

Equation 3. Cumulative Wind Multiplier

At our selected spatial resolution of approximately 0.5 km (1,141 sample locations within the prescribed 10 km radius), we found cumulative results converged at approximately n = 10 (Fig. 6).

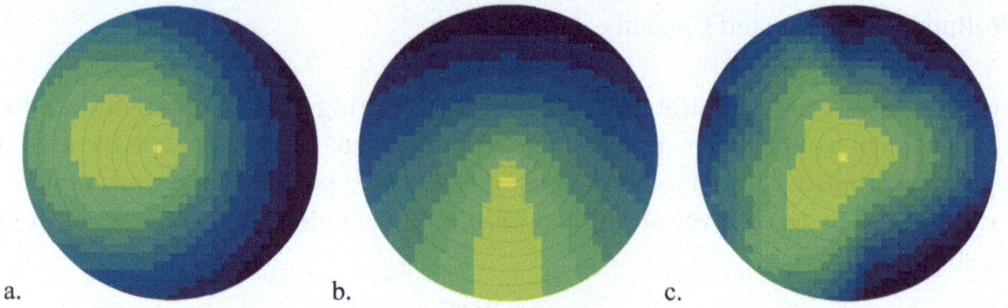

Fig. 6. a. Emission concentration decile maps for wheeling, OH, at 8 AM on July 2 when wind speeds are 1.1 m/s, b. July 4 when wind speeds are 5.7 m/s, and c. the cumulative emissions concentration map (2,543 h) approximates the pollutant plume.

1. Pollution Index

The pollution index vector is calculated through a combination of the EPD Pollution Quantity *(EPD)*, Wind Multiplier value *(WM)*, and Pollution Capacity *(PC)*. These are crosswise multiplied at the level of the census tract to create the Pollution Index vector *(PI)*:

$$PI = EPD \odot WM \odot PC$$

Equation 4. Crosswise Vector Multiplication to Generate the Pollution Index

2.5 EJ Impact Score

A single evaluative metric cannot be derived from a set of features on non-consistent scales in their raw form (Diez 2019). To create our EJ Impact Score, we propose a standard normal score (z-score) on a scale from 0 to 1 as the cumulative contribution of the three variable indices: Pollution Index, Demographic Index, and Vulnerability Index (Fig. 7).

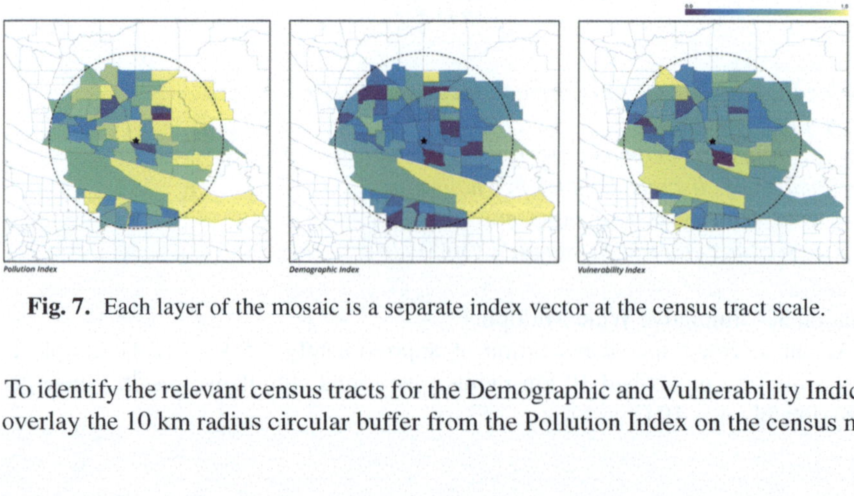

Fig. 7. Each layer of the mosaic is a separate index vector at the census tract scale.

To identify the relevant census tracts for the Demographic and Vulnerability Indices, we overlay the 10 km radius circular buffer from the Pollution Index on the census map and use the centroid containment method (Chakraborty et al. 2011). A sensitivity analysis of census tract sampling methods (weighted area containment, 50% containment, and centroid containment) found >99% similarity between methods. Where no centroid is contained within the buffer, the nearest centroid is used.

Each contributing index vector is averaged across the identified census tracts. Subsequently, we take the elementwise product of the vectors to reduce the data set to a single $[N \times 1]$ normalized vector to approximate how a manufacturer is performing relative to the mean (O'Neil 2011). Scores greater than zero indicate higher EJ impacts, whereas scores of less than zero indicate lower-than-expected effects.

3 Testing the Method

Two US-based architectural case studies demonstrate the gaps and strengths of the proposed methodology: OpenHome, a 500 m^2 prefabricated wood-frame house planned for Washington, and Loblolly House, a 205 m^2 prefabricated aluminium-frame house in Maryland. Each was modelled in Revit to a similar level of detail; then, we used Tally to generate and link a BOM directly to the EC3 EPD database. As several materials and classes of materials are not yet included in EC3, these could not be included in our calculations. This includes paints, resins, adhesives, and coatings; roofing products; door frames and hardware; plastics and composites; wood flooring products; heavy timber; and air barriers. For concrete, which is typically a locally sourced material, we limited the search range to within 160 km of the building site (Figs. 8 and 9).

Fig. 8. Case Study 1: OpenHome, a 500 m^2 prefabricated wood-frame single-family home located outside of Seattle, WA, USA

We found differences per category of material in reporting methods on the EPDs, so we applied a different assessment scope to match the PCR reporting requirements. Where fewer than two EPDs reported Module A3 separately within a product category, we expanded the search criteria to select EPDs that reported Module A1–A3 in aggregate, as noted in the data table (Table 1).

Fig. 9. Case Study 2: Loblolly House, a 205 m^2 prefabricated aluminium-frame single-family home located in Taylor's Island, MD, USA

For each case study, we tested the EJ Score method for the two product categories that contribute the highest percentage to the mass of the building and have two or more EPDs for comparison. In both cases, this is cast-in-place concrete and gypsum.

3.1 Case Study 1: OpenHome - Seattle, Washington, USA

Starting with a Revit model linked to Tally, we identified 44 unique products in the BOM. EC3 imported 93% of the Tally scope by mass but only 47% of the scope by the number of unique products. Within EC3, the opportunity for increased precision using EPDs shrank significantly, as only the insulation and gypsum categories separated environmental impacts associated with the product manufacturing stage A3. By expanding our LCA scope to include A1–A3, we had most of the remaining materials, representing 72% of the EC3 scope and 67% of the original Tally BOM, as measured by mass (Fig. 10).

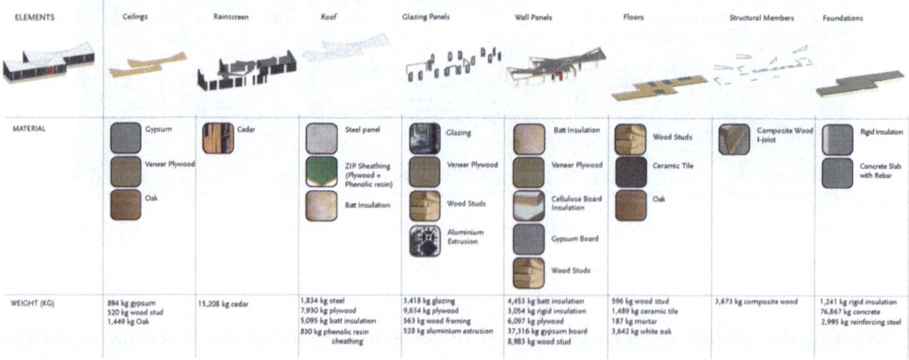

Fig. 10. BOM for OpenHome, showing quantities and location of each material or product included in the LCA.

Table 1. Availability of EPDs in the EC3 database for the building products in Case Study 1

Material	EPDs with EC	EPDs with POCP (* indicates scope A1–A3)	% of EC3 scope by mass
Rebar	38	6*	1%
Concrete	276	134*	24%
Self-levelling concrete	24	15*	1%
Structural steel	28	12*	1%
Structural wood products	12	7*	16%
Insulation (batt or blown)	91	3	3%
Glazing	15	4*	4%
Gypsum	38	29	20%
Floor tile	13	6*	<1%
Wood doors	10	10*	1%

We can minimise our EJ score (using EC as a tie-breaking metric when EJ scores from two sources are equivalent) by sourcing from widely distributed locations: rebar manufactured in Colorado, structural steel from California, gypsum from North Carolina, batt insulation and flooring tile from Texas, structural wood lumber and wood doors from Oregon, CLT from Washington, and glazing from Minnesota (Fig. 11).

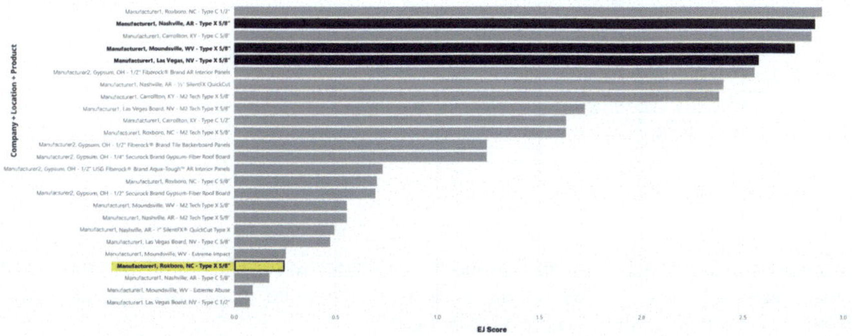

Fig. 11. Comparing the EJ score across options for Type X 5/8″ gypsum for Case Study 1. The best site of manufacture selected for the project is identified in yellow. The same product manufactured in other locations is called out in black.

Examining the supply chain for gypsum, we see the value in the EJ score. Although one can review different products manufactured at the same location (such as the five different gypsum products made at Manufacturer1's Nashville location) by simply comparing the declared POCP from the EPD, looking across products and manufacturing locations has a more complex story. Manufacturer1 produces an identical product at four different sites, with variation in the EJ score from 0.17 to 2.86. When we compare across manufacturers, the 29 options for Type X 5/8″ gypsum scores vary from 0.08 to 2.89 (Fig. 12).

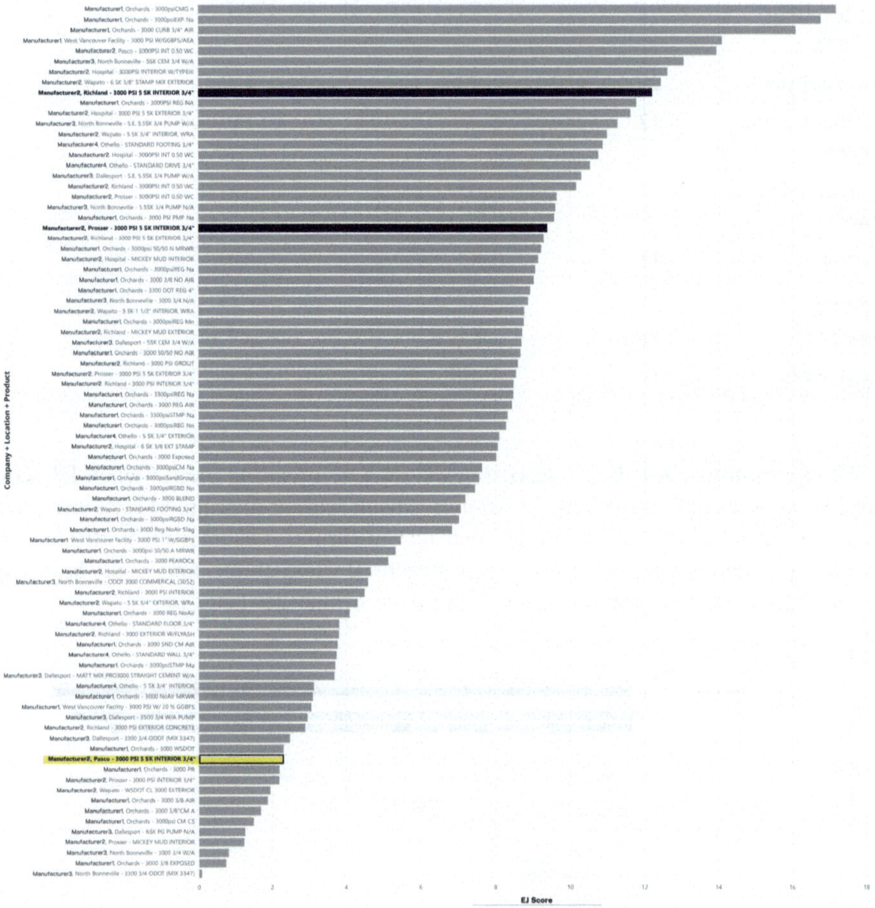

Fig. 12. Comparing the EJ score across options for concrete for Case Study 1. Mix and specific manufacturing sites selected for the project are identified in yellow. The same product manufactured in other locations is called out in black.

A similar story arises from the review of concrete. While a single producer location may have multiple options for mix designs meeting our structural criteria, based on additives and substitute cementitious material percentages, comparing across producer locations gives us a richer story, showing a variation of 557%. This implies that the site of the manufacturing plant matters, and the site of our material supply chains has a direct relationship to community health and wellness (Fig. 13).

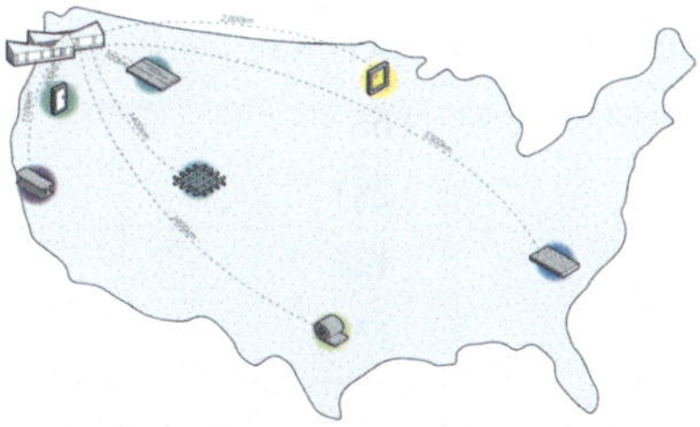

Fig. 13. Demonstration of supply chain mapping from the EJ score for a prefabricated wood structure in the US Pacific Northwest.

3.2 Case Study 2: Loblolly House – Taylor's Island, Maryland, USA

We identified 39 unique products in the BOM using Tally. EC3 imported 67% of the scope by mass and 47% by the number of unique products. This represents smaller building coverage by mass than Case Study 1, but an almost identical percentage of the number of unique products. Much of the unrepresented mass is in the structure: the aluminium scaffold, the engineered wood structural components, and the heavy timber foundation piles (Fig. 14 and Table 2).

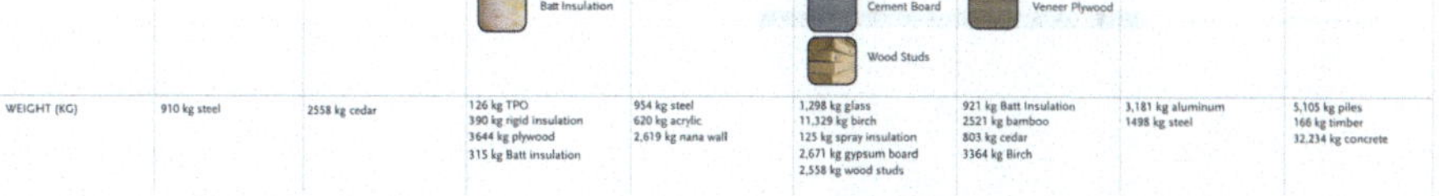

Fig. 14. BOM for Loblolly House, showing quantities and location of each material or product included in the LCA.

This is a problem in terms of the comprehensiveness of the scope of the data, as LCA practices assume a cut-off by mass or by the impact of no more than 5%, and these structural elements make up over 30% of the building by mass. However, we can still minimise our EJ score with the remaining elements. Although we select the same sources for gypsum, batt insulation, wood doors, and glazing as in Case Study 1, others are new due to the secondary selection criteria. Rebar is sourced from Tennessee, structural steel

Table 2. Availability of EPDs in the EC3 database for the building products in Case Study 2

Material	EPDs with EC	EPDs with POCP (* indicates scope A1–A3)	% of EC3 scope by mass
Rebar	38	6*	2%
Concrete	62	28*	58%
Structural steel	28	12*	2%
Plywood	5	1*	13%
Insulation (board)	39	7	<1%
Insulation (batt/blown)	91	3	<1%
Glazing	15	4*	3%
Gypsum	38	29	4%
Wood doors	10	10*	<1%

from North Carolina, board insulation from Ohio, and plywood from Alabama. Although some of the geographic distribution of this supply chain is artificially created by the limited number of EPD options from our source database, as fewer manufacturers in this region of the US produce EPDs, a further geographic distribution would be expected in the actual supply chain for the project, as many products in construction are sourced on the global market (Fig. 15).

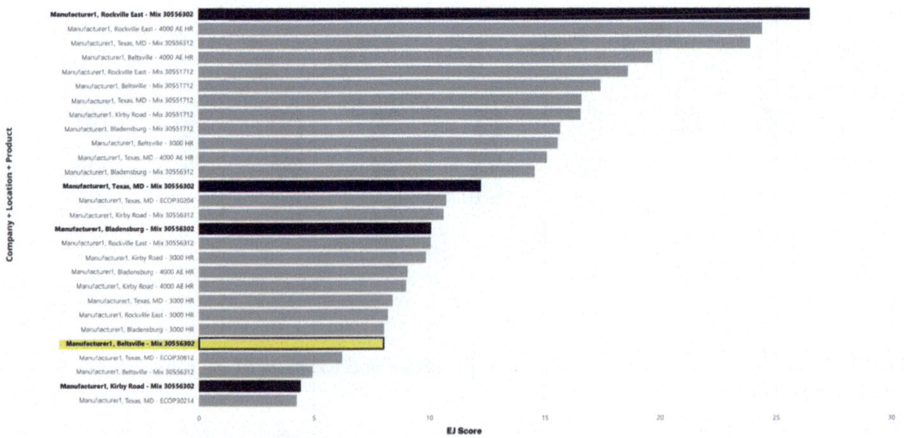

Fig. 15. Comparing the EJ score across options for concrete for Case Study 2. Mix and specific manufacturing site selected for the project are identified in yellow. The same product manufactured in other locations is called out in black.

We confront a significant constraint in the data for our concrete mix supplier for this project. While the Pacific Northwest has a culture promoting manufacturers participating in creating EPDs for the building design community, the East Coast does not have an equally substantial demand for manufacturer transparency documentation. However, even with only one company participating in the EPD process, we still have a selection of 28 mix designs from 10 plant locations, with the highest EJ scores for a single mix being more than five times the value of the lowest (Fig. 16).

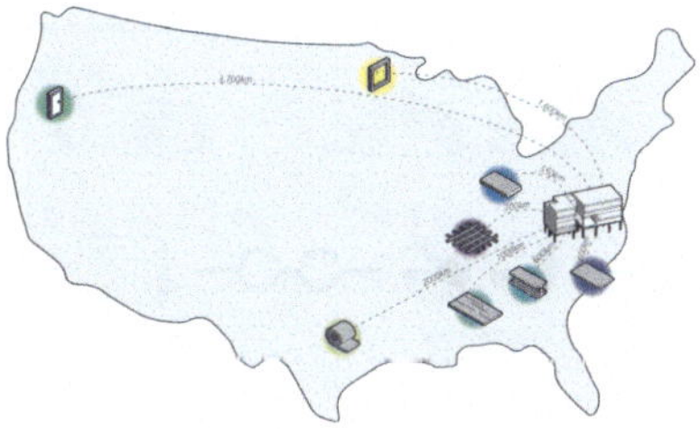

Fig. 16. Demonstration of supply chain mapping from the EJ score for a prefabricated aluminium structure in the Eastern US.

4 Findings

This proof-of-concept methodology demonstrates a modelling approach to evaluate building products through an EJ lens by mosaicking existing databases. It indicates that there could be significant differences in community health impacts for building product options. It also demonstrates the potential convergence between BIM-LCA and GIS-LCA emerging modelling practices. To use this workflow for decision-making, there needs to be increased confidence in the results by resolving issues within the underlying data. These issues fall into four categories: the resolution of LCI data, database API concerns, geographic coverage, and coverage by material.

- **Resolution of LCI data.** Using isolated A3 data published in an EPD reduces the analysis accuracy, as some impacts do not happen on-site, such as those created by power plants generating the power used. Combining A1–A3 impacts further obscures the location of emissions. Increasing the accuracy of the analysis requires access to the underlying LCI data.
- **Database API Weaknesses.** The EC3 database is focused on EC, which means many EPDs are not correctly parsed for localised impacts. Although our search identified 2,745 EPDs matching our needs, only 1,897 (69%) processed Module A (manufacturing) POCP impacts. Only 89 (3%) separate A3 (factory-based) impacts separately. Searches of the source EPDs revealed several cases where POCP A3 is reported but did not appear in the API.
- **Geographic Coverage.** Our data is limited to US sources for public health and community composition. Further issues in geographic representativeness relate to the uptake of EPD publication by AEC manufacturers in the US; coverage is biased biasing to the Pacific Northwest, especially in concrete. Additionally, some EPDs may not yet be in EC3.

- **Material and Product Diversity.** EC3 has limited coverage of material categories and many MasterFormat categories are unrepresented. Notably, for localised impacts, the percentage of a material by mass in a building is less likely to correlate with the percentage by impact, especially if categories like plastic are underrepresented (Fig. 17).

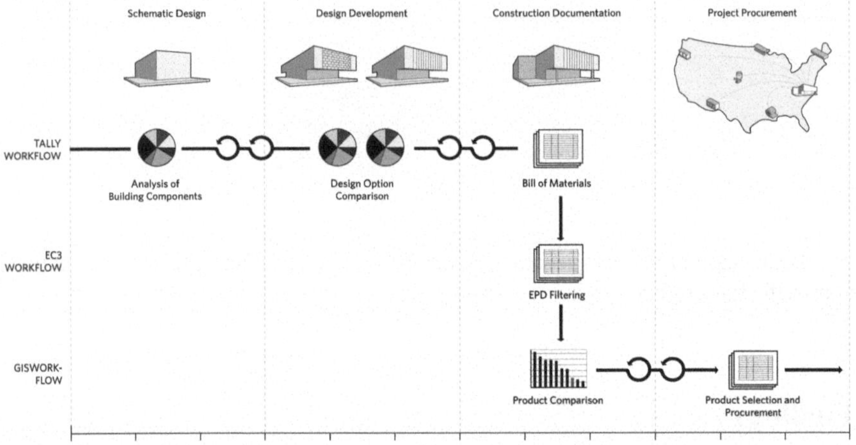

Fig. 17. Ideal workflow using the EJ score method, which is currently only valid for US manufacturing locations

5 Conclusion

While it is relatively simple to select products by comparing the reported values of potential pollutants in an EPD, for the local impacts such as POCP, an identical product manufactured at two or more sites can have vastly different EJ impacts. The proof-of-concept method we demonstrated successfully brings geographic differences to the forefront by including critical place-based metrics for environmental resilience, population vulnerability, and historic inequity. Providing visibility into these product differences could allow designers to select products more thoughtfully and advocate for specific actions within the supply chain. To increase the method's accuracy, designers should advocate for the EJ score to be included as a part of the EPD publication process or for manufacturers to release the underlying LCI data. For this methodology to be effective, the results, including the EJ Score and supply chain impact mapping, should be represented in the designer modelling environment for design decision-making, optioning, detailing, and specification writing.

References

Al-Ghamdi, S., Bilec, M.: Green building rating systems and whole-building life cycle assessment: comparative study of the existing assessment tools. J. Archit. Eng. **23**(1), 04016015 (2017)

Ash, M., Boyce, J.: Racial disparities in pollution exposure and employment at US industrial facilities. PNAS **115**(42), 10636–10641 (2018)

Azapagic, A., et al.: An integrated approach to assessing the environmental and health impacts of pollution in the urban environment: methodology and a case study. Process Saf. Environ. Prot. **91**(6), 508–520 (2013)

Bates, R., Carlisle, S., Faircloth, B., Welch, R.: Quantifying the embodied environmental impact of building materials during design. In: PLEA 2013 - 29th Conference, Sustainable Architecture for a Renewable Future, 10 September 2013 (2013)

Bulle, C., Margni, M., Patouillard, L., Boulay, A.-M., et al.: IMPACT World+: a globally regionalized life cycle impact assessment method. Int. J. Life Cycle Assess. **24**(9), 1653–1674 (2019). https://doi.org/10.1007/s11367-019-01583-0

Bruce-Hyrkäs, T., Pasanen, P., Castro, R.: Overview of whole-building life-cycle assessment for green building certification and ecodesign through industry surveys and interview. Procedia CIRP **69**, 178–183 (2018)

Castellani, V., Benini, L., Sala, S., Pant, R.: A distance-to-target weighting method for Europe 2020. Int. J. Life Cycle Assess. **21**(8), 1159–1169 (2016). https://doi.org/10.1007/s11367-016-1079-8

Chakraborty, S., Dutta, T., Sikdar, P.K.: Understanding the control of geology, geomorphology and landuse/landcover on arsenic distribution in groundwater of Bengal Basin using RS, GIS and PCA. Survey **52**(3&4), 70–85 (2011)

CDC: About the PLACES Project. Centers for Disease Control and Prevention, 18 October 2021. https://www.cdc.gov/places/about/index.html

Diez, D., Cetinkaya-Rundel, M., Barr, C.: OpenIntro Statistics, 4th edn. OpenIntro Inc., Boston (2019)

Environmental Protection Agency: EJScreen: Environmental Justice Screening and Mapping Tool. EPA (2022). https://www.epa.gov/ejscreen/ejscreen-map-descriptions#:~:text=The%20Demographic%20Index%20in%20EJScreen,numbers%20are%20simply%20averaged%20together. Accessed 21 Mar 2022

Federal Register Presidential Documents - Archives - Executive Order 12898 of February 11, 1994: Federal Actions To Address Environmental Justice in Minority Populations and Low-Income Populations. Presidential Documents (1994). https://www.archives.gov/files/federal-register/executive-orders/pdf/12898.pdf. Accessed 21 Mar 2022

Feng, H., Hewage, K.K.N., Sadiq, R.: Exploring the current challenges and emerging approaches in whole building life cycle assessment. Can. J. Civ. Eng. **49**, 149–158 (2021)

Huijbregts, M.A., Schöpp, W., Verkuijlen, E., Heijungs, R., Reijnders, L.: Spatially explicit characterization of acidifying and eutrophying air pollution in life-cycle assessment. J. Ind. Ecol. **4**(3), 75–92 (2000)

Ladybug-Tools: Ladybug-Tools/Epwmap: Map of Available .EPW Weather Files. GitHub. https://github.com/ladybug-tools/epwmap. Accessed 1 Apr 2022

Li, J., Tian, Y., Zhang, Y., Xie, K.: Spatializing environmental footprint by integrating geographic information system into life cycle assessment: a review and practice recommendations. J. Clean. Prod. **323**, 129113 (2021)

Megange, P., Ngae, P., Feiz, A.-A., Le, T.-P.: Dynamic site-dependent life cycle assessment for assessing impact of human toxicity of a double glazed PVC window. In: 27th CIRP Life Cycle Engineering (LCE) Conference, pp. 316–321. Elsevier B.V. (2020)

Nitschelm, L., Aubin, J., Corson, M.S., Viaud, V., et al.: Spatial differentiation in life cycle assessment (LCA) applied to an agricultural territory: current practices and method development. J. Clean. Prod. **112**, 2472–2484 (2016)

O'Neil, P.V.: Advanced Engineering Mathematics, 7th edn. Cengage Learning, Boston (2011). ISBN: 9781111427412

Seppälä, J., Posch, M., Johansson, M., Hettelingh, J.P.: Country-dependent characterisation factors for acidification and terrestrial eutrophication based on accumulated exceedance as an impact category indicator (14 pp). Int. J. Life Cycle Assess. **11**(6), 403–416 (2006)

Updates to the Air Quality Index (AQI) for Ozone and Ozone Monitoring Requirements. EPA (2015). https://www.epa.gov/sites/default/files/2015-10/documents/20151001_air_quality_index_updates.pdf

A Framework for Managing Data in Multi-actor Fabrication Processes

Lior Skoury[1,2](✉), Felix Amtsberg[1], Xiliu Yang[1], Hans Jakob Wagner[1], Achim Menges[1], and Thomas Wortmann[1,2]

[1] Institute for Computational Design and Construction, Cluster of Excellence Integrative Computational Design and Construction for Architecture, University of Stuttgart, Stuttgart, Germany
lior.skoury@icd.uni-stuttgart.de

[2] Chair of Computing, Institute for Computational Design and Construction, University of Stuttgart, Stuttgart, Germany

Abstract. This research proposes a design to fabrication data framework for multi-actor fabrication environments with robotic and human actors. The framework generates and exchanges fabrication data between the design elements and the fabrication environment. It features a uniform task data model to represent all processes in the fabrication procedure and link them to the design elements. The framework is demonstrated with a timber slab case study that shows the fabrication data generation, assignment to actors and ordering for execution with a multi-actor fabrication environment involving industrial robots and human workers. This framework opens new opportunities for continuous digital data exchange and feedback between the design and fabrication and allows for rapid changes in both the fabrication environment and the design.

Keywords: Design for humans & non-humans · Human-robot collaboration · Digital fabrication · Digital twin · Task-skill workflow

1 Introduction

Contemporary digital representations of a design or building contain data about many aspects of different disciplines in the architecture, engineering, construction (AEC) industry. The IFC schema is one of the main representations that describe building parts in detail [1]. Despite the benefits of using the IFC schema for building design, it falls short in generating, representing, and managing the data needed for digital fabrication processes. Nevertheless, IFC files can be converted to machine code after the model is done [2]. Another approach embeds the multidisciplinary aspect in the AEC industry through the notion of "Co-Design" [3]. This approach emphasises the importance of interdisciplinary research among the various stakeholders in the AEC industry through developing digitisation methods for planning, fabrication and construction. In contrast to the IFC schema that deals with the data as fixed data, a flexible data flow is needed to

C. Gengnagel et al. (Eds.): DMS 2022, *Towards Radical Regeneration*, pp. 601–615, 2023.
https://doi.org/10.1007/978-3-031-13249-0_47

accommodate a dynamic design and fabrication environment [4]. Approaching a design to fabrication process from Co-Design's perspective will allow for this flexible data flow.

A robotic fabrication environment can contain several robotic actors with different tools to achieve specific tasks [5]. Human actors work alongside the robots operating them and performing other tasks [6]. Industrial robots and their tasks are usually represented as machine-specific code. In contrast, other actors, such as human workers and their tasks, are described in analogue methods such as blueprints.

Digital representations of all actors and tasks in the fabrication process should be combined into one framework to facilitate the co-design approach, such that fabrication data can be generated and linked to the design geometry throughout the design process.

This research proposes a framework to manage design data, robotic fabrication data and human fabrication data in a continuous workflow that allows reciprocal feedback between design and fabrication and between different actors in the fabrication environment. To develop this framework, this research asks two main questions: (1) What is the relation between building data and fabrication data and (2) how can this data be generated and distributed to heterogenous fabrication actors? To answer these questions, the paper describes a framework to convert building data into fabrication data, analyse it, and assign it to the relevant actors in the fabrication environment.

2 Background

2.1 Relevance

Integrating information, communication technology, and industrial production technologies are crucial aspects of Industry 4.0 [7]. Cyber-physical systems (CPS) connect the design/product to the machine and the workers to keep companies flexible and relevant to the global market [8]. Industry 4.0, specifically CPS, are applied towards fully automated factories equipped with robots to execute production. Examples can be found in the automotive industry, where robotic production lines are the dominant actors in production [9]. Applying this approach to the AEC industry usually involves conceptualising buildings as products rather than project-based design challenges. This approach, guided by standardising building components [10], can cause design and architectural freedom constraints.

Integrating human actors in the automated prefabrication workflow will allow flexible and open design-fabrication processes. The workflow should consider all the actors in the fabrication process regardless of the level of automation [11].

The need to link design to fabrication has been described as a linear, continuous flow of data from the early design steps to the final building parts [12]. Instead of a linear workflow, the co-design methodology allows for a non-linear, reciprocal workflow for design to fabrication data flow. This workflow should represent the design alongside the fabrication environment and tasks and allow communication and feedback loops to support the co-design process.

2.2 State of the Art

An example of a project that deals with a design to fabrication multi-actor process is the research of large-scale spatial timber structure by Eversmann et al.[13]. Each element in the design is assigned with detailed fabrication data that will be translated and executed later by a robotic fabrication environment. This project shows high potential in robotic control, fabrication data communication strategies and feedback loops. Yet, the feedback loops are not involving the design. Furthermore, this project aims for a high level of automation using robotic arms to achieve all goals in the prefabrication process.

The BUGA wood pavilion is another project that deals with data workflow in the design-to-fabrication process [14]. This project suggests a comprehensive co-design approach, including the design, the structural model, and the detailing, which through robotic simulation, informs back the design. It also offers a machine control strategy and a fabrication workflow involving a robotic fabrication platform. However, the data workflow breaks between the simulation to the fabrication execution due to the need to translate it to machine code. Therefore, changes in the process can only occur before machine code generation. Additionally, it includes a high level of automation with industrial robotic actors working separately from human workers.

3 Methodology

The developed framework suggests a digital workflow that links the design data to the fabrication environment by a universal task data model. The task data model receives data from the design elements and the multi-actor fabrication environment.

A multi-actor fabrication environment consists of several actors performing the fabrication process. These actors can be industrial robots with various end-effectors, small autonomous robots or human workers. The framework explores the digital side of the design-to-fabrication process, where fabrication data is extracted, simulated and assigned to actors in the environments, alongside the physical aspect of communicating the tasks to the actors for the production. This research was developed in the Rhino-Grasshopper environment and implemented with C# and Python. The main considerations in the development of this framework consist of five parts described in the following section:

- Model Initiation
- Task Data Model
- Robotic Task Search
- Task Distribution and Ordering
- Fabrication Execution (Fig. 1)

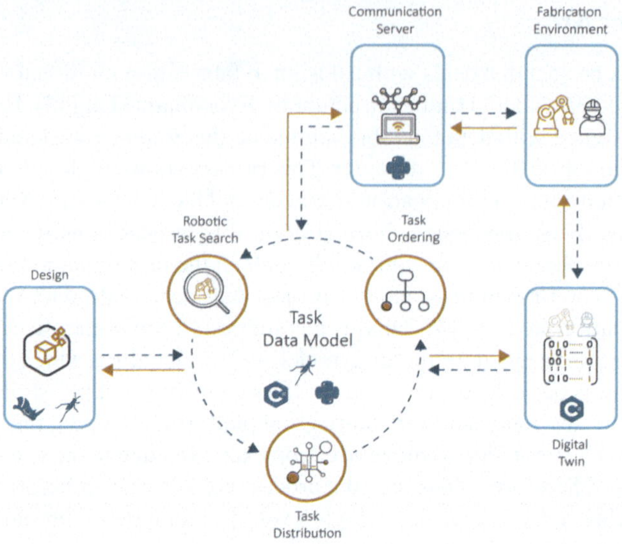

Fig. 1. Framework schema with the task data model as a link between design to fabrication.

3.1 Model Initiation

Analysing the design geometry and extracting fabrication tasks require two initial components to be defined: the actors' skills in the fabrication environment and the tasks for every design element. The fabrication environment's setup is built up from the actors' descriptions and the tools they can use. Every actor can obtain several tools allowing it to perform different tasks. The robotic actors and their tools (end effectors) are described as skills through a 3D model associated with the respective task they need to perform. Human actors and their tools are described as abstract (textual) objects containing data about the tasks they can execute. Both robotic and human actors' skill sets are determined according to the available robots, end effectors, workers and hand tools available in the physical fabrication environment. The setup of the fabrication environment outputs a list of all tasks that can be executed in the fabrication process. Accordingly, the design geometry is analysed and assigned abstract tasks needed for the fabrication of each element. Every design element is assigned a list of tasks that altogether allow for the fabrication of the element in the global design schema. Those tasks are modelled as objects using Object-Oriented Programming (OOP), allowing for fields in the object to be assigned and changed. The basic data of each task consist of a set of 3D planes, extracted from the geometry, representing the location or trajectory of the task.

3.2 Task Data Model

In a traditional digital fabrication process, each sub-process is translated to a specific machine language according to the designated actor after the digital generation and simulation of the process. This research proposes a universal data model to describe the fabrication tasks regardless of the actor executing them. This model is used for the digital side of planning, simulating the fabrication process, and communicating the tasks to the actors for the physical fabrication.

OOP allows assigning abstract tasks to the design elements, which later will be assigned with values to the different fields of the object. Furthermore, it allows for rapid changes in the fields for testing various options of fabrication actors.

The proposed data model consists of general data that each task should have and optional data according to the specific actor associated with the task. A detailed description of the task object fields is described in Fig. 2. This object is modelled as a dynamic OOP that allows it to be extended and assigned with more fields as the fabrication environment changes and expands to include more actors or tools requiring more information for the task execution. The task data model is actor independent, establishing a common language for robot and human tasks. This builds the framework for a flexible and resilient human-robot collaboration, task sharing and task allocation.

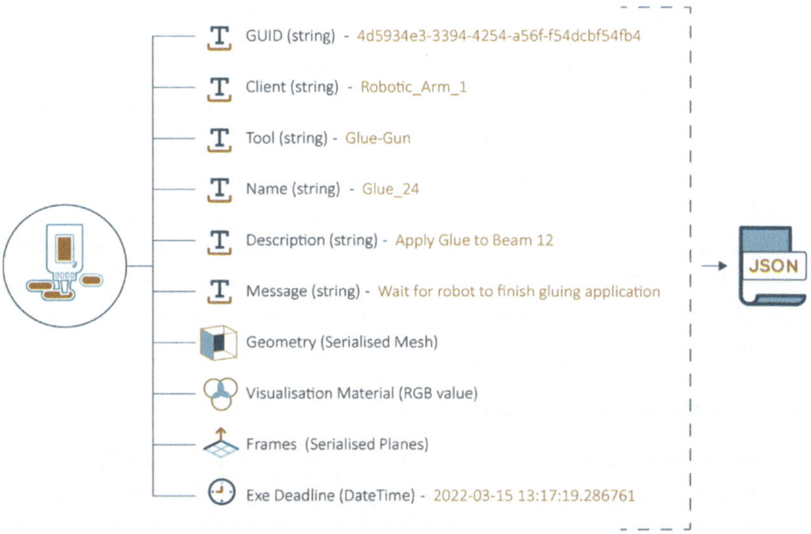

Fig. 2. Example of a robotic glueing task with basic and optional data in the data model.

3.3 Robotic Task Searching

Robotic tasks need to be tested for robot reachability before being approved and assigned to the robotic actors. Therefore, each task is tested for its specific robot and tool in the fabrication environment and adjusted if required. The robot object is defined using the existing robot library VirtualRobot for kinematic simulation and implemented in the Grasshopper environment.

Using the 3D planes and the tasks' data generated in Sect. 3.1, the algorithm matches tasks to a robot model and calculates the robot axis values using inverse-kinematic method. The axis values are tested to ensure they are not out of the robot's reachability. Additionally, the 3D model is tested to avoid self-collisions and robot-workpiece collisions. If a plane yields unreachability or collisions, the free axis parameters are adapted, and the plane is tested again. There are three options for tasks failing the search process: the design may be changed, the robot setup might be adapted, or the task can be allocated to another actor. Tested tasks that result in reachability for all planes in their set are flagged as possible robotic tasks (Fig. 3).

```
procedure Check_Robotic_Tasks(element El, robot R, tool T)
    foreach task t in EL tasks:
        foreach plane p in t.planes:
            flag = False
            while p can be adjusted:
                get axis values and collision of p(R,T)
                if axis value not exceeds R values and not collision:
                    flag = True
                    break
                adjest p
            if not flag:
                return El to design
                or
                change fabrication environment (R,T)
                or
                assign to other actor
                return
        task actor = R
        task tool = T
        task flag = True
```

Fig. 3. Robotic search algorithm. Given design elements with tasks and a robot with a tool, evaluate if a task is executable by the robot.

3.4 Task Distribution and Ordering

An entire fabrication process usually contains many tasks executed at different stages of the fabrication process by several actors, as illustrated in Fig. 4. The execution of tasks in groups is preferable to avoid unnecessary tool changes and allow the actors' continuous workflow in an overall fabrication process. For example, in a design containing ten beams, all glueing applications for the bottom surface will be executed, and only then all beams are picked and placed. Therefore, distributing the tasks to a multi-actor fabrication environment can become a complex process requiring a predefined task distribution and ordering strategy.

Task Actor Distribution
In this research, not all tasks are executed by the robotic actors, as this would have necessitated a disproportional increase in robot setup complexity and/or robotic fabrication time. Therefore, some tasks are actor-specific, while others are open to all actors.

For those tasks, the distribution strategy in this research prioritises the robotic actors over human actors whenever possible. Accordingly, a robotic actor and tool are assigned to tasks flagged for robotic execution, while the others are assigned to a human actor.

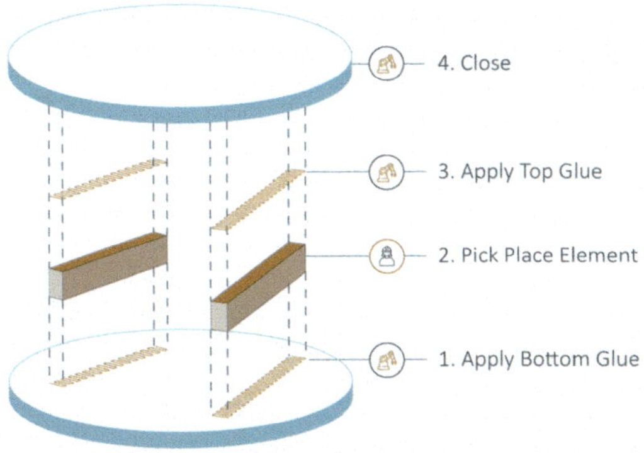

Fig. 4. An example of a simple fabrication involves top and bottom parts and two beams.

Task Ordering
The strategy used in this research groups tasks by process, actor and tool. For example, all possible robotic glueing tasks at a certain fabrication stage will be grouped and executed together. This approach prioritises task data over design data, which means that a fabrication process is not defined by the design elements as a linear sequence but by all the tasks involved in the fabrication process. However, the fabrication linearity of the design elements is still considered and influences the task order, such as some tasks have to be executed before others. Human and robotic groups are separated to allow a safe working environment. Subsequently, all groups are simulated. The task actor can be reassigned if needed, or the design elements can be modified to regenerate tasks.

3.5 Fabrication Execution

This research proposes a scalable communication strategy for the design to fabrication process to support tasks' execution in a multi-actor fabrication environment. The framework generates a JSON file containing all linked and ordered tasks for the fabrication execution. The centre part of this communication system is a Python server that can read, change, and send the tasks from the JSON file to clients for actor-specific execution. Each actor in the environment is modelled as an object (client) with a communication channel and a translator that converts tasks to actor language. The communication channel is two-sides, allowing the physical actor to receive and send data to the server. Since the central server only communicates tasks to the clients, this method allows more clients to be added to the system as the fabrication environment changes (Fig. 5).

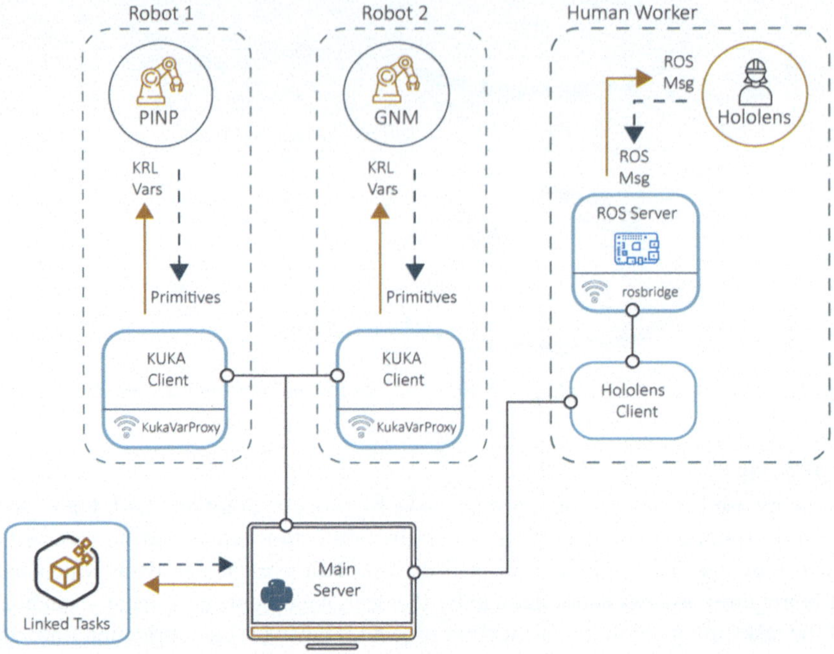

Fig. 5. The communication system in this process consists of two Kuka robot clients and one Hololens client. Kuka data transfer uses KukaVarProxy, while Hololens client uses a ROS server.

4 Case Study

4.1 Design

To demonstrate the framework, this research uses a timber slab based on a multi-directional composite timber slab system described in Orozco et al.[15]. The timber slabs in this system consist of 3 layers: bottom plate, mid-layer build-up from several elements, and top plate. The mid-layer elements consist of linear beams called webs that distribute the shear forces throughout the slab and solid timber parts called splotches dealing with high shear areas in the slab. In addition, the mid-layer is integrated with electrical wires that need to be placed before the slab is enclosed with the top plate. The slab explored in this research has a bounding box size of 2.1 × 6.9 m with a total height of 0.32 m, and it contains 16 shear webs and two splotches alongside the top and bottom plates (Fig. 6).

4.2 Fabrication Environment

The fabrication environment in this research consists of 3 actors: two industrial robots and one human worker. The two robots are part of a robotic platform with two industrial KUKA robotic arms (Wagner et al. 2020) [5]. One of the robots is defined as a Pick-and-Place robot (PINP), and it is equipped with a pneumatic gripper and a vacuum gripper. The second robot is defined as a Glue-Nail-Mill robot (GNM), and it is equipped with the

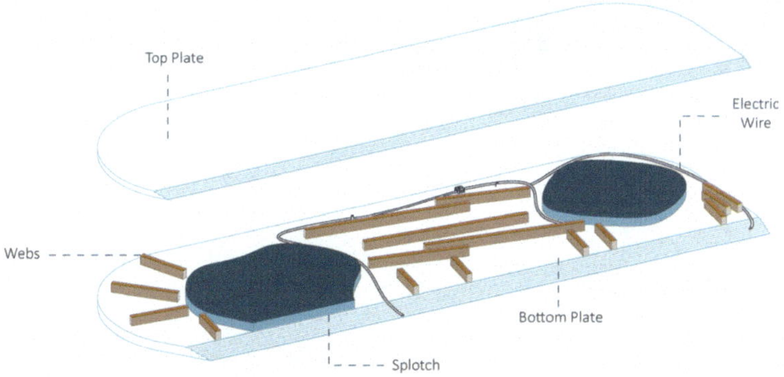

Fig. 6. Multi-directional timber slab and the different parts for fabrication.

corresponding tools, glue-gun, nail-gun and milling head. The tools used in this research are both grippers on PINP and the glue-gun on GNM.

The human worker is equipped with AR Hololens glasses that can visualise the task and the instruction for the task execution [16]. Since the workpiece is relatively large, an additional semi-tool is set in this environment: a linear table with wheels that the human worker can move. The table can be moved along one axis in the fabrication hall, allowing more flexibility for robotic tasks (Fig. 7).

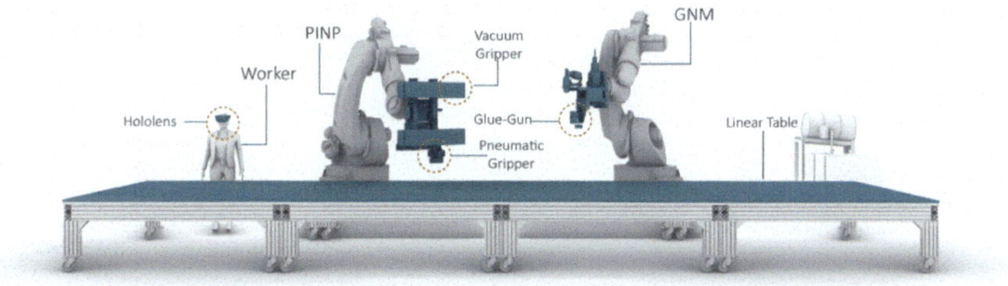

Fig. 7. Actors and tools in the fabrication environment.

4.3 Tasks

The fabrication tasks explored in this research are glue, pick-place, and electric wire installation. For each shear web, there are four tasks: apply bottom glue, pick, place, and apply top glue. The splotch tasks are similar to the webs except for the number of glueing tasks due to the splotch's interface area. Web tasks are not defined as actor-specific tasks, while splotch tasks are defined as robotic tasks due to the splotch size.

The electric installation and table position tasks are defined as human tasks since a robotic execution will require a bigger and more complex robotic setup. The total design-fabrication process contains 123 tasks in 15 groups, as described in Fig. 8.

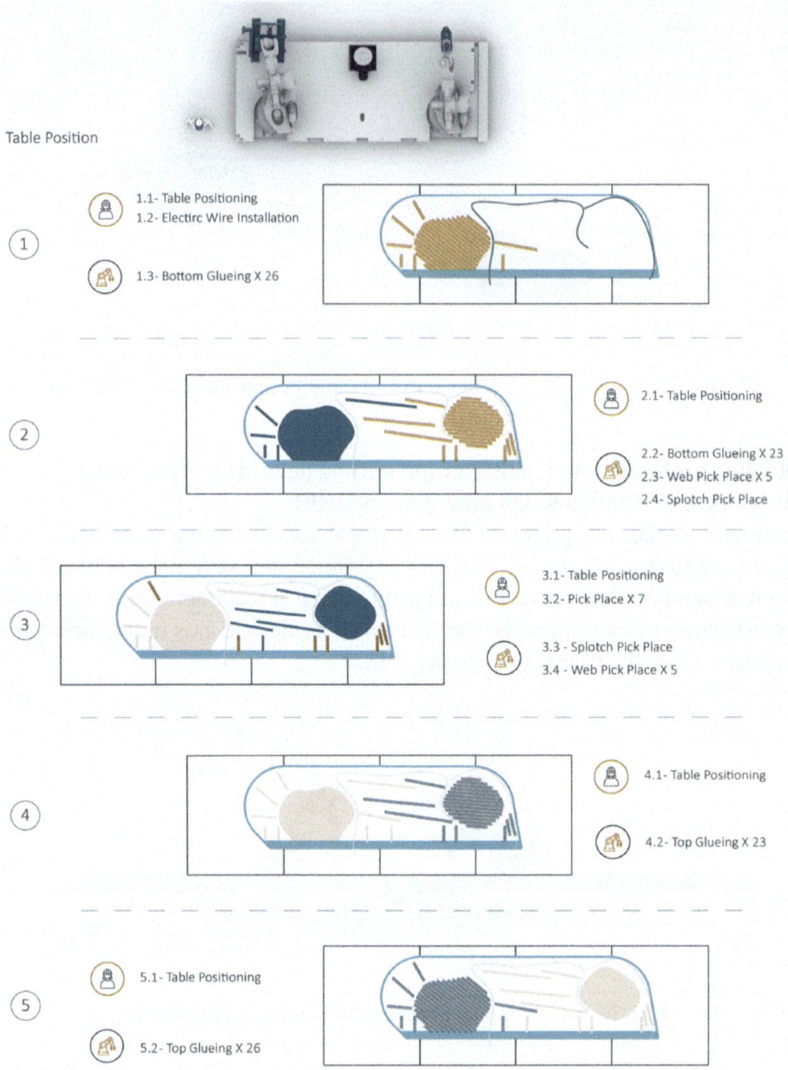

Fig. 8. Groups of tasks divided into table position, actor, and tool.

4.4 Physical Fabrication

The clients modelled in this research are a KUKA robot client and a Hololens client. The robot client uses a Kuka-Var-Proxy communication implemented in python. The client converter translates the tasks' set of planes to Kuka frames in KRL format for execution. The array is sent to the KUKA controller alongside the tool number and process number indicating the task. The human actor receives instructions from the Hololens client that uses a ROS server communication. For this client, the tasks are translated to visualise geometry alongside textual instructions and a message as AR objects on the Hololens.

The tasks are iterated through the server and sent to the relevant client. Subsequently to the task execution, the client sends a message to the server informing that the task is done (Figs. 9 and 10).

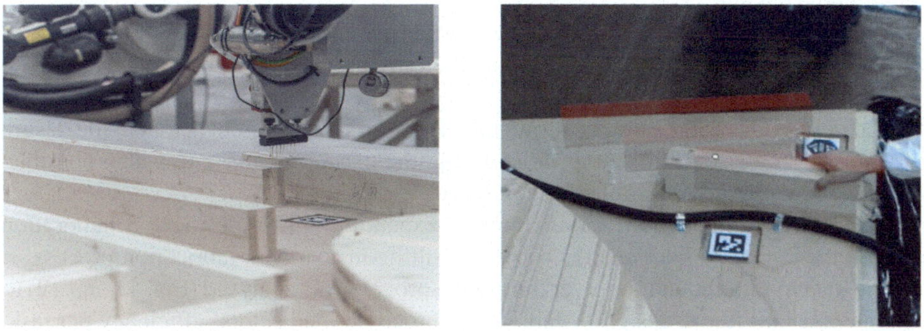

Fig. 9. Left - Robotic execution of a glueing task. Right - Worker execution of pick place task as viewed from the Hololens.

Fig. 10. Fabrication environment in the fabrication process showing all actors.

5 Results

Using the described framework results in a continuous data flow between design and fabrication. The tasks are linked to the design element from the beginning of the process and assigned with appropriate data for fabrication along the process.

The data backflow (from fabrication data to design data) indicated the out-of-reach tasks (for humans and robots) and the corresponding design parts to allow the designer to modify the relevant parts. Due to structural constraints, in this research's case study, the linear table positions were modified to allow the fabricability of all tasks instead of changing the design.

The constraints usually seen in design to fabrication processes that allow the generation of fabrication data as machine-specific only after the design is finalised are no longer applied with this framework.

The universal task data model allows allocating tasks to different actors in the design and simulation phases. Allocating tasks during the execution phase was also possible, although it caused a delay on the Python server.

Using the central Python server for task execution allows physical collaboration between the human worker and the robots, resulting in a smooth fabrication process.

The data monitored in the server during the process was analysed (Fig. 11 and 12) to gain insights into future design-to-fabrication processes and actors' efficiency per task. The data highlights robot efficiency over human actors in customising repetitive tasks (glue, pick and place). A more in detail look will discover the time needed for glueing procedures. This specific aspect can be crucial for time-sensitive glue, where a specific fabrication process must be executed within a time limit, informing the designer about the design element's sizes and fabrication batches.

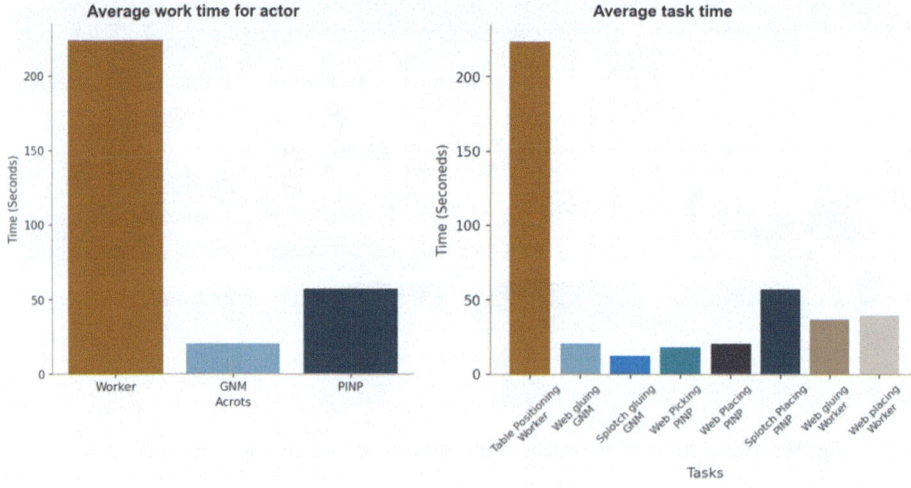

Fig. 11. (11) Comparison of fabrication time per actor (12) Comparison of fabrication time per task.

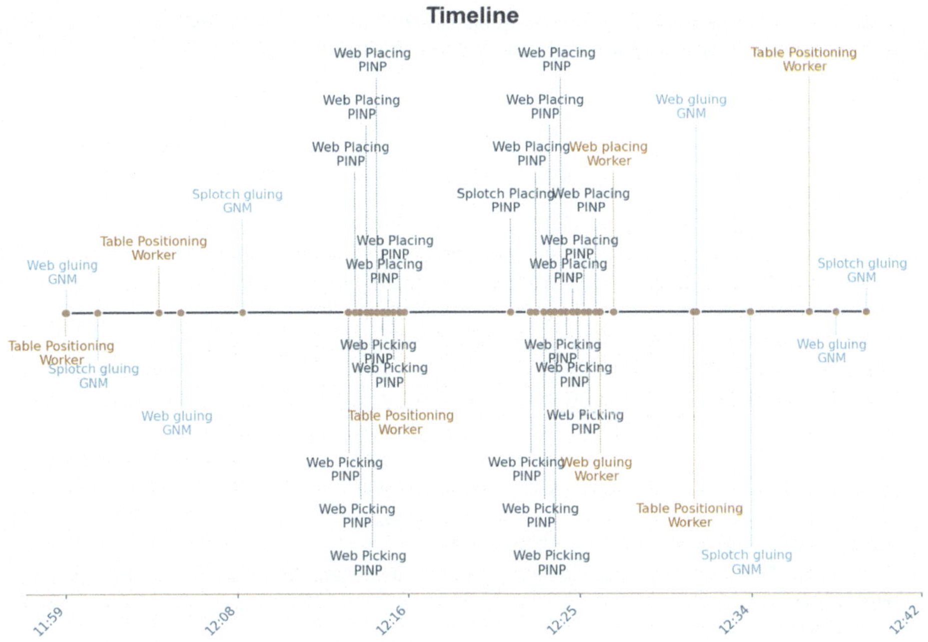

Fig. 12. Timeline for the fabrication process showing the overview of tasks progress during the fabrication.

6 Outlook

The proposed framework establishes a translation-free and universal task-based data flow between the design model and fabrication actors for instructive process description. Furthermore, the fabrication data is linked to the design data, allowing for testing and checking of fabrication alternatives, rapid prototyping and changing the design or the fabrication environment.

The task model can be extended to a fabrication ontology that represents more stakeholders in the fabrication process, such as sensors and smart tools, to validate further and improve the framework.

This research proposes a new data generation and representation framework that integrates the design data, fabrication task data, and fabrication environment by automation techniques and a new task data model that allows a direct link and continuous data flow between design and fabrication.

Acknowledgements. The research has been supported in part by the Deutsche Forschungsgemeinschaft (DFG, German Research Foundation) under the Excellence Strategy - EXC 2120/1 -390831618. The work presented in this paper was partially developed within the ITECH (Integrative Technologies and Architectural Design Research) Master's program at the University of Stuttgart.

Additionally, the authors would like to thank Long Nguyen for the development of the robot kinematics solver 'Virtual Robot' at the Institute for Computational Design and Construction (ICD).

References

1. Poerschke, U., Holland, R.J., Messner, J.I., Pihlak, M. (eds.): BIM collaboration across six disciplines. In: 17th International Workshop on Intelligent Computing in Engineering, EG-ICE 2010, Nottingham, UK, 30 June 2010–2 July 2010 (2010)
2. Correa, F.: Robot-oriented design for production in the context of building information modeling. In: Sattineni, A., Azhar, S., Castro, D. (eds.) Proceedings of the 33rd International Symposium on Automation and Robotics in Construction (ISARC), 33th International Symposium on Automation and Robotics in Construction, Auburn, AL, USA, 18–21 July 2016. International Association for Automation and Robotics in Construction (IAARC) (2016)
3. Knippers, J., Kropp, C., Menges, A., Sawodny, O., Weiskopf, D.: Integrative computational design and construction: rethinking architecture digitally. Civ. Eng. Des. **3**(4), 123–135 (2021). https://doi.org/10.1002/cend.202100027
4. Wortmann, T., Tunçer, B.: Differentiating parametric design: digital workflows in contemporary architecture and construction. Des. Stud. **52**, 173–197 (2017). https://doi.org/10.1016/j.destud.2017.05.004
5. Wagner, H.J., Alvarez, M., Kyjanek, O., Bhiri, Z., Buck, M., Menges, A.: Flexible and transportable robotic timber construction platform – TIM. Autom. Constr. **120**, 103400 (2020). https://doi.org/10.1016/j.autcon.2020.103400
6. Evjemo, L.D., Gjerstad, T., Grøtli, E.I., Sziebig, G.: Trends in smart manufacturing: role of humans and industrial robots in smart factories. Curr. Robot. Rep. **1**(2), 35–41 (2020). https://doi.org/10.1007/s43154-020-00006-5
7. Oesterreich, T.D., Teuteberg, F.: Understanding the implications of digitisation and automation in the context of Industry 4.0: a triangulation approach and elements of a research agenda for the construction industry. Comput. Ind. **83**, 121–139 (2016). https://doi.org/10.1016/j.compind.2016.09.006
8. Rauch, E., Vickery, A.R., Brown, C.A., Matt, D.T.: SME requirements and guidelines for the design of smart and highly adaptable manufacturing systems. In: Matt, D.T., Modrák, V., Zsifkovits, H. (eds.) Industry 4.0 for SMEs, pp. 39–72. Springer, Cham (2020). https://doi.org/10.1007/978-3-030-25425-4_2
9. KUKA: Automation solutions from KUKA. Case Studies for the Automotive Industry (2022). https://www.kuka.com/en-ca/industries/solutions-database. Accessed 13 Mar 2022
10. RANDEK: Zerolabor Robotic System (2022). https://www.randek.com/en/wall-floor-and-roof-production-lines/zerolabor. Accessed 13 Mar 2022
11. Shi, J., Jimmerson, G., Pearson, T., Menassa, R.:. Levels of human and robot collaboration for automotive manufacturing. In: Messina, E., Madhavan, R. (eds.) Proceedings of the Workshop on Performance Metrics for Intelligent Systems - PerMIS 2012, the Workshop, College Park, Maryland, 20–22 March 2012, p. 95. ACM Press, New York (2012)
12. Dohmen, P., Rüdenauer, K.: The digital chain in modern architecture. In: eCAADe 2007 Conference, eCAADe, pp. 801–804 (2007)
13. Eversmann, P., Gramazio, F., Kohler, M.: Robotic prefabrication of timber structures: towards automated large-scale spatial assembly. Constr. Robot. **1**(1–4), 49–60 (2017). https://doi.org/10.1007/s41693-017-0006-2

14. Wagner, H.J., Alvarez, M., Groenewolt, A., Menges, A.: Towards digital automation flexibility in large-scale timber construction: integrative robotic prefabrication and co-design of the BUGA Wood Pavilion. Constr. Robot. **4**(3–4), 187–204 (2020). https://doi.org/10.1007/s41693-020-00038-5
15. Orozco, L., et al. (eds.): Design Methods for Variable Density, Multi-Directional Composite Timber Slab Systems for Multi-Storey Construction (2021)
16. Amtsberg, F., Yang, X., Skoury, L., Wagner, H.J., Menges, A.: iHRC: an AR-based interface for intuitive, interactive and coordinated task sharing between humans and robots in building construction. In: 2021 Proceedings of the 38th ISARC, ISARC, pp. 25–32 (2021). https://doi.org/10.22260/ISARC2021/0006

Correction to: Hybrid Immediacy: Designing with Artificial Neural Networks Through Physical Concept Modelling

Mathias Bank, Viktoria Sandor, Robby Kraft, Stephan Antholzer, Martin Berger, Tilman Fabini, Balint Kovacs, Tobias Hell, Stefan Rutzinger, and Kristina Schinegger

Correction to:
Chapter 2 in: C. Gengnagel et al. (Eds.): *Towards Radical Regeneration*, DMS 2022, https://doi.org/10.1007/978-3-031-13249-0_2

The chapter "Hybrid Immediacy: Designing with Artificial Neural Networks Through Physical Concept Modelling", written by Mathias Bank, Viktoria Sandor, Robby Kraft, Stephan Antholzer, Martin Berger, Tilman Fabini, Balint Kovacs, Tobias Hell, Stefan Rutzinger, and Kristina Schinegger, was originally published electronically on the publisher's internet portal without open access. With the author(s)' decision to opt for Open Choice the copyright of the chapter changed on 04-September-2023 to © "The Author(s)" [2024] and the chapter is forthwith distributed under a Creative Commons Attribution 4.0 International License (http://creativecommons.org/licenses/by/4.0/).

The updated version of this chapter can be found at
https://doi.org/10.1007/978-3-031-13249-0_2

C. Gengnagel et al. (Eds.): DMS 2022, *Towards Radical Regeneration*, p. C1, 2024.
https://doi.org/10.1007/978-3-031-13249-0_48

Author Index

C. Gengnagel et al. (Eds.): DMS 2022, *Towards Radical Regeneration*, pp. 617–619, 2023.
https://doi.org/10.1007/978-3-031-13249-0

The manufacturer's authorised representative in the EU is Springer Nature Customer Service Centre GmbH, Europaplatz 3, 69115 Heidelberg, Germany. If you have any concerns regarding our products, please contact ProductSafety@springernature.com

Printed and bound by CPI Group (UK) Ltd, Croydon, CR0 4YY
01/10/2026
02225893-0016